Snack Foods

Snack Foods

Processing, Innovation, and Nutritional Aspects

Edited by
Sergio O. Serna-Saldivar

CRC Press
Taylor & Francis Group
Boca Raton London New York

CRC Press is an imprint of the
Taylor & Francis Group, an **informa** business

First edition published 2022
by CRC Press
6000 Broken Sound Parkway NW, Suite 300, Boca Raton, FL 33487-2742

and by CRC Press
2 Park Square, Milton Park, Abingdon, Oxon, OX14 4RN
CRC Press is an imprint of Taylor & Francis Group, LLC

This book provides a technical review of snack foods. Emphasis is made in flavored salty snacks commonly used as finger foods including popcorn, wheat-based snacks (crispbreads, pretzels, crackers), lime-cooked maize products (tortilla chips), potato chips, peanuts, almonds, and snacks from fruits/vegetables, milk, animal and marine sources.

Library of Congress Cataloging-in-Publication Data

Names: Serna Saldívar, Sergio R. Othón, editor.
Title: Snack foods: processing, innovation, and nutritional aspects /
edited by Sergio O. Serna Saldivar.
Description: First edition. | Boca Raton: CRC Press, [2022] | Includes
bibliographical references and index.
Identifiers: LCCN 2021040007 (print) | LCCN 2021040008 (ebook) | ISBN
9780367646875 (hardback) | ISBN 9780367653514 (paperback) | ISBN
9781003129066 (ebook)
Subjects: LCSH: Snack foods. | Food science.
Classification: LCC TX740.S6227 2022 (print) | LCC TX740 (ebook) | DDC
664/.6--dc23
LC record available at https://lccn.loc.gov/2021040007
LC ebook record available at https://lccn.loc.gov/2021040008

ISBN: 978-0-367-64687-5 (hbk)
ISBN: 978-0-367-65351-4 (pbk)
ISBN: 978-1-003-12906-6 (ebk)

DOI: 10.1201/9781003129066

Typeset in Times
by Deanta Global Publishing Services, Chennai, India

Contents

Preface

It has been almost two decades since Technomic Publishing Company produced the first edition of the *Snack Foods Processing* by my former professors Drs. Edmund Lusas and Lloyd W. Rooney. Unfortunately, both professors recently passed away and the much-needed and updated second edition became an orphan. It is for me a privilege to continue their legacy through the production and edition of this comprehensive book.

This book was conceptualized to thoroughly cover practical and scientific aspects related to the chemistry, technology, processing, functionality, quality control, analysis, and nutrition and health implications of the wide array of snacks derived from the major four groups of foods that sustain humankind (grains, fruits/vegetables, milk, and meat/poultry/seafood).

During the past 20 years, it has been massive improvements and advancements in the snack food industries that generate close to 520 billion dollars of annual sales. This new edition is based on the spirit of the original but with several new topics influencing food product development today: innovation, new emerging technologies, and nutritious and health-promoting snacks. The new volume is the result of a need to cover the intriguing world of snack food manufacturing in a deep format. The snack food industry has developed from a craft to technological industries around the globe. The mechanization of the food industry and the development of raw materials suited for these products have made it possible to produce a wide array of high-quality snacks with a high processing efficiency. Snack foods are commonly manufactured from low-cost raw materials such as cereal dry-milled fractions, starches, and oils, and the finished products are usually sold at a very high profit. The changes in trends in food consumption have always favored the snack food sector. For many years in developed countries, annual snack sales had grown more than the population increase, mainly due to convenience, long shelf life, and recently even through the COVID-19 pandemic. Nowadays, the consumer demands novel snacks with better nutritional attributes and processors have to innovate to keep or expand their markets. In addition, nutraceutical or functional additives are starting to form part of snack food developments. The industry has to adapt to these new developments and develop large-scale processes that increase manufacturing efficiency and reduce labor costs and risk of contamination, with pathogenic bacteria, mycotoxins, and toxicological compounds generated during deep-fat frying operations.

This book provides a technical review emphasizing flavored salty snacks commonly used as finger foods, including popcorn, wheat-based snacks (crispbreads, pretzels, crackers), lime-cooked maize products (tortilla chips and corn chips), extruded items (expanded and half products or pellets), potato chips, peanuts, almonds, tree nuts, and snacks from fruits/vegetables, milk, animal, and marine sources. The introductory chapter deals with the different categories of snacks and the statistics of volume, sales, and *per capita* consumption of these items. The following four chapters comprise the main or major unit operations used to process snacks: thermal processes, frying, seasoning, and packaging, including the novel and emerging technology of 3D printing. These sections accentuate emerging technologies and engineering and design aspects. Then, the book covers the most relevant raw materials used in snack operations: cereal-based refined grits, starches, and flours, followed by chapters for oils, seasoning formulations, and packaging materials. The core of the book is comprised of ten chapters, which describe the production and manufacturing of grain-based snacks including popcorn; wheat-based; alkaline or lime-cooked (corn and tortilla chips); extruded products; other grains such as legumes, pulses, oilseeds, and pseudocereals; and snacks derived from potatoes, almonds, tree nuts, fruits/vegetables, milk, and animal/poultry/seafood sources. Each of these manufacturing chapters includes flowcharts of the processes, relevant information in tables, and recent innovations and trends. The book is completed with chapters related to quality assurance and control procedures used by the various segments of the snack food industries and two chapters related to the nutritional and nutraceutical and health-promoting properties of all classes of snacks. At the end of each chapter, a set of updated references are included to provide key scientific material useful to get deeper into specific topics and expand knowledge.

This book was conceptualized and written to guide scientists, technologists, food engineers, technical salespeople, students, nutritionists, and even health practitioners interested in how the different classes of snacks are made and their relevant role in human nutrition and well-being.

The editor acknowledges the time, effort, and kind contributions of all authors and the Editorial Project Manager Stephen Zollo and Editorial Assistant Laura Piedrahita who dedicated many hours to this endeavor during the past two years. I especially recognize their tireless efforts through the completion of this volume.

Lastly, I certainly hope that this book will positively affect scientists, students, and food developers to upgrade snack food quality and health for the nearly eight billion people who presently inhabit the globe.

About the Editor

Tecnológico
de Monterrey

Sergio O. Serna-Saldivar, PhD
Professor and Head of CIDPRO
Research Areas

- Processing and biotechnology of cereal grains and legumes
- Nutraceutical properties of foods
- Production and isolation of novel food vegetable proteins

email: sserna@tec.mx

Sergio O. Serna-Saldivar is full professor of the School of Engineering and Sciences and head of CIDPRO at Tecnológico de Monterrey. Prior to this, he was a research scientist at the Soil & Crop Sci. Dept. at Texas A&M, consultant for EMBRAPA at Río de Janeiro, Brazil, and associate professor for the University of Sonora where he was recently awarded a *Doctor Honoris Causa*. He is currently the research chair leader regarding nutraceuticals associated with cereal and other grains. He has been a member of the American Association of Cereal Chemists for over 30 years and the Institute of Food Technologists and has acted as associate editor for the journals of *Cereal Chemistry* and *Cereal Science*. He was a member of the AACC International Board of Directors. He received his BS in Animal Science/Agricultural Engineering from ITESM and his MSc and PhD degrees in Scientific Nutrition and Food Sci. & Tech. from Texas A&M. He has published 12 books, 52 chapters, 255 refereed journal articles, and 16 encyclopedia articles and is the inventor of ten registered patents and four patent applications and codeveloper of the wheat variety TAM-202. His scientific and book publications have more than 10,000 citations with an H factor of 50 according to google scholar. He has more than 330 presentations in international and national conferences and symposiums. He has directed 70 MSc and 20 PhD students. His research interests focus on biotechnology of cereal grains, legume seeds, snack foods, mycotoxins, nutraceutical properties of grains and indigenous Mexican foods, and vegetable proteins. He belongs to the maximum category of the Mexican National Research System and the Mexican Academy of Sciences. In addition, he was granted the "Luis Elizondo" award in Agricultural and Food Industries, the 2004 AACC Excellence in Teaching award, the *Yum Kax* 2008 award for his contributions to the science and technology of nixtamalized products, and the AgroBio México 2012 national award in agrobiotechnology and six times the Teaching and Research Award at Tecnologico de Monterrey.

List of Contributors

Cristina Chuck-Hernandez
Tecnológico de Monterrey
Escuela de Ingeniería y Ciencias
Monterrey, México

Enrique Cuan-Urquizo
Tecnológico de Monterrey
Escuela de Ingeniería y Ciencias
Querétaro, México

Zamantha Escobedo-Avellaneda
Tecnológico de Monterrey
Escuela de Ingeniería y Ciencias
Monterrey, México

Johanan Espinosa-Ramírez
Tecnológico de Monterrey
Escuela de Ingeniería y Ciencias
Monterrey, México

Luis Eduardo García-Amezquita
Tecnológico de Monterrey
Escuela de Ingeniería y Ciencias
Zapopan, México

Tomás García-Cayuela
Tecnológico de Monterrey
Escuela de Ingeniería y Ciencias
Zapopan, México

Erick Heredia-Olea
Tecnológico de Monterrey
Escuela de Ingeniería y Ciencias
Monterrey, México

Carmen Hernandez-Brenes
Tecnológico de Monterrey
Escuela de Ingeniería y Ciencias
Monterrey, México

Daniel A. Jacobo-Velázquez
Tecnológico de Monterrey
Escuela de Ingeniería y Ciencias
Zapopan, México

Emilio Méndez-Merino
Sigma Alimentos
García, México

Esther Perez-Carrillo
Tecnológico de Monterrey
Escuela de Ingeniería y Ciencias
Monterrey, México

Ximena Quintero Fuentes
Private practice
Plano, Texas

Perla A. Ramos-Parra
Tecnológico de Monterrey
Escuela de Ingeniería y Ciencias
Monterrey, México

Sergio O. Serna-Hernandez
Graduate Program in Biotechnology
Tecnológico de Monterrey
Escuela de Ingeniería y Ciencias
Monterrey, México

Maria Soledad Tapia
Instituto de Ciencia y Tecnología de Alimentos
Caracas, Venezuela

Viridiana Tejada-Ortigoza
Tecnológico de Monterrey
Escuela de Ingeniería y Ciencias
Querétaro, México

Jorge Welti-Chanes
Tecnológico de Monterrey
Escuela de Ingeniería y Ciencias
Monterrey, México

1 Overview and State-of-the-Art of the Snack Food Industry

Sergio O. Serna-Saldivar

CONTENTS

1.1 INTRODUCTION

Snacks are defined as foods served or taken informally in small amounts between meals and are further described as light, casual, or hurried meals. Most snacks are ready-to-eat (RTE) items high in oil and generally flavored with salt or salt-based flavorings although many sweet fruit/vegetable-based snacks are growing in relevance throughout the globe. It is hard to establish the borderline between snacks, collations, and confectionary products, so in this book we will primarily stress the production and features of shelf-stable savory snacks containing low moisture. The major

DOI: 10.1201/9781003129066-1

distinction of snacks is that they are shelf-stable products at room temperature because most are subjected to dehydration by frying, toasting, baking, or salting. The packaging of these products is aimed to preserve freshness and avoid moisture uptake and oil rancidity. From the microbial viewpoint, most snacks do not pose health risks because pathogenic bacteria and molds require higher water activities (A_w) to prevail and grow. Most snack processors focus their technologies to enhance the shelf-life of their products from the rancidity or fat oxidation and texture standpoints. These critical parameters are mainly controlled by the selection of the proper frying oils, the frying process, the use of artificial or natural additives, and especially the packaging technology, which in most instances employ materials that exclude moisture, oxygen, and UV light to protect product crispiness and rancidity.

Today, consumers are seeking snacks that are aligned with their individual diet preferences, whether it's vegan, vegetarian, keto, or gluten free. The clean-label movement has created an array of products manufactured with clean ingredients and additives, and therefore more snack mixes or blends based on dry or intermediate moisture fruits, grains, legumes, and tree nuts are being developed. The other positive trend of the industry is the snacking of high protein and low-carb products mainly produced from dry and intermediate moisture meats and marine species.

Overall, the current trend of the snack industry is the development of nutritious snacks with lower amounts of fat and sodium, higher amounts of proteins and dietary fiber, with GRAS additives (green or clean label), and a better profile of fatty acids especially in terms of the ratio of unsaturated:saturated fatty acids.

The objective of this chapter is to introduce the reader to the scope of the snack industry including the historical preview of major snacks, statistics of volume and sales, and the importance, advancements, innovations, future trends, and constraints of the industry.

1.2 CLASSIFICATION OF SNACK FOODS

Snacks have always been an important part of life, and these foods represent a relevant segment of the food industry globally especially in developed countries. The snack food market is continually evolving and adapting to the new consumers' trends. During this century, the industry has gradually adapted to fulfill the demands of millennials and recently to forced patterns due to the COVID-19 pandemic. Snacking is increasing from factors such as increases in one-person households, a higher proportion of working parents and more school-age children obtaining their own meals, a highly mobile population, and the availability of RTE snacks in convenience stores and vending machines (Riaz 2004). From the processing viewpoint, the industry is continuously innovating and evolving to generate new foods packaged in attractive and functional materials to preserve their freshness for longer periods of time.

1.2.1 Current Commercial Products

Snacks are broadly classified according to their food source and to their manufacturing process (Tables 1.1 and 1.2). Practically all food groups have important examples of snacks. Undoubtedly, the most relevant and sold worldwide are those derived from grains, mainly cereals, followed by tubers. These products are fabricated taking advantage of the functionality of the gelatinized starch. Many other types of snacks are sourced from tree nuts, fruit/vegetables, milk, different meats, and marine products. Independently of the food source, practically all snacks are shelf stable at room temperature because of their low moisture content and water activity (A_w). The only exception is snacks derived from milk such as yogurt, cheeses, mousse, and flans, which commonly require refrigeration. Table 1.1 summarizes the major snack types according to the food source, characteristics, and general nutritional attributes. From the processing standpoint, most commercial snacks fit into the categories of first, second, or third generations (Table 1.2). First-generation snacks are the easiest to manufacture and consist of natural products such as roasted whole grains, peanuts,

TABLE 1.1
Classification of Snack Foods According to Their Food Source and Main Characteristics

Snack Food Source	Main Types of Snacks	Main Characteristics	Major Nutritional Attributes and Constraints
Grain Base			
Cereals	*Corn*: popcorn, parched corn, corn chips, tortilla chips, extruded puffs. *Wheat, oats, and/or rye*: toasted bread or crisps, crackers, pita bread chips, hard pretzels.	Shelf-stable products that contain low moisture and high-fat content commonly produced by conventional, lime, or extrusion cooking processes followed by baking, frying, and seasoning with salt-based flavorings.	Since most products are produced from refined milled fractions and flavored with fat they are considered high caloric and fat products. These snacks are rich in starch and oil and contain relatively low protein, dietary fiber, and micronutrients. The protein is medium to low quality because it lacks lysine and other essential amino acids. Wheat, rye, barley, and even oat products need to be declared on the food label as potential allergens for celiac or gluten intolerant people.
Legumes/oilseeds	Roasted peanuts, Japanese peanuts, soybeans, sesame, pumpkin, and poppy seeds.	Shelf-stable whole products that contain low moisture that in most instances are roasted or toasted and occasionally fried.	Considered nutritious snacks because of their high protein and dietary fiber contents and lipids rich in polyunsaturated fatty acids. Additionally, these grains are rich in phytochemicals that favor human health. Among these products, peanuts and soybeans should be declared as potential allergens in food labels.
Pseudocereals	Puffed and dried amaranth, chia, quinoa, and buckwheat.	Shelf-stable whole products that contain low moisture that in most instances are toasted or puffed for use in the preparation of granola bars.	Considered nutritious snacks because of their high protein and dietary fiber contents. Additionally, these grains are rich in relevant phytochemicals (antioxidants) that prevent oxidative stress and chronic diseases.
Tubers, Fruits, and Vegetable-Based			
Tubers	Regular and fabricated potato chips.	Shelf-stable savory snacks that contain low moisture and high lipid content that are commonly produced by deep-fat frying and salting or seasoning with salt-based flavorings.	Since most products are produced from sliced potato or potato flour and deep fat fried they are considered highly caloric due to their high-fat content. These snacks are rich in starch and oil and contain very low levels of protein, dietary fiber, and micronutrients. Besides, these snacks can develop significant levels of acrylamide, which is considered an important toxic compound.
Tree nuts	Toasted or roasted almonds, macadamia, pecans, pistachios, cashews, walnuts, and pine nuts.	Shelf-stable whole products that sustain minimum process because they are commonly only roasted. These products are usually sold in mixes with fruits or other nuts or employed for the production of snack bars.	Among snacks, tree nuts are considered one of the most nutritious because of their natural high fat, protein, and dietary fiber contents. Besides, their lipids are rich in polyunsaturated fatty acids, phospholipids, phytosterols and vitamin E, or tocopherols and other phytochemicals that favor human health mainly through the prevention of cardiovascular diseases. The main concern with almonds and tree nuts is their potential allergenicity.

(*Continued*)

TABLE 1.1 (CONTINUED)
Classification of Snack Foods According to Their Food Source and Main Characteristics

Snack Food Source	Main Types of Snacks	Main Characteristics	Major Nutritional Attributes and Constraints
Fruits	Dehydrated fruits, osmotic dehydrated foods, fruit bars.	Minimum processed snacks are usually obtained after dehydration with conventional means or novel technologies (osmotic, vacuum, or freeze-drying). The products have low Aw because of the moisture removal and concentration of natural sugars associated with the fruit matrix.	Snacks from fruits are rich in soluble sugars, dietary fiber, and an array of phytochemicals that favor human health. In addition, they are practically free of lipids and therefore are not as caloric dense as other snacks that are commonly fried.
Vegetables	Dehydrated vegetables, fried vegetables,	Minimum processed snacks are usually obtained after dehydration with conventional or novel technologies and alternatively by deep fat frying. The products have low Aw because of the moisture removal during drying or frying.	Snacks from vegetables are rich in dietary fiber, phytochemicals that favor human health, and are low in calories.
Milk-Based			
Dairy	Yogurts, cheeses, puddings, mousses, flans.	Snacks from dairy products generally require refrigeration and are produced after the coagulation of casein with enzymes or acidity created by fermenting lactic acid bacteria.	Dairy-based snacks are rich in good quality protein, soluble sugars, milk fat, and highly bioavailable calcium. However, the major constraints of these products are the potential hazards in terms of pathogenic bacteria and lactose intolerance.
Animal and Marine-Based			
Meats	Pork rinds, jerkies, dry meats, dry-cured hams, sausages.	Shelf-stable snacks from animal products are generally produced after solar or artificial dehydration, salting, or dry curing. The popular pork rinds are obtained after deep-fat frying of pork skin.	Meat-based snacks are rich in good quality protein and practically devoid of sugars and carbohydrates so they are demanded by people following the keto diets. In addition, they provide highly bioavailable iron and zinc. The major constraint of these products is their high susceptibility to pathogenic bacteria and their high contents of saturated fat and cholesterol.
Marine	Dehydrated fish, dry crustaceans.	Shelf-stable snacks from fish and crustaceans are generally produced after solar or artificial dehydration, salting, and pickling.	Marine-based snacks are rich in good quality protein and practically devoid of sugars and carbohydrates so they are demanded by people following the keto diets. In addition, they provide highly bioavailable iron and zinc. The major constrain of these products is their susceptibility to pathogenic bacteria contamination and for crustacean products like dehydrated prawns and shrimp their high cholesterol content.

TABLE 1.2
Classification of Snack Foods According to Conventional and Emerging Manufacturing Procedures

Category	Process	Products
Conventional Technologies		
First generation	Minimum processed snacks obtained from grains and fruits from natural products.	Popcorn, peanuts, other dehydrated whole grains, almonds, tree nuts, and dehydrated fruits and vegetables. Regular fried potato chips also belong to this category.
Second generation	Most popular category of snack foods. Snacks are obtained after simple shaping of doughs or dry milled fractions forced through a extruder (direct expanded products). Raw materials include various types of flours and grits.	Includes extruded corn chips, tortilla chips, and a wide array of direct extruded items such as corn puffs and curls. Toasted crisps, pita chips, hard pretzels, and fabricated potato chips or sticks are also included in this category.
Third generation	Most elaborated products also called half products or pellets generally produced by a two-step extrusion process (cooking and forming extruders). The formulation of these snacks uses a wide array of ingredients. These products required additional processes to reach the consumer such as baking and deep-fat frying. Most pellets are commonly processed via frying.	Imitation pork rinds, wheel-shaped puffs, or sticks obtained after thermal treatment or deep-fat frying of half products or pellets.
Novel Emerging Technologies		
Coextruded snacks	In this relatively new process two different types of raw materials are extruded from one die. These raw materials can come from two extruders or one extruder equipped with one pump. Coextruded snacks have two recognizable parts: an external cover or crust and an internal filling.	Cereal-based tubes filled with cheese or other fillings such as fruit pastes. Many commercial extruded cookies or crackers with salt-based or sweet fillings are produced with this technology.
Supercritical fluid extrusion	In this process, supercritical CO_2 is injected into the extruder barrel to create viscosity changes which affect the properties of second generation or puffed snacks especially in terms of crumb texture, radial expansion, color and flavor.	Extrusion with liquid CO_2 presents the main advantage of minimizing the loss of heat-labile nutrients that are usually lost during conventional thermoplastic extrusion processes.
3-D printed snacks	The 3-D printing process involves building 3-D forms from a drawing sample designed by a computer program. The potential snacks can be printed with special extruders or food-grade syringes that deposit the product through a nozzle. The printing is controlled by computers. Among the various 3-D printing technologies the most promising and practical is the extrusion-based printing that forces raw materials in the form of pastes through the specially shaped nozzle of the printer. Potential novel snacks that can be manufactured with this technology are dough-based bakery items, meat and fruit pastes, and cheeses.	The 3-D printing technology allows producers to tailor-make personalized snacks in terms of nutrition, shape, and organoleptic properties. The major constraints of this technology are production cost and large-scale production.

nuts, dried fruits and vegetables, popcorn, and regular potato chips. Most snacks consumed today belong to the category of second generation. These include simple formed products mainly obtained after direct extrusion (i.e., corn chips, expanded puffs, balls, and curls) or by sheeting/forming and baking/frying such as fabricated potato chips, tortilla chips, granola bars, pita chips, crisps/toasted breads, and hard pretzels. The most elaborated category in terms of ingredients and processing is the third-generation snacks commonly manufactured after two sequential extrusion steps known as cooking and forming that yield first pregelatinized dense pellets or half products which are further processed with deep-fat frying, baking, or microwave cooking to reach the consumer.

1.2.2 Emerging Products and Technologies

Novel processes that have been recently employed to produce innovative snacks are coextrusion, CO_2 extrusion, and 3D printing. Coextruded snacks are produced in a distinct type of extruder equipped with a special die that allows the coextrusion of two contrasting types of materials: an outer covering generally produced from starchy feedstocks and an inner filling consisting of salt-based flavorings or sweets.

Supercritical fluid extrusion, also known as SCFX, was patented by Rizvi and Mulvaney (1992) in the early 1990s. In this process, supercritical CO_2 is injected into the extruder barrel to create viscosity changes, which affect the properties of extrudates especially in terms of crumb texture, radial expansion, color, and flavor. Extrusion with liquid CO_2 presents the main advantage of coextruding with a nontoxic and relatively inexpensive supercritical carbon dioxide. SCFX has been applied for the manufacturing of novel puffed snacks at lower temperatures with improved texture, color, and taste minimizing the loss of heat-labile nutrients that are usually lost during conventional thermoplastic extrusion (Rizvi et al. 1995, Sharif et al. 2014, Panak-Balentić et al. 2017).

The novel and disruptive technology of 3D printing has been successfully used for the manufacturing of innovative snacks. Most frequently, different food products are printed with special extruders or food-grade syringes that deposit the product through a nozzle. The most advanced 3D printing technology allows producers to tailor-make personalized snacks in terms of shape, color, texture, and flavor on computers or related electronic devices. Among the 3D printing technologies, the most promising and practical is the extrusion-based printing that forces raw materials in the form of pastes through the specially shaped nozzle of the 3D printer. Potential novel snacks foods that can be developed using this technology are dough-based bakery items, meat pastes, and cheeses.

1.3 HISTORICAL PREVIEW AND GENERAL PROPERTIES OF SNACKS

Snack foods have their origin thousands of years ago likely since the men conceived agriculture, became sedentary, and developed simple processes to prolong the shelf-life of foods. The most antique snacks were grains, dehydrated meats, and marine products. However, the modern snacks available today became popular when they were first packaged for merchandising during the industrial revolution of the 19th century. It was until the first decades of the 20th century when that more snacks became accessible to the general public in Europe and the United States. The boom of the snack industry occurred after the Second World War when the postwar economic surge occurred and many novel snack types were launched into the market as convenient foods.

1.3.1 Popcorn

Popcorn (*Zea mays* L) is one of the oldest snacks and documented as a relevant source of nutrients to ancient American cultures. The special type of maize known as popcorn is regarded as one of the first landraces by many archeologists. It is quite possible that primitive Mesoamericans first discovered the usefulness of popcorn when this flinty wild kernel was unintentionally exposed to heat. To people with no grinding tools other than their own teeth realized that an unpromising food became tender, tasty, and attractive. The indigenous Mesoamericans knew popcorn before the Spaniards conquered America. The great Aztec culture honored one of the gods by scattering parched or popped corn, called *momochitl*, to emulate white flowers (Weatherwax 1954). The other great pre-hispanic Inca culture developed special pottery utensils for popping corn and remains of popped kernels have been often found in Peruvian graves. The oldest positively identified corn poppers date from the Mohican Indian culture of about 300 AD along the northeast coast of Peru. Christopher Columbus documented that when he arrived in San Salvador he observed natives wearing popcorn decorations. A Spaniard document dated 1650 quoted that the Mesoamericans "They toast a certain type of corn until it bursts. They call it pisancalla and use it as a confection." Popcorn also forms part of US history because it was present on the first Thanksgiving day when Quadequina, the brother of the Wapanong chief Massasoit, offered a deerskin bag full of popcorn as a gift to the celebration (Serna-Saldivar 2008a,b).

Historically, it is known that in India and neighboring countries other special types of popped kernels have been traditionally consumed. In this sense, a special type of hard endosperm sorghum (*Sorghum bicolor* L Moench) known as *Shallu*, and several types of millets have been traditionally used by Asian and African cultures (Malleshi and Desikachar 1985, Mishra et al. 2014).

Today, popcorn, especially the one suited for microwave popping (Chapter 9), continues to be one of the most popular snacks all over the globe with an astonishing market of about 10 billion USD in 2020.

1.3.2 Wheat-Based Snacks

There are many indigenous foods derived from ancestral wheat that we still widely consume today. The first sedentary cultures learn to stone-mill wheat and related cereals into flours, which upon hydration and mixing generated doughs that were adequate substrate for yeast and lactic acid bacteria. The unique combination of viscoelastic properties of wheat flour with the fermenting microorganisms resulted in the production of an array of traditional staples still produced by mankind today. The toasting of these primitive flat and leavened breads originated shelf-stable products that evolved into many wheat-based snacks. The major categories of wheat-based snacks thoroughly covered in Chapter 10 are crackers or hard biscuits, granolas, pretzels, pita chips, and toasted or crispbreads.

1.3.2.1 Crackers or Hard or Salt Biscuits

The preparation of biscuits or crackers dates to the early history of mankind and develop into an industry that today ranks among the leaders in the USA snack market. The first prototypes of crackers originated thousands of years ago when men stone-milled their grains and produced flours that were mixed with water and baked directly on fire or ashes to form small unleavened cakes. Remains of these primitive baked products have been found in Swiss dwellings dated more than 8,000 years old. The early Egyptian culture depicted illustrations of the baking operation and cookies served at the royal table in ancient tombs such as the Ramses III dating back to the year 1190 BC (Kulp 1994, Misra and Tiwari 2014).

According to Manley (2000) the word biscuit, which is the synonym for crackers and cookies, is derived from the Latin word *panis biscotus* meaning twice-cooked bread and refers to dried products originally made for mariners of the Middle Ages. The word biscuit is defined in the dictionary of Dr Samuel Johnson published in 1755 as "a kind of hard dry bread, made to be carried to sea," and a secondary one of "a composition of fine flour, almonds and sugar, made by the confectioners." William Shakespeare also refers to ships biscuits in his play *As You Like It* written more than four centuries ago. The early biscuits manufactured for long voyages were produced from just flour, salt, and water. In America, they were known as pilot biscuits or hardtack. Due to their hardness and low moisture content, biscuits were very firm to eat and normally had to be soaked in a beverage or soup to make them appetizing.

The category of crackers or salty biscuits is subdivided into cream, soda, and savory crackers (Chapter 10). Cream cracker biscuits were first introduced in about 1885 by the Irish firm Jacobs. Since then they have maintained a significant place in the sales of biscuits in Britain and have also spread to many other countries around the globe. They should not be confused with soda crackers, which are another traditional type of cracker. The soda cracker or saltines, derived from the cream cracker, is of American origin. The first record of this type of cracker is in 1840. The savory biscuits also known as sandwich crackers are more modern and are generally flavored with cheese-based fillings (Manley 2000).

1.3.2.2 Pretzels

A pretzel is a wheat-based snack food with 800 years of history. According to Terry Groff (2001), it was first produced by a monk in the 12th century as a reward to religious children. The typical pretzel configuration clearly shows a cross in the middle representing the crossed arms of a Christian in prayer. Pretzels are clearly represented on the coat of arms of the bakers of Vienna because these bakers in 1529 heard the digging of a tunnel that the Turks were building to conquer the city. Therefore, the pretzel bakers saved the city. Since then pretzels are the symbol of bakers throughout Europe especially in Germany, Austria, and Switzerland. According to Seetharaman (2014) hard pretzels were introduced to the United States in the 1860s at Lititz, Pennsylvania, and have since become a very popular and healthful snack item.

1.3.2.3 Granolas

Toasted granola pieces were first industrially manufactured in the late 19th century. The invention of the Granola is attributed to Dr James Caleb Jackson in the state of New York in 1863. The original granola was produced from Graham flour (coarsely ground whole wheat flour) and viewed as a healthy food. From the original granola, many other similar products developed mainly consisting of toasted whole grains, wheat-based bakery and toasted pieces, dehydrated fruits and nuts. Granolas were primarily considered as RTE breakfast cereals because they were generally consumed with milk. However, granola bars are viewed as one of the most healthy snacks available in the current market.

There is controversy regarding the invention of the first granola bars. Some credit the innovation to Stanley Mason and others to Herrick Kimball who also claimed the idea for this product in the mid-1970s. The key for the production of flattened bars is the addition of honey or other sweetened syrups, which are blended with many different toasted grains, nuts, and dehydrated fruits. Since then the granola bar market has increased to a multibillion-dollar industry with numerous different types. In the United States, Nature Valley followed by Nature Valley Sweet & Salty Nut and Quaker Chewy accounted for 23.4, 15.6, and 14.4% of the market share in 2017 (https://truenorthgranola.com/blog/what-is-granola-and-granola-history/).

1.3.3 Snack Bars

The category of snack bars is one of the fastest growing within the snack sector because customers, especially millennials and the health conscious perceive food bars as highly nutritional and convenient. The consumption of snack bars is influenced by age, gender, nutritional knowledge of the consumers, and convenience. As a result, the industry has developed numerous options with different nutritional features and numerous flavors and textures. Snack bars are widely available in multiple different flavors and textures depending on the types of cereals grains, other seeds (pulses and pseudocereals), proteins (vegetable and dairy), dehydrated fruits and vegetables, chocolate, and dairy products. The most important common characteristic is that all snack bars are shelf stable at room temperature because of their relatively low water activity (A_w) and in many instances fortified with micronutrients (vitamins and minerals) and nutraceuticals like fiber, antioxidants, and healthy fats such as omega-3 and DHA.

As a result of the great interest in snack bars, the industry is currently producing cereal-based, energy, protein, 40-30-30, gluten-free, fruit-mixed, nut-mixed, flavored with chocolate and other coatings with different nutritional profiles and functional properties (Estevez et al. 1995, Gonzales and Draganchuk 2003, Sun-Waterhouse et al. 2010, Aramouni and Abu-Ghoush 2011, Hogan et al. 2012, da Silva et al. 2014, Ferreira et al. 2015, Da Silva et al. 2016, Constantin and Istrati 2018, Munir et al. 2018, Ramirez-Jimenez et al. 2018). In recent years, the demand for high-protein snack bars has specially grown with some products containing up to 35% protein and bars are more commonly formulated with micronutrients (vitamins and minerals) and nutraceuticals like insoluble and soluble fiber sources, natural antioxidants, and healthy fats such as omega-3 and DHA (Aramouni and Abu-Ghoush 2011, Constantin and Istrati 2018). Table 1.3 depicts the main categories, features, and nutritional value of snack bars.

1.3.4 Alkaline-Cooked Corn Products (Corn and Tortilla Chips)

The alkaline or lime-cooked corn products have positioned themselves as the second most relevant savory snack food category. These unique products evolved from the nixtamalization process widely employed by the early Mesoamerican cultures about 10,000 years ago when corn (*Zea mays*) became a cultivated plant and the chief responsible for the initiation of formal agriculture in Mesoamerica. According to Garcia-Lara and Serna-Saldivar (2019), the origin of this crop occurred during the invention of the new world agriculture. The oldest paleoetnobotanic evidence of its domestication is in an archeological site named "Nac Neish" located northeast of Mexico. It consists of three primitive caves presenting data between 6,000 and 20,000 years BC (Serna-Saldivar 2015, Garcia-Lara and Serna-Saldivar 2019). In one of the caves it was found a tomb that contained human remains covered with reed mats (*petlatl*) and an offering consisting of a basket with corncobs. The American Indians created corn from a tiny two-rowed ear of teosintle of about 3 cm long into the first tiny corn ear with its four ranks of paired female spikelets. This transformation took perhaps only 100–200 years. Thereafter, two different corns classified as pyramidal or cylindrical originated the modern corns widely planted throughout the globe. The oldest remains of 8 and 12 rowed corns from Mesoamerica are dated 7,200 and 3,000 years old, respectively. These relevant developments originated different races varying in kernel morphology, color, endosperm texture, and other important features that make these types highly useful for snack processors (white, yellow, blue, and purple corns for corn and tortilla chips, cacahuacintle for parched corn, popcorn, etc.). More than 25 races have been used to produce high-yielding varieties and hybrids that make corn the most productive crop in the planet with yearly production exceeding 1.1 billion tons (FAO 2021).

After the discovery of the Americas, the crop eventually reached and adapted to the Iberia Peninsula and from there it was distributed to other parts of the globe. The Portuguese conquistadors and merchants were more responsible than the Spaniards for disseminating corn around the rest of Europe and Africa. However, the Spaniards catalyzed the Asian connection because they traded goods from the Pacific ports of Mexico to the Philippines. This occurred as early as 1565 but maize became important in Asia after the 17th century.

The ancient Mesoamerican cultures not only inherited regular corn and popcorn to the whole world but also the nixtamalization process that is still used to produce corn and tortilla chips. Historically, it is known that the Aztecs produced *totopochtli*, an ancient prototype of free-fat tortilla chips, by toasting tortillas on a hot griddle known as *comal*. This shelf-stable product was carried as a snack by travelers or used as a substance during wars (Serna-Saldivar 2008b, Serna-Saldivar 2015).

The dramatic increase in sales of lime-cooked snacks occurred during the past 50 years is attributed to Latin Americans inhabiting the southwestern states of the United States. They first introduced fresh and toasted tortillas (tostadas) in the market and then the industry evolved by designing high output industrial lines aimed toward the production of deep-fat fried corn and tortilla chips.

The first use of industrial equipment to produce corn chips occurred in San Antonio Texas in 1932. Elmer Doolin, founder of the Frito Company, bought rights from a Mexican who wanted to

TABLE 1.3
Characteristics and Nutritional Value of the Different Categories of Snack Bars

Snack Bar Type	Formulation	Nutritional Value
Cereal-based	These bars are the most popular and the base for blending with other ingredients like proteins, other grains, flavorings/coverings, dehydrated, or moisture intermediate fruits and nuts. The main cereals used are flaked oats, wheat, expanded rice, and maize. The preferred raw materials are whole grains, refined cereal flours, starches, and puffed grains, which are normally blended with honey, syrups, caramel, and in many instances covered with chocolate or vanilla-based flavorings.	Most bars are high in starch and energy and have relatively low levels of protein unless bars are fortified with protein sources from soybean, pulses, or dairy products. The use of whole grains or flours yields products with more dietary fiber, micronutrients, and phytochemicals that prevent oxidative stress.
Energy bars 40-30-30	These bars are balanced to fall into the 40-30-30 of total calories provided by carbohydrates, protein, and fat, respectively. The development of these bars is based on the principles of the Zone diet. These bars are higher in protein and fat and lower in total dietary fiber compared to the higher carbohydrate cereal-based energy bars. The key for the production of these bars is the intelligent combination of cereals, pulses, pseudocereals, and oils to balance the composition of macronutrients. The principles of production of bars are similar to energy or cereal-based snack bars, which after blending and forming or molding the shelf-stable product is baked, cooled, and packaged.	From the nutritional and dietary viewpoint, the 40-30-30 caloric ratio is recommended by many dieticians because the energy and protein are well balanced and create satiety. Therefore, these bars are known to provide energy, good quality protein, and micronutrients that favor human health. The bars low in soluble sugars promote a slow glycemic response.
Protein-enriched	Most of these bars are produced by commingling cereal meals or flours with other grains rich in protein or protein flours, concentrates, and isolates. The most popular protein sources are soybean (nuggets, texturized vegetable protein, defatted flour, concentrate, or isolates), other legumes (flours, concentrates, peanut butter), and dairy products (casein and whey protein concentrate). The combination of cereal proteins with legumes or dairy proteins improves both the protein concentration and protein quality of finished products. In terms of processing, most high protein bars are produced by first mixing raw ingredients (cereal flours, puffed wheat, or rice), protein ingredients (soybean, pulses, whey protein concentrate, casein, nuts) with water, syrup, oils, chocolate chips, and minor ingredients like glycerine, hydrocolloids, emulsifiers, and flavorings. Then, bars are formed, molded, and baked. After cooling, some bars are coated to improve organoleptic properties. Within this category, there are organic and gluten-free products.	These bars may contain up to 35% protein, which is generally provided by the cereal base and the supplemented protein ingredients. When produced from whole grains the bars also supply important amounts of dietary fiber, vitamins, minerals, phytochemicals, and antioxidants.
Fruit-based	Fruit-based bars are increasing in popularity due to the positive image of dehydrated fruits. Generally, dehydrated fruit pastes or pieces with *Aw* below 0.8 are bending with honey, glucose, or fructose syrups and other ingredients like rice crisps, vegetable oil, glycerol, maltodextrins, and pectins. Products are usually obtained by forming or molding with special extruders and in some instances baked.	Fruit bars are high in sugar, energy, dietary fiber, micronutrients (vitamins and minerals) and phytochemicals including polyphenols and flavonoids Most fruit bars are low in protein.
Nut-based	Nut-based bars are also popular due to the positive image of roasted nuts obtained from peanuts (roasted pieces and peanut butter), soybeans, and different types of tree nuts. Generally, the nut pieces are blended with honey, caramel, glucose, or fructose syrups and refined or whole cereal flours, vegetable oil, glycerol, maltodextrins, and hydrocolloids. Products are usually obtained by forming or molding with special extruders and in most cases baked to partially dehydrate the product and improve the flavor profile. Within this category, there are vegan nut-based snack bars.	Nut-based snack bars are high in fat, protein, energy, dietary fiber, micronutrients (vitamins and minerals), and phytochemicals associated with whole grain flours and roasted nuts. Generally, the lipids contain a favorable polyunsaturated fatty acid composition that prevents cardiovascular diseases.

return to his homeland. Mr. Dooling started the business in his kitchen and soon sales boomed and within a year moved to Dallas. When Mr. Dooling passed away, he had transformed the 100 USD loan from his mother into an industry with yearly sales of more than 50 million USD (Serna-Saldivar 2015). I. J. Filler was granted a trademark for Corn Chips in Texas in the early 1930s. That is probably the reason why Mr. Dooling chose the name Fritos™. One of the major advances for the industrial production of corn chips was the development of a mechanical screw press developed by Mr. Doolin and the Lawson Brothers Machine Shop, which extruded the bulk of dough or masa into ribbons, which fell directly into the deep-fat fryer (Chapter 11). The machine consisted of a vertically positioned barrel and a mechanically driven piston that pushed the masa into the die located at the end of the barrel. When the piston retracted a new load of masa was put into the machine. This press was improved when the piston was driven by compressed air and a 90° elbow tube was fitted to the discharge end of the barrel. A cutter was also driven by a pneumatic assembly with an air regulator to control speed. These new extruders were capable of producing 200 kg/hr. Hydraulics machines widely used nowadays replaced the pneumatic systems. The advantages of these corn chip extruders were increased production and a more consistent and homogenous chip length. Frito-Lay/L. C. Miller Co. first built these extruders in the early 1960s (Snack Food Association 1987, Burtea 2001).

The most popular tortilla chips developed about 20 to 30 years later than corn chips. A. West and G. Chesquiere of the newly formed Frito-Lay Incorporation got the idea of producing tortilla chips after seeing Mexican-style tostadas and USA taco shells. The trade name Doritos™ was selected for these new snacks. The first triangular-shaped Doritos were manufactured at Alex Foods in Anaheim, California. The tests went so well that soon the products were not only produced in California but also in the southwestern United States. Doritos™ went to the national market in 1966. Few years later, sales of tortilla chips surpassed those of corn chips. J. C. Ford Company, Casa Herrera, Electra Food Machinery, Lawrence Equipment, Maddox Metal Works, and Heat and Control are the major suppliers of tortilla chip and frying equipment and automated lines. The larger production lines are capable of producing 1.5 tons of chips/hr (Serna-Saldivar 2008b).

Nixtamalized or lime-cooked snacks have experienced one of the most relevant growth rates in terms of volumes and sales during the past five decades and nowadays are manufactured practically in all developed countries of the globe (i.e., European countries, Russia, China, Australia, South Korea). In fact, the current market of tortillas chips has increased above the rate of potato chips so tortilla chips are expected to surpass the market and sales of potato chips within the next five years.

1.3.5 Extruded Snacks

Historically, it is well known that the extrusion technology was first developed to produce clay pipe, tile, and soap. In 1867, two Englishmen developed the first known twin extruder to process sausage but it was not until the mid-1930s that the single screw continuous pasta extruder was developed (Riaz 2000). The history of extruded snacks is also traced back to this decade. Corn chips were introduced in the early 1930s while puffed fried and baked products just after the Second World War. In the late 1920s, C. B. Mathews developed in Wisconsin the first prototype of an extruder that pulverized, macerated, and partially cooked animal feed with the aim of improving digestibility and reducing intestinal problems. He collaborated with H. W. Adams to improve the method, launched a company named Flakall Corporation, and issued a US patent in 1938. Ed Wilson, an operator from Flakall Corp, while cleaning the apparatus with corn noticed the formation of puffs. Instead of throwing the puffs, he asked his wife to deep-fat fry and salt the product. His neighbors liked the product especially when flavored with cheese and soon the product was locally produced and sold with the trade name *Korn Kurls*. A US patent was issued for the machine and process in 1939 and granted in 1942. In 1946, Harry Adams, owner of Adams Corporation, started to commercially produce the original fried type corn snacks. The key processing step was the extrusion. The extruder used consisted of a rotor, stator, and a screw that produced irregularly shaped collets ready for frying.

The fried collet was commonly seasoned with a salted-cheese oil slurry. C. Heigl, an engineer from Adams Corporation, developed an extruder designed to produce less dense extrudates and produce puffs suitable for drying or baking. The lighter and smoother collets were produced by a short barrel, high shear extruder equipped with a die to form the gelatinized corn extrudate into various configurations. Resulting puffs were cut to a predetermined length by a rotating knife. After drying or baking the collets were flavored with a cheese slurry similarly to the deep-fat fried curls (Burtea 2001, Serna-Saldivar 2008b). Later on, larger and more sophisticated extruders were designed and built by Maddox, Wenger, Buhler, and other companies. These extruders are very versatile because they can be arranged to produce second-generation expanded products or third-generation pellets (Chapter 12). A key and major advancement in the extrusion field was the manufacturing of twin extruders that improved versatility and product consistency and quality (Serna-Saldivar 2008b).

The 80-year-old extruded snacks segment is expected to witness considerable growth over the next five years driven by category diversification coupled with equipment and ingredient innovation, especially in developed countries. Frito-Lay the main snack food company of PepsiCo will continue to dominate sales of these snacks.

1.3.6 Snacks from Legumes and Pseudocereals

Legumes, pulses, and pseudocereals have been part of the diet of men since prehistoric times and involved with the development of agriculture. These grains provide many nutrients required by humans especially in terms of proteins, carbohydrates, dietary fiber, vitamins, and minerals. Basically, all legume seeds contain antinutritional factors (Chapter 20) that protect seeds against insects, animals, and other biotic elements. Therefore, men learn to process these grains in order to deactivate antinutritional factors like trypsin inhibitors, urease, and hemagglutinins. The simple toasting or cooking of whole seeds or meals deactivates these factors and greatly enhances their nutritional value. In terms of snacks derived from legumes and pseudocereals covered in Chapter 13, peanuts have been positioned as the most relevant and sold. Recently, the use of other legume seeds or meals like soybean nuts, soybean meals and proteins, chickpeas, and green/yellow peas are increasingly being supplemented with nutritional bars and other snacks.

1.3.6.1 Peanuts

Peanuts or groundnuts (*Arachis hypogaea*) originated in Brazil or Peru. Historically peanuts are known as early as 950 BC. Explorers and missionaries to Africa introduced them to Africa. From Africa they returned to the American Continent by slave traders in the early colonial days. In the United States, peanuts started to be demanded and consumed after the Civil War. The Italian native, Amedeo Obici, was the founder of the Planters Peanuts Company in the United States. He was born in 1876 near Venice, Italy, and when he was a teenager migrated to New York without speaking any English. He started his career as a bellhop and fruit stand vendor in Scraton, Pennsylvania, and later on moved to Wilkes-Barre, Pennsylvania, and opened a fruit stand and invested in a peanut roaster. In a few years, Obici using a horse and wagon successfully sold peanuts and named himself "The Peanut Specialist." In 1906, Obici established a partnership with Mario Peruzzi and developed his own method of blanching whole roasted peanuts, doing away with the troublesome hulls and skins. Planters was founded, and soon the company was recognized as the benchmark in terms of quality and trade name (www.ideafinder.com/history/inventions/). In the year 1913, Amedeo Obici moved to Suffolk, Virginia, known for its peanuts, and then opened the first mass-production plant in this location. By the year 1930, Planters Peanut stores were open across the United States.

Nowadays, the Planters Company owned by Kraft Heinz processes and sells dry-roasted, honey-roasted, salted, non-salted, and hot peanuts. In 1998, Planters introduced the innovative product known as "trail mix" that evolved into an array of product blends with other nuts, granola, dehydrated fruits, chocolate candies, and other shelf-stable products that are viewed as healthy and convenient.

Another important historical development associated with peanuts and popcorn is the snack still known as Cracker Jack™. It consists of a unique mixture of popcorn, peanuts, and molasses first produced at the end of the 19th century by F. W. Rueckheim. This product was first commercialized at the World's Columbian Exposition at Chicago's First World Fair. Louis Rueckhein, brother and partner of the inventor, developed the process for keeping the molasses-covered popcorn morsels from sticking to each other (Serna-Saldivar 2008a).

1.3.7 Potato Chips

Potato chips are still the most consumed non-extruded snack in North America. Following a growing demand for potato chips across the region, a sizable number of market players are focusing on entering the tapping of the North American snacks market.

The potato (*Solanum tuberosum* L.) is indigenous to the Americas specifically the Andean Mountains of Peru, Ecuador, Bolivia, and Chile located in the western part of South America. Potatoes have been the staple of the inhabitants of these countries for more than 8,000 years (Gould 2001). Historically, the utilization of potato is traced to at least AD 200 where it was cultivated by the ancient Inca culture on the highlands of the Andes. The tubers were dried by allowing them to freeze overnight and thaw during the day. This cycle was repeated several times until lowering the moisture content enough to get a shelf-stable product. According to Gould (2001) and Talburt (1987), the Andean Indians mainly processed potatoes into dehydrated and convenient products named *Chuño* and *Tunta* used for the production of many indigenous foods. When the Spanish conquistadors arrived in the Americas, the wild potato crop was widely distributed throughout the continent. The Inca Indians cultivated potatoes intensively and were responsible for selecting a wide array of types varying in color (purple, pink, blue, yellow) and shape (round, oblong, cylindrical). Potatoes reached the Old World via ships that carried goods and people back to Europe. There are records that potatoes reached Sevilla, Spain, as early as 1573. From Spain, potatoes were carried to Italy and then disseminated throughout the rest of the old continent. Potatoes reached England at the end of the 16th century and soon were adopted as a staple by the Irish people. By the 18th and 19th centuries, the potato crop became the staple food in Europe, especially in the eastern countries. Interestingly, the North Americans introduced the crop via Irish settlers approximately in 1621 instead of obtaining the crop earlier from natives inhabiting the areas of New Mexico and Arizona. The Irish settlers of New Hampshire promoted the cultivation and acceptance of the potato crop.

In order to prolong the potato shelf-life and decrease its bulk the first documented production of potato meal was in 1786 whereas the first potato starch plant was created approximately 45 years later (Talburt 1987). The first potato chips were manufactured in 1853 by a New York chef named George Crum who was trying to develop a thinner version of French fries. The elegant restaurant menu included French fries prepared by Crum in the standard thick-cut French style that was popularized in France in the 1700s and enjoyed by Thomas Jefferson as ambassador to that country. Ever since Jefferson brought the recipe to America and served French fries to guests at Monticello. One guest found Chef Crum's fries too thick for his liking and rejected the order. Crum cut and fried a thinner batch, but these also were disapproved. Exasperated, Crum decided to rile the guest by producing fries too thin and crisp to skewer with a fork. The plan backfired because the guest was ecstatic over the first known prototype of potato chips. Since then paper-thin potato chips appeared on the menu as a house specialty. Soon chips were packaged and sold first locally then throughout the New England area and then disseminated to other parts of the United States. The invention of the mechanical potato peeler in the 1920s paved the road for potato chips to roar from a specialty item to a top-selling savory snack thoroughly described in Chapter 14. For several decades after their creation, potato chips were largely sold in Northern United States. In the 1920s Herman Lay was mainly responsible for popularizing potato chips in the southeast part of the United States. At the beginning of the 1960s he merged with Frito-Lay expanding even more volume sales (www.ideafinder.com/history/inventions/).

1.3.8 Almonds and Tree Nuts

The consumption of nuts probably dated back to the origin of the human species when men were tree dwellers and fruit gatherers and developed tools and skills to crush or break the hard shells of tree nuts. These products were preferred due to the high energy density and nutritional value.

The different types of tree nuts originated in diverse regions around the globe (Chapter 15). Almonds (*Amygdalus communis*), cashew (*Anacardium accidentale*), and macadamia (*Macadamia integrifolia*) trees originated in the Mediterranean, Brazil, and Australia, respectively. The first Macadamia plantations occurred in 1878 in the Hawaiian Islands. The Persian walnut (*Juglans regia L*) suggests Iraq as its native homeland but its original distribution was larger and extended from the Caucasus to Turkestan. On the other hand, walnuts were known to the ancient Greeks and from Europe reached North America with the early settlers during the 16th and 17th centuries. The term English walnut is also used because these nuts were transported in English ships.

The pecan tree (*Carya illinoensis*) is native to America. The pecan nuts were used for food for thousands of years before the discovery of the continent in 1492. The Spanish conquistador Hernan Cortes saw this indigenous tree during his journey to Tenochtitlan (today Mexico City). Lope de Oviedo, a member of the Spanish expedition, described for the first time in his journal the American pecan nut and described it as a much smaller nut compared to those commonly grown in Spain (walnut). In 1541, Cabeza de Vaca, a Spanish nobleman and treasurer, also wrote about pecans and mentioned that it was a subsistence crop for at least two months every other year for native Indians. The pecan name was adopted from the Algonquin Indian tribe language meaning that nuts were so hard that needed a stone for cracking. According to Brison (1974), the commercial pecan industry started about 1880. Therefore, the pecan industry is relatively new compared to other orchard crops.

The almond and tree nuts snack market is expected to observe significant demand in the upcoming years especially among millennials. The expected growth is due to their plant base nature and nutritional value in terms of protein, fatty acid profile, and micronutrients, which make them suitable for the surging market of organic/healthy snacks.

1.3.9 Fruits and Vegetables

Dehydrated fruits and vegetables initially obtained by solar drying have a long tradition of use dating back to the fourth millennium BC in the Middle Eastern region known as the Fertile Crescent where Mesopotamia (Iraq), Iran, southwest Turkey, Syria, Lebanon, and Israel are located. The earliest records of desiccated fruits can be found in Mesopotamian tablets dating to about 1500 BC, which depicted dietary patterns with fruits such as grapes, dates, figs, pomegranates, and apples. These early civilizations used dates, date juice evaporated into syrup, and raisins as sweeteners. They also included dried fruits in their breads for which they had more than 300 recipes. From the Fertile Crescent region, these fruits spread through Greece and Italy where they readily became a major part of the diet. The ancient Romans consumed raisins in impressive quantities and raisins-supplemented breads were part of the common diet. (https://en.wikipedia.org/wiki/Dried_fruit#History). The removal of most of the moisture associated with fruits and vegetables with the concomitant concentration of soluble carbohydrates is key to lower the *Aw* to levels which avoid the growth of bacteria, yeasts, and molds.

The intentional drying of raisins, dates, other fruits, and vegetables was practiced in order to prolong the shelf-life of these items without losing nutritional attributes and were fundamental for the prevention of scurvy among mariners that stayed long times offshore. There are several European and US documents that clearly depict the production of desiccated vegetables and the usefulness of these products to prevent scurvy and other nutrient deficiencies. As a result, The American Desiccating Company of New York was established in the middle of the 18th century.

Nowadays the consumption of dried fruits and vegetables is still widely practiced around the globe. Approximately half of the dried fruits produced today are raisins, followed by dates, prunes, figs, peaches, apricots, pears, mango, and apples (Chapter 16). During the past decades, high antioxidant fruits like blueberries, cranberries, and strawberries have gained market especially in the form of osmotic-dehydrated items (Chapter 16) offered in blends and nutritional bars.

1.3.10 Milk-Based Snacks

Humans have consumed dairy products from cows, sheep, and goats for at least 10,000 years. Archaeological evidence from the Neolithic revolution (8000 BC) points out the use of milk first in the Middle East and later on in regions of Europe and Africa. It is documented that 10,000 to 11,000 years ago goats were the first milk-producing domesticated mammals used by ancient cultures whereas cows were domesticated in eastern Sahara no later than 9,000 years ago. The emergence of dairying was a major innovation in prehistoric societies, enabling the supply of nutritious food without the slaughtering of prized livestock

The processing of milk into cheese and fermented products similar to yogurt would have been a critical development because it not only allowed the preservation of dairy products in a nonperishable and transportable form but also made milk a more digestible commodity for early prehistoric farmers. The key for the processing of these items is the natural fermentation of milk by lactic acid bacteria and the addition of salt to cheeses.

The first unequivocal evidence that humans in prehistoric Northern Europe made cheese more than 7,000 years ago is described by Salque et al. (2013) who found remains of fatty acids in pottery or cheese strainers especially made to produce the first prototypes of cheeses. The earliest perforated pottery sieves dated between 5200 and 4800 BC were found in the Linerbandkeramilk territory located in the strip of Hungary, Germany, Poland, France, Netherlands, and Ukraine/Russia (https://www.thoughtco.com/dairy-farming-ancient-history-171199).

Yogurt is also considered one of the oldest dairy foods produced by mankind. Most historians place its discovery somewhere between 9000 and 6000 BC in Central Asia (Mongolia Region) where men domesticated mares, cows, and camels. The history of the first fermented milk or yogurt goes back to Neolithic times when animal herders stored milk in the intestines of cows. Bacteria found in the intestines grew in the milk-yielding and acidic dairy product rich in probiotic bacteria. Then, the first fermented yogurts spread to the Middle East and Europe, and throughout the world. Yogurt is depicted in many old documents including the ancient Indian Ayurvedic scripts, the Bible, and historic texts. The great Mongol warrior and conquistador Genghis Khan is documented to have encouraged the drinking of fermented horse milk yogurt, known as *kumis*, to his warriors in order to maintain them healthy and brave. The Russian scientist Elie Metchnikoff was the first to record the health benefits of cultured milk on Bulgarian people and credited as the first one to indicate that the bacteria in fermented milk helped to diminish the amount of prejudicial bacteria in the gastrointestinal tract. His findings influenced the founder of the company Dannon Isaac Carasso to develop commercial products including yogurt supplemented with fruits. The expansion of yogurt sales started in the middle of the 20th century especially in the United States and Europe (http://www.indepthinfo.com/yogurt/history.htm).

1.3.11 Meat- and Marine-Based Snacks

Fresh meat, fish, and crustaceans rapidly deteriorate due to their high moisture and *Aw* contents. The use of sun-drying or camp fire-drying allowed primitive nomad races to desiccate meats for traveling and safe-keeping for future utilization. This easy-to-perform practice was key for the keeping of large game meats that otherwise rapidly deteriorate. The Egyptians made jerky from animal meat that was too big to eat all at once. Archeologists have found remnants of dried meat in Egyptian tombs dated 3200 BC.

The art of preservation by drying was not isolated to the African and European continents. In the Quechuan language of the Incas, the word *ch'arki* means dried meat. This product of the Andes was made of alpaca and llama meat debonded, pounded, and dehydrated in the arid climate of South America. The indigenous people shared this food with Spanish conquistadors who took the concept back to Europe. The Quechua word eventually evolved to jerky (https://delishably.com/meat-dishes/Exploring-Jerky-History-of-Preserving-Meat-Around-the-World#). Likewise, it is documented that the North American Indians mixed ground dried meat with dried fruits to make *pemmican*.

Practically all prehistoric cultures learn that the drying/smoking practice was key for their subsistence because it concentrated the meat nutrients while keeping the main nutritional attributes. The dehydration process evolved when meat and fish products were salted. Salting allowed a faster dehydration and significantly reduced microbial contamination especially during the first stage of drying. Men soon realized that the use of common salt contaminated with nitrates (Chapter 18) or pepper salt allowed the production of more stable products with a different flavor profile. This practice initiated the generation of dry-cured products. The nitrate salts avoid the growth of most pathogenic bacteria and practically assure the production of pathogen-free products (Chapter 18).

The use of freeze-drying or lyophilization was developed during the 20th century in order to improve the properties of dehydrated products. Freeze-drying removes the water by sublimation from the frozen meat. Freeze-dried products have a better texture, rehydration capacity, taste, color, and overall acceptability when compared with conventionally dehydrated counterparts. However, the excessive processing cost and the lack of continuous high output freeze-driers limit the use of this technology. Nowadays, the quantity of meat and seafood products processed by drying remains relatively small in comparison with the amounts of products consumed fresh, cured, or canned. However, the recent interest in low carbohydrate and keto diets and snacks rich in protein and other nutrients have gradually increased the demand for dehydrated animal-based snacks.

Fishing is a prehistoric practice dating back at least 40,000 years and is responsible for enriching the diet of our ancestors. Isotopic analysis of the skeleton of a 40,000-year-old modern human from eastern Asia has shown that he regularly consumed freshwater fish. Archaeological evidence of discarded fish bones and cave paintings show that seafoods were relevant for the survival of many ancient cultures. The Egyptians invented various implements and methods for fishing and these are clearly depicted in tomb scenes, drawings, and papyrus documents.

There is evidence of dehydration of fish and poultry meats 12,000 years BC by the ancient Egyptian culture. The drying operation of different types of fish evolved when products were assisted with salting and smoking. These three practices act synergistically to inhibit and prevent both the growth of spoilage and pathogenic microorganisms. Historically, the combination of salting and drying of cod has been practiced for at least five centuries. Desiccated fish has been a common snack and staple in Nordic cultures for centuries. The preferred species are white fish, usually cod or haddock, dried on wooden racks in the open air.

The simple solar drying technique used until the Middle Ages evolved when the Europeans constructed special structures, known as stillhouses, specifically designed to dehydrate foods via the heat generated by a fire.

1.4 STATISTICS OF SNACK FOOD PRODUCTION, SALES, AND CONSUMPTION

According to the global market of snack foods is approximately 520 billion USD in 2020 and expected to grow to more than 700 billion USD by 2025 (https://mordorintelligence.com/industry-reports/snack-food-market). Table 1.4 summarizes the global sales of the main sorts of snacks in the year 2018. The market is expected to grow at a compound annual growth rate of 6.2% from 2019 to 2025. The current market is dominated by nonextruded snacks because the consumer is demanding healthy products, such as granola and fruit bars, nuts, and high-protein low-carbohydrate items for special dietary regimes and prevent noncommunicable or chronic diseases like cardiovascular, hypercholesterolemia, and diabetes that are currently responsible for about 65% of the deaths.

The overall snack market in the United States, which comprises nonextruded and extruded snacks, exceeded 125 billion USD in 2019 and is expected to grow to 175 billion by the year 2025.

TABLE 1.4
Global Sales of the Main Types of Snacks in the Year 2018

Type of Product	Sales Billion USD (2018)
Chocolate/confectionary	110
Salty snacks	83
Ice cream	77
Sweet biscuits	73
Nuts, seeds, and trail mixes	27
Savory biscuits	23
Other savory snacks	18
Dairy and frozen desserts	15
Snack bars	14
Fruit snacks	11
Popcorn	5
Pretzels	3

Source: Chung (2019).

Nonextruded and extruded snacks represent about 85% and 15% of the current US market (www.grandviewresearch.com) and the salty snack category is expected to reach 109 billion USD by 2025. The savory snack food market in the United States estimated at 42 billion USD has grown since 2014 at an average annual rate of 1.4%. In this country, there are about 3,700 active businesses involved in the manufacturing of the different types of snacks providing employment opportunities for approximately 62,000 people. (IBIS World).

According to the packaging world, the global market for healthy snacks was worth 21.1 billion USD in 2016. The fruit snacks, nuts, and seeds accounted for approximately 25% of the market share in this particular category. Likewise, protein snacks are a significant driver for purchases in the snack food industry today. This category was up 6.6% from the year before, with some products that experienced 582% revenue growth in 2017 (IRI Market Advantage).

In the year 2018, the Americans, Netherlanders, Estonians, Canadians, and South Koreans had the highest annual *per capita* consumption of snack foods with an average annual intake of 22.4, 9.1, 8.7, 8.3, and 9.1 kg, respectively (Figure 1.1).

In 2019, the main companies that fabricated snacks and bakery items with yearly sales exceeding 10 billion USD were Nestle, PepsiCo, Mars, Mondelez International, Kraft-Heinz Co, General Mills, Kellogs, Ferrero, and the Mexican group of Bimbo. These nine companies sold nearly 310 billion USD (Table 1.5, Peckenpaugh 2020). Among savory snacks, potato and tortilla chips are the two most popular in terms of volume and sales. Frito-Lay North America accounts for 25% of the revenue at PepsiCo and is the leading producer of savory snack foods in the US market.

Retail sales of savory snacks in the United States projected to the year 2022 are expected to exceed 55.3 billion USD (Wunsch 2020). According to Mintel (Anonymous 2015) 95% and 50% of all Americans snack once and two to three times a day. The Americans also claim a preference for healthier snacking with 33% saying they are snacking with healthier foods specially formulated with simple ingredients and low-energy contents. However, 62% most often snack to satisfy a craving, highlighting the relevant role of taste. In fact, 63% of US consumers value the taste of salty snacks more than their nutrition. Importantly, the millennials are more likely to snack compared to older generations as a means to fulfill emotional and functional needs and added nutrition. As a result, they may be drawn to products with high fiber, energizing claims, or protein content to stay

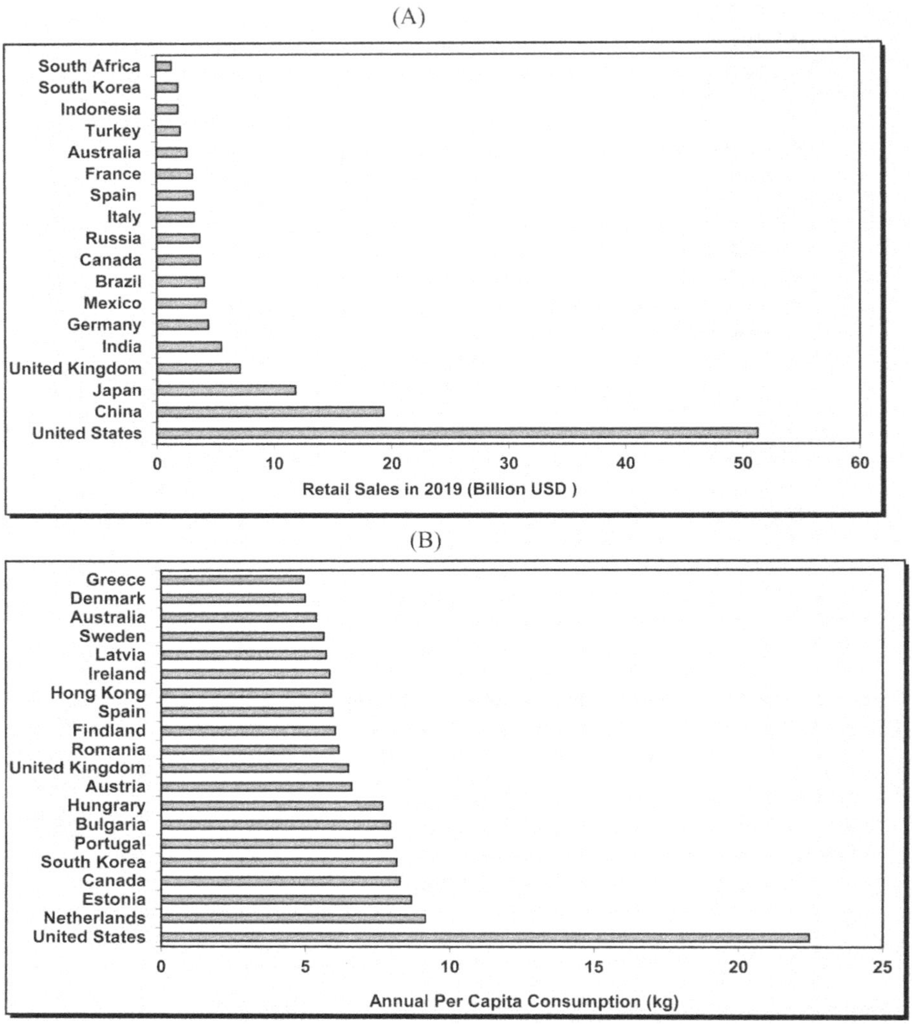

FIGURE 1.1 (A) Retail sales of savory snacks and (B) *per capita* volume consumption of snack foods in the leading countries of the world.

satiated, as well as bold flavors to help add variety to their frequent snacking occasions and eliminate boredom (Anonymous 2015).

North America and Western Europe are currently the largest snack markets, accounting for nearly 50%. According to Driscoll, the developed markets represent about 60% of total sales, while emerging markets represent the remaining 40%. Snacking in emerging markets has grown at a 7.3% compound annual growth rate over the past five years, three times faster than in developed markets (Chung 2019).

According to a contemporary survey performed by Datamonitor in Europe, the Britons munch their way through four times as many snacks a year as their Spanish cousins. In the snacking stake, in which they each shell out an average £204 a year on snack items. British consumers are situated far ahead of their nearest rivals in Sweden and the Netherlands. Residents of those countries spent £166 and £165, respectively, followed by the Germans who consumed £162 worth and the French £146. Taste was noted as the most influential factor on snack food product choice with consumers unwilling to compromise in exchange for added health benefits. Findings suggest new food product

TABLE 1.5
Top 20 Snack and Bakery Companies of the World in the Year 2019

Ranking	Company	Sales Billion USD (2019)
1	Nestle S.A.	101.5
2	PepsiCo	67.2
3	Mars, Inc.	35.0
4	Mondelēz International	25.9
5	Kraft Heinz Co.	25.0
6	General Mills	16.9
7	Kellogg Co.	13.6
8	Ferrero	12.6
9	Grupo Bimbo S.A.B. de C.V.	12.1
10	Conagra Brands	9.5
11	Arca Continental S.A.B. de C.V.	8.8
12	Campbell Soup Co.	8.1
13	The Hershey Co.	8.0
14	The Wonderful Co.	6.7
15	Post Holdings	5.7
16	TreeHouse Foods	4.3
17	Flowers Foods	4.1
18	Rich Products Corp.	4.0
19	ARYZTA	3.7
20	Premium Brands Holding Corp.	3.6

Source: Peckenpaugh (2020).

development in the UK market will continue to focus on the convenience and comfort from eating while on the move.

1.5 NUTRITIONAL VALUE OF SNACKS

In general, snacks are categorized as empty foods because they contain a high amount of fats/oils and sodium and most contain low protein and essential amino acids. The dietary fiber, vitamin, mineral, protein content, and protein quality of most snacks are low because they are generally manufactured from refined cereal grits and/or starches. The nutritional value of snacks discussed in Chapter 20, highly depends on the amount of fat absorbed during the frying process or added as flavoring base. Snacks are one of the highest energy-dense foods available because they practically do not contain water and are rich in fats which provide 9 kcal/g. The higher the oil content the lesser the amount of essential micronutrients such as vitamins and minerals. The issue of *trans* fatty acids present in partially hydrogenated frying oils and shortenings still used by many snack processors has further damaged the image of this particular industry.

Other major drawback of cereal-based snacks is that they contain low quantities of protein and more importantly the protein lacks lysine. Most snacks are salted or flavored with salt-based flavorings therefore these products commonly contain large quantities of sodium, which is related to hypertension and cardiovascular disease (Chapter 21). This essential electrolyte is considered the single nutrient most abused by modern humans because an average diet usually provides at least ten times the recommended daily intake. For these multiple reasons, snacks are viewed as "empty foods" and have a bad image among dietitians and health practitioners.

Among snacks, popcorn, peanuts, and other legumes or pseudocereals, almonds and tress nuts are gaining popularity because they are viewed as natural and nutritious. These products are rich in dietary fiber and also contain significant amounts of micronutrients and phytochemicals (Chapter 21).

Among cereal-based snacks, popcorn (Chapter 9) is considered one of the best nutritionally recommended because it is rich in starch, dietary fiber, and B-vitamins (Table 1.1). Likewise, the wheat-based hard pretzels, crispbreads, pita chips, and biscuits or crackers discussed in Chapter 10 are relatively lower in calories compared to other snacks. The nutritional value of these products resembles more bakery products than snacks (Chapter 20). The chief difference is that since these products are low in moisture they are usually more energy dense than soft bakery counterparts (Table 1.1).

Corn and tortilla chips produced with the nixtamalization process (Chapter 11) are considered energy-dense snacks because they mainly constitute starch and oil. These snacks are rich in sodium and lack good quality protein due to the deficiency of the essential amino acids lysine and tryptophan. However, they contain adequate levels of calcium and B-vitamins (Chapter 20) and their nutritional value is improved when combined with other legume seeds and vegetable proteins. Corn chips are more energy dense compared to tortilla chips, they normally contain 36 and 24% oil, respectively.

Most second- and third-generation extruded snacks (Table 1.2) are normally produced from refined cereal fractions practically devoid of vitamins, minerals, and dietary fiber and are obtained after various heat treatments that compromise even more the bioavailability of relevant vitamins. Most extruded snacks covered in Chapter 12 are considered as high caloric and empty foods and mainly constitute starch and oil (Chapter 20).

The acceptance of legume nuts, almonds, and tree nuts has increased recently because many epidemiological and clinical studies have shown that consumption of these items are associated with reduced risk of cardiovascular diseases because they contain high amounts of proteins, complex carbohydrates, polyunsaturated fats, both liposoluble and hydrosoluble vitamins, and antioxidants. These raw materials are blended with dehydrated fruits, concentrated forms of vegetable proteins, and sweeteners for the manufacturing of many new nutritional bars.

Likewise, the popularity of dehydrated fruits and/or vegetable-based snacks has increased because of their low oil content and high dietary fiber, micronutrient composition, and antioxidants.

Potato chips, which are still considered the most popular savory snack across the globe, are highly caloric because of their high starch and oil contents that provide more than 500 kcal/100 g serving. These snacks are usually devoid of dietary fiber, protein, and micronutrients and can contain relevant amounts of the potentially carcinogenic compound of acrylamide (Chapter 20).

Dairy and meat-based snacks contain more protein and the protein is much better in terms of essential amino acid balance. The consumption of meat/poultry/fish/crustaceans and milk-based snacks discussed in Chapters 17 and 18 are also gradually gaining market especially among consumers who would like to adopt a low-carbohydrate and high-protein diets (Table 1.1).

Nowadays, food scientists are including in their research and product development equations on the production of snacks with minimal fat calories and fat decomposition products known to have adverse effects on human health and produce snacks with similar organoleptic properties (flavor, texture, and overall acceptability) compared to regular counterparts. A healthful balance of fatty acids (saturated:monounsaturated:polyunsaturated and omega 6:omega 3) and the production of snacks free of *trans* fatty acids are into consideration.

The image of snacks has worsened because of the higher indexes of overweight and obesity in people inhabiting developed countries around the world. All chronic diseases (hypercholesterolemia, hyperlipidemia, cardiovascular disease, diabetes, hypertension, and cancer) are highly correlated with the body mass index and overconsumption of calories (Chapter 21). In order to overcome the bad reputation, the industry faces big challenges such as the production of low-calorie, high-fiber, and high protein items and snacks produced from natural products like whole grains high in nutraceuticals that diminish the incidence of noncommunicable diseases. In order to be successful in the market, new products should combine excellent organoleptic properties, high nutritional value, shelf-life, convenience, and marketing strategies.

1.6 MAJOR CONSTRAINTS OF SNACKS

The snack food industry faces important challenges in the near future. The first and most important is to reverse the bad image as producers of junk foods rich in calories, fat, and sodium and low in essential micronutrients and phytochemicals. In addition, processors should produce products that meet the minimum standards in terms of mycotoxins, microbial counts, trans fats, and toxic compounds like acrylamide and nitrosamines.

1.6.1 Major Nutritional and Health Concerns

As mentioned earlier most snacks as viewed as junk foods and processors should adapt products and processes in order to overcome the major constraints of new customers. Undoubtedly, the major goals should be to lower the amounts of calories, sodium, soluble sugars, trans fats, saturated fats, and cholesterol and increase levels of both insoluble and soluble and prebiotic dietary fiber, proteins, micronutrients, and phytonutrients so to produce novel products with enhanced nutritional value and health-promoting properties. In addition, processors should modify processing conditions in order to minimize the loss of important nutrients and the generation of harmful and toxic compounds.

1.6.2 Potential Toxicological Compounds

One of the main constraints of snack foods is related to public health in terms of affecting the health of consumers via pathogenic bacteria, toxigenic molds, and several toxic chemical compounds generated during deep-fat frying.

1.6.2.1 Pathogenic Bacteria

There are several categories of snack foods that might cause food-borne illnesses due to cross-contamination with pathogenic bacteria, although generally speaking these products rarely pose public health hazards due to their low moisture and *Aw*. The two most susceptible categories are snacks derived from meat/fish/crustacean and dairy products (Table 1.1) that are not thermally treated at temperatures that inactivate pathogenic microorganisms. These products that are increasingly being consumed due to their high nutrition profiles should be processed to meet the maximum allowed populations of pathogenic bacteria usually assessed by coliform counts. The most important pathogenic bacteria that are known to cause serious food-borne illnesses and even in some cases death are *Listeria*, *Salmonella*, *Clostridium*, *Shigella*, and *Vibrio*. These important bacteria can be also controlled with dehydration, use of curing salt, supplementation of fermenting microorganisms like lactic acid bacteria and yeast, and the intentional addition of organic acids and preservatives like propionates, sorbates, or benzoates.

1.6.2.2 Mycotoxins

Snacks based on peanuts, corn, and tree nuts might contain significant amounts of mycotoxins that can have acute and chronic toxicological effects. Unfortunately, most mycotoxins are heat resistant and therefore persist in finished snacks. Corn, peanuts, and tree nuts are susceptible to contamination to different mycotoxicogenic molds belonging to the *Aspergillus* and *Fusarium* genus when stored at relatively higher moistures of *Aws*. The most relevant and worrisome mycotoxins are aflatoxins, ochratoxins, and fumonisins that are known to cause cancer, neurological disorders, renal failure, and suppression of the immune system. In nature, aflatoxins are considered among the most powerful hepatocarcinogenic agents and most regulatory agencies regulate maximum permitted levels in foods (Generally < than 20 ppb or 0.02 ppm). These toxic compounds can be generated at the field and during faulty storage. The processes of dry milling for the production of refined fractions and nixtamalization or lime-cooking for the production of corn and tortilla chips are known to significantly reduce mycotoxins.

1.6.2.3 Toxic Compounds from Degraded Oil

One of the major public health concerns of the snack industry is related to toxicogenic compounds generated during frying and lipid oxidation. These main documented harmful reactions are autoxidation at temperatures higher than 100°C, polymerization in the absence of oxygen at temperatures ranging from 200 to 300°C, and thermic oxidation reactions at 200°C. According to Hosseini et al. (2016) approximately 400 types of decomposition products have been identified in fried products although about 50 are potentially carcinogenic. These compounds are additionally known as mutagenic, hyperlipidemic, and cholesterolemic, increasing the risk of cardiovascular diseases. The most relevant and worrisome toxic compounds are acrylamide, cyclic fatty acid monomers, aldehydes, acrolein, heterocyclic aromatic amines (HAA), and polycyclic aromatic hydrocarbons (PAH) discussed in Chapter 20 (Tareke et al. 2000, Mucci et al. 2003, Kim et al. 2005, Romero et al. 2000, Hosseini et al. 2016). For instance, the mutagenic potency of HAA is about 100-fold stronger than that of aflatoxins. As a result, the industry has focused on diminishing the formation and concentration of these harmful compounds in finished products. Several research fronts around the globe have investigated the use of additives or novel frying technologies to minimize the generation of these toxic compounds.

1.7 FUTURE OUTLOOK

The main elements driving the global snack food market include the growing trend of nuclear families, the adoption of Western eating habits among residents in emerging countries, busy lifestyles, diversity of snacks even within categories, and convenience. The increasing number of companies and the prompt influx of novel products in the market have enhanced consumer interest. The market is expected to witness healthy growth especially in developing countries where higher population growth is expected. The forecast of population growth in the next 30 years clearly indicates that the world will have approximately between 2 and 2.5 more billion people. The industry should adapt to these challenges by developing and launching new and innovative snacks with improved nutritional profiles and green labels in terms of usage of natural and GRAS ingredients to favor nutrition, health, and quality of life.

REFERENCES

Anonymous. 2015. A snacking nation: 94% of Americans snack daily. Food and Drink. Mintel Press Office. https://www.mintel.com/press-centre/food-and-drink/a-snacking-nation-94-of-americans-snack-daily.

Aramouni, F.M., and Abu-Ghoush, M.H. 2011. Physicochemical and sensory characteristics of no-bake wheat–soy snack bars. *Journal of the Science of Food and Agriculture*, 91(1):44–51. DOI: 10.10002/jsfa.4134.

Brison, F.R. 1974. *Pecan culture*. Austin: Capital Printing.

Burtea, O. 2001. Snack foods from formers and high shear extruders. In: *Snack foods processing*, eds. E.W. Lusas and L.W. Rooney, 281–314. Lancaster: Technomic Publishing Co.

Chung, H. 2019. Here's what the $605 billion global snack market looks like. https://finance.yahoo.com/news/citi-global-snack-market-110553461.html.

Constantin, O.E., and Istrati, D.I. 2018. Functional properties of snack bars. In: *Functional foods*, ed. V. Lagouri, 1–14. London: IntechOpen. DOI: 10.5772/intechopen.81020.

da Silva, E.P., Siqueira, H.H., Do Lago, R.C., Rosell, C.M., Boas, V., and De Barros, E.V. 2014. Developing fruit-based nutritious snack bars. *Journal of the Science of Food and Agriculture*, 94(1):52–56. DOI: 10.1002/jsfa.6282.

da Silva, E.P.D., Siqueira, H.H., Damiani, C., Boas, V., and de Barros, E.V. 2016. Physicochemical and sensory characteristics of snack bars added of jerivá flour (*Syagrus romanzoffiana*). *Food Science and Technology*, 36(3):421–425. DOI: 10.1590/1678-457X.08115.

Estevez, A.M., Escobar, B., Vásquez, M., Castillo, E., Araya, E., and Zacarías, I. 1995. Cereal and nut bars, nutritional quality and storage stability. *Plant Foods for Human Nutrition*, 47(4):309–317. DOI: 10.1007/BF01088268.

FAO. 2021. *Statistical data base*. Rome, Italy: Food and Agriculture Organization of the United Nations. http://www.fao.org/faostat.

Ferreira, M.S., Santos, M.C., Moro, T.M., Basto, G.J., Andrade, R.M., and Gonçalves, É.C. 2015. Formulation and characterization of functional foods based on fruit and vegetable residue flour. *Journal of Food Science and Technology*, 52(2):822–830. DOI: 10.1007/s13197-013-1061-4.

Garcia-Lara, S., and Serna-Saldivar, S.O. 2019. Corn history and culture. In: *Corn: chemistry and technology*, ed. S.O. Serna-Saldivar, 1–19. Cambridge: Elsevier (Woodhead Publishing), AACC-International.

Gonzales, E., and Draganchuk, M. 2003. Flavoring nutrition bars. *Cereal Foods World*, 48:250–251.

Gould, W.A. 2001. Potatoes and potato chips. In: *Snack foods processing*, eds. E.W. Lusas and L.W. Rooney, 227–259. Lancaster: Technomic Publishing Company.

Hogan, S.A., Chaurin, V., O'Kennedy, B.T., and Kelly, P.M. 2012. Influence of dairy proteins on textural changes in high protein bars. *International Dairy Journal*, 26(1):58–65. DOI: 10.1016/j.idairyj.2012.02.006.

Hosseini, H., Ghorbani, M., Meshginfar, N., and Mahoonak, A.S. 2016. A review on frying: Procedure, fat, deterioration progress and health hazards. *Journal of the American Oil Chemists' Society*, 93(4):445–466.

Kim, C.T., Hwang, E.S., and Lee, H.J. 2005. Reducing acrylamide in fried snack products by adding amino acids. *Journal of Food Science*, 70(5):354–358.

Kulp, K. 1994. *Cookie chemistry and technology*. Manhattan: American Institute of Baking.

Malleshi, N.G., and Desikachar, H.S.R. 1985. Milling, popping and malting characteristics of some minor millets. *Journal of Food Science and Technology*, 22:400–403.

Manley, D. 2000. Setting the scene history and position of biscuits. In: *Technology of biscuits, crackers and cookies*, ed. D. Manley, 1–8. Cambridge: Woodhead Publishing Company.

Mishra, G., Joshi, D.C., and Panda, B.K. 2014. Popping and puffing of cereal grains: A review. *Journal of Grain Processing and Storage*, 1:34.e46.

Misra, N.N., and Tiwari, B.K. 2014. Biscuits. In: *Bakery products science and Technology*, eds. W. Zhou, Y.H. Hui, I. DeLeyn, M.A. Pagani, C.M. Rosell, J. Selman, and N. Therdthai, 585–601. Chichester, West Sussex: Wiley Blackwell.

Mucci, L., Dickman, P., Steineck, G., Adami, H.O., and Augustsson, K. 2003. Dietary acrylamide and cancer of the large bowel, kidney and bladder: Absence of an association in a population-based study in Sweden. *British Journal of Cancer*, 88(1):84–89.

Munir, M., Nadeem, M., Qureshi, T.M., Qayyum, A., Suhaib, M., Zeb, F., and Ashokkumar, M. 2018. Addition of oat enhanced the physico-chemical, nutritional and sensory qualities of date fruit based snack bars. *Journal of Food and Nutrition Research*, 6(4):271–276.

Panak Balentić, J., Ačkar, D., Jozinović, A., Babić, J., Miličević, B., Jokić, S., Pajin, B., and Šubarić, D. 2017. Application of supercritical carbon dioxide extrusion in food processing technology. *Hemijska Industrija*, 71(2):127–134.

Peckenpaugh, D.J. 2020. The top 50 snack & bakery companies of 2020. Snack Food and Wholesale Bakery. https://www.snackandbakery.com/articles/95820-the-top-50-snack-bakery-companies-of-2020#top-50-list.

Ramirez-Jimenez, A.K., Gaytan-Martinez, M., Morales-Sanchez, E., and Loarca-Piña, G. 2018. Functional properties and sensory value of snack bars added with common bean flour as a source of bioactive compounds. *LWT- Food Science and Technology*, 89:674680. DOI: 10.1016/j.lwt.2017.11.043.

Riaz, M.N. 2000. Introduction to extruders and their principles. In: *Extruders in food applications*, ed. M.N. Riaz, 1–24. Lancaster: Technomic Publishing Co.

Riaz, M.N. 2004. Snack foods, processing. In: *Encyclopedia of grain science*. 1st edition, eds. C. Wrigley, C. Walker, and H. Corke, 98–108. Oxford: Elsevier.

Rizvi, S.S.H., and Mulvaney, S.J. 1992. Extrusion processing with supercritical fluids. *United States Patent* 5,120,559 A.

Rizvi, S.S.H., Mulvaney, S.J., and Sokhey, A.S. 1995. The combined application of supercritical fluid and extrusion technology. *Trends in Food Science and Technology*, 6(7):232–240.

Romero, A., Cuesta, C., and Sanchez-Muniz, F.J. 2000. Cyclic fatty acid monomers and thermoxidative alteration compounds formed during frying of frozen foods in extra virgin olive oil. *Journal of the American Oil Chemists' Society*, 77(11):1169–1175.

Salque, M., Bogucki, P.I., Pyzel, J., Sobkowiak-Tabaka, I., Grygiel, R., Szmyt, M., and Evershed, R.P. 2013. Earliest evidence for cheese making in the sixth millennium BC in northern Europe. *Nature*, 493(7433):522–525.

Seetharaman, K. 2014. Pretzel production and quality control. In: *Bakery products science and technology*, eds. W. Zhou, Y.H. Hui, I. DeLeyn, M.A. Pagani, C.M. Rosell, J. Selman, and N. Therdthai, 611–618. Chichester, West Sussex: Wiley Blackwell.

Serna-Saldivar, S.O. 2008a. Popcorn. In: *Industrial manufacture of snack foods*, 105–124. London: Kennedys Publications Ltd.

Serna-Saldivar, S.O. 2008b. The snack food industry. In: *Industrial manufacture of snack foods*, 1–24. London: Kennedys Publications Ltd.

Serna-Saldivar, S.O. 2015. History of wheat and corn tortillas. In: *Tortillas - Flour and corn products*, eds. L.W. Rooney and S.O. Serna-Saldivar, 1–28. St. Paul: AACC International Press.

Sharif, M.K., Rizvi, S.S.H., and Paraman, I. 2014. Characterization of supercritical fluid extrusion processed rice-soy crisps fortified with micronutrients and soy protein. *LWT – Food Science and Technology*, 56(2):414–420.

Snack Food Association. 1987. *50 years. A foundation for the future*. Alexandria: Snack Food Association.

Sun-Waterhouse, D., Teoh, A., Massarotto, C., Wibisono, R., and Wadhwa, S. 2010. Comparative analysis of fruit-based functional snack bars. *Food Chemistry*, 119(4):1369–1379. DOI: 10.1016/j.foodchem.2009.09.016.

Talburt, W.F. 1987. History of potato processing. In: *Potato processing*, eds. W.F. Talburt and O. Smith, 1–9. New York: Van Nostrand Reinhold Company.

Tareke, E., Rydberg, P., Karlsson, P., Eriksoon, S., and Tornqvist, M. 2000. Acrylamide: A cooking carcinogen? *Chemical Research in Toxicology*, 13(6):517–522.

Terry Groff, E. 2001. Perfect pretzel production. In: *Snack foods processing*, eds. E.W. Lusas and L.W. Rooney, 369–384. Lancaster:Technomic Publishing Co.

Weatherwax, P. 1954. *Indian corn in old America*. New York: The MacMillan Co.

Wunsch, N.-G. 2020. Savory snack sales in the United States from 2013 to 2022. *Statista*. https://www.statista.com/statistics/1026979/savory-snack-sales-in-the-us/.

2 Application of Traditional and Emerging Processes

Jorge Welti-Chanes, Viridiana Tejada-Ortigoza, and Luis Eduardo García-Amezquita

CONTENTS

2.1 INTRODUCTION

One of the most relevant food sectors within the ready-to-eat (RTE) category is the snack industry. This industrial sector began its development in the United States at the end of the 19th century, using cereals as the main ingredient. Nowadays, snack processing is still an important field within the food industry (Fast 1999). Several snack classifications depend on the feedstock and the manufacturing process (Chapter 1). Some snack-type foods are raw cut vegetables, potato-based products, corn-based products, half products or pellets, directly expanded extruded snacks, and corn pops, among others (Guy 2001). The elaboration of traditional snacks has three generalized

DOI: 10.1201/9781003129066-2

steps: (i) the first step implies the hydration of starch so that the formed dough can be manipulated; (ii) the next step includes the application of high temperatures to heat the dough, in this operation moisture is liberated as vapor from the inside of the material, resulting in the product expansion; (iii) finally, there is a drying step to remove the moisture excess, leaving a food product with low moisture content and with the desired sensory attributes, such as crispiness and flavor. In the second stage, the dough heating must be conducted with a fast temperature increment to a temperature that normally exceeds the boiling point. After reaching this temperature level, most of the water boils simultaneously and is released from the dough. This unit operation can be performed using several technologies such as extrusion, frying, drying, and baking. On the other hand, the selection of the manufacturing technology and the process variables are crucial to obtain the desirable sensory characteristics of these types of products.

This chapter first reviews some of the fundamental principles of the traditional processes used for the preparation and processing of various types of snacks and second covers some new ingredients and emerging technologies that are being increasingly adopted to improve processes and generate novel types of snacks.

2.2 TRADITIONAL PROCESSES APPLIED TO SNACK PRODUCTION

The following sections will cover mainly the processing conditions and currently used equipment for traditional processes employed by snack processors. In his chapter, the importance of extrusion is covered although the various types of extrusion processes used for snack production would be thoroughly studied in Chapter 12. A brief description of the principles of the process is described, while the focus of this section is the modification in the composition of fiber due to the thermoplastic process and its importance on the development of novel ingredients to improve the nutritional quality of snacks.

2.2.1 Extrusion

2.2.1.1 Principles of the Extrusion Process/Traditional Processes

Extrusion cooking is one of the most used technologies to produce snacks. In this process deeply described in Chapter 12, the raw material is converted into a food product through an efficient, rapid, and continuous operation. Food extruders are considered an efficient manufacturing equipment due to the thermic and mechanical energy produced to cause physicochemical modifications of the raw materials, resulting in complete homogenization of the extruded ingredients (Anton et al. 2009). This technology includes several unit operations, such as mixing, forming, texturizing, cooking, cutting, puffing, and drying, which occur almost simultaneously (Sumathi et al. 2007). The use of extrusion has continuously increased in the last 40 years to produce foods and food ingredients, mainly snacks, RTE breakfast cereals, baby foods, baking goods, supplementary foods, meat and cheese analogs, modified starch, among others (Ding et al. 2006, Shah et al. 2021). In its origins, an extruder basically formed by a rotor, stator, and a screw was used to produce an uncooked material that needed a subsequent frying process. Nevertheless, newer extruders designs constituted by a short-barrel and/or a high-shear mechanism with a plate die, allow the production of less dense materials suitable for drying or toasting (Serna-Saldivar 2010). Today, the extrusion process is characterized by a rapid heating process and by the formation of an extrudate in a relatively short time. The extrudates emerge from the die with high expansion rates (Bredie et al. 1998, Harper 2019).

The conditions necessary to maintain extruded products with a vitreous structure are the low moisture and the temperature. Usually, a food with moisture lower than 5% $_{(wb)}$ preserves its crispy texture at no more than 40°C. After this temperature, the food changes into a plastic or viscoelastic material (Guy 2001). During the extrusion process, there are high temperature and humidity conditions, which generally result in several reactions that produce colors and aromas, such as Maillard reaction, lipids and vitamin degradation, and Strecker degradation of amino acids. However, due to the short residence time of the material inside the extruder and the rapid cooling of the extruded food emerged from the extruder die, several compounds responsible for those reactions are immobilized within the vitreous structure of the food (Bredie et al. 1998). For this reason, the adequate control of the process variables can enhance the production of desirable aromas in the food products, making the extrusion process a superior technology to produce snack foods.

Twin-screw extruders to produce snacks have been more extensively used mainly due to the capacity of this technology to create a great variety of shapes, textures, and even flavors. The barrel temperature, the melt's moisture, and the shear produced by the two screws are essential variables of the process that modify the characteristics and final quality of the product. The expansion of the melt in snacks occurs mainly when it emerges from the die or orifice and depends primarily on two mechanisms: (i) one is known as the die swell phenomenon, where the volume increment occurs by the pressure differential between the internal matrix inside the die with the outside pressure; and (ii) the expansion produced by the sudden evaporation of water within the melt, where small bubbles are made in a process known as nucleation. The former mechanism is considered more relevant in foods. The bubbles enlarge their volume depending on the extensional viscosity of the food matrix (Figure 2.1) (Fan et al. 1994, Pai et al. 2009). Both phenomena depend on both the transition temperature of the extruded materials and the shear and extensional rheological properties of the melt (Blake 2006).

The quality and the functional properties of the extruded snacks depend on the composition and features of the raw ingredients especially in terms of moisture content, water absorption index, starch, and dietary fiber properties, fat content, particle size, and the hardness of grains or their dry-milled fractions. From the processing viewpoint, the screw(s) configuration, screw speed, feed rate, conditioning water, die shape and diameter, and heating rate are of paramount importance. Furthermore, the operational variables like residence time within the barrel, temperature, and pressure profile in the barrel's sections affect shear and specific mechanical energy produced by the screws (Brennan et al. 2013).

The raw material typically utilized for the production of extruded snacks are dehulled grains or refined flours with a high starch content, therefore, with a low content of phytochemicals and dietary fiber. Refined fractions are preferred because of their higher content of starch that enhances functionality in snacks and ready-to-eat foods. Extrusion may be used to cook, form, and expand materials known as directly expanded snacks, which are ready-to-eat products; however, this technology can yield pellets or half products used for the production of third-generation snacks (Chapter 12). Half products are known to be compact or dense materials that need an additional unit operation, basically frying, to produce the expansion and desired characteristics in finished products (Serna-Saldivar 2010). The diagram flow depicted in Figure 2.2 shows the basic process to produce both directly expanded and third-generation extruded snacks.

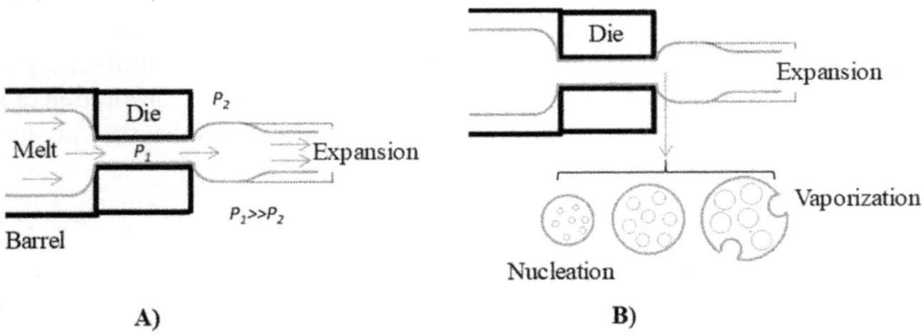

FIGURE 2.1 Phenomena involved in the expansion process during extrusion. (A) Die swell phenomenon and (B) vaporization of water in the extruded matrix.

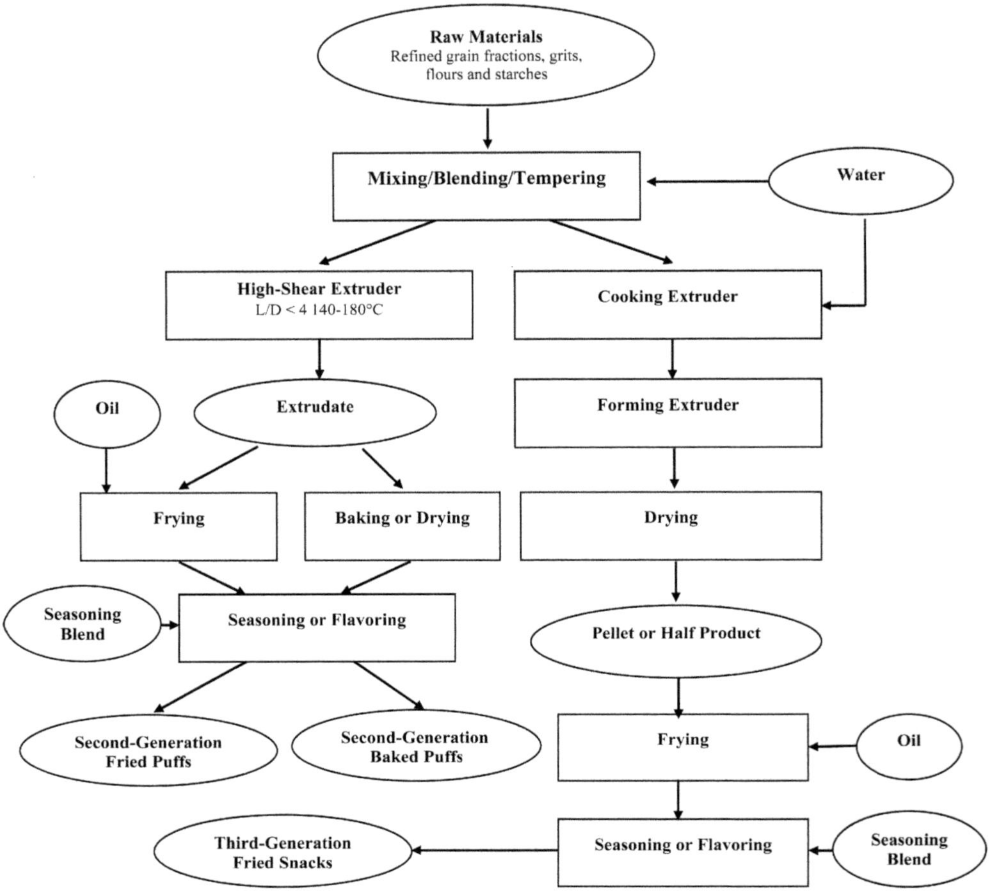

FIGURE 2.2 Flowchart of the manufacturing of directly expanded and third-generation snacks. Adapted from Serna-Saldivar (2010).

The use of non-starch carbohydrates like cellulose, guar gum, pectin, resistant starch, and β-glucans to manufacture directly expanded extruded foods has not been deeply explored. Nevertheless, during the last ten years, several research groups have studied the performance of non-starch polysaccharides on the quality of extruded snacks (Martinez-Bustos et al. 2011, Makowska et al. 2013, Paraman et al. 2015). One of the main reasons to use nonstarchy carbohydrates is to produce healthier foods. However, the search for novel ingredients with innovative characteristics is also studied. Although some ready-to-eat foods, mainly breakfast cereals, are available in the market, only a few examples can be considered low-glycemic index foods due to their high content of digestible starch and fat (Brennan et al. 2008). In this context, the partial substitution of starch-rich ingredients by noncaloric compounds, such as dietary fiber, is an option that may potentially result in healthier and improved products.

2.2.1.2 Current Trends

The demand of consumers for healthier ingredients, such as dietary fiber, has aroused the interest of food producers to develop and improve extruded snacks. Some in vivo studies have shown that the consumption of extruded ready-to-eat snacks with a high content (about 15%) of fiber substituting starch may decrease the plasma glucose level up to 43% (Brennan et al. 2012). However, due to the limited functional properties of the fiber and the undesirable texture that fiber produced in foods, there are some restrictions in its use in extrusion. Nevertheless, it has been suggested that with the control and correct manipulation of the extrusion process variables and the fiber sources, the fiber's functionality may be improved (Mendonça et al. 2000).

2.2.1.2.1 Effect of Process Conditions on the Fiber Fraction of Extruded Snacks

Nowadays, most determinations of the effects of processing conditions on the extruded snacks properties are mainly based on general models (Bisharat et al. 2013). However, due to the high fiber content in the feedstock, the temperature significantly increases because of the high increment of the extrusion torque and the specific energy. The variation of the process conditions has been studied as an alternative to decrease the undesirable effects resulting from the fiber addition in this technology (Jin et al. 1994, Mendonça et al. 2000). In this context, Bisharat et al. (2013) produced corn extrudates enriched with broccoli fiber. They observed that increasing both temperature and the screw speed reduced the viscosity of the melt and increased starch gelatinization, resulting in a porous product. On the other hand, increasing the tempering water showed a reverse effect. Similar observations were reported in a whole-wheat extrudate with a fiber content up to 12.4% (dwb) in the feedstock. In this study, an increment of the volumetric expansion and product porosity was obtained, increasing the temperature and the screw speed and decreasing the tempering moisture (Robin et al. 2012a). They also showed that increasing the barrel temperature, the screw speed, and the tempering moisture decreased the shear viscosity of the melt. The hardness of the extrudate could be modified by the adequate control of the screw speed and the tempering moisture.

The variation and control of the process variables affect the physical and sensorial attributes of the final product, but it has also been documented that may also modify the nutritional content of the extruded snacks. In this context, Stojceska et al. (2008a,b) produced extrudates using wheat flour and corn starch supplemented with brewer's sent grain and with red cabbage (Stojceska et al. 2009) with the aim to increase the dietary fiber content up to 14.5% (wb). Authors explained how the thermal cooking inside the barrel and increment of the tempering moisture (about 15%) significantly augmented the total dietary fiber, phenolic compounds, and antioxidant compounds in the extruded snacks with red cabbage. The same research group developed a gluten-free extrudate with high fiber content from several materials (e.g., potato and corn starch, wheat flour, milk, and soy). They observed that controlling the processing conditions, such as temperature, solids feeding rate, and screw speed, resulted in products with excellent textural properties and also a fiber content 60% higher compared to snacks produced under standard conditions (Stojceska et al. 2010). On the other hand, the extrusion process may modify certain carbohydrates' structures of the food system inducing to different degrees the polymerization of dietary fiber components. In a model system made of lactose, dextrose, and citric acid, about 60% of dietary fiber was obtained in an extruded product from the polymerization of lactose and low-molecular-weight oligosaccharides, which were soluble and nondigestible (Tremaine et al. 2014). These results suggest that controlling the conditions to produce fiber-rich snacks may result in foods with final quality similar to those made with starch-based ingredients.

2.2.1.2.2 Effect of Fiber Source and Composition on Extruded Snacks

Not only do the process conditions influence the physicochemical and functional characteristics of the final product, but also the composition and characteristics of the raw mixtures subjected to extrusion. The properties of the ingredients affect significantly process variables and quality of the finished product especially in terms of texture, structure, expansion, and sensory attributes. In this context, the high content of fiber negatively modifies the melt within the barrel. This effect has been tried to correct using additives such as glycerol monostearate (Mendonça et al. 2000) and dairy proteins (Onwulata et al. 2001), among others.

The inclusion of fibers or gums in formulations of extruded snacks modifies the pasting properties of the ingredients. This is reflected in the matrix forming nature and the water-binding capacity of the polysaccharides, changing the final viscosities of the melt. This inclusion also reduces the caloric content of the extrudates and decreases the digestibility of the starch (Brennan et al. 2008).

The composition of the fiber, which depends on the fiber's source and the process used to obtain them, greatly influences the selection of the extrusion conditions and the final quality of the snacks. Most studies have focused on the effect of the fiber type and composition of the extrudates. Stojceska et al. (2008b) obtained an extruded snack with 14% of fiber from brewer's spent grain with scarce effect on the color, although they also observed a reduction in the snack expansion. Nevertheless, the addition of starch and the adequate control of the processing parameters produced good quality snacks with up to 30% of brewer's spent grains. They also studied the addition of cauliflower (from 5.6 to 11.6%) as a fiber source in corn-, wheat-, and oat-based snacks. Their results showed that increasing the fiber content augmented the cell density and decreased the thickness of the walls and the moisture absorption index (Stojceska et al. 2008a). Likewise, Robin et al. (2011) evaluated the effect of wheat bran in starch extrudates. This research group showed that increasing the wheat bran at low concentrations increased the force of the extruded foam likely due to the fine structures obtained when low contents of wheat bran were supplemented. It should be noted that not all the added fiber ingredients yield acceptable snack products in terms of physical and sensory properties. The addition of fiber-rich legumes, such as common beans (*Phaseolus vulgaris*) in corn starch-based extrudates, although it resulted in a highly nutritious product, showed meager expansion due to the interaction of the fibers with the starch and besides extrudates showed undesirable color (Anton et al. 2009).

In general, fibers rich in soluble fraction yield extrudates with greater expansion than those rich in insoluble fiber. This improvement is mainly due to the low compatibility between insoluble fiber and starch (Robin et al. 2012a,b). The compatibility between fibers and starch may be improved by reducing the particle size of the fiber so that the contact surface with the starch increases. Blake (2006) decreased the particle size of corn fiber from 250 to 50 μm and concluded that this change considerably increase the sectional or radial expansion of extrudates. Other studies in which the solubility of the fiber was chemically modified using an alkaline treatment with NaOH showed an increase in solubility. The enhanced solubility made possible to obtain extrudates with higher expansion rates (Figure 2.3). The alkali modified fiber changes the apparent viscosity of the melt yielding produced higher crumb cells or bubbles in extrudates (Pai et al. 2009). Therefore, the effect of fibers on the texture of extruded snacks mainly depends on the interactions of the starch and the quantity and composition of the used fiber.

The use of insoluble fibers significantly reduces the expansion volume and increases the product density and hardness. The differences in the expansion volume of the fiber-rich extrudates can be explained by the water absorption, the effect of the viscoelastic properties of the melt when it emerges from the die, and the stabilization of the cell wall during the formation of internal or air cells. Some pretreatments, such as reducing the particle size and the increment of the fiber solubility, improve the properties of the final product (Robin et al. 2012b). However, it is not only the increment f soluble fiber that may improve the expansion and texture properties of extruded snacks. In this context, Peressini et al. (2015) elaborated on cereal-based extruded snacks with up to 7% of inulin (a well-known soluble fiber with prebiotic properties). In this research, inulins with two

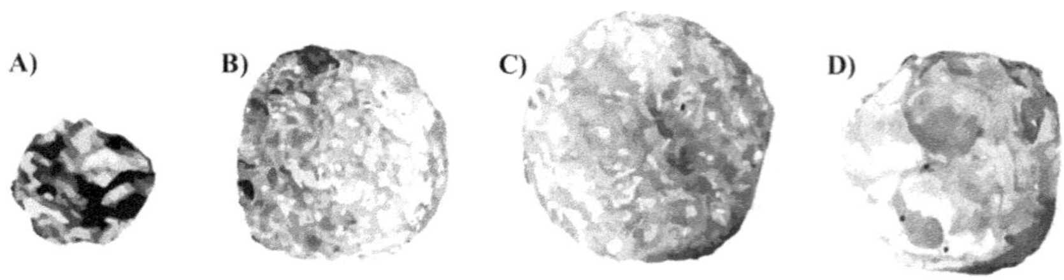

FIGURE 2.3 Extruded snacks with a high dietary fiber content (26%) with different soluble:insoluble dietary fiber weight-ratio. (A) 0.004:1, (B) 0.084:1, (C) 0.193:1, and (D) without fiber. Adapted and modified from Pai et al. (2009).

different degrees of polymerization (10 and 23) were used and compared. They observed that snacks produced with the lower molecular weight inulin showed better expansion and crispness compared to those elaborated with the high-molecular-weight counterpart.

The moisture content of extruded snacks containing fiber sources may modify the hygroscopic properties of the materials and therefore potentially cause textural problems during the shelf life. Tudorica et al. (2002) conducted SEM observations which showed that increasing the soluble fiber content resulted in changes in the hygroscopic properties due to the microstructural modifications. The results mentioned in this section suggest that the proportion of the different fiber fractions (soluble and insoluble) defines the capacity of these non-starch polysaccharides to be used in the manufacture of high-quality extruded snacks.

2.2.1.3 Modification of Snacks' Ingredients Using Extrusion

The technological functionality of the non-starch carbohydrates, such as fibers and gums, used in extrusion processes depends on their composition and the ratio between their soluble and insoluble fractions. Different technologies have been employed to modify these fractions and the fiber structures with the aim to improve functionality. The most utilized technologies are chemical and thermal modifications aimed to promote the partial hydrolysis of cell walls and fiber components and therefore their functional properties (García-Amezquita et al. 2020, Tejada-Ortigoza et al. 2020a). However, other technologies, such as drying (Amid and Mirhosseini 2012), milling (Chau et al. 2004), or even nonthermal processes, such as high hydrostatic pressure (Tejada-Ortigoza et al. 2017) and ultrasound (Martinez-Solano et al. 2020), also affect the composition and functionality of fiber sources. Thermoplastic extrusion effectively modifies non-starch polysaccharides. Several studies have shown that the thermal effect, the tempering moisture, and the shear produced by the screws work together to change these compounds (Larrea et al. 2005).

Studies conducted in cereals have shown that extrusion increases the fiber content and modifies the fiber profile. Østergård et al. (1989) reported that the fiber content in barley increased with extrusion while the starch content decreased with the same treatment due to the formation of nondigestible starch fragments. Similar observations were reported by Vasanthan et al. (2002), who significantly increased the soluble fraction and the type-3 resistant starch contents by transglycosylation.

On the other hand, the use of extrusion to modify the fiber content and composition of nonconventional sources, such as fruit and vegetable by-products, has been explored due to the high soluble fiber content. This alternative could be a solution for the production of high-quality fiber-rich extruded snacks (Table 2.1). Larrea et al. (2005) studied the effect of different extrusion conditions (83–167°C, 22–38% tempering moisture, 126–194 rpm) on the fiber fractions of orange pulp. Their results showed a decrement in the total fiber content accompanied by the increase of the soluble:insoluble fiber fractions ratio, from 0.37:1 to 1.15:1. These results agree with those reported by Mendez-García et al. (2011), who increased the soluble fraction (from 38.6 to 50%), decreasing the insoluble fraction (from 57.01 to 48%) in lemon peel. For their part, García-Amezquita et al. (2019) applied different extrusion processes (70–160°C, 24–58% tempering moisture, 100–200 rpm) to orange peel. They observed that in all treatments, the loss of insoluble fiber correlated to the increase in the soluble fraction, resulting in improved functional properties. The same correlation was reported by Jing and Chi (2013) in soy by-products processed with a twin-screw extruder at 110°C, 30% tempering moisture, and 180 rpm. These results suggest that the extrusion process may partially solubilize the insoluble fiber without the complete degradation of the polymeric structure.

2.2.2 Frying

2.2.2.1 Frying Principles and Fundamentals

Deep-fat frying is considered one of the most universal methods for cooking foods and is still widely used by practically all cultures on the planet. This process is greatly affected by the source,

TABLE 2.1
Substitution of Starch-Based Ingredients Using Fruit or Vegetable Fiber to Produce Extruded Snacks

Extruded Snack Base	Source	Cereal/Starch Substitution (%)	Effect of the Substitution	References
Barley flour	Tomato pomace	2–13	Similar hardness to control snack and higher sensory acceptability.	Altan et al. (2008)
Corn starch	Carrot pomace	5–15	Higher expansion and improved surface.	Kaisangsri et al. (2016)
Corn starch	Apple pomace	22	Density, hardness, and crispness were similar to control.	Paraman et al. (2015)
Rice grits and flour	fruits by-products blend (orange, grape, tomato)	3–7	Acceptable expansion and sensory attributes.	Yağcı and Göğüş (2008)
Wheat flour	Cauliflower	5–20	Higher water absorption index with scarce effect on hardness.	Stojceska et al. (2008a)
Wheat/corn flour	Orange bagasse	3–10	Higher expansion ratio and lower bulk density.	Pitts et al. (2016)

Source: Adapted from García-Amezquita et al. (2018).

type, and condition of the frying oil and processing parameters that are thoroughly discussed in Chapters 3 and 7. Frying consists of placing the food in hot oil during a certain period of time to achieve delectable texture and flavor particularly of fried foods (Banks and Lusas 2001). Several reactions occur in an instantaneous way in the frying oil and in the food being fried (Figure 2.4). When introducing a food into a fryer, oxygen comes in contributing to oxidation reactions; water also comes in, resulting in fat hydrolysis and in the increase in fatty acids, mono and diglycerides, and glycerin. Leachable metals and colored compounds remain in the oil with some lipids from the food. In frying, heat transfer and mass transfer are the transport phenomena leading the process.

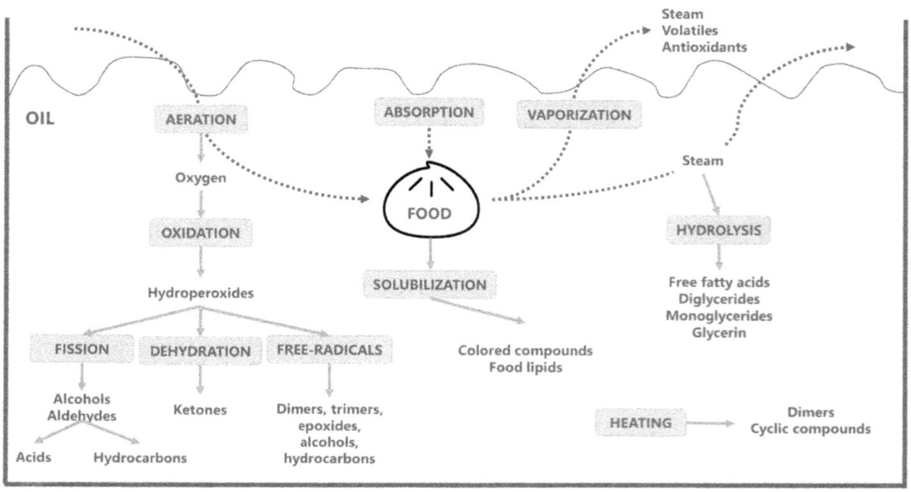

FIGURE 2.4 Reactions and chemical compounds produced during food frying. Adapted from Fritsch (1981).

Microstructure, molecular diffusion, capillary forces, and pressure gradients play an important role during the oil/water mass transfer when dealing with porous foods.

2.2.2.2 Frying Process

The general frying process described by Vitrac et al. (2000) relies strongly on the temperature. First, the initial heating where superficial boiling takes place and where the highest temperature difference between the oil and the product is observed. Then, the following period is related to the vaporization of free or capillary water. Here, free water is still being vaporized but also the boiling of free water moves toward the geometric center of the product. This water will move through the capillaries of the product to achieve the superficial region. Finally, the third stage has low drying rates and is when the product reaches the frying oil temperature that generally fluctuates from 140°C to 190°C. This process is detailed as follows (Banks and Lusas 2001).

2.2.2.2.1 Fryer Entry

Right after its immersion in hot oil, the starch contained in the food's surface is gelatinized and the product is immediately covered in bubbles as superficial boiling starts. The bubbling increases fast and avoids the sticking together of the product pieces. Simultaneously, a temperature front moves to the center by conduction. At this stage, the temperature difference between the oil and the product is the greatest, as well as the external heat flux (Vitrac et al. 2000). Steam evolution helps in limiting the product temperature to water's boiling point, also restricting the penetration of oil to the product.

2.2.2.2.2 Case Hardening

The cells situated on the surface of the product dehydrate and flatten. Only slight changes in the steam evolution can be observed while case hardening happens, and, in parallel, surface bubbling continues in some sites more rapidly than others. The initial moisture in the product is converted to steam, opening rupture channels in the structure of the product as surface moisture diminishes. Up to this point, dehydration has not allowed the formation of a crisp surface although the structure already shows some damage. Unlike other drying processes, during frying there is no true constant drying rate.

2.2.2.2.3 Surface Firming

The following layers of surface cells start a dehydration process, developing a crust structure. If the product is fried at high temperatures, then a thin and light-textured crust is developed; however, if the frying occurs at lower temperatures, the process is slower and results in a thicker crust and crunchier texture. Below the crust layer, the cell structure breaks leading to the formation of a hollow interior structure influenced by the frying temperature. At this point, both the crust and the internal structure are not yet terminated, but the elements affecting the texture in the finished product have been set.

2.2.2.2.4 Cooking/Moisture Reduction

The main objective of this phase is the penetration of heat and the reduction of moisture. It is important to highlight that uniformity in size of the products being fried is crucial for determining the cooking process and the time-temperature profile.

2.2.2.2.5 Finish Frying

During this stage, the temperature on the surface approaches the oil's temperature in a rapid manner. Flavor-producing reactions involving proteins and carbohydrates are produced by the low moisture content and the high temperatures. The increasing temperature in the oil allows a final moisture reduction resulting in a crust with a crispy texture. The oil intake increases during this stage, but it remains mostly on the surface of the product. To assure an optimum quality, the timing to remove the fried product must be precise.

2.2.2.2.6 *Oil Absorption*

Surface moisture, capillary action, and vacuum absorption are the factors affecting the oil intake by a product during frying. During the first stages, the moisture in the surface governs the absorption, followed by the capillary absorption. A large amount of oil associated with the final product is the one carried out of the fryer on the product's surface. During cooling, while water vapor contained by the product condenses, a partial vacuum effect occurs causing the absorption of oil on the surface of the product.

2.2.2.3 Current Trends in Frying

Recently, frying has experimented with novel solutions and alternatives that fit more the consumer's awareness on the impact of diet on health. The current strategies are focused on the reduction of oil uptake and on the increase in the stability of oil quality (Mellema 2003, Olayemi Oladejo et al. 2017, Pankaj and Keener 2017). The reduction of oil uptake targets the development or the use of alternative technologies for frying such as pressure frying, microwave frying, vacuum frying, radiant frying, or air frying. Besides, preprocessing and postprocessing treatments are suggested to achieve this goal (Figure 2.5).

2.2.2.3.1 *Preprocessing Strategies*

The use of batter and coatings is one of the strategies to reduce oil uptake since this is a function of the surface area of the material (Mellema 2003). Actually, this strategy has been reported as the most effective way to diminish oil absorption during frying according to Liberty et al. (2019). Commonly, coatings or batters have low moisture content, low moisture permeability, or thermogelling characteristics (Garcia et al. 2004). They also act as a barrier that avoids moisture loss (Liberty et al. 2019). For instance, the addition of hydrocolloids and gums (methylcellulose, carboxy methylcellulose or CMC, carrageenan, hydroxypropyl methylcellulose, pectin, guar gum, xanthan gum, locust bean gum, among others) into formulations helps in the reduction of oil uptake because they serve as a protective layer in the surface of the material (Garcia et al. 2004, Liberty et al. 2019). Recently, the effect of hydrocolloid addition on the batter of deep-fried banana fritters has been studied (Vengu et al. 2020). Hydrocolloids increased the batter pick-up and viscosity of the batter formulation resulting in a reduction of up to 55% oil content and equal sensory acceptance when

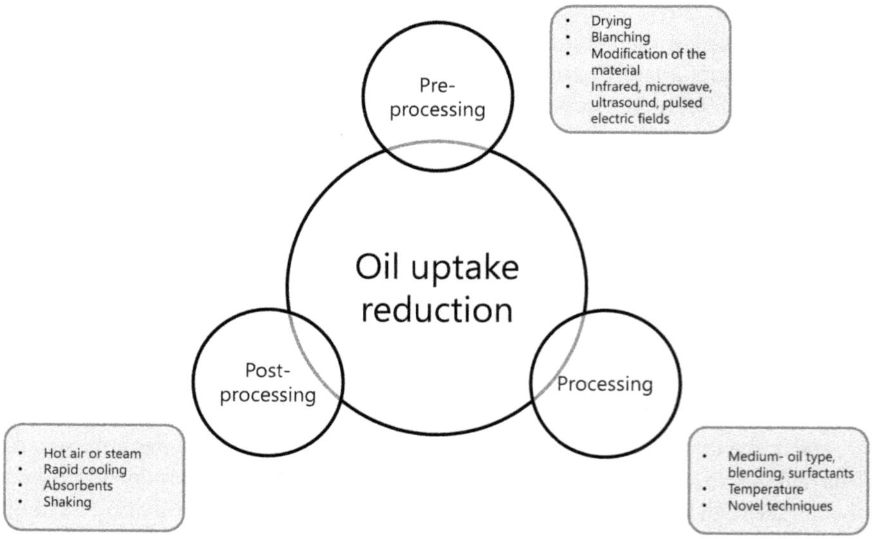

FIGURE 2.5 Strategies to reduce oil uptake during frying processing. Adapted from Pankaj and Keener (2017).

compared to the control. Fiber has also been added to a gluten-starch matrix to improve the quality attributes reducing the oil absorption (Dueik et al. 2014). In this work, the addition of wheat bran to the formulation achieved a reduction of up to 35% of oil uptake, resulting also in a high fiber content product. Wheat flour, a β-glucan source, has also been used in formulations to decrease the oil uptake of wheat chips (Yuksel et al. 2015).

2.2.2.3.1.1 Dehydration Drying is one of the most used strategies to reduce the absorption of oil during frying. The above because, in principle, if the layer at the surface has a low moisture content, less oil is retained. This reduction is also related to the shrinkage and plasticity that the food microstructure develops. In addition, when hot air drying is applied in starchy products, gelatinization might occur on their surface creating a barrier that avoids the penetration of oil (Dehghannya et al. 2016). Predrying has been applied in several products such as potato chips resulting in low-fat content and with the same level of sensory acceptance (Cruz et al. 2018). Ultrasound-assisted air drying has also been applied as pretreatment to potato strips before frying (Dehghannya et al. 2016). A combination of *novel technologies* like ultrasound and microwaves have been applied as pretreatment to improve the frying process. Fan et al. (2019) have evaluated the effect of osmotic-ultrasound dehydration as pretreatment to improve frying of purple-fleshed sweet potato. In addition, these authors applied microwave to assist vacuum frying of the material.

2.2.2.3.2 Processing Strategies

Several strategies have been recently proposed to improve frying processes. One of the most popular ones is air frying, due to the several domestic equipment currently available in the market. Some of the advantages of this technology are the reduction of expenses related to oil consumption, the reduction of the emission of contaminants to the environment, and the most attractive one is the offering of a healthier diet to the consumer (Zaghi et al. 2019). This technology uses small amounts of fat in a closed chamber. During this process, external deposition of oil droplets dispersed through a hot air stream is applied onto the product. Equipment design provides extremely high heat transfer rates between the product and the air in the chamber. In addition, the air is distributed uniformly through the product and thus a significantly lower amount of oil is used (Teruel et al. 2015). A comparative study between deep frying and air frying using different frying oils was performed by Santos et al. (2017). Air-fried potatoes contained about 70% less fat compared to the deep-fried counterparts and the novel technology also reduced the concentration of acrylamide (Chapter 20). Deep-fat frying, electrostatic frying, and vacuum frying were compared to air frying, being the last technology the one that yielded the lowest fat contents (9.8%) in fried fish skins (Fang et al. 2021).

Vacuum frying is a processing alternative for fried products which potentially yields products with high-quality attributes and low acrylamide content (Ravli et al. 2013). The use of vacuum conditions reduces the boiling point of water in the food, resulting in the elimination of moisture at lower temperatures thus frying is performed at lower oil temperatures. Because of these temperatures, nutrient preservation, oil quality protection, and toxic compound generation reduction can be achieved (Dueik et al. 2012). Moreover, the oil uptake in frying occurs during cooling and if vacuum conditions are applied, this cooling stage includes a period of vacuum break. This pressure recovery directly impacts the final oil content in the product. According to Mir-Bel et al. (2009), draining times and vacuum break velocity have a significant influence on oil uptake. Belkova et al. (2018) evaluated the use of vacuum frying achieving the minimization of acrylamide formation and the reduction of other undesirable chemical changes in frying oil that are discussed in Chapter 20.

Infrared radiation is also a promising technology to improve the frying process. Infrared radiation is transmitted as a wave that converts into heat once it gets to the water molecules and ions at the surface of the food. This technology provides less degradation of nutritional compounds when compared to convective or conductive mechanisms. It also has higher thermal efficiency with more uniform heating, impacting positively the energy costs and the processing time (Udomkun et al.

2019). Udomkun et al. (2019) tested how far infrared or conventional heaters affected the distribution of heat and the physicochemical properties of chicken nuggets. A more uniform distribution and a higher heating rate were observed with far infrared frying. No significant differences were observed in the physicochemical properties of the product. Additionally, infrared radiation saved energy and prolonged the shelf life of the frying oil.

Novel treatments have also been used in combination with vacuum frying to improve the process. The effect of ultrasound using microwave-assisted vacuum frying has been studied on the frying rate and quality of mushroom chips (Devi et al. 2018). These authors concluded that the combined technologies accelerated frying rate, diminished oil uptake, and produced mushroom chips with better texture and color when compared to vacuum-fried samples. Likewise, Sun et al. (2019) investigated the impact of this combination of processes on the frying oil quality and concluded that the assisted technologies delayed the deterioration of the frying oil. Sun et al. (2018 a,b) have also used the combination of these techniques to potato chips and concluded that the assisted methods decreased oil uptake and improved the texture and color properties of finished products. Additionally, these authors observed a decrease in energy consumption.

2.2.2.4 Postprocessing Strategies

The major portion of oil absorbed during frying occurs when the product is removed from the fryer and starts the cooling stage (Hoon 2010). Because of this, some strategies have also been proposed such as de-oiling the final product during the cooling period. De-oiling using a centrifuge helps in the removal of the surface oil resulting in a reduced oil content of the final product. Centrifugal forces facilitate the separation of the oil from the porous surface (Hoon 2010). These de-oiling mechanisms are commonly combined with other strategies to improve the frying process and the final product characteristics. Sothornvit (2011) proposed post-frying centrifugation of vacuum-fried banana chips. The combination of an edible coating and a high centrifugation speed leads to a reduction in the oil absorption of up to 33.7%. Moreira et al. (2009) explored the effect of centrifuging as a de-oiling mechanism of vacuum-fried potato chips. After centrifugation, the samples had a reduction of oil content of 77% when compared with the noncentrifuged ones. Post-frying vacuum applications have been applied by (Ahmad Tarmizi et al. 2013). Under vacuum conditions, draining of the final product reduces its moisture content. When compared to atmospheric drainage, vacuum drainage briefly diminished both the oxidization of the product and the generation of free fatty acids by 50%. Similarly, Ravli et al. (2013) studied a two-stage frying process (vacuum frying coupled to a de-oiling mechanism) to produce low-fat sweet potato chips. Their process reduced about 15% of the oil content when compared to their single-stage (only vacuum frying) process. More strategies during this critical stage are needed in order to reduce the oil content of the product without affecting the physicochemical and sensorial properties of the final snacks.

2.2.3 Drying Principles and Fundamentals

Drying is one of the oldest and most common unit operations applied to snacks. During drying, the water is eliminated to achieve a target content, resulting in a longer shelf life due to less water available for microorganisms and undesirable reactions. Because of this, it is a process commonly used in the production of many vegetable and animal-based snacks. The dehydration process is widely employed for the production of natural fruit rings or veggie chips and dehydrated animal/marine products discussed in Chapters 16 and 18. This unit operation involves heat and mass transfer that affect important chemical and physical transformations in the food matrix that undoubtedly affect end-product quality. Shrinkage, puffing, glass transitions, nutrient reduction, changes in color, odor, texture, are some of the changes that can occur during this process (Mujumdar and Devahastin 2004, Ibarz and Barbosa-Cánovas 2005).

The mechanisms of water transfer that occur during drying are: the water movement under capillary forces, diffusion due to concentration gradients, superficial diffusion, vapor diffusion in the

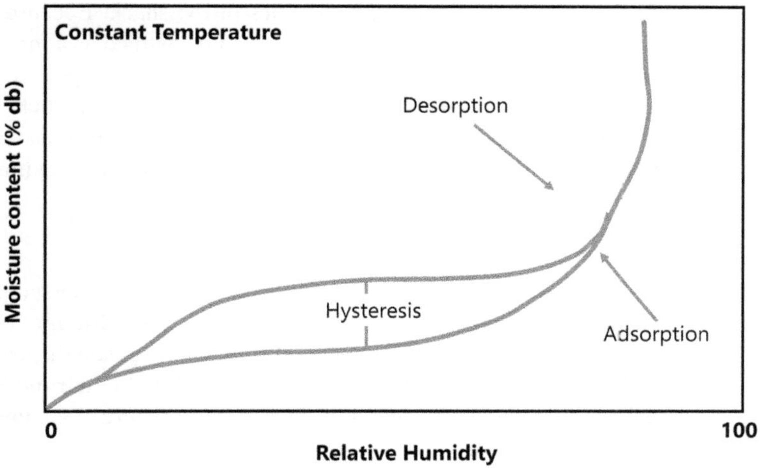

FIGURE 2.6 Typical adsorption and desorption isotherms.

pores filled with air, flow due to hydrostatic pressure gradients, and the combination of these phenomena. The material to be dried is also an important factor because of the possible changes that occur while eliminating the water contained in the food matrix (Ibarz and Barbosa-Cánovas 2005).

2.2.3.1 Drying Process

The equilibrium moisture content (EMC) is the moisture of a solid in equilibrium with air at a given relative humidity (or water activity, a_w) at a specific or fixed temperature. A sorption isotherm is the EMC (g water/100 g solids) of a product plotted against the relative humidity or a_w at a determined temperature (Chirife and Iglesias 1978, Al-Muhtaseb et al. 2002, Basu et al. 2006, Peleg 2020). Figure 2.6 shows typical adsorption and desorption isotherms, and their difference (hysteresis). The knowledge of this behavior, mainly the desorption one, is crucial when setting processing conditions for drying and selecting storage and packaging conditions of snack products.

Water sorption data is determined experimentally because, besides the temperature, the physical structure and the materials' composition and interactions affect the water sorption process. Physical and/ or chemical modifications such as the pore structure and size can cause important variations in the moisture-binding capacity of the material (Mujumdar and Devahastin 2004). Recently, a way to predict moisture sorption curves from three experimental data (initial moisture content and two more at different times) was proposed by means of a mathematical method by Tejada-Ortigoza et al. (2020b). During the drying process, *kinetics* is crucial to determine and predict the behavior of the material. Generally, once the material passes through an initial period of adjustment, the moisture content decreases linearly with time due to the evaporation of water. Then is followed by a nonlinear decrease in moisture content until the solid reaches an EMC.

Finally, another important term in drying processing is A_w, which is the availability of water for microorganisms' growth and for chemical reactions. It is defined as the ratio of the partial pressure and the equilibrium vapor pressure of water at the same temperature. Also, the amount of water and temperature defines the region in a state diagram at which a snack is, depending on its desired sensorial characteristics. Physical changes related to glass transition are a function of the moisture content. This will also have an impact on the stability and the shelf life of the product (Labuza and Hyman 1998, Payne and Labuza 2005). In low-moisture snacks, the loss of crispness due to water plasticization is an undesired phenomenon caused by the glass transition (Roos et al. 1998). The importance of this concept in snacks relies on its effect on crispness or gumminess as a textural characteristic, which is a major cause of product rejection by consumers (Katz and Labuza 1981). For instance, moisture content has an effect on the texture of most dry snacks by means of softening or plasticizing the matrix, altering the mechanical properties of the snack. Because of this, it has been suggested that textural acceptance in the snack industry can be defined as a function of A_w. For popcorn (Chapter 9), saltines (Chapter 10), puffed corn curls (Chapter 12), and potato chips (Chapter 14), the sensorial acceptability found ranged between A_w values of 0.35–0.50, where amorphous to crystalline transformations occur (Katz and Labuza 1981). Besides, it must be considered that at this A_w range lipid oxidation might also occur. Water might have a protective effect against oxidation at certain A_w (Labuza et al. 1972).

2.2.3.1.1 Equipment

The basic and general configuration of a dryer is a chamber where the food is introduced, equipped with a ventilator, and conducts that allow the circulation of the hot air through and around the product. Water is eliminated from the surface's product and conducted outside the dryer through an air flow. Several classifications of dryers have been proposed according to their mode of operation such as heat input, operating pressure, drying medium, drying temperature, and stages. Figure 2.7 depicts the most common drying systems.

Dryers are also classified based on their mode of thermal energy input into direct or convective and indirect dryers. Direct dryers are the most common ones in the industry despite their comparatively lower thermal efficiency. Indirect dryers are the ones that transfer the heat to dehydrate the material using a medium like steam, gas, or thermal fluids. These dryers commonly use vacuum or a gas flow to remove the moisture from the chamber.

2.2.3.2 Current Trends in Drying

The research in this area is focused on finding solutions to reduce the negative effects of the process, i.e., changes in cellular and chemical structures that led to physical variations and oxidative reactions. In addition, research is also aimed to study the addition of compounds with protective effects during the process, the development of structural elements for encapsulation before spray drying, and the promotion of reactions that result in positive functional changes (Betoret et al. 2015). Many of these studies converge in the development of healthy snacks based on fruits and vegetables (Chapter 16) due to their vitamin, minerals, and bioactive compounds with a positive impact on the human diet (Chen et al. 2018, Donno et al. 2019, Nowacka et al. 2019).

The use of freeze-drying is one of the most common methods to produce fruit and vegetable snacks and the development of healthy bars because, contrarily to the conventional hot air drying, yields crispy products with desirable color and texture (Chen et al. 2018, Ciurzyńska et al.

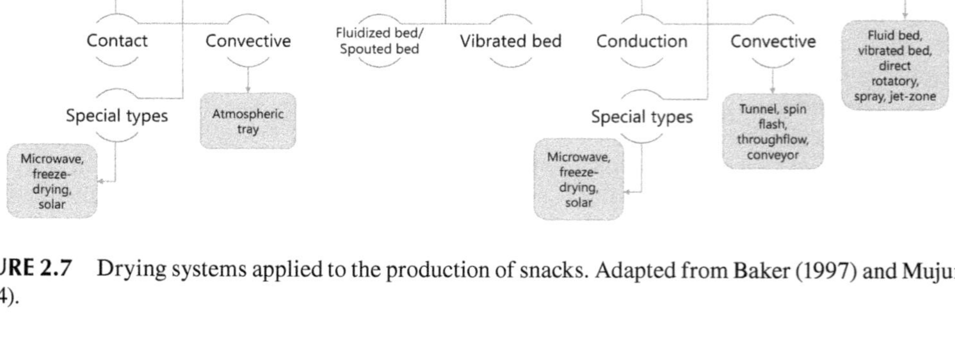

FIGURE 2.7 Drying systems applied to the production of snacks. Adapted from Baker (1997) and Mujumdar (2004).

2020). Apple, goji, kiwi, and kaki were freeze-dried to be consumed as snacks and the main health-promoting phytochemicals determined with the aim to stimulate large-scale exploitation of fruits (Donno et al. 2019). Likewise, probiotic enriched apple snacks have been dried by freeze-drying followed by microwave-vacuum drying (Li et al. 2018).

Freeze-dried vegetable snacks using hydrocolloids were developed by Ciurzyńska et al. (2020). Sodium alginate with calcium lactate and locust bean gum with xanthan gum was used to create a porous structure in the vegetable bar. The authors concluded that although a high-quality final product was obtained, freeze-drying is an energy-consuming process with high capital costs.

2.2.3.2.1 Microwave Drying

Microwave drying is not a common process applied in mass production in the food industry, although it has been studied as an alternative to hot air drying because it requires shorter drying times and lower temperatures (Hnin et al. 2019, Ciurzyńska et al. 2020). The development of equipment with combined methods to improve the quality of the final product is constantly studied. Microwave-vacuum drying is a variant of this method and it has been coupled with a freeze-drying system to shorten the processing time with results similar to the ones obtained in freeze-drying but with a reduction in the energy consumption due to shorter drying times (Hnin et al. 2019). These combined methods have been researched by Hnin et al. (2019) in the development of rose powder-yam snack chips resulting in high scores of sensory attributes and higher retention of total phenolics and anthocyanin pigments when compared to hot air or single microwave-vacuum drying. Similarly, combined freeze-drying and microwave-vacuum drying has been used to yield functional okra snacks with the advantages of significantly reducing the drying time and the energy consumption when compared to conventional hot air drying (Jiang et al. 2017). The effect of microwave-vacuum drying has also been compared to osmotic dehydration for the production of dried cranberries. Ultrasound and pulsed electric fields were used as pretreatments to improve the drying process (Nowacka et al. 2019).

2.2.3.2.2 Infrared Drying

Drying with *infrared* rays is a highly efficient method because energy is transferred directly to the product surface without heating the surrounding air (Chen et al. 2018). Consequently, it is an energy-saving method with a shorter processing time compared to convective drying. However, the infrared emitter might represent a comparatively higher investment cost (Ciurzyńska et al. 2020). Infrared has been successfully used as pretreatment to improve the conventional drying methods. Chen et al. (2018) evaluated the use of sequential infrared blanching and hot air drying to develop a crispy carrot snack resulting in a reduction of drying times (32, 41, and 45%) when applying 60°C, 70°C, and 80°C. The infrared pretreatment also led to crispier and more colorful products when compared to the noninfrared-treated samples.

2.2.3.2.3 Explosion Puffing Drying

This is an alternative to produce snacks with similar structural characteristics to those obtained by freeze-drying. It is an intermediate stage in the drying process caused by either a sudden increase in temperature or a sudden decrease in pressure that causes the water in the product to vaporize. The sudden water vaporization results in an expansion and creation of an internal porous crumb structure with less time and energy expended (Zou et al. 2013). Instant controlled pressure drop has also been used to produce crispy materials with a similar approach to puffing drying. Recently, Peng et al. (2018) evaluated the application of this method to fabricate dried carrot chips.

2.2.4 Baking

Various sorts of chips, crackers, pretzels, extruded snacks, and legume and tree nuts are produced by the application of heat in an oven (baking). The use of this thermal operation plays relevant roles in today's diet and consumers' interest due to the calories intake. Compared to frying, baking is an option to provide low-fat or fat-free snack foods (Oluwole et al. 2014). For the snack industry, baking enhances the reduction of moisture with the aim of getting desired organoleptic properties and longer shelf life. As in drying, concepts such as initial and final moisture contents, A_w, and water evaporation rate during the process are crucial to understanding the process. For instance, crispness and other mechanical parameters are closely related to A_w and glass transition (Oluwole et al. 2014, Roos et al. 1998). However, baked snacks not only look for the reduction of the water content of the product but also in the improvement of relevant sensorial features such as texture and flavor.

The concepts of "crispy" and "crunchy" are part of a complex of sensations that includes flavor, taste, odor, sight, texture, and sound. These firmness and brittleness are specially important because they affect how the snack fractures and disintegrates while being masticated (Dijksterhuis et al. 2007). The baking process is performed during a specific period of time and a particular temperature depending on the desired final characteristics. Temperature is one of the main parameters to consider because it controls Maillard reaction that leads to desirable sensorial qualities such as color, taste, and crispness (Nath and Chattopadhyay 2007, Oluwole et al. 2014). Because of this, baking is considered one of the most sensitive processes because it influences the chemical, physical, and sensory qualities of snacks and also affects production cost (Khatir et al. 2015).

Heat transfer during baking generally occurs in three major possible means: convection, conduction, and radiation. Besides the heating temperature and the time, the velocity of the heating medium and the nature of the food product are other factors affecting the final quality of the product. Thus, optimization of the oven parameters is crucial to assure temperature uniformity, efficiency, and humidity level inside the oven. The conduction mechanism governs during baking of snacks at low temperatures. Then, when the temperature is increased, simultaneous heat and mass transfer occur, where the evaporation-condensation mechanism dominates (Purlis and Salvadori 2009, Chhanwal et al. 2011, 2019). Here, water evaporates and vapor condenses at the colder side of the gas cells until the inside of the product reaches a temperature of 100°C (Chhanwal et al. 2011). The way the heat is provided plays an important role in the crust formation, browning, and flavor development of the baked snack. The industry constantly looks for ovens where the process can be accelerated, the yield increased, and the cost reduced while keeping or enhancing product quality (Chhanwal et al. 2019).

2.2.4.1 Current Perspectives on Ovens for Baking

Innovation and improvements in baking technology are mainly focused on final product quality, nutritional composition, and economic considerations (Mondal and Datta 2008). For small-scale production, electric ovens are commonly used; however, for large-scale operations, a light diesel oil-operated oven with digital temperature indicator and timer is normally used (Nath and Chattopadhyay 2007). These common convection ovens have the advantages of yielding good quality products and are relatively easy to design even though they are not as efficient and require longer baking times. Currently, hybrid heating techniques have proved to enhance the energy efficiency of the process maintaining the quality of the product. Besides convection, ovens using technologies like microwave, infrared, jet impingement have been recently studied (Chhanwal et al. 2019).

2.2.4.1.1 Microwave Oven

As previously reviewed, microwave technology has been widely used both in household and food industries not only for drying and baking, but also for thawing, heating, and cooking because this technology allows drying and precise control. It has the main advantages of eliminating pre-heating steps, instant start-up, space-saving, accuracy, and energy efficiency. However, quality parameters such as browning, starch gelatinization, and volume expansion ideal for snacks are affected by this technology. Along with these, nonuniform heating, increased cold and hot spots, and gummy texture are some other difficulties when baking using microwaves (Rakesh et al. 2009).

Microwave ovens generate the heat from inside the food, and then the heat is transferred by convection to the rest of the product. These patterns affect product characteristics. Moreover, due to the fast generation of heat, there is not sufficient time for gelatinization, expansion, and flavor development (Chhanwal et al. 2019). However, the combination of methods might be an option to improve the results obtained with this technology.

2.2.4.1.2 Infrared Oven

Infrared radiation is widely used because of its rapid surface heating capacity. The electromagnetic radiation has a wavelength ranging from 0.78 to 1,000 μm, being the far-infrared the most suitable for heating foods. Besides its enhanced efficiency, it also presents many advantages like reducing processing times, energy expenditure, uniform surface heating, and good sensorial and nutritional value of cooked foods. Some disadvantages of this technology are low penetration power inside the product and that heating is influenced by the characteristics of the snack surface and penetration depth of the radiation. For this technology, the penetration depth depends on the moisture content and it is affected by the wavelength. For this reason, the penetration depth varies significantly among food materials (Chhanwal et al. 2019, De Pilli and Alessandrino 2020).

Even though, infrared baking has been extensively studied for breads, cakes, or cookies but not for savory snack production. In general, in these studies, researchers have reported poor quality of the final products. For snacks, radiation has been mainly used as a roasting method for nuts (Chapter 13) because enhances flavor, color, texture, and appearance (Bagheri et al. 2019). Roasting nuts by infrared radiation has been extensively revised by Bagheri (2020).

2.2.4.1.3 Jet Impingement Oven

Jet impinging is a process where a jet of fluid impacts the surface of the product to achieve a fast heat and mass transfer. It is a forced convection heating through hot air applied at 10–100 m/s. Jet impingement has been used by many industries (textile, glass, metal, food) mainly for cooling. Nowadays, the food industry uses this technology for cooking, toasting, roasting, and baking with advantages such as high efficiency, short processing times, rapid moisture removal, uniform heating, reduced moisture loss, and most importantly, quality parameters like the ones obtained after usage of conventional ovens. This technology allows the reduction of about 25°C of the air temperature compared to conventional ovens which translates into 30% more efficiency. However, it might present disadvantages such as thick crust formation, low porosity, and darker colorations, besides that more energy is required (Supmoon and Noomhorm 2013, Chhanwal et al. 2019). Hot air impingement has been combined with infrared radiation to improve the quality of potato chips (Supmoon and Noomhorm 2013) and the preferred technology to process flash-fried savory snacks with reduced oil and caloric contents. Healthier snacks like apple and jujube slices have been also processed by using this technology as pulsed air (Wei et al. 2018), as a combination with microwave vacuum (Yin et al. 2019), or as hot-air assisted radiofrequency (Peng et al. 2019) with improved final product quality.

2.2.5 Specialized Processes: Principles and Equipment

2.2.5.1 Peelers and Slicers

One of the first stages in the production of snacks based on fruits and vegetables is the removal of skin or peel, as well as cutting the material to obtain the desired size and shape to subject the product to subsequent processes such as frying or drying. Advances in skinning and slicing equipment have been linked primarily to the potato-based industry (Chapter 14). The demand for potato (or other vegetables used in the snack industry) peeler and slicer equipment in many well-established markets is mainly driven by the need for replacements. Hence, most of the demand is to substitute old peeler and slicer equipment with new energy-efficient ones. The top three emerging market trends driving the global potato peeler and slicer equipment market according to Technavio food and beverage research analysts are: increasing focus on energy-efficient equipment, growing preference for automated peeler and slicer equipment, and applied new technological advancements (Technavio 2021).

Designers have introduced eco-friendly potato peelers and slicers, which optimize the efficiency of the production cycle in order to offer better alternatives to mitigate the negative effects of global warming and high energy consumption. This will encourage the adoption of new equipment and drive upgrades. Manufacturers are providing state-of-the-art components such as energy-efficient and electrically commutated motors that are used to exceed energy standards without forgoing performance and capability. The company provides stainless steel commercial peelers equipped with an auxiliary contact for an external electric valve that besides are energy efficient owing to engine optimization.

On the other hand, the market is witnessing an increasing demand for automated appliances, owing to their advantages such as lesser risk, efficiency, and quick return on investments. Many units of automated peelers and slicers come with improved features such as laser sorters and automatic defect removal (ADR) systems.

The quick adoption of technology and increasing consumer preference for innovative potato peelers and slicers provide high-quality potato peeling equipment with minimal skin loss, the design uses a combination of peeling and polishing mechanisms to provide a turnkey solution based on the specifications provided by end users. Other options are the water slicing methods, which are designed to apply high pressure to optimize cuts with a lower percentage of waste. All these advancements will likely increase the demand for new vegetable peeler and slicer equipment.

Some examples of industrial equipment used for the removal of skins and peels of fruits and vegetables are depicted in Figure 2.8. The brusher is a machine to remove skin and peel from food, gently remove loose peel, or aggressively product defects by using automatic brushing technology, dry peeler uses a specially designed stainless-steel drum with smooth perforated panels that adapt to each product. Inside the drum, there is an agitator that ensures the continuous mixing of the potatoes or the product to be treated, and the transfer of the remains of skin to the drum, which rotates at high speed using an external shaft so that the peel can thus pass to the pump of dry waste. The dry peeler manages to separate external residues (rest of skin, peel) without using water and with minimal losses of performance. On the other hand, the steam peeler achieves uniformity in the removal of the skin and has managed to reduce the peeling speed and energy consumption by up to 25%.

2.2.5.2 Poppers

A large category of starchy snacks is produced via indirect expansion, which happens if snack pellets are deep-fat fried or expanded with hot air or in a microwave oven (Moraru and Kokini 2003). Snacks like popcorn, popped and puffed rice, popped sorghum, popped and roasted wheat,

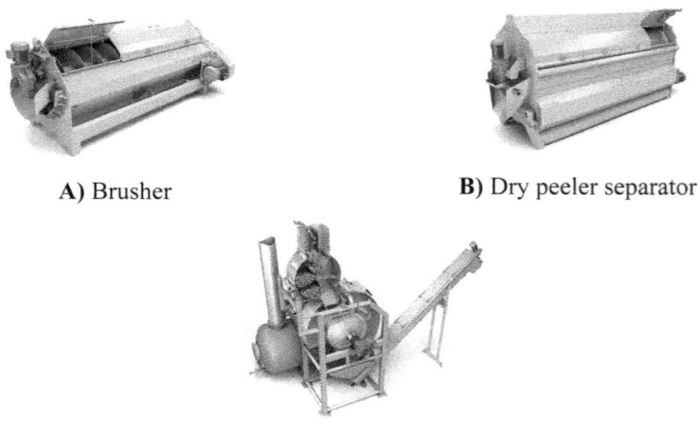

A) Brusher

B) Dry peeler separator

C) Steam peeler

FIGURE 2.8 Industrial equipment for the removal of skins and peels of fruits and vegetables (https://www.tomra.com/es-es)

and puffed soybean and other related legume seeds are examples of products obtained with these technologies.

One of the most popular and widely consumed snacks is popcorn thoroughly described in Chapter 9. Popped corn is made from a special type of maize which considerably differs in terms of physical characteristics and composition from other types used for human or animal nutrition. There are several varieties of popcorn, including rice popcorn, pearl popcorn, red, black, rainbow, yellow, white, and blue. These are known as gourmet specialty popcorn varieties. At present, work is being done to develop new varieties and hybrids with improved quality especially in terms of pop ability and stability of expanded products.

Popcorn kernels are normally environmentally dried to reach moisture levels between 13 and 14%. At this moisture level the popcorn kernels show a higher expansion rate (Metzger et al. 1989). When the popcorn temperature exceeds 100°C, its water content boils and reaches a thermodynamic equilibrium at the vapor pressure, as in a pressure cooker (Hoseney et al. 1983). Above a critical vapor pressure, the hull suddenly breaks. At the same time in the popcorn endosperm, the starch granules expand adiabatically and form a spongy flake of various shapes (Quinn et al. 2005). As shown in the Figure 2.1 the critical temperature (minimum to obtain a high percentage of popped corn) is around 180°C (Virot and Ponomarenko 2015) (Figure 2.9).

There are essentially two sorts of popcorn configurations: mushroom or butterfly. Popped kernels with a spherical shape are called a mushroom or ball type. Due to its configuration, it is preferred in the confection industry because it is less susceptible to breakage, more resistant to handling, and more efficiently coated with flavorings and confectionery syrups. The more popular butterfly type has a higher expansion or lower apparent bulk density, and better mouthfeel and it is preferred for on-premises popping, where it is sold by volume. For the production of mushroom or ball popcorn, more heat should be applied during popping. Temperatures of 215°C and 235°C are recommended to favor the production of butterfly- and mushroom-shaped flakes, respectively (Snack Food Association 1992, Serna-Saldivar 2008).

There are two main methods to popping popcorn: dry and wet methods. The dry method uses dry heat to pop the kernels, such as a fire or an air popper. The more popular and employed wet method uses oil to help distribute the heat and results in a more even popping. Microwave popcorn and most other commercial products are obtained using the wet method. Kernels with the right moisture content will be fluffy and light whereas counterparts with higher moisture will be more dense and gummy. Unpopped kernels have most likely dried out or were mechanically damaged especially in the outer cover or pericarp. On average, a grain pops when it reaches a temperature of 175–190°C (Virot and Ponomarenko 2015). The detailed dry and wet popping processes are deeply covered in Chapter 9.

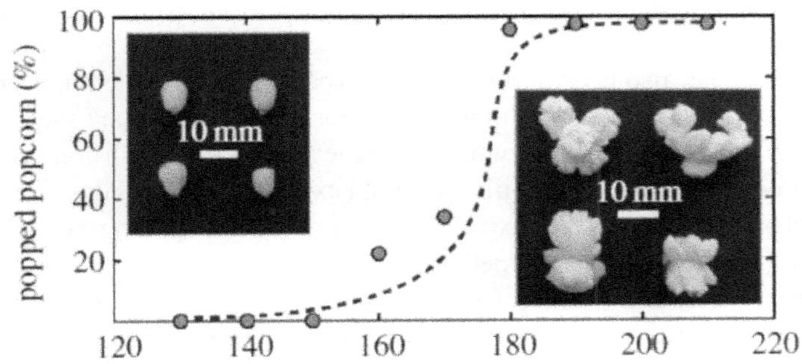

FIGURE 2.9 Percentages of popped popcorn in an oven at increasing temperature; the dashed line is a guide to the eyes. The critical temperature is approximately 180°C (insets). Snapshots of unpopped corn (kernels, left) and popped popcorn (flakes, right) (Virot and Ponomarenko 2015).

2.2.5.2.1 *Wet Poppers*

Normally, popcorn, oil, and salt are placed into a heated pan for the production of wet popped products. An agitator on the bottom constantly stirs the mixture, allowing better and even heat distribution among the kernels. When the temperature of the popcorn is high enough, the kernels gradually burst. At the moment of expansion, the oil is absorbed, and the salt sticks to the flakes. Poppers are either heated by gas or electricity, the first being the most preferred in the industry. The most frequent popcorn-to-oil ratio is 3:1 (w/w). The popping cycle varies from 2.5 to 3 min. In commercial operations, the wet poppers are usually arranged in tandem with 4 to 8 machines, which dump the popped corn onto a belt conveyor which transports the popcorn to a sifter, which removes unpopped kernels and other fines. After sifting, the popcorn should be immediately packaged to avoid moisture absorption and loss of texture (Snack Food Association 1992, Serna-Saldivar 2008, Serna-Saldivar 2010).

2.2.5.2.2 *Dry Poppers*

Dry poppers come in two basic types: the older rotary wire drum and the modern steel auger (Chapter 9). The rotary type heats popcorn kernels over an open flame, whereas in the steel auger kernels are thermally treated while traveling along the screw. The heated air is injected with enough pressure to force its way through the kernels, enhancing the transfer of heat. This action also stirs kernels in a fluidized bed fashion (Snack Food Association 1992, Serna-Saldivar 2010). The fluidized bed process has four basic variables: feed rate, temperature (210–230°C), popping time (affected by the rpm of the screw), and heated air pressure. The normal residence times vary from 65 to 105 s. Dry poppers are designed for large-scale operations and can pop up to 70–450 kg/h. When popcorn is processed, it loses 12–13% moisture and 8–10% in scrap (Serna-Saldivar 2008). Dry poppers are usually accompanied by a sifter and a coater or tumbler. The popper should have exhaust ventilation to remove the water vapor generated during processing. The sifting operation should efficiently remove unpopped kernels and broken flakes. The clean popped flakes are then flavored in a tumbler. Oil, salt, and flavorings are directly applied or sprayed onto the product. The main advantage of dry poppers is the possibility of manufacturing reduced or even fat-free popped corn snacks. The final oil content of snacks is controlled during the seasoning operation.

2.2.5.2.3 *Microwave Popping*

Microwave popping is today considered as a vast and substitute technology for popping grains such as maize, rice, and sorghum. Basically, microwave processing is being performed by electromagnetic radiations with a microwave frequency of approximately 2,450 megahertz (MHz) and the main principle behind food heating is dielectric heating. Besides processing conditions, the moisture content of the incoming food material is highly associated with expansion rate. This is the only food processing operation which transfers the energy from inside to outside of the food products (Solanki et al. 2018).

Microwave popping has also been applied for the expansion of extruded half products or pellets for the production of snacks with lower amounts of calories. The important steps in the microwave expansion of starchy pellets are: (i) the drying before the start of the expansion process, (ii) the start of the expansion process via nucleation, (iii) the actual expansion process due to the growth of individual steam bubbles, (iv) the end of the expansion process via cell opening, (v) the possible (partial) rupture of the cell, (vi) shrinkage of the pellet due to additional drying especially when the pellet is still in the rubbery state, and (vii) the fixation of the expanded structure if the expanded pellet goes into the glassy state. At the end of the heating process the expanded pellets can be (i) burnt, (ii) expanded, (iii) not expanded, or (iv) expanded, but clumped (van der Sman and Bows 2017). For this reason, the microwave process parameters should be carefully controlled especially in terms of power level, power density, and residence time (Sharma et al. 2014). The other critical factor is the design of the pillow-style microwave bag which normally has a susceptor plate that concentrates microwave heat onto the bed of the popcorn kernels blended with oil and flavorings (Chapter 4).

For microwave popping, the recommended power output in conventional microwaves is 660 W for 2 min on high power (Serna-Saldívar 2008).

2.2.5.3 Coating, Flavoring, and Battering

Oil, powder, and granule applicators, including oil and cheese sprayers, powder dispensers, electrostatic salters, and coating tumblers are normally used for snack flavoring, The detailed description of the equipment and flavoring technologies are discussed in Chapter 3.

The snack industry typically employs batch or continuous coaters. Batch coating for salty products is typically performed with an inclined enrobing pan that turns on an axis inclined at 30°. The base product is placed in the pan, and the oil coating and seasonings are introduced as the drum rotates. The coating is either poured as solid blends or pumped with an oil-based flavoring mix. The batch pan is allowed to work until the coating is fully spread and evenly distributed onto the product. The most common way to apply coatings is with a continuous machine consisting basically of a horizontal stainless-steel cylinder that rotates on its axis. The speed of rotation and the angle of the cylinder are adjusted to control the product flow rate or residence time. The coater is equipped with either a dry applicator or a pump system that sprays the coating in the first part of the inclined drum. The disadvantage of the dry powder is that it frequently comes off on the consumer's hands. The most common method is to spray cheese slurry onto the snacks. The cheese flavoring is produced by mixing oil with powdered cheese and other ingredients (Chapter 8). In order to apply the coating, the mixture should be heated to around 50°C to melt the cheese and prevent plugging the spray system. For instance, cheese-flavored popcorn normally contains 15% cheese, 28% oil, and 57% popcorn (Snack Food Association 1992, Serna-Saldívar 2008, Serna-Saldivar 2010).

A special type of coater is used to produce sugar-coated popcorn. The batch system consists of a copper or stainless-steel kettle in which sugar, glucose, and water solution are boiled to make caramel. The boiling solution is then gradually poured into a blender that is made of a rotating drum and a rotating auger inside at one edge. The resulting action lifts and mixes the popcorn with the sugar coating. A small amount of sodium bicarbonate can be added to the syrup to make it foam and double its volume. After the foamed sugar and popcorn are well coated, a small amount of oil/lecithin is sprayed to enhance the separation of the popcorn (Snack Food Association Association 1992, Serna-Saldivar 2008).

For caramel corn, coatings are generally composed of sugar syrup, vegetable oil, and other additives (coloring/flavoring compounds) that are heated at 144–149°C to produce a free-flowing solution (Serna-Saldivar 2008, Serna-Saldivar 2010).

2.2.5.4 Additional Processes

There are additional processes and equipment that are important for the operation of snack production lines, such as the following.

2.2.5.4.1 Transfer and Storage Equipment

In particular, storage of raw materials is of high importance, given that raw materials are natural products that have limited life and can be seriously affected by storage conditions. For instance, the sugar content of potatoes may be higher than desired when potatoes are stored at low temperature (2–4°C) due to lowing of sugars breakdown to carbon dioxide and water (Chapter 14). The sugar content can be lowered back to an acceptable level (< 0.2–0.4%) by reconditioning at > 13°C for several weeks prior to use. The right sugar content in potatoes ensures that Maillard (browning) reactions that result in undesirable dark spots on fried potato chips are avoided.

2.2.5.4.2 Packaging

A primary concern for packaging is to ensure a long shelf life of the finished product (Chapter 4). From a process control viewpoint, ensuring that the weight of packaged snacks is as close to but certainly above the value specified on the package has obvious economic implications.

2.3 APPLICATION OF EMERGING TECHNOLOGIES TO IMPROVE SNACK PRODUCTION

A lot of effort is put to decrease the energy consumption of drying, frying, or other thermal processes applied in the snack industry. This effect might be achieved by shifting traditional technologies like hot-air to microwave-vacuum, high hydrostatic pressure (HHP), pulsed electric fields (PEF), or ultrasound (US). These emerging technologies modify different cellular tissues and food structures that influence the process efficiency and quality of snacks. Dziki (2020) recently reviewed various pretreatments before the application of freeze-drying and concluded that mainly PEF and US presented advantages in terms of accelerating the drying process, generating products with improved sensory qualities and the retention of physical properties and relevant bioactive compounds in final products. Table 2.2 shows specific examples for freeze-dried snacks obtained from fruits and vegetables, confirming the beneficial effect of different pretreatments (HHP, PEF, and US) in terms of reduction of process times and improvement of both sensory and functional qualities. In the case of snacks and other fried foods, a review of Zhang et al. (2020) summarizes the analysis of novel alternative frying technologies, such as microwave-assisted vacuum frying (MVF), pulse-spouted microwave vacuum frying and power ultrasound, and microwave-assisted vacuum frying. Ultrasound microwave-assisted vacuum frying (USMVF) has proven to be efficient to produce products with lower oil content and better texture characteristics. In addition, different pretreatment technologies, such as HHP, PEF, US, and infrared radiation employed as post-frying processing, can be applied with positive synergistic effects. The operation and application of HHP, PEF, and US as pretreatment in the production of snacks are reviewed below.

2.3.1 High Hydrostatic Pressure (HPP) as Pretreatment in Snacks Elaboration

HPP is a nonthermal technology that has been adopted by various segments of the food industry in the past 30 years. The potential of maintaining nutritional, sensory, and quality while inactivating pathogenic and deteriorative microorganisms and modifying enzymes activities makes this technology very appropriate for meeting increasing consumer demand for minimally processed, healthier, and safer products. HPP is a postpackaging process in which a product kept in a flexible package is placed inside a compression chamber. The chamber is sealed and filled with a pressure transmission fluid (commonly water and oil) and the system is pressurized by a pumping and pressure intensification mechanism. HPP has been shown to provide efficient inactivation of vegetative microbial cells. Commonly, pressures between 400 and 800 MPa for 1 and 15 min have turned out to be effective for preserving quality and achieving microbial stability, thus providing prolonged shelf life under refrigerated conditions (Rosenthal et al. 2018). This technology has been mainly used for food preservation, although it has possibilities for other applications, such as texture modification, shift or assisted freezing, and extraction of food components (Knorr et al. 1998, Naik et al. 2013, Rastogi and Knorr 2013). Besides being used for the preservation of fresh-cut fruits, HPP has other potential applications such as pretreatment before drying and frying vegetable snacks. According to Oey et al. (2008), HPP can partially modify the cell permeability of fruits and vegetables benefiting the quality of osmotic-dried, air-dried, freeze-dried, or fried snacks. Rastogi et al. (2000) observed a higher diffusion coefficient in high pressure-pretreated pineapple cubes, showing that HHP pretreatment of vegetables can minimize the time of drying while maintaining the quality of the final products. Similarly, a study on textural changes and drying rates of high pressure-pretreated pineapple slices was conducted by Kingsly et al. (2009). These authors observed that hardness, springiness, and chewiness of pineapple slices were reduced by high pressure whereas cohesiveness was not significantly affected. On the other hand, in another comparative study of the effects of freezing or HHP as a pretreatment before vacuum frying of carrot-based snacks, Albertos et al. (2016) showed that both pretreatments caused cell modification, as observed from microstructural analyses. Pretreatment application helped in maintaining phenolic content and antioxidant

TABLE 2.2
Methods of Pretreatment of Fruits and Vegetables before FD, Drying Conditions and Effect of Pretreatment

Raw Material	Method and Conditions of Pretreatment	FD Conditions	Main Effect	References
Guava fruit	BL and ultrasounds (37 kHz, 240 W, 65°C, 5 and 10 min)	40°C, 75 Pa	Reduced DT, better color preservation, decreased TPC and AA	Alvarez et al. (2019)
Red bell pepper and strawberries	PEF, the field strength 1.0 kV/cm, the number of pulses was 20 and 200, treatment time 2.0–28.6 ms	−4±2°C and −40±2°C, 12,500 Pa, time 72 h	Reduced firmness, higher rehydration capacity	Fauster et al. (2020)
Apple	PEF, pulse duration of 40 ms and pulse width of 10 ms. The interval between the pulses 0.5 s (2 Hz). Energies of 0.5 and 1 kJ/kg and a field strength of 1.07 $kV \cdot cm^{-1}$	40°C, 100 MPa	Reduced DT, increased AA	Lammerskitten et al. (2019)
	PEF, pulses strength 1,000, 1,250, 1,500 $kV \cdot cm^{-1}$ Pulse duration 60, 90, 120 µs; pulses number 15, 30, 45	I step 70°C, 40~45 Pa, II step 90°C, 30~35 Pa	Reduced DT, increased hydration capacity	Wu and Guo (2010)
	PEF, 800 $V \cdot cm^{-1}$, 0.1 s	40°C, 1,000 Pa	Better preservation of shape, increased porosity	Parniakov et al. (2016)
Potato	PEF, electric field strength 1,000, 1,250; pulse width, pulses number 60, 90, and 120 µs, and 1,500 $V \cdot cm^{-1}$; pulses duration 500 ms	75°C, 40–45 Pa	Reduced DT	Wu and Zhang (2014)
Okra	US, 40 kHz, the power density is 25 W/L, treatment time 30 min, samples were frozen at −20°C for 24 h, then thawed by using US	Main drying at 18°C, final drying at 20°C, vacuum pressure of 52 Pa	Reduced FD time and total energy consumption, increased AA activity and lower degradation of chlorophyll	Xu et al. (2021)
Carrot	US, 45 kHz, the ultrasound power 150, 240, and 300 W, treatment time 30 min	60°C, 80 Pa	Reduced DT	Fan et al. (2020)
Red bell pepper	US, 76, 90, and 110 W	Drying temperature 50°C, 46 Pa	Reduced DT	Schössler et al. (2012)
Apple, carrot, eggplant	US, 25 and 50 W, 21.9 kHz, US 0.3, and 20.5 $kW \cdot m^{-3}$	−5, −7.5, −10°C, air velocity 2.5 $m \cdot s^{-1}$, atmospheric FD	Reduced DT, nondestructive effect on AA	Merone et al. (2020)
Sweet potato	US at 25°C, 30 kHz, power of 200, 400, and 600 W, duration 30 min	Drying temperature was 50°C, pressure 80 Pa	Reduced DT and improved color and texture	Wu et al. (2020)

(Continued)

TABLE 2.2 (CONTINUED)
Methods of Pretreatment of Fruits and Vegetables before FD, Drying Conditions and Effect of Pretreatment

Raw Material	Method and Conditions of Pretreatment	FD Conditions	Main Effect	References
Strawberry chips	US, 200 W, 40 kHz, duration 25 min	4°C, 10 Pa, 20 h	Reduced DT and enhanced antioxidant properties	Zhang et al. (2020)
Quince	US, 28 kHz, 50 W, time 10, 20, and 30 min	–25°C, 48 Pa	Reduced hardness and shrinkage, increased rehydration	Yildiz and Izli (2019)
Button mushrooms	US, 12.3 and 24.6 kW·m^{-3}	–10°C, 2 m·s^{-1}	Reduced DT and lightness, decreased hardness and chewiness, or rehydrated samples	Carrión et al. (2018)
Strawberry chips	HPP, from 0 to 250 MPa, 10 min	–50°C, 10 Pa	Increased redness, lightness, and total content of anthocyanin, reduced DT	Zhang et al. (2020)

Source: Adapted from Dziki (2020).
AA – antioxidant activity; BL – blanching; DT – drying time; FD – freeze-drying; HHP – high hydrostatic pressure; OD – osmotic dehydration; PEF – pulsed electric field; TPC – total phenolics content; US – ultrasound

capacity of the samples, an effect that could be observed throughout storage. Freezing pretreatment increased crispness and oil absorption values of the samples, as compared to HPP-pretreated or control samples. The authors concluded that the use of the proposed pretreatments (HPP or freezing) resulted in a feasible approach for improving the nutritional and organoleptic properties of vacuum-fried carrot snacks. Recently, Zhang et al. (2020) showed that HHP (50, 100, 150, 200, and 250 MPa) pretreatments significantly increased water mobility of strawberry slices which resulted in a significant reduction of drying time by 9–24%. Furthermore, the extent of HPP pretreatment increased redness values and anthocyanin contents (Figure 2.10 and Table 2.2). Higher HPP pressures increased and then decreased soluble pectins, while the contents of protopectin and cellulose decreased. After the HHP pretreatment, chromoplasts and moisture were more evenly distributed in the strawberry slices. Microscopy images showed the formation of microscopic holes or channels in the matrix and the breakdown of tissue structure after the application of HHP. Results suggested that this pretreatment disrupted the integrity of the fresh strawberry, which enhanced the drying efficiency and migration of the chromoplasts during the subsequent vacuum-freeze drying step.

Al-Khusaibi and Niranjan (2012) evaluated the application of HHP as a pretreatment prior to deep-fat frying of potato slices. The authors investigated the impact of HPP on the frying time and oil content of potato chips. The pressure ranged from low (200 MPa) to high pressure (800 MPa) for up to 5 min. It was found that high-pressure pretreatment could reduce the frying time necessary to obtain a final water content of ≤ 2%, but marginally increased the oil uptake. This might be related to higher water permeability caused by the high pressure, which enhanced oil penetration into the fried slices. Slices treated at 200 MPa for 5 min contained the highest oil content. Albertos et al. (2016) reported that high-pressure processing combined with freezing prior to vacuum frying had remarkable impacts on the final quality of fried products especially in terms of crispiness of carrot chips, susceptibility to oxidation, and retention of flavor and phenolics. The same authors explain

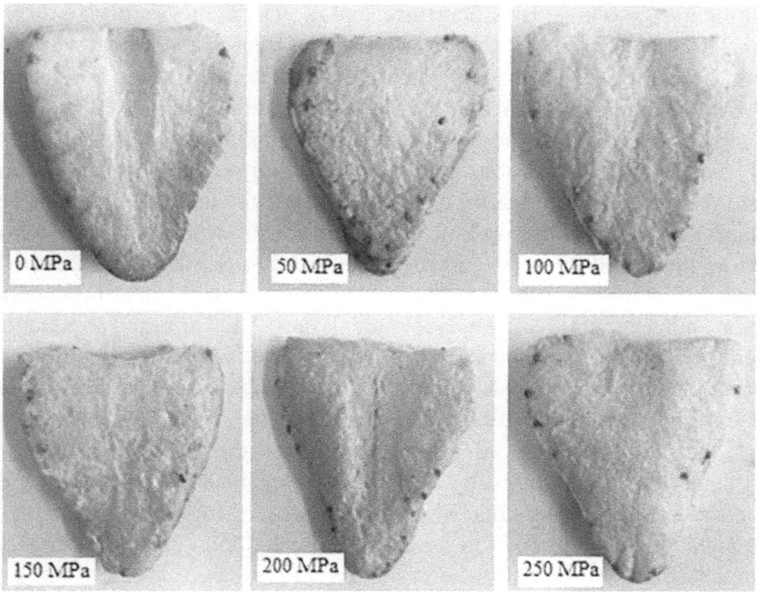

FIGURE 2.10 Effect of HHP pretreatment on color of freeze-dried strawberry chips (Zhang et al. 2020).

that HPP may probably affect color, oxidation, and enzymatic and non-enzymatic reactions during the frying process (Albertos et al. 2016).

2.3.2 Pulsed Electric Field (PEF) as Pretreatment in Snacks Elaboration

PEF processing involves the application of short pulses (μs) of electric fields intensities (E) varying from 0.1 to 40 kV/cm, depending on the pursued objective. The increase of product shelf life by the inactivation of microorganisms and enzymes is usually achieved with HIPEF (20–40 kV/cm). MIPEF (1–20 kV/cm) are commonly applied as pretreatments for the improvement/optimization of drying, frying, freezing, and extraction, among others processes (Morales-de la Peña et al. 2021, Wang et al. 2018). Likewise, MIPEF processing at low E (0.1–3 kV/cm) has demonstrated to be able of inducing stress reactions in horticultural crops generating secondary metabolites (Toepfl et al. 2006, Oey et al. 2016). During the last decades, the application of PEF has attracted large interest not only as cold pasteurization technology but also as an assisting process of well-established treatments in the food industry. Different research studies have demonstrated that the application of MIPEF as pretreatment enhances the efficiency of different processes by reducing operation time with an overall increment in extraction yields as well as nutritional and nutraceutical properties of extracted food products (Traffano-Schiffo et al. 2016, Lammerskitten et al. 2019). It has been observed that under the effects of MIPEF treatments, cell membranes may be irreversibly or reversibly permeabilized (Barba et al. 2015). According to Soliva-Fortuny et al. (2009), the effects of MIPEF have been well correlated to the more efficient recovery of high-added value compounds from different matrices, to the improvement of mass transfer rates, to the enhancement of osmotic dehydration, drying, frying, and freezing efficiency, and to the increase in oil, juice, and bioactive compounds extraction yields from different sources. In this sense, MIPEF processing has the capability of being implemented as an assisting process for diverse unit operations, thus allowing the development of high-quality products. An interesting application of MIPEF processing is to assist drying processes applied in snacks elaboration; Traffano-Schiffo et al. (2016) used kiwifruit as a model to investigate the transport mechanisms as affected by PEF pretreatments which are known to improve osmotic dehydration (OD). Kiwifruits were treated at 0.10, 0.25, and 0.40

kV·cm^{-1} evaluating mass, volume, Brix, moisture, and A_w. The authors reported an enhancement of water mass transfer and a decrease of the sugar content of treated kiwifruits, which were related to the E magnitude. These results were attributed not only to electroporation but also to a loss of key electrolytes involved in the active transport of the cell membrane, namely Ca^{+2} and Na^{+2}. In a further study, the authors examined microstructural changes and internal water transport of kiwifruits pretreated by PEF at the same conditions. In addition to altering active transport mechanisms, Ca^{+2} absence affected the mechanical integrity of the cell membrane by bond breakage between microtubules and membrane. Moreover, apoplastic and symplastic transport occurred in kiwifruit treated by PEF increasing the dehydration rate during OD (Traffano-Schiffo et al. 2016). Tylewicz et al. (2016) used several E values prior to freeze-drying of sliced apples to determine MIPEF effects on water distribution and water-solid exchange of apple tissues. They reported a reduction in the initial freezing temperature and freezable water content of pretreated apples relating it to an increment of cell disintegration index, which increased along with E. Moreover, the authors observed improvements in their stability throughout storage due to a decrement in A_w and an increment in solutes concentrations. Later on, Tylewicz et al. (2019) investigated the metabolic changes of strawberry and kiwifruit treated by PEF followed by OD. The application of an E of 0.2 kV·cm^{-1} produced irreversible electroporation of kiwifruits, while treatments of 0.1 kV·cm^{-1} maintained cell viability evidenced by a metabolic heat production similar to untreated kiwifruit. Additionally, the PEF-OD combination improved water loss in strawberry and kiwifruit treated at 0.2 kV·cm^{-1} regardless of the sugar employed for OD. In another research, the thermophysical and water adsorption properties of freeze-dried apples pretreated at 0.5 and 1.0 kJ·kg^{-1} were evaluated by Lammerskitten et al. (2019). At 1.0 kJ·kg^{-1}, pretreated apples showed minimal changes in *Aw* throughout storage with no differences in their hydroscopic properties compared to untreated counterparts. As shown in Table 2.2, the application of PEF as a pretreatment in different products processed by freeze-drying (apple, red pepper, potato, strawberry), not only achieved the reduction of the processing time, but also improved sensory characteristics (texture, porosity, color, rehydration capacity, etc.) so the combined technologies helped to develop snacks with improved functional and antioxidant properties.

PEF pretreatments (field strengths of E = 1.5–2.5 kV/cm) of potato slices were applied before their frying for 13 min at 190°C, the PEF treatment allowed a significant reduction of oil uptake that was more effective than hot water blanching (Janositz et al. 2011). Ignat et al. (2015) evaluated PEF-assistance prior to partial frying the potato cubes, freezing and final frying, potatoes were treated by PEF (E = 0.75 and 2.5 kV/cm), then fried (190°C for 1 min), and stored at −18°C for 7 days. Finally, potato cubes were finish-fried for up to 4 min at 190°C. It was demonstrated that PEF treatment results in improving product color and reducing oil uptake. The effects were attributed to the structural modification of potato tissue upon PEF-induced electroporation. Liu et al. (2016) studied the effect of PEF on the microstructure of "kumara" sweet potato (*Ipomoea batatas* cv. Owairaka) and its quality after frying. Whole sweet potato tubers were treated at different electric field strengths ranging from 0.3 to 1.2 kV/cm with specific energy levels between 0.5 and 22 kJ/kg, samples pretreated with PEF at electric field strength of 1.2 kV/cm and fried at 190°C had an 18% lower oil content than nonPEF treated. PEF pretreatment allowed frying the potato chips at lower temperature in order to achieve the same brown color intensity as the nonPEF treated tubers. This study showed clearly that PEF could reduce the energy required for cutting and frying kumara. Liu et al. (2018) evaluated the effects of PEF and preliminary convective air (CA) drying on the water loss and oil uptake during frying of potato disks. The PEF treatment was done using a monopolar generator with electric field strength of 600 V/cm and pulsed protocol that allows obtaining the high level of electroporation. Results obtained evidenced that PEF treatment accelerates both processes of preliminary CA-drying and posterior frying, the PEF treatment also resulted in smaller oil uptake after the frying for 600s. It was observed synergetic effects of preliminary PEF-treatment and CA-assisted dehydration on the frying time, oil uptake and texture changes evidently reflected impact of electroporation on the development of different processes on potato tissue during posterior frying. Acceleration of the effective moisture diffusivity can influence the distribution of water, surface porosity and cell wall softening during the preliminary CA-drying. This conclusion was supported by the observed differences in potato texture for the PEF treated and untreated dehydrated tissues. The authors speculated that the stronger texture for the PEF treated potato in the intermediate period of frying (100–500 s) reflects formation of a denser and less oil penetrable skin on the surface tissue. Recently the effects of preliminary vacuum drying (VD) and pulsed electric field (PEF) treatment on characteristics of fried potatoes were studied by Liu et al. (2020), the effects of PEF included significant shortening the VD time and absence of starch gelatinizing during the VD. For PEF treated samples the moisture contents were significantly smaller than for untreated ones and the PEF treatment also resulted in significant decrease in oil contents, for example, with potato (VD at 40°C and with a moisture reduction higher than 20%) the oil uptakes were 0.26 ± 0.02 and $0.17 \pm 0.01\%$ for untreated and PEF treated samples, respectively.

2.3.3 Ultrasound (US) as Pretreatment in Snacks Elaboration

Ultrasound is defined as sound waves with a frequency (20 kHz) that exceeds the hearing limit of the human ear (Awad et al. 2012). The generation of these waves is driven by the transformation of electrical energy into mechanical energy and the formation of mechanical oscillations. Ultrasound is a form of energy generated by acoustic waves and can be propagated in gases, liquids, or solids. Ultrasound processing can be divided into two groups according to frequency: low intensity, with low energy provided by frequencies higher than 100 kHz, and high intensity, with high energy provided by frequencies between 20 and 100 kHz (Nowacka and Wedzik 2016). Low-intensity ultrasound is used mainly in the medical field as a scanning and imaging tool. However, in the food industry, it serves as a physical tool for quality assurance. Low-intensity ultrasound can provide a rapid, accurate, inexpensive, simple, and nondestructive method to analyze the physiochemical properties of food and monitor the changes during food production (Chandrapala et al. 2012, Nowacka and Wedzik 2016). On the other hand, high-intensity ultrasound is used for disrupting and breaking up cellular structures, which later can activate or inhibit chemical or physical phenomena in food, leading to the intensification of heat and mass transfer-based processes (Nowacka and Wedzik 2016).

Ultrasound is based on acoustic cavitation; during its application, cavitation bubbles are generated inside the sample and around the sonotrode. These thousands of fast-moving bubbles undergo irregular oscillation and brutally collapse. As a consequence, high local temperature and pressure are produced leading to the disintegration of biological cells, denaturation of enzymes, formation of microscopic channels, and production of free radicals (Chemat et al. 2011, Nowacka and Wedzik 2016). Ultrasound, alone or combined with mild thermal preservation technologies, is receiving great interest because microbial inactivation can be achieved and the nutritional and organoleptic characteristics (texture, color, taste) of foods can be retained in some of the tested products. Moreover, ultrasound reduces the processing time, has a higher throughput and is environmentally safe (Rawson et al. 2011). Ultrasound has also been used in food processing to enhance operations such as filtration, defoaming, cooking, degassing, cutting, drying, meat tenderization, homogenization, and crystallization, among others (Chemat et al. 2011). Moreover, ultrasound has been studied as an emerging technology for application in food processing and preservation, but it has also been applied to improve mass transfer processes, to assist in thermal treatments and manipulation of texture (Knorr et al. 2011, Welti-Chanes et al. 2017).

Ultrasound has been applied to strengthen the productivity of osmotic dehydration prior to frying. Oladejo et al. (2017) investigated the feasibility of ultrasound-assisted osmotic dehydration (UOD) to reduce the moisture vapor content and oil absorption of deep-fried sweet potato, results showing that ultrasound-assisted osmotic dehydration is a good preprocessing method that can be applied to frying industry for reducing water vapor and oil content. Karizaki et al. (2013) investigated the feasibility of applying ultrasound-assisted osmotic dehydration as a preprocessing ahead of frying. Ultrasound-assisted osmotic dehydration is estimated to lower the initial water vapor content of fried chips, thus bringing about reduced oil content.

Su et al. (2018) combined power ultrasound with microwave-assisted vacuum frying (USMVF) to investigate drying kinetics, dielectric properties, and quality characteristics of fried purple-fleshed potato. The application of USMVF was found to achieve a higher moisture evaporation rate and therefore drying efficiency. The highest water evaporation rate was obtained n 600 W of ultrasound power and 800 W of microwave power combined with vacuum frying with about 82% of moisture diffusivity, and more importantly, the oil content diminished by 34%. The higher power level of ultrasound and microwave also created improvements in texture (crispness) and color. In the USMVF, the acoustic waves spread through the frying medium and the food and the pressure as well as oil viscosity could be affected by the alternation of diffusion boundary layer with the applied ultrasound.

Ostermeier et al. (2020) evaluated the influence of PEF and ultrasound (US) treatment on the frying behavior and quality of potato chips, some of the relevant results showed that both applied technologies impacted the water removal because PEF increased diffusion whereas ultrasound heat and mass transfer changes during frying. Additionally, the continuous water removal throughout the whole frying process (constant counter-current process) resulted in the lowest fat content (24.7%) for the PEFUS samples and 66% reduction of acrylamide content of 220 μg/kg (Table 2.3). The significant reduction of both fat and acrylamide can be beneficial for industrial potato chips producers as it enables better control within the production and product specification. It also allows using of raw materials with higher sugar contents that tend to develop more acrylamide.

2.4 3D PRINTING

Additive manufacturing, commonly called 3D printing, is a fabrication technology that has raised interest in many areas such as the food industry. These technologies are thoroughly covered in Chapter 5. In food, 3D printing can be used to personalize nutrition and to customize the design of the final product, among many other applications. The possibility of customizing food has become a necessity for some population groups such as children, elderly, and athletes (German and Watzke 2004).

In addition, because of its flexibility, this technology allows the use of a wide range of food materials options, while allowing the development of nutritious snack formulations. For example, novel fruits and vegetables have been developed as healthy snacks for children (Derossi et al. 2018, Severini et al. 2018). The use of novel materials has also been extended to formulations using insects and by-products for the printing of new snacks (Lupton and Turner 2018, Feng et al. 2020).

The most-used fabrication technique in snack foods is extrusion based. Once the material mixture is loaded, it is passed through a nozzle by using a mechanism controlled by the printer. Then, the material is deposited layer by layer in order to form a 3D structure (Sun et al. 2018a,b, Le-Bail

TABLE 2.3
Acrylamide Content for Chips Samples after 180 s of Frying

Processing Method	Acrylamide Content (μg/kg)
Untreated (NT)	651 ± 51[a]
Ultrasound (US)	431 ± 52[b]
Pulsed electric field (PEF)	542 ± 41[a,b]
Pulsed electric field and ultrasound (PEFUS)	220 ± 26[c]

Source: Adapted from Ostermeier et al. (2020).
Values with different letter(s) indicate statistically significant differences ($p = 0.05$).

et al. 2020). The formulations with the best potential are constituted by materials with rheological properties so to make them printable and capable of flowing through the nozzle. The printed material has to maintain the established design and structure after depositing. To achieve the specified figure, structure, and infill, parameters such as print speed, extrusion rate, layer height, and flow rate must be established (Sun et al. 2018a,b). Finally, depending on the material and the final product characteristics, postprocessing treatments can be applied. For instance, drying, baking, and frying with all their variants can be applied to improve the sensorial properties and overall features of printed snacks (Keerthana et al. 2020).

Even though 3D printing is a novel technique for snack production, one must remain aware that its use for large-scale production is still uncertain due to the manufacturing time and cost. Eventually, the development of cheaper and faster printers will be crucial for the future implementation of this technique in the production of snacks at home or at an industrial scale. Besides, there is still a lot to do related to the legal framework, labeling, safety, consumers acceptance for 3D printed snacks (Baiano 2020).

2.5 CONCLUSIONS

The increased consumer's need for healthier products has affected the snack industry. Traditional processes have evolved to enhance and assure the nutritional quality of the products. In addition, new technologies, also called emerging, have come out with the aim of developing new snack products. Both, traditional and emerging processes have the aim to maintain the organoleptic properties of the current ones, but with better nutritional profiles. These processes have broadened the possibility of developing snacks from more natural sources, and developing new ingredients for the production of novel snacks.

This chapter revised the effect of the extrusion process on the development of fiber-rich snacks and how the products are affected by the fiber's composition and source. An increase in the use of dietary fiber sources for the production of snacks would lead to the opportunity of using new ingredients with healthier characteristics. The thermal traditional processes of frying, drying, and baking, presented herein, have also changed due to the consumer's awareness on health. Alternate technologies such as microwave frying, air frying, or vacuum frying are presented as improvements for conventional deep-fat frying, although pretreatments or posttreatments are also being used to achieve the goal of reducing the oil uptake of snacks. Microwave drying, instant controlled pressure drop, and the use of infrared or jet impingement technologies are some of the improvements for drying and baking to reduce processing times and temperatures.

Finally, emerging technologies that have been commonly used in the recent years for microbial and enzyme inactivation have also been applied by the snack industry considering similar objectives: better nutritional and sensorial quality. Even some newer technologies, such as 3D printing have come to the interest of the consumers because they can offer the possibility of a personalized nutrition. The food industry will keep on moving toward the improvement of traditional and new technologies in the next future to suffice the consumer's demands. Research is needed, not only for the development and implementation of new technologies but also for the bettering of traditional processes.

REFERENCES

Ahmad Tarmizi, A.H., Niranjan, K., and Gordon, M. 2013. Physico-chemical changes occurring in oil when atmospheric frying is combined with post-frying vacuum application. *Food Chemistry*, 136(2): 902–908. https://doi.org/10.1016/j.foodchem.2012.08.001

Al-Khusaibi, M.K., and Niranjan, K. 2012. The impact of blanching and high-pressure pretreatments on oil uptake of fried potato slices. *Food and Bioprocess Technology*, 5(6): 2392–2400.

Al-Muhtaseb, A.H., McMinn, W.A.M., and Magee, T.R.A. 2002. Moisture sorption isotherm characteristics of food products: A review. *Transactions IChemE*, 80(June): 118–128.

Albertos, I., Martin-Diana, A.B., Sanz, M.A., Barat, J.M., Diez, A.M., Jaime, I., and Rico, D. 2016. Effect of high pressure processing or freezing technologies as pretreatment in vacuum fried carrot snacks. *Innovative Food Science and Emerging Technologies*, 33: 115–122. https://doi.org/10.1016/j.ifset.2015.11.004

Altan, A., McCarthy, K.L., and Maskan, M. 2008. Evaluation of snack foods from barley-tomato pomace blends by extrusion processing. *Journal of Food Engineering*, 84(2): 231–242. https://doi.org/10.1016/j.jfoodeng.2007.05.014

Alvarez, C., Ospina, S., and Orrego, C.E. 2019. Effects of ultrasound-assisted blanching on the processing and quality parameters of freeze-dried guava slices. *Journal of Food Processing and Preservation*, 43(12): 14288.

Amid, B.T., and Mirhosseini, H. 2012. Influence of different purification and drying methods on rheological properties and viscoelastic behaviour of durian seed gum. *Carbohydrate Polymers*, 90(1): 452–461. https://doi.org/10.1016/j.carbpol.2012.05.065

Anton, A., Fulcher, R.G., and Arntfield, S.D. 2009. Physical and nutritional impact of fortification of corn starch-based extruded snacks with common bean (*Phaseolus vulgaris* L.) flour: Effects of bean addition and extrusion cooking. *Food Chemistry*, 113(4): 989–996. https://doi.org/10.1016/j.foodchem.2008.08.050

Awad, T.S., Moharram, H.A., Shaltout, O.E., Asker, D., and Youssef, M.M. 2012. Applications of ultrasound in analysis, processing and quality control of food: A review. *Food Research International*, 48(2): 410–427. https://doi.org/10.1016/j.foodres.2012.05.004

Bagheri, H. 2020. Application of infrared heating for roasting nuts. *Journal of Food Quality*, 2020: 8813047. https://doi.org/10.1155/2020/8813047

Bagheri, H., Kashaninejad, M., Ziaiifar, A.M., and Aalami, M. 2019. Textural, color and sensory attributes of peanut kernels as affected by infrared roasting method. *Information Processing in Agriculture*, 6(2): 255–264. https://doi.org/10.1016/j.inpa.2018.11.001

Baiano, A. 2020. 3D printed foods: A comprehensive review on technologies, nutritional value, safety, consumer attitude, regulatory framework, and economic and sustainability issues. *Food Reviews International*, 1–31. https://doi.org/10.1080/87559129.2020.1762091

Baker, C.G.J. 1997. Dryer selection. In: *Industrial drying of foods*, ed. C. Blake, 255–268. London: Blackie Academic and Professional.

Banks, D., and Lusas, E. 2001. Oils and industrial frying. In: *Snack foods processing*, eds. E.W. Lusas and L.W. Rooney, 137–204. Boca Raton: CRC Press. https://doi.org/10.1201/9781420012545.ch6

Barba, F.J., Parniakov, O., Pereira, S.A., Wiktor, A., Grimi, N., Boussetta, N., Saraiva, J.A., Raso, J., Martin-Belloso, O., Witrowa-Raichert, D., Lebovka, N., and Vorobiev, E. 2015. Current applications and new opportunities for the use of pulsed electric fields in food science and industry. *Food Research International*, 77: 773–798.

Basu, S., Shivhare, U.S., and Mujumdar, A.S. 2006. Models for sorption isotherms for foods: A review. *Drying Technology*, 24(8): 917–930. https://doi.org/10.1080/07373930600775979

Belkova, B., Hradecky, J., Hurkova, K., Forstova, V., Vaclavik, L., and Hajslova, J. 2018. Impact of vacuum frying on quality of potato crisps and frying oil. *Food Chemistry*, 241: 51–59. https://doi.org/10.1016/j.foodchem.2017.08.062

Betoret, E., Betoret, N., Rocculi, P., and Dalla Rosa, M. 2015. Strategies to improve food functionality: Structure-property relationships on high pressures homogenization, vacuum impregnation and drying technologies. *Trends in Food Science and Technology*, 46(1): 1–12. https://doi.org/10.1016/j.tifs.2015.07.006

Bisharat, G.I., Oikonomopoulou, V.P., Panagiotou, N.M., Krokida, M.K., and Maroulis, Z.B. 2013. Effect of extrusion conditions on the structural properties of corn extrudates enriched with dehydrated vegetables. *Food Research International*, 53(1): 1–14. https://doi.org/10.1016/j.foodres.2013.03.043

Blake, O.A. 2006. *Effect of Molecular and Supramolecular Characteristics of Select Dietary Fiber on Extrusion Expansion*, PhD Dissertation. Purdue University, 1–118.

Bredie, W.L.P., Mottram, D.S., and Guy, R. 1998. Aroma volatiles generated during extrusion cooking of maize flour. *Journal of Agricultural and Food Chemistry*, 46(4): 1479–1487. https://doi.org/10.1021/jf9708857

Brennan, M., Derbyshire, E.J., Brennan, C.S., and Tiwari, B.K. 2012. Impact of dietary fibre-enriched ready-to-eat extruded snacks on the postprandial glycaemic response of non-diabetic patients. *Molecular Nutrition and Food Research*, 56(5): 834–837. https://doi.org/10.1002/mnfr.201100760

Brennan, M., Derbyshire, E., Tiwari, B.K., and Brennan, C.S. 2013. Ready-to-eat snack products: The role of extrusion technology in developing consumer acceptable and nutritious snacks. *International Journal of Food Science and Technology*, 48(5): 893–902. https://doi.org/10.1111/ijfs.12055

Brennan, M., Merts, I., Monro, J., Woolnough, J., and Brennan, C.S. 2008. Impact of guar gum and wheat bran on the physical and nutritional quality of extruded breakfast cereals. *Starch/Staerke*, 60(5): 248–256. https://doi.org/10.1002/star.200700698

Carrión, C., Mulet, A., García-Pérez, J.V., and Cárcel, J.A. 2018. Ultrasonically assisted atmospheric freeze-drying of button mushroom. Drying kinetics and product quality. *Drying Technology*, 36(15): 1814–1823.

Chandrapala, J., Oliver, C., Kentish, S., and Ashokkumar, M. 2012. Ultrasonics in food processing: Food quality assurance and food safety. *Trends in Food Science and Technology*, 26(2): 88–98.

Chau, C.F., Chen, C.H., and Lee, M.H. 2004. Comparison of the characteristics, functional properties, and in vitro hypoglycemic effects of various carrot insoluble fiber-rich fractions. *LWT – Food Science and Technology*, 37(2): 155–160. https://doi.org/10.1016/j.lwt.2003.08.001

Chemat, F., Zill, E., H., and Khan, M.K. 2011. Applications of ultrasound in food technology: Processing, preservation and extraction. *Ultrasonics Sonochemistry*, 18(4): 813–835.

Chen, J., Venkitasamy, C., Shen, Q., McHugh, T.H., Zhang, R., and Pan, Z. 2018. Development of healthy crispy carrot snacks using sequential infrared blanching and hot air drying method. *LWT – Food Science and Technology*, 97: 469–475. https://doi.org/10.1016/j.lwt.2018.07.026

Chhanwal, N., Bhushette, P.R., and Anandharamakrishnan, C. 2019. Current perspectives on non-conventional heating ovens for baking process - A review. *Food and Bioprocess Technology*, 12(1): 1–15. https://doi.org/10.1007/s11947-018-2198-y

Chhanwal, N., Indrani, D., Raghavarao, K.S.M.S., and Anandharamakrishnan, C. 2011. Computational fluid dynamics modeling of bread baking process. *Food Research International*, 44(4): 978–983.

Chirife, J., and Iglesias, H.A. 1978. Equations for fitting water sorption isotherms of foods: Part 1. A review. *International Journal of Food Science and Technology*, 13(3): 159–174. https://doi.org/10.1111/j.1365-2621.1978.tb00792.x

Ciurzyńska, A., Marczak, W., Lenart, A., and Janowicz, M. 2020. Production of innovative freeze-dried vegetable snack with hydrocolloids in terms of technological process and carbon footprint calculation. *Food Hydrocolloids*, 108: 105993. https://doi.org/10.1016/j.foodhyd.2020.105993

Cruz, G., Cruz-Tirado, J.P., Delgado, K., Guzman, Y., Castro, F., Rojas, M.L., and Linares, G. 2018. Impact of pre-drying and frying time on physical properties and sensorial acceptability of fried potato chips. *Journal of Food Science and Technology*, 55(1): 138–144. https://doi.org/10.1007/s13197-017-2866-3

De Pilli, T., and Alessandrino, O. 2020. Effects of different cooking technologies on biopolymers modifications of cereal-based foods: Impact on nutritional and quality characteristics review. *Critical Reviews in Food Science and Nutrition*, 60(4): 556–565. https://doi.org/10.1080/10408398.2018.1544884

Dehghannya, J., Naghavi, E.A., and Ghanbarzadeh, B. 2016. Frying of potato strips pretreated by ultrasound-assisted air-drying. *Journal of Food Processing and Preservation*, 40(4): 583–592. https://doi.org/10.1111/jfpp.12636

Derossi, A., Caporizzi, R., Azzollini, D., and Severini, C. 2018. Application of 3D printing for customized food. A case on the development of a fruit-based snack for children. *Journal of Food Engineering*, 220: 65–75. https://doi.org/10.1016/j.jfoodeng.2017.05.015

Devi, S., Zhang, M., and Law, C.L. 2018. Effect of ultrasound and microwave assisted vacuum frying on mushroom (*Agaricus bisporus*) chips quality. *Food Bioscience*, 25(February): 111–117. https://doi.org/10.1016/j.fbio.2018.08.004

Dijksterhuis, G., Luyten, H., de Wijk, R., and Mojet, J. 2007. A new sensory vocabulary for crisp and crunchy dry model foods. *Food Quality and Preference*, 18(1): 37–50. https://doi.org/10.1016/j.foodqual.2005.07.012

Ding, Q.B., Ainsworth, P., Plunkett, A., Tucker, G., and Marson, H. 2006. The effect of extrusion conditions on the functional and physical properties of wheat-based expanded snacks. *Journal of Food Engineering*, 73(2): 142–148. https://doi.org/10.1016/j.jfoodeng.2005.01.013

Donno, D., Mellano, M.G., Riondato, I., De Biaggi, M., Andriamaniraka, H., Gamba, G., and Beccaro, G.L. 2019. Traditional and unconventional dried fruit snacks as a source of health-promoting compounds. *Antioxidants*, 8(396): 1–15. https://doi.org/10.3390/antiox8090396

Dueik, V., Moreno, M.C., and Bouchon, P. 2012. Microstructural approach to understand oil absorption during vacuum and atmospheric frying. *Journal of Food Engineering*, 111(3): 528–536. https://doi.org/10.1016/j.jfoodeng.2012.02.027

Dueik, V., Sobukola, O., and Bouchon, P. 2014. Development of low-fat gluten and starch fried matrices with high fiber content. *LWT – Food Science and Technology*, 59(1): 6–11. https://doi.org/10.1016/j.lwt.2014.04.036

Dziki, D. 2020. Recent trends in pretreatment of food before freeze-drying. *Processes*, 8(12): 1661.

Fan, D., Chitrakar, B., Ju, R., and Zhang, M. 2020. Effect of ultrasonic pretreatment on the properties of freeze-dried carrot slices by traditional and infrared freeze-drying technologies. *Drying Technology.* https://doi.org/10.1080/07373937.2020.1815765

Fan, J.T., Mitchell, J.R., and Blanshard, J.M.V. 1994. A computer-simulation of the dynamics of bubble-growth and shrinkage during extrudate expansion. *Journal of Food Engineering*, 23(3): 337–356.

Fan, K., Zhang, M., and Bhandari, B. 2019. Osmotic-ultrasound dehydration pretreatment improves moisture adsorption isotherms and water state of microwave-assisted vacuum fried purple-fleshed sweet potato slices. *Food and Bioproducts Processing*, 115: 154–164. https://doi.org/10.1016/j.fbp.2019.03.011

Fang, M., Huang, G.J., and Sung, W.C. 2021. Mass transfer and texture characteristics of fish skin during deep-fat frying, electrostatic frying, air frying and vacuum frying. *LWT – Food Science and Technology*, 137:110494. https://doi.org/10.1016/j.lwt.2020.110494

Fast, R.B. 1999. Origins of the U.S. breakfast cereal industry. *Cereal Foods World*, 44(6): 394–397.

Fauster, T., Giancaterino, M., Pittia, P., and Jaeger, H. 2020. Effect of pulsed electric field pretreatment on shrinkage, rehydration capacity and texture of freeze-dried plant materials. *LWT – Food Science and Technology*, 121: 108937.

Feng, C., Zhang, M., Bhandari, B., and Ye, Y. 2020. Use of potato processing by-product: Effects on the 3D printing characteristics of the yam and the texture of air-fried yam snacks. *LWT – Food Science and Technology*, 125: 109265. https://doi.org/10.1016/j.lwt.2020.109265

Fritsch, C.W. 1981. Measurements of frying fat deterioration: A brief review. *Journal of the American Oil Chemists' Society*, 58(3): 272–274.

Garcia, M.A., Ferrero, C., Campana, A., Bértola, N., Martino, M., and Zaritzky, N. 2004. Methylcellulose coatings applied to reduce oil uptake in fried products. *Food Science and Technology International*, 10(5): 339–346. https://doi.org/10.1177/1082013204047564

García-Amezquita, L.E., Tejada-Ortigoza, V., Perez-Carrillo, E., Serna-Saldívar, S.O., Campanella, O.H., and Welti-Chanes, J. 2019. Functional and compositional changes of orange peel fiber thermally-treated in a twin extruder. *LWT – Food Science and Technology*, 111: 673–681. https://doi.org/10.1016/j.lwt.2019.05.082

García-Amezquita, L.E., Tejada-Ortigoza, V., Serna-Saldívar, S.O., and Welti-Chanes, J. 2018. Dietary fiber concentrates from fruit and vegetable by-products: Processing, modification, and application as functional ingredients. *Food and Bioprocess Technology*, 11(8): 1439–1463.

García-Amezquita, L.E., Tejada-Ortigoza, V., Torres, J.A., and Welti-Chanes, J. 2020. Extraction and modification of dietary fiber applying thermal processes. In: *Science and technology of fibers in food systems*, eds. J. Welti-Chanes, S.O. Serna-Saldívar, O.H. Campanella, and V. Tejada-Ortigoza, 329–342. Cham: Springer.

German, J.B., and Watzke, H.J. 2004. Personalizing foods for health and delight. *Comprehensive Reviews in Food Science and Food Safety*, 2(3): 145–151.

Guy, R. 2001. Snack food. In: *Extrusion cooking technologies and applications*, ed. R. Guy, 161–181. Boca Raton: CRC Press.

Harper, J.M. 2019. *Extrusion of foods* (Vol. 1). Boca Raton: CRC Press.

Hnin, K.K., Zhang, M., Wang, B., and Devahastin, S. 2019. Different drying methods effect on quality attributes of restructured rose powder-yam snack chips: Drying of restructured rose powder-yam chips. *Food Bioscience*, 32: 100486. https://doi.org/10.1016/j.fbio.2019.100486

Hoon, K.T. 2010. *De-oiling and pre-treatments for high-quality potato chips.* Master of Science Thesis. Texas A&M University.

Hoseney, R.C., Zeleznak, K., and Abdelrahman, A. 1983. Mechanism of popcorn popping. *Journal of Cereal Science*, 1(1): 43–52.

Ibarz, A., and Barbosa-Cánovas, G.V. 2005. Deshidratación. In: *Operaciones Unitarias en la ingeniería de alimentos*, eds. A. Ibarz and G.V. Barbosa-Cánovas, 583–629. Madrid: Ediciones Mundi-Prensa.

Ignat, A., Manzocco, L., Brunton, N.P., Nicoli, M.C., and Lyng, J.G. 2015. The effect of pulsed electric field pre-treatments prior to deep-fat frying on quality aspects of potato fries. *Innovative Food Science and Emerging Technologies*, 29: 65–69.

Janositz, A., Noack, A.-K., and Knorr, D. 2011. Pulsed electric fields and their impact on the diffusion characteristics of potato slices. *LWT – Food Science and Technology*, 44(9): 1939–1945.

Jiang, N., Liu, C., Li, D., Zhang, Z., Liu, C., Wang, D., Niu, L., and Zhang, M. 2017. Evaluation of freeze drying combined with microwave vacuum drying for functional okra snacks: Antioxidant properties, sensory quality, and energy consumption. *LWT – Food Science and Technology*, 82: 216–226. https://doi.org/10.1016/j.lwt.2017.04.015

Jin, Z., Hsieh, F., and Huff, H.E. 1994. Extrusion cooking of corn meal with soy fiber, salt, and sugar. *Cereal Chemistry*, 71(3): 227–234.

Jing, Y., and Chi, Y. 2013. Effects of twin-screw extrusion on soluble dietary fibre and pysicochemical properties of soybean residue. *Food Chemistry*, 138(2–3): 884–889. https://doi.org/10.1016/j.foodchem.2012.12.003

Kaisangsri, N., Kowalski, R.J., Wijesekara, I., Kerdchoechuen, O., Laohakunjit, N., and Ganjyal, G.M. 2016. Carrot pomace enhances the expansion and nutritional quality of corn starch extrudates. *LWT – Food Science and Technology*, 68: 391–399. https://doi.org/10.1016/j.lwt.2015.12.016

Karizaki, V.M., Sahin, S., Sumnu, G., Mosavian, M.T.H., and Luca, A. 2013. Effect of ultrasound-assisted osmotic dehydration as a pretreatment on deep fat frying of potatoes. *Food and Bioprocess Technology*, 6(12): 3554–3563.

Katz, E.E., and Labuza, T.P. 1981. Effect of water activity on the sensory crispness and mechanical deformation of snack food products. *Journal of Food Science*, 46(2): 403–409. https://doi.org/10.1111/j.1365-2621.1981.tb04871.x

Keerthana, K., Anukiruthika, T., Moses, J.A., and Anandharamakrishnan, C. 2020. Development of fiber-enriched 3D printed snacks from alternative foods: A study on button mushroom. *Journal of Food Engineering*, 287: 110116. https://doi.org/10.1016/j.jfoodeng.2020.110116

Khatir, Z., Taherkhani, A.R., Paton, J., Thompson, H., Kapur, N., and Toropov, V. 2015. Energy thermal management in commercial bread-baking using a multi-objective optimisation framework. *Applied Thermal Engineering*, 80: 141–149.

Kingsly, A.R.P., Balasubramaniam, V.M., and Rastogi, N.K. 2009. Effect of high-pressure processing on texture and drying behavior of pineapple. *Journal of Food Process Engineering*, 32(3): 369–381.

Knorr, D., Froehling, A., Jaeger, H., Reineke, K., Schlueter, O., and Schoessler, K. 2011. Emerging technologies in food processing. *Annual Review of Food Science and Technology*, 2: 203–235.

Knorr, D., Schlüter, O., and Heinz, V. 1998. Impact of high hydrostatic pressure on phase transitions of foods. *Food Technology*, 52(9): 42–45.

Labuza, T.P., and Hyman, C.R. 1998. Moisture migration and control in multi-domain foods. *Trends in Food Science and Technology*, 9(2): 47–55. https://doi.org/10.1016/S0924-2244(98)00005-3

Labuza, T.P., McNally, L., Gallagher, D., Hawkes, J., and Hurtado, F. 1972. Stability of intermediate moisture foods. 1. Lipid oxidation. *Journal of Food Science*, 37(1): 154–159.

Lammerskitten, A., Mykhailyk, V., Wiktor, A., Toepfl, S., Nowacka, M., Bialik, M., Czyzewski, J., and Witrowa-Raichert, D. 2019. Impact of pulsed electric fields on physical properties of freeze-dried apple tissue. *Innovative Food Science and Emerging Technologies*, 57: 102211.

Larrea, M.A., Chang, Y.K., and Martinez-Bustos, F. 2005. Some functional properties of extruded orange pulp and its effect on the quality of cookies. *LWT – Food Science and Technology*, 38(3): 213–220. https://doi.org/10.1016/j.lwt.2004.05.014

Le-Bail, A., Maniglia, B.C., and Le-Bail, P. 2020. Recent advances and future perspective in additive manufacturing of foods based on 3D printing. *Current Opinion in Food Science*, 35: 54–64. https://doi.org/10.1016/j.cofs.2020.01.009

Li, C., Niu, L.Y., Li, D.J., Liu, C.Q., Liu, Y.P., Liu, C.J., and Song, J.F. 2018. Effects of different drying methods on quality, bacterial viability and storage stability of probiotic enriched apple snacks. *Journal of Integrative Agriculture*, 17(1): 247–255. https://doi.org/10.1016/S2095-3119(17)61742-8

Liberty, J.T., Dehghannya, J., and Ngadi, M.O. 2019. Effective strategies for reduction of oil content in deep-fat fried foods: A review. *Trends in Food Science and Technology*, 92: 172–183. https://doi.org/10.1016/j.tifs.2019.07.050

Liu, C., Grimi, N., Lebovka, N., and Vorobiev, E. 2018. Effects of preliminary treatment by pulsed electric fields and convective air-drying on characteristics of fried potato. *Innovative Food Science and Emerging Technologies*, 47: 454–460.

Liu, C., Grimi, N., Lebovka, N., and Vorobiev, E. 2020. Impacts of preliminary vacuum drying and pulsed electric field treatment on characteristics of fried potatoes. *Journal of Food Engineering*, 276: 109898.

Liu, T., Dodds, E., Leong, S.Y., Eyres, G.T., Burritt, D.J., and Oey, I. 2016. Effect of pulsed electric fields on the structure and frying quality of "kumara" sweet potato tubers. *Innovative Food Science and Emerging Technologies*, 39: 197–208.

Lupton, D., and Turner, B. 2018. Food of the future? Consumer responses to the idea of 3D-printed meat and insect-based foods. *Food and Foodways*, 26(4): 269–289. https://doi.org/10.1080/07409710.2018.1531213

Makowska, A., Mildner-Szkudlarz, S., and Obuchowski, W. 2013. Effect of brewer's spent grain addition on properties of corn extrudates with an increased dietary fibre content. *Polish Journal of Food and Nutrition Sciences*, 63(1): 19–24. https://doi.org/10.2478/v10222

Martinez-Bustos, F., Viveros-Contreras, R., Galicia-García, T., Nabeshima, E.H., and Verdalet-Guzmán, I. 2011. Some functional characteristics of extruded blends of fiber from sugarcane bagasse, whey protein concentrate, and corn starch. *Ciência e Tecnologia de Alimentos*, 31(4): 870–878. https://doi.org/10.1590/S0101-20612011000400007

Martinez-Solano, K.C., Garcia-Carrera, N.A., Tejada-Ortigoza, V., García-Cayuela, T., and García-Amezquita, L.E. 2020. Ultrasound application for the extraction and modification of fiber-rich by-products. *Food Engineering Reviews*. https://doi.org/10.1007/s12393-020-09269-2

Mellema, M. 2003. Mechanism and reduction of fat uptake in deep-fat fried foods. *Trends in Food Science and Technology*, 14(9): 364–373. https://doi.org/10.1016/S0924-2244(03)00050-5

Mendez-Garcia, S., Martínez-Flores, H.E., and Morales-Sánchez, E. 2011. Effect of extrusion parameters on some properties of dietary fiber from lemon (Citrus aurantifolia Swingle) residues. *African Journal of Biotechnology*, 10(73): 16589–16593. https://doi.org/10.5897/AJB11.1582

Mendonça, S., Grossmann, M.V.E., and Verhé, R. 2000. Corn Bran as a fibre source in expanded snacks. *LWT – Food Science and Technology*, 33(1): 2–8. https://doi.org/10.1006/fstl.1999.0601

Merone, D., Colucci, D., Fissore, D., Sanjuan, N., and Carcel, J.A. 2020. Energy and environmental analysis of ultrasound-assisted atmospheric freeze-drying of food. *Journal of Food Engineering*, 283: 110031.

Metzger, D.D., Hsu, K.H., Ziegler, K.E., and Bern, C.J. 1989. Effect of moisture content on popcorn popping volume for oil and hot air popping. *Cereal Chemistry*, 66: 247–248.

Mir-Bel, J., Oria, R., and Salvador, M.L. 2009. Influence of the vacuum break conditions on oil uptake during potato post-frying cooling. *Journal of Food Engineering*, 95(3): 416–422. https://doi.org/10.1016/j.jfoodeng.2009.06.001

Mondal, A., and Datta, A.K. 2008. Bread baking – A review. *Journal of Food Engineering*, 86(4): 465–474.

Moralesde la Peña, M., RábagoPanduro, L.M., SolivaFortuny, R., MartínBelloso, O., and Welti-Chanes, J. 2021. Pulsed electric fields technology for healthy food products. *Food Engineering Reviews*. https://doi.org/10.1007/s12393-020-09277-2

Moraru, C.I., and Kokini, J.L. 2003. Nucleation and expansion during extrusion and microwave heating of cereal foods. *Comprehensive Reviews in Food Science and Food Safety*, 2(4): 147–165.

Moreira, R.G., Da Silva, P.F., and Gomes, C. 2009. The effect of a de-oiling mechanism on the production of high quality vacuum fried potato chips. *Journal of Food Engineering*, 92(3): 297–304. https://doi.org/10.1016/j.jfoodeng.2008.11.012

Mujumdar, A.S. 2004. Classification and selection of industrial dryers. In: *Guide to industrial drying: Principles, equipment and new developments*, ed. A.S. Mujumdar, 1–22. Mumbai: Colour Publications Pvt. Ltd. https://doi.org/10.1615/hedhme.a.000331

Mujumdar, A.S., and Devahastin, S. 2004. Fundamental principles of drying. In: *Guide to industrial drying: Principles, equipment and new developments*, ed. A.S. Mujumdar, 1–22. Mumbai: Colour Publications Pvt. Ltd. https://doi.org/10.1007/BF00826517

Naik, L., Sharma, R., Rajput, Y.S., and Manju, G. 2013. Application of high pressure processing technology for dairy food preservation-future perspective: A review. *Journal of Animal Production Advances*, 8: 232–241.

Nath, A., and Chattopadhyay, P.K. 2007. Optimization of oven toasting for improving crispness and other quality attributes of ready to eat potato-soy snack using response surface methodology. *Journal of Food Engineering*, 80(4): 1282–1292. https://doi.org/10.1016/j.jfoodeng.2006.09.023

Nowacka, M., and Wedzik, M. 2016. Effect of ultrasound treatment on microstructure, colour and carotenoid content in fresh and dried carrot tissue. *Applied Acoustics*, 103: 163–171.

Nowacka, M., Wiktor, A., Anuszewska, A., Dadan, M., Rybak, K., and Witrowa-Rajchert, D. 2019. The application of unconventional technologies as pulsed electric field, ultrasound and microwave-vacuum drying in the production of dried cranberry snacks. *Ultrasonics Sonochemistry*, 56: 1–13. https://doi.org/10.1016/j.ultsonch.2019.03.023

Oey, I., Lille, M., Van-Loey, A., and Hendrich, M. 2008. Effect of high-pressure processing on colour, texture and flavour of fruit- and vegetable-based food products: A review. *Trends in Food Science and Technology*, 19(6): 320–328.

Oey, I., Roohinejad, S., Leong, S.Y., Faridnia, F., Lee, P.Y., and Kethireddy, V. 2016. Pulsed electric field processing: Its technological opportunities and consumer perception. In: *Food processing technologies*, ed. A.K. Jaiswal, 447–516. Boca Raton: CRC Press.

Oladejo, A.O., Ma, H., Qu, W., Zhou, C., Wu, B., Yang, X., and Onwude, D.I. 2017. Effects of ultrasound pretreatments on the kinetics of moisture loss and oil uptake during deep fat frying of sweet potato (*Ipomea batatas*). *Innovative Food Science and Emerging Technologies*, 43: 7–17.

Olayemi Oladejo, A., Ma, H., Qu, W., C., Z., Wu, B., Benjamin Bernard, U., Onwude, D.I., and Yang, X. 2017. Application of pretreatment methods on agricultural products prior to frying- a review. *Journal of the Science of Food and Agriculture*, 98(2): 456–466. https://doi.org/10.1002/jsfa.8502

Oluwole, O., Kosoko, S.B., Owolabi, S.O., Olatope, A.O.A., Alagbe, G.O., Ogunji, O.A., Jegede, A.A., and Elemo, G.N. 2014. Effect of baking temperature on the quality of baked sweet potato crisps. *British Journal of Applied Science and Technology*, 4(23): 3419–3429. https://doi.org/10.9734/bjast/2014/10432

Onwulata, C.I., Konstance, R.P., Smith, P.W., and Holsinger, V.H. 2001. Co-extrusion of dietary fiber and milk proteins in expanded corn products. *LWT – Food Science and Technology*, 34(7): 424–429. https://doi.org/10.1006/fstl.2000.0742

Østergård, K., Björck, I., and Vainionpää, J. 1989. Effects of extrusion cooking on starch and dietary fibre in barley. *Food Chemistry*, 34(3): 215–227. https://doi.org/10.1016/0308-8146(89)90142-8

Ostermeier, R., Hill, K., Dingis, A., Töpfl, S., and Jägerb, H. 2020. Influence of pulsed electric field (PEF) and ultrasound treatment on the frying behavior and quality of potato chips. *Innovative Food Science and Emerging Technologies*, 67: 102553. https://doi.org/10.1016/j.ifset.2020.102553

Pai, D.A., Blake, O.A., Hamaker, B.R., and Campanella, O.H. 2009. Importance of extensional rheological properties on fiber-enriched corn extrudates. *Journal of Cereal Science*, 50(2): 227–234. https://doi.org/10.1016/j.jcs.2009.05.007

Pankaj, S.K., and Keener, K.M. 2017. A review and research trends in alternate frying technologies. *Current Opinion in Food Science*, 16: 74–79. https://doi.org/10.1016/j.cofs.2017.09.001

Paraman, I., Sharif, M.K., Supriyadi, S., and Rizvi, S.S.H. 2015. Agro-food industry byproducts into value-added extruded foods. *Food and Bioproducts Processing*, 96: 78–85. https://doi.org/10.1016/j.fbp.2015.07.003

Parniakov, O., Bals, O., Lebovka, N., and Vorobiev, E. 2016. Pulsed electric field assisted vacuum freeze-drying of apple tissue. *Innovative Food Science and Emerging Technologies*, 35: 52–57.

Payne, C.R., and Labuza, T.P. 2005. Correlating perceived crispness intensity to physical changes in an amorphous snack food. *Drying Technology*, 23(4): 887–905. https://doi.org/10.1081/DRT-200054237

Peleg, M. 2020. Models of sigmoid equilibrium moisture sorption isotherms with and without the monolayer hypothesis. *Food Engineering Reviews*, 12(1): 1–13. https://doi.org/10.1007/s12393-019-09207-x

Peng, J., Yi, J., Bi, J., Chen, Q., Wu, X., Zhou, M., and Liu, J. 2018. Freezing as pretreatment in instant controlled pressure drop (DIC) texturing of dried carrot chips: Impact of freezing temperature. *LWT – Food Science and Technology*, 89(2): 365–373. https://doi.org/10.1016/j.lwt.2017.11.009

Peng, J., Yin, X., Jiao, S., Wei, K., Tu, K., and Pan, L. 2019. Air jet impingement and hot air-assisted radio frequency hybrid drying of apple slices. *LWT – Food Science and Technology*, 116(May): 108517. https://doi.org/10.1016/j.lwt.2019.108517

Peressini, D., Foschia, M., Tubaro, F., and Sensidoni, A. 2015. Impact of soluble dietary fibre on the characteristics of extruded snacks. *Food Hydrocolloids*, 43: 73–81. https://doi.org/10.1016/j.foodhyd.2014.04.036

Pitts, K.F., McCann, T.H., Mayo, S., Favaro, J., and Day, L. 2016. Effect of the sugar replacement by citrus fibre on the physical and structural properties of wheat-corn based extrudates. *Food and Bioprocess Technology*, 9(11): 1803–1811. https://doi.org/10.1007/s11947-016-1764-4

Purlis, E., and Salvadori, V.O. 2009. Simulation, bread baking as a moving boundary problem. Part 2: Model validation and numerical. *Journal of Food Engineering*, 91(3): 434–442.

Quinn, P.V., Hong, D.C., and Both, J.A. 2005. Increasing the size of a piece of popcorn. *Physica A: Statistical Mechanics and Its Applications*, 353: 637–648. https://doi.org/10.1016/j.physa.2005.02.013

Rakesh, V., Datta, A.K., Amin, M.H.G., and Hall, L.D. 2009. Heating uniformity and rates in a domestic microwave combination oven. *Journal of Food Process Engineering*, 32(3): 398–424.

Rastogi, N.K., Angersbach, A., and Knorr, D. 2000. Synergistic effect of high hydrostatic pressure pretreatment and osmotic stress on mass transfer during osmotic dehydration. *Journal of Food Engineering*, 45(1): 25–31.

Rastogi, N.K., and Knorr, D. 2013. *Recent developments in high pressure processing of foods*. New York: Springer.

Ravli, Y., Da Silva, P., and Moreira, R.G. 2013. Two-stage frying process for high-quality sweet-potato chips. *Journal of Food Engineering*, 118(1): 31–40. https://doi.org/10.1016/j.jfoodeng.2013.03.032

Rawson, A., Patras, A., Tiwari, B.K., Noci, F., Koutchma, T., and Brunton, N. 2011. Effect of thermal and non thermal processing technologies on the bioactive content of exotic fruits and their products: Review of recent advances. *Food Research International*, 44(7): 1875–1887. https://doi.org/10.1016/j.foodres.2011.02.053

Robin, F., Dubois, C., Curti, D., Schuchmann, H.P., and Palzer, S. 2011. Effect of wheat bran on the mechanical properties of extruded starchy foams. *Food Research International*, 44(9): 2880–2888. https://doi.org/10.1016/j.foodres.2011.06.041

Robin, F., Dubois, C., Pineau, N., Labat, E., Theoduloz, C., and Curti, D. 2012a. Process, structure and texture of extruded whole wheat. *Journal of Cereal Science*, 56(2): 358–366. https://doi.org/10.1016/j.jcs.2012.02.014

Robin, F., Schuchmann, H.P., and Palzer, S. 2012b. Dietary fiber in extruded cereals: Limitations and opportunities. *Trends in Food Science and Technology*, 28(1): 23–32. https://doi.org/10.1016/j.tifs.2012.06.008

Roos, Y.H., Roininen, K., Jouppila, K., and Tuorila, H. 1998. Glass transition and water plasticization effects on crispness of a snack food extrudate. *International Journal of Food Properties*, 1(2): 163–180. https://doi.org/10.1080/10942919809524574

Rosenthal, A., Pokhrel, P.R., da Rocha-Ferreira, E.H., Tiburski, J.H., Barbosa-Cánovas, G.V., and Welti-Chanes, J. 2018. High pressure processing of fruit products. In: *Fruit preservation: Novel and conventional technologies*, eds. A. Rosenthal, R. Deliza, G.V. Barbosa-Cánovas, and J. Welti-Chanes, 351–398. New York: Springer Science+ Business Media Inc.

Santos, C.S., Cunha, S.C., and Casal, S. 2017. Deep or air frying? A comparative study with different vegetable oils. *European Journal of Lipid Science and Technology*, 119(6): 1600375.

Schössler, K., Jäger, H., and Knorr, D. 2012. Novel contact ultrasound system for the accelerated freeze-drying of vegetables. *Innovative Food Science and Emerging Technologies*, 16: 113–120.

Serna-Saldivar, S.O. 2008. *Industrial manufacture of snack foods*. London: Kennedys Publications Ltd.

Serna-Saldivar, S.O. 2010. Manufacturing of cereal-based snacks. In: *Cereal grains: Properties, processing, and nutritional attributes*, ed. S.O. Serna-Saldivar, 355–394. Boca Raton: CRC Press.

Severini, C., Derossi, A., Ricci, I., Caporizzi, R., and Fiore, A. 2018. Printing a blend of fruit and vegetables. New advances on critical variables and shelf life of 3D edible objects. *Journal of Food Engineering*, 220: 89–100. https://doi.org/10.1016/j.jfoodeng.2017.08.025

Shah, M.A., Mir, S.A., and Dar, B.N. 2021. Advances in extrusion technologies. In: *Food formulations: Novel ingredients and processing techniques*, eds. S.Pathnia and B.K. Tiwari, 147–163. Hoboken: John Wiley and Sons.

Sharma, V., Champawat, P.S., and Mudgal, V.D. 2014. Process development for puffing of sorghum. *International Journal of Current Research and Academic Review*, 2(1): 164–170.

Snack Food Association. 1992. Popcorn manufacturing. In: Snack Food Association, *Corn quality assurance manual*, 59–79. Alexandria: Snack Food Association.

Solanki, C., Indore, N., Mridula, D., and Nanda, K. 2018. Microwave vs conventional popping system: A comparative evaluation for maize popping. *International Journal of Chemical Studies*, 6(6): 176–181.

Soliva-Fortuny, R., Balasa, A., Knorr, D., and Martín-Belloso, O. 2009. Effects of pulsed electric fields on bioactive compounds in foods: A review. *Trends in Food Science and Technology*, 20(11–12): 544–556.

Sothornvit, R. 2011. Edible coating and post-frying centrifuge step effect on quality of vacuum-fried banana chips. *Journal of Food Engineering*, 107(3–4): 319–325. https://doi.org/10.1016/j.jfoodeng.2011.07.010

Stojceska, V., Ainsworth, P., Plunkett, A., Ibanoğlu, E., and Ibanoğlu, Ş. 2008a. Cauliflower by-products as a new source of dietary fibre, antioxidants and proteins in cereal based ready-to-eat expanded snacks. *Journal of Food Engineering*, 87(4): 554–563. https://doi.org/10.1016/j.jfoodeng.2008.01.009

Stojceska, V., Ainsworth, P., Plunkett, A., and Ibanoğlu, S. 2008b. The recycling of brewer's processing by-product into ready-to-eat snacks using extrusion technology. *Journal of Cereal Science*, 47(3): 469–479. https://doi.org/10.1016/j.jcs.2007.05.016

Stojceska, V., Ainsworth, P., Plunkett, A., and Ibanoğlu, Ş. 2009. The effect of extrusion cooking using different water feed rates on the quality of ready-to-eat snacks made from food by-products. *Food Chemistry*, 114(1): 226–232. https://doi.org/10.1016/j.foodchem.2008.09.043

Stojceska, V., Ainsworth, P., Plunkett, A., and Ibanoğlu, Ş. 2010. The advantage of using extrusion processing for increasing dietary fibre level in gluten-free products. *Food Chemistry*, 121(1): 156–164. https://doi.org/10.1016/j.foodchem.2009.12.024

Su, Y., Zhang, M., Bhandari, B., and Zhang, W. 2018. Enhancement of water removing and the quality of fried purple-fleshed sweet potato in the vacuum frying by combined power ultrasound and microwave technology. *Ultrasonics Sonochemistry*, 44: 368–379.

Sumathi, A., Ushakumari, S.R., and Malleshi, N.G. 2007. Physico-chemical characteristics, nutritional quality and shelf-life of pearl millet based extrusion cooked supplementary foods. *International Journal of Food Sciences and Nutrition*, 58(5): 350–362. https://doi.org/10.1080/09637480701252187

Sun, J., Zhou, W., Huang, D., and Yan, L. 2018a. 3D food printing: Perspectives. In: *Polymers for food applications*, ed. T. Gutierrez, 725–755. Cham: Springer. https://doi.org/10.1007/978-3-319-94625-2_26

Sun, J., Zhou, W., Yan, L., Huang, D., and Lin, L.Y. 2018b. Extrusion-based food printing for digitalized food design and nutrition control. *Journal of Food Engineering*, 220: 1–11. https://doi.org/10.1016/j.jfoodeng.2017.02.028

Sun, Y., Zhang, M., and Fan, D. 2019. Effect of ultrasonic on deterioration of oil in microwave vacuum frying and prediction of frying oil quality based on low field nuclear magnetic resonance (LF-NMR). *Ultrasonics Sonochemistry*, 51: 77–89. https://doi.org/10.1016/j.ultsonch.2018.10.015

Supmoon, N., and Noomhorm, A. 2013. Influence of combined hot air impingement and infrared drying on drying kinetics and physical properties of potato chips. *Drying Technology*, 31(1): 24–31. https://doi.org/10.1080/07373937.2012.711792

Technavio. 2021. Top 3 trends impacting the global potato peeler and slicer equipment market through 2021. Technavio. https://www.businesswire.com/news/home/20170307005135/en/Top-3-Trends-Impacting-Global-Potato-Peeler

Tejada-Ortigoza, V., García-Amezquita, L.E., Serna-Saldivar, S.O., Martín-Belloso, O., and Welti-Chanes, J. 2017. High hydrostatic pressure and mild heat treatments for the modification of orange peel dietary fiber: Effects on hygroscopic properties and functionality. *Food and Bioprocess Technology*, 11(1): 110–121. https://doi.org/10.1007/s11947-017-1998-9

Tejada-Ortigoza, V., García-Amezquita, L.E., Torres, J.A., and Welti-Chanes, J. 2020a. Chemical processes for the extraction and modification of dietary fiber. In: *Science and technology of fibers in food systems*, eds. J. Welti-Chanes, S.O. Serna-Saldivar, O.H. Campanella, and V. Tejada-Ortigoza, 343–361. Cham: Springer.

Tejada-Ortigoza, V., Welti-Chanes, J., Campanella, O.H., and Peleg, M. 2020b. Estimating equilibrium moisture content from relatively short sorption experiments. *LWT – Food Science and Technology*, 132(July): 109832. https://doi.org/10.1016/j.lwt.2020.109832

Teruel, M.del R., Gordon, M., Linares, M.B., Garrido, M.D., Ahromrit, A., and Niranjan, K. 2015. A comparative study of the characteristics of french fries produced by deep fat frying and air frying. *Journal of Food Science*, 80(2): E349–E358. https://doi.org/10.1111/1750-3841.12753

Toepfl, S., Heinz, V., and Knorr, D. 2006. Applications of pulsed electric fields technology for the food industry. In: *Food engineering series*, eds. J. Raso and V. Heinz, 197–221. New York: Springer.

Traffano-Schiffo, M.V., Tylewicz, U., Castro-Giraldez, M., Fito, P.J., Ragni, L., and Dalla Rosa, M. 2016. Effect of pulsed electric fields pre-treatment on mass transport during the osmotic dehydration of organic kiwifruit. *Innovative Food Science and Emerging Technologies*, 38: 243–251.

Tremaine, A.J., Reid, E.M., Tyl, C.E., and Schoenfuss, T.C. 2014. Polymerization of lactose by twin-screw extrusion to produce indigestible oligosaccharides. *International Dairy Journal*, 36(1): 74–81. https://doi.org/10.1016/j.idairyj.2013.12.013

Tudorica, C.M., Kuri, V., and Brennan, C.S. 2002. Nutritional and physicochemical characteristics of dietary fiber enrihed pasta. *Journal of Agricultural and Food Chemistry*, 50(2): 347–356.

Tylewicz, U., Aganovic, K., Vannini, M., Toepfl, S., Bortolotti, V., Dalla Rosa, M., Oey, I., and Heinz, V. 2016. Effect of pulsed electric field treatment on water distribution of freeze-dried apple tissue evaluated with DSC and TD-NMR techniques. *Innovative Food Science and Emerging Technologies*, 37(C): 352–358.

Tylewicz, U., Tappi, S., Genovese, J., Mozzon, M., and Rocculia, P. 2019. Metabolic response of organic strawberries and kiwifruit subjected to PEF assisted-osmotic dehydration. *Innovative Food Science and Emerging Technologies*, 56: 102190.

Udomkun, P., Niruntasuk, P., and Innawong, B. 2019. Impact of novel far-infrared frying technique on quality aspects of chicken nuggets and frying medium. *Journal of Food Processing and Preservation*, 43(5): 1–9. https://doi.org/10.1111/jfpp.13931

Van der Sman, R.G.M., and Bows, J.R. 2017. Critical factors in microwave expansion of starchy snacks. *Journal of Food Engineering*, 211: 69–84.

Vasanthan, T., Gaosong, J., Yeung, J., and Li, J. 2002. Dietary fiber profile of barley flour as affected by extrusion cooking. *Food Chemistry*, 77(1): 35–40. https://doi.org/10.1016/S0308-8146(01)00318-1

Vengu, S., Ravandran, H.P., Gannasin, S.P., and Muhammad, K. 2020. Effect of hydrocolloid addition on batter properties and quality of deep-fried banana (*Musa* spp.) fritters. *British Food Journal*, 122(10): 3227–3238. https://doi.org/10.1108/BFJ-02-2020-0114

Virot, E., and Ponomarenko, A. 2015. Popcorn: Critical temperature, jump and sound. *Journal of the Royal Society. Interface*, 12(104): 20141247. https://doi.org/10.1098/rsif.2014.1247

Vitrac, O., Trystram, G., and Raoult-Wack, A.L. 2000. Deep-fat frying of food: Heat and mass transfer, transformations and reactions inside the frying material. European. *Journal of Lipid Science and Technology*, 102(8–9): 529–538. https://doi.org/10.1002/1438-9312(200009)102:8/9<529::aid-ejlt529>3.0.co;2-f

Wang, Q., Li, Y., Sun, D.W., and Zhu, Z. 2018. Enhancing food processing by pulsed and high voltage electric fields: Principles and applications. *Critical Reviews in Food Science and Nutrition*, 58(13): 2285–2298.

Wei, S.Y., Wang, G., Huang, Y., Zhu, R.G., Wang, Q., Zhang, X.N., Cao, Y.X., and Yao, X.D. 2018. Pulsed air jet impingement drying characteristics of Winter Jujube slices. *Czech Journal of Food Sciences*, 36(4): 324–332. https://doi.org/10.17221/258/2017-CJFS

Welti-Chanes, J., Morales-de la Peña, M., Jacobo-Velázquez, D.A., and Martín-Belloso, O. 2017. Opportunities and challenges of ultrasound for food processing: An industry point of view. In: *Ultrasound: Advances for food processing and preservation*, ed. D. Bermudez-Aguirre, 267–292. London: Academic Press an imprint of Elsevier Inc.

Wu, X.F., Zhang, M., Ye, Y., and Yu, D. 2020. Influence of ultrasonic pretreatments on drying kinetics and quality attributes of sweet potato slices in infrared freeze drying (IRFD). *LWT – Food Science and Technology*, 131: 109801.

Wu, Y., and Guo, Y. 2010. Experimental study of the parameters of high pulsed electrical field pretreatment to fruits and vegetables in vacuum freeze-drying. In: *International conference on computer and computing technologies in agriculture*, 691–697. Nanchang: Springer-Heidelberg.

Wu, Y., and Zhang, D. 2014. Effect of pulsed electric field on freeze-drying of potato tissue. *International Journal of Food Engineering*, 10(4): 857–862.

Xu, X., Zhang, L., Feng, Y., Zhou, C., Yagoub, A.E.G.A., Wahia, H., Ma, H., Zhang, J., and Sun, Y. 2021. Ultrasound freeze-thawing style pretreatment to improve the efficiency of the vacuum freeze-drying of okra (*Abelmoschus esculentus* (L.) Moench) and the quality characteristics of the dried product. *Ultrasonics Sonochemistry*, 70: 105300.

Yağcı, S., and Göğüş, F. 2008. Response surface methodology for evaluation of physical and functional properties of extruded snack foods developed from food-by-products. *Journal of Food Engineering*, 86(1): 122–132. https://doi.org/10.1016/j.jfoodeng.2007.09.018

Yildiz, G., and Izli, G. 2019. The effect of ultrasound pretreatment on quality attributes of freeze-dried quince slices: Physical properties and bioactive compounds. *Journal of Food Process Engineering*, 42(5): e13223.

Yin, X., Jiao, S., Sun, Z., Qiu, G., Tu, K., Peng, J., and Pan, L. 2019. Two-step drying based on air jet impingement and microwave vacuum for apple slices. *Journal of Food Process Engineering*, 42(5): 1–9. https://doi.org/10.1111/jfpe.13142

Yuksel, F., Karaman, S., and Kayacier, A. 2015. Barley flour addition decreases the oil uptake of wheat chips during frying. *Quality Assurance and Safety of Crops and Foods*, 7(5): 621–628. https://doi.org/10.3920/QAS2014.0472

Zaghi, A.N., Barbalho, S.M., Guiguer, E.L., and Otoboni, A.M. 2019. Frying process: From conventional to air frying technology. *Food Reviews International*, 35(8): 763–777. https://doi.org/10.1080/87559129.2019.1600541

Zhang, L., Liao, L., Qiao, Y., Wang, C., Shi, D., An, K., and Hu, J. 2020. Effects of ultrahigh pressure and ultrasound pretreatments on properties of strawberry chips prepared by vacuum-freeze drying. *Food Chemistry*, 303: 125386.

Zhang, L., Qiao, Y., Wang, C., Liao, L., Shi, D., An, K., Hu, J., Wang, J., and Shi, L. 2020. Influence of high hydrostatic pressure pretreatment on properties of vacuum-freeze dried strawberry slices. *Food Chemistry*, 331: 127203.

Zhang, X., Zhang, M., and Adhikari, B. 2020. Recent developments in frying technologies applied to fresh foods. *Trends in Food Science and Technology*, 98: 68–81.

Zou, K., Teng, J., Huang, L., Dai, X., and Wei, B. 2013. Effect of osmotic pretreatment on quality of mango chips by explosion puffing drying. *LWT – Food Science and Technology*, 51(1): 253–259. https://doi.org/10.1016/j.lwt.2012.11.005

3 Industrial Frying and Seasoning Operations

Sergio O. Serna-Saldivar

CONTENTS

3.1 INTRODUCTION

The process of frying is the most noteworthy and conventional operation in the production of many savory snacks. Needless to say, it is extremely pertinent to understand the quality and characteristics of frying oil sources thoroughly described in Chapter 7, the features of the product to be fried and the frying technology. The sort of frying oil also affects organoleptic and shelf-life expectations, particularly from the oxidative rancidity perspective.

According to Moreira et al. (1999), the operation of frying is one of the oldest processes of food preparation. Historically, the first frying process originated around the Mediterranean where olive (*Olea europaea*) oil was utilized. At present, deep-fat fried snacks are common in practically all cultures around the globe. Frying is defined as a process in which the food is cooked in hot oil with the aim of modifying its flavor, texture and overall organoleptic properties. Nutritionally, fried products are more energy dense than unfried counterparts because of the significant loss of water during the process and mainly due to the oil uptake during the operation. In addition, the ratio of saturated to unsaturated fatty acid composition of the frying oil and presence of trans fats greatly affects the nutritional profile and concurrent health implications (Chapters 20 and 21). From the

DOI: 10.1201/9781003129066-3

stability viewpoint, the loss of water improves the microbial stability of the product by lowering the water activity (*Aw*) but the oil absorption enhances the susceptibility of the product to oxidative rancidity. The selection of the frying oil, frying conditions and other related processing parameters such as packaging materials and technology (Chapter 4), storage temperature, and addition of artificial or natural antioxidants are important strategies to decrease the rate of rancidity and preserve freshness.

Deep-fat frying of foods carried out under atmospheric conditions at temperatures near 190°C darkens and scorches products even before they are completely fried. In addition, conventional frying generates oil decomposition products associated with adverse health effects (Chapter 20). Vacuum frying is a technology that can counteract these negative effects. This process is carried out under pressures below atmospheric levels (< 6.65 kPa). According to Moreira (2014) vacuum-fried products have higher retention of phytochemicals, better color scores and are less prone to oxidation and formation of harmful oil degraded compounds.

According to Tizhe-Liberty et al. (2019) there are three major strategies to diminish the amount of oil in fried foods based on modifications of product surface, frying technique and frying medium. Owing to the fact that frying is majorly a surface phenomenon, modification of product surface has proven to be the most effective in reducing oil uptake. Within this strategy, coating with soy protein isolate, methylcellulose, whey protein isolate and wheat gluten protein has been effectively used to modify product surface and reduce oil uptake by 35 to 40%.

The application of seasonings has become a vital element in the growth and success of the snack food industry. The development of many new products is directly related to the use of the appropriate seasoning and the desired percentage of covering.

The interaction between the snack and the seasoning dictates the final flavor profile and consumer acceptance. Seasonings for salty snacks are mixtures of salt, spray-dried cheese or sour cream, dehydrated onion, garlic and spices, flavor enhancers, coloring agents and fillers such as maltodextrins and/or flours (Chapter 8). These seasonings can be applied in powders or in oil-based liquid mixes or slurries that are directly sprayed onto the snack units. The main goal of the coating operation is to apply the seasoning in a uniform and consistent manner so all sides of the product should have the same appearance and amount of cover. Normally, the seasonings formulations are in dry form, but they can alternatively be delivered in oils or two-phase slurries. Slurries are produced by blending dry flavors with a liquid carrier that in most instances is oil for savory snacks or water for sweetened products. As not all dry flavorings are soluble in the liquid carrier, a two-phase mixture is in some cases applied. The foremost function of the seasoning device is to correctly deliver the desired amount of seasoning to the incoming product flow to achieve the desired proportion of seasoning in the finished snack. The correct and even application of the seasonings that are recognized as one of the most expensive raw ingredients should be carefully controlled because the over-application represents relevant economic losses each year to snack processors.

This chapter describes traditional and novel frying operations in terms of processing conditions and fryers and degradative changes that occur in frying oils during this thermal operation. Moreover, it describes and covers the most used devices for the delivery of powdered and/or liquid seasonings widely applied by snack food processors.

3.2 FRYING PROCESS

Most industrial snack processes involve frying in which heat energy is transferred from the frying oil to the food. Most operations employ deep-fat frying that is considered when the food is submerged at least 2 cm in the oil (Moreira et al. 1999). During frying, the snack is cooked and dried through contact with the heated oil and therefore the process involves both heat and mass transfer mechanisms (Hosseini et al. 2016). During frying, water vapor is liberated from the product and transferred to the air and part of the heated oil enters the product at different rates. These phenomena play critical roles in terms of quality of the finished products. Deep-fat frying of snacks is typically made at 165 to 195°C for several seconds to several minutes depending on the type of snack food and its dimensions. When the product is immersed in the heated oil, the surface temperature of the food rapidly augments enhancing water evaporation. During this first stage of the frying process, a thin layer of water vapor is formed in the interphase between the frying oil and the product. Importantly, the steam acts as a partial insulator because its heat conductivity is lower compared to the frying oil. Nevertheless, after the subsequent steps of the frying, the steam layer eventually breaks enhancing the direct contact of the oil with the product surface. The disruption of the steam layer can be boosted by the use of emulsifiers contained in the food matrix or the frying oil. As steam migrates out of the snack it creates voids or cavities and the frying oil flows into the product. This greatly affects oil uptake and the internal porosity or texture of the snack food. When larger volumes of steam become trapped between intracellular spaces larger pores, known as blisters or pillows, are commonly generated. These large blisters are considered as a serious defect in most snacks and are created in higher extents when the incoming food product contains more moisture and when the frying oil is heated to higher temperatures (Moreira and Barrufet 1998, Moreira et al. 1999, Kawas and Moreira 2000, Saguy and Dana 2003, Moreira 2007, Bouchon 2009, Bouchon and Dueik 2018).

Most of the water:oil exchange takes place during the initial stage of the deep-fat frying process. For instance, when raw tortilla pieces (Chapter 11) are fried at 175°C most of the moisture is evaporated and replaced by the frying oil during the first 10 seconds (Serna-Saldivar et al. 1990). During this brief period, the moisture content of tortilla chips diminishes from 34% to 10–15% and the oil content increases from 2% to close to 20%. Independent of the type of snack, the oil uptake is mostly dictated by intrinsic features of the unfried product especially in terms of initial moisture and starch contents, degree of starch gelatinization (Moreira and Barrufet 1998, Moreira et al. 1999, Kawas and Moreira 2000, Saguy and Dana 2003, Moreira 2007, Bouchon 2009, Bouchon and Dueik 2018).

3.2.1 Conventional Frying Operations

During deep-fat frying operations, oils are repeatedly used at high temperatures in the presence of oxygen and therefore undergo oxidative and thermal abuse. Throughout the prolonged use of frying oils, many chemical reactions occur like generation of free fatty acids (FFA), volatile and nonvolatile decomposition products and known toxicogenic and polymerized compounds (Chapter 20). Figure 3.1 summarizes the most common and relevant changes that occur during prolonged use of heated oils. The operating conditions and the kind of material being fried affect the shelf life of the oil (Warner 2004, Serna-Saldivar 2008).

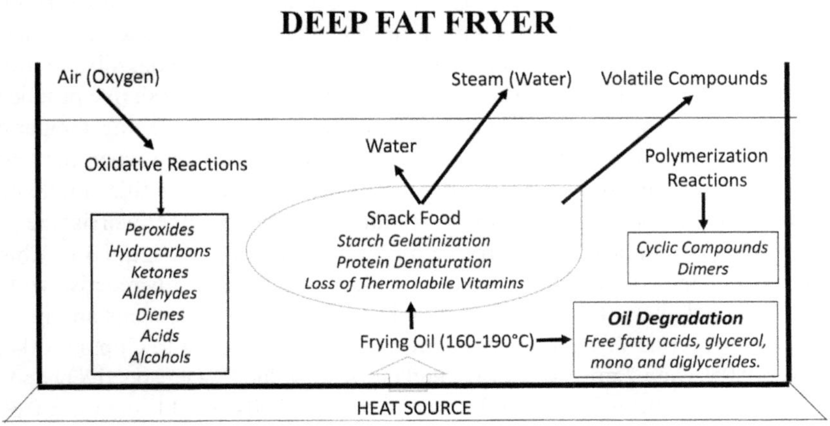

FIGURE 3.1 Relevant chemical changes which occur during prolonged used of frying oils.

The oil turnover rate, which is associated with the time it resides in the fryer, is of utmost importance in frying operations. Gupta et al. (2004) define this term as the ratio of the weight of the oil in a system to the hourly rate of oil picked up by the snack. This is one of the major factors in designing the frying system and dictates oil quality parameters especially in terms of FFA, peroxide value (PV) and smoke point. These quality control assays are discussed in Chapter 19. A comparative example of the calculation of oil turnover time for potato chips and tortilla chips is as follows: Oil content in the frying system = 2,000 kg, production throughput = 1,000 kg/hr, oil uptake by the product = 36% for potato chips and 24% for regular tortilla chips, respectively. With these numbers, the oil carried out by the two snacks is 360 and 240 kg/hr. Therefore, the comparative theoretical oil turnover rates for potato chips = 2,000 kg oil in the fryer/360 kg oil uptake = 5.55 hr whereas for tortilla chips = 2,000 kg oil in the fryer/240 kg oil uptake = 8.33 hr. Dunford (2012) suggests that the turnover time should be adjusted to compensate times required to start-up and shut down the fryer and death times for common interruptions due to mechanical failures. Therefore, the same author suggests dividing the theoretical oil turnover rate by 0.8 or 0.9 (average 0.85) so the adjusted values for potato chips and tortilla chips specified above are 5.55/0.85 = 6.53 hr and 8.33/0.85 = 9.8 hr, respectively. Dunford (2012) further indicates that factors such as types of oil and product used, frying temperature, fryer design, oil heating system and operating conditions determine the oil turnover rate for a given frying operation. High oil turnover rates are undesirable because oil degradation increases and product shelf life diminishes. For industrial fryers, oil turnover rates are recommended to fall between 5 to 10 hr. Gupta et al. (2004) indicate that the desired oil turnover rates for potato chips, tortillas chips and extruded products are 8, 6 and 4 hr, respectively.

Table 3.1 summarizes the main causes and suggested troubleshoot adjustments to control the quality of deep-fat fried snacks.

3.2.1.1 Operational Temperature of Frying Oil

The rate of oil degradation is highly associated with the degree of unsaturation or iodine value and the rate of FFA formation. The higher the iodine value the more susceptible the oil to oxidation and the higher the acid or FFA value the higher the formation of undesirable volatile and oxidative compounds. High acid or FFA values are strongly related to the oil smoke point temperature. As the FFA value of a given frying oil increases the smoke and ignition points decrease, therefore these two quality control parameters are commonly used to determine the condition of the oil (Chapter 19). For instance, oils containing 1% FFA have at least 50°C less smoke and ignition points compared with the fresh counterparts containing practically zero FFA. The iodine values of commercial oils and quality parameters to determine rancidity and stability are discussed in Chapters 7 and 19, respectively.

The two most relevant operating conditions to maintain the quality of frying oils are operating oil temperature and filtration. Table 3.1 summarizes the troubleshooting and recommendations to adjust the quality of fried snack foods. In terms of the temperature of frying oils, the rule of thumb is that the higher the temperature the faster the formation of FFA and oxidative products. The rate of these reactions generally doubles for each 10°C of increase in the frying temperature. From the operational viewpoint, one of the most relevant considerations is that when the fryer is not in operation the oil temperature should be lowered to at least 30°C because this simple practice will decrease eightfold the rate of degradation (Serna-Saldivar 2008). The other worrisome reaction that normally occurs during prolonged use of frying oils is polymerization (Figure 3.1). This particular phenomenon is associated with the generation of epoxides which are highly reactive with each other forming dimers and polymers that usually accumulate in the heating elements and interior surface contact areas of the fryer as varnish (Serna-Saldivar 2008). According to Gupta (2004), fresh oils should contain $< 0.05\%$ and 2 mEq/kg of FFA and peroxide value (PV) respectively and oils should be bland, sweet or with buttery flavor. On the other hand, used oils should contain $< 0.5\%$ FFA and < 5 mEq/kg of PV at shutdown. During normal operation of fryers, the FFA rises, reaches a steady-state value within 24 hr of frying, and remains essentially unchanged unless process interruptions

TABLE 3.1
Trouble Shooting Causes and Suggested Adjustments to Improve Quality of Deep-Fat-Fried Snacks

Problem	Main Reasons	Adjustments
Excessive oil absorption or uptake	Low frying oil temperature	Decrease snack food load, increase capacity of the heat elements of the fryer, clean heat elements from polymerized fat.
	Use of degraded oil with high acid and peroxide values	Increase turnover rate of the oil, increase frequency of removal of food particles by filtration, lower the temperature of the frying oil when fryer is not operating.
	Improper unfried food properties	Control degree of starch gelatinization, control the particle size of unfried product and diminish the use of emulsifiers added to the formulation.
Low-fat absorption	High oil temperature	Increase snack food load and slightly decrease frying temperature.
	Use of fresh oil	Adjust oil turnover to equilibrate the oil acid value.
	Improper unfried food properties	Control degree of starch gelatinization, control the particle size of unfried product and increase the use of emulsifiers added to the snack food formulation.
Development of undesirable darkness in final products	High frying oil temperature	Lower the temperature of the frying oil.
	Degraded oil	Totally or partially replace degraded oil or prevent oil degradation by filtering and lowering temperature between uses.
	Improper unfried food properties	Reduce the use of reducing sugars in the formulation and excessive alkaline conditions (high pH).
Snack product overly light	Low frying oil temperature	Increase the operational oil temperature, decrease snack food load and clean heating elements to improve heat exchange.
	Short frying time	Adjust time-temperature profile in the fryer.
Foaming	High frying temperature	Lower operational temperature of the fryer.
	Inadequate turnover	Increase snack food load.
	Presence of food particles or leftovers	Increase the frequency of filtration.
	High level of polymerized oil	Increase filtration, lower fat temperature when fryer is not in use, use of chelating agents in the oil to lower oxidation rate.
Oil smoking	High frying temperature	Lower operational temperature.
	High acid or free fatty acid value	Lower oil temperature when fryer is not in use, increase oil turnover and remove food particles via filtration.

Source: Adapted from Serna-Saldivar (2008).

occur. The steady-state value for FFA depends on the following factors: (i) type of product being fried; (ii) temperature of frying; (iii) type of fryer; and (iv) impurities in the incoming oil (Gupta et al. 2004).

The generation of FFA increases when water associated with the food matrix transfers into the frying oil. The water increases the rate of hydrolysis of the ester bonds of the triglycerides yielding FFA and mono and diglycerides. The polar mono and diglycerides act as emulsifiers because they possess one or two polar sides (OH groups) which further aggravates the rate of hydrolysis. The oils rich in polyunsaturated FFA can form cyclic fatty acid monomers, which are considered highly toxic and carcinogenic (Chapter 20). Furthermore, FFAs are more prone to oxidation of their double bonds which leads to the formation of many polar compounds such as aldehydes, ketones, peroxides, acids and alcohols (Hosseini et al. 2016). These surface-active materials affect fryer performance and more importantly the sensory properties and rancidity of fried snacks. It is recognized that the presence of metal ions such as iron and copper at concentrations of ppm in the frying oil enhances oxidative reactions. This is the main reason why copper is avoided in the fabrication of commercial fryers and ethylenediamine tetra-acetic acid (EDTA), phosphoric acid or other metal sequestrants are normally employed as oil additives (Serna-Saldivar 2008).

3.2.1.2 Filtration

Most snacks undergoing deep-fat frying lose solids during the operation as food particles. The pH of snacks affects the rate of degradation because both strong acidic or alkaline conditions speed up the generation of FFA. The alkaline pH associated with certain snacks like corn and tortilla chips (Chapter 11) and high operational temperatures increases the amount of FFA due to saponification reaction. Furthermore, the food crumbs tend to collect on heat exchange surfaces and carbonize further catalyzing undesirable reactions. Therefore, the elimination of these particles by filtration with a steel mesh, cloth or filter paper is highly recommended in both continuous and batch operations. Practically all continuous fryers are equipped with an integrated filtration device whereas the oil of batch fryers are commonly hand-filtered after a shift (Serna-Saldivar 2008).

3.2.2 Vacuum Frying

3.2.2.1 Conventional Vacuum Frying

Vacuum frying is a modern technology suited for the production of low-fat healthy snacks with lower amounts of toxic compounds that originate from thermal degradation and oxidation. This novel technology is performed in a sealed system (Figure 3.2) at pressures below atmospheric levels so as to enhance the removal of product moisture in a low-oxygen environment at comparatively lower temperatures (Garayo and Moreira 2002, Bouchon and Dueik 2018). The application of this technology is mainly used nowadays to dehydrate heat-sensitive food products. As a result, vacuum-fried snacks maintain their natural color, flavor, have comparatively better sensory properties and more importantly maintain key essential nutrients that play key roles in human nutrition and health (Shyu and Hwang 2001, Da Silva and Moreira 2008, Dueik and Bouchon 2011a, Dueik and Bouchon 2011b, Bouchon and Dueik 2018). Moreover, vacuum frying comparatively degrades frying oils to

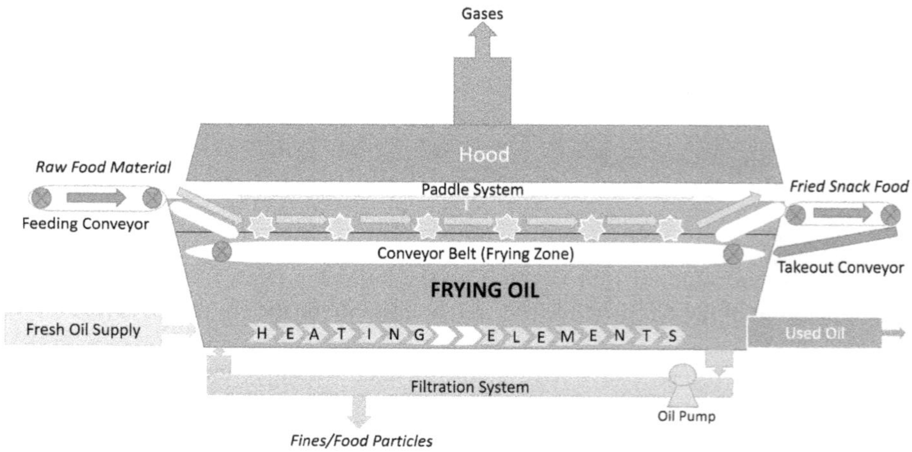

FIGURE 3.2 Illustration of the conventional continuous equipment used for deep-fat frying of snacks.

lesser extents improving their oxidative stability (Shyu et al. 1998) and diminishes the generation of relevant toxic compounds like acrylamide (Granda et al. 2004, refer to Chapter 20).

According to Garayo and Moreira (2002) vacuum-fried snacks normally contain lower amounts of oil compared to counterparts fried regularly. For example, the oil content of vacuum-fried potato chips was about 44% less compared to atmospheric fried counterparts. Likewise, Da Silva and Moreira (2008) and Dueik et al. (2010) reported that vacuum-fried sweet-potato slices and carrot chips absorbed approximately 24% and 50% less oil, respectively, compared to counterparts fried at atmospheric conditions. The significant differences in oil uptake are attributed to the lower vapor pressure of water during vacuum frying and to the higher temperatures reached during regular frying, which increases structural changes and tissue degradation. Nevertheless, a comparison of the oil uptake of vacuum and regular fried mango slices showed similar results indicating that the food matrix and its internal structure play a key role. According to Moreira et al. (2009) the use of centrifugal force (40 sec @ 750 rpm) before pressurizing the vessel further diminishes the oil content of vacuum-fried products like potato chips that only contained 23% oil compared to regular counterparts. Likewise, the use of post frying centrifugation (2 min @ 400 rpm) significantly decreased the oil content by approximately 65% of vacuum-fried potatoes and carrots and 24% in apples.

It is well documented that the quality or oxidative stability of frying oils used in vacuum-frying operations is maintained for longer periods favoring also the oxidative stability of fried snacks. The longer oxidative shelf life is attributed to the lower oil uptake or content and the lower FFA value and PV of the frying oil.

One of the major advantages of vacuum-frying technology is the production of healthy products containing higher amounts of beneficial compounds like vitamins and phytochemicals such as vitamin C, carotenoids and betalains that are normally lost during conventional frying (Yen et al. 2002, Herbach et al. 2008, Dueik et al. 2010). Dueik and Bouchon (2011b) compared ascorbic acid retention in vacuum and atmospheric fried potato and apple slices and concluded that the retention of this important thermal labile vitamin was 95 and 50% in potato and apple slices vacuum and conventionally fried at 98 and 160°C, respectively.

Granda et al. (2004) studied the color development and acrylamide formation during vacuum and atmospheric frying using different potato varieties. The authors observed that in potato chips of different varieties fried under vacuum conditions (118°C, 1.3 kPa for 8 min), the acrylamide contents were always significantly lower compared to atmospheric frying (165°C for 4 min), with reductions ranging from 85 to 99%. In relation to color, the same study revealed that for all varieties L, a and b color values were lower in the case of vacuum-fried potato chips due to less browning. From the organoleptic viewpoint, vacuum-fried products are normally rated with higher scores compared to regular fried counterparts, which usually have darker colorations, more intensity of bitter off-flavors, and lower overall acceptability especially in terms of flavor (Warner 2004).

Crosa, et al. (2014) studied changes produced in oils during vacuum and traditional frying of potato chips. Two different refined sunflower oils were used one classified as high oleic (Chapter 7) and the other supplemented with synthetic antioxidant TBHQ. Oil degradation was monitored by measuring the FFA, PV, p-anisidine (p-AV),) total polar compounds (TPC) and oxidative stability. Oil samples were taken every 4 h of frying for ten consecutive days. Values of FFA, p-AV, and TPC using TBHQ with traditional frying were (0.201, 207.0, 25.0) significantly higher compared to values with vacuum frying (0.073, 25.8, 11.2). The same parameters by using the high oleic oil were (0.327, 82.0, 21.9) with traditional frying and (0.099, 33.3, 6.4) with vacuum frying. These notorious differences affected the oil stability with values of 2.44 and 7.95 with the oil supplemented with TBHQ used for traditional and vacuum frying respectively. The same trend was observed when the high oleic oil with values of 0.65 and 2.67, respectively. At the end of the frying processes, the percentages of alpha-tocopherol reduction ranged from 4.9% for the TBHQ oil used in vacuum frying to > 97% for both oils used in traditional frying. Therefore, vacuum frying achieved a significant decrease in the rate of degradative reactions in oils.

Shyu et al. (1998) also evaluated frying oil quality with different assays during vacuum frying of carrot slices. Palm oil, lard, and soybean oil were fried under vacuum at 105°C for 20 min each hour in an 8-hr shift for six consecutive days. PV, acid value, carbonyl value, total polar components, dielectric constant, viscosity and fatty acid composition were used to evaluate the quality of these frying oils. Results showed that palm oil and lard possessed greater thermal stability than soybean oil. Of the methods used, PV, carbonyl value, total polar components, and dielectric constant showed good correlations with frying time and between each other.

3.2.2.2 Vacuum Frying Assisted with Other Novel Technologies

According to Su et al. (2018) the main disadvantages of vacuum frying are low efficiency especially with low-temperature frying and relatively higher oil uptake in certain types of products. One of the recommended methods to overcome these limitations is the usage of other methods that act synergistically especially in terms of enhancement of energy transfer. The most promising mechanisms are microwave radiation and ultrasound. The combination of vacuum frying with these two technologies leads to significant improvements in energy efficiency and quality attributes of fried fruits and vegetables. The combination of ultrasound and microwave exerted a synergistic effect on the energy efficiency and quality attributes of vacuum-fried potato chips. The ultrasound applied at 300 W and 600 W decreased frying time by 36.4 to 54.5% and improved moisture diffusivity from 60.5 to 144.5%, respectively. The novel ultrasound-microwave-assisted processes consumed about 34.9 to 48.3% less energy compared to sole vacuum frying. More importantly, the oil uptake of vacuum-fried potato chips was reduced by 27.4–32.3% whereas the texture and color significantly improved. In addition, scanning electron micrographs revealed a more porous microstructure in samples processed with ultrasound-microwaves. The moisture loss kinetics and quality parameters improved markedly when higher ultrasound power was used. The authors concluded that the vacuum-frying technology assisted with both ultrasound and microwaves greatly improved energy efficiency and product quality. These experimental results have encouraged processors to recently adopt novel vacuum-frying technologies of fruits and vegetables because resulting snacks contain less oil and retain their quality and from the processing viewpoint are manufactured with less time and energy.

Moreira et al. (2009) studied ways of removing adhered surface oil on vacuum-fried potato chips because this problem exacerbates after the pressurization step. The authors designed experiments to establish the effects of de-oiling and frying temperature on potato chips oil absorption and their quality attributes. Potato slices were fried for 360 sec in a vacuum fryer operating at pressures of less than 1.33 kPa at 120, 130, or 140 C. An integrated centrifuging system, which operated at 750 rpm for 40 sec, was used before pressurizing the vessel with the aim to reduce surface oil. Chips fried at 120°C for 360 sec without centrifugation contained 43% oil compared to 9.77 % oil observed in counterparts that were centrifuged. Most of the oil was absorbed during the first 180 sec of vacuum frying and by the end of the frying schedule only 14% of oil was internal whereas the rest surface oil. At the end of frying, bulk density and porosity values of the noncentrifuged samples were around 800 kg/m^3 and 0.36 and 564 kg/m^3 and 0.61 for the centrifuged samples, respectively. Therefore, the use of the proposed centrifugation step yielded vacuum-fried potato chips with lower oil content and superior quality.

3.3 OIL DEGRADATION

Oil oxidation or rancidity is an undesirable series of chemical reactions involving oxygen that end in the generation of off-flavors and odors. When oil oxidizes, it produces sequentially a series of degraded products starting with primary oxidation like FFA and peroxides followed by secondary products such as carbonyls, aldehydes and trienes and finally tertiary products. The rate of oxidation is affected by many factors like type of oil especially in terms of its iodine value or degree of unsaturation and external factors like availability of oxygen, temperature, light and presence of trace metals like copper and iron.

TABLE 3.2
Stages of the Oxidative Reactions That Lead to Oil and Snack Food Rancidity

Stage of Oxidative Reaction	Main Chemical Mechanisms
Initiation	It was thought in the past that oxidation occurred directly on the double linkages between two carbon atoms of unsaturated fatty acids. However, today it is known that a double linkage between two carbon atoms favors the oxidation of the neighboring carbon linkage. The key reaction is the formation of a highly reactive alkyl radical from fatty acids by removing hydrogen mainly from neighboring double bonds. Through this process, an oxygen ion will likely replace the hydrogen ion yielding hydroperoxides. The heat, ultraviolet light, metals, and reactive oxygen species facilitate this reaction. The main prooxidant metals are Cu2+ and Fe3+.
Propagation	Also known as a free radical chain reaction. The peroxy radical abstracts a hydrogen atom from another fatty acid and forms new hydroperoxides and new alkyl radicals. The chain reactions of free alkyl radicals and peroxy radicals accelerate the thermal oxidation of oil exponentially.
Termination	The termination reactions occur when the alkoxy radicals react with other radicals or are decomposed to form nonradical volatile and nonvolatile compounds that produce off-odors and flavors. Hydroperoxides are extremely prone to oxidation and breakdown into numerous unreactive components like aldehydes, hydrocarbons and ketones but also new highly reactive radicals. The main aldehydes produced are malonaldehyde and its isomeric combinations, hydroxyacrolein and ephidrinaldehyde. The aldehydes and ketones are the main ones responsible for the rancid flavor and odor of snacks.

The oxidation of oils is divided into three phases: initiation, propagation and termination (Table 3.2). During the initiation stage free radicals are generated whereas in the propagation these free radicals generate more free radicals and peroxides. The final stage is when these peroxides are broken down into nonradical products that affect the flavor and odor of the oil and fried products.

Blumenthal (1991) described five sequential phases that oil passes through during the degradation process: break-in, fresh, optimum, degrading and runaway oils. Break-in and fresh oils produce slight browning, relatively little crisping of the surface and lower oil pickup. The optimum stage is when the food acquires a golden brown color with crisp and rigid surfaces and optimal oil uptake. Products fried with degrading and/or runaway oils have darkened and/or spotty surfaces, excess oil pickup and off-odors and flavors. From a health perspective, these oils are considered harmful because the polar and polymerized compounds are known to produce cancer in laboratory animals (Chapter 20). Figure 3.1 summarizes the major factors that affect decomposition of frying oils and the products formed after their continuous usage. The measurement of heat abuse of oils is the most relevant quality control parameter in deep-fat frying operations. Many methods of analysis like acid value, peroxide value and smoke point (Chapter 19) correlate with the condition of the abused oil.

Deep-fat frying produces both desirable and undesirable flavor compounds and changes the stability, color, properties and overall quality of oils. Tree types of major deteriorative reactions occur to frying oils: hydrolytic, oxidative and polymerization. These reactions generate both nonvolatile and volatile compounds that modify the physical and chemical properties of oils. The frying decreases the unsaturated fatty acids of oil and increases foaming, color, viscosity, density, specific heat and contents of FFA, polar materials, and polymeric compounds. In addition, the nonvolatile compounds affect flavor stability and quality and texture of snacks and their shelf stability especially in terms of oxidative rancidity.

Tsuzuki et al. (2010) evaluated the impact of thermal treatments of oils on the generation of *trans* fatty acids (TFA). Slices of raw potatoes (10% of the frying oil (w/w)) were fried in commercially available canola oil at 160, 180 and 200°C during ten frying cycles. The oil and TFA contained in raw potatoes was about 0.1% (w/w) and practically zero, respectively. The TFA of potatoes fried by the tenth frying operation was between 0.99 to 1.05 g/100 g lipids. Therefore, the intake of 100 g of fried potatoes provided approximately 0.1 g of TFA. These authors further evaluated the generation of TFA in six types of commercially available vegetable oils that were heated to 180°C. The authors concluded that the slight changes in TFA concentrations were observed after several hours of heating but the ordinary frying process with nonhydrogenated oils did not impact TFA from the practical viewpoint.

Besides the intake of harmful TFA, one of the major concerns in the snack food industry is the generation of toxic compounds during frying like acrylamide, aldehydes, acrolein, heterocyclic amines (HCA) and polycyclic aromatic hydrocarbons (PAH). The toxicology and adverse health effects of these compounds are covered in Chapter 20.

3.3.1 Oil Hydrolytic Reactions

The snacks undergoing deep-fat frying release their moisture as steam creating a bubbling action that incorporates air (oxygen) in the frying oil. These components are greatly associated with the deteriorative chemical reactions in the frying oil and snack. The presence of small amounts of water hydrolyze ester bonds of triglycerides yielding FFA, mono and diglycerides and glycerol. All these hydrolyzed compounds have certain polarity and affinity for water and their higher concentrations accelerate further the hydrolytic reactions. The glycerol is known to evaporate at 150°C or higher and the leftover molecules are also known to promote the release of FFA (Naz et al. 2005). The main indicator of this deteriorative reaction is the oil acid or FFA value (Chapter 19) which increases in function of the oil usage and operation time. The suggested maximum FFA content of commercial fresh frying oils should be 0.05% to 0.08% (Choe and Min 2007).

Other relevant hydrolytic reaction is the thermal hydrolysis that mainly occurs within the oil phase rather than the water–oil interface. Hydrolysis is more preferable in oils containing short and unsaturated fatty acids because they are more hydrosoluble compared to long and saturated-containing oils. The moisture content associated with the unfried snack matrix is easily accessible to short-chain oils (Choe and Min 2007). Large contact between the oil and the aqueous phase of food increases the rate of hydrolysis and formation of FFA. According to Houhoula et al. (2003) the mono- and diglycerides in cottonseed oil during frying of potato chips at 155°C–195°C increase initially and then reach a plateau. In terms of oil management, the frequent replacement of the frying oil with fresh oil diminishes the hydrolytic rate.

3.3.2 Oil Oxidative Reactions

The oil oxidative reactions are considered the principal related to organoleptic properties of snack foods. These reactions occur differently according to the available oxygen in the oil, the time-temperature of the frying oil and the type of oil especially in terms of iodine value and presence of natural or artificial antioxidants. The oxidative mechanism of thermal oxidation is very similar to autoxidation although these reactions occur at different rates. The thermal oxidation rate is faster and involves the initiation, propagation and termination sequential deteriorative steps. Table 3.2 summarizes the chemical reactions that occur in each of these stages. The initiation reaction starts with the formation of a radical which is normally formed in the hydrogen with the weakest bond. For instance, the energy required to break the carbon-hydrogen bonds of the 8 and 11 carbons of linoleic acid is 75 and 50 kcal/mole (Choe and Min 2007). The double bonds at carbons 9 and 12 decrease the carbon-hydrogen bond at carbon 11 by withdrawing electrons. In comparison, the carbon-hydrogen bond on the saturated carbon without any double bond next to it is approximately 100 kcal/mol. These differences in strengths of hydrogen-carbon bonds of fatty acids explain the differences in oxidation rates among stearic, oleic, linoleic, and linoleic acids during oxidation. As mentioned above the weakest carbon-hydrogen bond of linoleic acid is the one on carbon 11 and this will be preferably removed to yield a free radical. On the other hand, the alkyl radical of saturated fatty acids normally forms at the α-position of the carboxyl group owing to electron-withdrawing property.

The alkyl radicals with a reduction potential of 600 mV react rapidly with diradical triplet oxygen and produce peroxy radicals at the rate of 109/M/sec. The peroxy radical with a reduction potential of 1,000 mV abstracts hydrogen from oleic acid and linoleic acid and produces hydroperoxide at a rate of approximately 1 × 100/M/sec and 1 × 10/M/sec, respectively (Choe and Min 2005). The rate of oxidative degradation increases as the concentration of oxygen and free radicals increases. The amount of mono- and diacylglycerols is small at the beginning of the frying process. The high interfacial tension in the frying system breaks steam bubbles and forms a steam blanket over the oil surface. The steam blanket reduces the contact between the oil and oxygen, and lowers the oil oxidation (Blumenthal 1991).

The amount of volatile compounds present in the oil varies widely depending on the type of oil, snack food and frying conditions. Most of the volatiles generated during the terminal step of the oxidation process are removed from the frying oil by steam. The disappearance or volatile compounds during frying is due to the combination of evaporation and decomposition, and the reaction of volatile compounds with other food components. Volatile compounds are extremely important to the flavor qualities of frying oil and fried foods, although volatile contents in total decomposition products of frying oil are present at concentrations of only part per million.

The naturally present or intentionally added antioxidants in oils and foods influence oil quality during deep-fat frying. Tocopherols, butylated hydroxyl anisole (BHA), butylated hydroxyl toluene (BHT), propyl gallate (PG) and tert-butyl hydroquinone (TBHQ) slow down the oxidation of oil at room temperature. However, all these antioxidants become less effective when exposed to frying temperatures due to losses through volatilization or decomposition. Tyagi and Vasishtha (1996) reported the ineffectiveness of 0.01% BHA and TBHQ during deep-fat frying of potato chips in soybean oil. However, Choe and Lee (1998) indicated that the degradation of tocopherols in soybean oil, beef tallow and palm oil after 8 hr frying of steamed noodles at 150°C were 12.5, 100 and 100%, respectively. Kim and Choe (2004) concluded that lignans (sesamol, sesamin and sesamolin,) were stable during heating and contributed to improving the oxidative stability of roasted sesame oil subjected to heating at 170°C. Likewise, the extracts of rosemary (*Rosmarinus officinalis*) and sage (*Salvia officinalis*) diminished oil deterioration during a 30-hr intermittent deep-fat frying of potato chips (Che Man and Jaswir 2000). Jaswir et al. (2000) further reported the synergistic antioxidant effects of rosemary, sage and citric acid on palm olein used to fry potato chips.

3.3.3 Oil Polymerization Reactions

Besides the hydrolytic and oxidative reactions discussed before, frying oils also degrade into non-volatile polar compounds and triacylglycerol dimers and polymers (Choe and Min 2007). Dimers and polymers are large molecules with a molecular weight range of 692 to 1,600 daltons and formed by a combination of –C–C–, –C–O–C–, and –C–O–O–C-bonds (Choe and Min 2007). Dimers derived from linoleic and oleic acids such as dehydroxydimer, ketodehydrodimer, monohydrodimer and dehydrodimer are generated during the frying at 195°C with soybean oil (Christopoulou and Perkins 1989). The dimers and polymers have hydroperoxy, epoxy, hydroxy and carbonyl groups, and –C–O–C– and –C–O–O–C-linkages. These compounds are either acyclic or cyclic depending on the reaction process and fatty acid composition of the oil (Choe and Min 2007). The dimers are formed from the reactions of allyl radicals by C–C linkage whereas the formation of acyclic polymers results from the reactions of triglycerides with oxygen via alkyl hydroperoxides (ROOH) or dialkyl peroxides (ROOR). These compounds are promptly decomposed to alkoxy and peroxy radicals by RO–OH and ROO–R scission, respectively. The alkoxy radicals are known to produce hydroxy compounds, or alternatively combine with other alkyl radicals to produce oxydimers. On the other hand, the peroxy radicals can react with other alkyl radicals and yield peroxy dimers (Choe and Min 2007). The formation of dimers and polymers is mainly affected by continuous usage of the oil, the type of frying oil and the operational oil temperature. In terms of the type of oil, the ones containing high amounts of linoleic acid (Chapter 7) are more prone to polymerization compared to counterparts rich in oleic acid like olive and sesame oils (Tompkins and Perkins 2000, Bastida and Sanchez-Muniz 2001, Choe and Min 2007). Cyclic polymers are produced within or between triglycerides by radical reactions and the Diels-Alder reaction. The formation of cyclic is also associated with the degree of unsaturation or iodine value and the frying oil temperature The formation of cyclic monomers and polymers is higher as the amount of linolenic acid increases in the oil (Rojo and Perkins 1987, Tompkins and Perkins 2000). In fact, the generation of cyclic monomers was negligible until the linolenic acid content exceeds 20% of the total fatty acid composition. Cyclic compounds are not formed to a significant extent until the oil temperature exceeds 200°C. Generally, most snacks are fried at temperatures ranging from 160 to 190°C. Polymers formed in deep-fat frying are known to accelerate oil oxidation, oil foaming and oil viscosity (Yoon et al. 1988, Choe and Min 2007) and the deposited brown varnish reduces the transfer of heat from the heating elements to the oil.

3.4 FRYING EQUIPMENTS

The proper operation of a fryer requires a uniform feed rate, the control of the oil temperature and the replenishment of the oil at a uniform rate as it is depleted by snack food. This concept suggests that fryers should be designed to maintain the maximum designed capacity, the inlet and outlet temperatures of the oil and an efficient filtration system to remove fines and prevent oil degradation. The types of fryers can be divided into two broad categories: batch and continuous. These fryers generally operate at atmospheric pressure although vacuum fryers are gaining popularity.

3.4.1 Batch Fryers

Batch fryers are generally small sized and used in fast-food restaurants and small establishments. These fryers are generally built with stainless steel to protect the oil stability and the product is positioned in one or two wire mesh baskets in a battery that automatically moves up or down so the deep-fat frying time is controlled (Serna-Saldivar 2008). In addition, they are equipped with heating elements and an oil-draining valve normally located at the bottom of the pan. The heat is either applied by gas flames or with electrical elements. The filtering of the oil in most batch units is performed manually although some models are furnished with an integrated system. The main controls in these fryers are oil temperature (temperature range from 150 to 200°C), frying time and the switch for downing or lifting the wire baskets. The small, intermediate and high capacity units are designed to work with 8–11 L, 17–22 L and 22–28 L of oil, respectively. There are also versions of high-pressure fryers developed to speed up the process and improve the texture of products like fried chicken. Generally, these fryers work at pressure ranges of 62,052 to 96,526 N/m^2 (9–14 psi) or 193,053 to 220,632 N/m^2 (28–32 psi). The main disadvantages of the high-pressure fryers are that oils tend to degrade quicker and therefore require more frequent filtration to compensate their higher degradation (Serna-Saldivar 2008).

3.4.2 Continuous Fryers

Continuous fryers typically process snack products in a moving belt or conveyor and are equipped with a tank that contains the frying oil, a heating unit with a temperature control system, a moving belt, a filtering system and an exhaust system to remove water vapor and other vapors emerging from the frying oil and/or product (Figure 3.2). Equipment are usually built from stainless steel, furnished with hoods to remove heat and gases and are insulated to save energy.

Several important factors come into play when deciding the optimum fryer for a given operation. The most critical issues are the sort of snack being produced, the equipment cost, space availability, the manufacturer's experience and environmental issues or regulations. Gupta et al. (2004) indicate that the fryer capacity is mainly selected based on production volume, the number of operating hours per day, frying time and frequency of fryer sanitation. The above information is critical to determine the dimensions of the fryer, which are dictated by the production rate, frying time and product loading (product weight/area, product weight/length or product pieces/length) which are known as "belt loading." The cook area (effective cooking area available in a fryer) and fryer length are calculated as follows: Cook area required = [production rate (product weight/time) × frying time]/[product loading (product weight/area)] whereas the fryer length = cook area required/fryer width (Dunford 2012).

Continuous fryers are mainly classified according to the heating system. The most popular direct heating system contains heating elements immersed in the cooking oil and the heat is mainly transferred by conduction. Direct heating is usually achieved by burning natural or propane gas. The temperature control is achieved by modulating the fuel input to the burners or switching the electric elements on and off. A thermocouple or temperature probe immersed in the frying oil is used to sense the temperature that is regulated to slightly overheat the frying oil.

These fryers are generally less expensive than those equipped with indirect heating systems but have disadvantages that the oil temperature is higher close to the heating elements and is not evenly distributed throughout the oil tank and that heating elements need more frequent cleaning to remove the varnish or gummy filth that lower the heat transfer efficiency.

The indirect or external heating systems use a heated thermal fluid that increases the temperature of the frying oil. The heat source is located outside the frying tank and pumps the oil to a heat exchanger and the fryer. The external heating systems commonly employ either high-pressure saturated steam or natural gas. In a steam heater, considered an excellent heat exchange system, the oil is pumped through a bank of heat exchange tubes with steam condensing on the outside. This is considered an exceptional system because it prevents overheating of the oil and therefore oil degradation. A modified design for the gas-fired system is designed to reduce overheating of the oil. In this design, the gas burner heats inducted air that is forced to pass across the tubes through which the fryer oil is recirculated. This indirect heating system is superior to direct heating due to the better oil flow and distribution through the fryer (Gupta el al 2004).

Continuous fryers are supplied with different conveying systems therefore the selection of the conveyor will vary according to the type of snack. These conveyor belts should push the product uniformly so as to control the residence time. For those snacks that float during frying, the conveying system should turn the snack to attain uniform frying. Turning is generally achieved by a rotating paddling system (Figure 3.2) that submerges and controls the movement of the snack through the frying zone. Another popular system is the one that holds the frying product submerged under the oil surface with a couple of conveyors. The conveyor system generally ends on a slop for draining the residual oil. Output capacities of continuous fryers vary from 100 to 2,500 kg/hr.

Fryers are also classified as single or multizone. For instance, in a single-zone fryer for potato chips these are fried with an inlet temperature of 177–180°C and a temperature drop of 4.4 to 7.2°C, between the inlet and the outlet of the equipment. In a multizone fryer, the oil enters the fryer at different locations helping to retain a more even temperature along the fryer and also significantly increase the oil volume unless the design is done with utmost care. Frying times for tortilla chips (Chapter 11), extruded products (Chapter 12) and potato chips (Chapter 14) are generally 1 min, < 60 sec and 2–3 min, respectively.

One of the most relevant parts of commercial fryers is the filtration system aimed to remove charred batter commonly referred to as cracklings. The fines liberated from the feedstocks should be removed to minimize darkening of the oil and prevent burnt flavors in finished products. If the oil is heated through the bottom of the fryer the cracklings also impede heat transfer. Various types of screens, cartridges and paper filters can be used to eliminate fines or cracklings. The first filtration step is the pass of the frying oil through screens that will take out large particles. The filter cartridges or paper filters are used to remove fine particles and some have alkaline properties so to neutralize FFA and produce fatty acid salts or soaps. In many applications, the system employs a

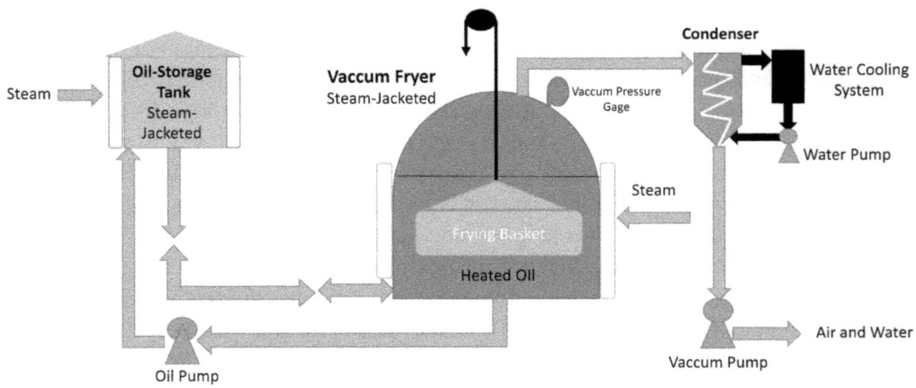

FIGURE 3.3 Illustration of the vacuum frying technology.

combination of centrifugal or rotary drum filters. New adsorbents are being employed to preserve the quality of the oils within the preset limits of polar compounds and FFA (Gupta et al. 2004, Serna-Saldivar 2008).

3.4.3 Vacuum Fryers

The main inconvenience of vacuum frying is the comparatively higher capital investment compared to regular operations. The first prototype of vacuum fryers were developed by Florigo B.V. about 60 years ago to produce high-quality potato chips.

Vacuum fryers are comparatively more expensive compared to regular fryers and most are designed for relatively smaller operations. The most commonly used fryers of this category are batch fryers (Figure 3.3). Continuous vacuum fryers are extremely costly and are rarely used. These fryers have several distinctive features including the following: (1) The fryer is operated under vacuum. Typical operating pressure is <100 mmHg. (2) The frying is conducted at about 120°C instead of > 165°C. The food can be dehydrated at this temperature under vacuum. (3) The food is placed in a basket. (4) The basket is inserted and placed in the vacuum chamber above the surface of the oil. (5) The vacuum is applied. (6) The basket is lowered into the hot oil at approximately 120°C. (7) Frying begins and the oil temperature drops (as described in a batch fryer). (8) The oil is circulated continuously through an external heater. (9) The oil regains the temperature when the moisture content of the product reaches the predetermined value. (10) The vacuum in the fryer is broken slowly. (11) The fryer is opened and the product basket is taken out. The excess oil is allowed to drain and the product is cooled before packaging (Bouchon and Dueik 2018).

Today there are commercial batch and continuous state-of-the-art vacuum fryers. The comparatively less expensive batch vacuum fryers (Figure 3.3) are designed to hold the incoming raw material inside a basket whereas the continuous systems are normally equipped with a frying pan in a stainless steel vacuum tube, where the raw product is fed through a rotary airlock feeder and a conveyor belt takes the product along the vacuum tube. These continuous units are equipped with centrifuges located in the vacuum chamber attached to the vacuum tube in order to remove surface oil (Bouchon and Dueik 2018).

3.5 SEASONING OPERATIONS

3.5.1 Seasonings

Most savory snacks are supplemented after baking or frying with seasonings with the aim to enhance their sensory properties and overall acceptance. The seasoning is designed to affect color, texture and mainly flavor. Traditionally, the most popular flavorings have been cheese, barbecue, sour cream-onion and ranch (Chapter 8). In Mexico and Latin America, chili pepper/citric acid–based flavorings are extremely accepted. The seasoning mixes are complex systems and produced from a wide array of ingredients. The seasoning mix generally consists of salt, spices, carriers or fillers, flavor enhancers, coloring agents, sweeteners and other minor additives such as antioxidants and anticaking agents (Chapter 8). Most snack food processors rely on external suppliers of seasoning mixes (Seighman, 2001, Serna-Saldivar 2008).

Flavor is defined as the property of a substance that causes a sensation of taste. Four basic flavors are perceived by taste buds on the tongue: sweet, salty, sour and bitter. Salt is one of the main ingredients in seasonings because it strengths or enhances the overall sensation of flavor. Several snacks are only flavored with salt. Powdered and oil-based seasonings are usually added from 5 to 10% of the snack base weight and are normally formulated with 15 to 25% of fine salt. Coarse and flaked salts are normally used for the production of pretzels and crackers discussed in Chapter 10. Most formulations contain flavor enhancers such as monosodium glutamate, hydrolyzed yeast, hydrolyzed vegetable protein or other less common such as guanylate and inosinate. Some producers avoid the use of monosodium glutamate because it can cause allergic reactions to a small segment of the population and in most instances, it should be declared on the label. A new trend of the industry is the formulation of reduced sodium snacks. The flavorings with reduced sodium chloride content are usually supplemented with potassium chloride and hydrolyzed yeast or vegetable proteins. Most powdered seasonings contain high quantities of carriers or fillers being the most common maltodextrin, whey solids and various kinds of cereal flours such as refined wheat flour (Chapter 6). These ingredients have a bland taste and small particle size and play an important role in the even distribution of the flavorings. The storage of seasoning mixes should always be in a cool place at a temperature of around 15°C and preferably avoid direct contact with light. These conditions are ideal to retard rancidity and undesirable flavor changes (Seighman, 2001, Serna-Saldivar 2008).

3.5.2 Industrial Application of Seasonings

There are several different systems utilized to apply diverse kinds of seasonings to savory snacks. According to the type of seasoning, there are basically two ways of application: liquid or slurry and dry or powdered. The industrial delivery of powdered seasonings known as "dusting" is normally performed inside a revolving tumbler or coating drum. On the other hand, the slurry application uses oil as the main carrier of the flavorings whereas powder application uses fillers or carriers. Most snack seasonings are applied in powdered forms although several types of snacks such as directly extruded snacks are alternatively seasoned with oil-based slurries. From the technological viewpoint, the seasoning systems are divided into conveyor-based for dry seasoning application, drum-based single-stage, drum-based slurry and drum-based two stage. One additional variant of the one-phase technology is the use of electrostatic seasoning. Independent of the type of system, the main goal of these technologies is to apply the seasoning in a uniform and consistent matter in such a way that flavored snacks have the same amount of coverage and appearance.

3.5.2.1 Conveyor Coating

The application of seasonings on moving conveyors is normally employed for the manufacturing of fabricated potato chips (Chapter 14), crackers or saltines (Chapter 10) and cookies. The major advantage of the conveyor belt system is that it requires less floor space in processing plants. The conveyor can be a vibratory, open-wire belt, closed fabric belt, or other selected types. Several principles should be considered when designing a conveyor-based coating system. The first is that, without some device for turning the product over, only the top surface or face of the snack will be coated like pretzels and crackers. For this reason, processors should avoid the formation of several layers of the unseasoned snack units on the seasoning belt in order to avoid the over and under flavoring of the units positioned on top and bottom, respectively. However, this may be desirable in some instances, for example, the top coating of crackers. Also, because the base product seldom covers 100% of the conveyor, a means for collecting and (if possible) recycling excess seasoning is required. Separate conveyor sections should be used for each material if more than one is applied, especially for liquids. This will reduce the amount of excess carried forward to the succeeding application and onto downstream equipment (Hanify 2001). The percent covering is carefully adjusted according to the flow of the dry ingredient dispenser, the speed of the conveyor belt and the oncoming product feed rate. Dispensers are normally built from stainless steel and contain a hopper with the dispenser unit designed to even seasoning distribution with special systems to avoid bridging.

3.5.2.2 Drum Coating

These devices are the most commonly used for flavoring snacks. They are also known as coating reels, tumblers or enrobers and are normally manufactured of stainless steel. The purpose of a coating drum is to expose all surfaces of the base product to the various seasonings applied. In their simplest form, coating drums are inclined cylinders that rotate to lift the product. The drum inner wall includes product-lifting flights to create the desired exposure (Figure 3.4). The product is picked up by the flights near the bottom of the drum and lifted as the drum rotates. Once the product reaches a critical height, usually 90–120° from the bottom, it begins to turn or roll down to the bottom of the drum where the lifting process begins again. The product is transported forward because the drum is inclined downward from entrance to exit. Each time the product completes a cycle of lifting and falling, it moves closer to the discharge end of the drum. The extent of product exposure is determined by drum design, rotation speed and angle of drum inclination. Some seasoning lines are furnished with tumblers consisting of a perforated drum designed to remove chaff and fines of the seasoned snack before packaging. Drum coaters are divided into single-stage slurry, single-stage dry seasoning and two phase.

3.5.2.2.1 Single Stage-Slurry

These tumblers apply the seasoning blend through an oil spray system. The single stage slurry application normally consists of a tumbler furnished with a spray system, pump and a holding tank with temperature controls where the oil-based flavoring slurry is contained and agitated. The flow rate of the spraying system plus the feed rate of the incoming snack units and inclination of the revolving tumbler dictates residence time and the amount of covering. The seasoning slurry normally contains about 60% oil and is usually delivered at temperatures above 100°C in order to enhance

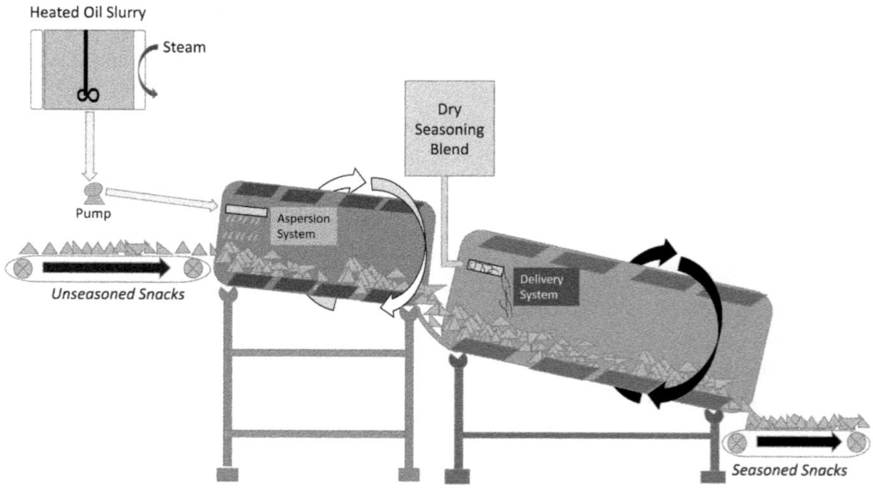

FIGURE 3.4 Illustration of the two-stage drum-based seasoning system commonly employed to flavor snack foods.

the adherence of the flavorings. Slurries are commonly added at target covering levels of 10 to 25%. The liquid slurry technology is commonly used for the flavoring of air-popped popcorn (Chapter 9) and second-generation puffs and other extruded items (Chapter 12).

3.5.2.2.2 Single-Stage Dry Seasoning

Single stage tumblers are the most popular in the industry because they are very effective for the application of seasonings onto fried snacks. The technology is more effective when seasonings are applied immediately after deep-fat frying and the snack base still contains hot oil for the more effective adherence of the flavoring mix. The single stage tumblers are equipped with an auger feeder that delivers a given or preset amount of seasoning per unit time. The use of vibrating trays helps to better dispense the mix onto the snacks. Another alternative is the use of a long distribution tube that dispenses the free-flowing mix as a curtain. These devices are normally used on products that have natural tack so the seasoning easily adheres to the product such as oil or wet popped popcorn (Chapter 9), corn chips (Chapter 11), third-generation extruded snacks (Chapter 12) and potato chips (Chapter 14).

3.5.2.2.3 Electrostatic Seasoning

Electrostatic application of dry flavorings or liquid slurries is a subset of the single-stage seasoning technologies described above. The electrostatic technology reduces the amount of overspray that might otherwise escape the tumbler, reduces waste and improves product consistency. The liquid electrostatic slurry technology is based on the delivery of the slurry with a peristaltic pump into an electrostatic spray nozzle assembly. The liquid seasoning is sprayed into a negatively charged field toward the drum wall, which is grounded causing the attraction of the fine spray onto the surface of the snack units. Likewise, the electrostatic coating of dry seasonings follows the same principles because both the tumbler is grounded and the base product is grounded by contact with the drum.

These devices normally consist of a flex wall screw feeder, vibratory pan feeder, fluidizing air system and the electrostatic probe. The seasoning is delivered into a mixing area, which has an energized electrode. The seasoning receives the electrostatic charge within the mixing area and is carried onto the base product by air. The seasoning picks up a charge as it passes the charged area and then falls onto the snack units (Hanify 2001). The electrostatic seasoning technology has been successfully used for the flavoring of peanuts, cashews and other tree nuts (Chapters 13 and 15).

3.5.2.2.4 Two-Stage Seasoning

The dual-stage system is recommended when there is not enough oil on the surface of the unflavored snack. Therefore, the product is first covered with oil or other adhering liquid so the following delivery of the dry seasoning mix sticks to the product. For snacks, the first coating is generally oil although other liquid mixes can be alternatively used. One possible problem of these systems is the buildup of the seasoning in overlap areas of the two-stage coating drum (Hanify 2001). This technology is preferred when the unflavored snack units do not have natural or sufficient tack and therefore need an oil coating in order to enhance the application of the dry flavoring. The technology is used for air-popped popcorn (Chapter 9), baked tortilla chips (Chapter 11) and second-generation extruded snacks (Chapter 12).

REFERENCES

Bastida, S., and Sanchez-Muniz, F.J. 2001. Thermal oxidation of olive oil, sunflower oil and a mix of both oils during forty discontinuous domestic fryings of different foods. *Food Science and Technology* 7: 15–21.

Blumenthal, M.M. 1991. A new look at the chemistry and physics of deep fat frying. *Food Technology* 45(2): 68–71.

Bouchon, P. 2009. Understanding oil absorption during deep-fat frying. In: *Advances in Food and Nutrition Research*, vol. 57, ed. S.L.Taylor, 209–223. Amsterdam: Academic Press.

Bouchon, P., and Dueik, V. 2018. Frying of foods. In: *Fruit Preservation. Novel and Conventional Technologies*, eds. A. Rosenthal, R. Deliza, J. Welti-Chanes, and G.V. Barbosa-Cánovas, 275–310. New York: Springer.

Che Man, Y.B., and Jaswir, I. 2000. Effect of rosemary and sage extracts on frying performance of refined, bleached and deodorized (RBD) palm olein during deep-fat frying. *Food Chemistry* 69(3): 301–307.

Choe, E., and Lee, J. 1998. Thermo oxidative stability of soybean oil, beef tallow, and palm oil during frying of steamed noodles. *Korean Journal of Food Science and Technology* 30: 288–292.

Choe, E., and Min, D.B. 2005. Chemistry and reactions of reactive oxygen species in foods. *Journal of Food Science* 70(9): R142–R159.

Choe, E., and Min, D.B. 2007. Chemistry of deep-fat frying oils. *Journal of Food Science* 72(5): 77–86.

Christopoulou, C.N., and Perkins, E.G. 1989. Isolation and characterization of dimmers formed in used soybean oil. *Journal of the American Oil Chemists' Society* 66(9): 1360–1370.

Crosa, M.J., Skerl, V., Cadenazzi, M., Olazabal, L., Silva, R., Suburu, G., and Torres, M. 2014. Changes produced in oils during vacuum and traditional frying of potato chips. *Food Chemistry* 146: 603–607.

Da Silva, P.F., and Moreira, R.G. 2008. Vacuum frying of high-quality fruit and vegetable-based snacks. *LWT - Food Science and Technology* 41(10): 1758–1767.

Dueik, V., and Bouchon, P. 2011a. Development of healthy low-fat snacks: Understanding the mechanisms of quality changes during atmospheric and vacuum frying. *Food Reviews International* 27(4): 408–432.

Dueik, V., and Bouchon, P. 2011b. Vacuum frying as a route to produce novel snacks with desired quality attributes according to new health trends. *Journal of Food Science* 76(2): E188–E195.

Dueik, V., Robert, P., and Bouchon, P. 2010. Vacuum frying reduces oil uptake and improves the quality parameters of carrot chips. *Food Chemistry* 119(3): 1143–1149.

Dunford, N. 2012. Industrial deep fat frying. FAPC-176. Robert M. Kerr Food & Agricultural Products Center, Oklahoma Cooperative Extension Service, Division of Agricultural Sciences and Natural Resources. Stillwater: Oklahoma State University.

Garayo, J., and Moreira, R. 2002. Vacuum frying of potato chips. *Journal of Food Engineering* 55(2): 181–191.

Granda, C., Moreira, R.G., and Tichy, S.E. 2004. Reduction of acrylamide formation in potato chips by low temperature vacuum frying. *Journal of Food Science* 69(8): E405–E411.

Gupta, M.K. 2004. Procedures of oil handling in a frying operation. In: *Frying Technology and Practices*, eds. M.K. Gupta, K. Warner, and P.J. White. Champaign: AOCS Press.

Gupta, M.K., Grant, R., and Stier, R.F. 2004. Critical factors in the selection of an industrial fryer. In: *Frying Technology and Practices*, eds. M.K. Gupta, K. Warner, and P.J. White. Champaign: AOCS Press.

Hanify, D.E. 2001. Snack seasonings application. In: *Snack Foods Processing*, eds. E.W. Lusas and L.W. Rooney, 517–527. Lancaster: Technomic Publishing Co.

Herbach, K., Stintzing, F., and Carle, R. 2008. Betalain stability and degradation-structural and chromatic aspects. *Journal of Food Science* 71(4): R41–R50.

Hosseini, H., Ghorbani, M., Meshginfar, N., and Mahoonak, A.S. 2016. A review on frying: Procedure, fat, deterioration progress and health hazards. *Journal of the American Oil Chemists' Society* 93(4): 445–466.

Houhoula, D.P., Oreopoulou, V., and Tzia, C. 2003. The effect of process time and temperature on the accumulation of polar compounds in cottonseed oil during deep-fat frying. *Journal of the Science of Food and Agriculture* 83(4): 314–319.

Jaswir, I., Che Man, Y.B., and Kitts, D.D. 2000. Synergistic effects of rosemary, sage, and citric acid on fatty acid retention of palm olein during deep-fat frying. *Journal of the American Oil Chemists' Society* 77(5): 527–533.

Kawas, M.L., and Moreira, R.G. 2000. Characterization of product quality attributes of tortilla chips during the frying process. *Journal of Food Engineering* 47(2): 97–107.

Kim, I., and Choe, E. 2004. Oxidative stability and antioxidant content changes in roasted and bleached sesame oil during heating. *Food Science and Biotechnology* 113: 762–767.

Moreira, R.G. 2007. Deep fat frying. In: *Heat Transfer in Food Processing: Recent Developments and Applications*, eds. S.Yanniotis and B. Sunden, 209–237. Ashurst: WIT Press.

Moreira, R.G. 2014. Vacuum frying versus conventional frying—An overview. *European Journal of Lipid Science and Technology* 116(6): 723–734.

Moreira, R.G., and Barrufet, M.A. 1998. A new approach to describe oil absorption in fried foods: A simulation study. *Journal of Food Engineering* 35(1): 1–22.

Moreira, R.G., Castell Perez, M.E., and Barrufet, M.A. 1999. *Deep Fat Frying. Fundamentals and Applications*. Gaithersburg: Aspen Publishers, Inc.

Moreira, R.G., Da Silva, P.F., and Gomes, C. 2009. The effect of a de-oiling mechanism on the production of high quality vacuum fried potato chips. *Journal of Food Engineering* 92(3): 297–304.

Naz, S., Siddiqi, R., Sheikh, H., and Sayeed, S.A. 2005. Deterioration of olive, corn, and soybean oils due to air, light, heat and deep-frying. *Food Research International* 38(2): 127–134.

Rojo, J.A., and Perkins, E.G. 1987. Cyclic fatty acid monomer formation in frying fats. I. Determination and structural study. *Journal of the American Oil Chemists' Society* 64(3): 414–421.

Saguy, S.I., and Dana, D. 2003. Integrated approach to deep-fat frying: Engineering, nutrition, health and consumer aspects. *Journal of Food Engineering* 56(2–3): 143–152.

Seighman, J. 2001. Snack food seasoning. In: *Snack Foods Processing*, eds. E.W. Lusas and L.W. Rooney, 495–516. Lancaster: Technomic Publishing Co.

Serna-Saldivar, S.O. 2008. The process of frying and seasoning. In: *Industrial Manufacture of Snack Foods*, ed. S.O. Serna-Saldivar, 329–352. London: Kennedys Publications Ltd.

Serna-Saldivar, S.O., Gomez, M.H., and Rooney, L.W. 1990. The technology, chemistry and nutritional value of alkaline cooked corn products. In: *Advances of Cereal Science and Technology*, vol. X, ed. Y. Pomeranz, 243–307. Saint Paul: American Association of Cereal Chemists.

Shyu, S., Hau, L., and Hwang, L. 1998. Effect of vacuum frying on the oxidative stability of oils. *Journal of the American Oils Chemical Society* 75(10): 1393–1398.

Shyu, S.L., and Hwang, L.S. 2001. Effects of processing conditions on the quality of vacuum fried apple chips. *Food Research International* 34(2–3): 133–142.

Su, Y., Zhang, M., Adhikari, B., Mujumdar, A.S., and Zhang, W. 2018. Improving the energy efficiency and the quality of fried products using a novel vacuum frying assisted by combined ultrasound and microwave technology. *Innovative Food Science and Emerging Technologies* 50: 148–159.

Tizhe-Liberty, J., Dehghannya, J., and Ngadi, M.O. 2019. Effective strategies for reduction of oil content in deep-fat fried foods: A review. *Trends in Food Science and Technology* 92: 172–183.

Tompkins, C., and Perkins, E.G. 2000. Frying performance of low-linolenic acid soybean oil. *Journal of the American Oil Chemists' Society* 77(3): 223–229.

Tsuzuki, W., Matsuoka, A., and Ushida, K. 2010. Formation of trans fatty acids in edible oils during the frying and heating process. *Food Chemistry* 123(4): 976–982.

Tyagi, V.K., and Vasishtha, A.K. 1996. Changes in the characteristics and composition of oils during deep-fat frying. *Journal of the American Oil Chemists' Society* 73(4): 499–506.

Warner, K. 2004. Chemical and physical reactions in oil during frying. In: *Frying Technology and Practices*, eds. M.K. Gupta, K. Warner, and P.J. White, 16–28. Champaign: AOCS Press.

Yen, G.C., Duh, P.D., and Tsai, H.L. 2002. Antioxidant and pro-oxidant properties of ascorbic acid and garlic acid. *Food Chemistry* 79(3): 307–313.

Yoon, S.H., Jung, M.Y., and Min, D.B. 1988. Effects of thermally oxidized triglycerides on the oxidative stability of soybean oil. *Journal of the American Oil Chemists' Society* 65(10): 1652–1656.

4 Packaging of Snack Foods

Sergio O. Serna-Saldivar

CONTENTS

DOI: 10.1201/9781003129066-4

4.1 INTRODUCTION

Packaging plays a key role in the appearance and quality attributes of snacks and significantly affects production cost. Conventionally, a package is designed to protect foods from environmental conditions, such as light, oxygen, moisture, microbes, and mechanical stresses allowing producers to extend shelf life. The chief functions of packaging are containment, protection, convenience, and communication through the food label. Historically, the packaging of potato chips with sheets of waxed paper is attributed to a Californian woman about 100 years ago. The wax paper was ironed into bags that were filled with fried products. Polyethylene, polystyrene, and polyvinylchloride rapidly replaced waxed paper because of their higher resistance to moisture transfer and processing into preprinted formed bags suitable for highly productive packaging lines (Lusas 2001).

Nowadays, the competitive snack market is evolving into intensive packaging innovation. The overwhelming majority of snacks are packaged in multilaminated flexible bags commonly made of petroleum-based polymers. The future of packaging might be in the development of biodegradable multilaminated functional films that will be friendlier with the environment and the adoption of intelligent packages that will show whether the product is still in good condition. Starch is the preferred material for the production of biodegradable plastics whereas thermoplastic extrusion is the most used manufacturing technology.

The key to packaging requirements for snacks and particularly among the wide array of deep-fat fried items is the prevention of oil oxidation and loss of crispiness. The latter resulting from moisture gain from the environment because most snacks are very hygroscopic and consequently prone to texture changes. In addition, most packaging systems should be designed to avoid or minimize product breakage during handling and distribution and attract consumers through the mandatory labeling information that includes in most instances the nutrition information and list of ingredients (Chapter 20).

Oils associated with snack foods change rapidly due to oxidation that leads to undesirable flavor changes. Many factors related to the intrinsic characteristics of the frying oil (Chapter 7) and deep-fat frying conditions (Chapter 3) affect the rate of lipid oxidation and flavor deterioration. The main factors are temperature, ultraviolet light, presence of oxygen and salt, and nature of the fried snacks. In general, the higher the degree of oil instauration the more possibilities of oxidation. Some raw materials contain natural antioxidants such as tocopherols, flavonoids, and phenolics that protect or retard oxidation. Natural or artificial antioxidants intentionally added to the products or packaging materials help although none of these compounds can totally impede all oxidative rancidity (Serna-Saldivar 2008).

Millennials are the main responsible for the downsizing trend and on-the-go, convenience-style packaging. Moreover, the industry is evolving into the trend of resealable packaging and easy-to-open/close features. Another trend is the increased desire to diminish the environmental impacts of packaging materials by focusing on issues such as lightweight containers, reusability, and material selection. This is highly relevant because packaging materials constitute as much as 65% of the global solid waste (Azzi et al. 2012, Grace 2016).

The presence of large amounts of fats/oils is relevant to prevent unsightly staining of the package, seepage, and grease package feel (Serna-Saldivar 2008). The oil can also in some instances lead to splitting of lamination by dissolving the laminate Most snacks are dried-crisp foods and easily gain moisture when exposed to the air or packed with materials that are a low barrier against water vapor. As they absorb moisture they gradually increase their water activity (*Aw*) until a level where the snack losses crispness and overall acceptability. Normally this occurs usually in an *Aw* range of 0.35 to 0.5 under regular storage conditions. There is an optimum *Aw* in which dehydrated snacks exhibit their longest shelf life. As moisture is gained or lost from this critical *Aw* range, the rate of undesirable reactions like lipid oxidation, browning, bleaching, and nutrient degradation increases. For some critical reactions such as lipid oxidation and nonenzymatic browning the rate doubles or triples for every 0.1 increase above the *Aw* of 0.4. Moreover, the temperature also plays a critical role because for dry foods for an increase of 10°C the rancidity rates augment from 1.5 to 2 whereas rates of nonenzymatic browning from 5 to 8 (Labuza and Contreras-Medellin 1981).

The synthetic petroleum-based polymers have been extensively used for multipurpose food applications due to their satisfactory mechanical and thermal properties although they are known to harm both land and aquatic ecosystems due to their low biodegradability. The extra durability of synthetic polymeric films beyond the expected product shelf life is not essential and these materials normally persist in the environment for several decades. To bring this waste disposal under control, integrated waste management practices including the use of biodegradable green plastic materials and recycling of petroleum-based films are increasingly being implemented. Biopolymers possess inferior mechanical properties and processability but they are progressively being used due to their biodegradability features at the end of their life cycle (de Vileger 2003).

In most developed countries, intensive research and development work on active and intelligent packaging are being performed although there are only a few commercially significant systems on the market. The idea of intelligent packaging is to monitor food quality and safety of food products and inform customers of the condition of the packaged food. Active, intelligent, and green packaging technologies can work synergistically to yield novel multipurpose packaging systems with no negative interactions between components (Han et al. 2018).

This chapter covers characteristics and properties of petroleum-based and green packaging materials, active and intelligent packaging technologies, and their use in different types of snacks. In addition, commercial packaging equipment commonly employed in snack food settings is described.

4.2 SYNTHETIC PLASTIC POLYMERS

Polymers and plastics are typical materials of the past century and have made a tremendous growth of some hundreds of tons/year at the beginning of the 1930s to nearly 300 million tons today. Nowadays, the per capita yearly use of plastic in Europe and the United States is 60 and 80 kg, respectively (de Vlieger 2003). Most snacks are still packaged with synthetic or petroleum-based single or multilaminated films or plastics. Homo or heteropolymers are by far the most common materials. Homopolymers have the same repeating building block unit throughout the molecule whereas heteropolymers are constituted with two or more different units. Table 4.1 summarizes the types and features of plastic films approved and used by the different segments of the food industry. Needless to say, these polymers possess different physical properties such as light transmission rate, resistance to tearing and puncture, haze, vapor water transmission rate, oxygen permeability, fat permeation, and thermo-sealability (Moreira et al. 1999, Serna-Saldivar 2008, Robertson 2012, Ebnesajjad 2013).

From the common modes of deterioration of snacks, the package should provide good barriers against oxygen, light, and moisture as they are typically placed inside multilaminated films with modified atmospheres although spiral wound, paperboard cans lined with aluminum foil and sealed under vacuum technologies (i.e. potato chips or sticks, see Chapter 14) are also used. Many sorts of flexible films are employed to package deep-fat fried items, such as potato chips (Chapter 14), corn and tortilla chips (Chapter 11), and third-generation snacks (Chapter 13). The popular metallized flexible films are extensively used by the snack industry because they are normally less expensive and are considered the best barriers against ultraviolet light, moisture, and oxygen. Oriented polypropylene (OPP), oriented polyethylene (OPE), and biaxial oriented nylon (BON), described in Table 4.1, are used as base films to apply a thin layer of aluminum on the outside and produce the aluminized bags with different barrier properties. The popularity of snacks packaged in recloseable bags is growing because the opened product could be protected from humidity before consumption resumes (Moreira et al. 1999, Serna-Saldivar 2008, Robertson 2012).

TABLE 4.1
Types and Characteristics of Plastic Films Widely Used for Packaging

Type of Polymer	Abbreviation	Properties and Features
Low-density polyethylene	LDPE	Most widely used food-packaging material. It is tough, slightly translucent, an excellent barrier against water and water vapor, and poor barrier against gases. It has excellent chemical resistance to chemicals such as acids and alkalis. It is sensitive to hydrocarbons and lipids. It has excellent sealability.
Linear low-density polyethylene	LLDPE	Improved chemical resistance, higher surface gloss, higher strength, and puncture resistance than LDPE.
High-density polyethylene	HDPE	HDPE is stiffer, harder, with more tensile and bursting strengths than LDPE but has lower impact and tear strengths than LDPE. It has better resistance to chemicals, oils, and greases and offers excellent moisture protection and a much lower gas permeability compared to LDPE. Heat sealing is more difficult than LDPE.
Polypropylene	PP	Has lower density than polyethylene, low water vapor transmission, medium gas permeability, good resistance to lipids and chemicals, good gloss, and high clarity.
Oriented and biaxially oriented PP	OPP, BOPP	Has higher clarity, tensile strength, and better moisture barrier than PP. OPP and BOPP are usually coated with PVdC and/or acryl polymers to improve water vapor and gas permeabilities.
Polystyrene, oriented and biaxially oriented polystyrene	PS, OPS, BOPS,	Polystyrene is a synthetic aromatic hydrocarbon polymer made from the monomer styrene (C8H8). Polystyrene can be solid or foamed. General-purpose polystyrene is clear, hard, and rather brittle whereas the high-impact polystyrene (HIPS) that comes in foam or solid forms is considered as one of the most common plastics on the planet. PS possesses a poor barrier to both oxygen and water vapor and has a relatively low melting point. PS can be naturally transparent, but can be colored and are used for a variety of packaging, laminating, and printing applications. Some grades of these films are also approved for direct food contact. BOPS films and sheets are mainly used for thermoforming and shrink film products. They have excellent transparency, high thermoformability, and good printability.
Polyvinyl chloride	PVC	PVC can be compounded to produce a wide array of physical properties in plastic films. PVC can yield materials ranging from rigid, crisp to limp, tacky, and stretchable films. High glossy and clear films can be produced. PVC has high water permeability but low gas permeability and excellent resistance to oils, fats, acids, and alkalis.
Poly vinylidene chloride	PVdC	Pure PVdC is rarely used for packaging; however, is usually copolymerized with vinyl chloride to yield a soft, tough, and relatively impermeable film known under the trade name of Saran. PvdC copolymer films have low permeability to water vapor, gases, and odors. PVdC copolymers can be sealed to themselves or other materials.
Ethylene vinyl acetate	EVA	EVA copolymers with vinyl acetate to form films similar in flexibility to plasticized PVC. EVA copolymers have better heat sealing and stretch properties than LDPE. Some EVA is used in coextrusion for manufacturing multilaminated materials.
Ethylene vinyl alcohol	EVOH	EVOH copolymers have better barriers to gases, odors, and solvents. EVOH resins have high mechanical strength, elasticity, surface hardness, high gloss, low haze, and excellent resistance to oils. It is one of the best materials against the penetration of oxygen to the package enhancing flavor and quality retention. EVOH films are ideally suited for gas fill packaging.

(*Continued*)

TABLE 4.1 (CONTINUED)
Types and Characteristics of Plastic Films Widely Used for Packaging

Type of Polymer	Abbreviation	Properties and Features
Polyethylene terephthalate	PET	Biaxially oriented PET is widely used because of its great tensile strength, excellent chemical resistance, lightweight, and elasticity. To improve barrier properties, PET is coated with LDPE or PVdC copolymer. Coextruded heat-sealable films are frequently metalized and used as an inside layer of snacks food packages. Rigid grades of metalized PE can be used in thermoformable applications.
Nylon or synthetic polyamides	PA	Specialty films that have a carbon-carbonyl-amido-carbon chemical configuration are used especially in the textile industry. Several types of nylons exist with use in the food-packaging industry. The most popular are the nylon 6 and nylon 11 films. Nylon 6 has higher temperature, grease, and oil resistance than nylon 11. In general, nylons are highly permeable to water vapor. PVdC copolymer-coated nylons offer improved oxygen, moisture vapor, and UV light barrier properties.
Biaxial oriented nylon	BON	Two direction-oriented nylon films with improved strength and durability. BON films provide improved flex-crack resistance, mechanical properties, and barrier characteristics. These films have applications in fresh and processed meats.
Aluminum foil		Aluminum foil is flexible and has one of the best barriers against moisture, gases, and light. However, foil packages are comparatively expensive in comparison to plastic films and preclude the visibility of the packed product.
Metalized films		OPP, OPET, and BON are used as base films to deposit a thin layer of aluminum on the outside. The composite films are one of the best barriers against light, moisture, and oxygen and are generally less expensive than aluminum foil.

Source: Moreira et al. (1999), Serna-Saldivar (2008), Robertson (2012), and Ebnesajjad (2013).

As mentioned before, the most relevant features for polymeric materials used to package snacks are water vapor transmission, light transmission, and oxygen permeability rates. The package should avoid or minimize the entrance of water vapor because the products are typically highly hygroscopic because of their low moisture and *Aw*. If products adsorb environmental moisture, they will progressively lose texture and sensory attributes. Table 4.2 summarizes and compares vapor water transmission rates of different packaging films. The polyethylene terephalate or PET and polyesters are the films that possess the best barriers against water vapor. In addition, a low gas transmission rate is required in order to reduce the entrance of oxygen and rancidity and maintain inert gases flushed and contained into the package headspace. Nitrogen is the inert gas most commonly used to prevent rancidity in sealed containers. Additionally, the flushed gas produces a pillow that protects the product from breakage during the critical stages of distribution and handling (Serna-Saldivar 2008, Robertson 2012).

Plasticizers play a critical role in the design of synthetic polymeric films because they greatly affect functional and mechanical properties. In terms of plasticizers, Vieira et al. (2011) indicate that these low-molecular-weight and nonvolatile additives are widely used in the polymer industries to improve the flexibility and processability of films by lowering the second-order transition temperature or the DSC glass transition temperature (T_g). Plasticizers fill intermolecular spaces between polymer chains changing their three-dimensional structure and therefore diminishing the energy required for molecular motion and the formation of hydrogen bonding between chains.

TABLE 4.2
Comparison of Water Vapor Transmission and Water Vapor Permeability Rates of Different Packaging Films

Sort of Film	Film Thickness (mm)	Water Vapor Transmission Rate (g water/day/100 sq. inches) (645.2 cm²)	Water Vapor Permeability (g water/m² * day * mm Hg)
Polyethylene (PE)	0.031	1.10	0.178
PE-nitrocellulose-coated cell	0.076	1.40	0.279
Polyethylene terephalate (PET)	0.021	0.40	0.133
PET/adh/PE	0.064	0.50	0.058
PET Saran coated/adh/PE	0.089	0.40	0.027
Polypropylene (PPE)	0.032	0.70	0.148
Polyester	0.051	0.40	0.079
Polyester	0.061	0.35	0.043
Teflon (Polytetrafluoroethylene)	0.025	0.90	0.055

Source: Data obtained from Labuza and Contreras-Medellin (1981).
Note: Parameters determined at 21°C and 65% relative humidity.

There are several plasticizers that are normally used for the manufacturing of different plastics. Most of them are based on carboxylic acid esters with linear or branched aliphatic alcohols of moderate chain lengths (predominantly C6–C11). The most commonly used are phthalate esters, adipates, citrates, acids esters, alkane-dicarboxylic, glycols, and phosphates. The esters of phthalic acid constitute more than 85% of the total plasticizer consumption (Vieira et al. 2011). The degree of plasticity of polymers is largely dependent on the chemical structure of the plasticizer, molecular weight, and functional groups affecting chiefly the flexibility of the material. The selection and concentration of the plasticizer are commonly based on the compatibility, processing characteristics, desired thermal and mechanical properties of the packaging material. In addition, plasticizers are selected based on the material resistance to water, chemicals, and solar radiation.

Plasticizers are broadly divided into two categories: internal or external. External plasticizers are low-volatile compounds, which interact with polymer chains but do not react chemically to primary bonds and therefore can be lost by evaporation, migration, or extraction. On the other hand, internal plasticizers co-polymerize or react chemically with the polymers and become part of the film. Plasticizers diminish both the T_g and elastic modulus of plastic films. Plasticizers are also classified as primary and secondary. Primary plasticizers are soluble in the polymer at relatively high concentrations and are utilized as sole sources or as the main element of the plasticizer. They should gel the polymer rapidly under regular processing temperatures and should not exude from the film. Secondary plasticizers possess lower gelation capacity and limited compatibility with the polymer and therefore are regularly mixed with primary counterparts in order to enhance product properties. Industrially, the most common plasticized polymers are PVC, poly(vinyl butyral) (PVB), poly(vinyl acetate) (PVAc), acrylics, cellulose molding compounds, nylon, polyamides, and certain copolyamides. PVC processing is by far the most relevant use of plasticizers, consuming over 80% of production (Vieira et al. 2011).

4.3 BIOPOLYMERS (GREEN PLASTICS) AND REGENERATED CELLULOSE

The development of biodegradable packaging materials has been the subject of extensive research particularly with regard to renewable alternatives to traditional petrochemical-derived polymers discussed earlier. Biopolymers are defined as those synthesized by nature such as starch and other polysaccharides whereas regenerated fiber cellulose (RFC) also known as cellophane is derived from renewable fiber sources like cotton and wood pulp.

According to Gadhave et al. (2018), the interest in the development and usage of green or bioplastics has grown due to the growing concern over environmental problems. Green plastics are normally derived from renewable agro/food resources, mainly starch. However, these natural polymers on their own do not demonstrate the same material properties as traditional plastic polymers thereby limiting their potential for food applications. According to de Vlieger (2003) there are two major categories of biodegradable commercial pseudoplastics: starch and polylactic acid (PLA)-based polymers. The main advantages of these new polymers are biodegradability and renewability. On the other hand, Vieira et al. (2011) categorized biodegradable polymers into four groups: (i) Polymers from agro-resources biomass like polysaccharides (starches, lignocellulosic materials, pectins, gums, and chitosans); (ii) Polymers produced from proteins like zeins, soybean, gluten, casein, whey, and collagen or obtained by microbial production like polyhydroxyalkanoates; (iii) Polymers chemically synthesized using monomers like PLA; and (iv) polymers whose monomers and polymers are obtained by chemical synthesis from fossil resources like polycaprolactones, polyesteramides, and both aliphatic and aromatic copolyesters.

4.3.1 Starch-Based Biopolymers

Starch-based biofilms and related materials are extensively produced on an industrial scale and are the earliest commercialized source of bio-based plastics (Shen et al. 2010). In 2018, the total volume of starch-based biomaterials reached 7.5 million tons representing about 2% of the production volume of petrochemical polymers. The capacities and production of these polymers will continue to grow with an expected CAGR of about 4% until 2023 (Chinthapalli et al. 2019). Flexible packaging accounts for about half of the starch plastic market, the remainder is used in agriculture, rigid packaging, and consumer goods.

Starch is the most consumed single product of the average human diet and the main way of storing energy of cereals, pulses, roots, and tubers. The starch is stored in granules in two major polymeric forms: linear amylose (AMY) and branched amylopectin (AMP). AMY is constituted by glucose units joint by α-1,4 glycosidic bonds whereas AMP contains branches that originate from α-1,6 bonds. Regular starch granules normally contain 25% AMY and 75% AMP whereas waxy starches contain more than 95% AMP and less than 5% AMY. The main sources of commercial starches are maize or corn followed by wheat (Chapter 6), potato (Chapter 14), and tapioca. By far, the major source of starch is maize that is wet milled using sulfur dioxide as explained in Chapter 6. This source of starch is highly refined (< 0.3% protein), extensively available, and compared to other sources of starch with a relatively lower cost. In addition, the major source of waxy or high amylopectin starch, which possesses different viscoamylograph properties especially in terms of retrogradation, is from waxy maize (Chapter 6).

According to Broeren et al. (2017) native starch cannot be used directly in biodegradable plastics due to its granular or crystalline structure that needs to be disrupted via gelatinization and also with the use of plasticizers such as glycerol. The most common way to achieve gelatinization is via thermoplastic extrusion similar to the typical processing of petroleum-based plastics. The sole use of extrusion yields materials with poor mechanical properties and a relatively high affinity for water, which limits potential applications. The main challenge for researchers and the packaging industry in terms of producing starch-based blends with commercial utility is overcoming miscibility problems at high starch contents (Gadhave et al. 2018).

Therefore, starches are commonly composited with other polymers, typically aliphatic and hydrophobic polyesters, to improve mechanical, functional, and barrier properties (Yu et al. 2006, Gadhave et al. 2018). The intelligent combination of starch, petrochemical-based polymers, and additives like plasticizers generate a wide range of innovative packaging materials with adequate technical properties.

The starch-based plastics are complex blends of starch with compostable plastics such as PLA, polybutylene succinate, polybutylene adipate terephthalate polycaprolactone, and phthalates. The last category of compounds tends to bioaccumulate and persist in organisms with potential toxicological effects (Gadhave et al. 2018).

For the production of green plastics with 100% natural ingredients, the starch is commonly mixed with water and glycerol, which act as plasticizers. In terms of processing, most films are obtained after thermoplastic extrusion that has advantages from the labor, cost, and energy usage viewpoints. The starch-based films can be tailor-made to have a very low oxygen permeability rate, making them suitable for the packaging of most snacks. Nevertheless, the water vapor permeability is comparatively higher compared to synthetic polymeric materials (Tables 4.1 and 4.2) so in order to maintain texture, it needs to be composited with other films. Due to this drawback, there are no products available at the moment made from pure thermoplastic treated starch and producers overcome this problem by blending starch with hydrophobic synthetic polymers discussed in Table 4.1. Unfortunately, all these production processes make the starch-based products rather expensive in comparison to common plastic alternatives (de Vlieger 2003).

Blends of biopolymers and synthetic materials are promising because they profit from the advantages of both types of materials especially in terms of water resistance, mechanical properties, and processing properties in highly automated packaging machines. For instance, the low-density polyethylene (LDPE)/starch blends have shown favorable functional physical properties and biodegradability and found commercial use. Apart from disturbing the continuity of the LDPE network and adding to the heterogeneity of the composite film, the presence of starch particles as fillers within the LDPE matrix enhances gas permeability (de Vlieger 2003, Gadhave et al. 2018).

Broeren et al. (2017) assessed the environmental impacts of six different starch plastic compositions and concluded that the use of starch plastics diminish greenhouse gas emissions and nonrenewable energy use but production of these films had higher eutrophication potential and required more agricultural land use compared to common petrochemical counterparts. The savings of greenhouse gas emissions were strongly associated with the biomaterial formulation with some grades offering an 85% reduction and others an 80% increase compared to the petrochemical counterpart. Interestingly, the additives can account for up to 40% of the gas emissions of starch plastics. The highest savings of gas emissions were obtained when components such as polybutylene adipate-co-terephthalate (PBAT) and polybutylene succinate (PBS) were minimized and the starch or natural fibers were maximized.

4.3.2 Polylactic Acid (PLA)-Based Materials

Synthetic polymeric films can be replaced with green plastics by converting plant sugars to PLA with specific microorganisms or by synthesis from hydrogen cyanide and acetaldehyde. Both high- and low-molecular-weight polymers of PLA are currently produced for different applications (de Vlieger 2003). PLA biopolymers have similar properties and features compared to polyolefines and have the major advantage that they can be processed with the same procedures and equipment like injection molding and thermal extrusion. The main disadvantage of the PLA-based plastics is their comparatively higher price compared to the petroleum-based counterparts. The polymer of choice for most applications is produced from 90% L-lactide and 10% racemic D-L-lactide. Both lactides readily polymerize and melt producing oriented films. Its DSC glass transition temperature or T_g is 60°C, and its melting temperature is around 155°C. These films are reported to possess similar properties and appearance compared to oriented polystyrene films. Residual lactide is not a concern since it hydrolyzes harmless lactic acid. Therefore, the GRAS PLA polymers can be applied in foods that have direct contact with the film (de Vlieger 2003).

4.3.3 Regenerated Fiber Cellulose (RCF)

RFC is manufactured from highly purified cellulose extracted from bleached or sulfite-treated pulp derived from wood or cotton linters. Cellulose is considered the most abundant natural polymer on earth, and, needless to say, is attractive due to its renewability, wide availability, low-cost, biocompatibility, and biodegradability. RCF is produced via physical dissolution and regeneration with an environmentally friendly process that uses other green reagents. The widely used cellulose dissolution viscose process, developed over 100 years ago, still dominates the RCF products market.

For the production of RFC, the sheet of pulp is steeped in sodium hydroxide to enhance the depolimerization of the long cellulose molecules after which carbon disulfide is added to yield sodium cellulose xanthane. The reaction takes place in the hydroxyl group of carbon 6. The resulting high viscous solution is continuously extruded through a slot die into a sulfuric acid-sodium sulfate coagulating vat at 40°C. The regenerated cellulose is washed and bleached to yield a transparent film that is further plasticized in a bath containing glycerol or ethylene glycol before drying. The plain RFC is permeable to steam and impervious to gases, fats, and organic solvents. However, RFC is rarely sued alone so in most food-packaging applications film is coated with nitrocellulose, PVC, PVC/PVdC (copolymer-coated cellophane), or PET (Robertson 2012, Wang et al. 2016). According to Kalka et al. (2014) all-cellulose composites possess the intriguing combination of high strength and exceptional biodegradability with mass losses of up to 73% following a soil burial of 70 days.

4.3.4 Protein-Based Materials

Several types of proteins have been researched for the production of films and coatings. Casein, sodium caseinate, and other derivatives have been extensively used for film formation. Likewise, collagen and hydrolyzed gelatin have been used for the manufacturing of edible sausage casings. Protein films are effective as gas barriers but their water vapor transmission rate is high (Arvanitoyannis and Gorris 1999).

4.4 ACTIVE PACKAGING MATERIALS

The development of active packaging materials is aimed to purposely incorporate additives or components that would release or absorb substances into or from the packaged food or the surrounding environment. Therefore, active packaging materials are deliberately added to extend the shelf life or to maintain or improve the condition of packaged food. Although extensive research on active packaging technologies is being conducted, many of these innovations have not yet been implemented in commercial systems (Yildirim et al. 2018). Nevertheless, the outlook of active packaging is promising because it effectively extends shelf life and keeps organoleptic properties of snacks especially in terms of preventing rancidity, texture loss, and microbial growth. The last issue is critically relevant for intermediate moisture snacks derived from dairy (Chapter 17) and meat/fish products (Chapter 18).

The more significant challenges in developing these new packaging materials are to find secure absorbers and/or releasers that can be safely incorporated into packaging materials. These active materials should be effective during the expected shelf life of the product, and at the same time, they should keep good appearance, permeability, and mechanical properties at reasonable prices. Moreover, the new film materials should be suitable for packaging machines already used in the food industry especially in terms of sealability.

Active packaging techniques for preservation and improving, the quality and safety of foods can be divided into three categories: absorbers or scavengers, releasing systems, and others (Ahvenainen 2003). Absorbing or scavenging systems remove compounds such as O_2, CO_2, excessive water, taints, and other undesired compounds. Oxygen absorbers have been successfully applied for cheeses (Chapter 17), ready-to-eat items, nuts, and bakery goods whereas CO_2 absorbers for beef jerky and dehydrated poultry products (Chapter 18). Other relevant compounds are absorbers of UV light, off-flavors, amines, and aldehydes (Gomez-Estaca et al. 2014). These are recommended for oil-containing cereal-based snacks (Chapters 10–12) and potato chips (Chapter 14).

Oxygen is one of the most important substrates for many mechanisms involved with food-product degradation. This essential gas is involved in anabolic reactions such as respiration and is responsible for many catabolic or deteriorative reactions. In terms of snacks, the oxidative rancidity of unsaturated and polyunsaturated fatty acids (Chapters 3 and 7) leads to off-flavors and even in extreme circumstances to the generation of toxic compounds (Chapter 20). In dehydrated fruit and vegetable snacks (Chapter 16), the scavenging of oxygen helps to prevent discoloration, darkening, and browning reactions (Brody et al. 2001).

4.4.1 Moisture Scavenger

The moisture content and *Aw* of snack foods are critical factors, which affect texture, quality, and safety. Most snacks contain moisture contents lower than 5% and therefore are very sensitive to environmental air moisture during storage and distribution. For other types of snacks such as intermediate moisture like meats, fruits, and vegetables, the control of the relative humidity level inside the package is desirable in order to prevent moisture loss or grain. There are several strategies to control moisture in packaged products that are thoroughly detailed by Yildirim et al. (2018). Two main active mechanisms are used as moisture scavengers: desiccants placed in the headspace or removers of liquids or air moisture contained in the package. Examples of desiccants are silica gel, clays, molecular sieves from zeolite, sodium, potassium, and calcium alumina silicates. Desiccants are normally placed inside sachets, microporous bags, or pads.

4.4.2 Oxygen Scavengers

The application of oxygen scavengers aims to remove residual O_2 present in the food package. Most snacks are sensitive to oxygen because they contain high amounts of lipids highly susceptible to oxidation whereas some intermediate moisture products are susceptible to the growth of aerobic putrefactive and pathogenic microorganisms. Independently of the type of snack food, the oxygen scavengers also protect products from color, nutritional and organoleptic changes. Most oxygen-scavenging systems are iron based. Other systems that scavenge oxygen are chemical or biocatalytic. The first includes the use of photosensitive dyes, ascorbic acid, gallic acid, or even unsaturated fatty acids whereas the second utilizes enzymes. Oxygen scavengers are used to protect and extend the shelf life of seeds (Chapter 13), nuts (Chapter 15), dehydrated and spread cheeses (Chapter 17), meats (Chapter 18), and fried snacks mainly through the prevention of rancidity, mold growth, and discoloration (Yildirim et al. 2018).

Cichello (2015) concluded that oxygen absorbers have a relevant role in the removal of dissolved oxygen, preserving the color, texture, and aroma of different products and that ferrous iron oxides are the most reliable sources. These oxygen absorbers can be applied as self-adhesive patches-sachets of dried iron powder and are known to effectively diminish oxygen to about 100 ppm. This type of active package has a wide range of food applications including for the prevention of lipid oxidation of roasted and ground soybeans (Takenaga et al. 1987).

Ahn et al. (2016) prepared and characterized LDPE oxygen scavenger films containing from 1 to 20% gallic acid and potassium chloride and compared them with pure LDPE. The surface morphology, gas barrier, and thermal properties indicate that the different concentrations of the oxygen scavenger were well incorporated into the LDPE film structure. The surface roughness of the film increased with the amount of oxygen-scavenging material. The oxygen and water vapor permeability of the developed film also increased with the amount of oxygen-scavenging material, though its elongation significantly decreased. The oxygen-scavenging capability of the prepared film was analyzed at different temperatures. The initial O_2 (20.90%) in the headspace of films containing 20% gallic acid decreased after seven days to 16.6% at 4.8°C, 14.6% at 23.8°C, and 12.7% at 50.8°C. The film impregnated with 20% gallic acid showed an effective oxygen-scavenging capacity of 0.709 mL/cm^2 at 23.8°C.

4.4.3 Carbon Dioxide (CO_2) Emitters

As same as oxygen scavengers, the CO_2 emitters are mainly used due to their antimicrobial and antioxidant effects. The antimicrobial outcome has been found to be proportional to the partial pressure of the gas inside the package. The emitters usually come in the form of a pad or sachet. There are commercial products that combine oxygen scavengers and CO_2 emitters based on either ferrous carbonate or a mixture of ascorbic or citric acid and sodium bicarbonate. Most of their applications are for the extension of the shelf life of fresh meats and fish fillets (Yildirim et al. 2018).

4.4.4 Antioxidant Releaser

The most promising active packaging materials for snacks are the ones that release or contain synthetic antioxidants like butylated hydroxytoluene (BHT) and butylated hydroxyanisole (BHA) or natural antioxidants like phenolics, resveratrol, tocopherols, plant extracts from rosemary, oregano, and essential oils (Yildirim et al. 2018). Products high in fats and oils are particularly prone to oxidation including nuts, crisps, and processed meats. Antioxidants can be incorporated into the packaging to reduce the degree of oxidation. BHA and BHT have been incorporated into the lining of the packaging material from which they are released during storage by diffusion. In this regard, several investigators have reported the release of synthetic and natural antioxidants added to packaging films (LDPE and HDPE). However, the presence of synthetic antioxidants in food packages is questioned due to their potential risks and regulatory controls.

A benefit of antioxidants associated with packaging films compared to the direct addition to the snack food is the release of the active compound at controlled rates. Independent devices like sachets, pads, or labels that contain the antioxidant separate from the food have been the first approach to produce antioxidant active packaging. These are separated from the food product and are added to a conventional "passive" package. Fine powders of iron and ferrous oxides are the most common oxygen scavengers, although ascorbic acid, sulfites, catechols, ligands, and enzymes such as glucose oxidase have also been utilized (Fang et al. 2017).

Natural alpha-tocopherol can also be extruded onto high-density polyethylene (HDPE) with antioxidant properties. It has a comparatively lower release rate compared to BHT. A large proportion of the antioxidant was lost during release but the remainder was absorbed by the food, offering some protection.

According to Gomez-Estaca et al. (2014) there are basically two ways to produce these active packaging systems: independent devices in which sachet, pad, or labels containing the agent separately from the food is supplemented to conventional packages or the manufacturing of materials containing antioxidants. In the last system, antioxidants are commonly incorporated via coating or included in the formulation before the fabrication of films. The active antioxidant packaging materials should release antioxidants at controlled rates so a predetermined concentration of these compounds is maintained in the food. In terms of the mode of action, antioxidant packages are designed to release antioxidants to the food or to scavenge undesirable compounds such as oxygen, radical oxidative species, or metal ions from the food or headspace (Gomez-Estaca et al. 2014, Sanches-Silva et al. 2014).

4.4.5 Antimicrobial Releaser

Antimicrobial active packaging consists of a system designed to inhibit the growth of spoilage and pathogenic microbes. This type of active packaging has two main categories, namely migratory and nonmigratory. The first system allows the reversible release of nonvolatile or volatile active components from polymer matrix to food constituents or packages headspace by diffusion and/or partition at interfaces. On the other hand, the active components in nonmigratory systems are irreversibly tied to the package's surface and no diffusivity of the antimicrobial agents occurs.

The most studied antimicrobial packaging systems are based on the use of organic acids (citric and sorbic acids), essential oils (cinnamon, carvacrol, and thymol), antimicrobial polymers (chiotosan), bacteriocins, and enzymes and independently of the type of system they should possess a wide spectrum against most pathogenic and food spoilage microorganisms (Gadhave et al. 2018, Yildirim et al. 2018).

Essential oils with antimicrobial properties have been incorporated into starch biofilms using the casting method and have improved the microbial shelf life of food products, water vapor permeability, and mechanical features. Gadhave et al. (2018) indicate that polyphenolic compounds also exert antioxidant properties.

Recently, several antimicrobial nanoparticles have been investigated. The main ones are derived from titanium, magnesium, copper, silver, platinum, and zinc (Gadhave et al. 2018, Yildirim et al. 2018). The goal of these novel systems is to extend the microbial shelf life of fresh and processed meats (beef, pork, and poultry), fish products, and cheeses discussed in Chapters 17 and 18.

4.5 INTELLIGENT PACKAGING

The snack packaging industry will keep evolving likely adopting novel technologies such as intelligent packaging, which includes external or internal indicators used for quality control and freshness of the packaged food mainly related to time temperature, concentration of oxygen or carbon dioxide, and microbial growth.

These innovative systems will allow customers to perceive changes in the condition of packed snacks throughout their shelf life and to assure they are safe and still maintain desired organoleptic properties. In relation to snacks, one of the key design factors is the assessment of the condition of the product in terms of oxidative rancidity and generation of off-flavors. Intelligent packages that will inform customers of the freshness and oxidative stability have been developed but unfortunately at a relatively higher cost. The implementation of these new technologies is likely to occur in the near future because most snacks are sold with a good profit and new customers are concerned about the quality and wholesomeness of their foods (Yam et al. 2005, Han et al. 2018).

Besides the research and development of novel active packaging materials, Han et al. (2018) categorized other novel packaging technologies comprising intelligent or smart packaging (IOSP), time-temperature indicators (TTIs), gas indicators (GIs), and other indicators for microwavable foods. These innovations further improve food quality, food safety, and shelf life.

4.5.1 Intelligent or Smart Packaging

IOSPs are designed to supervise the condition of packaged food or its environment by using chemical, electronic, and/or mechanical sensors with the aim to alert customers to possible problems related to the packaged food (Yam et al. 2005). IOSP schemes possess intelligent devices like bar codes, small labels, or tags, which are included in packaging materials to obtain information about the food's quality and transfer it in a simple way to retailers and consumers. The most regularly used devices are broadly categorized into direct and indirect sensors. The direct indicators are designed to display information to buyers about the wholesomeness, quality, and safety of foods. These smart devices are placed inside the primary packaging so they are in direct contact with the food or surrounding environment. On the other hand, indirect sensors, which do not provide direct information to consumers, are designed to indirectly evaluate the effects of the encompassing environment of the packed foods (Han et al. 2018).

Today, powerful and affordable scanner and wireless technologies have provided more favorable environments for companies to develop advanced time-temperature indicators or TTI systems for tracking and controlling the quality of perishable food products (Bhushan and Gummaraju 2004). For example, Bioett (Lund, Sweden) has developed a TTI/barcode system in which data may be read by a handheld scanner, displayed on a computer monitor, and downloaded into a database for analysis (Anonymous 2004). KSW Microtec (Dresden, Germany) and Infratab (Oxnard, California) have developed battery-powered TTI/RFID tags using a technology in which thin-film batteries are printed onto a flexible substrate (Giaglis et al. 2003). Unlike the traditional TTI that is based on diffusion or a biochemical reaction, the TTI/RFID tags use microchips to sense and integrate temperature over time to determine the shelf life of a product (Nitzan 1999, Collins 2003).

4.5.2 Time-Temperature Indicators

Temperature is the most relevant factor influencing the kinetics of physical, chemical, and microbiological deteriorations of foods. The shelf lives of many food items, especially those belonging to perishable meats, fish, and dairy products, are very sensitive to temperature abuse, which is the major responsible factor for the deterioration, food-borne illnesses, and economic losses. TTIs are designed to assess temperature abuse during transportation, distribution, storage, and expenditure to alert customers and improve safety measurements. Therefore, temperature monitoring is vital to provide consumers with necessary information about food quality and safety throughout the process of food circulation (Han et al. 2018). The TTIs are typically small self-adhesive labels or tags also known as "freshness indicators" attached to shipping containers or individual food packages. These labels provide visual indications of temperature history during distribution, storage, and circulation mainly through changes in color intensity. There are three basic types of commercially available TTIs: critical temperature indicators, partial history indicators, and full-history indicators (Yam et al. 2005, Han et al. 2018). The critical temperature indicators show whether a product has been exposed to a temperature above and in some instances below the reference temperature. The second category is designed to indicate the cumulative or historical effect of the time-temperature changes related to food quality and safety in products exposed to a temperature above the reference value. Lastly, the full history sensors are designed to continuously monitor temperature variations throughout time.

The fundamental principles of TTIs are the identification of permanent reactions in the form of enzymatic, electronic, chemical, biological, and nanoparticle modifications after a product has been exposed to temperature abuse (Han et al. 2018). Electronic devices are designed to warn retailers and consumers about the quality of a product by using a thermal sensor that converts temperature signals to electrical signals into a visual output. These TTIs are generally expensive and inconvenient and require some training on the part of consumers limiting their commercial usage.

Other types of TTIs that have been developed are based on nanoparticles, enzymes, and chemical and biological changes, which generally yield an irreversible color modification related to the thermal history of the food. These types of TTIs are generally stuck on or incorporated into the product packaging by printing or coating, and they have a comparatively lower cost, compared to electronics-based counterparts (Han et al. 2018).

According to Yam et al. (2005), the most used TTIs are powerful and affordable scanners and wireless technologies mainly designed to track and control the quality of perishable foods. The most utilized devices are barcode and RFID tag systems in which data may be read by a hand scanner, which displays and saves information on a computer system. The RFID tag consists of a microchip designed to sense and integrate temperature over time highly associated with the shelf life of the perishable food product.

4.5.3 Gas Sensors Indicators (GIs)

The gas composition in the headspace of sealed packages is very much related to the respiratory activity of the food product and microorganism types and counts and also to relevant oxidative reactions related to rancidity. Needless to say, the most relevant for snacks are reactions that lead to lipid oxidation. The internal gas composition can also indicate the effectiveness of the film material to gas transmission rate so it determines whether the package ruptured or was improperly sealed.

These GIs are included in the film or package label in order to inform retailers and consumers of the quality and safety of the packaged food products (Yam et al. 2005).

Different types of GIs are utilized in food packaging in the form of package labels or package printing films that detect oxygen, ethanol, hydrogen sulfide, water vapor, carbon dioxide, or other gas components (Han et al. 2018).

Undoubtedly, oxygen indicators are the most common ones because this gas enhances oxidative rancidity, color and flavor changes, and microbial spoilage. The oxygen GIs are mainly designed to show color changes due to leaking, improper sealing, or tampered packages mainly produced with modified atmosphere packaging systems. The recent advances in smart ink and printing technologies (Chapter 5) will also allow GIs to be read automatically with optical systems (Yam et al. 2005). Wilhelm and Wolfbeis (2011) designed a couple of absorption-based opto-chemical oxygen indicators consisting of leuco indigo or leuco thioindigo dyes that were combined into two sorts of synthetic polymers (copolymer of polystyrene-*co*-acrylonitrile and a highly O_2-permeable polyurethane hydrogel). These produced irreversible and detectable color changes caused by chromogenic oxidation of the colorless dyes after exposure to oxygen. The authors concluded that these sensors represent a cost-effective means to visually detect and monitor oxygen ingress with potential applications for both the modified-atmosphere food and pharmaceutical industries.

4.6 PACKAGING OF SNACK FOODS

The different sorts of snacks foods differing in chemical and physical properties especially in terms of moisture, *Aw*, fat content and composition, friability, texture, and color normally require different packaging materials, technologies, and systems. The next section describes the most common packaging requirements for most snacks described in Chapters 9 to 18.

4.6.1 Popcorn

Popcorn is one of the most hygroscopic of the crispy snack foods. In addition, pre-popped popcorn for retail distribution often contains salt, flavorings, and oil or glazed sugar-based coatings, which also tend to absorb moisture from the air. Therefore, popcorn-packaging materials should be an excellent barrier against moisture or water vapor. When popcorn is not flavored with an oil base such as sugar-coated popcorn the packaging film should not provide a good oxygen barrier because the product is low in fat content. Freshly popped popcorn containing 1.7% moisture (Chapters 9 and 20) possesses an *Aw* of only 0.062. Consequently, the product is very susceptible to absorbing water vapor from the atmosphere with the consequent loss of crispness (Katz and Labuza 1981). According to Robertson (2012) the crispness intensity of popcorn starts deteriorating at the critical moisture of 6.1% corresponding to an *Aw* of 0.49. The low density of the popcorn results in a product that can actually cushion a compressing force. As a result, puncture and tear requirements are low allowing the utilization of low gauge films. PVdC coated or coextruded heat-sealable OPP is typical plies for the inside face of packaging films. Metalized OPP and polyester better protect the popcorn from environmental moisture and therefore are used when long shelf life is required (Serna-Saldivar 2008, Robertson 2012).

In recent years the market and utilization of microwavable popcorn for in-home preparation has rapidly increased (Chapter 9). The microwave energy is generated by electromagnetic radiation with a characteristic wavelength, amplitude, and frequency produced by a magnetron. According to Decareau (1985) most commercial microwaves work at a frequency of 2.45 billion cycles per second or 2450 MHz and wavelength of 3×10^8 m/sec. In the microwave oven, the oven cavity is three dimensional and waves are scattered in all directions. For this reason, the electric field varies along the different sections of the oven and most ovens are furnished with a turntable, which rotates the food. Dipole rotation and ionic polarization are the two main mechanisms by which microwaves heat foods being the first comparatively more relevant than ionic polarization. Ionic polarization occurs when ions in solution move in response to the electric field and the kinetic energy is given up by the field to the ions enhancing the coalition with other ions converting energy into heat.

Most microwavable pillow style and multilaminated bags are packaged with 75 g popcorn plus 25 g of an oil cake that in most instances is blended with flavorings and salt. The most used ones are bags built from PET-coated paperboard that are placed inside cellophane bags in order to prevent environmental moisture uptake. The microwavable bags contain a 15 cm susceptor plate, commonly produced from vacuum-deposited aluminum, which absorbs microwave energy into the center of the package where the popcorn kernels and oil cake are mainly localized (Robertson 2012). The package is designed with inner- and outer-paper plies with the susceptor positioned between them. Upon microwave heating, the package is designed to expand due to the volume of the popped corn and the steam produced by the thermal operation. In addition, the container is designed to prevent heat loss in order to optimize popping and at the same time vent the water vapor generated during the operation. The microwavable container must not absorb or reflect significant amounts of the microwave radiation, and should not soften and release toxic compounds like per-and polyfluoroalkyl surfactants (Zabaleta et al. 2017) and objectionable odors at the conclusion of the popping cycle.

4.6.2 Wheat-Based Products

4.6.2.1 Toasted Bread, Crispbreads, and Crackers

According to Robertson (2012) four modes of deterioration are usually associated with crackers and related products: loss of crispness, development of rancidity, and product breakage. Since most crackers or soda biscuits have moisture contents of about 2% and *Aw* of approximately 0.1 they are normally packaged inside moisture-proof materials. If the packaging material is permeable to air moisture or if the seal is not perfect the product will tend to equilibrate with the surrounding moisture with the resulting loss of crispness. According to Labuza and Contreras-Medellin (1981), crackers lose crispness when they absorb moisture to increase their *Aw* to 0.43. The moisture sorption isotherm at 20°C for crackers at this particular *Aw* is about 8%. The packaging material should also protect crackers from the entry of oxygen and make a perfect seal in order to delay the initiation of reactions leading to rancidity. Consumers regard broken crackers as unacceptable and therefore the package should provide mechanical protection.

The traditional material used for the packaging of crackers has been regenerated cellulose fiber (RSF) coated with either LDPE or PVC/PVdC copolymer and often with a layer of glassine in direct contact with fat-containing crackers. Recently, this material has been replaced by OPP, coextruded OPP containing acrylic coatings on both surfaces of the film (i.e. acrylic coating/OPP/PVdC). Coextruded OPP films provide better sealability. Mechanical protection is provided by placing the individual packages inside a rigid carton container (Robertson 2012). Broady (2017) suggested the use of multilaminates with aluminum foil to protect these crispy products from moisture gain. Particularly the aluminum foil and PVdC are suited for use in VFFS packaging machines.

The effects of two innovative multilayer packaging materials (metalized PLA/paper and metalized OPP with EVA and pro-oxidant additive/paper) were compared with a control (OPP)/paper) using crackers or biscuits subjected to accelerated storage at 25, 35, or 45 °C and 50% relative humidity for 92 days. Results indicated that there were no remarkable differences in the evolution of primary and secondary lipid oxidation compounds during storage. The crackers stored in flexible metalized PLA showed significantly higher moisture and *Aw* levels, but the package did not adversely affect other important quality features. Therefore, the two novel packaging materials were suitable for commercial use (Romani et al. 2015).

4.6.2.2 Pretzels and Pita Chips

Both pretzels and pita chips described in Chapter 10 are generally packaged with the same multilaminate films employed for tortilla chips, potato chips, and extruded snacks (refer to next sections).

4.6.2.3 Cereal-Based Snack Bars

This category is comprised of snacks based mainly on flaked, puffed, and/or cereal flours or meals commonly mixed with syrups, caramel, dehydrated fruits (Chapter 16), toasted nuts, and/or pseudocereals (Chapters 13 and 15). Most of these snacks have a chewy texture and an *Aw*, which places them in the intermediate moisture class. Packaging materials for these snack bars should provide a barrier to water vapor ingress in order to avoid the development of stickiness due to moisture uptake by the sugar in the product. In addition, the package should prevent the exchange of oxygen so as to retard rancidity especially in bars containing high-fat content.

Senhofa et al. (2015) studied the effects of three packaging materials on the quality of muesli consisting of a mixture of grain flakes chocolate and dried fruits. The packaging materials tested were paper bags, paper tubes, and a stand-up pouch. Products were stored for nine months at 20 ± 2°C and 55 ± 3% RH. During the storage moisture content, water activity, total plate count of mesophilic aerobic and facultative anaerobic microorganisms mold and yeast counts, volatile compounds, and sensory parameters were evaluated. After nine months' storage, the muesli samples packaged in stand u pouches had the best sensory properties and overall quality. The worst results were observed in muesli stored in paper bags.

Similar research was undertaken to test consumer acceptability of puffed glutinous rice bars supplemented with Malaysian fruits (date, figs, and raisins) and syrups (honey and glucose syrup) that were packed with four different materials (PA, metalized polyester, PET, and aluminum foil). The cereal-based bars retained their properties when packed with metalized polyester or aluminum foil.

An oat and wheat-based cereal bar was packed in PP, paper, aluminum foil, PET, and metalized polyester LDPE/LLDPE with and without vacuum and stored under ambient temperature conditions and 37°C. It was found that the bar remained acceptable during the entire storage period irrespective of packaging material at ambient temperature. At 37°C, the bar packed in PP film remained acceptable for 6 months while bars packed in PET and metalized polyester remained stable and acceptable for 12 months (Padmashree et al. 2013).

4.6.3 Nixtamalized Snacks (Corn and Tortilla Chips)

The relatively higher bulk density of both corn chips and tortilla chips compared with other snacks demands different packaging requirements and technologies. The high density allows the packaging at faster speeds. Corn and tortilla chips require the typical moisture barrier for most savory snacks. Oxygen and light barriers are not as critical as required in other types of snacks. Therefore, glossy, low-haze transparent packaging materials are regularly utilized (Serna-Saldivar 2008). Corn chips are not likely to cause punctures or tears to packaging films because of their uniform shape and size. Contrarily, the sharp edges of triangular tortilla chips can cause punctures to the body of the package, especially on the bottom seal. PVdC-coated PP or glassine, PP/PP, metalized PP or PE, and coextruded OPP are the most common materials used to pack corn and tortilla chips (Chapter 11). A good hot tack and initial seal strength, especially the bottom seal, are critical factors for packaging these chips. Overinflation of bags, a practice commonly employed to cushion the chips and prevent breaking, also puts a demand on seal strength during bag forming, storage, and distribution.

New bio-based materials combined with modified atmosphere technology have the potential for packaging tortilla chips and related types of snacks containing low moisture and high oil content. Peelman et al. (2016) demonstrated that the quality and crispness of tortilla chips are guaranteed when packed in multilayer biomaterials consisting of xylan and cellulose. Most of the bio-based packages showed sufficient barriers toward moisture and gases to serve as a food-packaging material suitable for nixtamalized chips.

4.6.4 Extruded Snacks

The major mode of deterioration for extruded and puffed snacks is loss of crispness (Robertson 2012). These snacks described in Chapter 12 are usually dried or baked to lower their moisture to 1.5%, which corresponds to an *Aw* of 0.08. These products lose texture and quality when the *Aw* exceeds 0.36, which corresponds to a moisture of 4.2%. Most extruded and puffed snacks are normally packaged in identical petroleum-based films discussed for corn and tortilla chips (Serna-Saldivar 2008, Min et al. 2010). Processors should make special emphasis on the prevention of loss of crispness so packages should offer an excellent water vapor barrier. Some extruded and puffed snacks are comparatively less sensitive to oxygen compared to fried snacks, and, therefore the barrier requirements of these films are consequently less stringent than fried snacks (Min et al. 2010).

4.6.5 Potato Chips

Potato chips or crisps are considered the most relevant savory snack food in terms of sales and volume. These products are divided into regular and fabricated potato chips and normally contain less than 3% moisture and fat contents that exceed 32% (Chapter 14). The type of oil used for frying purposes greatly affects their susceptibility to oxidative rancidity. According to Robertson (2012) potato chips packaged in bags manufactured of HDPE, LDPE coated with RCF or PVC/PVdC copolymer coated with RCF had a shelf life of about 15 days at 27°C and 60% relative humidity. The shelf life could be significantly extended to approximately ten weeks when chips were stored in an aluminum foil/LDPE construction that also had HDPE plus titanium dioxide and a brown light-absorbing pigment. This multilayered film had zero oxygen and water vapor transmission rates and protected potato chips from the harmful effects of the ultraviolet light partially responsible for initiating oxidative rancidity.

Abong et al. (2011) studied the effects of packaging and storage temperature on the shelf life of potato chips processed from four Kenyan cultivars. The chips were packaged inside aluminum foil packs or PE bags and then stored at 25, 30, or 35°C for 24 weeks. The results showed that the aluminum foil pack was the most effective in controlling increase in moisture content, peroxide values, and FFA levels. The type of potato cultivar and storage temperature significantly influenced the formation of peroxides and overall quality of chips. The potato chips stored at 35°C had shorter shelf life compared to counterparts stored at 25 and 30°C.

Fabricated potato chips (Chapter 14) are traditionally packaged in a single stack of saddle-positioned pieces in a cylindrical fiberboard laminated and with a reclosable plastic lid located at the top of the container. During packaging the cylindrical cans are first evacuated, and then the oval-shaped snacks placed inside the container. The last step is the flushing of nitrogen before closing the container. Alternatively, fabricated potato chips are packed in blow-molded plastic containers obtained by thermoplastic extrusion or injection. The last blowing technique is the most widely employed by the industry (Robertson 2012). As same as plastic flexible films, blow-molded containers can be made with multiple layers to impart different functionalities. For fabricated snacks with high oil content, the blow-molded container should present an excellent barrier against oxygen, water vapor, and UV light. Some containers are manufactured with corrugations to improve the sidewall strength and vacuum resistance. Most blow-molded containers for fabricated chips have a U-shaped bottom structure or package adaptation with the same shape to receive chips and protect them against breakage.

4.6.6 Nuts and Related Products

According to Min et al. (2010), lipid oxidation is the most relevant quality parameter in nuts and related products because this phenomenon yields undesirable off-flavors and odors. Many intrinsic and extrinsic factors affect the rate of lipid oxidation and rancidity. In terms of intrinsic factors, the oxidative shelf life is affected by variety, maturity, oil content, fatty acid composition, and presence of natural antioxidants like phenolics, flavonoids, and tocopherols (Chapter 20). On the other hand, the most relevant extrinsic factors are storage temperature, oxygen concentration, light, air relative humidity, and composition of the gas atmosphere. Processed nuts are optimally packed in multilayer structures that present good barriers against water vapor, oxygen, and light. These multilayered flexible films are alternatively manufactured with a UV light-absorbing additive. Fu et al. (2018) studied the effects of four types of packaging materials on the storage quality of peanut kernels. The results showed that the type of material had a major influence on peanut quality indexes. The color of the peanut seed coat packaged in the PET/aluminum/polyamide/PE composite bag did not change significantly during storage. The color deterioration was slower with polyamide/PE materials compared with the PE and woven bags. The use of PET/aluminum/polyamide/PE and polyamide/PE bags maintained kernel quality and freshness for more than one year and additionally prevented insect deterioration and aflatoxin contamination. According to Ucherek (2004) the packaging of peanuts in cans flushed with a gas mixture containing 99% nitrogen and the rest oxygen optimized shelf life. Mate et al. (1996) concluded that the storage with less than 2.5% oxygen diminished both peroxide and hexanal contents in shelled peanuts and walnuts and that products stored at 54% RH contained higher peroxide values than counterparts stored at 21% RH. Likewise, the replacement of atmospheric air with CO_2 extended the shelf life of raw peanuts and pecans from 2 to 27 weeks (Holaday et al. 1979).

Robertson (2012) indicates that legume nuts and tree nuts generate CO_2 during respiration and this gas is absorbed into the pores of these products resulting in the accumulation and formation of vacuum inside the pouches. With this packaging technique, salted in the shell peanuts, shelled raw and processed peanuts, and pecans are protected from any significant deterioration for 4, 12, and almost 7 months, respectively.

The shelf life of nuts and related products (Chapters 13 and 15) can be optimized by combining inert gas flushing (nitrogen and CO_2), vacuum packaging with bags resistant to oxygen, and CO_2 or fabricated with oxygen-absorbing materials (Min et al. 2010).

In another study conducted by Jensen et al. (2003), it was determined that the use of oxygen absorbers or packages flushed with nitrogen enhanced the preservation of walnuts. The use of both oxygen absorbers and high-oxygen-barrier materials (e.g. laminate with EVOH) acted synergistically to prolong shelf life and maintain quality of nut products (Jensen et al. 2003, Min et al. 2010).

Mexis and Kontominas (2010) investigated the effects of active packaging, nitrogen flushing, and storage conditions on quality of raw whole unpeeled almond kernels packed in PET/LDPE or LDPE/EVOH/LDPE pouches. Tests were conducted under N_2 with or without an oxygen absorber in heat-sealed bags stored for 12 months. The peroxide values of almonds ranged between 0.17 for fresh kernels to 9.22 meq O_2/kg oil for almonds packaged in PET/LDPE pouches under N_2 exposed to light at 20°C for 12 months whereas the respective hexanal values were <28.5 μg/kg and 4.88 mg/kg. These authors concluded that the oxygen absorber was effective to optimize shelf life for one year providing the use of pouches with low oxygen permeability. The almonds stored under optimum conditions showed lower peroxide and hexanal values and amounts of relevant volatile compounds like aldehydes, ketones, alcohols, alkanes, and aromatic hydrocarbons.

Likewise, Leufven et al. (2010) investigated the influence of packaging systems on the stability of dried raw pistachio nuts packed in multilayer pouches of PE/Polyamide/PE under atmosphere, vacuum, and oxygen scavenger conditions stored at 20, 35, or 50°C throughout 12 weeks. The results showed that the oxidative degradation reactions were best controlled by decreasing the oxygen concentration by using the oxygen scavenger system.

4.6.7 Fruit and Vegetable-Based Snacks

The chief use of fruits products in the snack industry is as semi-dried or intermediate moisture items with prolonged shelf life at room temperature. The packaging of partially or totally dehydrated fruits and vegetables covered in Chapter 16 is aimed toward the minimization of the entrance of moisture and in some instances oxygen especially when fruits have been previously treated with sulfur dioxide to retard nonenzymatic browning and increase the retention of vitamin C. In addition, fruits or vegetables, which contain significant amounts of carotenoids and xantophylls can undergo oxidative deterioration and loss of the characteristic yellow-orange coloration (Robertson 2012).

These products are stable mainly due to the combination of *Aw* and pH and require packaging technologies to preserve *Aw*, color, and sensory properties. Intermediate moisture fruits normally are dehydrated to lower their moisture and *Aw* between 20 to 30% and 0.70 and 0.85, respectively. The application of edible coatings is practiced in order to enhance the quality and shelf life of intermediate moisture fruits because the technology prevents the food products from mechanical damage, microbial contamination, loss of volatiles, and relevant organoleptic properties. These edible coating materials are normally applied as a thin layer onto the surface of foods in liquid forms after dehydration. The edible coating acts as a barrier to moisture loss and microbial contamination and also delays the transfer of moisture into other ingredients or materials of snack mixes. Lipids, hydrocolloids, and chitosan are the most commonly used coating materials (Qiu et al. 2019).

According to Gvozdenovic et al. (2007) packaging materials suited for the preservation of dried fruits should provide appropriate barrier characteristics for water, oxygen, nitrogen, and carbon dioxide, as well as for UV light. Dried apple pieces were packed in PE, OPP-metalized, OPP/PE, OPP-metalized/PE, or PET/OPP-metalized/PE and evaluated for 120 days at room temperature. The authors concluded that the type of package influenced the overall quality and nutritive properties of dried apple samples. The qualitative changes of color, polyphenolic compounds, hydroxymethyl furfural, and sensory characteristics indicated that apple pieces packed with the PET/OPP- metalized/PE packaging best preserved their quality attributes.

Many cereal products contain dried fruits. Packaging materials for these products must prevent the development of stickiness as a result of moisture uptake. The *Aw* of these products should be maintained below 0.5. The moisture equilibrium between dried grains and fruits in the package should also be considered. Most dried fruits mixed with other snacks are intermediate moisture foods that have relatively higher moisture contents so it is imperative to study moisture transfer mechanisms between the fruits and other ingredients (i.e. toasted grains, tree nuts) (Min et al. 2010).

Sergio et al. (2015) studied the postharvest performance of intermediate moisture peach and prune packed for 60 days in PPE trays sealed with a composite covering plastic film with partial vacuum stored under both refrigerated and unrefrigerated conditions. Semi-dried peach and prune pieces that were not stored under refrigeration showed diminished quality after 20 days of storage whereas counterparts stored under refrigeration preserved their physical, biochemical, organoleptic, and microbiological properties for 40 days. After 60 days, only the semi-dried peaches maintained the minimum acceptable quality properties.

A large variety of dried fruits commonly blended with nuts and sweeteners (honey, sugar, or syrups) are sold as fruit bar snacks. Their composition is extremely variable, leading to the development of bars with a wide variety of flavors and textures. Generally, these products are in the intermediate moisture category and have *Aws* in the range of 0.20–0.40. This is because the dehydrated fruits contain relatively high amounts of sugars (25–30%) that bind water. Packaging materials for fruit bars should provide an adequate barrier against water vapor to avoid moisture uptake and the development of stickiness (Robertson 2012). Recently, Pawase et al. (2019) determined the effects of different packaging materials on fruit bars produced from fig-mango or fig-papaya stored for 60 days in HDPE, LDPE, aluminum foil, or PET. These authors concluded that both aluminum foil and PET maintained for a longer time the quality of the fruit bars.

4.6.8 Dairy-Based Snacks

The wide array of dairy products available in the market are one of the most nutritious because they contain most of the essential nutrients required for growth and development. Unfortunately, milk products are highly perishable because of their high *Aw*, adequate pH, and composition that favors the growth of spoilage and pathogenic microorganisms. Nevertheless, humans have learned to process dairy foods with extended shelf life via the modification of *Aw*, pH, and application of thermal treatments aimed to pasteurize or even sterilize packed products. The different sorts of dairy products that are considered snacks are thoroughly described in Chapter 17. In this sense, the packaging requirements vary according to the type of dairy product and its expected shelf life at room or controlled temperature. The packaging systems practiced for most dairy products have been recently addressed by Sabikhi et al. (2018) and Mizra et al. (2020).

Of the total milk packed, it is estimated that flexible pouches dominate (92%), followed by bottles (5%) and aseptic packaging cartons (3%). The main advantages offered by the plastic packages include good barrier properties, visibility of the contents, container weight, convenience, cost, and attractiveness especially in terms of multicolor printing and ergonomic designs. Moreover, the use of plastic containers eliminates noise in the milk bottling plants and during delivery and also reduces water pollution caused by milk residues and detergents used in the bottle washing process (Sabikhi et al. 2018).

4.6.8.1 Fermented Milk and Yogurt

Among dairy products, yogurt is the most popular fermented practically consumed all over the globe. As thoroughly covered in Chapter 17, yogurts are classified as a set, stirred-type, drinking, frozen type, and concentrated or Labaneh. Even though these products are acidic they are considered perishable so their packaging is of utmost importance to preserve their physicochemical, nutritional, and organoleptic properties. The packaging materials should be acid-resistant, present a good barrier against oxygen to prevent the growth of aerobic yeast, molds, and bacteria. These products are commonly packed in rigid or semi-rigid plastic containers. The most common polymers are HDPE, PPE, PS, and PVC. These containers are usually covered with preformed lids of aluminum foil and PE or PS laminates of about 100 mm diameter. Shelf-stable yogurts like drinking, stirred, or set-types are aseptic packaged using six-layered laminates containing PE, aluminum foil, and PE/PE/aluminum foil/PE/paperboard/PE (Raju and Singh 2016, Sabikhi et al. 2018).

In view of the anaerobic and microaerophilic ecology of the strains normally used in the production of probiotic dairy products, the level of oxygen within the package should be as low as possible in order to avoid the death of these beneficial microorganisms. It is well known that the exposure to dissolved oxygen during processing and storage is highly detrimental to *Bifidobacterium bifidum* and *Lactobacillus acidophilus*. *B. bifidum* is more sensitive to oxygen compared to *L. acidophilus* (Da Cruz et al. 2007). A high oxygen barrier packaging materials such as HIPS-EVOH-PE and PS minimize the presence of oxygen in probiotic yogurts maintained for 42 days at refrigeration (Miller et al. 2002). The same research group (Miller et al. 2003) further investigated the incorporation of oxygen absorbers into PS and concluded that this combination created optimum conditions for the growth of anaerobic probiotic cultures.

Talwalkar et al. (2004) investigated the effect of HIPS, HIPS-EVOH-PE, and HIPS-EVOH-PE integrated with oxygen-scavenging material on the survival of *Bifidobacterium* spp. and *L. acidophilus* in yogurt by monitoring the oxygen concentration during storage and reported that the dissolved oxygen depended on the type of packaging material used. At the end of 42 days of storage, the dissolved oxygen content decreased in the HIPS-EVOH-PE and significantly increased in the HIPS packaged yogurt. The oxygen levels in the oxygen scavenger integrated HIPS-EVOH-PE container dropped to the lowest concentration indicating that the active packaging favored the needed anaerobic conditions.

4.6.8.2 Cheeses

According to Sabikhi et al. (2018) cheese packaging can be broadly categorized into wrapping and retail packaging. The first is performed in processing plants for storage and ripening whereas the second is for the retail market and consumers. The packaging systems should prevent loss of quality mainly induced by the degradation of lipids and milk proteins and additionally water loss and color changes. Light catalyzes the oxidation of lipids even at refrigeration and in unripened cheeses results in "cardboardy" or "metallic" off-flavors. The production of plastic materials with titanium dioxide diminishes the light transmittance rate. The selection of the most appropriate material for a particular cheese is affected by the type and its characteristic texture (hard or soft), the presence of desirable microorganisms and the permeability to water vapor, oxygen, carbon dioxide, and ammonia. Bulk cheeses are either paraffin wax coated or vacuum-packed in flexible films. Normally, low-moisture fresh cheeses are covered with impermeable paraffin or latex emulsions to protect them against moisture loss, molds, and other microorganisms. In addition, the protective cover reduces the penetration of air. The other more modern alternatives are the utilization of vacuum packaging, gas-flushing, overwrapping, or vacuum skin packaging machines using high barriers synthetic coextruded films such as LLDPE-nylon-LLDPE, LLDPE-EVOH-LLDPE, and LLDPE-PVdC-LLDPE.

The oxidative stability of cream cheese stored in thermoformed trays made of amorphous PET-PE, PS-EVOH-PE, and PP-PE with a transparent lid made of PET-aluminum oxide-PE film with different depth and color was studied by Pettersen et al. (2005). The authors concluded that cream cheese stored in PET-PE trays offered the best protection against fat oxidation, sensory properties, and color.

For fresh cheeses, thermoformed HIPS coextruded or extrusion coated with PVC or PVdC containing titanium dioxide are recommended because of their barrier properties and protection against light. In addition, thermoformed nylon-LDPE or injection molded HDPE containers with slits in the side are utilized to allow drainage of whey (Sabikhi et al. 2018).

Mozzarella, a fresh soft and unripened cheese (Chapter 17) is susceptible to microbial spoilage because of its high moisture > 50% and high *Aw*. The stability of sliced Mozzarella cheese packed under three different modified atmospheres was studied by Alves et al. (1996) who reported that the shelf life was 63 and 45 days when packed in 100% CO_2 and 50% CO_2/50% N_2, respectively compared to the 13 days observed when packed in normal air. In a similar investigation, semisoft-aged Danbo or Denmark cheese blocks were first packed in thermoformed trays with a lid made of PLA under a 70% N_2/30% CO_2 atmosphere and then stored in a display cabinet with exposure to fluorescent light that emulated conditions of retail stores. Authors concluded that PLA packages extended the shelf life of Danbo cheese for 56 days with limited loss of moisture and lipid oxidation (Holm et al. 2006).

Moreira et al. (2011) investigated the antimicrobial effectiveness of sodium caseinate, chitosan, and a combination of sodium caseinate-chitosan coatings produced by food immersion or as wrapper film or packages on the native microflora of cheese. Results revealed that chitosan and sodium caseinate-chitosan coatings and wrappers exerted significant bactericidal effects on mesophilic and psychotropic bacteria and diminished yeast and mold counts.

Interest in edible with or without antimicrobial coatings for extending the shelf life of food products including fermented dairy products is increasingly growing. Cerqueira et al. (2009) investigated different biopolymers (chitosan and two different galactomannans) as coating agents for the Saloio Portuguese full-fat cheese which is produced with a mixture of caprine, bovine, and ovine pasteurized milk. The aim was to select the best coating material especially in terms of lowering the gas transfer rate. The solutions of *Gleditsia triacanthos* (formulation: 1.5% of galactomannan, 2.0% of glycerol, and 0.5% of oil) presented the best properties to coat the Saloio cheese in terms of wettability, water vapor permeability, and both oxygen and carbon dioxide permeabilities. Visual evaluation also indicated that the uncoated cheese suffered from an extensive mold growth when compared with the coated counterpart.

Likewise, fresh Ricotta cheese coated with chitosan-whey protein composite film and stored at 4°C under 60% N_2/40% CO_2 showed reduced microbial counts and a shelf life extension of 30 days (Di Pierro et al. 2011).

For surface mold-ripened cheeses such as Camembert and Brie, packing should be performed after the mold has grown to a desirable extent with films that have limited oxygen permeability to prevent the growth of anaerobic proteolytic bacteria. For this purpose, perforated OPP, paper coated with wax, or laminated-perforated films are recommended (Robertson 2012).

Cheese is also available in slices individually wrapped in plastic films. About ten such slices are placed and sealed in an outer high barrier plastic pouch. Although used for a wide range of products in the dairy sector, modified atmosphere packaging (MAP) has been mainly applied to the packaging of cheese. Acceptable eating quality with shelf life trebled from around four weeks to up to three months was achieved by using MAP. Hard cheeses such as cheddar are commonly packed in 100% CO_2 using horizontal form-fill-seal pillow pack machines. The packaging materials used include PVdC-coated PE and PET-PE. Cheese spreads may also be packed in plastic PPE tubs commonly sealed with aluminum foil laminates and closed with a press-on plastic lid or plastic laminated or coextruded squeezable tubes or bottles (Sabikhi et al. 2018).

4.6.8.3 Ice Cream and Frozen Products

The oil-in-water emulsion of ice cream and related frozen products are commonly packaged in filling machines designed for handling paperboard cartons on a time cycle basis or on a volume/weight basis. Most of the folding cartons used today are paperboard laminates containing LDPE. Other containers used today include plastic cups and form-seal plastic containers produced from PS or HIPS (Sabikhi et al. 2018).

4.6.9 Meat-Based Snacks

There are many packaging options for processed meat products, which vary according to the type of product and desired shelf life during different types of storage conditions as affected by susceptibility to putrefactive and pathogenic bacteria, lipid oxidation, discoloration, and loss of organoleptic properties. The packaging of processed meats sold as snacks is usually performed with synthetic plastics that present adequate barriers against water vapor and gas permeability (Tables 4.1 and 4.2). Many products are vacuum or modified atmosphere packaged for long-term storage at refrigeration or even ambient temperatures. Processed cured meats are commonly packaged in materials which offer an anoxic environment, which maintains the nitrosylhemochrome pigments associated with these products. Common polymers for food packaging are LDPE, HDPE, PP, polytetrafluoroethylene (PTFE), PS, nylon, PVC, PVdC PVdC, PA, and EVA. Each type of packaging material has advantages and disadvantages in terms of consumer and marketing issues, environmental considerations, and different costs.

Today, meat processors are alternatively employing green materials with improved biodegradability, functional features, and convenience or novel technologies of smart packaging mainly aimed to prevent microbial growth and lipid oxidation. There are many antimicrobial and antioxidant agents that have potential in meat packaging including organic acids, bacteriocins, antibiotics, plant extracts, phenolics, and synthetic and natural antioxidants (Scetar et al. 2010, McMillin 2017, Shumann and Schmid 2018).

4.6.9.1 Dehydrated Meat Products

Dried meats or jerky products described in Chapter 18 are one of the most popular meat-based snacks in many places around the globe. This category of meat snack is stable at room temperature because is generally formulated with curing agents, salt and further subjected to dehydration aimed to diminish the *Aw* to levels in which putrefactive and pathogenic bacteria cannot grow ($Aw < 0.85$ and preferably around 0.75). However, lipid oxidation is the major cause of quality deterioration because rancidity lowers the flavor and overall acceptance of these products.

Gaikwad et al. (2020) tested novel polyisoprene-based UV-activated oxygen-scavenging film for the packaging of beef jerky. The polyisoprene was supplemented 5, 10, or 20% into LDPE to develop novel oxygen-scavenging films designed to be activated by UV light. The tensile strength and elongation break of the film manufactured with 20% polyisoprene was similar to the control whereas its thermal stability significantly improved. Furthermore, the barrier properties were higher compared to those of pure LDPE. The experimental films containing 20% polyisoprene had 16.72 mL O_2/g oxygen-scavenging capacity so beef jerky packed with this film had improved stability, color, and pH during storage. Furthermore, this experimental film was effective as BHT to maintain the oxidative stability of beef jerky and significantly decreased microbial counts after storage for 90 days due to the lower oxygen concentration in the headspace of the package.

4.6.9.2 Intermediate Moisture Meat Products

The main requirement for partially dehydrated meat packaging is the exclusion of water vapor and oxygen via vacuum packaging. Vacuum packaging involves enclosing meat products in flexible plastic containers to prevent moisture gain and exclude oxygen. The multilaminates used for vacuum packaging must have low moisture and gas permeabilities and be strong enough to maintain the vacuum during handling and distribution. Films are commonly constructed with PA-PET laminate or EVA copolymer PVC/PVdC copolymer laminate (Robertson 2012). The hot tack or seal strength is crucial for vacuum packaged food items because if vapor water permeates into the package it will gradually change the *Aw* and negatively affect microbial growth and rate of lipid oxidation. Most dehydrated or semi-dried meats are prone to mold growth because they are one of the few microorganisms that can grow at *Aws* lower than 0.8 and relatively low pH´s (chapter 18).

Cured meats also require vacuum packaging in order to keep the nitrosomyoglobin that imparts the desired pink coloration. During storage cured meats deteriorate in the first instance due to discoloration, secondly due to rancidity, and thirdly due to nitrate-resistant microorganisms. The pigment of cured meats named nitrosomyoglobin is stable in the absence of oxygen or under vacuum but its oxidation to metamyoglobin is very fast when oxygen is present. Ultraviolet light is another factor that exacerbates color deterioration of cured meats. The modifications of the film materials with coloring agents or ultraviolet light absorbers decrease light transmission rates and protect products against rancidity.

4.6.9.3 Pork Rinds

Korge and Laos (2019) studied the influence of different packaging materials and atmospheric conditions on the properties of pork rinds especially in terms of fat oxidation or rancidity. The properties of pork rinds packaged in four different packaging materials with and without nitrogen were stored for 120 days at 22°C, 60% RH, and absence of light. The PET/PE packaging material had lower barrier properties for the product in both atmospheric conditions compared with PP/metalized PP and PET/PET-metalized/PE, PP/metalized PP, and PET/PET metalized/PE flushed with nitrogen. The pork rinds packed in PET/PET metalized/PE, PP/metalized PP, and PP/metalized PP without N_2 started to develop detectable rancid off-flavors after 90 days. The packaging systems PET/PET-metalized/PE and PP/metalized PP with N_2 were found to be good methods for preventing rancidity and textural changes of pork rinds throughout 120 days of storage.

4.7 INDUSTRIAL SNACK PACKAGING EQUIPMENT AND TECHNOLOGIES

Packaging is one of the most expensive operations in the manufacturing of snacks and greatly affects shelf life and consumer acceptance. Needless to say, the types of synthetic or biodegradable packaging materials described before and the packaging lines affect profitability. Packaging production costs are affected by the selection and cost of the packaging materials, the equipment investment and cost of its maintenance, the floor space, labor necessary to operate this part of the process, and the waste of packaging films.

Packaging equipment used for snacks varies from the simplest and most inexpensive used for filling hand weighed predetermined portions into pre-made bags that are sealed with wire twists, glue, or heat to automated expensive packaging devices that use sequential weighing, fill the pouch formed from roll stock, and seal at the rate of several hundred per minute. The automation in the packaging of snack foods started in the decade from 1940 to 1950. Danil Woodman designed a semiautomatic system that pulled premade bags and clipped them beneath a series of funnels on a rotating turret where operators could easily fill predetermined amounts of product into the bags. The bags were detached from the funnels and then individually sealed with heat (Kuhr 2001).

One of the most relevant factors affecting the selection of the packaging technology is the nature of the food item, especially in terms of physicochemical features, and the number of packages produced per unit of time in full-scale production lines. It is also necessary to know the anticipated pack size or product variations and to plan for any changes that are likely to occur in the requirements of the marketplace (Matche 2006). Other factors that play a major role are the availability of space and utilities, safety requirements, and the level of skill of the operators involved in packaging operations (Matche 2006).

There are a large number of packaging options and machines used for snack foods. Undoubtedly the form-fill-seal (FFS) operation is the most extensively used for the packaging of savory snacks.

4.7.1 Cartoning

The semiautomatic and automatic vertical and horizontal cartoning machines are thoroughly described by Kuhr (2001) and Matche (2006). According to Matche (2006) the principle of the cartoning operation is to erect cartons from a flat condition, fill it with the product, and sealing, or closing. The machinery varies from simple hand-fed machines to automatic coupled with packing the cartons directly into cases for dispatch. The semi or fully automatic machines differ based on the filling of the preformed cartons. The carton style most commonly employed in snacks is produced from a blank that has been preglued into a tube or sleeve by the carton supplier who also dies cuts and prints the package. The sleeve style blanks are typically delivered to the user in corrugated cases also known as "end-load style" are subdivided into horizontal or vertical (Kuhr 2001). There are several factors to consider when selecting a cartoning system like variations in the product, post-packaging handling of the product, protective barriers requirement mainly in terms of barrier to moisture and oxygen, product handling, number of package sizes required, frequency of size changes, production rate, and labor requirement (Matche 2006)

4.7.2 Bag and Pouch Packaging

4.7.2.1 Premade Bags

Bag is one of the oldest forms of packaging and still the most popular way to pack snacks because it performs the fundamental functions of containment, protection, and communication at the lowest cost. The bag-manufacturing machines have two essential components: tube forming and bottoming equipment. Rolls of film or paper are formed into the tube. In case of multilaminated paper bags, spout, or adhesive are applied between the plies to hold them together. The material is then formed into the tube. The forming tube can be furnished with or without gussets and converts bags by closing one or both the ends of the tube (Matche 2006). Premade bags can be plain or unprinted to which labels can be affixed or they can be supplied preprinted. The main disadvantage of premade bags is the necessity of manual filling and handling (Kuhr 2001).

4.7.3 Form-Fill-Seal (FFS) Packaging Technologies

Today, most snacks are filled into pillow pouches that are not free-standing with highly automated and sophisticated FFS machines. These automatic machines are divided into two basic types: vertical and horizontal.

4.7.3.1 Vertical Form-Fill-Seal (VFFS) Packaging Technology

The VFFS is by far the most popular way to pack snacks. These pouches are formed simultaneously while wrapping the product. The VFFS consists of first making a pouch or bag from flexible material followed by inserting a preweighed amount of the snack food and closing the top of the pouch with heat. The VFFS machine is equipped with two sets of seal tooling. First, a vertical seal tooling seals the back of the bag in preparation for fitting the package material into the feed tube. The other tooling seals the top and bottom parts of the package. A knife or a hot wire cuts the bags at the same time of the closing-sealing operation. These machines capable of producing up to 200 bags per minute have sophisticated load cells, digital electronics and rely on gravity to deliver the product into the preformed containers (Kuhr 2001). Since gravity is the sole determinant of the rate of fill the product density and rate of fall determine the velocity of the equipment during the operation. The general form of the pillow pouch is two end seals and a back seal at the middle or along one side (Brody 2017).

VFFS machines usually employ two types of materials: thermoplastic and heat-sealable films. Most thermoplastic films are produced from polyethylenes and these require a special sealing technique. PET films must be heated under controlled conditions until the areas to be sealed to each other are melted and fused. Heat is applied to fuse or weld the materials followed by cooling that permits the seal to set. These sequential steps should be carefully controlled in order to minimize defects. The heat-sealable materials include paper, cellophane, metallic foils, and most coextruded laminated films. These materials do not melt at sealing temperatures: therefore, they require a special film for heat sealing. Figure 4.1 shows a diagram of the three-side seal pouch formed on vertical form, fill and seal packaging machine, and an illustration of a commercial equipment.

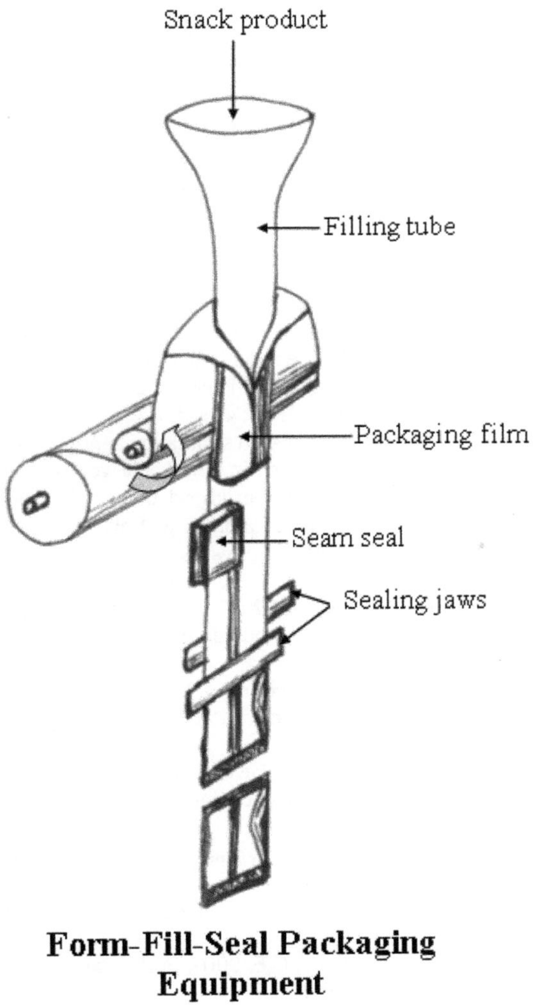

FIGURE 4.1 Illustration of the three-side seal pouch formed on vertical form-fill and seal packaging machine.

4.7.3.2 Horizontal Form-Fill-Seal (HFFS) Packaging Technology

HFFS are commonly employed to pack snacks consisting of individual items or items packaged by count—such as snack cakes, individually wrapped cookies, chocolate bars, ice lollies, and dried meat snacks. There are five types of HFFS which are able to continuously pack bags or pouches and thermoforms. The HFFS takes a web of material and forms it in a horizontal tube where the product is fed by conveyors. The long seal is made by passing the material through a heater and then through cold rolls. The motion is maintained by friction belt drives on the tube of material containing the product. The cross-seals are made by rotary crimping sealers, which also cut off the packages. The speed is mainly determined by how quickly the product can be fed and some models are capable of gas flushing (Kuhr 2001).

REFERENCES

Abong, G.O., Okoth, M.W., Imungi, J.K., and Kabira, J.N. 2011. Effect of packaging and storage temperature on the shelf life of crisps from four Kenyan potato cultivars. *American Journal of Food Technology*, 6(10): 882–892.

Ahn, B.J., Gaikwad, K.K., and Lee, Y.S. 2016. Characterization and properties of LDPE film with gallic-acid-based oxygen scavenging system useful as a functional packaging material. *Journal of Applied Polymer Science*, 133(43): 44138. DOI: 10.1002/APP.44138.

Ahvenainen, R. 2003. Active and intelligent packaging - An introduction. In: *Novel Food Packaging Techniques*, ed. R. Ahvenainen. Boca Raton: CRC Press.

Alves, R.M.V., Grigioli de Luca-Sarantopoulos, C.I., Fernandes Van Dender, A.G., and Fonseca Faria, J.de A. 1996. Stability of sliced mozzarella cheese in modified atmosphere packaging. *Journal of Food Protection*, 59(8): 838–844.

Anonymous. 2004. Monitor your cold chain. Lund, Sweden. Bioett. http://www.bioett.com/pdf/bioett.pdf, accessed 18 May 2004.

Arvanitoyannis, I., and Gorris, L.G.M. 1999. Edible and biodegradable polymeric materials for food packaging or coating. In: *Processing Foods - Quality, Optimization and Process Assessment*, eds. F.A.R. Oliveira, J.J. Oliveira, M.E. Hendrickx, D. Korr and L.G.M. Gorris. Boca Raton: CRC Press.

Azzi, A., Battini, D., Persona, A., and Sgarbossa, F. 2012. Packaging design: General framework and research agenda. *Packaging Technology and Science*, 25(8): 435–456.

Bhushan, N., and Gummaraju, K. 2004. A Petri net based simulation approach for evaluating benefits of time temperature indicator and wireless technologies in perishable goods retail management. Bangalore: Infosys Technologies Limited. http://www.infy.com/knowledge_capital/thoughtpapers/ENG-17.pdf.

Brody, A.L. 2017. *Flexible Packaging of Foods*. Boca Raton: CRC Press.

Brody, A.L., Strupinsky, E.P., and Kline, L.R. 2001. *Active Packaging for Food Applications*. Boca Raton: CRC Press.

Broeren, M.L., Kuling, L., Worrell, E., and Shen, L. 2017. Environmental impact assessment of six starch plastics focusing on wastewater-derived starch and additives. *Resources, Conservation and Recycling*, 127: 246–255.

Cerqueira, M.A., Lima, A.M., Souza, B.W.S., Teixeira, J.A., Moreira, R.A., and Vicente, A.A. 2009. Functional polysaccharides as edible coatings for cheese. *Journal of Agricultural and Food Chemistry*, 57(4): 1456–1462.

Chinthapalli, R., Skoczinski, P., Carus, M., Baltus, W., de Guzman, D., Kab, H., Raschka, A., and Ravenstijn, J. 2019. Bio-based building blocks and polymers – Global capacities, production and trends 2018 – 2023. *Industrial Biotechnology*, 15(4): 237–241.

Cichello, S.A. 2015. Oxygen absorbers in food preservation: A review. *Journal of Food Science and Technology*, 52(4): 1889–1895.

Collins, J. 2003. Active RFID labels for the masses. *Hauppauge, N.Y. RFID Journal*. http://www.rfidjournal.com/article/articleview/555/1/1/.

Da Cruz, A.G., Faria, J.D.A.F., and Van Dender, A.G.F. 2007. Packaging system and probiotic dairy foods. *Food Research International*, 40(8): 951–956.

Decareau, R.V. 1985. *Microwaves in the Food Processing Industry*. London: Academic Press.

de Vlieger, J.J. 2003. Green plastics for food packaging. In: *Novel Food Packaging Techniques*, ed. R. Ahvenainen. Boca Raton: CRC Press.

Di Pierro, P., Sorrentino, A., Mariniello, L., Giosafatto, C.V.L., and Porta, R. 2011. Chitosan/whey protein film as active coating to extend Ricotta cheese shelf-life. *LWT - Food Science and Technology*, 44(10): 2324–2327.

Ebnesajjad, S. 2013. *Plastic Films in Food Packaging: Materials, Technology and Applications*. Oxford: William Andrew.

Fang, Z., Zhao, Y., Warner, R.D., and Johnson, S.K. 2017. Active and intelligent packaging in meat industry. *Trends in Food Science and Technology*, 61: 60–71.

Fu, X., Xing, S., Xiong, H., Min, H., Zhu, X., He, J., Feng, J., and Mu, H. 2018. Effects of packaging materials on storage quality of peanut kernels. *PLOS ONE*, 13(3): e0190377.

Gadhave, R.V., Das, A., Mahanwar, P.A., and Gadekar, P.T. 2018. Starch based bio-plastics: The future of sustainable packaging. *Open Journal of Polymer Chemistry*, 8(2): 21–33.

Gaikwad, K.K., Singh, S., Shin, J., and Lee, Y.S. 2020. Novel polyisoprene based UV-activated oxygen scavenging films and their applications in packaging of beef jerky. *LWT- Food Science and Technology*, 117: 108643.

Giaglis, G.M., Minis, I., Tatarakis, A., and Zeimpekis, V. 2003. Real time decision support systems in urban distribution: Opportunities afforded by mobile and wireless technologies. In: *Conference Proceedings of 2nd International ECR Research Symposium—Consumer-Driven Electronic Transformation: Applying New Technologies to Enthuse Consumers*, 1–8. St. Gallen: ECR Europe Academic Partnership.

Gomez-Estaca, J., Lopez-de-Dicastillo, C., Hernandez-Muñoz, P., Catala, R., and Gavara, R. 2014. Advances in antioxidant active food packaging. *Trends in Food Science and Technology*, 35(1): 42–51.

Grace, R. 2016. The millennial impact on future packaging design. As a generation with broad life goals, Millennials are influencing the shape and style of packaging. *Plastics Engineering*, 72(6): 8–10.

Gvozdenovic, J.J., Aljilji, A.R., Lazic, V.L., Tepic, A.N., and Svrzic, G.V. 2007. Influence of protective characteristics of packaging material on packed dried fruits. *Acta Periodica Technologica*, 38(38): 21–28.

Han, J.W., Ruiz-Garcia, L., Qian, J.P., and Yang, X.T. 2018. Food packaging: A comprehensive review and future trends. *Comprehensive Reviews in Food Science and Food Safety*, 17(4): 860–877.

Holaday, C.E., Pearson, J.L., and Slay, W.O. 1979. New packaging method for peanuts and pecans. *Journal of Food Science*, 44(5): 1530–1533.

Holm, V.K., Mortensen, G., and Risbo, J. 2006. Quality changes in semi-hard cheese packaged in a poly (lactic acid) material. *Food Chemistry*, 97(3): 401–410.

Jensen, P.N., Sorensen, G., Brockhoff, P., and Bertelsen, G. 2003. Investigation of packaging systems for shelled walnuts based on oxygen absorbers. *Journal of Agricultural and Food Chemistry*, 51(17): 4941–4947.

Kalka, S., Huber, T., Steinberg, J., Baronian, K., Mussig, J., and Staiger, M.P. 2014. Biodegradability of all-cellulose composite laminates. *Composites – Part A: Applied Science and Manufacturing*, 59: 37–44.

Katz, F., and Labuza, T.P. 1981. Effect of water activity on the sensory crispness and mechanical deformation of snack food products. *Journal of Food Science*, 46(2): 403–409.

Korge, K., and Laos, K. 2019. The influence of different packaging materials and atmospheric conditions on the properties of pork rinds. *Journal of Applied Packaging Research*, 11(3): 1.

Kuhr, C. 2001. Snack foods filling and packaging. In: *Snack Foods Processing*, eds. E.W. Lusas and L.W. Rooney, 567–606. Lancaster: Technomic Publishing Co.

Labuza, T.P., and Contreras-Medellin, R. 1981. Prediction of moisture protection requirements for foods. *Cereal Foods World*, 26: 335–343.

Leufven, A., Sedaghat, N., and Habibi, M.B. 2010. Influence of different packaging systems on stability of raw dried pistachio nuts at various conditions. *American-Eurasian Journal of Agricultural and Environmental Science*, 8(5): 576–581.

Lusas, E.W. 2001. The snack food setting. Overview. In: *Snack Foods Processing*, eds. E.W. Lusas and L.W. Rooney, 1–29. Lancaster: Technomic Publishing Co.

Matche, R.S. 2006. Packaging machinery. In: *Handbook of Frozen Food Processing and Packaging*, ed. D.W. Sun, 679–711. Boca Raton: Taylor & Francis Group.

Mate, J.I., Saltveit, M.E., and Krocha, J.M. 1996. Peanut and walnut rancidity: Effect of oxygen concentration and relative humidity. *Journal of Food Science*, 61(2): 465–472.

McMillin, K.W. 2017. Advancements in meat packaging. *Meat Science*, 132: 153–162.

Mexis, S.F., and Kontominas, M.G. 2010. Effect of oxygen absorber, nitrogen flushing, packaging material oxygen transmission rate and storage conditions on quality retention of raw whole unpeeled almond kernels (*Prunus dulcis*). *Journal of the Science of Food and Agriculture*, 91(4): 634–649.

Miller, C.W., Nguyen, M.H., Rooney, M., and Kailasapathy, K. 2002. The influence of packaging materials on the dissolved oxygen content in probiotic yogurt. *Packaging Technology and Science*, 15(3): 133–138.

Miller, C.W., Nguyen, M.H., Rooney, M., and Kailasapathy, K. 2003. The control of dissolved oxygen content in probiotic yogurts by alternative packaging materials. *Packaging Technology and Science*, 16(2): 61–67.

Min, S.C., Kim, Y.T., and Han, J.H. 2010. Packaging and the shelf life of cereals and snack foods. In: *Food Packaging and Shelf Life. A Practical Guide*, ed. G.L. Robertson, 340–352. Boca Raton: CRC Press.

Mirza, A., Masoomian, A., Shakooie, M., Zabihzadeh Khajavi, M., and Farhoodi, M. 2020. Trends and applications of intelligent packaging in dairy products: A review. *Critical Reviews in Food Science and Nutrition*, 14: 1–15. DOI: 10.1080/10408398.2020.1817847.

Moreira, M.D.R., Pereda, M., Marcovich, N.E., and Roura, S.I. 2011. Antimicrobial effectiveness of bioactive packaging materials from edible chitosan and casein polymers: Assessment on carrot, cheese and salami. *Journal of Food Science*, 76(1): M54–M63.

Moreira, R.G., Castell Perez, M.E., and Barrufet, M.A. 1999. Packaging fried snacks. In *Deep Fat Frying. Fundamentals and Applications*, eds. M.D.R. Moreira, M. Pereda and N.E. Marcovich. Gaithersburgh: Aspen Publishers, Inc.

Nitzan, Z. 1999. Flexible thin layer open electrochemical cell and applications of same. United States Patent Office. US 5,897,522A.

Padmashree, A., Sharma, G.K., and Govindaraj, T. 2013. Development and evaluation of shelf stability of flaxoat nutty bar in different packaging materials. *Food and Nutrition Sciences*, 4(5): 31236.

Pawase, P.A., Veer, S.J., and Chavan, U.D. 2019. Studies on effect of different packaging materials on shelf life of mix fruit bar. *International Journal of Food Science and Nutrition*, 4(5): 156–162.

Peelman, N., Ragaert, P., Verguldt, E., Devlieghere, F., and De Meulenaer, B. 2016. Applicability of biobased packaging materials for long shelf-life food products. *Packaging Research*, 1(1): 7–20.

Pettersen, M.K., Eie, T., and Nilsson, A. 2005. Oxidative stability of cream cheese stored in thermoformed trays as affected by packaging material, drawing depth and light. *International Dairy Journal*, 15(4): 355–362.

Qiu, L., Zhang, M., Tang, J., Adhikari, B., and Cao, P. 2019. Innovative technologies for producing and preserving intermediate moisture foods: A review. *Food Research International*, 116: 90–102.

Raju, P.N., and Singh, A.K. 2016. Packaging of fermented milks and dairy products. In: *Fermented Milk and Dairy Products*, ed. A.K. Puniya, 637–671. Boca Raton: CRC Press.

Robertson, G.L. 2012. *Food Packaging - Principles and Practice*. 3rd edition. Boca Raton: CRC Press.

Romani, S., Tappi, S., Balestra, F., Rodriguez Estrada, M.T., Siracusa, V., Rocculi, P., and Dalla Rosa, M. 2015. Effect of different new packaging materials on biscuit quality during accelerated storage. *Journal of the Science of Food and Agriculture*, 95(8): 1736–1746.

Sabikhi, L., Khetra, Y., and Raju, P.N. 2018. Processing and packaging of dairy-based products. In: *Food Process Engineering and Quality Assurance*, eds. C.O. Mohan, E. Carvajal-Millan, C.N. Ravishankar and A.K. Haghi, 311–375. Boca Raton: CRC Press, Taylor and Francis Group.

Sanches-Silva, A., Costa, D., Albuquerque, T.G., Buonocore, G.G., Ramos, F., Castilho, M.C., Machado, A.V., and Costa, H.S. 2014. Trends in the use of natural antioxidants in active food packaging: A review. *Food Additives and Contaminants*, 31(3): 374–395.

Scetar, M., Kurek, M., and Galic, K. 2010. Trends in meat and meat products packaging–A review. *Croatian Journal of Food Science and Technology*, 2(1): 32–48.

Schumann, B., and Schmid, M. 2018. Packaging concepts for fresh and processed meat–Recent progresses. *Innovative Food Science and Emerging Technologies*, 47: 88–100.

Senhofa, S., Straumite, E., Sabovics, M., Klava, D., Galoburda, R., and Rakcejeva, T. 2015. The effect of packaging type on quality of cereal muesli during storage. *Agronomy Research*, 13(4): 1064–1073.

Sergio, L., Spremulli, L., Gatto, M.A., Pieralice, M., Linsalata, V., La Sala, G., and Di Venere, D. 2015. Postharvest performance of intermediate moisture peaches and prunes as affected by packaging and storage conditions. *Acta Horticulturae*, 1071(1071): 739–746.

Serna-Saldivar, S.O. 2008. Packaging of snack foods. In: *Industrial Manufacture of Snack Foods*, ed. S.O. Serna-Saldivar, 353–370. London: Kennedys Publications Ltd.

Shen, L., Worrell, E., and Patel, M. 2010. Present and future development in plastics from biomass. *Biofuels, Bioproducts and Biorefining: Innovation for a Sustainable Economy*, 4(1): 25–40.

Takenaga, F., Itoh, S., and Tsuyuki, H. 1987. Prevention of lipid oxidation in roasted and ground soybean with oxygen absorber during storage. *Journal of Japanese Society of Food Science and Technology*, 34(11): 705–713.

Talwalkar, A., Miller, C.W., Kailasapathy, K., and Nguyen, M.H. 2004. Effect of packaging materials and dissolved oxygen on the survival of probiotic bacteria in yogurt. *International Journal of Food Science and Technology*, 39(6): 605–611.

Ucherek, M. 2004. An integrated approach to factors affecting the shelf life of products in MAP. *Food Reviews International*, 20(3): 297–307.

Vieira, M.G.A., da Silva, M.A., dos Santos, L.O., and Beppu, M.M. 2011. Natural-based plasticizers and biopolymer films: A review. *European Polymer Journal*, 47(3): 254–263.

Wang, S., Lu, A., and Zhang, L. 2016. Recent advances in regenerated cellulose materials. *Progress in Polymer Science*, 53: 169–206.

Wilhelm, S., and Wolfbeis, O.S. 2011. Irreversible sensing of oxygen ingress. *Sensors and Actuators B: Chemical*, 153(1): 199–204.

Yam, K.L., Takhistov, P.T., and Miltz, J. 2005. Intelligent packaging: Concepts and applications. *Journal of Food Science*, 70(1): R1–R10.

Yildirim, S., Rocker, B., Pettersen, M.K., Nilsen-Nygaard, J., Ayhan, Z., Rutkaite, R., Radusin, T., Suminska, P., Marcos, B., and Coma, V. 2018. Active packaging applications for food. *Comprehensive Reviews in Food Science and Food Safety*, 17(1): 165–199.

Yu, L., Dean, K., and Li, L. 2006. Polymer blends and composites from renewable resources. *Progress in Polymer Science*, 31(6): 576–602.

Zabaleta, I., Negreira, N., Bizkarguenaga, E., Prieto, A., Covaci, A., and Zuloaga, O. 2017. Screening and identification of per-and polyfluoroalkyl substances in microwave popcorn bags. *Food Chemistry*, 230: 497–506.

5 3D-Printed Snacks

Viridiana Tejada-Ortigoza, Enrique Cuan-Urquizo, Luis Eduardo García-Amezquita, and Jorge Welti-Chanes

CONTENTS

5.1 INTRODUCTION

Additive manufacturing, or 3D printing, is a widely used fabrication technique with applications in several areas. Food is not an exception, taking advantages such as customization of food designs, personalized nutrition, simplifying the supply chain, and opening the possibility of using novel food materials (Liu et al. 2017). Customization of food design opens the possibility of creating attractive and novel 3D models with edible materials, an important criterion for consumers. Complex and artistic designs, that cannot be made manually or by molding, have been successfully printed, for example, by the confectionery industry or by chefs in restaurants (Sun et al. 2015). Additionally, the possibility of personalizing the nutrition requirements for individuals or specific population groups is impacting the food sector. According to their physical activities, ages, diseases, or any specific nutrimental requirement, 3D printing of foods has moved toward fulfilling the necessities of the elderly, children, or even space missions, just to name some (Baiano 2020, Derossi et al. 2018, Jędrusek-Golińska et al. 2020, Liu et al. 2017, Terfansky and Thangavelu 2013). This novel technology has triggered the development of conscious and healthier diets and functional foods.

3D printing also broadens the use of nonconventional food ingredients with high protein, fiber, or nutraceutical contents such as insects, fruit and vegetable by-products, animal by-products, among others (Severini and Derossi 2016, Liu et al. 2017). Besides these advantages and regarding the complex food design that can be achieved, 3D printing allows having full control of the internal structures or distribution of material, resulting in differentiated texture. The possibility of varying the internal structure of the printed food impacts not only the texture but also the control of the nutritional profile (Godoi et al. 2019). Besides, it might have sustainable implications due to the reduction of wastage and the usage of packaging materials. With respect to packaging, 3D printing has also been used for prototyping and producing biodegradable materials derived from agri-food by-products (Baiano 2020).

Recent research related to 3D printing of foods is vast. When looking for "3D AND printing AND *food*" in the Scopus database, one can find a total of 532 documents related to the keywords.

DOI: 10.1201/9781003129066-5

Since 2018, around 70% of the documents were published emphasizing the importance and the tendency of this novel technique. Although the trend is clear, and 3D printing of food can easily be adopted to the manufacturing of snacks, only a few works can be found in this field. When repeating the search but using the words "3D AND printing AND *snack*" only 13 documents from 2016 to date are found and not all of them refer to original research for snack production. There is an important gap in the use of this technology to produce snacks, especially considering the importance of personalized nutrition. The possibility of printing personalized snacks at home is not far from now because of the reduction of printers' costs, the development of easy-to-handle software, and the facility to operate printers by nonexperienced users. Nevertheless, there is still a lot of research to perform related to materials, rheology, correlation with printability, and postprocessing treatments.

This chapter starts with the discussion of personalized nutrition and its growing relevance in the food industry. Then, the 3D printing fabrication procedures are described including techniques currently used to print foods and their possible applications for the manufacturing of novel snacks. A description of the materials used for this purpose is also included, specifying the ones currently used to print snacks and the possibilities of broadening material sources for development of snacks. The printability and its relationship with the rheology of the materials are described. Finally, insights into the future perspectives of this manufacturing technique in the food area are discussed.

5.2 FUTURE CONTRIBUTIONS OF 3D SNACKS TO PERSONALIZED NUTRITION

Nowadays, health has become an important value for individuals and the society, and the prevention of chronic or noncommunicable diseases has become a priority. Consequently, research has focused on the study of the effect of genetic variants on the nutrients' tolerance (nutrigenetics) and the effect of food consumption on gene expression (nutrigenomics). The study of nutrigenetics and nutrigenomics established a clear relationship between food and health and strongly supported the development of the concept of functional foods. As a result, the market of functional foods has been growing along with the production of foods prepared with nutritional and functional compounds that are known to prevent diseases (Severini and Derossi 2016).

In addition, there is an increased consumer interest and demand for personalized food products. Customization of food implies connecting self-knowledge to food choices (German and Watzke 2004). Personalized nutrition has become a need for some segments of the population, such as children, diabetics, pregnant women, athletes, the elderly, patients with difficulties in eating, autistic, among many others. It is based on specific nutritional needs and calorie intake to fulfill their specific needs and the goal would be to satisfy the needs, not only for a specific population but for an individual person (Pérez et al. 2019). The importance of personalized nutrition is that the metabolic response to specific foods is individual, and alters the microbiota composition which plays an important role in the development of obesity and chronic diseases (Christensen et al. 2018). This interest has been reflected in the advent taken by research trends in the latest years. According to Scopus, the number of documents with the words "personalized nutrition" has increased almost exponentially since 2004, with a total of 3,126 documents up to date. From 2016 to 2020, more than 2,000 papers in this field have been published.

Recently, 3D printing has been proposed as a technology to approach the ambitions of personalized nutrition by using functional ingredients. This technology allows the use of materials that can fit the dietary needs and sensorial preferences of people of different ages, sex, lifestyle, allergies, among others, by changing the composition, structure, flavor, of the printed materials according to their selection (Dankar et al. 2018). The goal to achieve is that food printers can integrate nutritional models so that users can control their diet and caloric intake by selecting ingredients and the printing parameters (Baiano 2020). This would allow the consumer to customize the amount of protein, sugar, vitamins, minerals, dietary fiber, etc.

Some 3D printing projects are based on fulfilling the food necessities of particular groups. For example, Derossi et al. (2018) printed a fruit-based snack aimed to satisfy children's nutritional intake and to furnish a novel and attractive visual aspects highly appreciated by these consumers. The formulation was nutritionally tailored for 3–10 years old children in order to meet their energy, vitamin D, iron, and calcium requirements. More recently, the potential of 3D printing to meet individual needs and preferences was tested in an exploratory consumer study in a military setting (Caulier et al. 2020). In this study, the participants' attitudes toward 3D printing to personalize a snack selection were evaluated. The authors concluded that a desired degree of customization, consumers' empowerment, and sensory profile optimization influenced the acceptance of a 3D printed product.

Regarding customization, 3D printing also opens the possibility of creating novel food structures and the use of nonconventional food ingredients. The technology allows the printing of patterns, shapes, and structures that impact the texture and therefore the sensorial experience (Dankar et al. 2018). With this, the consumer could also be able to control not only the aesthetics of the food to be eaten but also the biting and chewing experience. The possibility of controlling the infill patterns impacts not only the texture but also the amount of material used and the uptake of oil when frying and therefore the calorie intake (Liu et al. 2020). Moreover, 3D printing grants the use of food ingredients that are not commonly used in snack formulations, such as algae, insects, fungi, among many others (Dankar et al. 2018, Severini et al. 2018).

In the near future, this technology will allow the production of snacks to satisfy specific individual needs by controlling the printed material, its amount, its distribution, its nutritional content, and even its sensorial (textural) properties. 3D printed foods will offer meals adapted to individual diets, allow the use of new and nonconventional ingredients, and customize sensorial and functional properties of foods (Baiano 2020).

5.3 FABRICATION PROCEDURES

The implementation of 3D food printing is based on the procedure summarized in Figure 5.1. The material, the technique and design used, the printing parameters, and postprocessing handling are the main and most relevant process components. Regarding *formulation* with a broad possibility of

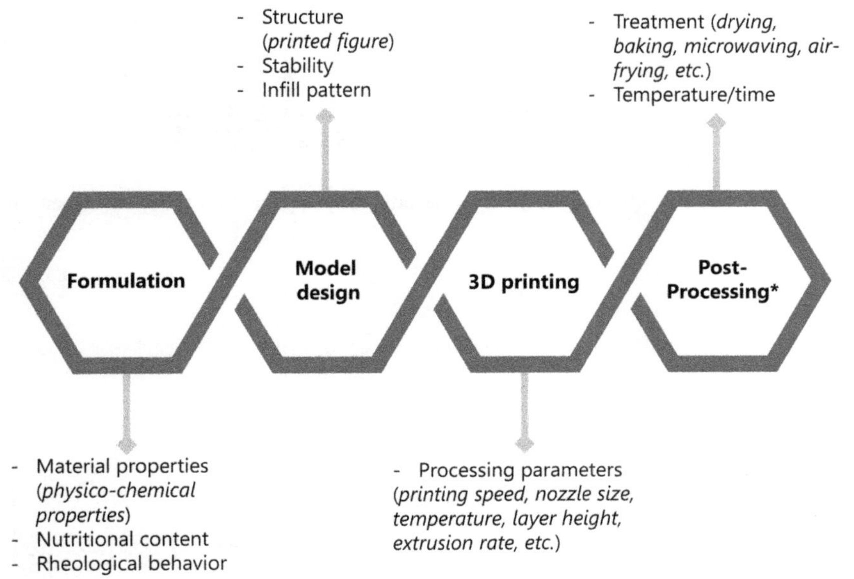

FIGURE 5.1 Stages of the food 3D printing process and their correspondent parameters. *Postprocessing treatment is not always required; it depends on the food material used for printing.

materials, the chief parameters to consider are the nutritional content, the interaction among materials, and especially the rheological behavior that will affect the printability of the blend. This will be deeply discussed in Sections 5.4 and 5.5. The importance of the *model design* relies on the structure and stability that the printed snack will have. The infill pattern is also relevant for the postprocessing and the desired texture characteristics. The *printing parameters* depend on the properties of raw materials, the printing technique, and the sort of printer (Sections 5.3.1 and 5.3.2). Finally, *postprocessing* is not always needed and depends on the printed food materials and the desired end-product characteristics (section 5.3.3).

Printing techniques commonly used in other additive manufacturing industries have been applied for foods. The most widely used in the food industry is extrusion, due to the range of materials that can be potentially used and its versatility. Ink powder-based (binder jetting and selective sintering) techniques are less common but also used to print food materials. They consist of a powder material bed constantly refilled and are based on the chemical and/or physical interaction between the powder and different agents. These techniques are depicted in Figures 5.2 and 5.3 and described as follows.

5.3.1 Food 3D Printing Techniques

5.3.1.1 Extrusion-Based

This technique works with a system composed of a vertical extruder on a three degrees-of-freedom mechanism that moves in the 3D space staying perpendicular to the base. After the material of interest is loaded and pushed out of the nozzle in a controlled way, it is deposited layer by layer to form a 3D solid structure (Sun et al. 2018, Le-Bail et al. 2020). The material can be extruded by three different pneumatic and/or mechanical methods: screw-based extrusion, syringe-based extrusion, and

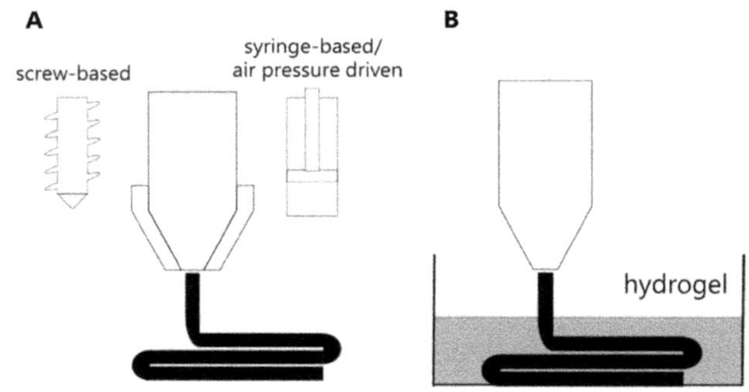

FIGURE 5.2 Extrusion-based printing. (A) Screw-based and syringe and air pressure-driven extruders, (B) hydrogel-forming extruder.

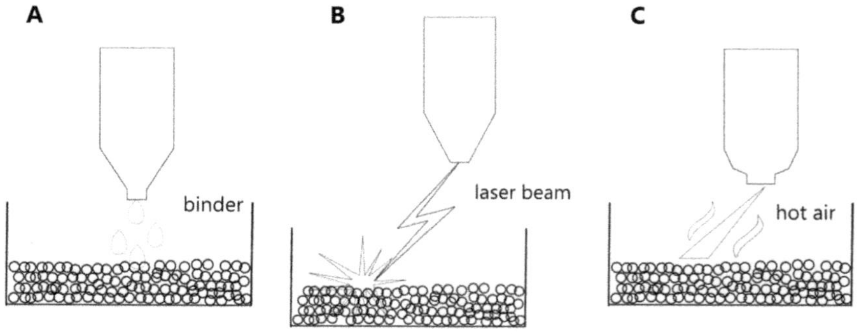

FIGURE 5.3 Powder-based printing. (A) Binder jetting, (B) laser sintering, and (C) hot air sintering.

air pressure-driven extrusion (Sun et al. 2018, Godoi et al. 2019) (Figure 5.2A). Pneumatic arrays commonly provide more control over the flow of viscous materials, while mechanical arrays help in the mixing of materials with higher viscosities (Godoi et al. 2019).

In addition, extrusion can be achieved with (melting) or without (non-phase change) high temperature depending on the material and its properties. Without the use of high temperatures, the extrusion process commonly depends on the rheological properties of the materials. On the counterpart, when a hot extrusion process is used, the material is at first in a solid state and then heated up to a semi-molten state to be extruded; once it's deposited, the material must solidify to assure its vertical assembly or bonding (Godoi et al. 2019, Le-Bail et al. 2020).

Hydrogel-forming extrusion depicted in Figure 5.2B is also classified by some authors as part of this group, which can also be carried out with or without using high temperatures, depending on the material. The gelation mechanisms can be induced by thermal treatment, chemical cross-linking, ionotropic cross-linking, coacervate formation, and enzymatic cross-linking (Godoi et al. 2019). Several materials used in this technique are hydrocolloids in solutions or dispersions which are loaded in a syringe pipette or similar device and printed into a settled bath with the conditions to harden or gelatinize. It can also be referred to as inkjet printing, where 2D and 3D patterns are deposited as liquid on a substrate (Sun et al. 2018, Godoi et al. 2019).

Regardless of the extrusion technique, rheology is one of the main parameters to be considered for extrusion-based printing. First, the material must have the capability to be extruded through the nozzle or die. Then, the printed layers must have enough mechanical integrity to support the deposited structure avoiding collapse, and finally, the structure must be stable and well defined. *Printability* of materials will be thoroughly defined and explained later in Section 5.4. Several types of food materials such as dairy, cereals, vegetables, meats, and confectionaries have been successfully printed using this technology (Periard et al. 2007, Lipton et al. 2015, Severini et al. 2016, Kim et al. 2018, Le Tohic et al. 2018, Liu et al. 2018, Karyappa and Hashimoto 2019, Kim et al. 2019).

5.3.1.2 Binder Jetting

This technology uses a liquid agent known as *binder* to solidify the powder particles contained in a so-called *bed*. These machines have two mechanisms, one for controlling the dropping of the binder (Figure 5.3A) and one for positioning the binder in the printing plane. The binder is sprayed in the powder bed to bind two consecutive layers and for each layer, a layer of powder is also spread. While the cost of the equipment is high and a poor definition of the printed objects, some advantages of using this technology are a fast fabrication and low material cost (Holland et al. 2018, Le-Bail et al. 2020). Hydrocolloids and gel-type materials such as cellulose and xanthan gum have been used with this procedure to produce cohesive geometric food structures (Holland et al. 2018).

5.3.1.3 Selective Sintering

Selective sintering is performed using a laser or hot air as a sintering source (Figures 5.3B and 5.3C) to fuse or melt the powder particles and make them a solid layer. After solidifying the first powder layer, a new powder layer is distributed on the top and the procedure is repeated to build up the multilayered object. Some advantages of this technique are a fast construction of complex food structures in short times without needing postprocessing, however, the fabrication might be complicated due to the materials and the several printing parameters involved (Le-Bail et al. 2020).

Compared to extrusion-based printing, these techniques are less used for processing complex mixtures of food materials although they have been employed with single food ingredients like sugar. CandyFab printers, by Evil Mad Scientist Laboratories, are well-known imprinters that use laser sintering to build up complex and precise sugar structures as shown in Figure 5.4. In addition, laser sintering uses maltodextrin and starch as structural components. The binders glucose, gluten, maltodextrin, whey protein, and soy protein were recently patented (Villarama Díaz et al. 2018). Some authors have also reported the use of electrostatic inkjet printing using chocolate with sufficient accuracy to make artistic and complicated figures (Suzuki et al. 2019).

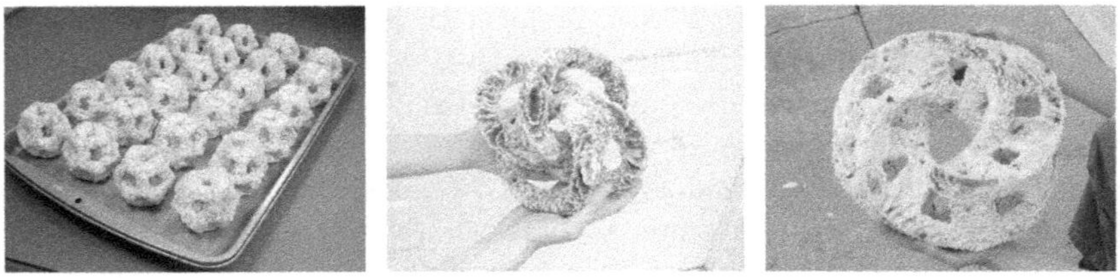

FIGURE 5.4 Pictures from the website: https://candyfab.org/. Courtesy of Windell H. Oskay and Lenore M. Edman www.evilmadscientist.com

5.3.2 3D Food Printers

Many printers have been successfully used for food processing although most of the reported studies are with extrusion-based printers using a mechanical actuator or air pressure to move the piston or syringe. In general, two types of food printers are found in the literature and the market: the adapted ones and the counterparts specifically fabricated to print foods. The first ones are common in scientific papers and were originally made to print plastic polymers (PLA- polylactic acid), or clay such as Delta WASP 2040, nScrypt, or Shinnove-D1 (Derossi et al. 2018, Lille et al. 2018). The modifications on these printers are mainly directed to the hardware, meaning the extrusion mechanisms of the printers so that the head used for fusing a plastic material is changed by a piston or syringe. Usually, these adapted printers assemble the food extruder on the three degrees-of-freedom mechanism that moves along the printer bed. Some advantages of these devices are that they can be used almost with any type of material by changing the printing parameters and that the extrusion process is not limited to a specific motor power. In addition, the printing inner structure (infill) is totally customizable, allowing the design of nonconventional patterns and textures (Figure 5.5). Some authors have even added an infrared lamp heating mechanism for simultaneous cooking of the printed material (Hertafeld et al. 2019), widening the possibilities. These printers, even after being adapted, might be considerably less expensive compared to commercial counterparts although knowledge especially related to mechanics, controlling, design, and food is needed. A complete report of the most relevant patents concerning 3D printed foods has been recently published by Baiano (2020). This author highlights that most related patents are Chinese.

On the other hand, the commercial food printers already found in the market (Figure 5.6) are furnished with a closed software, where the changes of printing parameters can become challenging. Moreover, most of these printers are extrusion based where not all types of materials can be printed, being an important limiting factor in the mechanical actuator system that pushes the food material from the extruder through the nozzle. Normally, the devices are not able to handle highly viscous materials. Also, they are programed to print figures that are already in the software, and the printing parameters are set for materials such as chocolate, potato puree, or similar paste-like foods with specific rheological properties. Furthermore, the infill cannot be controlled so the texture and mechanical properties cannot be regulated. The main advantage of these printers is their user-friendly interface that, without any knowledge related to rheology, allows 3D printing with practically any of the designs declared by the manufacturer. With an easy operation, one can also experience customization (at some level) using the several creative and attractive designs available on each printer software or on their webpages. Commercial printers have been mainly directed to commercial kitchens and chefs, bakeries, confectionaries, hotels, among others. In addition, their use has also been proposed at schools or hospitals for personalized nutrition with the aim of promoting their use later on at homes (Hamilton et al. 2017, Derossi et al. 2018, Jiang et al. 2019). Focus (byFlow), Mycusini (Procusini), Choc Creator V2.0 Plus (Choc Edge Ltd.), and 3D Food Printer (Createbot) are some examples of commercially available printers.

5.3.2.1 Printing Parameters

Once the printing mechanism has been set, the printing parameters are crucial in the process and they depend completely on the material and the printing form. The print speed, flow,

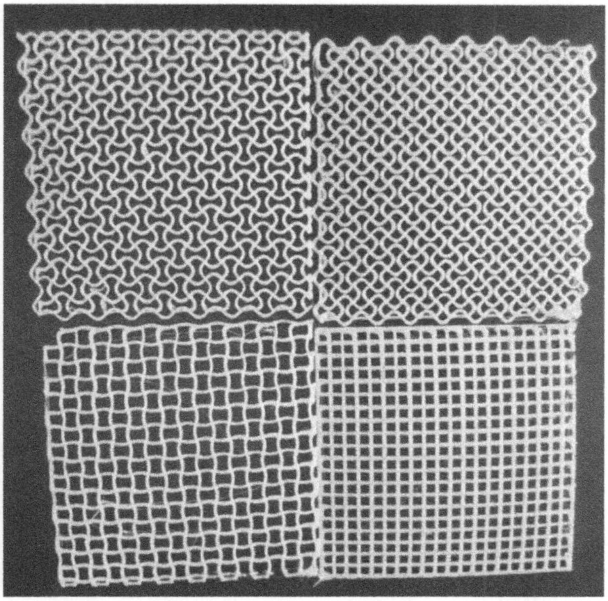

FIGURE 5.5 Infill patterns that would lead nonconventional texture in food printing.

FIGURE 5.6 Commercial 3D food printer. Foodini from Natural Machines®. (Picture taken from website: https://www.naturalmachines.com/store.)

extrusion rate, layer height, among many others, are some of the printing parameters that can be customized (Table 5.1). These parameters are commonly studied when an adapted printer is used, although in some commercial devices these parameters can be modified. Several studies have focused on the optimization of some of these printing parameters, based on the printer and on the material used. However, it is virtually impossible to generalize these parameters because many variables affect performance (the printer, the material, the figure, the infill, etc.). Not all the printing parameters can be changed and the recommendation as in every experimental procedure is to stay with few variables.

TABLE 5.1
Important Parameters Related to 3D Printing of Snacks

Parameter	Description
Print speed	Rate of the printer movements (mm/s)
Flow	Deposited material with respect to time (mm^3/s)
Extrusion rate	Speed at which the material is extruded (mm/s)
Layer height	Height of the deposited material (mm)
Nozzle size	Diameter of the extruder nozzle (mm)
Diameter of the filament	Diameter of the printed filament (mm)
Infill structure	Lattice or tessellated arrangement used to fill the 3D printed products
Infill density	Level of infill with respect to the total material that could have been printed in a layer (%)
Infill at the bottom or top layer	Level of infill at the first and last layer (%)
Printing temperature	Temperature of the extruder and nozzle (°C)
Bed temperature	Temperature where the material is deposited in the printer (°C)

Source: Adapted from Derossi et al. (2019).

5.3.3 Postprocessing Treatments

In some cases, the printed materials need to be postprocessed to yield edible foods. Postprocessing can cause relevant changes in the physicochemical characteristics, nutritional value, and the sensorial characteristics of the foods (Keerthana et al. 2020). Baking, drying, frying, to name some, are examples of postprocessing treatments. In general, these treatments require the handling of the printed material and its exposition to different temperatures inside and outside the matrix resulting in textural changes. Therefore, it is important to consider that the printed material must resist the manipulation and thermal treatment. Nevertheless, postprocessing has not been deeply studied and there are few investigations that include these treatments after the printing process.

Lille et al. (2018) compared freeze-drying and oven-drying after the 3D printing of healthy and customized snacks rich in protein and fiber. Freeze-drying preserved better the printed structure resulting in samples with higher values of hardness and reduced moisture content. On the other hand, oven-drying caused spread-ability of the samples when viscosity decreased due to an increase in the drying temperature. Recently, the texture of air-fried 3D printed yam snacks was evaluated (Feng et al. 2020). The authors printed various infill levels and structures to change the texture after air-frying. Postprocessing conditions ranged between 100°C and 160°C depending on the structure and the infill level. Once printed, a low infill level resulted in a low weight and highly porous material. After air-frying, the hardness of the final products decreased in relationship with the porosity. This finding opens the possibility of controlling the textural properties of printed materials.

Likewise, Keerthana et al. (2020) evaluated microwave cooking as a postprocessing treatment for the production of mushroom and fiber-enriched snacks. The authors optimized the parameters focusing on avoiding the shrinkage of the printed material while keeping positive sensorial features. The treatment resulted in a crispy texture without compromising the geometry of the printed figure. Similarly, other fiber-rich printed snacks based on legumes, cereals, and seeds were postprocessed with deep frying, hot-air drying, and microwave drying (Krishnaraj et al. 2019). The results showed that microwave drying enhanced the retention of key nutrients and minimized color and texture changes compared to the other frying schemes. There is still a lot of research to perform related to postprocessing of 3D printed materials especially in terms of development of novel sensorial experiences (texture, color, flavor) and nutritionally enhanced products.

5.4 PRINTING SNACKS: MATERIALS

One of the advantages of 3D printing technologies is the possibility of using a wide variety of materials from different sources. These materials are normally paste/gel-like or powder ingredients such as chocolate, fat, and dough (Feng et al. 2019). These materials are classified according to their rheological characteristics as natively printable and nonprintable (Table 5.2). The first ones can be easily extruded through the nozzle due to their viscoelastic properties whereas the second type of materials cannot be extruded through the nozzle in their raw state. Therefore, they often require additional postprocessing or the use of additives aimed to improve the flow behavior, sedimentation, and lubrication properties of the blends (Feng et al. 2019; Voon et al. 2019). Other classifications have also been proposed (Voon et al. 2019), such as traditional and nontraditional food materials

TABLE 5.2
Characteristics of Natively Printable and Non-printable Materials

Classification	Characteristics	Food Materials	References
Natively printable	• Materials easily extruded through a nozzle. • Generally, these materials have a high content of fat, denatured proteins, and carbohydrates (starch or soluble fibers) that give them self-support properties. • Commonly, gel and paste-type materials. • Stable to hold the shape once deposited or printed.	Chocolate	Lanaro et al. (2017), Karyappa and Hashimoto (2019), Mantihal et al. (2019), Mantihal et al. (2019), Mantihal et al. (2017, 2019), Suzuki et al. (2019)
		Dough	Hyun Woo Kim et al. (2019), Liu et al. (2019), Pulatsu et al. (2020)
		Dairy	Kern et al. (2018), Le Tohic et al. (2018), Liu et al. (2018), Liu et al. (2019)
		Hydrogels	Lipton et al. (2010), Vancauwenberghe et al. (2017), Vancauwenberghe et al. (2018), Gholamipour-Shirazi et al. (2019), García-Segovia et al. (2020)
Nonnatively printable	• Require the use of additives to improve their printability for extrusion through the nozzle. • With low-fat content but it might contain protein in its natural state (non-denatured). • Not gel or paste-type materials.	Fruits and vegetables	Azam et al. (2018), Derossi et al. (2018), Yang et al. (2018), Kim et al. (2018), Severini et al. (2018), Lee et al. (2019), Yang et al. (2019)
		Insect protein	Lupton and Turner (2018), Severini et al. (2018)
		Meat	Liu et al. (2018), Wang et al. (2018), Dick et al. (2019), Gudjónsdóttir et al. (2019)

(Feng et al. 2019) or according to the nature of the chemical component as carbohydrates, fats, or proteins (Jiang et al. 2019, Pérez et al. 2019).

Printable and nonprintable materials have been used for snack printing. Table 5.3 presents information on these materials, postprocessing treatment conditions, and the parameters evaluated from recent studies carried out aimed to obtain printed snacks. For instance, a healthy snack for children containing a mixture of fruits and vegetables (banana, canned beans, fat milk, lemon juice, and dried mushrooms) was successfully printed by Derossi et al. (2018). For this development, the authors employed a pectin solution (30% of the composition) to ease the printing properties of the formulation. Some authors have also printed formulations composed of carrots, pears, kiwi fruit, broccoli raab leaves, and avocado (Severini et al. 2018). A reduction of moisture content was needed in order to increase the viscosity and facilitate its printing by extrusion. Azam et al. (2018) compared the effect of using gum arabic, carrageenan, and xanthan gums in an orange concentrate and wheat starch blend as a prospective formulation for healthy snacks.

TABLE 5.3
Materials Commonly Used for the Production of 3D Snack Printing

Materials	Postprocessing	Evaluated Parameters	Reference
Wheat flour, olive oil, sodium chloride	Baking (150°C, 18 min)	Textural properties, microstructural analysis, and relative density	Derossi et al. (2020)
Potato flakes, wheat flour, bovine fat	Air-frying (140°C, 12–21 min)	Rheological properties, textural properties, oil content	Liu et al. (2020)
Cookie dough	Air-frying (170°C, 7–10 min), steamer (100°C, 10 min)	Consumers' acceptance, sensory evaluation	Caulier et al. (2020)
Mushrooms, wheat flour	Microwaving (450–900 W, 2–12 min)	Rheological properties, textural properties, color, water activity, shrinkage, sensory analysis	Keerthana et al. (2020)
Yam, potato	Air-frying (100–160°C, time depended on the infill level)	Rheological properties, textural properties, bending height	Feng et al. (2020)
Green gram, fried gram, barnyard millet, ajwain seeds	Deep frying (165°C), hot air drying followed by deep frying (100°C, 2 min followed by 165°C), and microwave drying (280–1,400 W, 60 s)	Rheological properties, textural properties, color, sensory analysis	Krishnaraj et al. (2019)
Carrot, pear, potato, kiwi fruit, broccoli raab leaves, avocado	Not required	Shelf-life, microstructure, antioxidant capacity and total phenolic content, microbiological analysis, sensory evaluation	Severini et al. (2018)
Banana, mushroom, canned white beans, dried nonfat milk, lemon juice, pectin	Not required	Rheological properties, microstructure, behavior after printing	Derossi et al. (2018)
Orange fruit concentrate, wheat starch, gums	Not required	Rheological properties, textural properties, FT-IR analysis, dimensional stability, microstructure	Azam et al. (2018)
Wheat flour	Baking (200°C, 15 min)	Microstructure, textural properties, appearance	Severini et al. (2016)

A printed snack produced from wheat flour dough enriched with edible insects was developed by Severini et al. (2018). The authors evaluated the quality and the nutritional attributes of the printed snacks after baking at 180–220°C for 14–22 min. Recently, yam powder rich in starch with appropriate viscoamylograph/gelling properties was successfully used to 3D print a snack that was air-fried after processing (Feng et al. 2020). Likewise, a potato-based snack with reduced oil was successfully produced (Liu et al. 2020). These authors evaluated variations on the internal structures of the printed figures and their effect on the oil uptake.

In contrast to Table 5.2, Table 5.3 shows that there are more possible raw materials (salty and sweeteners) that can be applied in the development of new snacks. More research is needed, to formulate healthy snacks using nonconventional materials such as insects' protein and setting the most desirable postprocessing treatments, keeping in mind the importance of personalized nutrition and the possibilities that will be open in the near future especially in terms of usage at home.

5.5 PRINTABILITY OF SNACKS: RHEOLOGY

A good understanding of materials' rheological properties is needed to improve the printability of foods (Liu et al. 2019). The main material features to be considered and studied to assure proper printability are: (i) materials should be easily extruded through a nozzle and then, once extruded, and (ii) materials should possess the mechanical strength to "self-support" the layer-by-layer deposition retaining the printed structure while minimizing the deformation. Pseudoplastic fluids with suitable shear-thinning behavior are the ones that, under an appropriate shear force, are easily extruded. Rheological properties and even some phase transition (melting point, glass transition) are crucial on this "self-supporting" ability (Godoi et al. 2016, Feng et al. 2019, Liu et al. 2019).

There are two strategies that have been devised to achieve the rheological behavior for 3D printing. The first is the preparation of food inks with low viscosity that can be easily extruded. These materials must pass through a gelation process once printed and additionally should possess high mechanical strength in order to resist deformation. The second strategy is the preparation of food materials with desired rheological properties such as shear-thinning behavior to facilitate the extrusion and after deposition with sufficient strength to maintain the printed structure (Liu et al. 2019). However, since printability is affected by several other factors like temperature, materials, and type of printer, these parameters should be carefully controlled so as to obtain homogenous and good quality products.

Almost all the studies performed for the production of 3D extruded foods report the rheological properties of the materials. Nevertheless, only a few of them calculated correlations of printability with the rheological behavior of the material. This information is crucial in order to improve printing processes and to generalize these properties for broader applications with other materials. Several authors have established assumptions related to rheological parameters and printing behavior, although only few have demonstrated it quantitatively. For example, a low value of yield stress is not recommended because the extruded material does not attain the mechanical strength to support the deposited layers causing deformation and poor resolution. Also, these materials are easily extruded when a low flow behavior index is shown (Liu et al. 2018). According to Jiang et al. (2019), rheological properties such as yield stress, storage modulus, loss modulus, flow behavior, and consistency indexes are effective to predict the effect of extrusion through the nozzle. Particularly, the yield stress and the storage modulus are related to the self-support ability of the matrix whereas the flow behavior and consistency indexes to extrudability and printability. These and some other rheological parameters correlated with extrusion behavior and printability as depicted in Table 5.4.

A first approach to establish the printability of foods by using hydrocolloids as referenced was reported by Kim et al. (2017). These authors evaluated printability after assessing dimensional stability, handling properties, and rheological properties. Interestingly, the shear modulus of samples corresponded to the simulations based on dimensional stability tests so the authors concluded that the food printability can be predicted and classified through a comparison to reference material.

TABLE 5.4
Rheological Parameters Correlated with Extrusion Behavior and 3D Printability

Parameter	Correlation with 3D Printing
Consistency index (K)	Low values are related to easily extruded materials through the nozzle.
Flow behavior index (n)	Low values indicate high shear-thinning properties. They can be easily extruded out of a nozzle when an increased shear stress is applied.
Loss modulus (G")	*G'* and *G"* values indicate the ability of the matrix to support itself once printed. These parameters are related to "self-supporting" structure. A strong frequency dependence might indicate a material that behaves like a solid at higher frequencies and like a liquid at lower frequencies.
Shear modulus	It might be used to quantify the printability. It also predicts and quantifies the deformation behavior after the printing process, so it is used to predict the shape deformation.
Storage modulus (G')	In combination with the yield stress, it can be used to predict the shape retaining of a printed material and a good resolution (printing fidelity).
Tan (δ) = G"/ G'	High values indicate a fluid-like behavior and low values of a solid-like behavior. If $G' \geq G''$, then it can be observed as resistance against collapse and a better holding shape after printing.
Viscosity	High viscosity materials are easy to stick on the extruder walls and block the nozzle output. Thus, an inaccurate production of the final shape of the product is obtained.
Yield stress	It is the minimum force required for material to be extruded and it is related to self-support. When this value is passed, the structure of the material collapses and flows because the internal structure cannot hold the pressure and store the energy.

Source: From Jiang et al. (2019), Kim et al. (2018), Liu et al. (2018), Pérez et al. (2019).

Even though this research categorized different grades of printability, the use of this system has not been extended to other works. Kim et al. (2018) valuated the effect of hydrocolloids on the rheological properties and printability of vegetable materials by measuring the shear modulus and extrusion force with a rheometer and a texture analyzer, respectively. Hydrocolloids improved the printability of the vegetable-based foods, particularly when combined with xanthan gum. This gum reduced differences in printability of several vegetable materials (carrot, broccoli, and spinach) so it is recommended to standardize recipes for extrusion-based printing. Moreover, xanthan gum can suppress the swelling capacity of the incorporated particles, improving the extrudability of novel snack food mixtures. Liu et al. (2019) devised a gel system based on carrageenan-xanthan-starch to improve rheology and printability. The authors divided the printing process into three stages (extrusion, recovery, and self-supporting) and evaluated each phase in terms of rheological properties (yield stress, viscosity, shear-thinning behavior, shear and temperature recovery properties, and complex modulus). Yield stress and shear-thinning behaviors were the parameters most closely related to the material extrusion behavior. In this work, printability was evaluated by means of the consistency of the printed lines, the average area of the interconnected channels, and the wall thickness of a hollow cylinder. Similar findings were obtained using lemon juice gels (Yang et al. 2018, 2019).

Zhu et al. (2019) found out quantitative correlations of printability with some rheological parameters like the stress at collapse with the flow stress, the shear viscosity, and the storage modulus. The authors defined stress at collapse as the relationship of the total sample weight at collapse and the bottom surface area of the printed figure. They reported a linear correlation between the printing stability and the rheological properties, proposing an algorithm to develop food recipes with desired printability. The algorithm is mainly based on the flow stress obtained by shear rheology. Even though this information is valuable and a quantitative manner to solve the limitations of 3D food printing, the stress collapse might not be the best or optimum parameter. Structural failure during the printing process also depends on the shape and mechanical performance of the printed material, and furthermore on the distribution of the infill material (Suiker 2018, Cuan-Urquizo et al. 2019, Wolfs and Suiker 2019).

5.6 FUTURE PERSPECTIVES

The printing of personalized foods will be possible at home in the near future when the cost of printers becomes competitive. This novel technique will expedite new developments, the printing of personalized meals, and the design of healthy foods. The use of new components, nonconventional textures with the concomitant customization of foods are the main advantages of 3D food printing, being different types of snacks one of the most approachable categories to develop. Nutritional customization will be accomplished through printers with integrated nutrition models, easily handled by non-expert users to manage their diet and the exact amount of calories by the selection of the amount and types of ingredients (Baiano 2020).

These novel technologies will further open the possibility of using nonconventional ingredients like fruit peels, pseudo cereals, and many other underutilized sources of proteins, fibers, and nutraceuticals. Tailored-made healthy formulations can be developed from these materials as previously reported (Severini et al. 2018). This might revolutionize the snack food industry especially in terms of developing new ingredients and recipes for printing. Sustainability is also a relevant term to consider due to the reduction of source consumption and food waste by using underutilized materials (Baiano 2020). This will inspire new research efforts toward new post-processing methods or the improvement of existing ones. Moreover, the novel technologies will allow the fabrication of less expensive and multi-material food printers, perhaps with the postprocessing treatment already integrated.

4D printing is a new concept recently explored by researchers. It originated from the 3D printing concept but also displays potential and attractive applications by changing chemically or physically through time. A mixture of anthocyanin-potato starch gel that changes its color over time due to a change in pH has been successfully printed (Ghazal et al. 2019). Similarly, a mixture of mashed potato/purple sweet potato puree with spontaneous color change was developed by He et al. (2020). 3D printing is also moving in this direction to produce even more attractive foods for customers.

Food packaging might also be an industry impacted by 3D printing technology because it has been successfully used for prototyping design. The use of this technology has considerably reduced prototyping to few weeks instead of the commercial usage of molds that commonly takes several months. As in food printing, customization of packaging applying stereographic designs is a possibility. The production of biodegradable packaging from food by-products is also a challenge to be pondered. Smart Cups of Mission Viejo introduced a printed and biodegradable cup with embedded microencapsulated compounds. The addition of water to this printed cup yielded, after mixing, a drink free of sugar, carbohydrates, and calories (Siegner 2017). Milling fractions of rice husks supplemented with guar gum have been used for printing a material for food packaging (Nida et al. 2021).

There are still several open issues to be addressed in this area. The most worrisome is the massive production of foods using printers. The application of 3D printing for the large-scale production of food is yet unsure. The key factor will be the development of advanced cheaper and faster printers. Other relevant issues to consider are associated with the legal framework, consumer acceptance, safety, and labeling (Baiano 2020). 3D printing technology will not replace the conventional methods for food production; however, it is an interesting way to create customized and novel products for small settings that sooner or later will be adopted in households.

REFERENCES

Azam, R. S. M., Zhang, M., Bhandari, B., and Yang, C. 2018. Effect of different gums on features of 3D printed object based on vitamin-D enriched orange concentrate. *Food Biophysics*, 13(3): 250–262.

Azam, S. M. R., Zhang, M., Mujumdar, A. S., and Yang, C. 2018. Study on 3D printing of orange concentrate and material characteristics. *Journal of Food Process Engineering*, 41(5): 1–10. https://doi.org/10.1111/jfpe.12689

Baiano, A. 2020. 3D printed foods: A comprehensive review on technologies, nutritional value, safety, consumer attitude, regulatory framework, and economic and sustainability issues. *Food Reviews International*, 1–31. https://doi.org/10.1080/87559129.2020.1762091

Caulier, S., Doets, E., and Noort, M. 2020. An exploratory consumer study of 3D printed food perception in a real-life military setting. *Food Quality and Preference*, 86: 104001. https://doi.org/10.1016/j.foodqual.2020.104001

Christensen, L., Roager, H. M., Astrup, A., and Hjorth, M. F. 2018. Microbial enterotypes in personalized nutrition and obesity management. *American Journal of Clinical Nutrition*, 108(4): 1–7.

Cuan-Urquizo, E., Barocio, E., Tejada-Ortigoza, V., Pipes, R., Rodriguez, C., and Roman-Flores, A. 2019. Characterization of the mechanical properties of FFF structures and materials: A review on the experimental, computational and theoretical approaches. *Materials*, 12(6): 895. https://doi.org/10.3390/ma12060895

Dankar, I., Haddarah, A., Omar, F. E. L., Sepulcre, F., and Pujolà, M. 2018. 3D printing technology: The new era for food customization and elaboration. *Trends in Food Science and Technology*, 75: 231–242. https://doi.org/10.1016/j.tifs.2018.03.018

Derossi, A., Caporizzi, R., Azzollini, D., and Severini, C. 2018. Application of 3D printing for customized food. A case on the development of a fruit-based snack for children. *Journal of Food Engineering*, 220: 65–75. https://doi.org/10.1016/j.jfoodeng.2017.05.015

Derossi, A., Caporizzi, R., Paolillo, M., and Severini, C. 2020. Programmable texture properties of cereal-based snack mediated by 3D printing technology. *Journal of Food Engineering*, 289: 110160. https://doi.org/10.1016/j.jfoodeng.2020.110160

Derossi, A., Caporizzi, R., Ricci, I., and Severini, C. 2019. Critical variables in 3D food printing. In: *Fundamentals of 3D Food Printing and Applications*, eds. F. C. Godoi, B. R. Bhandari, S. Prakash and M. Zhang. 41–92. London: Elsevier Inc. https://doi.org/10.1016/b978-0-12-814564-7.00003-1

Dick, A., Bhandari, B., and Prakash, S. 2019. 3D printing of meat. *Meat Science*, 153: 35–44. https://doi.org/10.1016/j.meatsci.2019.03.005

Feng, C., Zhang, M., and Bhandari, B. 2019. Materials properties of printable edible inks and printing parameters optimization during 3D printing: A review. *Critical Reviews in Food Science and Nutrition*, 59(19): 3074–3081. https://doi.org/10.1080/10408398.2018.1481823

Feng, C., Zhang, M., Bhandari, B., and Ye, Y. 2020. Use of potato processing by-product: Effects on the 3D printing characteristics of the yam and the texture of air-fried yam snacks. *LWT- Food Science and Technology*, 125: 109265. https://doi.org/10.1016/j.lwt.2020.109265

García-Segovia, P., García-Alcaraz, V., Balasch-Parisi, S., and Martínez-Monzó, J. 2020. 3D printing of gels based on xanthan/konjac gums. *Innovative Food Science and Emerging Technologies*, 64: 102343. https://doi.org/10.1016/j.ifset.2020.102343

German, J. B., and Watzke, H. J. 2004. Personalizing foods for health and delight. *Comprehensive Reviews in Food Science and Food Safety*, 3(4): 145–151.

Ghazal, A. F., Zhang, M., and Liu, Z. 2019. Spontaneous color change of 3D printed healthy food product over time after printing as a novel application for 4D food printing. *Food and Bioprocess Technology*, 12(10): 1627–1645. https://doi.org/10.1007/s11947-019-02327-6

Gholamipour-Shirazi, A., Norton, I. T., and Mills, T. 2019. Designing hydrocolloid based food-ink formulations for extrusion 3D printing. *Food Hydrocolloids*, 95: 161–167. https://doi.org/10.1016/j.foodhyd.2019.04.011

Godoi, F. C., Bhandari, B. R., Prakash, S., and Zhang, M. 2019. An introduction to the principles of 3D food printing. In: *Fundamentals of 3D Food Printing and Applications*, eds. F. C. Godoi, B. R. Bhandari, S. Prakash and M. Zhang, 1–18. London: Elsevier Inc. https://doi.org/10.1016/b978-0-12-814564-7.00001-8

Godoi, F. C., Prakash, S., and Bhandari, B. R. 2016. 3d printing technologies applied for food design: Status and prospects. *Journal of Food Engineering*, 179: 44–54. https://doi.org/10.1016/j.jfoodeng.2016.01.025

Gudjónsdóttir, M., Napitupulu, R. J., and Petty Kristinsson, H. T. 2019. Low field NMR for quality monitoring of 3D printed surimi from cod by-products: Effects of the pH-shift method compared with conventional washing. *Magnetic Resonance in Chemistry*, 57(9): 638–648. https://doi.org/10.1002/mrc.4855

Hamilton, C. A., Alici, G., and in het Panhuis, M. 2017. 3D printing vegemite and marmite: Redefining "breadboards". *Journal of Food Engineering*, 220: 83–88. https://doi.org/10.1016/j.jfoodeng.2017.01.008

He, C., Zhang, M., and Guo, C. 2020. 4D printing of mashed potato/purple sweet potato puree with spontaneous color change. *Innovative Food Science and Emerging Technologies*, 59: 102250. https://doi.org/10.1016/j.ifset.2019.102250

Hertafeld, E., Zhang, C., Jin, Z., Jakub, A., Russell, K., Lakehal, Y., Andreyeva, K., Nagaraj Bangalore, S., Mezquita, J., Blutinger, J., and Lipson, H. 2019. Multi-material three-dimensional food printing with simultaneous infrared cooking. *3D Printing and Additive Manufacturing*, 6(1): 13–19. https://doi.org/10.1089/3dp.2018.0042

Holland, S., Foster, T., MacNaughtan, W., and Tuck, C. 2018. Design and characterisation of food grade powders and inks for microstructure control using 3D printing. *Journal of Food Engineering*, 220: 12–19. https://doi.org/10.1016/j.jfoodeng.2017.06.008

Jędrusek-Golińska, A., Górecka, D., Buchowski, M., Wieczorowska-Tobis, K., Gramza-Michałowska, A., and Szymandera-Buszka, K. 2020. Recent progress in the use of functional foods for older adults: A narrative review. *Comprehensive Reviews in Food Science and Food Safety*, 19(2): 835–856. https://doi.org/10.1111/1541-4337.12530

Jiang, H., Zheng, L., Zou, Y., Tong, Z., Han, S., and Wang, S. 2019. 3D food printing: Main components selection by considering rheological properties. *Critical Reviews in Food Science and Nutrition*, 59(14): 2335–2347. https://doi.org/10.1080/10408398.2018.1514363

Karyappa, R., and Hashimoto, M. 2019. Chocolate-based ink three-dimensional printing (Ci3DP). *Scientific Reports*, 9(1): 14178. https://doi.org/10.1038/s41598-019-50583-5

Keerthana, K., Anukiruthika, T., Moses, J. A., and Anandharamakrishnan, C. 2020. Development of fiber-enriched 3D printed snacks from alternative foods: A study on button mushroom. *Journal of Food Engineering*, 287: 110116. https://doi.org/10.1016/j.jfoodeng.2020.110116

Kern, C., Weiss, J., and Hinrichs, J. 2018. Additive layer manufacturing of semi-hard model cheese: Effect of calcium levels on thermo-rheological properties and shear behavior. *Journal of Food Engineering*, 235: 89–97. https://doi.org/10.1016/j.jfoodeng.2018.04.029

Kim, H. W., Bae, H. J., and Park, H. J. 2017. Classification of the printability of selected food for 3D printing: Development of an assessment method using hydrocolloids as reference material. *Journal of Food Engineering*, 215: 23–32. https://doi.org/10.1016/j.jfoodeng.2017.07.017

Kim, H. W., Lee, I. J., Park, S. M., Lee, J. H., Nguyen, M. H., and Park, H. J. 2019. Effect of hydrocolloid addition on dimensional stability in post-processing of 3D printable cookie dough. *LWT- Food Science and Technology*, 101: 69–75. https://doi.org/10.1016/j.lwt.2018.11.019

Kim, H. W., Lee, J. H., Park, S. M., Lee, M. H., Lee, I. W., Doh, H. S., and Park, H. J. 2018. Effect of hydrocolloids on rheological properties and printability of vegetable inks for 3D food printing. *Journal of Food Science*, 83(12): 2923–2932. https://doi.org/10.1111/1750-3841.14391

Krishnaraj, P., Anukiruthika, T., Choudhary, P., Moses, J. A., and Anandharamakrishnan, C. 2019. 3D extrusion printing and post-processing of fibre-rich snack from indigenous composite flour. *Food and Bioprocess Technology*, 12(10): 1776–1786. https://doi.org/10.1007/s11947-019-02336-5

Lanaro, M., Forrestal, D. P., Scheurer, S., Slinger, D. J., Liao, S., Powell, S. K., and Woodruff, M. A. 2017. 3D printing complex chocolate objects: Platform design, optimization and evaluation. *Journal of Food Engineering*, 215: 13–22. https://doi.org/10.1016/j.jfoodeng.2017.06.029

Le-Bail, A., Maniglia, B. C., and Le-Bail, P. 2020. Recent advances and future perspective in additive manufacturing of foods based on 3D printing. *Current Opinion in Food Science*, 35: 54–64. https://doi.org/10.1016/j.cofs.2020.01.009

Le Tohic, C., O'Sullivan, J. J., Drapala, K. P., Chartrin, V., Chan, T., Morrison, A. P., Kerry, J. P., and Kelly, A. L. 2018. Effect of 3D printing on the structure and textural properties of processed cheese. *Journal of Food Engineering*, 220: 56–64. https://doi.org/10.1016/j.jfoodeng.2017.02.003

Lee, J. H., Won, D. J., Kim, H. W., and Park, H. J. 2019. Effect of particle size on 3D printing performance of the food-ink system with cellular food materials. *Journal of Food Engineering*, 256: 1–8. https://doi.org/10.1016/j.jfoodeng.2019.03.014

Lille, M., Nurmela, A., Nordlund, E., Metsä-Kortelainen, S., and Sozer, N. 2018. Applicability of protein and fiber-rich food materials in extrusion-based 3D printing. *Journal of Food Engineering*, 220: 20–27. https://doi.org/10.1016/j.jfoodeng.2017.04.034

Lipton, J., Arnold, D., Nigl, F., Lopez, N., Cohen, D., Norén, N., and Lipson, H. 2010. Mutli-material food printing with complex internal structure suitable for conventional post-processing. In *21st Annual International Solid Freeform Fabrication Symposium - An Additive Manufacturing Conference, SFF 2010*, 809–815.

Lipton, J. I., Cutler, M., Nigl, F., Cohen, D., and Lipson, H. 2015. Additive manufacturing for the food industry. *Trends in Food Science and Technology*, 43(1): 114–123. https://doi.org/10.1016/j.tifs.2015.02.004

Liu, C., Ho, C., and Wang, J. 2018. The development of 3D food printer for printing fibrous meat materials. *IOP Conference Series: Materials Science and Engineering*, 284(1). https://doi.org/10.1088/1757-899X/284/1/012019

Liu, Y., Liang, X., Saeed, A., Lan, W., and Qin, W. 2019. Properties of 3D printed dough and optimization of printing parameters. *Innovative Food Science and Emerging Technologies*, 54: 9–18. https://doi.org/10.1016/j.ifset.2019.03.008

Liu, Y., Liu, D., Wei, G., Ma, Y., Bhandari, B., and Zhou, P. 2018. 3D printed milk protein food simulant: Improving the printing performance of milk protein concentration by incorporating whey protein isolate. *Innovative Food Science and Emerging Technologies*, 49: 116–126. https://doi.org/10.1016/j.ifset.2018.07.018

Liu, Y., Yu, Y., Liu, C., Regenstein, J. M., Liu, X., and Zhou, P. 2019. Rheological and mechanical behavior of milk protein composite gel for extrusion-based 3D food printing. *LWT - Food Science and Technology*, 102: 338–346. https://doi.org/10.1016/j.lwt.2018.12.053

Liu, Z., Bhandari, B., Prakash, S., Mantihal, S., and Zhang, M. 2019. Linking rheology and printability of a multicomponent gel system of carrageenan-xanthan-starch in extrusion based additive manufacturing. *Food Hydrocolloids*, 87: 413–424. https://doi.org/10.1016/j.foodhyd.2018.08.026

Liu, Z., Dick, A., Prakash, S., Bhandari, B., and Zhang, M. 2020. Texture modification of 3D printed air-fried potato snack by varying its internal structure with the potential to reduce oil content. *Food and Bioprocess Technology*, 13(3): 564–576. https://doi.org/10.1007/s11947-020-02408-x

Liu, Z., Zhang, M., Bhandari, B., and Wang, Y. 2017. 3D printing: Printing precision and application in food sector. *Trends in Food Science and Technology*, 69: 83–94. https://doi.org/10.1016/j.tifs.2017.08.018

Liu, Z., Zhang, M., Bhandari, B., and Yang, C. 2018. Impact of rheological properties of mashed potatoes on 3D printing. *Journal of Food Engineering*, 220: 76–82. https://doi.org/10.1016/j.jfoodeng.2017.04.017

Lupton, D., and Turner, B. 2018. Food of the future? Consumer responses to the idea of 3D-printed meat and insect-based foods. *Food and Foodways*, 26(4): 269–289. https://doi.org/10.1080/07409710.2018.1531213

Mantihal, S., Prakash, S., and Bhandari, B. 2019. Texture-modified 3D printed dark chocolate: Sensory evaluation and consumer perception study. *Journal of Texture Studies*, 50(5): 386–399. https://doi.org/10.1111/jtxs.12472

Mantihal, S., Prakash, S., Godoi, F. C., and Bhandari, B. 2017. Optimization of chocolate 3D printing by correlating thermal and flow properties with 3D structure modeling. *Innovative Food Science and Emerging Technologies*, 44: 21–29. https://doi.org/10.1016/j.ifset.2017.09.012

Mantihal, S., Prakash, S., Godoi, F. C., and Bhandari, B. 2019. Effect of additives on thermal, rheological and tribological properties of 3D printed dark chocolate. *Food Research International*, 119: 161–169. https://doi.org/10.1016/j.foodres.2019.01.056

Nida, S., Anukiruthika, T., Moses, J. A., and Anandharamakrishnan, C. 2021. 3D printing of grinding and milling fractions of rice husk. *Waste and Biomass Valorization*, 12(1): 81–90. https://doi.org/10.1007/s12649-020-01000-w

Pérez, B., Nykvist, H., Brøgger, A. F., Larsen, M. B., and Falkeborg, M. F. 2019. Impact of macronutrients printability and 3D-printer parameters on 3D-food printing: A review. *Food Chemistry*, 287: 249–257. https://doi.org/10.1016/j.foodchem.2019.02.090

Periard, D., Schaal, N., Schaal, M., Malone, E., and Lipson, H. 2007. Printing food. In *Proceedings of the 18th Solid Freeform Fabrication Symposium*, 564–574, Austin, TX.

Pulatsu, E., Su, J. W., Lin, J., and Lin, M. 2020. Factors affecting 3D printing and post-processing capacity of cookie dough. *Innovative Food Science and Emerging Technologies*, 61: 102316. https://doi.org/10.1016/j.ifset.2020.102316

Severini, C., Azzollini, D., Albenzio, M., and Derossi, A. 2018. On printability, quality and nutritional properties of 3D printed cereal based snacks enriched with edible insects. *Food Research International*, 106: 666–676. https://doi.org/10.1016/j.foodres.2018.01.034

Severini, C., and Derossi, A. 2016. Could the 3D printing technology be a useful strategy to obtain customized nutrition? *Journal of Clinical Gastroenterology*, 50(2): 175–178. https://doi.org/10.1097/MCG.0000000000000705

Severini, C., Derossi, A., and Azzollini, D. 2016. Variables affecting the printability of foods: Preliminary tests on cereal-based products. *Innovative Food Science and Emerging Technologies*, 38: 281–291. https://doi.org/10.1016/j.ifset.2016.10.001

Severini, C., Derossi, A., Ricci, I., Caporizzi, R., and Fiore, A. 2018. Printing a blend of fruit and vegetables. New advances on critical variables and shelf life of 3D edible objects. *Journal of Food Engineering*, 220: 89–100. https://doi.org/10.1016/j.jfoodeng.2017.08.025

Siegner, C. 2017. A new dimension: The 3-D-printed cup embedded with energy drink ingredients. *FoodDive*. https://www.fooddive.com/news/a-new-dimension-the-3-d-printed-cup-embedded-with-energy-drink-ingredients/512887/. Accessed 11 October 2020.

Suiker, A. S. J. 2018. Mechanical performance of wall structures in 3D printing processes: Theory, design tools and experiments. *International Journal of Mechanical Sciences*, 137: 145–170. https://doi.org/10.1016/j.ijmecsci.2018.01.010

Sun, J., Zhou, W., Huang, D., Fuh, J. Y. H., and Hong, G. S. 2015. An overview of 3D printing technologies for food fabrication. *Food and Bioprocess Technology*, 8(8): 1605–1615. https://doi.org/10.1007/s11947-015-1528-6

Sun, J., Zhou, W., Huang, D., and Yan, L. 2018. 3D food printing: Perspectives. In: *Polymers for food applications*, ed. T. Gutierrez, 725–735. Cham: Springer.

Sun, J., Zhou, W., Yan, L., Huang, D., and Lin, L. Y. 2018. Extrusion-based food printing for digitalized food design and nutrition control. *Journal of Food Engineering*, 220: 1–11. https://doi.org/10.1016/j.jfoodeng.2017.02.028

Suzuki, Y., Takagishi, K., and Umezu, S. 2019. Development of a high-precision viscous chocolate printer utilizing electrostatic inkjet printing. *Journal of Food Process Engineering*, 42(1): 1–6. https://doi.org/10.1111/jfpe.12934

Terfansky, M. L., and Thangavelu, M. 2013. *3D Printing of Food for Space Missions, AIAA SPACE 2013 Conference and Exposition*. 5346. https://doi.org/10.2514/6.2013-5346

Vancauwenberghe, V., Baiye Mfortaw Mbong, V., Vanstreels, E., Verboven, P., Lammertyn, J., and Nicolai, B. 2017. 3D printing of plant tissue for innovative food manufacturing: Encapsulation of alive plant cells into pectin based bio-ink. *Journal of Food Engineering*: 1–11. https://doi.org/10.1016/j.jfoodeng.2017.12.003

Vancauwenberghe, V., Delele, M. A., Vanbiervliet, J., Aregawi, W., Verboven, P., Lammertyn, J., and Nicolaï, B. 2018. Model-based design and validation of food texture of 3D printed pectin-based food simulants. *Journal of Food Engineering*, 231: 72–82. https://doi.org/10.1016/j.jfoodeng.2018.03.010

Vancauwenberghe, V., Katalagarianakis, L., Wang, Z., Meerts, M., Hertog, M., Verboven, P., Moldenaers, P., Hendrickx, M. E., Lammertyn, J., and Nicholai, B. 2017. Pectin based food-ink formulations for 3-D printing of customizable porous food simulants. *Innovative Food Science and Emerging Technologies*, 42: 18–150. https://doi.org/10.1016/j.ifset.2017.06.011

Villarama Díaz, J., Cornelis van Bommel, K. J., Willem-Jan Noort, M., Henket, J., and Briër, P. 2018. Method for the production of edible objects using SLS and food products. US Patent No. 10,092,030. 9 Oct. 2018.

Voon, S. L., An, J., Wong, G., Zhang, Y., and Chua, C. K. 2019. 3D food printing: A categorised review of inks and their development. *Virtual and Physical Prototyping*, 14(3): 203–218. https://doi.org/10.1080/17452759.2019.1603508

Wang, L., Zhang, M., Bhandari, B., and Yang, C. 2018. Investigation on fish surimi gel as promising food material for 3D printing. *Journal of Food Engineering*, 220: 101–108. https://doi.org/10.1016/j.jfoodeng.2017.02.029

Wolfs, R. J. M., and Suiker, A. S. J. 2019. Structural failure during extrusion-based 3D printing processes. *International Journal of Advanced Manufacturing Technology*, 104(1–4): 565–584. https://doi.org/10.1007/s00170-019-03844-6

Yang, F., Guo, C., Zhang, M., Bhandari, B., and Liu, Y. 2019. Improving 3D printing process of lemon juice gel based on fluid flow numerical simulation. *LWT - Food Science and Technology*, 102: 89–99. https://doi.org/10.1016/j.lwt.2018.12.031

Yang, F., Zhang, M., Bhandari, B., and Liu, Y. 2018. Investigation on lemon juice gel as food material for 3D printing and optimization of printing parameters. *LWT - Food Science and Technology*, 87: 67–76. https://doi.org/10.1016/j.lwt.2017.08.054

Zhu, S., Stieger, M. A., van der Goot, A. J., and Schutyser, M. A. I. 2019. Extrusion-based 3D printing of food pastes: Correlating rheological properties with printing behaviour. *Innovative Food Science and Emerging Technologies*, 58: 102214. https://doi.org/10.1016/j.ifset.2019.102214

6 Production of Cereal-Based Raw Materials for the Snack Industry

Sergio O. Serna-Saldivar

CONTENTS

6.1 INTRODUCTION

The features of most snacks deeply depend on the quality of the raw materials used for their manufacturing. Most second- and third-generation extruded snacks (Chapter 12) employ refined fractions from cereals because they are highly available, relatively cheap, and rich in starch and low in fiber and oil. Cereal grains are broadly divided into naked and covered caryopses. The covered caryopses of paddy rice, barley, and oats tightly retain their glumes after harvesting whereas the rest of the cereals lose these hulls during harvesting and threshing. Therefore, the hulls (lema and palea) of covered caryopses are removed by special equipment. The naked caryopses of all cereals are covered by the fruit coat or pericarp and contain a germ that holds most of the fat. The dry-milling industry has existed for centuries in order to provide materials for the baking, breakfast cereals, and snack industries. Today industrial milling processes are modern, highly automated, and very versatile in terms of the arrays of products generated. The resulting products have a long shelf life due to their low moisture, *Aw*, and oil contents. Most dry-milled products are highly refined due to the removal of the husks in the case of covered caryopses, pericarp, and germ tissues. Dry milling consists of the separation of the different grain anatomical parts with the aim of obtaining the endosperm as a whole, in pieces or grits or as meals and flours. These raw materials, especially grits, are highly in demand for production of a wide array of savory snacks. The dry-milling processes for corn, wheat,

DOI: 10.1201/9781003129066-6

and rice have different aims and product yields. Dry millers of rice produce white-polished kernels and various types of grits and flours. White rice is commonly expanded for the production of puffed kernels demanded by producers of granola bars whereas broken kernels are normally reduced in particle size for the production of grits used for the manufacturing of extruded snacks.

On the other hand, the major cereal grains are alternatively wet milled into refined undamaged starches that meet specifications for maximum amount of protein. These starches differ in terms of functionality, viscoamylograph behavior, and expansion rate when extruded. Particularly, cornstarch is highly demanded because of its relatively low cost, high availability as raw and modified versions, and versatility. Refined starches are modified in order to change their functional properties like solubility in cold water, water retention, gelatinization temperature, gel clarity, viscosity, and retrogradation. The use of these modified starches allows food product developers to modify the bulk density, texture, crispness, and structure of snacks. This chapter describes industrial dry and wet-milling processes of cereal grains for the production of the wide array of products highly demanded by snack food processors.

6.2 DRY-MILLING PROCESSES

6.2.1 Dry Milling of Corn (*Zea mays*)

It is estimated that the 12 largest US corn dry-milling facilities had a combined daily grind capacity of 13,335 tons, equivalent to 525,000 bushels and that this capacity represented 83% of the total volume. Assuming an average five-day-per-week "grind schedule," this would equate to an annual volume consumed of about 4.165 million tons (Anderson and Almeida 2019). These millers supply most of their refined products to the brewing, breakfast cereal, and snack industries.

Dry millers process corn either by stone grinding or the highly efficient degerming tempering (DT) milling procedure. Stone grinding yields hominy grits and whole meals rich in pericarp and germ whereas DT highly refined grits, meals, and flours with extended shelf life. Whole corn meals have rich flavor because of their high oil and germ content and nutritionally are higher in total dietary fiber, minerals, and lipids compared to counterparts obtained by the DT milling procedure. From the functionality viewpoint, refined fractions are preferred because they favor expansion rate, color, shelf life, and organoleptic properties, especially when processing second-generation extruded snacks (Chapter 12). However, the new trend of the industry is the production of snacks with better nutritional and nutraceutical profiles so dry millers are increasing the production of whole meal products and meals.

The detailed corn DT milling process has been recently reviewed and thoroughly described by Anderson and Almeida (2019). US No. 2 yellow dent maize is the type most frequently used by millers. Test weight (72.1–73.4 kg/hL), kernel density (1.27 g/cm^3) or endosperm texture (intermediate), and freedom from stress cracks, aflatoxins, and heat-damaged kernels are the major grain quality criteria. High-temperature grain drying creates stress cracks, which leads to breakage during handling and milling (Serna-Saldivar 2010a, Anderson and Almeida 2019).

The aim of the DT milling procedure described in Figure 6.1 is to produce the maximum percentage of refined grits, containing minimum fat, fiber, and specks from the hilum or tip cap and to recover the maximum percentage of clean germ with maximum oil content. This coproduct is generally used by the oil crushing industry to obtain refined corn oil (Chapter 7).

A rigorous cleaning of the kernels precedes the milling process. The typical cleaning operation starts with air aspiration to remove fines, cob pieces, and other light foreign material followed by size separation to remove coarse impurities and broken kernels. The cleaning equipment are normally provided with magnets to remove metals and other ferrous particles. Many dry-milling plants also use destoners to remove stones, glass, and nonmagnetic metals with similar dimensions of the corn kernels (Serna-Saldivar 2010a, Anderson and Almeida 2019). All mills carefully examine incoming lots of maize for mycotoxins, chiefly aflatoxins, and grain molds (Chapters 19 and 20).

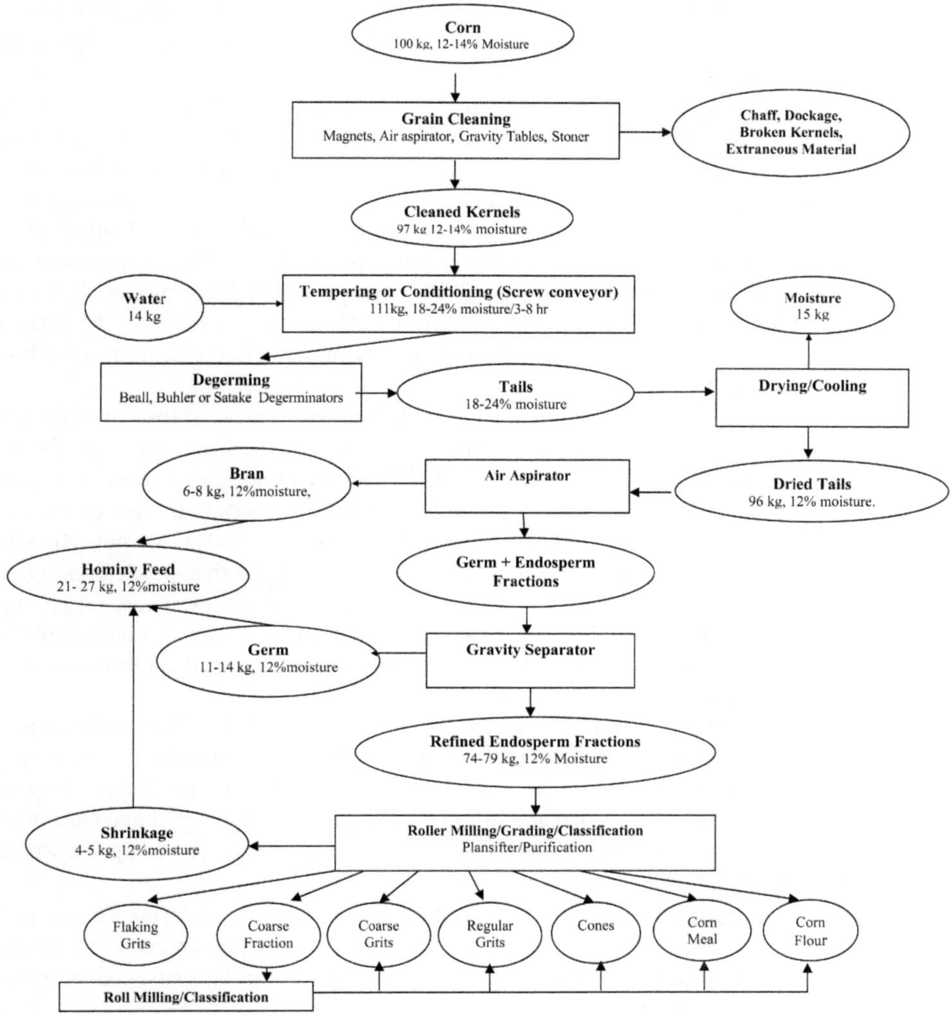

FIGURE 6.1 Diagram of the DT dry-milling process of corn.

The clean corn kernels are commonly tempered to 20 to 23% moisture using screw conveyors equipped with water sprayers and placed in a tempering bin for 1 to 3 hrs. The aim of tempering is to toughen the bran and germ to facilitate subsequent separations. Tempering also hydrates the endosperm so maximum grit yields with minimum flour are achieved.

Degermination is the most critical step for efficient milling. The most usual degerminator is the Beall although recently the Buhler degerminator has gained acceptance. The Beall equipment consists of a conical knobbed rotor and a stator shell that is also knobbed on the lower convex surface and slotted on the upper surface. The degerminator is set to produce large, clean pieces of endosperm known as hominy tails where most of the germ and pericarp are liberated. Likewise, the Buhler degerminator removes most or the bran and germ leaving intact the endosperm fractions. The Buhler technology utilizes a cylindrical design, with a central rotor running horizontally the length of the machine, to which rows of spiked fingers known as "beater bars," are attached. This assembly is enclosed in a perforated screen housing or designed to enhance friction between kernels and the beater bars (Anderson and Almeida 2019). The Buhler equipment requires uniform tempering to the optimal moisture and time to achieve best results in terms of pericarp removal and germ release. Other less utilized degerminators are the Satake vertical, the impact entoleter, and the Ocrim degerminators (Anderson and Almeida 2019).

Normally, tempering is done via a two-stage procedure. The first step incorporates a hold phase of 1–10 hr and is intended to optimize germ release whereas the second of shorter duration is aimed to toughen the bran and help to release it in cleaned flakes. The final moisture content upon entering the decorticator is typically in the 18–20% range. The cleaned and tempered kernels processed in the Beall degerminator are subjected to a combination of physical and mechanical abrading resulting in the release and separation of the various components. This initial separation results in two streams exiting the degerminator: "tail stock," which is primarily large pieces of endosperm that exit the tail end of the degerminator; and "through stock," which is made up of the germ, pericarp, and smaller endosperm pieces that pass through the perforations in the degerminator housing (Anderson and Almeida 2019).

Independently of the type of degerminatior, the additional processing steps comprise drying, cooling, aspirating, density separating, roller milling, and sizing or classification. With the Buhler decorticator, the tail stock tends to be composed of slightly smaller endosperm pieces compared to the ones produced by the Beall counterpart, and pericarp removal is generally less extensive. For these reasons, flaking grits widely demanded by the breakfast cereal industries are not typically one of the prime products recovered from this system. On the other hand, the thru-stock stream tends to contain larger and more intact germ pieces as well as larger pericarp pieces. These larger coproducts are easier to remove downstream. Finally, the Buhler decorticating system is comparatively less energy intensive and commonly yields higher prime products such as brewers and snack grits, corn meal, and corn flour (Anderson and Almeida 2019).

Reduction of the large endosperm pieces is done with roller mills followed by sifting and aspiration to remove bran particles from the grits. Then, gravity tables separate low-density germ pieces from endosperm particles with higher density. Endosperm fractions are finished in purifiers designed to remove fine pieces of pericarp and hilum (black specks) and are packaged at a moisture content of approximately 12%. The mean chemical composition of maize kernels, grits, whole and refined meals, and bran are summarized in Table 6.1.

The typical yields of dry-milled fractions vary according to the quality of corn kernels especially in terms of endosperm hardness or texture, type of degerminator used, and objective of the milling industry. However, the average yields of flaking grits, coarse grits, regular grits, coarse meal, and corn flour are 12, 15, 23, 6, and 4%, respectively (Brekke 1970).

Corn grits are low in fiber and contain less than 1.2% oil. Most manufacturers of extruded corn puffs (Chapter 13) demand highly refined grits with less than 1% oil. Grits of various granulations are widely utilized by snack food processors. Refined grits, meals, or flours are cooked to gelatinize the starch, denature the protein, and produce a dough, which can be processed into directly extruded puffs or third-generation pellets or half products (Chapter 12). The extrudate expansion rate is closely related to the quality of the grits especially in terms of the amount and condition of the starch (Serna-Saldivar and Perez-Carrillo 2019).

Refined maize meals and flours are used to a lesser extent by snack food processors mainly due to their finer particle size or granulation that makes their flow difficult through thermoplastic extruders (Serna-Saldivar and Perez-Carrillo 2019). However, their chemical composition is similar to grits (Table 6.1).

Zhang and Hoseney (1998) assessed the performance of two contrasting corn meals in terms of expansion properties and concluded that the cornmeal with lower expansion rate had a slightly larger particle size and proportion of opaque particles. Water diffusion tests showed that the sample containing more opaque particles was more competitive for water. The underplastisized vitreous particles remained glassy (unmelted) after thermoplastic extrusion resulting in reduced expansion of extrudates.

TABLE 6.1
Chemical Composition of Maize and Its Dry-Milled Products Commonly Used for Production of Snack Foods

Nutrient per 100 g	Raw Grain	Refined Dry Milled White Grits	Yellow Grits	Hominy Grits	Degermed Meal	Whole Meal	Bran
Water, g	10.37	10.00	10.00	10.48	11.59	10.91	4.71
Energy, kcal	365	371	371	347	366	361	224
Protein, g	9.42	8.80	8.80	8.53	8.48	6.93	8.36
Lipids, g	4.74	1.20	1.20	1.36	1.65	3.86	0.92
Carbohydrates, g	74.26	79.60	79.60	79.16	77.68	76.85	85.64
Dietary fiber, g	7.3	1.6	1.6	4.8	7.4	9.6	79.0
Calcium, mg	7	2	2	4	5	7	42
Iron, mg	2.71	1	1	3.52	1.10	2.38	2.79
Sodium, mg	35	1	1	2	3	5	7
Thiamine, mg	0.385	0.130	0.130	0.580	0.140	0.246	0.01
Riboflavin, mg	0.201	0.040	0.040	0.330	0.050	0.080	0.10
Niacin, mg	3.627	1.200	1.200	4.810	1.000	1.900	2.735
Folate mcg	19	–	–	N/A	48	25	4
Type of fat							
Saturated	0.667	0.155	0.155	0.260	0.225	0.543	0.13
Monounsaturated	1.251	0.300	0.300	0.200	0.412	1.018	0.243
Polyunsaturated	2.163	0.516	0.516	0.680	0.710	1.759	0.421

Source: USDA (2021).

6.2.2 Dry-Milling of Wheat (*Triticum aestivum, durum*)

Worldwide, most wheat is milled into flours for different applications using the conditioning roller procedure (Posner 2009, Serna-Saldivar 2010a). The aim of millers is to obtain the maximum flour yield when soft and hard wheats are processed or semolina when durum wheat is used. The coproducts of these industries are bran and shorts and sometimes germ. The conventional milling procedure starts when selected wheat is thoroughly cleaned to remove metals, chaff, glumes, stones, and other extraneous materials including other grains like oats (Figure 6.2). The cleaning operations include use of magnets, air aspiration, gravity tables, and disk separators. The last step of the cleaning process is the pass of the wheat kernels through a scourer. In this device, the wheat is subjected to friction in order to separate impurities adhering to the kernel's surface. In other words, it has a polishing effect. The scourer consists of a rotor and a cylindrical woven steel wire. The rotor bounces the kernels against the perforated cylindrical wall causing the small impurities to pass through the screen (Posner 2009, Serna-Saldivar 2010a). Cleaning is critical to protect the milling equipment and to produce the best quality flour. The cleaned grain is conditioned or tempered before milling with the aim of separating more cleanly the bran from the endosperm, to soften the endosperm and enhance its gradual reduction in particle size, and to improve the sieving efficiency. Wheat is usually tempered with water sprayed on the grain being conducted in a screw conveyor. The water is evenly distributed on the surface of kernels by the blending action of the screw conveyor. Then the wheat is transported into a tempering bin for conditioning for the optimum time that generally fluctuates from 12 to 18 hr. During this time, kernels absorb the water that concentrates in the external part of the grain (germ and pericarp). When kernels are hard or vitreous such as durum or kernels

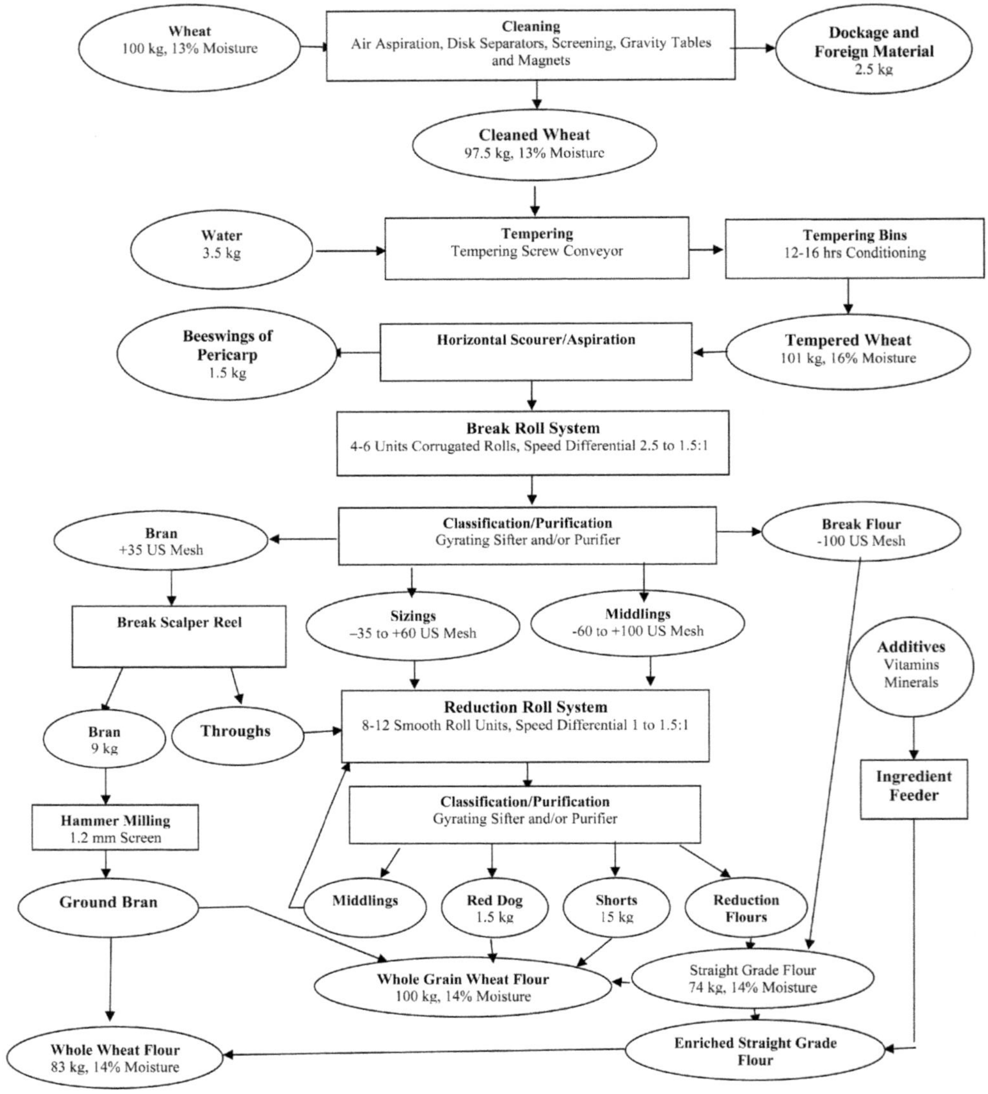

FIGURE 6.2 Diagram of the typical dry-milling process of wheat for production of refined flour, whole wheat flour, or whole grain wheat flour.

contain low moisture, the tempering operation is preferably performed in two stages. The different classes of wheat are tempered for different times and final moisture contents. Hard wheats require more moisture and conditioning time than soft wheats. Generally, hard wheats are conditioned for 12–24 hr to 16.5% moisture whereas soft counterparts for 5–15 hr and 15–15.5% moisture. Durum wheats require the highest tempering moisture (17–17.5%) and conditioning time of 12–24 hr.

After tempering and equilibration, kernels are first passed through an abrasive machine equipped with an air aspiration system with the aim of eliminating impurities located on the pericarp and breaking damaged kernels (i.e., insect-damaged kernels). The kernels are transported in between a pair of inverted conic-shaped device. One of the cones is static and the other moves creating the abrasive action. The sound and damaged exiting grains are separated in a cylindrical metal sieve. This operation allows millers to decrease the number of insect fragments in flours (Posner and Hibbs 2005, Posner 2009, Serna-Saldivar 2010a).

Milling is accomplished in two types of roller mills: break and reduction roller mills. These mills consist of pairs of horizontal, parallel iron cylinders rotating in opposite directions. They pull the stock down between the rolls into the nip, which is considered the grinding zone. The separation or classification of resulting particles is normally performed with plansifters and purifiers. The aims of break rolls are to open kernels and remove the endosperm and germ from the pericarp with the least amount of contamination and obtain a particle size distribution of maximum large middlings with a minimum of flour. The bran should be detached from the kernel in flakes and without any adhered endosperm pieces. The break rolls consist of a pair of corrugated metal rolls that counter-rotate at different speeds. The rpm differential (1.5–3.0 to 1) and corrugation enhances the abrasive action and removal of the bran in large pieces. A typical mill has between 4 and 6 break rolls operating on battery. The break material is classified into bran (US mesh +35), sizings (US mesh +70), middlings (US mesh + 100), and break flour (US mesh –100). Sizings are the coarsest separation of the endosperm whereas middlings are finer particles of endosperm that require further reduction in order to yield flour. The bran obtained from the break system is usually passed through a bran finisher consisting of a beater that frees endosperm from the pericarp by impact and friction. The beaters are steel bars on a rotating shaft enclosed in a cylindrical perforated screen cage (Posner and Hibbs 2005).

The two most common ways to separate stocks from the break and reduction rolls are the plansifter and purifiers. The plansifter consists of a gyrating bolting device consisting of a set of sieves (up to 30) positioned parallelly that gyrate in a horizontal plane. It is the primary sifter used to classify particles by size. The purifier consists of a reciprocating sieve enclosed in an airtight container with controlled airflows passing through the screen. It is ideally suited for the removal of unattached bran from middlings and at the same time classifies the middlings into several fractions. Purifiers can remove bran and classify middlings, which is not possible with the plansifter alone. The less dense particles of about the same size are removed by controlled air current. Therefore, purifiers are used to produce more refined flours with lower ash contents and better color scores (Posner and Hibbs 2005).

The intermediate fractions are usually purified before further milling in the reduction rolls. The objective of the reduction system is to reduce sizings and middlings into flour minimizing the starch damage. The reduction roll system consists of smooth rolls that generally operate at the same velocity or with a small speed differential. The reduction roll system consists of 6–10 mills in battery. These mills work with an integrated particle classification system that segregates flours that meet the desired particle size (US mesh -100) and coarser particles that are recycled for further milling. At the end of the operation, several break and reduction flours are obtained. These flours have different color, degree of damaged starch, and ash contents. The mixture of all flours yields straight-grade flours whereas the combination of approximately 80% of the best quality flours yields patent flours that are generally manufactured for the production of yeast-leavened breads (Chapter 10). The worst quality flours taken away from patent flours are usually denominated clear flours. These flours contain more fat and ash and starch damage and are easily recognized because of their lower color scores. The term extraction rate commonly used by millers refers to the amount of flour produced from a given amount of wheat. Generally, the extraction rate for refined flours varies from 72 to 78% (Figure 6.2) whereas the extraction rate of patent flours is normally from 60 to 64% due to the extra removal of the clear flours steams. The coproducts of the milling system are bran, shorts (inseparable mixture of bran endosperm and germ that remains after flour extraction), germ, and red-dog flour (a low-grade flour consisting of endosperm and bran taken from the tail of the mill).

The basic principles of flour milling are applied to the production of semolina. However, durum milling is different from conventional wheat milling especially in terms of the usage of more break roll passes and number of purifiers used because kernels are harder and the semolina is coarser than flour. The average extraction rate is in the range of 65–70 and particle size distribution is in the range of -60 to +100 US mesh. No more than 3% of the particles should pass a No. 100 sieve. The semolinas are classified depending on average particle size distribution into coarse, middle, or fine semolina (Serna-Saldivar 2010a).

There are several types of commercial refined wheat flours widely used for the preparation of baked (Chapter 10) and extruded (Chapter 12) snacks. These are normally classified as hard, all-purpose, soft, and chlorinated. The main differences among these flours are protein content (Table 6.2) and dough rheological properties. The hard wheat flour usually contains from 10.5 to 12.5% protein and the dough has strong gluten suitable for production of yeast-leavened breads whereas the soft flours contain 7–10% protein and the dough has a weak but extensible gluten appropriate for the production of cookies and related chemical-leavened products. The chlorinated flours are obtained from soft flours treated with chlorine gas, which bleaches the flour, lowers the pH, and weakens the gluten structure due to the disruption of disulfide groups. The chlorinated flours are tailor-made for cake and related products production. The all-purpose or family flours are usually produced from a mixture of hard and soft wheats and contain about 10–10–5% protein and intermediate dough rheological properties. These flours are used for production of donuts, crackers (Chapter 10), crisps, and most wheat-based snacks and nutritional bars. Durum wheats are usually processed into refined semolina for pasta production. Semolina is coarser than flour and usually has a particle size distribution of US mesh -60 to +100. In addition, they are yellow colored because the endosperm has significant amounts of carotenes and xantophylls.

6.2.3 Dry Milling of Rough Rice (*Oryza sativa*)

Rice is the cereal grain that is consumed more frequently as a dehulled, decorticated, and degermed product. Yearly, nearly 780 million metric tons of paddy rice is transformed into white rice by the milling industry. Most rice varieties are composed of roughly 20% husks or glumes, 11% pericarp and aleurone layers, and 69% starchy endosperm. Rough rice milling consists of a series of operations highly mechanized and complex that has the aim of producing high yields of unbroken kernels that meet particle size, color, and shelf-life specifications. The milling process mainly consists of six major operations: rough rice drying, grain cleaning, dehulling, decortication, polishing, and classification (Bond 2006, Serna-Saldivar 2010a, Bao 2019).

Yields of unbroken white rice are affected mainly by management practices during harvesting and drying. Paddy or rough rice is normally harvested at moistures ranging from 20 to 24% and therefore it should be carefully dehydrated to lower its moisture to around 12%. This moisture is optimum for prolonged storage and milling. Harvesting and drying should be performed with the aim of minimizing mechanical damage to the endosperm structure that results in fissures that weaken the grain structure increasing the amount of broken kernels. Rice is either solar or artificially dehydrated. The natural sun drying is the least expensive operation and produces the best quality rice. However, it takes a long time and depends on the environmental conditions so it is less flexible especially during the rainy season. In addition, solar-dried paddy rice is usually more prone to infestation with insects, rodents, and other pests. Nowadays, most paddy rice is artificially dried in continuous driers with strict controls of temperature, relative humidity, and airflow rate. When artificially dried, rough rice is normally dehydrated to lower its moisture to 12% at temperatures not higher than 39°C. Drying also allows millers to preserve the rough rice in grain elevators for prolonged periods.

The milling operation starts when the rough rice is thoroughly cleaned with air aspiration to remove light contaminants, such as free glumes, empty kernels, and extraneous light materials. Then the precleaned rice is further cleaned in cylindrical or on flat sieves. The equipment continuously separates foreign material larger than paddy rice and broken kernels and small contaminants on a second stream. The cylindrical sieves are placed in tandem whereas the most common sifter tables are normally equipped with at least two sieves and in some instances with an integrated air aspiration system. Ideally, the air aspirated and classified rough rice is subjected to the last cleaning stage in gravity tables designed to effectively remove foreign material like stones with the same

TABLE 6.2
Chemical Composition of the Different Classes of Wheats and Their Dry-Milled Products Widely Used for Production of Snacks

	Whole Wheat Grains			Refined Products				
Nutrient per 100 g	**Soft[a]**	**Hard[b]**	**Durum Semolina**	**Hard Flour**	**All-purpose Flour[c]**	**Soft Flour**	**Durum Semolina**	**Wheat Bran**
Water, g	11.3	12.9	10.94	13.36	11.92	12.51	12.67	9.89
Energy, kcal	335	328	339	361	364	362	360	216
Protein, g	10.50	14.0	13.68	11.98	10.33	8.20	12.68	15.55
Lipids, g	1.79	1.73	2.47	1.66	0.98	0.86	1.05	4.25
Carbohydrates, g	74.75	68.53	71.13	70.31	76.31	78.03	72.83	64.51
Dietary fiber, g	12.6	12.2	–	2.4	2.7	1.7	3.9	42.8
Calcium, mg	31	27	34	15	15	14	17	73
Iron, mg	4.31	3.40	3.52	0.90	1.17	7.32	1.23	10.57
Sodium, mg	2	2	2	2	2	2	1	2
Thiamine, mg	0.40	0.56	0.419	0.080	0.120	0.892	0.280	0.523
Riboflavin, mg	0.01	0.11	0.121	0.060	0.040	0.430	0.080	0.577
Niacin, mg	4.79	5.65	6.738	1	1.250	6.790	3.310	13.578
Folic acid mcg	41	41	43	-	-	138	72	79
Type of fat, g								
Saturated	0.33	0.29	0.454	0.244	0.155	0.127	0.150	0.630
Monounsaturated	0.20	0.26	0.344	0.140	0.087	0.073	0.124	0.637
Polyunsaturated	0.80	0.71	0.978	0.727	0.413	0.379	0.430	2.122

Source: USDA (2021).

[a] Average composition of red and white soft wheats.

[b] Average composition of hard spring and hard winter wheats.

[c] All-purpose flours obtained by blending soft and hard wheats prior to milling.

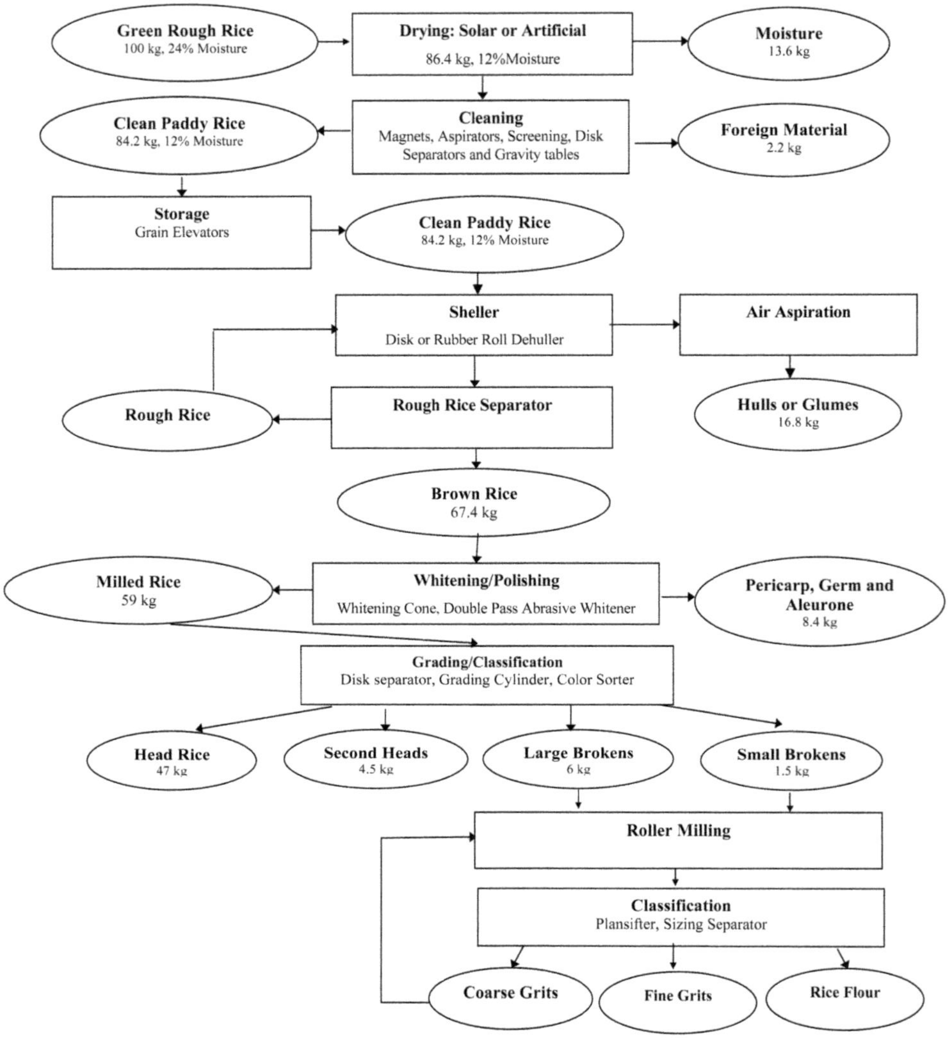

FIGURE 6.3 Diagram of the typical dry-milling process of rough rice and production of refined grits.

size but different density than the clean rough rice. This equipment is utilized to remove damaged-unfilled kernels and stones that have lower and higher densities than sound paddy rice. During these cleaning steps, grains are usually passed through magnets integrated into the cleaning devices in order to trap ferrous metals that can damage the milling equipment and compromise the quality of end-products (Bond 2006, Serna-Saldivar 2010a, Bao 2019).

Paddy rice is dehulled to remove the glumes or husks yielding brown rice. During this procedure, approximately 20% of the paddy rice weight is lost (Figure 6.3). Dehulling is accomplished by passing the cleaned paddy rice through a couple of plastic-covered milling rolls that counterrotate at different speeds or rpm. The gap between the rolls is carefully adjusted so as to dehull about 90–95% of the kernels and minimize broken kernels. The speed differential creates an abrasive mechanical action that removes most of the glumes or husks in the first pass. The other type of dehulling device is the one consisting of two abrasive discs horizontally positioned. The lower disk rotates below the upper stationary disk. Rough rice flows vertically through an orifice located in the central part of the upper stationary disk and is dehulled by the abrasive action of the surfaces of the discs. Since these dehullers are adjusted to an efficiency of 90–95% the output is first passed through an

TABLE 6.3
Chemical Composition of Rough Rice and Its Dry-Milled Products Commonly Used for Production of Snack Foods

	Type of Rice				Refined Products		Bran[c]
Nutrient per 100 g	**Rough**	**Brown**	**White Regular[a]**	**White Glutinous**	**Grits**	**Flour**	
Water, g	14.00	10.37	11.6	10.5	11.62	11.89	6.13
Energy, kcal	379	370	365	370	365	366	316
Protein, g	6.75	7.94	7.13	6.81	7.13	5.95	13.35
Lipids, g	1.90	2.92	0.66	0.55	0.66	1.42	20.85
Carbohydrates, g	68.40	77.24	80.0	81.7	79.95	80.13	49.69
Dietary fiber, g	nd[d]	3.50	1.3	2.8	1.3	2.4	21.0
Calcium, mg	45	23	28	11	28	10	57
Iron, mg	3.70	1.47	4.31	1.6	4.31	0.35	18.54
Sodium, mg	43	7	5	7	5	–	5
Thiamine, mg	2.9	0.401	0.576	0.180	0.576	0.138	2.753
Riboflavin, mg	0.9	0.093	0.049	0.055	0.049	0.021	0.284
Niacin, mg	42	5.091	4.192	2.140	4.192	2.590	33.995
Folic acid mcg	300	20	223	7	223	–	63
Type of fat, g							
Saturated	–	0.584	0.180	0.11	0.180	0.386	4.17
Monounsaturated	–	1.056	0.206	0.20	0.206	0.442	7.55
Polyunsaturated	–	1.044	0.177	0.20	0.177	0.379	7.46

Source: USDA (2021) and Serna-Saldivar (2010a).

[a] Regular long and enriched raw rice.

[b] Glutinous unenriched raw rice.

[c] The rice bran consists of the abraded pericarp, germ, and aleurone tissues.

[d] The rough rice contains from 7.2 to 10.4 g/100 g of crude fiber, or 16.4 to 19.2 g/100 g of neutral detergent fiber.

aspiration system to remove the detached glumes and then through a gravimetric table to separate the less dense paddy rice from the more dense brown kernels. The paddy rice is recirculated while the brown rice is sent to the decorticator and polishing machines.

Decortication/polishing is after drying the most critical rice-milling step because most of the broken kernels are generated during this unit operation. During this step, the pericarp, germ, and the multialeurone layer are removed. There are two types of systems generally used by the milling industry. In both, the outer layers are progressively removed by abrasion to yield white rice. These tissues, commonly named rice bran, make about 12% of the brown rice weight and are rich in protein, fat, and minerals (Table 6.3). The abrasive system is the oldest one and consists of a vertical conical stone that is encased inside a perforated metal frame. Brown rice is fed through the gap existing between the rotating stone and the static metal frame. Decortication is mainly achieved by the abrasive action of the external part of the stone and the metal mesh. The stone generally rotates at 800 m/min. The finely ground outer tissues pass through the metal screen whereas the white rice exits at the bottom of the milling machine. Rice bran has little use as a source of dietary fiber unless the fat is removed with hexane. Rice bran contains approximately 20% oil rich in monounsaturated and polyunsaturated fatty acids (Table 6.3).

The cylindrical decortication device consists of a horizontally positioned cylindrical roll encased inside a metal mesh. Brown rice is gradually decorticated by the abrasion of the rotating roll and the friction among kernels. The efficiency of the mill can be improved when calcium carbonate (3.3 g/kg brown rice) is used. The degree of decortication is adjusted by controlling the gap between the rotating roll and the fixed metal mesh. Rice polishing is accomplished with the same cylindrical mill but the rotating roll is usually covered with leather. The decorticated white kernels are mixed with magnesium silicate in order to produce a bright coloration and a shiny gloss (Serna-Saldivar 2010a).

After decortication, a mixture of white rice and broken kernels are obtained. Ideally, the yield of refined fractions should be approximately 68–70% (Figure 6.3). The higher the yield of white rice the better the quality of the rough rice and the processing equipment and adjustments. The broken kernels are classified as second heads, half kernels, and rice grits. Second head rice is the one that loses the tip of the kernel and is approximately ¾ of the total whole white grain length. These refined particles are classified according to their particle size with flat sieves or with a cylindrical sizing system.

Rice grits alone or blended with other raw materials are commonly used to manufacture various types of snack foods. The advantage of rice grits is that they are light colored, high in starch, low in fat, and practically free of fiber (Table 6.3). These grits have a high expansion rate so are ideally suited for production of direct-extruded puffs (Chapter 12). For production of rice grits, the whole or broken white rice is gradually reduced in particle size with the use of roller mills. The resulting particles are classified to meet particle size specifications. The most demanded rice grits have particle sizes from US mesh 40–60.

6.2.4 Dry Milling of Oats (*Avena sativa*)

Among cereals, oats are considered the best in terms of nutritional value and positive health effects. This is the reason why snack food processors are developing new products like nutritional bars based on this cereal. This cool seasoned crop yields a covered caryopsis rich in protein and both soluble and insoluble dietary fiber. The processing of oats for human foods commands dehulling to produce naked caryopses regularly known as groats, which are considered by regulatory agencies as whole grain. During a typical oat milling procedure, covered or hulled caryopses are thermally treated before the mechanical removal of the husks or glumes which are largely constituted by lignin and cellulose. This thermal treatment is carried out in order to deactivate lipases and lipoxygenases, which are known to generate off-flavors and odors. Practically all oat-based-foods are manufactured from whole, rolled, or ground groats, which contain between 11.9 and 14% (dwb) of total dietary fiber and up to 14% protein. Research by Manthey et al. (1999) reported that the dietary fiber of six groats obtained from different cultivars contained from 10.2 to 12.1% (dwb) with soluble and insoluble fiber contents ranging from 6.0 to 7.1% and 4.1 to 4.9%, respectively. Oats are particularly rich in soluble dietary fiber that is mainly constituted by β-glucans. In addition, groats are high in ferulic acid and other phenolic antioxidant compounds. Therefore, its consumption is linked to gastrointestinal health maintenance and reduction or prevention of constipation, colon cancer risk, blood cholesterol, and cardiovascular diseases (Bleidere et al. 2014, Serna-Saldivar and Sanchez-Hernandez 2020).

The typical stages of the oat milling process summarized in Figure 6.4 are grain cleaning, heat treatment, dehulling, cutting, flaking, and roll milling to produce meals or flours. The cleaning operation involves the utilization of magnets, air aspiration, disc separators, and gravity tables (Serna-Saldivar 2010a). The aim is to remove dust, chaff, extraneous seeds, and other impurities. Then oats are classified into thinner or plumper kernels according to width or thickness in a width sizer, which consists of a rotating horizontal cylinder with round or rectangular perforations and a vibrating conveyor located underneath the drum (Serna-Saldivar 2010a). After cleaning and sizing, the oats are heat treated in order to inactivate lipases and lipoxygenases which enhance oxidation or rancidity and formation of off-flavors and odors. Heating is regularly performed in pan dryers or radiator vertical columns. The thermal treatment is usually performed at temperatures varying from 88 to 93°C in order to enhance the posterior removal of the glumes and the inactivation of lipases (Serna-Saldivar 2010a, Girardet and Webster 2011).

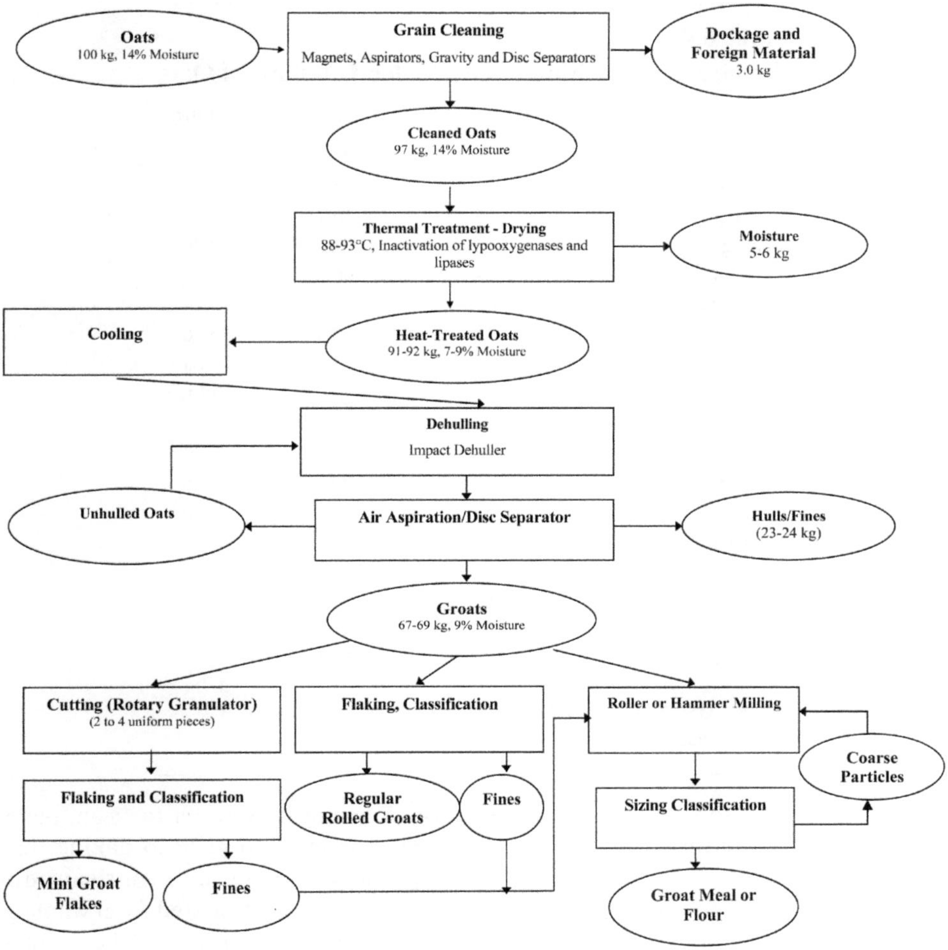

FIGURE 6.4 Diagram of the typical dry-milling process for the production of rolled and oat meals used for the production of snacks.

The critical step of dehulling is commonly performed in impact dehullers equipped with a high-speed rotor that expels the oats against a carborundum ring attached to the machine housing. The speed of the rotor is tuned to detach the hulls from the grain and minimize breakage of groats. Normally, the rotor speed is adjusted between 1,400 and 2,000 rpm in order to achieve a dehulling efficiency of 90–95%. The dehulling apparatus produces a mixture of groats, free hulls, unhulled oats, broken kernels, and fines. The fines and hulls are separated by air aspiration whereas the unhulled oats by sizing in disk or gravity separators. The recovered whole oats are sent again to the dehullers in order to increase yield and efficiency. Optimally, the yield is 75% groats and 25% hulls. The sorted naked groats are a then cut and flaked for the production of rolled oats or sent to a milling unit to be ground into a coarse meal or fine flour. Rolled oats constitute the main finished product followed by meals or flours obtained after grinding of fines. Whole groats yield large flakes that are difficult to handle, store, and package. Therefore, groats are usually cut into uniform pieces before flaking and precooking. The groat is cut with rotary granulators into two to four pieces. This highly efficient equipment, which should yield less than 2% fines, consists of rotating drums perforated with countersunk round holes through which the groats align themselves endwise and fall against stationary knives that are

TABLE 6.4
Chemical Composition of Dry-Milled Fractions of Oats

Nutrient per 100 g	Oats	Groats	Hulls
Water, g	14.0	8.2	8.0
Energy, kcal	286	389	–
Protein, g	14.7	16.7	5.7
Lipids, g	5.50	6.90	2.30
Carbohydrates, g	53.32	66.3	–
Dietary fiber, g	nd[a]	10.6	–
Ash, g	2.75	1.72	6.90
Calcium, mg	60.0	54.0	130.0
Iron, mg	6.2	4.7	12.8
Sodium, mg	2.0	2.0	–
Thiamine, mg	0.66	0.76	0.60
Riboflavin, mg	0.12	0.14	0.15
Niacin, mg	0.83	0.96	0.70
Folic acid mcg	52	56	–
Type of fat, g			
Saturated	0.97	1.22	–
Monounsaturated	1.73	2.18	–
Polyunsaturated	2.02	2.54	–

Source: USDA (2021) and Serna-Saldivar (2010a).
[a] Hulled oats contain 10.4 g/100 g of crude fiber.

arranged around the bottom and outside of the drum. Some millers prefer to steam the groats immediately ahead of cutting rather than after granulation and just prior to flaking. In other mills, groats are tempered 15–30 min before cutting. The cut groats are then steamed in the traditional way immediately ahead of flaking. The sized groats are separated from whole groats by length on a disk separator that lifts the cut pieces and reject whole kernels. The uncut groats are then recycled back to the granulator. In order to produce instant flakes, the groats are conditioned and steamed before rolling or flaking. Oat meal or flour can be alternatively produced by reducing the particle size of the cut and steamed pieces of groats with break roller mills and/or hammer mills. The resulting meal is usually sifted and coarse particles recycled for further grinding until the desired particle size distribution is achieved (Girardet and Webster 2011). The average chemical compositions of groats and milled products are detailed in Table 6.4.

6.2.5 Dry Milling of Other Cereal Grains

Barley (*Hordeum vulgare*), rye (*Secale cereale*), and triticale (*Triticum secale*) are usually dry milled into flours following the same principles described for wheat milling described above. In addition, these cereal grains can be pearled and used for the production of nutritional bars or other snacks.

On the other hand, sorghum (*Sorghum bicolor*) and some types of millets are commonly abrasively decorticated and in some cases degermed. A wide array of refined or semi-refined milled fractions could be obtained varying from decorticated kernels, grits, meals, and fine flours. In recent years, mechanized milling processes have been developed especially for grain sorghum. The most common is by abrasive decortication followed by reduction of particle size via roller or hammer milling (Serna-Saldivar 2010a). Particularly, decorticated sorghum and refined flours and meals are increasingly being used for the production of extruded gluten-free snacks (Alavi et al. 2019).

6.3 WET-MILLING PROCESSES FOR PRODUCTION OF CEREAL STARCHES

Starch is the most abundant chemical component associated with cereal grains and considered the most functional for the production of snacks. Historically starch has been extracted for many centuries due to its functionality and economic value and has found wide use for the production of many types of snacks.

The wet-milling industries have learned to optimally mill the different sorts of cereals that differ in pericarp and endosperm structures and association of starch granules to the endosperm protein matrix. The various types of wet-milling industries aim toward the extraction of the maximum possible amount of native or undamaged starch granules. By far, maize or corn is the main source of refined starch in the globe because of its availability, cost, high starch content (> 65%), and value of its coproducts (germ and pericarp). Globally, more than 85 million tons of refined starches are produced. Of this output about 53, 36, and 11% are used for sweeteners, native starch, and modified starch production, respectively. The market for modified starch reached a value of USD 11.1 billion in 2020 (https://www.beroeinc.com/category-intelligence/starch-market/). Of this output, about 77–80% is obtained from corn followed by potato and wheat.

The objective of wet millers is to obtain the highest possible yield of undamaged starch that meets specifications for color, chemical, and functional properties. In the particular case of wheat, the industry aims toward the production of refined starch and vital gluten.

6.3.1 Wet Milling of Maize

Most of the industrially produced starch is obtained from yellow dent maize. The optimum grain physical properties for milling are undamaged and soft-textured dent kernels with test and thousand kernels weights of 67.5 kg/hL and greater than 300 g, respectively. These kernels contain a favorable ratio of endosperm to germ and pericarp, higher amounts of starch, and their starch granules are easier to extract. Furthermore, the softer endosperm structure hydrate faster during the critical step of steeping, considered the bottleneck of the process (Eckhoff and Watson 2009, Serna-Saldivar 2010b, Serna-Saldivar 2013, Raush et al. 2019). The typical process starts when shelled maize containing pieces of cob, chaff, foreign seeds, other extraneous material, and broken kernels is cleaned with air aspirators followed by sifters and gravity separators. Aspiration is used to remove dust and light chaff, the reciprocating screening cleaner is commonly equipped with a sieve of 4.8 mm orifices to remove broken kernels and smaller pieces of foreign material. The gravity separators are designed and calibrated to remove ferrous contaminants and stones of approximately the same size as the corn kernels. These cleaners usually contain magnets to trap metal trash (bolts, tools, scrap iron, etc.) and protect milling equipment from damage (Eckhoff and Watson 2009, Serna-Saldivar 2013).

Figure 6.5 summarizes the sulfur dioxide wet-milling process of corn. The most critical step is the steeping of the kernels in a sulfur dioxide solution. Kernels are normally steeped for 24–48 hr at a temperature of 48–52°C. Generally, softer kernels require shorter steeping times compared to harder counterparts. At the end of the steeping period, the kernels should contain approximately 45% moisture and release about 6.0 to 6.5% of their dry matter weight into the steep liquor and absorb about 0.2 to 0.4 kg of sulfur dioxide per ton (Eckhoff and Watson 2009, Serna-Saldivar 2013). The typical countercurrent steeping can be divided into three sequential stages known as fermentation, sulfur dioxide diffusion, and dominated sulfur dioxide. The first stage that normally lasts from 8 to 12 hr is when kernels absorb the steep solution which contains approximately 2% lactic acid produced by *Lactobacillus*. The length of the fermentation phase is controlled by the sulfur dioxide concentration in the steep liquor. The growth of this bacteria stops when the sulfur dioxide concentration exceeds 100 ppm. The second stage of steeping also lasts from 8 to 12 hrs and is characterized by the sulfur dioxide diffusion into the kernels. The reducing action of the sulfur dioxide that disrupts disulfide bonds enhances the reactions that lead to the release of starch granules and the weakening of the protein matrix. The sulfur dioxide acts synergistically with the proteolytic action of the *Lactobacillus* breaking endosperm cell walls and hydrolyzing the protein matrix that

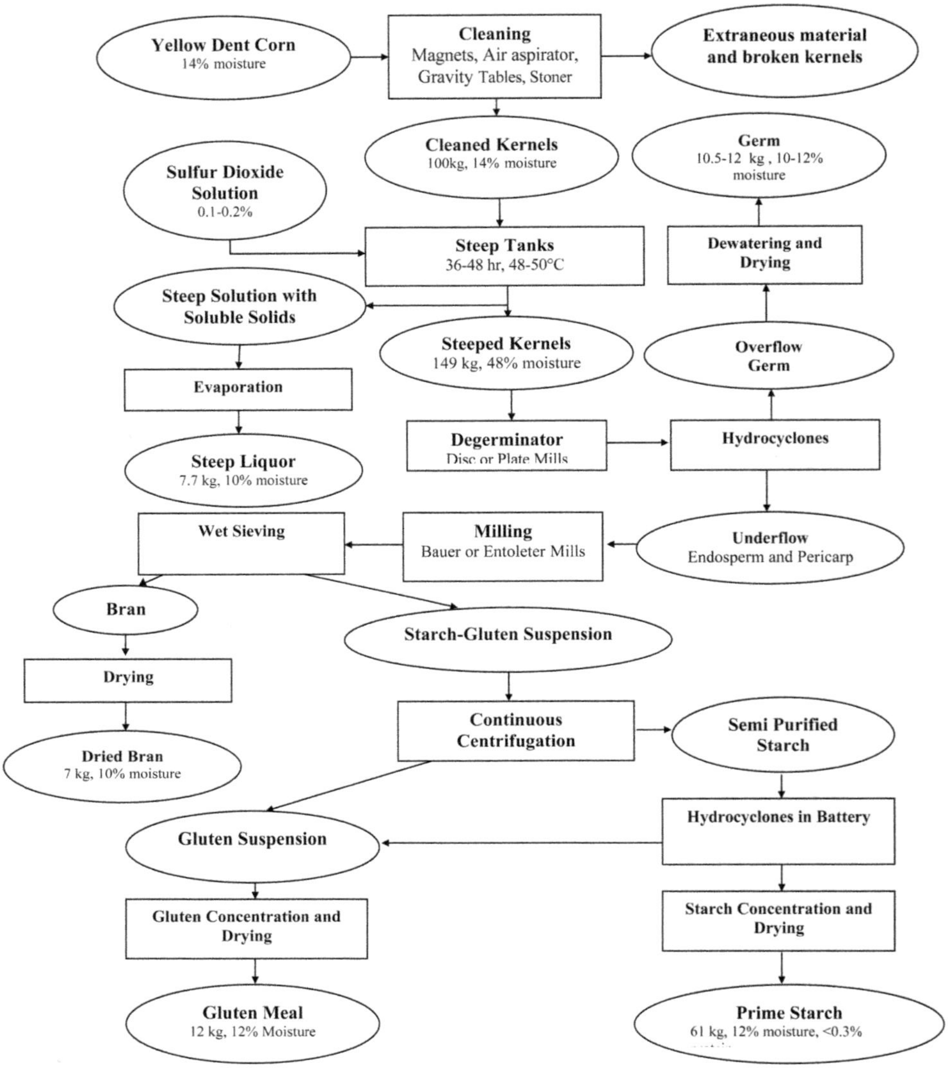

FIGURE 6.5 Diagram of the wet-milling process commonly used for the production of corn starch.

engulfs starch granules. The final stage is characterized by the addition of high concentrations of sulfur dioxide (up to 2,500 ppm) to the oldest steeped kernels. Besides softening the endosperm structure, the sulfur dioxide and lactic acid avoid germination and the growth of undesirable microorganisms (Eckhoff and Watson 2009, Serna-Saldivar 2010b, Serna-Saldivar 2013, Raush et al. 2019). The steeping operation is conducted in 6 to 12 large stainless steel tanks arranged in batteries that each have capacity to hold 250 to 600 tons. In addition, the battery of tanks is furnished with heat exchangers to maintain the desired temperature and pumps to move steep solutions from one tank to another or to withdraw it from the system (Eckhoff and Watson 2009).

The steeping step increases the original grain moisture from 14% to around 48%. The steep solution enters the kernel typically at the tip cap and moves quickly to the germ and the pericarp´s tube cells where it is gradually distributed to the endosperm. The steep solution moves quickly up the sides of the kernel and enters through the dent region into the porous floury endosperm region. According to Eckhoff and Watson (2009), the germ and endosperm tissues become wet in about 4–8 hrs, respectively. In addition, the steep solution uptake causes kernels to expand 55–65%.

Approximately 5 to 7% of the corn solids, mainly from the germ, solubilize and lixiviate during steeping. These are chiefly constituted by albumins and globulins, lactic acid, minerals, phytic acid, and B-vitamins (Eckhoff and Watson 2009, Serna-Saldivar 2010b, Serna-Saldivar 2013). Thereafter, the conditioned kernels are milled in plate or disk attrition mills in preparation for degermination. These mills consist of plates covered with pyramidal knobs and the impact ring is absent. The bulk of the germ is liberated in the first pass, but a second pass is usually provided after the free germ has been removed in hydroclones. The mill gap is adjusted to give the maximum free germ with minimum tissue breakage because the mechanical damage disrupts *scutellum* cells that will liberate oil, which is mostly absorbed by the gluten. Interestingly, about half of the total starch and gluten, especially associated with the hydrated chalky endosperm, is also liberated during this milling phase. The ground particles diluted in water are transported into a series of hydroclones to separate the less dense germ from the rest of the kernel components. The germ has a lower density because it contains most of the oil (density = 0.9) and does not contain any starch (density = 1.5 g/cm^3). The ground maize with 12–14% solids is forced into hydroclones that effectively and continuously separate heavier endosperm and fiber particles from the lighter germ stream which exits the equipment from the top (Eckhoff and Watson 2009, Serna-Saldivar 2010b, Serna-Saldivar 2013, Raush et al. 2019).

The germ stream is dewatered to 45–50% solids by passing it through mechanical squeezers prior to dehydration. This coproduct is generally used for the mechanical, solvent, or dual extraction of the oil (Chapter 7). The free starch and protein, are separated from the unmilled endosperm and fiber by screening over a screen bend device to reduce the load of solids and water to subsequent milling operations (Eckhoff and Watson 2009). The second milling stage starts when the underflow from the hydrocyclones is further milled to release the starch granules. The operation is normally performed with Bauer or Entoleter mills. The latter is the most preferred by corn wet millers. These mills release the starch granules from the protein matrix causing minimal mechanical damage (Eckhoff and Watson 2009, Serna-Saldivar 2013).

The next step consists of the separation of the starch and gluten released by milling from the larger pieces of pericarp. This is best achieved by pumping the slurry with considerable force in a fixed concave wedge bar screen surface arranged in a vertical position. The screens have slit widths of 50, 75, 100, and 150 μm. Generally, a 120° pressure-fed screen with a bar screen slit width of 50 μm is employed by the industry. The separation screen is designed to minimize clogging and for continuous processes. The pericarp mat discharging from the screen is reslurried and reprocessed over other units in order to recuperate as much endosperm as possible. This coproduct is first dewatered in a centrifugal screen, mechanical squeezers, and/or a continuous discharge centrifuge to a moisture content of 65–75% and normally blended with concentrated steep liquor, maize cleanings, spent or defatted germ, and other coproducts for drying to produce corn gluten feed that contains 18–22% protein (Eckhoff and Watson 2009).

The following steps are designed to separate the more dense starch granules from the less dense gluten. This is accomplished by first passing the starch–gluten slurry in a series of continuous centrifuges followed by a secondary purification using a series of hydrocyclones. The starch slurry is the first vacuum filtered to concentrate solids from 40 to 50% and then in perforated belt type dryers at 65–150°C (Raush et al. 2019). The gluten stream is first dewatered in basket centrifuges, followed by solid concentration with vacuum filters and dehydration to 12% moisture in flash, rotary, or drum driers. The purified starch is refined in order to reduce the protein to approximately 0.3% (dwb) whereas the dry gluten to 68 to 75% protein (Eckhoff and Watson 2009, Serna-Saldivar 2010b, Serna-Saldivar 2013). The yield of prime starch and gluten meal from clean maize is from 61 to 65% and 12 to 13%, respectively (Figure 6.5).

Some wet-milling industries obtain identity-preserved waxy and high-amylose starches for niche markets. By much, the mutant maize most widely processed is the waxy type that contains more than 95% amylopectin. More than one million tons of this particular genotype are annually wet milled in the United States. The recommended steeping temperature of the waxy and high-amylose kernels is 3°C lower and higher compared to the temperature commonly used for regular maize (Raush et al. 2019).

6.3.2 Wet Milling of Wheat

Historically it is known that the starch from wheat was extracted by the ancient Egyptian culture for the preparation of glue of papyrus and as an aid to solidify the cloth that covered the mommies. Nowadays, about 4.7 million tons of wheat starch are annually produced in the world. This represents about 7.8% of the total starch production worldwide (Maningat et al. 2009, Serna-Saldivar 2013). More than half and 17% of the wheat starch are produced in the United States and Europe, respectively. The modern wet-milling plants process refined flours with two major goals: the production of vital gluten and native starch. The vital gluten is a food additive used in the baking and meat processing industries whereas the starch has analogous uses as corn starch.

The wheat endosperm possesses two distinctive types of starch granules, which differ in size, shape, and protein content. These starches are classified as A or B types. The A-starch granules are large and lenticular with a mean diameter of 14.2 μm whereas the B counterparts are small and spherical with a mean diameter of only 4.1 μm. Quantitative separations of wheat starch granules indicated that wheat starch contains approximately 70% by weight of A granules. The analysis of 130 lines of Australian wheat showed that the different genotypes ranged from 23 to 50 by weight of B granules and that soft wheats normally contained the lowest levels. The physical, chemical characteristics, and susceptibility to amylase hydrolysis of these types of starches are described by Maningat et al. (2009). The A and B pure starches normally contain 0.3% and up to 1% protein, respectively, and both types are recognized to trigger allergic reactions associated with gluten intolerance or celiac disease.

There are more than a dozen ways to extract starch from whole wheat or refined flours, However, most commercial wheat starches are currently obtained from three manufacturing procedures that employ refined flours as raw materials. These are known as dough-Martin, dough-batter, and batter procedures (Maningat et al. 2009, Serna-Saldivar 2010b, Serna-Saldivar 2013).

6.3.2.1 Martin Process

This process was originally proposed in Paris about 1835 and is still considered the most practiced worldwide. This method summarized in Figure 6.6 employs refined wheat flour and water as major feedstocks. The process comprises first the blending of the wheat flour with water in a 2:1 ratio to yield a uniform and properly developed dough. The starch is simply extracted by washing the dough placed in a large reactor with a water spray system. The water and mechanical action gradually release starch granules without breaking the gluten matrix. The industry uses several washing devices such as a rotary drum, mixers with twin screws, and blenders equipped with a couple of sigma blades that counterrotate at different speeds. Regardless of the equipment, the water extracts most of the starch granules along with some soluble compounds like proteins (albumins and globulins) and soluble sugars. The insoluble part, consisting of concentrated wet gluten (gliadins and glutelins) left behind on the screens are subsequently compressed through a pair of rolls to remove excess water in preparation for further dehydration in vacuum or drum driers. On the other hand, the starch slurry containing A- and B-starch granules with approximately 10% solids is first passed through a set of vibrating screens in order to remove small pieces of gluten and other contaminants and then separated with continuous centrifuges and/or hydrocyclones. Both the gluten and starch are dehydrated to reduce their moisture to about 10%. The industrial specification for prime wheat starch does not allow more than 0.3% residual protein. The Martin process yields about 65% A- and B-starch and 14% vital gluten (Figure 6.6). The rest are other solids that are separated during the washing and centrifugation steps (Maningat et al. 2009, Serna-Saldivar 2010b).

6.3.2.2 Dough-Batter Process

This industrial process also named "hydroclone" consists of two basic stages known as gluten agglomeration and starch separation. The gluten is extracted with high-shear homogenizers whereas the starch with hydrocyclones or decanters. The main advantage of this method is that it requires less water compared to the Martin procedure. The process starts when refined wheat flour normally

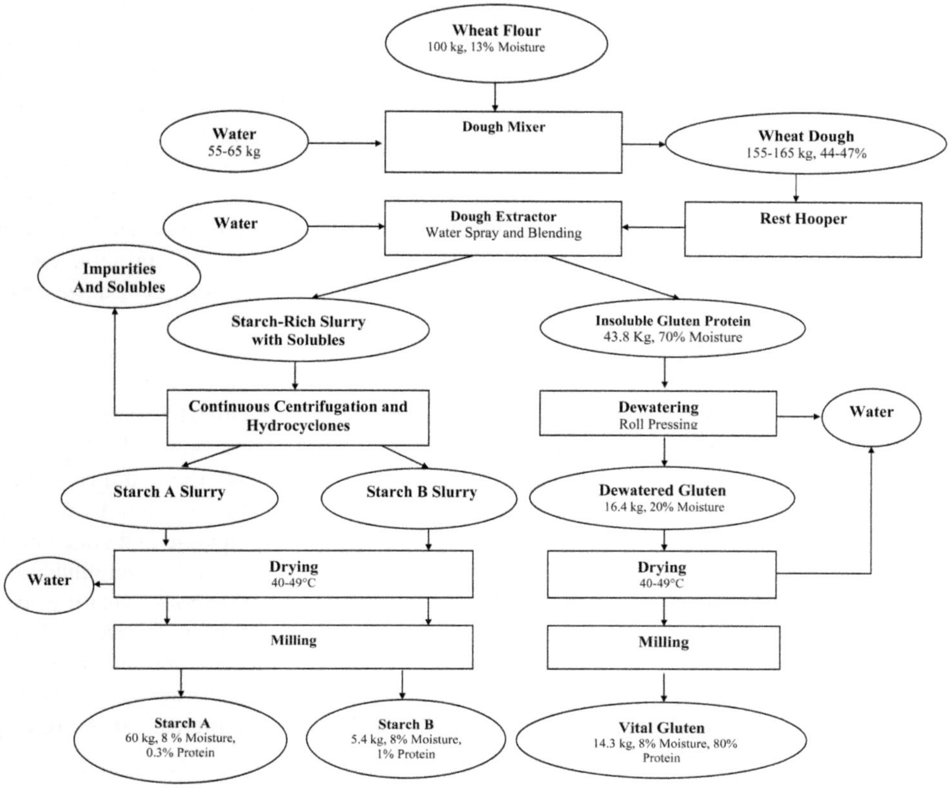

FIGURE 6.6 Diagram of the dough-Martin process commonly used for the production of wheat starch and vital gluten (adapted from Serna-Saldivar 2013).

obtained from the dry-milling process described before is kneaded with warm water to form a slack dough. After a 10– to 12-min resting, the dough is strongly agitated with water to form gluten threads and a starch-rich slurry. The slurry is filtered through a strainer and then pressure pumped to a set of hydrocyclones where the gluten migrates in the light stream leaving the units. The gluten agglomerates in strands at the exit port due to its concentration and intense mixing. Then, the gluten is screened, washed to remove B-starch, bran, and cell-wall debris, and then dewatered prior to drying. On the other hand, the starch-rich slurry is sent to a set of hydrocyclones that effectively separate, wash, and concentrate the large A-starch granules in the underflow fraction. The A-starch is further separated from fiber in static and rotary screening machines. The remaining starch slurry separated in the overflow fraction is screened first to remove fine fibers and then separated into A-starch and B-starch fractions. The advantages of the dough-batter process compared to the dough-Martin counterpart include better yield of vital gluten, lower water, and energy requirements, and lower volume of effluents (Maningat et al. 2009).

6.3.2.3 Batter Processes

These main batter processes are known as Fesca, Raisio, and high-pressure disintegration. The Fesca method was first used about 100 years ago to obtain both wheat starch and vital gluten. This process was later on modified and improved in order to obtain better yields of high protein gluten and refined starch. However, the principles of the old Fesca process were upgraded to develop the Alfa-Laval/Raisio and high-pressure disintegration processes. Independently of the type of process, the technologies use refined wheat flour with water to yield a thin batter that is washed and later on subjected to centrifugal forces in order to separate vital gluten and the A and B starches from other contaminants.

In the Fesca process, the wheat flour is mixed with water to form a thin batter which is dispersed by shearing to prevent gluten development. The batter is centrifuged in order to separate the starch from a protein concentrate. The starch is reslurried, refined, and dried whereas the protein concentrate usually spray dried. The main advantage of this process is that water usage is minimal and solids recovery is nearly 100% (Maningat et al. 2009, Serna-Saldivar 2013).

The Raisio process, also known as Alfa-Laval, significantly saves water. The batter is prepared by blending three parts water per one part flour with approximately 30% solids. The dispersion is mixed at 30°C with a high-shear mixer and the batter is further separated by a decanter into a starch fraction containing approximately 1% protein and a gluten fraction with 40% protein. The starch is further refined with conical screens and decanters whereas the gluten screened, dewatered, and dehydrated (Maningat et al. 2009).

The latest batter technology of high-pressure disintegration is mainly employed in Europe, Australia, and Canada and is based on starch extraction and separation by centrifugal forces from a highly sheared batter. The procedure starts when wheat flour is mixed with 0.85–0.95 parts water at 35°C in a continuous dough mixer to achieve a smooth batter. The batter is pumped at high pressure (100 bar) through a homogenizer in order to release starch granules from the hydrated gluten that is also disrupted into small particles. Thereafter, the homogenized batter is diluted with water to 30% solids and fed into a three-phase decanter centrifuge designed to separate and concentrate the denser A-starch from the intermediate density stream rich in B-starch granules and gluten and the least dense viscous stream containing pentosans (arabinoxylans) and other contaminating solubles. The medium-dense stream is pumped into a sieve drum where B-starch granules are washed from the fine gluten particles. The A-starch slurry is also screened and then refined either with multistage hydrocyclones or with decanters. The major advantages of the process are reduction in water consumption to as low as three parts per part of wheat flour and a 10% increase in A-starch yield (Maningat et al. 2009, Serna-Saldivar 2013).

6.3.3 Wet Milling of Rice

It is estimated that only 1% of the total starch produced worldwide is rice starch. The commercial production of this starch is limited due to the relatively high cost of the rice and the difficulty to extract the starch due to the hard endosperm structure. However, rice starch is unique in terms of properties and niche functionalities because it synthesizes compound starch granules that upon milling release tiny starch angular-shaped subunits that only measure 2–9 μm (Table 6.5).

Rice starch is normally obtained after an alkali extraction that is required to optimize yields and starch purity. The main disadvantage of this process is that yields effluents that are difficult to treat. In order to obtain prime starch, preferably broken kernels are treated with sodium hydroxide in order to soften the vitreous endosperm via solubilization of glutelins. Normally the rice is stepped in a 0.3 to 0.5% NaOH solution adjusted to 50°C for up to 24 hrs.

The steeped stock is fed to disc or attrition mills and the slurry is allowed to rest for 10–24 hrs to further dissolve the proteins. The fiber is removed by filtration and the starch slurry recovered by centrifugation. The starch is washed with water to remove excess alkali and then dehydrated to 10% moisture (Mitchell 2009). The rice starch normally contains less than 1% protein (Mitchell 2009, Serna-Saldivar 2010b, Serna-Saldivar 2013).

6.4 PROPERTIES OF MAIN CEREAL STARCHES

Starch granules contain the linear and branched starches named amylose and amylopectin, respectively. Amylose is essentially a linear polymer composed of glucose units linked with alpha 1-4 glycosidic bonds whereas amylopectin has also several alpha 1-6 glycosidic bonds. In amylopectin only 4 to 5% of the total glycosidic bonds are alpha 1-6. Amylose is usually

TABLE 6.5
Morphology, Physical, and Viscoamylograph Properties of Main Cereal Starches

	Corn		Wheat	Rice	
	Regular	Waxy		Regular	Waxy
Type of granule	Single	Single	Single	Compound	Compound
Granule shape	Round, Angular	Round, Angular	Round, Angular	Polyhedral, pentagonal	Polyhedral, pentagonal
Granule size, µm	5–30	5–25	5–35	1.6–8.7	1.9–8.1
Granule diameter, µm	13.8–14.5	13.9–14.2	16.4–16.6	–	5.4–5.6
AMY:AMP	25:75	1:99	25:75	19:81	1:99
Paste temp. °C	73	69	77	79	78.5
Peak viscosity, BU	470	1,000	65	420	---
Viscosity at 95°C, BU	470	400	60	400	---
Viscosity hold 95°C/30 min, BU	350	250	60	350	---
Viscosity at 50°C, BU	830	390	300	840	---
Viscosity hold at 50°C/30 min, BU	760	370	270	760	
Retrogradation	High	Low	High	Medium	Low

Note: Tests were conducted with 35 gr starch/500 ml.

composed of approximately 1,500 glucose units whereas amylopectin is a larger molecule containing up to 600,000 units. The ratio of amylose:amylopectin in regular starches is generally 25:75 whereas waxy starch granules contain more than 95% amylopectin. Although amylose is called linear starch the molecule form helixes, which allows these molecules to complex with iodine, free fatty acids, alcohols, and emulsifiers. One important characteristic of amylose is that stains purple with iodine and is the most important molecule related to retrogradation and formation of cohesive gels. The native starch granule is water insoluble and swells reversibly when placed in water. The most important feature of native starches is that they show birefringence when exposed to the polarized light plane. This phenomenon occurs due to the high structure organization within the granule that makes these structures behave like pseudocrystals. As indicated before, amylopectin is the predominant starch molecule and is much larger than amylose. Due to its larger size and more complex structure, amylopectin stains brownish red with iodine. Amylopectin has a low retrogradation and forms weak and sticky gels (Fuentes-Zaragoza 2010, Serna-Saldivar 2010c).

The comparative physical and viscoamylograph properties of various types of cereal starches are summarized in Table 6.5. Properties of wheat starch are similar to corn starch although the latter is a slightly better thickener. Rice has one of the smallest starch granules of the cereal starches varying in size from 2 to 9 µm. The forms of the granules are polyhedral or pentagonal dodecahedron. The main industrial use of rice starch is in the cosmetic industry. However, the rice starch has unique viscoamylograph properties that make it ideal for use in second-generation extrusion processes (Chapter 12). Potato starch differs from maize and wheat starches and therefore has other applications. Potato produces the largest starch granules and when heated produces the highest peak viscosity. The most important characteristics of potato starch are: low gelatinization temperature, high consistency or viscosity on pasting, a high decrease in viscosity when further heated and stirred, excellent flexible film formation, and binding capacity. The production of potato starch is covered in Chapter 14 (Serna-Saldivar 2010c).

TABLE 6.6
Chemical and Functional Properties of Main Types of Modified Starches

Type of Modified Starch	Reactant and Production	Functionality and Application
Pregelatinized	Heat treatment in the presence of water to enhance starch gelatinization.	Instant starches that disperse in cold water. Used in beverages and puddings.
Resistant Starch type 4	Chemically modified starches known as RS4 produced by cross-linking to form products like distarch phosphate ester, starch ethers, or starch esters.	Foods in which these low caloric modified starches have less susceptibility to digestion due to cross-linking. RS4 is used to impart crispness to snacks and breakfast cereals. These properties have been successfully used in a range of reduced-calorie baked and extruded products.
Acid-treated	Treatment with acids in order to randomly hydrolyze starch molecules into dextrins. Reduces viscosity (acid thinning) and produces firmer gels due to the higher rate of retrogradation.	Confectionary industry (fillings for candies).
Dextrinized	Dry heat with acid or alkaline catalyst that disrupts the starch granules.	Improved water solubility and reduced viscosity.
Oxidative	Treatment with sodium hypochlorite in order to oxidize hydroxyl groups to carboxyl or carbonyl groups. Reduces viscosity, gelling, and retrogradation and yields clear pastes.	Food applications for high solids and lower viscosity. Use as an additive for batters and breading.
Bleached	Treatment with hydrogen peroxide, peracetic acid, ammonium persulfate, potassium permanganate, and sodium hypochlorite to oxidize pigments such as carotenoids and xantophylls.	Used as fluidizing agent for dry powders such as confectioner's sugar.
Acetate derivative	Starches are reacted with acetic anhydride in order to substitute acetate groups. The acetate derivative starch has reduced gelatinization temperature, higher hot peak viscosity, and lower cold paste viscosity and retrogradation.	Food thickeners and stabilizers in gravies, fruit pies, salad dressings, and filled cakes.
Succinate derivatives (OSA starches)	Starch suspensions are treated with octenyl succinic anhydride under alkaline conditions to impart surface activity with both hydrophilic and lipophilic groups.	Used as food thickener for encapsulation purposes and mainly to produce stable emulsions for the beverage and meat products.
Phosphated	Treatment with monosodium orthophosphate or sodium tripolyphosphate in order to produce stable and clear gels.	Used as flocculants, emulsifiers, and adhesives in the food industry.
Hydroxypropylated	Starch is treated with propylene oxide to substitute with hydroxypropyl groups. The modification reduces gelatinization temperature and produces more clear and stable pastes.	Hydroxyethyl starches are mainly used as surface sizing whereas hydroxypropylated are used in food gravies, pie fillings, and salad dressings.
Cross-bonded	Also known as distarch phosphate. This type of modification is made after reaction with phosphorus oxychloride or trimetaphosphate, which enhances cross-linking of glycosidic chains.	Reduce swelling and loss of viscosity due to temperature, shear, and acid. The cross-bonded starches are used in the canning industry as a stable thickener, for salad dressings and baby foods. In addition, they are used as a functional ingredient in the production of extruded snacks.

Source: Johnson (2000), Sajilata et al. (2006), Fuentes-Zaragoza et al. (2010), Serna-Saldivar (2010c), BeMiller (2019).

6.5 PROPERTIES OF MODIFIED STARCHES

Native starches, obtained by the various wet-milling processes described above, are industrially transformed by physicochemical and enzymatic processes into modified or functional starches for special applications in the various segments of the food industry. Modified starches are industrially manufactured to fulfill special functionalities or uses. These starch modification operations are normally integrated with the wet-milling operations described before because most reactions are aqueous and take place in the wet starch slurries saving the energy-demanding dehydration step. Starch modifications are aimed toward altering the starch granule features and the susceptibility of starch granules to gelatinization, cooking, retrogradation, and/or gelling. Chemical modifications can be hydrolytic, oxidative, ester-forming, or ether forming. The most wide types of modified starches are acid thinning, oxidized/bleached, molecular cross-bonded, and chemically substituted or derivatized. Some starches are manufactured by combining the methods mentioned before. Most manufacturing technologies consist of treating wet slurries of native starches, chiefly corn starch, in a reactor with temperature and agitation controls. Generally, a starch slurry with approximately 35% solids and adjusted to a certain pH is treated with chemicals to promote the desired reactions (Table 6.6). In most instances, gelatinization and starch solubilization is avoided so to prevent the disruption of the starch granule.

Recently the intended starch modification to lower its digestion in the gastrointestinal tract has resulted in the development of resistant starch 4 or RS4. These particular chemically modified starches (Table 6.6) are physiologically similar to soluble dietary fiber with similar prebiotic properties that lower the colonic pH and promote the growth of microbiota which exerts many positive health implications described in Chapter 21 (Sajilata et al. 2006, Fuentes-Zaragoza et al. 2010). Therefore, the usage of RS4 for the production of healthy and reduced-calorie snacks is expected to increase.

REFERENCES

Alavil, S., Mazumdar, S.D., and Taylor, J.R.N. 2019. Modern convenient sorghum and millet food, beverage and animal feed products, and their technologies. In: *Sorghum & Millets: Chemistry, Technology and Nutritional Attributes*, eds. J.R.N. Taylor, and K.G. Duodu, 293–329. Duxford: Woodhead Publishing, AACC-International.

Anderson, B., and Almeida, H. 2019. Corn dry milling: Processes, products, and applications. In: *Corn Chemistry and Technology*, ed. S.O. Serna-Saldivar, 405–433. Duxford: Woodhead Publishing, AACC International.

Bao, J. 2019. Rice milling quality. In: *Rice Chemistry and Technology*, ed. J. Bao, 339–369. Saint Paul: AACC International.

BeMiller, J. 2019. Corn starch modification. In: *Corn Chemistry and Technology*, ed. S.O. Serna-Saldivar, 537–550. Duxford: Woodhead Publishing, AACC-International.

Bleidere, M., Timbare, R., Bobere, N., Paklone, M., Jākobsone, I., and Zute, S. 2014. Characterisation of hulled oat grain physical and biochemical parameters significant for dietary products. *Proceedings of the Latvian Academy of Sciences. Section B. Natural, Exact, and Applied Sciences*, 68(3–4): 166–173.

Bond, N. 2006. Rice milling. In: *Rice Chemistry and Technology*, ed. E.T. Champagne, 283–300. St. Paul: AACC.

Brekke, O.L. 1970. Corn dry milling industry. In: *Corn Culture, Processing, Products*, ed. G.E. Inglett, 269–291. Westport: AVI Publishing Co.

Eckhoff, S.R., and Watson, S.A. 2009. Corn and sorghum starches. In: *Starch Chemistry and Technology*, eds. J. Bemiller, and R. Whistler, 373–440. New York: Academic Press.

Fuentes-Zaragoza, E., Riquelme-Navarrete, M.J., Sanchez-Zapata, E., and Perez-Álvarez, J.A. 2010. Resistant starch as functional ingredient: A review. *Food Research International*, 43(4): 931–942.

Girardet, N., and Webster, F.H. 2011. Oat milling: Specifications, storage and processing. In: *Oats: Chemistry and Technology*, eds. F.H. Webster, and P.J. Wood, 301–320. Saint Paul: AACC-International.

Johnson, L.A. 2000. Corn: The major cereal of the Americas. In: *Handbook of Cereal Science and Technology*, eds. K. Kulp, and J. Ponte. New York: Marcel Dekker.

Maningat, C.C., Seib, P.A., Bassi, S.D., Woo, K.S., and Lasater, G.D. 2009. Wheat starch: Production, properties, modification and uses. In: *Starch Chemistry and Technology*, eds. J. BeMiller, and R.Whistler, 441–510. New York: Academic Press.

Manthey, F.A., Hareland, G.A., and Huseby, D.J. 1999. Soluble and insoluble dietary fiber content and composition in oat. *Cereal Chemistry Journal*, 76(3): 417–420.

Mitchell, C.R. 2009. Rice starches: Production and properties. In: *Starch Chemistry and Technology*, eds. J. BeMiller, and R. Whistler, 570–578. New York: Academic Press.

Posner, E.S. 2009. Wheat flour milling. In: *Wheat: Chemistry and Technology*, eds. K. Khan, and P.R. Shewry, 119–152. Saint Paul: AACC International.

Posner, E.S., and Hibbs, A.N. 2005. *Wheat Flour Milling*. Saint Paul: American Association of Cereal Chemists.

Rausch, K.D., Hummel, D., Johnson, L.A., and May, J.B. 2019. Wet milling: The basis for corn biorefineries. In: *Corn Chemistry and Technology*, ed. S.O. Serna-Saldivar, 501–536. Duxford: Woodhead Publishing, AACC-International.

Sajilata, M.G., Singhal, R.S., and Kulkarni, P.R. 2006. Resistant starch–A review. *Comprehensive Reviews in Food Science and Food Safety*, 5(1): 1–17.

Serna-Saldivar, S.O. 2010a. Dry-milling operations. In: *Cereal Grains: Properties, Processing and Nutritional Attributes*, ed. S.O. Serna-Saldivar, 177–223. Boca Raton: CRC Press (Taylor & Francis Group).

Serna-Saldivar, S.O. 2010b. Wet-milling operations. In: *Cereal Grains: Properties, Processing and Nutritional Attributes*, ed. S.O. Serna-Saldivar, 225–239. Boca Raton: CRC Press (Taylor & Francis Group).

Serna-Saldivar, S.O. 2010c. Industrial production of modified starches and syrups. In: *Cereal Grains: Properties, Processing and Nutritional Attributes*, ed. S.O. Serna-Saldivar, 395–416. Boca Raton: CRC Press (Taylor & Francis Group).

Serna-Saldivar, S.O. 2013. Wet milling and starch extraction. In *Engineering Aspects of Cereal and Cereal-Based Products*, eds. R.P. Ferreira Guine, and P.M. dos Reis Correia, 241–270. Boca Raton: CRC Press (Taylor & Francis Group).

Serna-Saldivar, S.O., and Perez-Carrillo, E. 2019. Food uses of corn and dry milled fractions. In: *Corn Chemistry and Technology*, ed. S.O. Serna-Saldivar, 435–468. Duxford: Woodhead Publishing, AACC International.

Serna-Saldivar, S.O., and Sanchez-Hernandez, D. 2020. Dietary fiber in cereals, legumes, pseudocerals and other seeds. In: *Science and Technology of Fibers in Food Systems*, eds. J. Welti-Chanes, S.O. Serna-Saldivar, O. Campanella, and V. Tejada-Ortigoza, 87–122. New York: Springer.

USDA, United States Department of Agriculture. 2021. Nutrient data laboratory. *Agricultural Research Service*. Electronic page: http://www.nal.usda.gov/fnic/foodcomp.

Zhang, W., and Hoseney, R.C. 1998. Factors affecting expansion of corn meals with poor and good expansion properties. *Cereal Chemistry Journal*, 75(5): 639–643.

7 Production and Properties of Fats and Oils

Sergio O. Serna-Saldivar

CONTENTS

7.1 INTRODUCTION

Fats and oils have always been an important part of the human diet. Among all nutrients, lipids provide the highest amount of energy per unit weight (9 kcal/g). In addition, some lipid materials carry important fat-soluble vitamins such as A, D, E and K and liposoluble nutraceuticals like phospholipids and phytosterols (Chapter 21). Most lipids consumed are triacylglycerol or triglycerides composed of three fatty acids (FA) esterified to a molecule of glycerol. These are considered the main reserve lipids in plant and animal tissues where they are stored in spherosomes and adipocytes, respectively. Fats and oils differ in properties and applications due to their FA composition consisting of an aliphatic chain containing from 4 to 26 carbons with a terminal carboxy group (COOH). The carboxyl group esterifies with the glycerol forming mono, di or triglycerides, and these molecules are broken down by lipases during digestion. All FA are produced from acetyl Co-A and therefore the molecule is built by adding two carbons. That is the reason why most FA contain an even number of carbons. The FA can be saturated, monounsaturated or polyunsaturated. The saturated does not contain double bonds whereas the monounsaturated and polyunsaturated contain one and two or more double bonds, respectively. According to the Geneva system of nomenclature, widely used among food scientists and technologists, carbons are numbered starting from the carboxyl end and ending with the methyl termination. Polyunsaturated FA contains their double

DOI: 10.1201/9781003129066-7

bonds following the methyl-interrupted rule. This means that three carbons separate the double bond positions being the middle carbon a methyl group. For every double bond the FA molecule has two less hydrogen atoms. Unsaturated and polyunsaturated FA can be also named according to the omega (ω) nomenclature widely used by medical doctors, dieticians and physiologists. The number of carbons existing between the last double bond and the terminal methyl group is designated as ω. Most unsaturated or polyunsaturated FA are ω9, ω6 or ω3. For instance, linolenic acid is denominated by the Geneva nomenclature 18:3 Δ 9, 12, 15 and 18:3 ω3 by the omega system. Interestingly, the omega system only specifies the number of carbons separating the last double bond from the methyl end. Milk oil contains short-chain FA such as butyric (C4:0), caproic (C6:0), capric (C8:0) and caproic (C10) whereas vegetable oils from tropical sources such as coconut, palm and palm kernel and cocoa are rich in medium-chain triglycerides containing lauric (12:0), myristic (14:0), palmitic (16:0) and palmitoleic (16:1 Δ 9) FA. The other vegetable oils are rich in oleic (18:1 Δ 9), linoleic (18:2 Δ 9, 12) and linolenic (18:3 Δ 9, 12, 15). These 18 C FA are ω9, ω6 and ω3, respectively. Animal fats such as lard and tallow contain significant amounts of saturated stearic (18:0) and palmitic (16:0) FA (Juttelstad 2004, List 2004, Erickson 2007, Serna-Saldivar 2008). Marine oils including nutraceutical oils from algae contain high amounts of high molecular weight polyunsaturated FA such as arachidonic (20:4 Δ 5, 8, 11, 14), eicosapentaenoic (20:5 Δ 2, 5, 8, 11, 14) most frequently called EPA and docosahexenoic (22:6 Δ 4, 7, 10, 13, 16, 19) commonly denominated DHA. The DHA is highly nutraceutical because its deficiency is associated with neurological, macular, high cholesterol and cardiovascular diseases (Chapter 21).

The FA composition of the oil greatly affects the functionality for frying (Chapter 3) and the susceptibility to oxidative rancidity. In general, the degree of unsaturation of FA lowers the melting point of the oil and increases susceptibility to oxidation. For instance, stearic acid has a higher melting point than oleic, oleic than linoleic and linoleic than linolenic. Contrarily, linolenic is more prone to oxidation compared to counterparts with lower degree of unsaturation. On the other hand, the fewer the carbons or molecular weight the lower the melting point. For instance, palmitic has lower melting point than stearic (Juttelstad 2004, List 2004, Serna-Saldivar 2008).

Fats/oils used in the preparation of snack foods can be divided into two classes based on their consistency at room temperature: (i) liquid oils such as most vegetable oils like soybean, cottonseed, sunflower, safflower, peanut, olive, canola and (ii) solids and semisolids such as vegetable shortenings, lard, butter, margarine, coconut and palm kernel. The main difference between these two categories is the degree of instauration related to the iodine value. The triglycerides of liquid oils are mainly composed of polyunsaturated and monounsaturated FA whereas solid, crystallized or plastic fats are high in saturated FA. Crystalized fats can exist in two major polymorphic forms named β and β´ that affect the appearance and texture. Liquid oils are the most commonly used or suitable for frying purposes. The characteristics of the frying oil/fat are critically important in the final quality of snacks because it is highly associated with flavor, eating quality and mouth feel.

Unsaturated FA can exist in either the *cis* or *trans* form depending on the geometric configuration of the hydrogen atoms bound to the carbons shearing the double bonds. If the hydrogen atoms are on the same side of the carbon chain the isomer is *cis* and if they are located on opposite sides the configuration is *trans*. The *cis* isomers exist in most naturally occurring FA whereas the *trans* counterparts are mainly formed when oils are hydrogenated to produce partially hydrogenated frying oils and plastic shortenings. The *cis* and *trans* FA have different melting points and crystallization properties. An unsaturated *trans* FA containing equal molecular weight has a higher melting point compared to its *cis* isomer. Recently, *trans* FA have received special attention from regulatory agencies because of their proven negative effects on human health (Chapters 20 and 21). In some countries around the world, labeling of the amount of *trans* fats is mandatory whereas in others products containing significant amounts of this harmful FA are prohibited. The trend of the industry today is to produce snacks and other products free of *trans* FA (Juttelstad 2004, List 2004).

Oils are also used to formulate seasonings and flavorings (Chapter 8), which undoubtedly play an important role in the consumer acceptance of savory snacks. The aim of this chapter is to cover the main oilseeds used to extract commercial oils widely used by snack food processors and to describe their comparative chemical, physical and functional properties. In addition, the chapter depicts processes commonly employed to mechanically or chemically extract crude oils followed by the steps of refining to yield commercial oils and the alternative modification of their physical properties via fractionation, hydrogenation, and intra or inter esterification. The new emerging technology for the production of oleogels as an alternative for plastic shortenings and animal fats is also discussed.

7.2 OILSEEDS AND OIL-BEARING FRUITS

For several thousand years, fats and oils have been recovered from oil-bearing fruits and seeds. In 2019, 1.1 billion tons of oilseeds were harvested from 324.5 million hectares with a yield of approximately 3.4 ton/Ha (FAO 2021). More than several hundred varieties of plants produce oil in sufficient quantities to warrant recovering but less than 15 sources are commercially significant. Table 7.1 lists the most important oilseeds along with their average oil content and most practiced method of oil extraction. Oil-bearing materials are divided according to growth habit into perennial and annual crops and growth ecosystem into tropical and temperate crops. Tropical sources are all-perennial and normally extracted using mechanical presses. These include copra (*Cocus nucifera*), African palm (*Eleais guinensis*) and Cocoa (*Theobroma cacao*). The African palm is the fastest-growing oil-bearing crop in the world with an estimated production of fruits and oil in the year 2019 of 410 and 71.47 million tons, respectively. More than 28 million Ha are planted with an average yearly yield of 14.5 Ton/Ha. Indonesia and Malaysia are the leading producers with 60 and 24% of the total global production (FAO 2021). The fruit yields two different types of oil: palm and palm kernel. The first is widely used for frying, whereas the second is for confectionary purposes. Triglycerides from tropical oils are rich in 12 C, 14 C and 16 C saturated FA and are solid at room temperature (Table 7.2). Their melting point (solid at room temperature and liquid at 37°C) makes these oils suitable for confectionary purposes. This is the reason these oils are considered premium (high market value). The other perennial crop of economic importance is olive (*Olea europea*)

TABLE 7.1
Growth Habit and Amount of Oil in the Main Oilseed Crops Planted throughout the World

Oilseed or Oil-Bearing Fruit	Type of Crop	Maximum Oil Yield, %	Method of Oil Extraction
Copra	Perennial	60–63	Mechanical Pressing
African Palm	Perennial	Kernel (48) and Pulp (< 20)	Mechanical Pressing
Cacao	Perennial	50	Mechanical Pressing
Olive	Perennial	18	Mechanical Pressing
Soybean	Annual	17–20	Direct Solvent Extraction
Cottonseed	Annual	18	Dual (Mechanical-Solvent)
Sunflower	Annual	25	Dual (Mechanical-Solvent)
Colza/Canola	Annual	35	Dual (Mechanical-Solvent)
Safflower	Annual	28	Dual (Mechanical-Solvent)
Peanut (shelled)	Annual	48	Dual (Mechanical-Solvent)
Sesame	Annual	47	Dual (Mechanical-Solvent)
Flax or linseed	Annual	34	Dual (Mechanical-Solvent)
Corn (germ)*	Annual	26–36	Dual (Mechanical-Solvent)

* Germ obtained from wet-milling generally contains higher amounts of oil compared to germ obtained from dry-milling operations (refer to Chapter 6).

TABLE 7.2
Classification of Major Oilseeds According to the Type of Oil and Uses

Classification	Main Fatty Acid Composition	Main Oils	Characteristics and Uses
Lauric acid or medium-chain triglyceride group	Lauric, myristic and palmitic acids	Coconut, palm oil, palm olein, palm stearin and cocoa butter	Low iodine value, solid at room temperature, good oxidative stability, most are solid at room temperature.
High Oleic	Oleic acid	Olive, sesame, high-oleic soybean, high-oleic corn, high-oleic cottonseed, regular and high-oleic canola, high-oleic sunflower, high-oleic safflower	Lower iodine value compared to high linoleic counterparts. Adequate oxidative stability. Source of omega 9 fatty acids.
Linoleic group	Linoleic acid	Soybean, corn, cottonseed, peanut, sunflower, safflower	High iodine value, liquid at room temperature and excellent source of omega 6 fatty acids. Low oxidative stability.
Linolenic group or drying oils	Linolenic acid	Linseed or flax	Very high iodine value, very low oxidative stability, excellent source of omega-3 fatty acids.
Milkfat or short-chain triglyceride group	Short-chain saturated fatty acids butyric, caproic, caprilic and capric.	Milkfat (bovine, ovine, caprine and others)	Low iodine value, high saponification value, solid at room temperature, high oxidative stability.
Animal fats	Saturated stearic and palmitic acids followed by oleic acid	Lard, beef tallow	Low iodine value, solid at room temperature, high oxidative stability.

grown in areas around the Mediterranean especially in Spain, Greece, Italy and Tunisia. The olive oil is the only source that is consumed as unrefined or crude oil and is the one that has the highest market value. Virgin olive oil is rich in tocopherols, phospholipids, chlorophyll and other liposoluble compounds that are considered relevant nutraceuticals (Chapter 21). The main FA present in olive oil is oleic acid (Table 7.2).

The most relevant annual oilseeds in terms of production and utilization are soybean (*Glycine max*), cottonseed (*Gossypum hirsitum*), canola (*Brassica campestris*) and sunflower (*Helianthus annus*). These are abundantly grown in temperate climates around the globe. Soybeans are the most relevant annual oilseed in terms of production and utilization. The estimated production in the year 2019 was 333.7 million tons harvested from 120.5 million hectares mainly located in the United States, China, Brazil and Argentina (FAO 2021). Soybeans are highly valued because the extraction procedure yields a highly unsaturated oil rich in linoleic acid and a defatted protein meal that contains nearly 50% protein. These two products are highly used by snack processors in frying operations and for the production of nutritional bars and other high-protein items.

Canola and sunflower have gained market during the past years due to their favorable FA profile. Peanut (*Arachis hypogea)* and sesame (*Sesame indicum*) oils are considered premium because the first is generally blended with other oils to improve flavor of processed snacks and the second is demanded as salad oil. The corn (*Zea mays*) germ fraction from dry and wet-milling processes described in the preceding Chapter 6 are used for the production of corn oil. The triglycerides of these vegetable oils are mainly composed of 18C FA. Canola and sesame are rich in 18:1 oleic acid whereas the rest of the annual crops in 18:2 linoleic FA. Flax (*Linumusitatissimum*) is the only crop that yields an oil that contains more than 50% 18:3 linolenic acid but the oil is not commercially extracted because is highly prone to oxidation. However, whole flax seeds are increasingly being used for production of nutritional bars.

Table 7.2 summarizes the classification of oils according to their fatty acid composition depicted in Table 7.3. The main categories are tropical oils that belong to the lauric o medium-chain triglyceride group that are generally solid at room temperature due to the comparatively higher melting points (Table 7.4) and lower iodine value (Table 7.5). The annual oilseed crops are generally

TABLE 7.3
Typical Fatty Acid Composition of Vegetable Oils and Animal Fats

Type of oil	8:0	10:0	12:0	14:0	16:0	16:1	18:0	18:1	18:2	18:3
Vegetable Oils from Perennial Crops										
Palm kernel	3	4	47	16	8		3	11	2	
Palm oil				1	44	1	4	37	9	
Palm olein	4.3	3.7	45.9	14.2	8.2		2.2	18.3	2.9	
Palm stearin			0.4	1.2	52.6		4.7	32.7	7.6	0.4
Coconut	8.2	6.5	47.4	18.3	9.3		2.7	5.2	1.5	
Cocoa butter	1	13	3	3.1	27.7	2.6	12	28.2	7.5	
Olive					11.6	1.0	2.4	76.0.	6.8	0.6
Vegetable Oils from Annual Crops										
Soybean					11		4	24	54	7
High-oleic soybean					7.3		3.4	85.1	1.3	2.0
Cottonseed				0.9	25.5	0.6	2.5	17.7	52.2	0.1
High-oleic cottonseed					15		2	78	4	
Sunflower					7		5	19	68	1
High-oleic sunflower				0.04	3.6		4	81.5	8.9	0.1
Canola					4.3		1.7	60.5	20.8	10.1
Low-linolenic canola					5.1		1.9	65.6	24.0	1.7
High-oleic canola				0.1	3.0	0.3	2.0	73.7	14.4	2.9
Corn					11.1		1.7	23.7	62.6	1.0
High-oleic corn					9.1		2.3	65.4	22.7	0.6
Peanut				1	45		4	56.1	24.2	
High-oleic peanut								75.6	4.7	
Safflower					7		2	13	78	
High-oleic safflower					7		6	81	8	
Olive					11	1	2	72	9	1
Sesame					9		5	39	41	
Flaxseed or linseed					5		4	20	13	53
Animal Fats										
Lard				1	24	3	14	41	10	1
Beef Tallow			1	4	25	4	19	36	3	1
Milkfat*	1.4	2.7	3.3	10.9	30.6	1.0	12.2	22.8	1.6	0.7

Source: Warner and Knowlton (1997), Kochhar (2001), Liu et al. (2002), Ericsson (2007), Lindmark Månsson (2008), Serna-Saldivar (2008), Marina et al. (2009), O´Brien (2009), Shafer de Martini Soares et al. (2009), Hosseini et al. (2016), Ollivier et al. (2006).

* Contains 4.4% C4 butyric acid and 2.4% C6 caproic acid.

TABLE 7.4
Comparison of the Major Physical Properties of Vegetable Oils and Animal Fats

Type of Fat or Oil	Specific Gravity, g/ cm³ @ 25°C Unless Otherwise Specified	Refractive Index @25°C Unless Otherwise Specified	Melting Point °C	Natural Crystal Habit or After Hydrogenation
Vegetable Oils from Perennial Crops				
Palm	0.889	1.455	37.5	β′
Palm kernel	0.867 @30°C	1.451 @ 40°C	28.3	β′
Palm olein	0.912@40°C	1.459@40°C	21.4	β
Palm stearin	0.887 @40°C	1.448@60°C	50.2	β
Coconut	0.917 @30°C	1.449 @ 40°C	24.5	β′
Cocoa	0.864 @15°C	1.456@	32.5	β and β′
Olive	0.912	1.470	0	β
Vegetable Oils from Annual Crops				
Soybean	0.918	1.473	0.6	β
Cottonseed	0.917	1.470	13	β′
Peanut	0.913	1.468	–2	β
Corn	0.919	1.472	–9.5	β
Sunflower	0.920	1.464	–19	β
High-oleic sunflower	0.914	1.468	5.8	β
Safflower	0.922	1.474	–17	β
High-oleic safflower	0.915	1.468	–17	β
Canola	0.917	1.466	–9	β
Sesame	0.916	1.472	–6	
Flax	0.934	1.481	–22	
Animal Fats				
Lard	0.898 @ 50°C	1.454 @ 50°C	33.5	β
Beef tallow	0.893 @40°C	1.454 @40°C	46.5	β′
Milkfat	0.907 @40°C	1.465 @ 60°C	35	β′

categorized as high oleic, linoleic or linolenic groups. These 18C rich oils clearly differ in their melting points (Table 7.4), iodine value and oxidative stability (Table 7.5).

Most frying operations (Chapter 3) select their oils based on the oxidative stability and employ natural oils rich in oleic acid and low in linolenic acids. The other alternative is the utilization of hydrogenated vegetable oils belonging to the linoleic group but this practice is currently avoided because partially hydrogenated oils commonly contain significant amounts of trans fats. For this reason, many commercial frying oils used today consist of blends of oils so to reduce the iodine value and improve oxidative stability. The palm and palm stearin oils are the preferred sources for blending purposes.

Most annual oilseed crops have been manipulated by plant breeding to significantly modify their FA profiles. Most efforts have been toward decreasing linolenic and linoleic FA in favor of monounsaturated oleic acid (Merill et al. 2008). These crops yield oils that have improved frying stability. The most relevant high-oleic crops that have been successfully manipulated are soybeans (Mounts et al. 1998), canola (Wilson et al. 1989), corn (Warner and Knwolton 1997, Barrera et al. 2019), peanuts (O'Keefe et al. 1993), sunflower (Purdy 1986, Flagella et al. 2002), palm (Ramli et al. 2009) and safflower (Merill et al. 2008, Hamdan et al. 2009). The low-linolenic and high-oleic acid soybean crops were developed as alternatives to produce oils with better oxidative stability

TABLE 7.5
Comparison of Oxidative Stability, Resistance to Rancidity and Iodine and Saponification Values of Vegetable Oils and Animal Fats

Type of Oil	Iodine Value	Saponification Value	Relative Resistance to Rancidity	Oxidative Stability Index, OSI @ 110°C
Vegetable Oils from Perennial Crops				
Palm kernel oil	19.3	248	Good	26.8
Palm oil	52.6	198	Good	20
Palm olein	56.7	198	Good	25.5
Palm stearin	28.5	199	Excellent	
Coconut	4.7	260	Excellent	28.1
Cocoa butter	35.5	191	Excellent	23.0
Olive	86.4	190	Excellent	40.9
Vegetable Oils from Annual Crops				
Soybean	131	190	Poor	5.8
High-oleic soybean	82.1		Good/Excellent	25
Cottonseed	109	192	Fair	5.6
Sunflower	140	191	Poor	4.5
High-oleic sunflower	87.8	191	Good	14.3
Canola	118	173	Fair	7
High-oleic canola	97.7		Fair/Good	12
Corn	131	190	Poor/Fair	9.0
High-oleic corn	97.2		Good	16
Peanut	96	190	Poor/Fair	5.1
Safflower	145	191	Poor	2.3
High-oleic safflower	92	192	Good	10.9
Olive	85.5	190	Good	4.9
Sesame	109	192	Good	33.2
Flax	187		Poor	1.2
Animal Fats				
Lard	61.5	193	Excellent	9.1
Beef tallow	46.5	195	Excellent	
Milk fat	34.5	227	Excellent	

Source: Tan et al. (2002), Rotondi et al. (2004) and Merrill et al. (2008).

and less flavor reversion. Rapeseed is one of the oilseeds that has received more attention by plant biotechnologists and was manipulated to lower the erucic acid (22:1 ω9) content to practically nil in favor of oleic and linoleic acids. Erucic acid is known to cause heart lesions and cardiovascular problems in laboratory animals. The improved oilseed free of erucic and rich in oleic acid known as Canola (registered trademark of the Canadian Council) is today considered within the five more important oilseeds in the world because the public perceives it as healthful due to the high amounts of unsaturated fatty acids. The crop has been successfully modified with modern techniques of haploid tissue culture, genetic transformation and hybridization using microspore tissue culture. These DNA transfer techniques have been feasible due to *Agrobacterium tumefaciens* infection (Rattray 1991). Special types of canola crops have been developed in order to yield oils rich in oleic acid, low-linolenic acid and even high-lauric acid. The high-oleic canola oil has excellent potential as frying oil because it is comparatively less prone to oxidation. Likewise, the development of new sunflower varieties obtained by traditional plant breeding contains between 55 and 75% of oleic acid, which allows the production of more stable oils suited for deep-fat frying operations.

In terms of perennial crops, efforts in African palm have focused on increasing the amount of oleic and linoleic acids with the consequent decrease in saturated palmitic acid (Ramli et al. 2009, Cuellar-Sánchez 2016).

7.3 ORGANOGELS

An organogel or oleogel is defined as a self-assembled or crystalline particle structure formed by the entanglement of one or more structural units such as crystals, fibrillar networks or suspended polymer strands. Organogels impart the desired functionality to foods without generating *trans* fatty acid (TFA) and with reduced saturated FA when unsaturated oils are used. Oleogelation has several advantages compared to hydrogenation and interesterification, the most relevant being that no biochemical processes are involved (Hughes et al. 2009, Co and Marangoni 2012, Sintang et al. 2017, da Silva et al. 2018).

Oloegels are regularly formed with three different systems consisting of (i) crystalline particles and self-assembled structures of low molecular weight, (ii) self-assembled structures of polymers or polymeric strands, and (iii) miscellaneous. At first formation, gelation occurs by trapping the liquid oil phase into triglycerides. Also, both mono and diglycerides and FA are capable of the formation of similar structures like triglycerides. The self-assembled fibrous networks are formed by using a low molecular weight felling agent like phytosterols and ricinoleic acid. Helical and twisted crystalline ribbons are obtained by this system (Rogers et al. 2008, Co and Marangoni 2012, Pehlivanoglu et al. 2018).

Structuring mechanisms involved in these systems can be divided into a dispersion of the fat phase as crystallized or uncrystallized solid particles or self-assembled complex structures, which are held together by specific supramolecular interactions (Pernetti et al. 2007). It is essential to optimize and design the textural and rheological properties when developing a new organogel. However, during storage, organogels may change their rheological properties and appearance (phase separation) due to oil exudation, which is the major cause of rejection (Sintang et al. 2017).

Practically any oil can be used as a base to produce organogels with different consistencies for diverse applications. The key is the selection of one or more structuring agents to change the physical properties of the selected vegetable oil that facilitate the formation of semi-solid or solid plastic materials. Oleogels with different consistencies have been successfully produced with structurants or oleogelators like waxes (candelilla, carnauba and rice bran wax), emulsifiers like sorbitan tristearate, polyglycerol esters and mono- and diglycerides generally added at concentrations of less than 5% of the total oleogel weight. Oleogels are produced by simply mixing the base oil with the oleogelator(s) followed by heating and cooling. Organogels rich in unsaturated FA based on soybean, sunflower, canola, olive and other vegetable oils have been successfully produced to replace trans-rich plastic shortenings in several food applications like margarines, bakery items, and coatings.

Recently, oleogels rich in nutraceutical compounds have been produced with omega-3 FA, phytosterols, phospholipids, fat-soluble antioxidants like tocopherols and many other bioactive compounds (Pernetti et al. 2007, Rocha et al. 2013, Ogutcü and Yılmaz 2014, Palla et al. 2017, Sintang et al. 2017, Chaves et al. 2018, da Silva et al. 2018, Pehlivanoglu et al. 2018).

7.4 ANIMAL FATS

The animal rendering industry recovers fats from fatty tissues not suitable for meat products. These tissues free of muscle and bone and rich in adiposities normally contain from 70 to 90% fat that is rendered to produce oleostocks like bleachable tallow and lard. Dry rendering is the simplest method of animal fat extraction. The fat-rich tissues are simply heat treated in order to release the fat that is recovered by draining. The residue may be screw pressed to further reduce its fat content. The recovered fat is allowed to settle, centrifuged and/or filtered and in some instances bleached before marketing. Tallow and lard are rich in saturated FA especially stearic (18:0) and palmitic (16:0) (Tables 7.2 and 7.3) and therefore have higher melting points (Table 7.4) and excellent oxidative stability due to the comparatively lower iodine values (Table 7.5). Beef tallow contains higher amounts of saturated FA and lower iodine value compared to lard because the dietary fat is saturated by the ruminal microflora (Table 7.3). On the other hand, the FA composition and consistency of the lard rendered from the monogastric pig varies according to the composition of the dietary fat. The fats associated with animal tissues are virtually devoid of free FA and therefore are not commonly refined. Edible tallow often contains 0.5–1.5% of free fatty acids (FFA) whereas lard from 0.2 to 0.5%. In addition, they have low iodine values and high melting points (Tables 7.4 and 7.5). Unfortunately, the consumption of large quantities of these fats rich in saturated FA has been associated with cardiovascular disease.

7.5 OIL EXTRACTION AND REFINING

Vegetable oil crushing and extraction is the most critical and relevant operation of the oil industry (Figure 7.1). During this process, the oil is extracted from its natural matrix. The extraction consists of the separation of the liposoluble compounds from the oilseed or oil-bearing fruits using mechanical, chemical or a combination of these two processes known as a dual system. The aim of the industry is to obtain the maximum yield of crude oil at the minimum possible cost. Quality wise, the industry aims toward the extraction of oils with the minimum amount of impurities and damage so to minimize refining losses. The vegetable oil crushing and extraction industries also produce a range of value-added coproducts such as partially or totally defatted protein meals and hulls (Serna-Saldivar 2008).

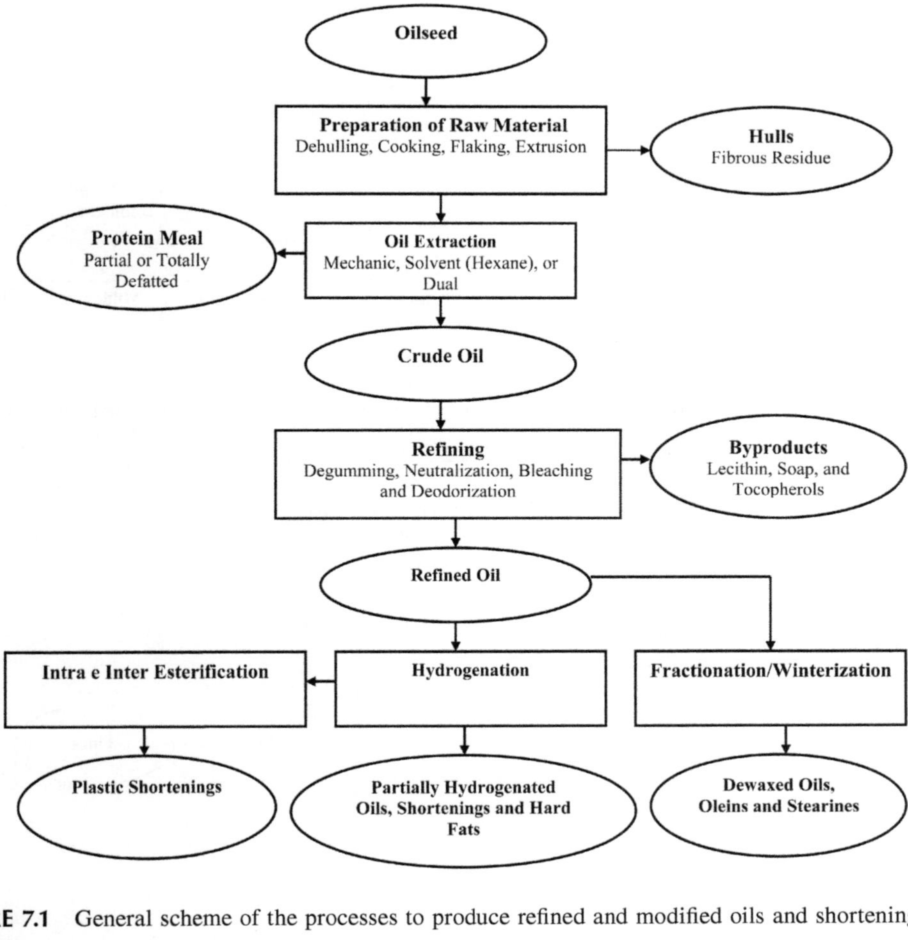

FIGURE 7.1 General scheme of the processes to produce refined and modified oils and shortenings from oilseeds.

Extractions of oils via mechanical means have been performed since the epoch of the Egyptian pharaohs. Nowadays, practically all the tropical oilseeds palm, copra and cacao and olive fruits are mechanically extracted (Table 7.1). The mechanic presses are less expensive but yield a partially defatted protein meal containing 7–12% residual oil that needless to say is prone to oxidation. This limits the value of the partially defatted meals or byproducts. Soybeans, due to their comparatively low oil content, are generally solvent (hexane) extracted allowing the almost complete oil removal and the production of a protein meal with less than 1% residual oil that is in many cases used for the production of texturized vegetable protein, concentrates and isolates. The main disadvantages of solvent extraction processes are the high investment in equipment and the danger of using highly flammable solvents. The rest of the vegetable oils are normally extracted using the dual system consisting of first extracting the oil with mechanic presses or expellers followed by treating the resulting partially defatted meal with hexane. The advantage of the dual extraction system is that it takes advantage of the goodness of both processes that translate into a more economically extraction yielding defatted meals practically free of residual oil (< 0.7% oil).

7.5.1 Oil Extraction Processes

Figures 7.2 and 7.3 detail general diagrams of the oil mechanical, solvent and dual extraction processes that are commonly used to extract most of the edible crude oils available worldwide.

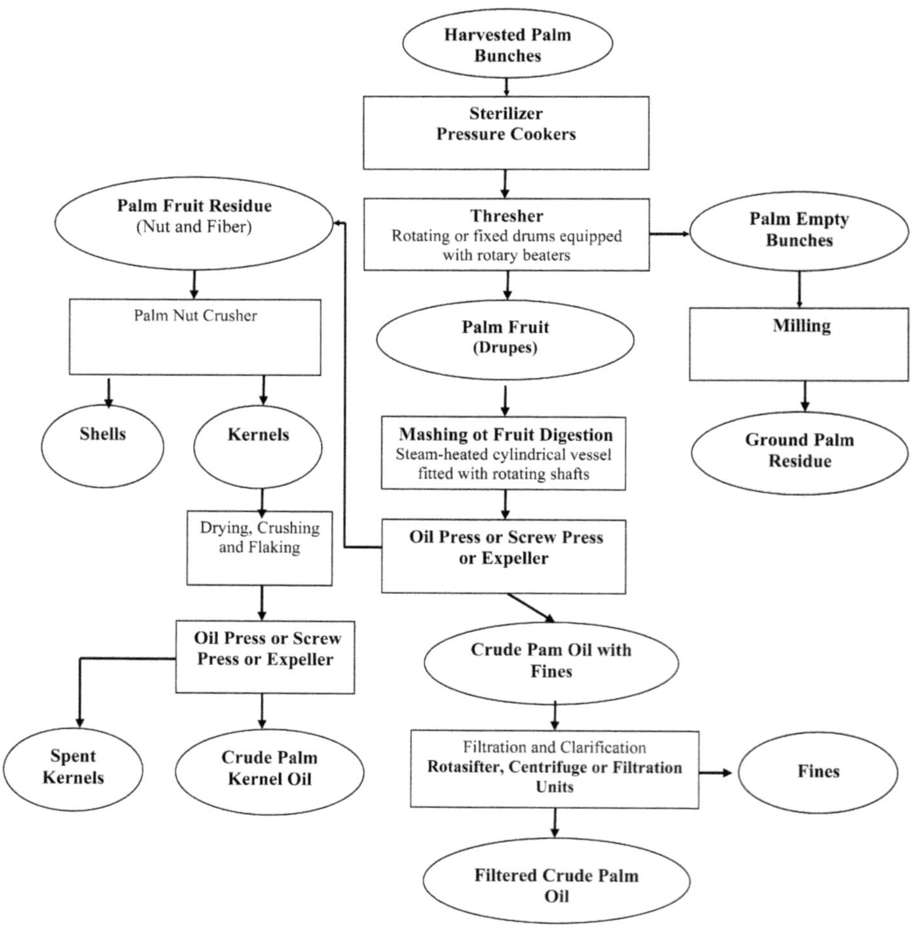

FIGURE 7.2 General diagram of the mechanical extraction process of crude palm oils.

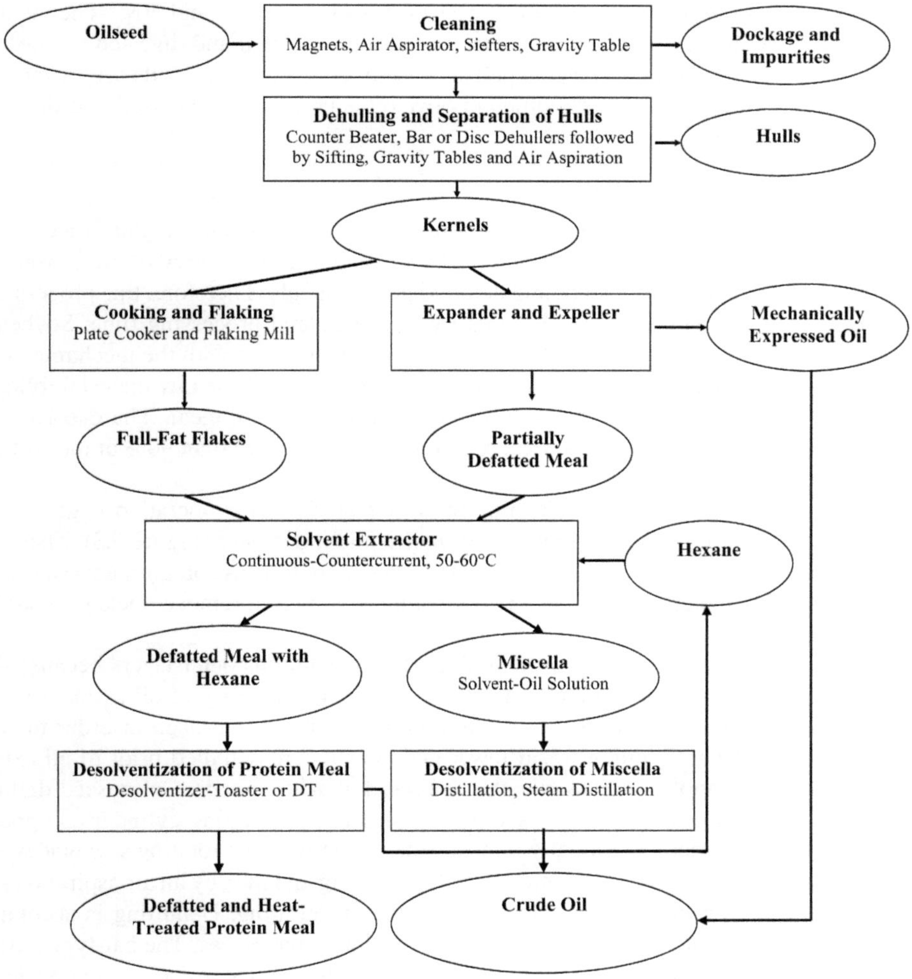

FIGURE 7.3 General diagram of the dual (mechanical-solvent) extraction process of crude vegetable oils.

7.5.1.1 Mechanical Extraction

Practically all oils from tropical bearing fruits and olive are extracted with mechanical presses or continuous expellers. Among these oil-bearing materials, the process of extraction of palm and palm kernel oil is the most practiced especially in Malaysia and Thailand. Figure 7.2 depicts the sequential steps to obtain palm oils. The key steps are the heat treatment aimed to inactivate enzymes and microorganisms and rupture organelles that store fat and the mechanical or screw presses that free the oil from fruits. The small palm oil processors still utilize batch presses although most of the palm and palm kernel oils are extracted in continuous processing plants equipped with large screw presses or expellers. There are different types of batch presses still used especially in developing countries. One of the most common ones is the box hydraulic press. The oilseed feedstock is placed in boxes and gradually pressed. The loading, pressing and unloading cycle regularly takes 30 min. The more efficient expellers are extruders equipped with special screws that apply high pressure to the oilseed cake inside a barrel that is ribbed at the terminal part so to allow the flow of the free oil into a collection pan or container. The advantage of these continuous equipment is that they defat the oilseed cake in less than one minute. Independently of the type of mechanical extraction process, the leftover oil in the fruits or spent kernels ranges between 10 and 30%. The yield of the crude oil significantly increases when the fruits are thermally treated and digested or mashed and when fruits are extracted with proficient expellers. Copra, cocoa and olive oils are extracted with similar systems described in Figure 7.2 although olive oil is usually cold-pressed in order to fulfill regulatory labeling issues.

7.5.1.2 Solvent and Dual Extraction

The most effective way to extract oils is via solvent extraction but this particular process requires more investments in special equipment necessary for extraction and recovery of the solvent associated with the miscella (solvent oil blend) and defatted protein meals. Therefore, this process is regularly combined with mechanical extraction that saves energy and processing time. Soybeans are commonly only extracted with solvent whereas the rest of the oilseeds with the mechanical-solvent process known as dual. These processes start with the preparation of the raw material followed by extraction and recovery of the solvent from the miscella and protein meal. The desolventization steps are the most energy demanding and are required to recover more than 90% of the solvent that is quite expensive.

The first step of these processes is cleaning the raw material. This operation is generally performed with an air aspirator, followed by gravity tables and a destoner (Figure 7.3). These equipment are usually provided with magnets to assure the removal of ferrous contaminants such as nails, bolts and other ferrous materials. The cleaning operation is aimed to remove dockage, chaff, other seeds and broken seeds.

Most oilseeds require the removal of the hull, pericarp or fibrous outer layers because they do not contain significant amounts of oil and the fiber hinders the efficiency of oil extraction. For the specific case of cottonseed, the whole seed is first delinted and then broken in order to separate hulls and kernels. Soybeans, sunflower, safflower and peanuts are dehulled prior to oil extraction whereas linseed, colza/canola, and sesame are processed as whole seeds. Cottonseed delinting is accomplished with the use of delinting machines consisting of a rotating cylinder equipped with circular saw blades. The friction gradually removes the cotton with the rotating saw blades and the detached lints are removed from the machines first by brushing and then by an air aspiration system. The distance between the blades determines the degree of delinting. Dehulling is accomplished with impact, pin, roll or plate mills that break the oilseed into large pieces. The bar-type hullers are widely used especially to break cottonseed, sunflower and safflower. It consists of a cylindrical rotor equipped with pins encased inside another static cylinder with pins on the internal surface. The pins of the rotating cylinder break the oilseed into large pieces and release the oil-bearing kernels. The universal roll mill is also widely used. The gap between the rolls is controlled as well as the speed differential. The performance of these mills is greatly affected by the moisture content of the feedstock. If the moisture is too high, the hulls are difficult to separate and if it is extremely low it may cause excessive rupture of the oilseed affecting oil yield and extraction performance. The output of these mills is classified or separated into kernels and hulls with the use of gravity tables, air aspiration or a combination of both.

The resulting kernels are heat treated or cooked in order to coagulate proteins, break cell walls and spherosomes (organelles that store oil in the cells), lower the viscosity of the oil, lower microbial counts and denature enzymes that increase refining losses, denature antinutritional factors and insolubilize the phospholipids. The spherosomes break because their membranes are constituted of proteins, which denature due to the thermal treatment. The resulting oil droplets coalesce and are easier to extract by mechanical or chemical means (Allen et al. 1982). The thermal treatment also imparts plasticity to the oilseed material improving its flaking performance. The heat treatments are commonly applied by the use of continuous plate cookers or special types of extruders known as expanders in the industry. The plate cooker consists of a series of plates adjusted to a certain temperature. Pieces of tempered kernels (12–15% moisture) are fed on the top part of the cooker and the oilseed material treated with heat while traveling down the plates. The top plate temperature averages 88–90°C in order to assure the inactivation of lipases and other enzymes and prevent the formation of free FA. The other plates are adjusted to higher temperatures (115–130°C) depending on the kind of oilseed. The average residence time is approximately 90 min. Plate cookers are generally teamed with flaking mill units. The aim of flaking is to further damage the oilseed cells and augment the surface area so to improve efficiency of oil extraction especially when processing only with hexane. The critical control point in this operation is the thickness of the flake. Generally speaking, the thinner the flake the lower the solvent extraction time and the higher savings in energy and labor. Commercial flakes are usually flaked to thicknesses varying from 0.25 to 0.5 mm. There are basically two types of flakers used by the industry the single flaker consisting of a pair or rolls and the multi roller system consisting of five rolls (Allen et al. 1982). The latter is widely used to produce cottonseed flakes.

Expanders are gaining popularity because the heat treatment is applied in less than 1 minute instead of the more than 60 min treatment of the plate cooker. This saves considerably processing time and energy. The expander consists of an extruder that houses a screw that applies more pressure or mechanical shear as it advances through the barrel and nears the die. The barrel acts as a heat exchanger and heat is normally applied with vapor. The pressure differential causes the extrudate or collet to expand immediately as soon as it exits the die orifice and equilibrates with the atmospheric pressure. The heat treatment and pressure causes protein and enzyme denaturation, breakdown of spherosomes and inactivation of antinutritional factors such as trypsin inhibitor and urease. The expansion of the resulting low-density collet favors the extraction of the oil.

The key step of the industry is the extraction that is accomplished by the use of mechanic presses, solvent extractors or a combination of both. Most mechanical presses used today are continuous equipment, named expellers. These special types of extruders press the oilseed cake at high temperatures and pressures to release the oil. The pressed cake usually contains from 8 to 12% residual oil. The expeller is a special type of extruder consisting of a heated barrel that houses a screw with threads that increases pressure as the material travels along the extruder. The last part of the barrel has grooves so the oil drains into a collecting pan. The extruded cake is partially defatted and commonly contains from 8 to 12% residual oil (Allen et al. 1982).

Hexane extraction constitutes the most efficient process for the recovery of oil from oil-bearing materials. After solvent extraction protein meals normally contain less than 1% residual oil. The disadvantages of this process are that is more expensive and uses a flammable and potentially explosive solvent. The process is more expensive because of extraction equipment and especially due to the equipment and energy required to recover the hexane. Soybeans, due to their relatively low oil content, are usually solvent extracted directly without a previous mechanic extraction. The dual extraction process generally processes most other vegetable oils, excluding tropical crops and olive.

The principle of solvent extraction is passing the hexane through a bed of the oil-bearing material that was previously prepared to optimize processing time and extraction rate. The percolation is performed by gravity until the removal of the oil is substantially complete. Industrially, this operation is performed with counter currently batch or continuous systems. The incoming defatted oil-bearing material is first extracted with half-miscella and as oil is gradually removed the partially defatted protein meal extracted with a mixture richer in hexane. The goal is to remove the oil as soon as possible with the least possible solvent:meal ratio.

Most continuous extractors work with percolation although there are some extractors that employ the immersion principle. Both systems use a principle of liquid-solid extraction. In percolation systems, hexane or half-miscella is distributed over a bed of flakes or cake and percolates down and exits the bed at the bottom through some sort of a filtering device (i.e., perforated plate, mesh screen etc.). Total extraction times vary according to the oilseed but generally ranges from 40 to 90 min. Between 1 to 1.5 L hexane per kg of the material is required by most extractors. Extractors most commonly commercially used are the basket type, rotary type (Rotacel) or crown consisting of a continuous loop. The first consists of a series of baskets with perforated bottoms placed like in a fortune wheelchair encased in a frame that avoids the loss of solvent. The baskets move slowly so it takes about 60 min to make a turn or revolution. The certain amount of the oil-bearing material is placed on the top basket and this gradually descends. When descending, half-miscella is sprayed on top of the oilseed material and percolates through the bed. The half-miscella then percolated to the contents of the next baskets. At the bottom of the extractor the full-miscella is pumped out for the selective recovery of the crude oil and hexane. When baskets start ascending, they will receive pure hexane that percolates from the top so to remove efficiently the most difficult to extract residual oil. At the bottom of this side of the extractor the pure hexane with some oil (half-miscella) is recovered and pumped to treat the fresh lot of oilseeds being treated in the descending baskets. The Rotacel uses a similar principle than the basket type of extractors. It contains baskets that rotate in a horizontal plane and miscella percolates through the baskets and falls into bottom compartments where it is pumped and recirculated counter currently to the flakes. The Crown extractor combines percolation and immersion and consists of a conveyor in which a drag chain moves inside a stationary casing conveying the solids over two sections of wedge wire screens (Allen et al. 1982). Regardless of the type of extractor, at the end of the extraction cycle two main materials are obtained: defatted solvent-rich protein meal and a full-miscella. The first contains about 30% hexane and the second consists of a mixture of about 70–75% hexane and 25–30% oil. The miscella is first filtered to remove fines and then distilled to enhance a liquid (hexane)-liquid (oil) separation. Distilling hexane from crude oil is relatively easy because the solvent has a boiling point of 68–69°C whereas crude oils higher than 190°C. The operation is commonly performed in three sequential steps: the first two consist of regular distillation that recover most of the hexane. The last is a steam distiller or stripper aimed toward the removal of residual hexane (Allen et al. 1982). This is critically important because regulatory agencies do not allow practically any residual hexane in the oil.

Desolventization of the protein meal is accomplished in a desolventizer-toaster commonly named D-T by the industry. The D-T consists of a series of stack plate cookers in which both live and indirect steam is applied. The material enters on the top of the machine and a rotating arm pushes the material toward a triangular hole that feeds the next plate down. Live steam is normally injected into the top plates where most of the solvent is stripped from the meal. However, the wet thermal treatment increases the moisture of the meal. Dry heat is usually applied in the bottom plates or toasting zone in order to remove residual hexane and moisture. The hexane-rich vapors are impelled to a condenser located on the top part of the D-T. Besides removing the hexane from the protein meal, the D-T conditions (residence time and temperatures) are regulated to heat-treat meals with the aim to achieve a certain degree of protein denaturation and inactivation of antinutritional factors associated with legume oilseeds such as trypsin inhibitor, urease, and hemaglutinins. The exiting defatted meal is usually heat treated to achieve a certain PDI value.

The recovery of the hexane of the desolventization steps of the miscella and protein meals is more than 90% efficient.

7.5.2 Oil Refining

With the exception of olive oil, the rest of vegetable oils are chemically or physically refined. The aim of refining is to remove natural contaminants that lower the color and sensory properties of the oil. At the end of the refining process, the refined oil practically consists of pure triglycerides. Many color, flavor and breakdown products are selectively removed. The compounds mainly removed are phospholipids known as lecithins in the industry, FFA, color compounds such as phenolics, flavonoids, carotenoids and chlorophyll, tocopherols, waxes, monoglycerides, diglycerides, aldehydes, ketones and alcohols (Dijkstra and Segers 2007, Serna-Saldivar 2008).

There are two commercial kinds of refining procedures: chemical and physical. Chemical refining is the most widespread used and involves four sequential steps: degumming, neutralization, bleaching and deodorization (Figure 7.4). On the other hand, physical refining consists of three major stages: degumming, bleaching and a special kind of deodorization. It is called physical

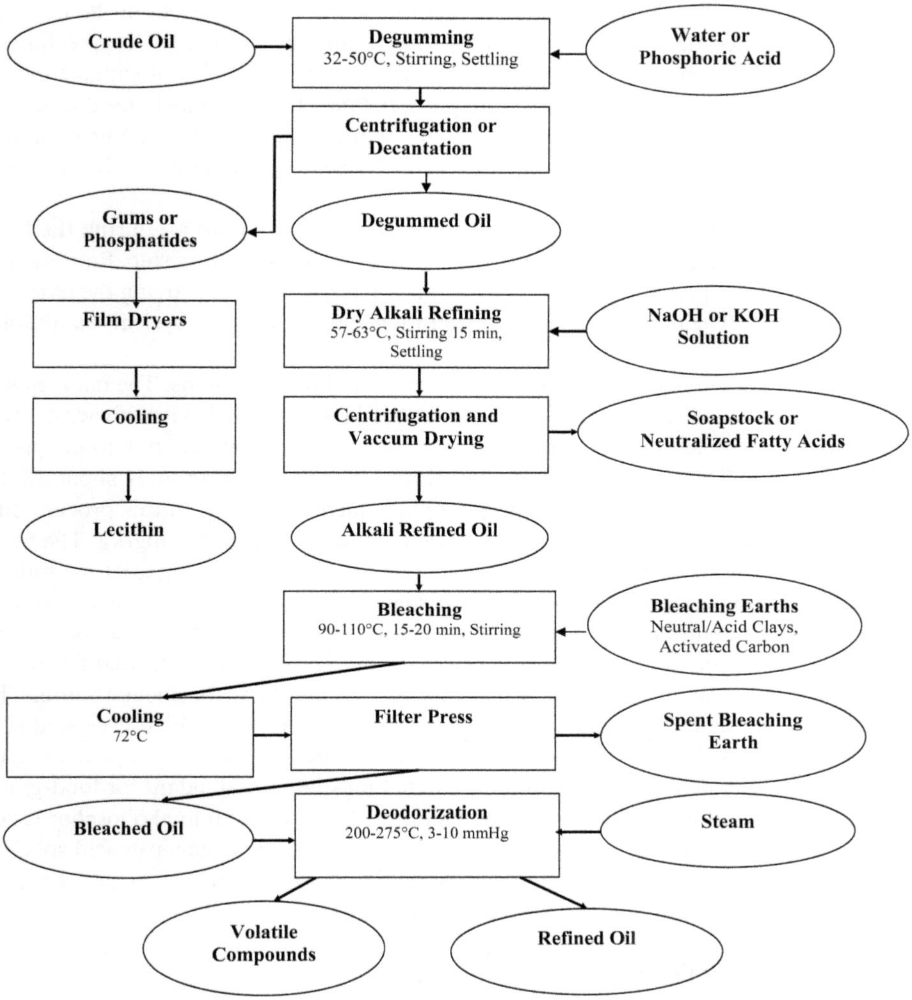

FIGURE 7.4 General diagram of the sequential steps of the chemical refining process of vegetable oils.

refining because the key operation is the deodorization or distillation of FFA, monoglycerides and diglycerides from the triglycerides (Dijkstra and Segers 2007, Serna-Saldivar 2008).

7.5.2.1 Chemical Refining

It is still the most widely used refining process although physical refining is gaining in popularity. Chemical refining consists of four sequential steps in which selective chemical compounds are removed.

7.5.2.1.1 Degumming

The first step is aimed toward the removal of phospholipids although some FFA and other contaminants are also partially removed. The most relevant quality control parameter for this specific step is the determination of phosphorus, which is an indicator of the total amount of phospholipids. The phospholipids become oil insoluble when water or a phosphoric acid solution is mixed with the crude oil. Generally, 3% water is mixed with the crude oil, heated to about 65°C and mixed to enhance the contact between the water and phospholipids. Water binds to the polar side of the phospholipids and upon the discontinuation of blending settles on the bottom of the mixing tank. It is commercially called degumming because of the gummy appearance of the precipitate. The degummed oil is removed by suction or by continuous centrifugation. The precipitate can be further processed in order to isolate and purify the phospholipids that are widely used as emulsifiers by the food industry. Commercial lecithin is normally obtained from the precipitate obtained after degumming crude soybean oil (Dijkstra and Segers 2007, Serna-Saldivar 2008, Barrera-Arellano et al. 2019).

The phospholipids are removed in order to prevent the loss of neutral oil during the following neutralization process and the darkening of the oil during the deodorization step. Removal of these polar lipids results in the parallel removal of iron and copper, thereby improving the oxidative stability of the refined oil. The process is based on hydration and subsequent physical separation of the phospholipids.

Several schemes are industrially utilized for the removal of phospholipids. The main approaches are either classical or acid degumming. The classic method is carried out with water, with the purpose of making possible the transportation and storage of the oil, in addition to the production of food-grade lecithin. The water commonly added for effective degumming is about 2% in relation to the crude oil or about 75% of the phospholipid content of the oil. In this process, the final phosphorus content in the degummed oil should range between 80 and 200 mg/kg. The process is based on the addition of water to the preheated oil, blending contents for a fixed time followed by centrifugation to obtain the denser lecithin from the degummed oil. The degumming with water can be carried out in either continuous or batch modes, where the main controlled process variables are the mixing time and temperature. In the batch process, 60–70°C is recommended for 30 minutes whereas in continuous operations temperatures between 70 and 80°C for about 15 min. The acid degumming step consists of a variant of the classic or regular process, which uses acid solutions for greater effectiveness. In the process, food-grade acids, such as citric or phosphoric are used. However, the gums obtained by acid degumming do not make up the standard for food-grade lecithin. Acid degumming allows the production of degummed corn oil with final phosphorus contents between 5 and 30 mg/kg. This process is usually continuous and supplements an acid solution with a concentration between 50 and 85%, in the proportion of 0.2 to 1% in relation to the oil content and is carried for 30 minutes at 65 to 70°C (Barrera-Arellano et al. 2019).

7.5.2.1.2 Neutralization or Saponification

The most relevant phase of the chemical refining process is neutralization consisting of the addition of alkali to neutralize FFA and yield salts known as soaps. The alkali is commonly added as an aqueous sodium hydroxide solution. Caustic soda is more effective in its decolorization action than its weaker counterparts are. The strength and concentration vary according to the acid value of the crude oil. Undoubtedly, the fat acidity or FFA of the crude oil is the most relevant quality control parameter for the neutralization step. The higher the FFA or acid value the higher the alkali solution needed. The degummed-neutralized oil is separated from the more dense soap solution by suction or centrifugation. Alkali refining with caustic soda reduces the FFA content to 0.01–0.03% and further removes residual phosphatides or gums present in degummed oils. The soaps formed in the reaction also adsorb natural pigments, gums and mucilages that were not removed by the prior step of degumming. The separation of the soapstock is done by continuous centrifuges.

The two ways of expressing the concentration of the alkaline solution of NaOH for use in the neutralization are Baumé degrees (°Bé) and common concentration (g/L). Common parameters for the neutralization of degummed oils refer to excess between 0.12 and 015%, with an alkaline solution of 18° Bé or 145 g/L. These indications minimize the loss of neutral oil and help to bleach the oil. After incorporation of the NaOH solution, the oil is slightly agitated to ensure uniformity of reaction, but the formation of emulsion or soap aggregates should be avoided. After a residence time of 15 minutes, the oil is pumped to a heat exchanger, which raises its temperature to 70–75° C, and then pumped to the continuous centrifuge. The oil soaps content after separation should be less than 200 mg/kg. These soaps are removed in single or double wash operations. The amount of water required to wash the neutralized oil is 12–15% relative to the oil weight, with the best results obtained with 15%. The quality of the water, free of calcium and magnesium salts, is essential, and the use of demineralized or condensed water is recommended. For more effective removal of soaps from the oil, the water/oil mixture is heated between 80 and 90°C prior to entering the washing centrifuge. Finally, the neutralized oil is dried in vacuum dryers at about 50 mBar. The level of residual soaps considered ideal after drying is a maximum of 50 mg/kg (Dijkstra and Segers 2007, Barrera-Arellano et al. 2019).

7.5.2.1.3 Bleaching or Clarification

Clarification is defined as the treatment that aims at partial or complete removal of the pigments that give color to the oils. It consists of an adsorption process, characterized as a physicochemical phenomenon involving mass transport of an adsorbate (solute adsorbed) from the solution phase to the surface of the adsorbent.

Bleaching is aimed to selectively remove pigments and oxidation products with bleaching earths (natural neutral or acid-treated clays or bentonites) and/or activated carbon. The main pigments associated with crude oils are gossypol in cottonseed, carotenoids in palm and corn oils, chlorophyll in most vegetable oils and phenolic acids and flavonoids. In addition to the removal of the pigments, the clarification step is essential in the removal of various undesirable compounds, such as peroxides, aldehydes, ketones, phospholipids, traces of pro-oxidants (iron and copper), soaps and other contaminants such as pesticides and polycyclic aromatic hydrocarbons (Taylor 2010, Barrera-Arellano et al. 2019).

Bleaching earths or clays normally used for clarification are broadly classified into two groups: naturally activated clays; and acid-activated clays by treatment with hydrochloric or sulfuric acids. The choice of clay depends on the quality of the oil, clarification requirement in terms of color score and the intrinsic bleaching powder. In general, the clarification of most oils requires acid-activated clays because they are comparatively more effective compared to neutral clays. Other adsorbents such as alumina, silicic acid and aluminum and magnesium silicates can be used with clays in combined processes. The use of activated charcoal for the clarification of vegetable oils is not very common due to its high cost, high oil retention and difficult handling because of its fine powder form, but may be advantageous in some combinations with acidic clays.

The clarifying process is preferably carried out under vacuum conditions (20–30 mmHg). The preheated oil is pumped into the slurry tank, which refers to the mixture of oil and clarifying clay. After stirring, the mixture is transferred to the clarification reactor under reduced pressure conditions. The time of the clarification process is from 15 to 30 minutes, with a temperature range of 80–120° C. Although the increase in temperature contributes to the increase in adsorption efficiency, the use of high temperatures may provide conditions for the development of undesirable reactions. The temperature should be such as to allow low viscosity of the oil, which increases the rates of mass transfer and diffusion. The water content is an important factor to avoid the formation of FFA during the process. The moisture content of the adsorbent and the oil should be less than 15% and 0.1%, respectively. The amounts of bleaching earths depends on the composition and quality of the oil to be clarified, type of clay and the effectiveness of previous steps of degumming and saponification. The dosage range of the adsorbent is between 1 and 2%, but in some cases may be greater than 5%. The Lovibond color or crude and bleached oils are widely used to determine the effectiveness of the clarification process. Another important quality control parameter is bleaching loss, which determines the amount of oil loss during the sequential steps. Generally, bleaching earths adsorb from 15 to 20% of the oil (Taylor 2010).

7.5.2.1.4 Deodorization

The last refining step is the deodorization that is aimed toward the removal of undesirable odors and flavor chemical compounds (Serna-Saldivar 2008, Barrera- Arellano et al. 2019). Deodorization consists of a steam distillation whereby volatile odoriferous and flavored substances are stripped from the relatively nonvolatile oil at temperatures below those damaging to the oil (Figure 7.4). The application of reduced pressure during the operation protects the heated oil from oxidation (Barrera-Arellano et al. 2019). The main objectives of deodorization are the removal of oxidized compounds with undesirable taste and odor in order to produce oils with better organoleptic features. The process further diminishes FFA and removes peroxides and odoriferous organic compounds like ketones, aldehydes and toxic contaminants such as pesticides and polycyclic aromatic hydrocarbons (Chapter 20). The oil color improves due to the destruction of thermosensitive pigments like carotenoids and xantophylls. However, this process also removes tocopherols and phytosterols that are considered as relevant nutraceuticals associated with the prevention of oxidative stress and hypercholesterolemia (Chapter 21).

The optimization of stripping parameters (temperature, time, vacuum and gas pressure) is defined according to the properties of the oil at the beginning of the process, specifications of the final product, limitations of the equipment and maximum preservation of compounds of nutritional or nutraceutical interest. The temperature of deodorization has important effects on the flavor, odor and color of refined oils. In this operation, color components may be removed by thermal degradation or clarification Most deodorization systems operate at temperatures between 230 and 260°C and a maximum pressure of 3 mBar, with steam consumption of approximately 10 kg per ton of processed oil (Dijkstra 2007a, Barrera-Arellano et al. 2019). The deodorizer should be able to deaerate the oil before exposure to the atmosphere. Normally, the process is considered finished when the FFA content is reduced to 0.01–0.03% and the peroxide value is reduced to practically zero (Serna-Saldivar 2008).

7.5.2.2 Physical Refining

Physical refining of oils is gaining popularity because it avoids the free FA neutralization and high refining losses. Instead, the FFA are removed by steam distillation. The physical refining steps start with acid (phosphoric acid) degumming so as to produce an oil low in phosphatides followed by bleaching and the critical step of steam distillation-deodorization.

Physically refined corn oil requires a greater amount of clarifying clay compared to chemical refining oil due to the high residual content of phospholipids. In this case, the silica can be used as a coadjuvant because it presents selectivity for phospholipids and metals (Dijkstra 2007a, Dijkstra and Segers 2007).

The last stage of the physical refining is known as distillation and differs from deodorization because it is conducted at higher temperatures and vacuum conditions in order to effectively remove FFA. In addition, the distillation removes volatiles and odoriferous compounds, tocopherols and phytosterols. Generally, the physically refined oils should end the distillation process with < 0.05% FFA (De Greyt 2013).

7.6 MODIFICATION OF FUNCTIONAL PROPERTIES OF OILS

The processes of blending, fractionation, hydrogenation, interesterification and combination of these are employed by the industry to modify FA composition and improve functionality and oxidative stability. These processes changed the physical, chemical and functional features of modified oils especially in terms of iodine value or degree of unsaturation, melting point and crystallinity or fat texture (Dijkstra 2007b, Serna-Saldivar 2008). However, these modifications not only alter the functionality but also the nutritional properties. The current trend is to generate functional fats and oils free of *trans* fats and with reduced levels of saturated fats (Duijn et al. 2006). In terms of snack foods, these technologies generate partially hydrogenated oils for frying purposes (Chapter 3), seasonings (Chapter 8) and shortenings and confectionary fats for bakery items (Chapter 10) and enrobing purposes.

7.6.1 Blending

Natural or modified refined oils are blended to meet specific requirements of consistency and oxidative stability especially for application as general-purpose fats and more functional and stable frying oils (Chapter 3). The lipid bases for obtaining mixtures may also include hydrogenated, interesterified and fractionated fats (Ghotra et al. 2002). The mixing process is relatively simple because it consists of first blending two or more oils followed by heating at a temperature above the melting point of the components and the subsequent stirring for approximately 20 min in order to homogenize the new blend. The necessary equipment consists of jacketed vessels and dosifiers for accurate measurement of the amount of each lipid fraction. Mixing tanks shall be equipped with mechanical stirrers and heating coils to ensure uniformity and complete homogenization of the product. Nitrogen atmosphere must be used to minimize lipid oxidation (Dijkstra 2007b, Gupta 2008, Barrera-Arellano et al. 2019). Blending allows processors of deep-fat fried snack foods the selection of oils with better oxidative stability and in order to avoid *trans* FA the blending with common vegetable oils like palm oil, palm olearin or palm stearin. Generally, these blends substitute partially hydrogenated oils with iodine values ranging from 90 to 110.

7.6.2 Winterization and Fractionation

There are two major physical processes that aim to separate undesirable compounds or at least two distinctive fractions from oils. The first and most common one is named winterization that is used to remove waxes from vegetable oils like corn, sunflower, safflower and rice. The second process known as fractionation or crystallization is performed to obtain two contrasting fractions especially in terms of melting points from crystalized oils subjected to low temperature. The higher iodine value fraction is known as olein whereas the one containing more saturated fats with higher melting point and lower iodine value than stearin. These products are mainly obtained from the fractionation of pam oil.

7.6.2.1 Winterization or Dewaxing

The term "winterization" evolved from the observation that refined and bleached vegetable oil stored in outside tanks during the winter months physically separated into hard and liquid-clear fractions. The decanted liquid fraction provided oils without clouding for long periods at refrigeration temperatures. These oils readily found usage as salad oil and for the production of mayonnaise (O´Brien 2009). The classic winterization process is designed to remove higher melting point triglycerides and mainly waxes from the winterized oil. Waxes are considered undesirable compounds because they tend to crystallize at room temperature due to their low solubility, causing turbidity in refined oils. The critical factors that affect this process are the nature of the oil especially in terms of iodine value and presence of waxes, the rate of cooling and the temperature of crystallization. The amount of wax associated with corn, safflower, sunflower and canola oils ranges from 0.2 to 1%. The common process consists of chilling heated refined oils to about 5–8°C without or minimum agitation followed by heating to about 18°C and slow filtering with plate and frame presses to separate the wax crystals from the liquid oil. The analytical methods most used to determine the efficiency of dewaxing or winterization are the chill and cold tests (O´Brien 2009).

7.6.2.2 Fractionation

Fractionation is defined as a thermomechanical process that separates oil into two fractions, each with a specific melting behavior caused by modification of the triglyceride composition. This process is the modern way to obtain oils with different physicochemical and functional properties because when applied to natural or unhydrogenated oils generally yields two contrasting fractions with 100% of their FA in the cis configuration. The principal goals of fractionation are to generate a liquid oil with improved properties (e.g., cloud point) or to produce a fraction with a narrower composition and melting behavior. The former may find use as frying or a salad oil while the latter, depending on its composition, may be used as shortening, margarine hard stocks or in confectionary fats. Fractionation has gained relevance especially in the palm oil industry because the palm oil rich in medium-chain triglycerides containing about equal amounts of saturated and unsaturated fatty acids yields two or more fractions with different FA composition, iodine values, oxidative stability and functionalities. The physical nature of palm oil allows its separation into a low-melting fraction known as olein and a high-melting fraction commercially named stearin.

The fractionation process is considered a thermomechanical separation scheme where triglycerides are separated according to differences in crystallization. There are three fractionation processes used to fractionate palm oil: dry fractionation, detergent fractionation and solvent fractionation. Dry fractionation, carried out without solvent, is relatively cheap and extensively utilized to tailor the properties of the oil. Detergent fractionation is a modification of the dry method whereas the solvent or wet fractionation is the least applied and most expensive (Smith 2001). The most practiced one is dry fractionation where two or more components are obtained from the oil involving two major stages: selective crystallization and filtration.

The fractionation generally processes refined oils although crude palm oil can be alternatively fractionated. Most processes are batch although several industries are adopting semi-continuous methods. Fractionation consists of first heating the oil to around 70°C min order to completely melt all solid fat and remove all nuclei followed by gradual cooling to a temperature between 15 and 20°C while agitating to encourage selective crystallization to enhance nucleation crystal growth to a size and shape that allows efficient separation via slow filtration. The most utilized filtration systems are Florentine, rotary drum and membrane filters. The efficiency of filtration and composition of oleins and stearins are clearly affected by the structure and size of the crystals of the solid fat.

Today, an array of palm oil-derived products are obtained from single, double and even triple-stage dry fractionation. The single-stage dry fractionation commonly yields palm oleins with iodine values of 56 and 62. The saturation content of palm olein can be further reduced by using multiple-stage fractionation in which double- and triple-stage fractionation can be used to produce oleins with iodine values of 65 (super olein) and 70 (top olein), respectively (Smith 2001, Calliauw et al. 2007, Kellens et al. 2007, Huey et al. 2015).

7.6.3 Hydrogenation

Hydrogenation was first applied to vegetable oils at the beginning of the 19th century and is practiced toward the selective hydrogenation of double bonds of monounsaturated and polyunsaturated FA in order to change their melting point and crystallization properties (Allen 1982). There are three types of hydrogenation processes: partial, elevated and full. The first is normally applied to frying oils in order to improve its shelf stability and the oil keeps its liquid form at room temperature (Chapter 3). The elevated and full hydrogenation processes yield vegetable shortenings that are generally solid at room temperature and have plastic properties due to the formation of crystals. Elevated hydrogenated fats are processed to lower their iodine values to around 80 whereas full-hydrogenated fats to values lower than 10. The last type is commonly known as hard fats.

The three major ingredients for hydrogenation to take place are the refined oil, the catalyst and hydrogen that are blended under controlled conditions of temperature and pressure in a closed reactor. Edible oils are usually hydrogenated with the catalyst Ni-Raney, which has a porous structure. Contents are very well agitated to enhance the distribution of the hydrogen gas and catalyst in order to promote the saturation of double bonds. The rate of hydrogenation highly depends on the temperature and pressure, the nature of the oil especially in terms of iodine value and FA composition and the activity and concentration of the Ni-Raney. The temperature and pressures commonly employed vary from 160 to 220°C and 69 to 414 kilopascals (10 to 60 psi). The Ni-Raney concentration generally added varies from 0.03 to 0.08%. The rate of agitation of 600–1,000 rpm plays an important role in the rate of hydrogenation and selectivity. This catalyst usually selectively saturates double bonds starting with FA with a higher degree of unsaturation (Allen 1982).

In partial hydrogenation processes, the aim is to lower the degree of instauration of linolenic acid because this particular FA is more prone to flavor reversion and oxidation (Chapter 3). On the other hand, the full hydrogenation to produce hard fats practically free of *trans* FA is gaining acceptance for the formulation of blends and oleogels. The palm kernel, palm, soybean, cottonseed, sunflower and canola oils have been successfully hydrogenated into hard fats. These high-melting-point fats are used as alternatives to improve the functional characteristics of lipid systems and also for intra and interesterification with other oils (DeMan et al. 1989, Ribeiro et al. 2013).

For successful hydrogenation, the oil should be refined or free of impurities that can inhibit or poison the catalyst. The hydrogenation process is carried in special types of batch reactors equipped with a high-speed agitator and rate of injection of the hydrogen gas. The process is ended when the iodine value or refractive index (Chapter 19) of the oil decreases to a certain target value. The feedstock is then chilled down under controlled conditions to allow crystallization.

The main advantages of selective hydrogenation processes are that the melting points of partially or elevated hydrogenated oils can be customized especially in terms of iodine value, refractive index and consistency. The main advantage of hydrogenation is the improvement in the oxidative stability and shelf life of the products and the production of solid fats with different consistencies and textures.

However, the main withdraw of hydrogenation is the formation of significant amounts of *trans* FA that are known to harm health worse compared to saturated fats. For this reason, the hydrogenation process is losing acceptance especially in countries where the production and use of *trans* oils and shortenings are prohibited or highly regulated. These isomers are not present in nature and have been associated with cardiovascular disease and cancer (Chapters 3 and 20).

7.6.4 Intra e Interesterification

The esterification of FA with glycerol was known as early as 1844 (Sonntag 1982). Esterification is defined as the movement or shuffle of FA to different positions in the triglycerides in order to change crystallization and the plasticity of the shortening whereas transesterification characterized by the interchange of FA between or among triglycerides. Generally, the result of esterification is the modification of the melting point and dilatometry of the oil or fat material (Sonntag 1982, Karabulut et al. 2004) and the production of smoother shortenings with excellent creaming properties.

The interesterification is currently the main method for the preparation of plastic fats with low levels of *trans* isomers or even the absence of these harmful compounds. In contrast to hydrogenation, interesterification does not promote the isomerization of double bonds of FA nor the degree of saturation; however, it allows the modification of the behavior of oils and fats, offering important contributions to optimize their use and functionality. Most interesterification processes start with the blending of two or more oils differing in melting points and FA profiles. Tropical oils like palm oil, palm stearin, palm olein, coconut and cacao butter are generally preferred due to their high melting point (Shafer de Martini Soares et al. 2009).

The rearrangement of the FA within or among triglycerides does not change the iodine value or the isomeric state of the FA. Therefore, in order to produce a shortening free of *trans* FA a hydrogenated hardfat or a stearine that does not contain any *trans* isomers should be used. For instance, the random interesterification of 80% liquid soybean oil with 20% completely hydrogenated soybean oil yields a shortening with minimum amounts of *trans* FA (List 2004, O´Brien 2009). The process is performed in agitated reactors operating under vacuum (20 mm pressure) at a high temperature with sulfonic acid or sodium methoxide. Commercially, interesterification is mainly used to produce coating fats, shortenings, confectionary fats and margarine oils. Interesterified fats generally change their texture, spreadability, smoothness, creaming and appearance. The primary use of interesterification in the United States is for the production of confectionary and coating fats known as hard butters that possess a sharp melting point. These fats are generally obtained from palm kernel, coconut oil with small amounts of vegetable oils (Sonntag 1982).

Interesterification is performed with three distinctive processes known as direct, random and enzymatic. Direct interesterification is carried out at relatively low temperatures to allow the liquid portion of the fat to shuffle fatty acids. In random interesterification the FA rearrange within the glyceride or from one glyceride to another in a predictive way. Enzymatic interesterification is becoming popular to yield trans-free fats for confectionary use. The main advantage of this technology is the specificity and prediction of the reaction mediated by lipases under anhydrous conditions at comparatively lower temperatures (70°C) compared to chemical interesterification (O´Brien 2009). Chemical interesterification is performed in agitated reactors operating under vacuum (20 mm pressure) at temperatures of 250°C or higher with sulfonic acid, sodium methylate, sodium-potassium alloy or sodium methoxide (Sonntag 1982, O'Brien 2009).

7.7 TRANS FATS

Trans fats (TFA) are the major concern of the oil industry because they negatively affect human health (Juttelstad 2004). Recent studies indicate that TFA increase LDL and lower HDL-cholesterol blood levels and the risk of developing cardiovascular diseases is considered the leading global cause of death (Chapter 20). The consumption of TFA is considered more harmful compared to consumption of saturated fats and has also been linked to a higher risk of developing type-2 diabetes (de Souza et al. 2015).

In several countries the presence of TFA is prohibited in human foods and in many others are mandatory in the nutritional label of processed foods. For this reason, the hydrogenation process is losing acceptance in favor of fractionation and interesterification (Hosseini et al. 2016). Linolenic acid (18:3 Δ9,12,15) has been recognized as the main responsible for the formation of TFA during the harsh thermal treatment of hydrogenation. Moreover, the frying process is known to generate TFA at temperatures above 200°C in a time-dependent and oil-turnover relationships (Chapter 3). The primary dietary TFA are vaccenic (18:1, trans-11) and elaidic (18:1, trans-9) being the second the main isomer in industrially hydrogenated oils and shortenings (Dhaka et al. 2011).

REFERENCES

Allen, R.R. 1982. Hydrogenation. In *Baileys industrial oil and fat products*, Volume 2, Fourth Edition, ed. D. Swern, 1–95. New York: John Wiley and Sons.

Allen, R.R., Formo, M.W., Krishnamurthy, R.G., McDermott, G.N., Norris, F.A., and Sonntag, N.O. 1982. Extraction of fats and oils. In *Baileys industrial oil and fat products*, Volume 2, Fourth Edition, ed. D. Swern. New York: John Wiley and Sons.

Barrera-Arellano, D., Badan-Ribeiro, A.P., and Serna-Saldivar, S.O. 2019. Corn oil: Composition, processing and utilization. In *Corn: chemistry and technology*, ed. S.O. Serna-Saldivar, 593–614. Cambridge: Elsevier (Woodhead Publishing), AACC-International.

Calliauw, G., Gibon, V., De Greyt, W., Plees, L., Foubert, I., and Dewettinck, K. 2007. Phase composition during palm olein fractionation and its effect on soft PMF and superolein quality. *Journal of the American Oil Chemists' Society*, 84(9): 885–891.

Chaves, K.F., Barrera-Arellano, D., and Ribeiro, A.P.B. 2018. Potential application of lipid organogels for food industry. *Food Research International*, 105: 863–872.

Co, E.D., and Marangoni, A.G. 2012. Organogels: An alternative edible oil-structuring method. *Journal of the American Oil Chemists' Society*, 89(5): 749–780.

Cuellar-Sánchez, M. 2016. Experiencias en la producción y mercadeo del aceite de palma alto oleico. *Revista Palmas*, 37: 322–330.

da Silva, T.L., Barrera-Arellano, D., and Martini, S. 2018. Physical properties of candelilla wax, monoacylglycerols, and fully hydrogenated oil oleogels. *Journal of the American Oil Chemist Society*, 95: 797–811.

De Greyt, W.F.J. 2013. Edible oil refining: current and future technologies. In *Edible oil processing*, eds. W. Hamm, R.J. Hamilton and G.H. Calliauw, 127–151. Chichester: John Wiley & Sons.

De Souza, R.J., Mente, A., Maroleanu, A., Cozma, A.I., Ha, V., Kishibe, T., Ulelyk, E., Budylowski, P., Shunemann, H., Beyene, J., and Anand, S.S. 2015. Intake of saturated and trans unsaturated fatty acids and risk of all cause mortality, cardiovascular disease, and type 2 diabetes: Systematic review and meta-analysis of observational studies. *The British Medical Journal*, 351: h3978.

DeMan, L., DeMan, J.M., and Blackman, B. 1989. Polymorphic behavior of some fully hydrogenated oils and their mixtures with liquid oil. *Journal of the American Oil Chemists' Society*, 66(12): 1777–1780.

Dhaka, V., Gulia, N., Ahlawat, K.S., and Khatkar, B.S. 2011. Trans fats - Sources, health risks and alternative approach - A review. *Journal of Food Science and Technology*, 48(5): 534–541.

Dijkstra, A.J. 2007a. Vacuum stripping of oils and fats. In *The lipid handbook*, Third Edition, eds. F.D. Gunstone, J.L. Harwood, and A.J. Dijkstra, 235–253. Boca Raton: CRC, Taylor & Francis Group.

Dijkstra, A.J. 2007b. Modification processes and food uses. In *The lipid handbook*, Third Edition, eds. F.D. Gunstone, J.L. Harwood, and A.J. Dijkstra, 263–353. Boca Raton: CRC, Taylor & Francis Group.

Dijkstra, A.J., and Segers, J.C. 2007. Production and refining of oils and fats. In *The lipid handbook*, eds. F.D. Gunstone, J.L. Harwood, and A.J. Dijkstra, 143–262. Boca Raton: CRC, Taylor & Francis Group.

Dujin, G.V., Dumelin, E.E., and Trautwein, E.A. 2006. Virtually trans free oils and modified fats. In *Improving the fat content of foods*, eds. C. Williams and J. Buttriss, 490–506. Boca Raton: Woodhead Publishing Limited and CRC Press LLC.

Erickson, D.R. 2007. Production and composition of frying oils. Chapter 1. In *Deep frying: Chemistry, nutrition, and practical applications*, Second Edition, ed. M. Erickson. Urbana, IL: Academic Press and AOCS Press.

FAO. 2021. Statistical databases. http://apps.fao.org/. February 28, 2021.

Flagella, Z., Rotunno, T., Tarantino, E., Di Caterina, R., and De Caro, A. 2002. Changes in seed yield and oil fatty acid composition of high oleic sunflower (*Helianthus annuus* L.) hybrids in relation to the sowing date and the water regime. *European Journal of Agronomy*, 17(3): 221–230.

Ghotra, B.S., Dyal, S.D., and Narine, S.S. 2002. Lipid shortenings: a review. *Food Research International*, 35(10): 1015–1048.

Gupta, M.K. 2008. *Practical guide to vegetable oil processing*, Second Edition. American Oil Chemists' Society, London: AOCS Press – Academic Press.

Hamdan, Y.A., Perez-Vich, B., Velasco, L., and Fernandez-Martínez, J.M. 2009. Inheritance of high oleic acid content in safflower. *Euphytica*, 168(1): 61–69.

Hosseini, H., Ghorbani, M., Meshginfar, N., and Mahoonak, A.S. 2016. A review on frying: Procedure, fat, deterioration progress and health hazards. *Journal of the American Oil Chemists' Society*, 93(4): 445–466.

Huey, S.M., Let, C.C., and Beng, Y.C. 2015. New developments in palm oil fractionation. *Palm Oil Developments*, 62: 4–14.

Hughes, N.E., Marangoni, A.G., Wright, A.J., Rogers, M.A., and Rush, J.W. 2009. Potential food applications of edible oil organogels. *Trends in Food Science & Technology*, 20(10): 470–480.

Juttelstad, A. 2004. Trans fats: Status and solutions. *Food Tech.* 58(1): 20–22.

Karabulut, I., Turan, S., and Ergin, G. 2004. Effects of chemical interesterification on solid fat content and slip melting point of fat/oil blends. *European Food Research and Technology*, 218(3): 224–229.

Kellens, M., Gibon, V., Hendrix, M., and De Greyt, W. 2007. Palm oil fractionation. *European Journal of Lipid Science and Technology*, 109(4): 336–349.

Kochhar, S.P. 2001. The composition of frying oils. In *Frying: Improving quality*, ed. J.B. Rosell, 87–114. Boca Raton: Wood Head Publishing.

Lindmark Mansson, H. 2008. Fatty acids in bovine milk fat. *Food & Nutrition Research*, 52(1): 1821. DOI: 10.3402/fnr.v52i0.1821.

List, G.R. 2004. Decreasing trans and saturated fatty acid content in food oils. *Food Technology*, 58(1): 23–31.

Liu, Q., Singh, S., and Green, A. 2002. High-oleic and high-stearic cottonseed oils: nutritionally improved cooking oils developed using gene silencing. *Journal of the American College of Nutrition*, 21(sup3): 205S–211S.

Marina, A.M., Che Man, Y.C., Nazimah, S.A.H., and Amin, I. 2009. Chemical properties of virgin coconut oil. *Journal of the American Oil Chemists' Society*, 86(4): 301–307.

Merrill, L.I., Pike, O.A., Ogden, L.V., and Dunn, M.L. 2008. Oxidative stability of conventional and high-oleic vegetable oils with added antioxidants. *Journal of the American Oil Chemists' Society*, 85(8): 771–776.

Mounts, T.L., Warner, K., List, G.R., Kleiman, R., Hammond, E.G., and Wilcox, J.R. 1998. Effect of altered fatty acid composition on soybean oil stability. *Journal of the American Oil Chemists' Society*, 65: 624–628.

O'Brien, R.D. 2009. *Fats and oils. Formulating and processing for applications*, Third Edition. Boca Raton: CRC Press.
Ogutcü, M., and Yılmaz, E. 2014. Oleogels of virgin olive oil with carnauba wax and monoglyceride as spreadable products. *Grasas y Aceites*, 65(3): 040.
O'Keefe, S.F., Wiley, V.A., and Knauft, D.A. 1993. Comparison of oxidative stability of high- and normal-oleic peanut oils. *Journal of the American Oil Chemists' Society*, 70(5): 489–492.
Ollivier, D., Artaud, J., Pinatel, C., Durbec, J.P., and Guérere, M. 2006. Differentiation of French virgin olive oil RDOs by sensory characteristics, fatty acid and triacylglycerol compositions and chemometrics. *Food Chemistry*, 97(3): 382–393.
Palla, C., Giacomozzi, A., Genovese, D.B., and Carrín, M.E. 2017. Multi-objective optimization of high oleic sunflower oil and monoglycerides oleogels: Searching for rheological and textural properties similar to margarine, *Food Structure*, 12: 1–14.
Pehlivanoglu, H., Demirci, M., Toker, O.S., Konar, N., Karasu, S., and Sagdic, O. 2018. Oleogels, a promising structured oil for decreasing saturated fatty acid concentrations: Production and food-based applications. *Critical Reviews in Food Science and Nutrition*, 58(8): 1330–1341.
Pernetti, M., van Malssen, K., Kalnin, D., and Floter, E. 2007. Structuring edible oil with lecithin and sorbitan tri-stearate. *Food Hydrocolloids*, 21: 855–861.
Purdy, R.H. 1986. High oleic sunflower: physical and chemical characteristics. *Journal of the American Oil Chemists' Society*, 63(8): 1062–1066.
Ramli, M.R., Siew, W.L., and Cheah, K.Y. 2009. Production of high oleic palm oils on a pilot scale. *Journal of the American Oil Chemists' Society*, 86(6): 587–594.
Rattray, J. 1991. *Biotechnology of plant fats and oils*. Champaign: American Oil Chemists Society.
Ribeiro, A.P.B., Basso, R.C., dos Santos, A.O., Andrade, G.C., Cardoso, L.P., and Kieckbusch, T.G. 2013. Hardfats as crystallization modifiers of cocoa butter. *European Journal of Lipid Science and Technology*, 115(12): 1462–1473.
Rocha, J.C.B., Lopes, J.D., Mascarenhas, M.C.N., Barrera-Arellano, D., Guerreiro, L.M.R., and da Cunha, R.L. 2013. Thermal and rheological properties of organogels formed by sugarcane or candelilla wax in soybean oil. *Food Research International* 50: 318–323.
Rogers, M.A., Wright, A.J., and Marangoni, A.G. 2008. Engineering the oil binding capacity and crystallinity of self-assembled fibrillar networks of 12-hydroxystearic acid in edible oils. *Soft Matter* 4: 1483–1490.
Rotondi, A., Bendini, A., Cerretani, L., Mari, M., Lercker, G., and Toschi, T.G. 2004. Effect of olive ripening degree on the oxidative stability and organoleptic properties of cv. Nostrana di Brisighella extra virgin olive oil. *Journal of Agricultural and Food Chemistry*, 52(11): 3649–3654.
Serna-Saldivar, S.O. 2008. Properties and manufacturing of oils. In *Industrial manufacture of snack foods*, ed. S.O. Serna-Saldivar, 73–103. London: Kennedys Publications Ltd.
Shafer de Martini Soares, F.A., da Silva, R.C., Gilmaraes da Silva, K.C., Lourenço, M.B., Ferreira Soares, D., and Gioielli, L.A. 2009. Effects of chemical interesterification on physicochemical properties of blends of palm stearin and palm olein. *Food Research International*, 42(9): 1287–1294.
Sintang, M.D.B., Danthine, S., Brown, A., Van de Walle, D., Patel, A.R., Tavernier, I., Rimaux, T., and Dewettinck, K. 2017. Phytosterols-induced viscoelasticity of oleogels prepared by using monoglycerides. *Food Research International*, 100: 832–840.
Smith, K.W. 2001. Crystallization of palm oil and its fractions. In *Crystallization processes in fats and lipid systems*, eds. N. Garti and K. Sato, 357–380. New York: Marcel Dekker.
Sonntag, N.O.V. 1982. Fat splitting, esterification and interesterification. In *Baileys industrial oil and fat products*, Volume 2, ed. D. Swern. New York: John Wiley and Sons.
Tan, C.P., Man, Y.C., Selamat, J., and Yusoff, M.S.A. 2002. Comparative studies of oxidative stability of edible oils by differential scanning calorimetry and oxidative stability index methods. *Food Chemistry*, 76(3): 385–389.
Taylor, R.D. 2010. Adsorbents. In *Bleaching and purifying fats and oils: Theory and practice*, ed. G.R. List, 69–85. Urbana: AOCS Press.
Warner, K., and Knowlton, S. 1997. Frying quality and oxidative stability of high-oleic corn oils. *Journal of the American Oil Chemists' Society*, 74(10): 1317–1322.
Wilson, R.F., Burton, J.W., and Kwanyuen, P. 1989. *Edible fats and oils processing: Basic principles and modern practices*, ed. D. Erickson, 355–359. Champaign: American Oil Chemists' Society.

8 Seasonings for Snack Foods

Johanan Espinosa-Ramírez

CONTENTS

8.1 INTRODUCTION

The market of savory snacks has registered a strong growth during the last years, expecting an annual growth of 4% in the following years (MarketLine 2019). This market includes a wide variety of products where potato chips (Chapter 14), corn chips (Chapter 11) and extruded snacks (Chapter 10) are the most popular (Buechler and Lee 2019). Other savory snack products include popcorn (Chapter 9), pretzels, crackers (Chapter 10), peanuts (Chapter 13), tree nuts (Chapter 15), among others.

The most popular snacks are produced with bland flavor ingredients (i.e., potato or cereals), which make essential the addition of seasonings to increase their acceptability by consumers (Ainsworth and Plunkett 2007). Seasoning results in a highly relevant operation (Chapter 3) during the production of snacks, especially when the snack flavor determines the selection of a product among others of the same category.

The ingredients selected during formulation of a snack seasoning must present both sensory and technological properties. This is required to obtain adequate acceptance by the consumer but also ease

DOI: 10.1201/9781003129066-8

of processing by the snack producer. Among the sensory properties, seasonings may affect the mouthfeel, flavor and appearance of snacks. On the other hand, the ingredients added to improve technological properties contribute to the seasoning adhesion, flowability, storage stability and to achieve adequate snack coverage. Formulations may be applied on powdered or liquid basis. Then, selection of seasoning ingredients should also consider the application method. Moreover, seasoning ingredients are selected to satisfy the increasing demand of consumers for more natural and healthier foods, and with lower contents of fat and sodium. Adequate formulation of seasonings is key to achieving these goals, especially since they are the main contribution to the salt or sodium content of the snack. Also, depending on the application method, seasonings may increase significantly the fat content of the product. Finally, each country may prefer certain flavor profiles according to the culinary practices, which must be considered when seasonings are formulated. For instance, the preferred flavors in the United States are cheese, barbecue, sour cream-onion and ranch. In Latin America, chili pepper/citric acid and cheese/chili pepper-based flavorings are extremely accepted (Serna-Saldivar 2008). European countries have a wider variety of popular flavors including sweet chili pepper, cider vinegar and spices, cheese, sour cream and onion, paprika and more exotic flavorings such as prawn, oyster or meat.

The aim of this chapter is to describe the functionality of the different ingredients used in the formulation of savory seasonings applied in the diversity of snack products. Alternatives to reduce salt, natural ingredients to substitute flavor enhancers, antioxidants and coloring agents, and options to reduce or eliminate the use of oil during seasoning application are also addressed.

8.2 SEASONING INGREDIENTS

Snack seasonings are constituted by a combination of ingredients with different technological and sensory functions. In general, formulations contain flavoring materials including sweeteners and acidulants, salt, carriers and fillers, flavor enhancers, mouthfeel components, coloring agents and minor ingredients including anticaking agents and antioxidants (Serna-Saldivar 2008). These ingredients will be described in this section. The selection of ingredients during seasoning formulation is highly relevant since they will determine how the flavor will release upon consumption, what the aftertaste will be, and if the seasoning can be applied evenly on the snack surface (Seighman 2001). Moreover, seasoning ingredients may help to improve the stability of the seasoning and to increase the appeal of the snack.

8.2.1 Flavorings

Flavoring materials are ingredients incorporated into the seasoning to provide or reinforce its characterizing flavor profile. Flavor selection is the most important step when developing a seasoning. Typically, two components are added to construct the flavor profile: primary and middle part flavors. The primary flavors, also called top notes, are savored first and are responsible for the initial flavor of the snack (Seighman 2001). The primary flavors come from compounds that dissolve and release flavor rapidly due to their volatile nature and may include smoke, Worcestershire, natural flavors as oleoresins and essential oils of herbs and spices, or artificial flavors (Seighman 2001, Hall 2012). The burst of primary flavors dissipates quickly, leading to flavors perceived from ingredients that solubilize slower such as ground spices/herbs and dehydrated vegetable powders as onion and garlic. These flavors are used to fill the middle of the taste experience (Seighman 2001). To achieve the flavor profile in savory seasonings, the flavorings used are: (i) compounded flavors which are concentrated forms of meaty, dairy or vegetable tastes plus their top note flavors, and (ii) base flavors including spices or dairy ingredients. Compounded flavors provide stronger long flavors compared to base flavors and may be obtained after extraction, thermal or fermented processes (Hall 2012).

The flavor industry has several materials to create flavorings. Snack seasonings may contain natural ingredients including plant and animal materials, products of fermentation and enzymatic reactions, and also synthetic chemicals (Reineccius 2005). The flavors most commonly used to season savory snacks are spices and herbs and their essential oils/oleoresins (Section 8.2.1.1), dehydrated vegetables (Section 8.2.1.2) and dairy ingredients (Section 8.2.1.3) which are summarized in Table 8.1.

TABLE 8.1
Flavorings Commonly Used in Snack Seasonings

Category and Flavor	Description
Spices and Oleoresins	
Black, White and Red Pepper	Obtained from the berry of the *Piper nigrum* plant which produces both black and white pepper. Dried berries are called peppercorn. May come whole, cracked or ground (coarse or fine). The essential oil gives the aroma, whereas the non-volatiles such as piperine, give the pungency. **Black pepper** berries are picked immature, fermented and dried, and have a hot, biting taste with pungent and woody aromas. **White pepper** berries are picked near ripe, soaked or steamed to soften their skin, stripped of the outer hull by rubbing and sundried or prepared from black peppercorn by decortication. They have a sharp, winey and biting taste with little aroma. **Red peppers** are ripe matured berries, dried at high temperature and have a delicate sweet taste.
Clove	Dried unopened flower buds of the *Syzyium aromaticus* tree which are sold whole or ground. They have a spicy, woody, burning, sweet and musty aroma with a taste sharp, bitter and pungent with a numbing feeling. The essential oil from buds is yellowish-green and contains eugenol (80%) that gives the sharp burning flavor. The oleoresin is brownish-green.
Cumin	Dried, brown, ripe fruit or seed of *Cuminium cyminum* and is available whole or ground. It has a nutty earthy, spicy and bitter taste with a penetrating, warm and slightly lemony aroma. It presents a more intense aroma when it is dry roasted.
Mustard	There are three types of seeds from the genus *Brassica*: pale yellow, brown and black mustards. The smaller and darker the seed, the mustard is more intense and pungent. Ground yellow mustard acts as a physical emulsifier, flavor enhancer and water binder.
Oregano	Dried leaves from *Origanum vulgare*, available whole, flaked or ground. Can be mild or strong depending on its origins. It has a phenolic, herbaceous, slightly floral and bitter taste with lemony and pungent notes. Oleoresin oregano (made from Spanish oregano) is a dark brownish-green liquid.
Parsley	Dried leaves of *Petroselinum satiuum* (flat leaves). It comes whole, flaked or minced. When it is consumed fresh, parsley has a grassy, green taste with an herbaceous and slight lemony aroma. Dry parsley has no flavor and is commonly used in flakes to make snacks look more appetizing. Essential oil is produced from leaves or seeds while oleoresins are manufactured from seeds and blended with leaf oil.
Paprika	In most of the world, paprika refers to the ground or powdered form of the dried peppers from the nonpungent to slightly pungent red varieties of *Capsicum annum* with most of the seeds and veins removed. Depending on its origin and grade, it can be sweet to pungent. Paprika loses its aroma and develops bitterness when exposed to high heat and its color when exposed to light and air.
Dehydrated Vegetables	
Onion Powder	It is an indispensable flavoring from *Allium cepa*. The common onion is a white, yellowish-white or brownish-red bulb and are mild or pungent depending on the variety and source. The flavor-contributing chemicals are sulfide compounds that are released through enzymatic action when the cells are damaged.
Garlic Powder	It is one of the most world's popular spices and is obtained from the cloves of *Allium sativum*. Dried garlic used in seasonings is powdered and has a very strong, persistent aroma and taste. The oils come in many dilutions for easier blending and handling. Garlic salt consists of garlic powder, salt and anticaking agent.
Chili Pepper	Belong to the genus *Capsicum* with five species: *C. annum*, *C. frutescens*, *C. chinense*, *C. pubescens* and *C. baccatum* and more than 3,000 known varieties differing in shape, color, size, flavor and degree of pungency, the latter depending on the capsaicinoids content ranging from 0.2 to 2%. Dried chilies add flavor and heat to foods, providing a background note for spices and other flavorings and are available whole, ground, crushed or powdered. Dried chili peppers can be roasted to develop more intense flavors.
Tomato Powder	Tomato powder may be used to impart color and flavor to the seasoning. Its flavor is acidic and can be found in barbeque and nacho cheese seasonings.

(*Continued*)

TABLE 8.1 (CONTINUED)
Flavorings Commonly Used in Snack Seasonings

Category and Flavor	Description
Dairy Products	
Cheese Powder	Spray-dried cheese is one of the most used flavorings in salty snack foods. Cheddar cheese is the most popular, followed by mozzarella and parmesan. Cheese seasonings are obtained by blending cheese and flavors with water, dairy-based ingredients, color emulsifiers and melting salts to obtain a flavor concentrate, which is then spray-dried. This slurry contains approximately 35-45% solids. The advantage of spray drying is that the product is immediately dried so undesirable color changes are minimized.
Sour Cream	Obtained after spray drying of a slurry containing approximately 35–45% solids. The cream is the portion of the milk rich in fat that rises to the top of undisturbed milk and generally contains 18 to 40% fat.
Whey Powder	Whey is the portion of the milk remaining after coagulation and removal of the curd. It is rich in lactose and lactoalbumin. Whey powder has a clean, mild flavor and is essential as a building block in seasonings.

Source: Raghavan (2000), Ainsworth and Plunkett (2007), Serna-Saldivar (2008), Chandan (2011).

Basic flavorings that compliment and/or boost seasoning flavors are sweeteners (Section 2.1.4) and acids (Section 2.1.5). Artificial flavorings are mixtures of synthetic compounds recognized as safe for use in foods and may also be found in snack seasonings. They are added in a liquid form (solubilized in food-grade solvents) or in solid carriers. Liquid flavorings may be water-soluble (dissolved in alcohol or propylene glycol), or fat-soluble (dissolved in triacetin, benzyl alcohol or vegetable oil). Artificial flavorings are used when the natural counterpart is not available (i.e., meat top notes or fruit flavors) or also to reduce costs. However, the market for synthetic flavor chemicals has significantly decreased over time (Reineccius 2005).

8.2.1.1 Spices, Herbs and Oleoresins

Spices and herbs are available in several presentations. They are ground into fine powders to blend them with fine salt, garlic and onion powders or are added in coarsely ground, whole or flake forms to also contribute to the appearance of the seasoning as occurs with parsley and oregano (Seighman 2001). Spices and herbs need to be treated to reduce microbial loading with techniques that minimize flavor losses. To do so, irradiation was the most common method; however, since it is not currently permitted in the European Union (EU), other methods have been explored and optimized. Whole/ground spices or herbs are treated with steam during a short period of time which represents a good alternative to radiation (Hall 2012). A safe alternative to the use of whole or ground spices and herbs is the use of their essential oils or oleoresins due to their significantly low microbial risk after the extraction processing (Seighman 2001).

Spices and herbs that are commonly used in snack seasonings include pepper (black, red or white), clove, cumin, mustard, oregano, parsley and paprika (Table 8.1). However, depending on the world region, other spices may be preferred by the consumers and incorporated into the flavor profile of snacks. Ground spices are used at levels ranging from 0.25 to 2.00% in snack seasonings. Since flavors are concentrated in spice extracts and spray-dried essential oils, these ingredients are used in lower quantities. The replacement level would be determined based on the manufacturer's instructions. Spices are expensive ingredients, but their use is cost effective due to the strengths of their flavor (Seighman 2001).

During spice milling, oil glands are ruptured causing evaporation of essential oils and therefore, loss of aromas (Serna-Saldivar 2008). To overcome this problem, essential oils and oleoresin extracts of spices and herbs, which have more flavor impact compared to ground fresh counterparts, may be incorporated into the seasoning (Ainsworth and Plunkett 2007). Oleoresins contain both the essential oil and the nonvolatile constituents of a given spice and while they are more expensive than spices, they are used in lower amounts due to their concentrated flavorful compounds (Serna-Saldivar 2008). For instance, 1.8 kg of oregano oleoresin is equivalent to 45.3 kg of freshly ground oregano (Raghavan 2000). Besides their larger amount of flavor, essential oils and oleoresins are readily released when consumed in contrast to the whole spice. This is useful to develop the flavor profile. However, the negative aspect is that they are heat-sensitive (Reineccius 2005).

The essential oils of spices and herbs are obtained by distillation which lead to accumulation of volatile compounds that give the flavor a top note with a very strong aroma. On the other hand, oleoresins are typically produced by solvent extraction although they can be obtained after cold pressing or cryogenic supercritical carbon dioxide extraction. The oleoresins have a very similar flavor compared to the whole profile of the original material but with the advantage of having less color, sharper top note flavors, and low microbiological load (Hall 2012). Essential oils and oleoresins are typically spray-dried to accentuate their flavor or encapsulated to increase their shelf stability and microbial risk (Ainsworth and Plunkett 2007). Spray drying of essential oils and oleoresins also helps to accelerate the flavor release (Seighman 2001). For their application, they are commonly mixed with carriers in adequate concentrations to emulate the intensity of the flavor in the original spice or herb (Hall 2012).

8.2.1.2 Dehydrated Vegetables

Dehydrated vegetables are also used as flavoring agents in snack seasonings. They are produced by drying the vegetable slurry under heat and vacuum to reach moisture contents lower than 5% (Seighman 2001). Heat-vacuum drying presents the advantage of reducing flavor losses and color changes of dried vegetables (Chapter 16). Also, the combination of microwave and vacuum drying presents a great potential to improve the flavor preservation (Serna-Saldivar 2008). To increase the drying efficiency when microwave is used to dehydrate, the use of fluidization drying devices is recommended. Microwave fluidization drying avoids local overheating and uneven drying, which also reduces nutrition losses compared to hot air drying and static microwave drying (Lv et al. 2019).

Vegetables may also be toasted or roasted prior to drying to modify the flavor profile. The most common dehydrated vegetables used in seasonings are onion, garlic, chili pepper and tomato powders (Table 8.1). These ingredients are relatively inexpensive and have the advantage of containing concentrated flavor. For instance, onion and garlic bring depth to the middle part of the seasoning flavor profile making the taste profile more complex and prolonging its sensation. Onion powders may be added at levels 1–10% and garlic powder is used in lower concentrations (0.5–5%) (Seighman 2001). Garlic may also be found as oil and oleoresin forms. Garlic oil presents 200 times the strength of dried garlic and 900 times the strength of fresh garlic. For garlic oleoresin, 3.6 kg are equivalent to 45.3 kg of dehydrated garlic, and 0.9 kg is equivalent to 45.3 kg of fresh garlic (Raghavan 2000).

8.2.1.3 Dairy Ingredients

Dairy-derived ingredients are very popular as flavorings for snack products. Moreover, they are also used in seasonings to enhance color in snack items and as mouthfeel components which will be discussed in Section 8.2.5. Dairy-derived ingredients include dry cheese, cheese flavors and sour cream which are used in coated nuts, crackers, potato chips, extruded and puffed snacks. Enzyme-modified cheese and butter flavor concentrates are also used in snack seasonings as flavor boosters and to round flavors. These ingredients give intense flavor notes at low usage levels and are used along with whey powder to replace a fraction of cheese solids. Additionally, in nacho, ranch and barbeque flavors, whey and other dairy ingredients are essential building blocks. The addition of whey powder which contains high protein quality and calcium may represent an advantage from the nutritional viewpoint. However, this would depend on the specific seasoning formulation and its application level (Chandan 2011).

Cheese seasonings may contain different cheese types and levels. Cheese powders come in a variety of flavors including cheddar, blue, Colby, Monterrey jack, mozzarella, Romano and parmesan, being cheddar cheese the most popular alternative (Chapter 17). Cheese powders may contain more than 90% cheese while blends contain less than 50–70% cheese (Chandan 2011). Dairy powders may be used in seasoning formulations at different levels giving different properties to the flavor: at low levels, they help to smooth out the flavor, especially when a high level of flavors and spices are present, and at high levels, they contribute to the mouthfeel and flavor of the seasoning (Seighman 2001). However, using high concentrations of cheese increases the cost of the seasoning. This may be reduced by substituting cheese solids with enzyme-modified cheese and butter flavor concentrates along with whey solids, buttermilk, skim milk solids and/or vegetable oil (Chandan 2011). The price of cheese ingredients will also depend on whether the product is kosher or not (Seighman 2001).

The most common process to obtain cheese seasonings involves blending cheese and flavors with water, dairy-based ingredients, color, emulsifiers and melting salts to obtain a flavor concentrate which is then spray-dried. Herbs and spices may be blended with the resulting cheese powders to complement the flavor and appearance (Chandan 2011).

8.2.1.4 Sweeteners

Sweeteners are used to stimulate the basic taste sensations. The most common sweeteners used in seasoning formulations are sugar, brown sugar, dehydrated honey solids, spray-dried molasses, dextrose and fructose. These ingredients must have a low particle size distribution as the rest of the ingredients in the formulation. The selection of the specific sweetener will depend on the desired sweetness profile. These are inexpensive components of the seasoning. However, a significant problem associated with their use is their high hygroscopic nature, which may lead to flowability problems, especially when processed under hot conditions. Therefore, when sweeteners are used in seasoning formulations, flowing agents must be added (Seighman 2001). Sweeteners including dextrose, sucrose and fructose are also used as carriers in snack seasoning formulations (Ainsworth and Plunkett 2007).

8.2.1.5 Acids

These ingredients are used in seasoning formulations to impart acidic flavors and also to enhance or boost the overall flavors. They may also function to inhibit color losses and as preservatives (Ainsworth and Plunkett 2007). The most used acids in seasonings are citric acid, lactic acid, vinegars and acetic acid. Citric acid is found in powder or fine granulated forms and is used as a flavor enhancer or to sharpen the overall flavors due to its immediate impact on the palate (Hall 2012). This acid accentuates the hot or pungent sensation of seasonings which may help to reduce the salt required in the blend. It also acts as a chelating agent of oxidant metals (Serna-Saldivar 2008). Lactic acid is used as flavoring or preservative. It boosts dairy flavors such as cheese and sour cream-type flavorings. Vinegars may be found in spray-dried powder versions using maltodextrin as a carrier and containing up to 10% acetic acid which makes them comparable to liquid alternatives (Hall 2012). Acetic acid or sodium diacetate may be used to mimic the vinegar flavor (Seighman 2001, Serna-Saldivar 2008). Lemon juice may also be used as an acid source and flavoring in snack seasonings. To achieve a higher stability, lemon juice is spray-dried (Hall 2012).

Citric or acetic acids are usually used jointly with flavorings that contain chili pepper, powdered garlic, black pepper, basil, clove, coriander and/or oregano (Serna-Saldivar 2008). Popular flavors in potato chips and other snacks are simply pepper and vinegar or salt and vinegar.

8.2.2 Salt

Salt is one of the main ingredients in seasonings since it strengthens the global flavor and it is even the only seasoning used in some snacks (Serna-Saldivar 2008). Salt refers to sodium chloride (NaCl) and is probably the cheapest ingredient in seasoning formulations. It may be used in snack foods in different particle sizes, from fine to coarse, and also different shapes of granules including flaked particles (Ainsworth and Plunkett 2007). The most common salt used in snack seasoning formulations is flour salt, which is a fine granular material that contains 96% of its particles with sizes lower than 178 μm. Granulated salt or fine flake salt may be used for snacks where only salt is added, but are not recommended for seasoning blends. This is because the larger salt particles adhere to the snack differently than the fine-particle-size ingredients in the seasoning blend, resulting in salt falloff and uneven distribution on the snack surface (Seighman 2001). However, these larger salt crystals are appreciated in some snacks such as pretzels or crackers, since they dissolve slowly giving a sharper lingering salt taste compared to smaller crystals and also contribute positively to the snack appearance (Chandan 2011).

From a flavor point of view, the salt is formulated to be within two outer limits. The upper limit where the salty flavor becomes overpowering and unpalatable, and the lower limit where salt still has a flavor. The latter vary among individuals that have different thresholds for tasting salt although it is considered to be around 0.3–0.5%. Therefore, for most savory products, salt levels ranging from 0.8 to 1.0% in RTE products are considered adequate as starting points during seasoning formulation (Hall 2012). The snack type also influences the salt perception and therefore, its application level. For instance, potato chips and extruded snacks contain 1–2.25% salt, tortilla crisps and pretzels contain 1.5–2.5% salt and popcorn may contain up to 4% salt in a snack basis (Ainsworth and Plunkett 2007). Determination of the salt level must also take into consideration its presence in other ingredients such as spray-dried dairy powders, hydrolyzed vegetable proteins (HVPs), autolyzed yeast extracts or compounded flavors included in the seasoning formulation. Finally, the selection of adequate salt levels must be confirmed through consumer testing (Seighman 2001).

8.2.2.1 Salt Substitutes and Reduction Strategies

Although salt is essential for health, high sodium diets are associated with increased blood pressure which increases the risk for cardiovascular disease and strokes (Hall 2012). Therefore, reducing the salt content of many food products is practiced worldwide. Still, salt reduction in snacks is difficult since it may affect the product acceptance by the consumers. Moreover, the reduction levels are dependent on the salt consumption/requirements of the population that will consume the product, which complicates the scenario for international producers. One of the main challenges when replacing salt is to find substitutes with the same functionality at comparable prices. Gradual reductions in salt over time, allowing the consumer to become habituated to lower levels has been performed by companies and is recognized as one of the simplest methods (Ainsworth and Plunkett 2007). A study performed by Mitchell et al. (2011) found that it was possible to diminish up to 40% the initial salt content in a ready-to-eat meal, without affecting taste by panelists. However, authors indicated that the level of reduction depends on the product and the processing conditions. Therefore, food product development studies need to be performed for each specific product to validate the maximum level of reduction. To implement this gradual reduction strategy, it is recommended to reduce the salt content in increments of approximately 10% to guarantee that the salt taste threshold of consumers is reduced over time (Mitchell et al. 2011). Therefore, the limitation of this strategy is that only limited amounts of salt can be reduced immediately, and significant reductions would be achieved over a long period of time (Israr et al. 2016). Moreover, even when consumers may adapt to the flavor change, the salt reduction could also be accompanied by changes in the seasoning flowability, shelf life and texture of the snack food, which may be unacceptable by consumers and would require reformulation of the seasonings (Ainsworth and Plunkett 2007).

Strategies for salt reduction in short periods include: (i) salt substitutes such as potassium chloride and other ingredients containing mineral salts, (ii) usage of flavor enhancers to increase the salt perception, (iii) increase the intensity of flavors in the snack base raw materials or in the seasoning, (iv) reduction in the salt particle size, (v) contrast of flavors and microencapsulation of salt, and (vi) use of mock salts. These strategies are discussed in the following paragraphs. It is important to mention that in most cases, a combination of several strategies is practiced to achieve optimal results.

One of the most extended strategies to reduce sodium in seasonings at an industrial level is the use of potassium chloride (KCl) as a substitute. However, since potassium ion has a less salty flavor compared to sodium, KCl must be added in higher concentrations producing noticeable bitter metallic flavors. To mask this flavor, bitter blockers or masking agents such as L-lysine hydrochloride may be used (Hall 2012). The use of umami taste promoters such as monosodium glutamate (MSG), disodium inosinate, disodium guanylate and kokumi aroma, also helps to reduce the bitter taste of KCl (Tavares Filho et al. 2020). In a study performed by Tavares Filho et al. (2020), authors found that the combination of KCl with MSG resulted in similar sensory perceptions to the use of NaCl. These authors further concluded that this combination was an adequate strategy to reduce sodium contents in foods without affecting negatively the consumer acceptance. The use of other mineral salts such as ammonium, calcium and magnesium chlorides have limited applications due to their unpleasant flavors. Whey permeates and spray-dried plant salt substitutes produced with saltwort, sea tangle and kukoshi, are ingredients that naturally contain mineral elements, which contribute to a salty taste (Israr et al. 2016). These ingredients have been tested to replace salt in baked products and may also be used as natural alternatives to diminish sodium in seasoning formulations.

Previous studies have also demonstrated that the addition of flavor enhancers help to reduce the salt content, without affecting the acceptability by consumers. Flavor enhancers compensate for the loss of flavor caused by the salt reduction, being an adequate strategy for immediate modifications in the salt content of food formulations (Mitchell et al. 2011). It is important to consider that flavor enhancers only boost the salty flavor of the product, but cannot be used as a salt substitute. The ingredients contained in the seasoning formulation may contribute to modifying the maximum level of reduction of salt achievable by the addition of flavor enhancers. Therefore, the reformulation of snack seasonings to lower their salt content should be accompanied by sensory analyses to validate the results. Flavor enhancers used to achieve sodium reductions include MSG or disodium salts of inosinate and guanylate, which are discussed in Section 8.2.4. Yet, the use of natural flavor enhancers to reduce the salt content in food has been an increasing trend due to the consumer interest in consuming foods with less additives. Jimenez-Maroto et al. (2013) used a natural flavor enhancer to replace salt in white bread. This flavor enhancer was produced after fermentation of a solution containing water, soybeans, wheat and sodium chloride in a procedure similar to the production of soy sauce but using different processing parameters. After fermentation, the ingredient was spray-dried using maltodextrins as a carrier. The resulting product was suitable for powdered snack seasonings. This fermented soy-based flavor enhancer improved the saltiness perception of low-salt products being possible to replace up to 25% salt without negative effects in consumer acceptance. Natural flavor enhancers have also been tested in snack products. Harada-Palermo et al. (2021) evaluated a flavor enhancer rich in umami compounds to reduce the sodium content of extruded corn puffs. This new umami ingredient was obtained from shiitake mushroom (*Lentinula edodes*) by-products and was free of MSG. Their study proved that this new umami ingredient was an efficient flavor enhancer that imparted similar results compared to MSG in features such as seasoning flavor, saltiness and umami tastes. Therefore, it could be used as a substitute for MSG in low-sodium formulations contributing to the development of snacks with less additives (Harada-Palermo et al. 2021).

Increasing the intensity of other flavors in the snack base raw materials or in the seasoning may also help to reduce their salt content. This is achieved by increasing flavors such as pepper and spicy or by the addition of small amounts of acids in the formulation. The increase of these flavors causes overwhelming in taste receptors helping to reduce the salt threshold and making possible its reduction (Hall 2012).

The reduction in salt particle size contributes to the in-mouth delivery of sodium, modifying the saltiness perception. Therefore, the use of salts with lower particle size is an alternative to reduce the sodium content of snacks. Rama et al. (2013) used sliced potato chips as a snack model to evaluate the impact of different salt crystal sizes on the delivery rate of sodium to the tongue and the saltiness taste. Results indicated that small salt crystals (< 106 μm) dissolved and diffused quicker in the saliva and throughout the mouth and tongue leading to a higher sensation of saltiness in comparison to medium (106–425 μm) and large (425–710 μm) crystal fractions. Smaller crystals also presented better coating efficiency and broader distribution compared to the larger particles that showed an inhomogeneous distribution on the crisp surface. However, authors recommended consumer insight studies prior to the implementation of this strategy since the use of smaller crystals may also lead to a quicker loss of saltiness after chewing which may not be desirable in all products.

The creation of contrasting salt levels throughout the product and use of encapsulation to increase the saltiness perception with lower salt concentrations are techniques that have been recently evaluated. Contrast flavors are achieved by inhomogeneous distribution of salt in different layers of the product. This generates saltiness peaks during chewing, increasing the perception of the salty taste (Israr et al. 2016). Microencapsulation of salt is another technique that contributes to the inhomogeneous distribution of salt and may help to control the sodium release and therefore, the saltiness perception. Cai and Lee (2020) altered the sodium release dynamics after salt microencapsulation by means of spray drying with a three-fluid nozzle and carriers such as modified starch or maltodextrin. However, authors declared that sensory analyzes are still required to determine the release profile most suitable for snack products.

Finally, the usage of mock salts could also help to reduce sodium contents in seasonings. Mock salts consist of starch particles coated with a thin layer of salt (Ainsworth and Plunkett 2007).

8.2.3 Carriers and Fillers

Carriers and fillers are ingredients of bland taste and small particle size used to carry flavorings and ensure the even distribution of the seasoning over the snack surface (Serna-Saldivar 2008). Formulators use fillers to adjust the application level of the seasoning ensuring the desired coverage and flavor impact since changes in the proportion of the bland filler will have little impact on the overall flavor of the seasoning (Seighman 2001). These ingredients, also called extenders, are one of the main ingredients in snack seasonings accounting for 20–50% of its composition depending on the type of seasoning and its level of application (Seighman 2001, Ainsworth and Plunkett 2007).

Examples of fillers in snack seasonings include grain flours, maltodextrins, corn starch, corn syrup solids, lactose, dextrose or whey solids (Seighman 2001, Ainsworth and Plunkett 2007, Serna-Saldivar 2008). Among cereal flours, wheat and maize (Chapter 6) are the most widely employed. Refined wheat flour functions as bulking agent or carrier of flavorings and is appreciated due to its bland flavor, and since its starch and gluten function as adhesives. Maize meal or flour is used as a body and binder agent. Other flours that are less frequently used include soybean, barley, oats, rice and rye. Maltodextrins are the product obtained after partial acid or enzymatic hydrolysis of starch. In snack seasonings, they are widely used as bulking agents and carriers for active ingredients such as flavors (Serna-Saldivar 2008). Cornstarch has a bland taste and functions as a carrier but also helps with the adhesion of the seasoning. Corn syrup solids are used as carriers and also impart flowability properties to seasoning powders. Lactose serves as a filler, flavor carrier and improves the coating properties of seasonings. It also provides dairy flavor and relatively low sweetness. Dextrose is a carrier that also imparts sweetness, which is desired in some seasoning formulations. Whey powders are dairy powders that have a clean, mild flavor and may be used to carry cheese, oleoresin and other savory flavors helping to coat snacks uniformly and giving an appealing appearance (Chandan 2011).

8.2.4 Flavor Enhancers

Flavor enhancers increase the overall flavor impact and give a mouth-watering sensation attracting the consumer to eat more. Therefore, as salt, flavor enhancers are key ingredients in seasoning formulations to avoid bland or flat tastes and to develop salty snacks with better taste (Seighman 2001). The most widely used flavor enhancer is monosodium glutamate which is the sodium salt of glutamic acid. This salt does not contribute a flavor of its own which may represent an advantage in several applications (Serna-Saldivar 2008). MSG is most effective in the pH range from 5.5 to 8.0 and is unstable under heat processing where it may participate in Maillard reactions. These factors would lead to loss of the flavor potentiating activity (Reineccius 2005). Hydrolyzed vegetable proteins are also commonly used as flavor enhancers. HVPs are produced after acidic hydrolysis of proteins using HCl and heating conditions. The hydrolyzed proteins are neutralized with NaOH, bleached and aged for around 8 months before drying. HVPs contain up to 12% MSG, which is naturally present in the protein source and may contain up to 45% salt, depending on the neutralization conditions. The sensory properties depend on the source of proteins and the production conditions (Reineccius 2005). The most common raw material to produce HVP is defatted soybean flour (Serna-Saldivar 2008). However, HVPs may be also produced from maize. Besides its potential as a flavor enhancer, HVPs are also added to seasonings to make the overall flavor meatier in character (Seighman 2001). Other flavor enhancers that are used in snacks include the 5'-nucleotides disodium inosinate and disodium guanylate (Serna-Saldivar 2008). The 5'-nucleotides are more resistant to short-term heat treatments and only minimal destruction occurs during deep-fat frying. MSG and 5'-nucleotides act synergistically. A 1:1 ratio may result in up to 30 times more flavor intensity. However, due to the higher price of inosinate and guanylate salts, they are typically blended in a 1:1 ratio and this preparation is used in an MSG:inosinate/guanylate salts ratio of 95:5 (Reineccius 2005).

The level of flavor enhancers used in snack seasonings may vary depending on the seasoning profile, but common levels are MSG, 1–5%; HVP, 1–5% and disodium salts of inosinate and guanylate, 0.01–0.05%. The cost also differs being MSG and HVP cheaper compared to disodium inosinate and disodium guanylate that may have a cost up to 4 times higher (Seighman 2001).

Flavor enhancers are used to reduce the salt concentration in snack seasoning formulations. These compounds do not replace salt but increase the stimulation of the salt receptors (Ainsworth and Plunkett 2007). In a recent study performed by Buechler and Lee (2019), authors proved that MSG and disodium inosinate and guanylate combinations could maintain or improve the overall liking of reduced-sodium potato chips (up to 51% reduction) and puffed rice (up to 57% reduction). Moreover, these authors found that the consumer acceptability of the reduced-sodium snacks was benefited when combined with ingredient labeling and sodium reduction education.

Even though the inclusion of flavor enhancers in seasoning formulations results in better overall snack taste, sometimes they are not used by snack producers due to the consumers' increasing interest in clean food labels and natural products (Seighman 2001). Flavor enhancers that are considered natural alternatives are discussed in Section 8.3.1.

8.2.5 Mouthfeel Components

The coating of snacks modifies their texture giving a smooth mouthfeel (Chandan 2011). This mouthfeel is appreciated by consumers, who generally prefer fatty and smooth textures instead of dry sensations. Thus, ingredients that add a fatty mouthfeel are typically added in seasoning formulations. The fattiness of the mouthfeel components also helps to blend the harsh flavors with background flavors providing smooth transitions from taste to taste. Seasonings without any fat tend to "clean up" more rapidly even if the flavor profile is well formulated. Spray-dried dairy powders are normally used as mouthfeel components and to help blend all the flavorings. The reason why dairy fat is the preferred alternative for this application is its melting point, which is below 37.8°C. Therefore, it is readily melted when eaten, coating the mouth. Furthermore, the lipophilic flavors solubilize in the fat while it is melting, creating a long-lasting flavor sensation in the mouth. Due to this behavior, the aftertaste of snack seasonings may be manipulated by adjusting the fat-soluble flavor components present in the formulation (Seighman 2001). Examples of mouthfeel components used in seasonings include cheese powders, dry buttermilk, shortening powder and whey powders (Ainsworth and Plunkett 2007, Chandan, 2011).

8.2.6 Coloring Agents

The function of coloring agents is to improve the appearance of the snacks. It is relevant since it complements the flavor and texture properties by making the foodstuff more appealing to consumers. Colorants may be classified as (i) naturally derived also called pigments, and (ii) synthetic colorants or dyes (Pasias et al. 2015). The seasoning color must resemble its flavor. For instance, the color in cheese powder ranges from white to deep orange depending on the manufacturer and the regional preferences (Chandan 2011). Both natural and synthetic colors are used in relatively low levels, therefore the contribution to the price is relatively low (Pasias et al. 2015).

The advantage of using synthetic colorants is that they produce strong, consistent and stable colors in food formulations (Pasias et al. 2015). The artificial colors used in snack seasonings are FD&C alumina lakes due to their high stability, nonreactivity, nonflavor contribution and compatibility with oil-soluble formulations (Seighman 2001, Pasias et al. 2015). Most FD&C colors are controlled by regulatory agencies since they have demonstrated harmful effects on laboratory animals and pose health risk to humans (Serna-Saldivar 2008). The colorants most widely used in snacks include Yellow #5, Yellow #6, Red #40, Blue #1 and Blue #2. The combination of two or more of these colors practically makes any other color. Colorants can be directly applied in seasoning formulations or can also be added indirectly by using colored spray-dried ingredients where the dye is added to the slurry before drying (Seighman 2001).

In the European Union, the coloring agents are almost exclusively natural driven by health concerns and also by the increasing tendency of consumers to eat more natural products (Hall 2012). Natural colorants are discussed in Section 8.3.2.

8.2.7 Anticaking Agents and Processing Aids

Anticaking agents are considered processing aids that absorb available moisture and physically separate moisture donors and hygroscopic materials (Hall 2012). These ingredients coat the particles in the seasoning blend with fine powders helping them to resist agglomeration and allowing the seasoning to flow freely during the application procedure. This is especially important to avoid the clogging of the nozzles that are used to spray the seasoning onto the snack surface ensuring even application and avoiding seasoning falloff (Seighman 2001). The most common anticaking agents are silicon dioxide, tricalcium phosphate, magnesium carbonate, magnesium or calcium silicates and calcium stearate (Serna-Saldivar 2008, Chinwan and Castell-Perez 2019). The preferred free-flow agent for seasonings is silicon dioxide. It is a powdery material with small particle size and high surface area. Its functionality is to coat ingredients and reduce their trend of agglomeration helping to achieve an even application of the seasoning to the snack surface (Seighman 2001). Silicon dioxide can absorb 120% of its weight and remain free flowing. Tricalcium phosphate is a white powdered anticaking agent with very low water solubility. It may also serve as a calcium source. Magnesium carbonate or silicates (magnesium or calcium salts) are used to avoid agglomeration in salt and also in dry seasoning blends (Serna-Saldivar 2008). Calcium stearate is used as a flow conditioner in food products. It reduces the water absorption of powders during storage, avoiding their agglomeration and further improving their flow behavior. Anticaking agents must be effective at low concentrations (around 1%) and their use is limited to levels up to 3% (Chinwan and Castell-Perez 2019).

Vegetable oil is also used as a processing aid to prevent the agglomeration of ingredients and formation of lumps in the blend. It prevents hydrophilic ingredients to absorb moisture. To obtain the desired effect, the vegetable oil must be blended with the hydrophilic components promoting a complete coating of particles. Vegetable oil is also useful to agglomerate seasoning components with different particle sizes avoiding their segregation, for example, when whole herbs are used. However, when used in powdered formulations, the seasoning flowability must be adjusted according to the operating characteristics. To do so, agents such as silicon dioxide must be added when the rest of the ingredients were blended (Seighman 2001).

8.2.8 Antioxidants

Antioxidants are used to inhibit the loss in the seasoning quality in terms of color, flavor or nutritional properties (Ainsworth and Plunkett 2007). They are typically added to oil-based raw materials to protect them during storage (Seighman 2001). The most important synthetic antioxidants used in snack seasonings are butylated hydroxyanisole (BHA), butylated hydroxytoluene (BHT) and tert-butyl hydroquinone (TBHQ) (Ainsworth and Plunkett 2007, Serna-Saldivar 2008). BHA is more soluble in fats compared to BHT. Both antioxidants are more effective in less unsaturated animal fats compared to polyunsaturated vegetable oils. On the other hand, TBHQ presents higher effectivity in polyunsaturated vegetable oils and is more stable for heat processing. However, studies have concluded that TBHQ might trigger allergic reactions, negatively affect the immune system and increase the risk of tumor development in laboratory animals (Van Esch 1986).

Antioxidants react with free radicals to slow down the rate of autooxidation of fats and oils (Chapter 3) and retarding their rancidity (Frankel 2005). Therefore, to increase the stability they should be added before starting the oxidation. The maximum concentration allowed for synthetic antioxidants is 0.02% based on the fat content (Serna-Saldivar 2008). Antioxidants can be used alone or as antioxidant mixtures where they show useful synergistic activity, improving their effectivity. The synergism has been related to an increased oxidative stability of antioxidants caused by mutual protection of the antioxidant radicals. Formulations containing antioxidants commonly include carriers to increase their solubility and facilitate the uniform dispersion and incorporation into the oils and food products. Carriers that are appropriate for this function include propylene glycol, mono- and diglycerides and vegetable oils (Frankel 2005).

Natural alternatives to replace synthetic antioxidants have also been explored. These natural antioxidants are discussed in Section 8.3.3.

8.2.9 Other Ingredients

Typically, preservatives are not used in seasonings. However, cheese seasonings require the addition of these compounds to inhibit fungal growth. Preservatives that may be found on cheese seasonings are potassium sorbate and sodium benzoate.

Also, emulsifiers are not extensively used in all types of seasonings but are commonly found in cheese flavorings. Among the emulsifiers used in seasonings are soy lecithin and mono- and diglycerides, which are considered natural additives. Mono- and diglycerides are also used as carriers of antioxidants since due to their emulsifying properties, help to improve the interactions among the antioxidants and the oily fraction of the formulation.

Additives that serve as pH modifiers can also be found in seasoning formulations. Sodium bicarbonate is used to increase the pH while phosphate salts serve as acidulants, decreasing the pH. Phosphate salts include calcium, potassium or sodium phosphates and disodium phosphate. Sodium citrate is also used as acidulant and pH regulator. Citrates and phosphates also serve as cheese-emulsifying salts and are usually found in cheese and hot-spicy seasonings (Serna-Saldivar 2008, Chandan 2011).

Gums are included in seasoning formulations as tackifier agents. In low-fat seasonings, gums are used to form a film in the snack surface where the powdered seasoning will adhere. Examples of gums used in snack seasonings are acacia, guar, xanthan or Arabic. Modified starches are also added to snack seasonings to accomplish tackiness functionalities. The advantage of using modified starches over gums is that they are typically acid-thinned starches with low viscosities even at high solid concentrations. Starches that are modified to function as tackifiers present better properties compared to vegetable oils. Additionally, there are commercial tackifier starches that can be used in their solid forms, without the previous solubilization in water (Sajilata and Singhal 2005). Gums and modified starches may also be included in seasoning formulations due to their emulsifying properties.

Finally, some ingredients are added to improve the nutritional profile of the seasonings. Shaviklo et al. (2014) evaluated the addition of fish protein powder or omega -3 fish oil to a cheese-flavored seasoning applied on extruded corn snacks (chapter 12). Results indicated an adequate overall acceptability of the snack by children in the range of 7–12 years old. The fish-derived ingredients did not influence the sensory attributes of extruded puffs after development and within 12 week of storage. However, fish flavor and off-odor were developed after 16 weeks of storage, especially when fish oil was added.

8.3 NATURAL ALTERNATIVES FOR SEASONING INGREDIENTS

Although natural ingredients have always been used in the production of savory food products, and thus, in the formulation of snack seasonings, its usage to exclude the non-natural ingredients, is recent. The movement toward the change to natural ingredients is worldwide, with Europe in the lead. Accordingly, this trend started with movements in the UK and Iceland where food retail brands started to remove artificial colors and flavors in the late 1990s. Nowadays, worldwide consumers are more aware looking for more healthy and natural foods and reading nutritional and ingredient declarations (Hall 2012). Therefore, this trend increases the necessity of the study of natural alternatives to replace functional additives generally used to prepare food formulations.

Natural ingredients that are already widely used in snack seasonings include dry vegetables, dairy-based powders such as sour cream or cheddar cheese, spices and herbs which were treated in Section 8.2.1. Although salt is considered as a natural ingredient, the use of sea salt gives this ingredient a more natural connotation especially when fine salt contains anticaking agents. Moreover, the usage of sea salt reduces the sodium content since besides NaCl it also contains potassium chloride and other chlorides. The negative aspect is that the presence of these salts can produce bitter/metallic flavors (Hall 2012).

In this section are discussed natural alternatives for flavor enhancers (Section 8.3.1), colorants (Section 8.3.2) and antioxidants (Section 8.3.3).

8.3.1 Natural Flavor Enhancers

Although MSG, HVP and 5'-nucleotides may be obtained from natural sources, they have an unfavorable perception with consumers (Reineccius 2005). Consumption of HVP declined in the UK since 2000 due to the discovery of carcinogenic chloropropanol compounds produced when HVPs were produced by acid hydrolysis. However, current manufacturing practices remove nearly all the chloropropanol formed after acid hydrolysis. Yet, HVPs continue to be excluded from UK products (Hall 2012). Therefore, natural alternatives have been evaluated to substitute these flavor enhancers, especially for consumers that search for clean labels.

A natural alternative to replace flavor enhancers could be by enhancing the overall flavor levels. This could be achieved by the addition of citric acids or also of some herbs and spices. Citric acid is considered a natural flavor enhancer. It causes an immediate impact on the palate, which sharpens the overall flavors. Citric acid can be found as a powder or fine granulated (Hall 2012). On the other hand, blends of spices, herbs and seafood have also resulted useful to replace flavor enhancers. Badejo (2016) used combinations of dried bonga fish (*Ethmalosa fimbriata*), dried catfish (*Caprius gariepinus*), shrimp, garlic, ginger, onion, moringa, shrimp, salt and basil to develop seasonings that could serve as natural flavor enhancers. Sensory results showed that in general, no significant differences were found in the aroma and taste of foods seasoned with the developed mix, compared to commercially available seasonings that contained MSG. Moreover, the natural mixes used to enhance flavors were high in calcium and presented antioxidant properties.

Autolyzed yeast and yeast extracts are also used as natural flavor enhancers. Yeast additives contain high levels of 5'-nucleotides, which are responsible to potentiate savory flavors in seasonings and cause the mouth-watering response (Seighman 2001). The high levels of RNA contained in yeasts serve as a source of nucleotides when it is hydrolyzed. Hydrolysis may be achieved by autolysis where the yeast nucleases act upon RNA, or via acid hydrolysis using concentrated HCl (Reineccius 2005). Yeast additives produced with different processes and raw materials and diverse features are found in the market. Dried and autolyzed yeasts resulting as by-products of alcohol manufacture have light brown colors and are used to give or enhance cheese-like flavors. Autolyzed yeasts are also used as flavor enhancers due to their high glutamic acid contents and contain salt levels lower than 10%. Yeast extracts are normally derived from bakery or lactic acid yeasts and are found in light to dark brown colors. These forms have beefy or chicken-like flavors, and are additionally savory and enhancing flavor properties due to their high level of glutamate and nucleotides. Yeast extracts can be found in low-salt (< 2%) or salty (~ 40%) forms (Hall 2012). Yeast extracts also contain peptides, polypeptides and amino acids which contribute to the flavor profile and enhancing properties (Serna-Saldivar 2008). Yeast ingredients are incorporated in formulations at levels ranging from 0.5 to 5%, depending if they are used alone or in combination with other enhancers (Seighman 2001, Serna-Saldivar 2008). The disadvantage of yeast additives that serve as enhancers is that their cost is comparatively higher than both MSG and HVP (Seighman 2001).

8.3.2 Natural Colorants

Natural colorants are pigments extracted from natural sources. The major challenges while formulating with these additives include their higher and fluctuating price compared to artificial coloring agents and also their relatively lower stability to pH, temperature and light. The stability of natural colorants may be improved by including ingredients to buffer the pH or by including antioxidants joint with the natural color agent. The use and levels of most of these natural colorants are regulated by the European Union (Hall 2012). Natural colorants that may be used in snacks and seasoning are described in Table 8.2.

8.3.3 Natural Antioxidants

Several authors have highlighted the use of herb extracts such as sage or rosemary extracts to give protection against oxidation and rancid off notes in snack foods and snack seasonings (Hall, 2012, Berdahl and McKeague 2015). Rosemary (*Rosmarinus officinalis* L.) and sage (*Salvia officinalis* L.) belong to the family Laminacae. Their antioxidant potential is associated with the high variety of phenolic compounds contained in these herbs. Moreover, carnosic acid which is a phenolic compound found in both rosemary and sage is a radical scavenger classified as a super-stoichiometric antioxidant. Unlike synthetic antioxidants such as BHA or BHT, it has the ability to act repeatedly as a reducing agent by donating hydrogen atoms through rearrangement of the initially oxidized species. Since rosemary contains a higher concentration of carnosic acid in comparison to sage, rosemary extracts with higher antioxidant capacity are more widely used in food formulations. Due to the flavor profile of snack foods, the sage or rosemary extracts used as antioxidants must be reduced-flavored forms. To obtain extracts with reduced flavors, producers decrease their concentration of essential oils and therefore their flavor strength and aroma. Commercially, rosemary/sage extracts can be found in liquid (oily or aqueous) or powder forms. When possible, it is preferred to add directly the herb extracts to the snack dough. However, in snacks where it is not possible, the antioxidant could be added to the frying oils or mixed with the seasoning and sprayed on the surface during the seasoning operation (Berdahl and McKeague 2015).

Tocopherols are also natural antioxidants that have gained popularity since they are also essential nutrients and protect humans against oxidative stress (Serna-Saldivar 2008). α-Tocopherol has a synergic effect with ascorbic acid explained by the regeneration of the intermediate tocopherol radical into the initial α-tocopherol and its subsequent recycling. Rosemary extracts may also have been used in combination with tocopherol giving better results probably due to their mutual reinforcing activity (Frankel 2005). Tocopherols are also combined with synthetic antioxidants such as BHT in some snack seasoning formulations.

TABLE 8.2
Natural Colorants Commonly Used for Snack Products

Colorant	Description
Curcumin (E100) or Turmeric	Oil-soluble, orange-yellow pigment, that is present in turmeric. It must be mixed with salt, sugar or maltodextrin to help distribution of color in the seasoning blend. This pigment is unstable in alkaline pH and also light-sensitive, and is stable to heat. It can be added to snacks in levels up to 200 mg/kg.
Riboflavin (E101)	Yellow pigment that is considered one of the most important natural food colors. It can be naturally found in eggs, milk, yeast or cheese and is commercially available from bacterial sources. This colorant is soluble in aqueous solvents and is stable at pH 2–6 but it is susceptible to oxidation/reduction processes and unstable to light. Therefore, requires addition of antioxidants or protection with packaging.
Carminic acid (E120)	Red pigment with a pink shade found naturally in cochineal. It is stable to heat, light and oxygen. Cochineal is insoluble at pH 3 and is sensitive to pH changes. It can be applied in concentrations ranging 50–500 mg/kg.
Chlorophyll (E140)	Oil-soluble, green pigment that can be found in green plants. There are different types which present different green tones. They can be converted to water-soluble molecules by saponification (called clorophyllins). It is unstable to light.
Caramel or melanoidins	Brown pigment found in caramels and mellanoids. Only plain or spirit caramel (E150a) is considered natural. It has adequate stability.
Carotenoids (E160, E161)	Oil-soluble pigments that generate colors from yellow and orange to red. They can be naturally found in carrots, oranges, red pepper, saffron, annatto or marigolds. They are unstable to light and oxygen. Therefore, antioxidants must be included in the seasoning formulation.
Annatto (E160b)	Pigments that generate red/orange colors obtained from the seed *Bixa orellana*. The main two pigments associated to annatto are bixin and norbixin. Annatto precipitates at low pH and is not stable at high temperatures (130 to 250 °C) but its stability may be increased when it is bound to starch or proteins.
Paprika (E160c)	Oil-soluble, orange-yellow pigments. It must be mixed with salt, sugar or maltodextrin to help distribute the color in the seasoning blend. It has good heat and pH stability but is sensible to oxidation and light and will fade if it is not properly stabilized with antioxidants. Due to its flavor, its use is limited to spicy savory products.
Betanin (E162)	Red pigment obtained from beetroot. It is water-soluble and unstable to heat and oxidation which modify its color to brown. It is stable in the pH range of 3 to 7 and is used in snacks at levels ranging from 0.1 to 2.0%.
Anthocyanins (E163)	Water-soluble pigments that generate colors from red to blue. They are naturally present in blackcurrant, black grapes, red cabbage, cherries or strawberries, among other fruits and vegetables. These pigments are unstable to oxygen, light, pH > 4 and metal complexation. Acetylated forms present higher stability to heat, light and pH.

Source: Seighman (2001), Hall (2012), Pasias et al. (2015).

Citric and ascorbic acids can also be used as antioxidants in seasonings. They are considered preventive antioxidants due to their metal chelating properties. However, ascorbic acid may behave either as an antioxidant or prooxidant, depending on its molar ratio in relationship with the metal. It can promote oxidation at ratios 1:1 but may inhibit oxidation when the molar ratio of ascorbic acid is increased (Frankel 2005). These compounds are used to improve the antioxidant effect of tocopherols. The recommended ratio to improve the antioxidant potential is one part tocopherol and three parts of ascorbic acid (Hall 2012). For citric acid, better results are obtained with one part citric acid and two parts tocopherol (Frankel 2005).

To reduce or even eliminate the necessity of using antioxidants in the seasoning formulations, high-barrier packaging films combined with gas flushing of modified atmospheres in packaged snacks, can be used (Seighman 2001, Hall 2012). For instance, Agarwal et al. (2018) found that gas flushing with nitrogen significantly increased the shelf life of potato chips (unsalted or with a cheese and onion seasoning) in terms of the stability of volatile compounds and sensory perception. Interestingly, these authors also found that the addition of cheese and onion seasoning improved the stability of the potato chips compared to the unsalted counterpart, probably due to the antioxidant properties of milk proteins.

8.4 SEASONING FORMULATIONS

The formulation and development of snack seasonings is a highly specialized area where the flavor, granulation and flow characteristics are relevant to guarantee the uniform appearance and taste of the finished product (Hall 2012). Most snack seasonings share a series of ingredients, which can be divided into the following categories: (i) defining flavor ingredients, (ii) top notes and acids, (iii) appearance improvers, (iv) flavor enhancers, (v) carriers and fillers and (vi) functional ingredients. These categories are described in Table 8.3. Seasonings are generally formulated considering the blend as a 100%, although salt is formulated considering the desired level in the finished snack (Hall 2012). The formulation development is a trial and error process where the formula starts with salt, fillers, enhancers and flavors. To this blend, formulation developers add other additives to adapt the seasoning for the desired flavor profile and application (Seighman 2001).

Examples of snack seasoning formulations are presented in Table 8.4. Importantly, seasonings must be stored in tightly closed containers at cool temperatures, avoiding light and oxygen to retard rancidity and undesirable flavor changes (Serna-Saldivar 2008).

Snack seasonings can be formulated as powdered-based (Section 8.4.1), or liquid-based. The latter may be based on oil (Section 8.4.2) or water if it is a low-fat flavoring (Section 8.4.3). Seasonings added to the snack doughs are called background formulations (*Section 4.4*). Each type of seasoning

TABLE 8.3
Seasoning Ingredient Categories and Their Functionality

Category	Function	Examples
Defining Flavor Ingredients	These ingredients impart flavor to the final products.	Flavorings including cheese ingredients and dehydrated vegetables/ spices/herbs.
Top Notes and Acids	These ingredients are added to restore the flavors lost while the products are cooked. Include volatile components that primarily hit the nose and then dissipate.	Essential oils, oleoresins, acids and vinegar.
Appearance Improvers	Make the product more appealing and attractive to consumers. Enhance visual attraction and give the impression of a premium product.	Colorants, cracked spices or flaked salt.
Flavor Enhancers	These ingredients boost the flavors already present and provide fullness of flavor to the product.	Monosodium glutamate, 5'-nucleotides, hydrolyzed vegetable proteins, yeast enhancers.
Carriers and Fillers	Aid to dilute the flavor components and extend seasoning mix to ensure even application.	Maltodextrin, flours, starches, lactose, whey powder, salt, dextrose, sugar.
Functional Ingredients	Processing aids that help to maintain the flowability of the blend avoiding agglomeration of particles. Also includes ingredients that improve adhesion and help maintaining organoleptic properties of the seasoning.	Anticaking agents, vegetable oil, hydrocolloids, mouthfeel components (fat powders), antioxidants.

Source: Ainsworth and Plunkett (2007), Hall (2012).

TABLE 8.4
Examples of Snack Seasoning Formulations

Ingredient Category	Ingredient	Sour Cream-Onion Seasoning[1]	Barbecue[2]	Nacho Flavoring for Corn and Tortilla Chips and Popcorn[3]	Seasoning for Extruded Snack[3]	Sea Salt and Black Pepper for Potato Chips (Natural Seasoning)[4]
Defining Flavor Ingredients	Mix of cheeses, dried	–	–	50.0	68.0	–
	Sour cream powder	20.0	–	–	–	–
	Onion powder	5.0	4.0	1.5	–	3.0
	Toasted onion powder	15.0	–	–	–	–
	Garlic powder	–	1.0	0.5	–	–
	Tomato powder	–	3.0	5.0	–	–
	Red/black pepper, ground	–	0.5	0.3	–	4.0
	Mustard	–	10.0	1.0	–	–
	Compounded flavor	1.0	–	–	–	–
–	Non-fat dry milk	5.0	-	–	–	–
	Dry chili pepper		4.0			
Top Notes and Acids	Black pepper oil	–	–	–	–	0.5
	Black pepper extract	–	–	–	–	0.16
	Enzyme-modified cheese powder	–	–	–	1.0	–
	Enzyme-modified butter oil	–	–	–	1.0	–
	Citric acid	0.8	1.0	1.0	–	0.4
	Lactic acid	2.0	–	–	0.3	–
	Natural spray-dried lemon flavor	–	–		–	0.3
	Oleoresin paprika		0.5			
Flavor Enhancers	Salt	22.0	20.0	18.9	6.2	17.7
	MSG	2.0	2.0	3.0	4.0	–
	HVP	3.0	–		–	–
	Yeast extract	-	15.0	–	–	4.0
Appearance Improvers	Parsley rubbed or flakes	3.0	–	–	–	2.0
	Artificial colorants	–	–	0.7	–	–
	Caramel powder	–	–	0.5	–	–

(Continued)

TABLE 8.4 (CONTINUED)
Examples of Snack Seasoning Formulations

Ingredient Category	Ingredient	Sour Cream-Onion Seasoning[1]	Barbecue[2]	Nacho Flavoring for Corn and Tortilla Chips and Popcorn[3]	Seasoning for Extruded Snack[3]	Sea Salt and Black Pepper for Potato Chips (Natural Seasoning)[4]
Carriers and Fillers	Maltodextrin	14.7	–	17.6	–	37.14
	Sugar	–	18.0	–	–	14.0
	Dextrose	5.0	–	–	–	–
	Lactose	–	–	–	–	16.0
	Soybean flour (defatted)		20.0			
	Dry sweet whey	–	–	–	19.5	–
Functional Ingredients	Silicon dioxide	1.0	1.0	–	–	0.3
	Sunflower oil	0.5	–	–	–	0.5
	Total	100%	100%	100%	100%	100%

Notes: Data is presented in percentage in dry seasoning. MSG = monosodium glutamate; HVP= hydrolyzed vegetable proteins. *Source:* [1]Seighman (2001), [2]Serna-Saldivar (2008), [3]Chandan (2011), and [4]Hall (2012).

requires different application methods. For better results, formulations must be adapted to the method by which the seasoning will be applied (Hall 2012). Also, depending on the snack, the seasoning formulation may vary slightly to be adequate for the surface characteristics. Table 8.5 summarizes the characteristics of ideal seasonings for different types of snacks and their application levels.

TABLE 8.5
Common Characteristics, Usage Levels and Application Techniques According to the Snack Type

Snack	Seasoning Characteristics and Usage Levels
Popcorn	Seasonings for popcorns may be powdered-based, which are applied at levels ranging from 6 to 12% after spraying oil, or oil based which are used at 22–29% on snack basis. The salt levels range from 1.8 to 4% on snack basis. Salted and cheese seasonings are common flavors for this snack, but popcorn can also be coated with caramel which is produced with sugar, water and dextrose at temperatures around 150 °C.
Potato Chips	A two-stage seasoning is done for potato chips. In the first stage, chips are salted directly out of the fryer to a salt level around 1–2%. Some chips are packaged after this step whereas flavored chips are subjected to a second stage where a powdered seasoning is added in a tumbler, at levels ranging from 5 to 8% on snack weight basis. The overall surface must be considered to set the application level (thin chips require lower use levels). Since salt is added to the base, potato chip seasonings contain lower salt content compared to counterparts designed for other snacks. Seasonings must be formed by coarse particles (from 40 to 100 mesh sizes) for optimum adhesion.
Corn Collets	Collets are extruded "half product pellets" that are dried and can be puffed later by oil frying, rapid hot air heating, or microwave heating in ovens. Corn collets may be flavored using powdered-based seasonings at usage levels of 6–12% or using oil-based seasonings at levels ranging from 22 to 29%, expressed on a snack basis. These snacks may also be seasoned using background flavors, which are incorporated into the doughs. In this case, the seasoning is added at a level around 10%, expressed on a flour weight basis. It is common to add coloring agents in the background formulation.
Extruded Snacks	These snacks are commonly flavored using an oil-based seasoning slurry. Due to the porous surface of extruded snacks, the oil absorbs readily, making additional seasoning necessary. The usage level for seasoning extruded snacks ranges from 10 to 15% on a snack basis. All the salt is contained in the seasoning, being common values from 10 to 15% on seasoning basis (1–2.25% on snack basis). Background flavorings including corn syrup and citric acid may be added in this type of snacks. It is also common to add coloring agents in the dough prior to extrusion.
Tortilla Chips	Seasonings should be designed with enough flavor impact to overcome the taste of the corn base. The usage level of tortilla chip powdered seasoning ranges 8–10% on snack basis, and are applied after spraying the snack with 3–4% oil to help adhesion of the seasoning to the surface. Seasonings must be formed by fine particles. If only salt is used, it is applied downstream after the frying step. The salt levels range from 1.5 to 2.5% on snack basis. Sometimes annatto rich in bixin and norbixin is used in the background formulation as a coloring agent.
Corn Chips	As for tortilla chips, seasonings should be designed to overcome the taste of fried corn flavor, which may result in overwhelming in some consumers. Therefore, strong flavors must be used in the seasoning formulation. The usage level of powdered seasoning is 8–10% on snack weight basis which is applied in a tumbler downstream from the fryer. The salt levels range from 2 to 2.5% on snack basis. Sometimes annatto is used in the background formulation as a coloring agent.
Pretzels	This snack is difficult to flavor due to the smooth and crusty surface which makes it difficult for the seasoning to adhere. Then, the seasoning can be applied whether by using a sticky and adhesive base that sets on the snack surface or by exposing the internal porous structure of the pretzel. The seasoning level ranges 2–2.5%. The use of background flavors (incorporated into the dough) is not common since the flavor release is slower than flavors found on the surface. The salt level in pretzels is 2–2.5% and coarse or flaked forms are preferred.

Source: Burtea (2001), Cretors (2001), Seighman (2001), Ainsworth and Plunkett (2007), Chandan (2011).

8.4.1 Powdered-Based Formulations

Powdered-based seasonings are the most commonly used. The application method takes advantage of the remnant oil on the surface of the snack after frying to achieve the seasoning adhesion. Yet, for snacks where less oil is found on the surface after production (i.e., baked snacks, popcorn or nuts), warm vegetable oil is first sprayed on the surface to assist and enhance adhesion. Sometimes the dry powder is electrostatically charged to enhance adhesion (Hall 2012). Seasonings applied by this dual technology are added at values fluctuating from 6 to 15% based on the snack weight. Powder seasonings are normally applied to the snacks on inclined drum tumblers described in Chapter 3. It is important that powder formulations contain adequate amounts of free-flowing or anticaking agents to achieve uniform application (Seighman 2001).

To produce powdered-based seasonings, the different ingredients can be dry blended or mixed to form a slurry, which is subsequently spray-dried. The latter offers uniformity on the flavor and application since the ingredients cannot fractionate during storage or processing. However, this form of production is more expensive compared to the dry blending process (Reineccius 2005). In these formulations, liquid flavors should be encapsulated to avoid loss of volatile compounds during application at high temperatures (Seighman 2001).

The use of powdered-base seasonings presents the advantages of ease of application, production of one base snack that could be seasoned with different flavors and also the immediate flavor impact since it is concentrated on the surface. Yet, the disadvantages include the use of anticaking agents that could add dry mouth sensations, presence of hygroscopic ingredients that could difficult the seasoning application, fractionation of seasonings during storage or handling, difficulty to achieve uniform application, and comparatively higher usage levels required to obtain desired organoleptic properties (Reineccius 2005).

8.4.2 Oil-Based Formulations

In oil-based formulations, the snack seasoning is mixed with vegetable oil to form a slurry which is sprayed onto the product. These oil slurries normally contain 55–60% oil and 40–45% seasoning. Oil-based seasonings are used in snacks that have a very dry surface and are applied at levels up to 29% on snack basis (Chandan 2011). During formulation, particular attention should be given to obtaining fine granulated seasonings to avoid nozzle blockages during spraying. To obtain the desired result, fine presentations of ingredients (i.e., fine salt or spices) should be used (Hall 2012). The seasoning base of these formulations should be prepared by spray drying and flavors must be encapsulated. This helps to increase the stability of the seasoning components during application at high temperatures. Oil-based seasonings can also be found as a paste, called flavor concentrate, which contains flavors, flavor enhancers, acids and colors, but does not contain fillers. This paste is oil-soluble and is diluted with oil previous to its application (Seighman 2001).

The advantage of applying oil-based seasoning on the snack surface is that it reduces dry sensations in the mouth. However, the disadvantage is that oxidation or light-induced undesirable reactions occur during storage (Reineccius 2005). This could be solved by including antioxidants in the formulation or with the use of proper packaging laminates that block oxygen and UV light (Chapter 4).

8.4.3 Low-Fat Formulations

For powdered-based formulations, a surface film is needed on the snack to facilitate the adhesion of the seasoning. In low-fat formulations, this film may be generated by spraying acacia gum, corn syrup solids or maltodextrins onto the surface of snacks and subsequent drying (Chandan 2011). Recently, the use of xanthan gum aqueous solutions to coat extruded snacks prior to the application of seasoning was investigated. Results indicated that xanthan gum solutions yielded good coating properties and did not damage the texture of the extrudates that showed good overall acceptability scores (Graça et al. 2020). Also, dry tack special starches may be found commercially. These starches adhere to snacks without the use of oil or water. Seasonings containing dry tack starch must be applied while the snack surface is hot (100 to 150°C) which causes melting of the powder in a short time (~30 s). Dry tack starches present a high degree of tackiness which helps to keep the seasoning attached to the snack (Sajilata and Singhal 2005).

Liquid-based low-fat seasonings can also be found being slurries where water, instead of oil, is used as the carrier. The seasoning formulation should be adjusted since ingredients must be water soluble and contains maltodextrins and starches that will serve as adhesion aids. After application, a drying step must be applied to remove the excess water from the snack surface (Seighman 2001). Dextrin coatings present adequate tackiness properties, comparable to vegetable oils. To produce dextrin coatings, these compounds are dissolved in water at concentrations of around 30%. They are easy to spray at high solid contents (30–40%) due to their low viscosity (Sajilata and Singhal 2005).

The major disadvantage of low-fat formulations is that after their application, the snack typically requires a drying step that causes volatilization of flavor compounds, resulting in loss of balance of the flavor profile. An alternative to avoid this problem is the use of encapsulated flavors on water-insoluble excipients that do not melt during the drying step (Seighman 2001).

8.4.4 Background Formulations

Addition of background seasonings may be desirable to provide a boost to the flavor or to make the flavor more uniform in the snack product (Reineccius 2005). With this type of seasoning application, the flavor is not only on the surface but in the snack itself. Clearly, adding background flavors to the snack base is not possible for all types of snacks. The most suitable are extruded snacks where seasonings may be added to the feedstock prior to extrusion or frying. However, during the formulation of this type of seasonings, the sensibility of the ingredients to the high temperatures of extrusion or frying must be considered. Volatile compounds should be avoided since they would be lost during processing, but may be included as top flavor note components in the seasoning applied on the snack surface (Hall 2012). Losses during processing increase the cost of this type of flavoring strategy. Moreover, when the flavor is released into the frying oil, the oil use would be limited only to that type of snack. This could affect manufacturing operations and production logistics (Reineccius 2005).

REFERENCES

Agarwal, D., Mui, L., Aldridge, E., Mottram, R., McKinney, J., and Fisk, I. D. 2018. The impact of nitrogen gas flushing on the stability of seasonings: Volatile compounds and sensory perception of cheese and onion seasoned potato crisps. *Food and Function* 9(9):4730–4741. doi.org/10.1039/c8fo00817e

Ainsworth, P., and Plunkett, A. 2007. Reducing salt in snack products. In: *Reducing Salt in Foods: Practical Strategies*, eds. D. Kilcast and F. Angus, 96–315. Cambridge: Woodhead Publishing. doi.org/10.1533/9781845693046.3.296

Badejo, A. A. 2016. Nutrient composition and antioxidative potential of seasonings formulated from herbs, spices, and seafood. *Journal of Culinary Science and Technology* 14(3):222–233. doi.org/10.1080/15428052.2015.1103675

Berdahl, D. R., and McKeague, J. 2015. Rosemary and sage extracts as antioxidants for food preservation. In: *Handbook of Antioxidants for Food Preservation*, ed. F. Shahidi, 177–217. Cambridge: Woodhead Publishing. doi.org/10.1016/B978-1-78242-089-7.00008-7

Buechler, A. E., and Lee, S. Y. 2019. Consumer acceptance of reduced sodium potato chips and puffed rice: How does ingredient information and education influence liking? *Journal of Food Science* 84(12):3763–3773. doi.org/10.1111/1750-3841.14907

Burtea, O. 2001. Snack foods from formers and high-shear extruders. In: *Snack Foods Processing*, eds. E. Lusas and L. W. Rooney, 281–314. Boca Raton: CRC Press.

Cai, J., and Lee, Y. 2020. Controlling sodium release using maltodextrin and octenyl-succinic-anhydride-modified starch with two types of spray-dryer nozzles. *Journal of Food Process Engineering* 43(5):e13238. doi.org/10.1111/jfpe.13238

Chandan, R. C. 2011. Dairy ingredients in bakery, snacks, sauces, dressings, processed meats, and functional foods. In: *Dairy Ingredients for Food Processing*, eds. R. C. Chandan and A. Kilara, 473–500. Iowa: Blackwell Publishing Ltd. doi.org/10.1002/9780470959169.ch18

Chinwan, D., and Castell-Perez, M. E. 2019. Effect of conditioner and moisture content on flowability of yellow cornmeal. *Food Science and Nutrition* 7(10):3261–3272. doi.org/10.1002/fsn3.1184

Cretors, C. 2001. Popcorn products. In: *Snack Foods Processing*, eds. E. Lusas and L. W. Rooney, 385–420. Boca Raton: CRC Press.

Frankel, E. 2005. *Lipid Oxidation*. Cambridge: Woodhead Publishing. doi.org/10.1533/9780857097927.209

Graça, C., Marques, D., Sousa, I., and Monteiro, A. R. G. 2020. Xanthan gum as an alternative to replace the fat for coating and flavoring the extruded snacks. *Journal of Food Science and Technology* 57(8):3151–3156. doi.org/10.1007/s13197-020-04542-y

Hall, R. H. 2012. Applications of natural ingredients in savoury food products. In: *Natural Food Additives, Ingredients and Flavourings*, eds. R. Seal and D. Baines, 281–317. Oxford: Woodhead Publishing. doi.org/10.1016/B978-1-84569-811-9.50013-X

Harada-Padermo, S. dos S., Dias-Faceto, L. S., Selani, M. M., Conti-Silva, A. C., and Vieira, T. M. F. de S. 2021. Umami ingredient, a newly developed flavor enhancer from shiitake byproducts, in low-sodium products: A study case of application in corn extruded snacks. *LWT- Food Science and Technology* 138:110806. doi.org/10.1016/j.lwt.2020.110806

Israr, T., Rakha, A., Sohail, M., Rashid, S., and Shehzad, A. 2016. Salt reduction in baked products: Strategies and constraints. *Trends in Food Science and Technology* 51:98–105. doi.org/10.1016/j.tifs.2016.03.002

Jimenez-Maroto, L. A., Sato, T., and Rankin, S. A. 2013. Saltiness potentiation in white bread by substituting sodium chloride with a fermented soy ingredient. *Journal of Cereal Science* 58(2):313–317. doi.org/10.1016/j.jcs.2013.06.001

Lv, W., Li, D., Lv, H., Jin, X., Han, Q., Su, D., and Wang, Y. 2019. Recent development of microwave fluidization technology for drying of fresh fruits and vegetables. *Trends in Food Science and Technology* 86:59–67. doi.org/10.1016/j.tifs.2019.02.047

MarketLine. 2019. *Savory Snacks Industry Profile: Global*. Manchester: MarketLine.

Mitchell, M., Brunton, N. P., and Wilkinson, M. G. 2011. Current salt reduction strategies and their effect on sensory acceptability: A study with reduced salt ready-meals. *European Food Research and Technology* 232(3):529–539. doi.org/10.1007/s00217-010-1420-6

Pasias, I. N., Asimakopoulos, A. G., and Thomaidis, N. S. 2015. Food colours for bakery products, snack foods, dry soup mixes, and seasonings. In: *Colour Additives for Foods and Beverages*, ed. M. J. Scotter, 211–226. Cambridge: Woodhead Publishing. doi.org/10.1016/B978-1-78242-011-8.00008-8

Raghavan, S. 2000. *Handbook of Spices, Seasonings, and Flavorings*. Boca Raton: CRC Press. doi.org/10.1201/9781420012552.ch4

Rama, R., Chiu, N., Carvalho Da Silva, M., Hewson, L., Hort, J., and Fisk, I. D. 2013. Impact of salt crystal size on in-mouth delivery of sodium and saltiness perception from snack foods. *Journal of Texture Studies* 44(5):338–345. doi.org/10.1111/jtxs.12017

Reineccius, G. 2005. *Flavor Chemistry and Technology*. 2nd ed. Boca Raton: CRC Press.

Sajilata, M. G., and Singhal, R. S. 2005. Specialty starches for snack foods. *Carbohydrate Polymers* 59(2):131–151. doi.org/10.1016/j.carbpol.2004.08.012

Seighman, J. 2001. Snack food seasonings. In: *Snack Foods Processing*, eds. E. W. Lusas and L. W. Rooney, 495–527. Boca Raton: CRC Press.

Serna-Saldivar, S. O. 2008. The process of frying and seasoning. In: *Industrial Manufacture of Snack Foods*, ed. S. O. Serna-Saldívar, 329–352. London: Kennedys Publications Ltd.

Shaviklo, A. R., Kargari Dehkordi, A., and Zangeneh, P. 2014. Interactions and effects of the seasoning mixture containing fish protein powder/omega-3 fish oil on children's liking and stability of extruded corn snacks using a mixture design approach. *Journal of Food Processing and Preservation* 38(3):1097–1105. doi.org/10.1111/jfpp.12068

Tavares Filho, E. R., Almeida Esmerino, E., de Almeida Santos-Junior, V., Cazzelato Lins da Silva, A., and André Bolini, H. M. 2020. Dynamic aspects of salt reduction in tomato sauce by use of flavor enhancers and a bitter blocker. *Food Science and Technology International* 26(6):549–559. doi.org/10.1177/1082013220913361

Van Esch, G. J. 1986. Toxicology of tert-butylhydroquinone (TBHQ). *Food and Chemical Toxicology* 24(10–11):1063–1065.

9 Popcorn and Other Puffed Grains

Sergio O. Serna-Saldivar

CONTENTS

9.1 INTRODUCTION

Popcorn (*Zea mays everta*) has been a favorite traditional snack in the United States for more than 100 years (Ziegler 2000). Its popularity is rapidly increasing in other areas of the world because it is perceived as a nutritious snack. The global market for popcorn products was valued at over 9 billion USD in 2016 and is projected to reach 15 billion USD by 2023. More importantly, the Compound Annual Growth Rate (CAGR) of this category is estimated at 7.6% from 2017 to 2023**.** According to the 2020 State of the Industry Report of the Snack Industry the ready-to-eat popcorn and caramel corn segment in the United States contributed 1.5 billion USD to the 26.7 billion USD salty snacks category. This represented about 5% increase compared to the previous year. The overall popcorn/popcorn oil category saw a 13.4% increase to nearly 1 billion USD in sales whereas the microwave popcorn segment accounted for the bulk of this explosive growth, rising 13.3% to 848.5 million in sales (Finkel 2020).

Americans consume 17 billion quarts of popcorn yearly and 70% of this volume is eaten at home. This volume represents the *per capita* consumption of more than 60 quarts per year. In 2020 more than 233 million of the total US population estimated at 327 million are reported to consume popcorn. For the health-conscious consumer, a diverse array of new popcorns have been launched into the market. The occupied lifestyle of new consumers and its convenience has motivated the market especially in developed countries around the globe. Today, microwavable popcorn dominates sales because of its convenience and that is commercially available in an extensive variety of flavors including cheese, lemon, low fat and butter. Besides, it is popped at sporting events, amusement parks and movie theaters. The development, production, processing and marketing of popcorn

DOI: 10.1201/9781003129066-9

encompasses many disciplines such as genetics, agronomy, food science, engineering and logistics in order to provide a safe and healthy snack to consumers (Scott et al. 2019).

There are three major market classes of popcorn, ready-to-eat (RTE), microwavable and whole kernel popcorn (Scott et al. 2019). The market segments of raw or whole kernel, RTE and microwave popcorn are estimated in 5.3, 59 and 35% of the total market. Genetics, production, technology and processing of popcorn are described by Ziegler (2000, 2003), Serna-Saldivar (2008, 2010b) and Scott et al. (2019).

The technologies of oven and gun puffing aimed to expand whole grains from cereals, pseudocereals and legumes are used to generate expanded grains with different texture, aroma, flavor and functionality especially in terms of liquid retention. These highly nutritious grains are employed for the production of bars and other snacks.

The aims of this chapter are to describe the main physical grain properties of popcorn related to its popping ability, the mechanism of popping and the main industrial and in-home processes to obtain popped corn. In addition, the chapter covers oven, microwave, and gun-puffing technologies used to expand whole grains.

9.2 HISTORICAL OVERVIEW OF POPCORN

Popcorn is perhaps the oldest snack and considered a relevant source of nutrients to ancient Mesoamerican and Andean cultures. It is regarded as the first type of the corn we know today by many archeologists. It is believed that primitive Mesoamericans first discovered the usefulness of popcorn when a flinty wild popcorn kernel was accidentally exposed to heat. To people with no grinding tools other than their own teeth realized that an unpromising food became tender, delicious and attractive. The oldest ears of popcorn ever found, dated about 5,600 years old, were discovered in the Bat Cave of west central New Mexico at the end of the 1940s. These ears ranged in size from 2.5 to 5 cm long (1 to 2 inches). On the coast of Peru, archeologists have found kernels of popcorn more than 1,000 years old. These kernels have been so well preserved that they kept their popping capability. Likewise, 1,000 year-old popped kernels were found in a dry cave inhabited by predecessors of the Pueblo Indian culture in the southwestern region of Utah (http://specialcollections.nal.usda.gov/popcorn-exhibit), which is nearly 7,000 km from Peru indicating the widespread of this type of corn in the American continent (Serna-Saldivar 2008).

Popcorn was used by the ancient Inca civilization for both food and decoration in many religious ceremonies. The oldest positively identified corn poppers date from the Mohican Indian culture of about 300 AD along the northeast coast of Peru. These pottery utensils as well as specimens of prehistoric grains are frequently found in Peruvian graves. The natives of many American cultures knew popcorn before the Spaniards arrived in the new world. In a fiesta that honored one of the Aztec gods, the Indians scattered parched or popcorn, called *momochitl*, which look like very white flowers (Weatherwax 1954). Christopher Columbus documented that when he arrived in San Salvador he saw natives wearing popcorn garlands. The oldest written record is from a Spaniard in 1650 who quoted "They toast a certain type of corn until it bursts. They call it pisancalla and they use it as a confection" (Serna-Saldivar 2008).

The North American natives introduced popcorn to the first Europeans who explored the current territories of the United States and Canada. Around 1612, early French explorers through the Great Lakes region noted that the Iroquois popped kernels with heated sand in a pottery vessel and used it to make soup and other indigenous foods. Popcorn also forms part of the US history because it was present at the first Thanksgiving when Quadequina, the brother of the Wapanong chief Massasoit, brought a deerskin bag full of popcorn to the celebration as a gift (Serna-Saldivar 2008).

The invention of the popcorn-based confection known as Cracker Jack™ is attributed to F.W. Rueckheim in 1896 in the city of Chicago USA. The mixture of popcorn, peanuts and molasses was released at the World's Columbian Exposition at Chicago's First World Fair. Louis Rueckhein, the brother and partner of the inventor, developed the process for keeping the molasses-covered popcorn morsels from sticking (Serna-Saldivar 2008). Popcorn is viewed as one of the most nutritious snacks especially when popped with hot air. The nutritional value of various popcorn products is covered in Chapter 20.

9.3 POPCORN PRODUCTION, BREEDING AND KERNEL QUALITY

According to the United States Department of Agriculture (USDA) Farm Service Agency (FSA) nearly 100,000 hectares of popcorn are planted in the United States (Scott et al. 2019) with a yield of approximately 3.4 tons/Ha. Needless to say, popcorn yields are lower compared to field corn due to its relatively smaller cob and the agronomic qualities of popcorn such as disease resistance, stalk strength, root development and overall yield are comparatively inferior (Johnson and Eldredge 1953).

Popcorn production greatly increased when seed companies released hybrids resulting in the crossing of selected maternal and parental lines. The major advantages of growing hybrids are improved yields, better plant standing ability, better popping expansion and more uniform kernel type and maturity (Ziegler 2000). The first commercial popcorn hybrid named Minhybrid 250 was released in 1934 by the Minnesota Agricultural Experiment Station and was readily adapted to grow along the northern edge of the US corn belt. Thereafter, more hybrids were released in the early 1940s by the Indiana and Kansas Agricultural Stations (http://specialcollections.nal.usda.gov/popcorn-exhibit). Although the breeding of popcorn has significantly improved yields and quality, the plant breeding effort continues especially in terms of expansion volume, flake type and eating quality (Scott et al. 2019).

Popcorn is a special kind of maize, which is similar to flint types in that almost all of the endosperm consists of tightly, packed starch granules giving the typical glassy or vitreous and hard appearance. The major traits that distinguish popcorn from other kinds of corn genotypes are the size and shape of the kernels and the ability of the sound grains when heated to explode and produce large puffed flakes (Ziegler 2003). The popcorn kernel possesses a significantly higher test weight or apparent density due to the vitreous endosperm texture, a 1,000 kernel weight of approximately half the value of regular corn and the characteristic round shape that clearly differs from dent corn.

Table 9.1 summarizes the physical properties of popcorn kernels as described by several authors. The regular dent maize kernels weigh at least twice as much as the flint popcorns used nowadays but in contrast, the test or bushel weight expressed in kg/HL of popcorn kernels is at least 10–12 kg higher compared to regular dent corns. Approximately 97.5% of the total popcorn belongs to the yellow class.

The popcorn kernel is a caryposis constituted by a single seed and a fruit coat botanically named pericarp. The pericarp is fused to the true seed formed by the germ and endosperm, which comprise 7, 12 and 81% of the kernel weight, respectively. The most relevant seed trait that distinguishes popcorn from other types of corn is its hard pericarp, which minimizes the exchange of moisture with the surrounding environment (Pordesimo et al. 1990, 1991, Scott et al. 2019). The endosperm has three distinctive anatomical parts: the aleurone, the soft or chalky starchy endosperm and the hard or vitreous endosperm. The aleurone is a single layer of cells rich in oil, protein, minerals (mainly phosphorus from phytic acid), vitamins (mainly B-vitamins) and enzymes. One of the major differences between popcorn and dent regular corns is that popcorn hybrids possess practically 100% vitreous or corneous endosperm although some bisected kernels show a tiny area of floury endosperm contiguous to the germ. The hard endosperm cells contain tightly packed small starch granules with polyhedral or angular forms interacting with protein bodies and protein matrices enclosed in rectangular-shaped fibrous cell walls. The tightly packed starch granules have indentations that look like small craters from the protein bodies and the protein matrix holds the internal structure. The protein bodies are rich in prolamins or zeins whereas the protein matrix is in glutelins (Serna-Saldivar 2008).

Popping expansion volume (PEV) is the most critical quality factor for popcorn (Crumbaker et al. 1949, Hoseney et al. 1983, Robbins and Ashman 1984, Dofing et al. 1990, Pordesimo et al. 1990,

TABLE 9.1
Physical Characteristics of Popcorn

Physical Properties	Popcorn	Reference
Test weight		Serna-Saldivar (2008), Scott et al. (2019)
Lb/bu	65–68	
Kg/HL	81–84.8	
Kernel density, ml/g	1.37–1.39	Mohamed et al. (1993)
Kernel volume, ml	0.093–0.129	Mohamed et al. (1993)
1000 Kernel weight, g	130–142	Mohamed et al. (1993)
Large	149–152	
Medium	133–147	
Small	95–131	
Pericarp thickness μm	98–138	Mohamed et al. (1993)
Kernel size, kernels/10 g	52–105	Scott et al. (2019)
Large	52–67	
Medium	68–75	
Small	76–105	
Kernel length, mm	8.06–8.59	Pordesimo et al. (1990)
Kernel width, mm	5.25–5.96	Pordesimo et al. (1990)
Kernel thickness, mm	3.99–4.42	Pordesimo et al. (1990)
Kernel sphericity	0.66–0.75	Pordesimo et al. (1990)
Endosperm texture	Hard or Vitreous	Serna-Saldivar (2010b)
Popping expansion volumeb	36–55 cm^3/g	Sweley et al. (2013)

[1] Kernel sphericity = geometric mean diameter/major diameter (Mohamed et al. 1993).
[2] Measured preferably with the metric weight volume tester or MWVT.

1991, Song et al. 1991, Song and Eckhoff 1994a, b, Ziegler 2000, Karababa 2006, Serna-Saldivar 2008, Scott et al. 2019). The popped volume is important because the commercial buyer purchases the popcorn by weight and sells it by volume. In popcorn breeding, the most relevant quality performance trait is PEV, a quantitative trait controlled by many genes (Crumbaker et al. 1949, Robbins and Ashman 1984, Dofing et al. 1990, Ziegler 2000, Lu et al. 2003, Karababa 2006, Scott et al. 2019). A recent study conducted by Jele et al. (2014) concluded that the best performing popcorn hybrids in terms of PEV and flake volume were strongly regulated by genetics and that additive gene effects were more prominent than non-additive counterparts were. A significant negative relationship was observed between flake volume and kernel size whereas the critical number of unpopped kernels was positively correlated with kernel size.

Sakin et al. (2005) compared the yield and quality features of single-cross and three-way cross popcorn genotypes with open-pollinated cultivars. There were significant differences among genotypes for yield, popping volume, and percentage of unpopped grains in favor of hybrid genotypes. Likewise, Sweley et al. (2012) studied the effect of hybrid and environment on physical and chemical characteristics of three popcorn hybrids grown in three different environments. Hybrid had a significant effect on kernel sphericity, time to grind, dietary fiber, sugars, and starch whereas the environment only affected total mineral content. The interaction between hybrid and environment influenced test weight, hardness as measured with the tangential abrasive dehulling device, thousand-kernel weight, total carbohydrates and kernel protein content. Oil adherence to the bag averaged 15.8% and was proportional to the oil amount added prior to microwave popping. PEV was 44.7 ± 3.7 and 47.3 ± 6.4 cm^3/g, depending on the method of determination whereas the unpopped kernels averaged 11.4 ± 5.3%. PEV was significantly correlated with several kernel physiochemical parameters that were influenced by the hybrid effect. Sphericity, thousand-weight, and total fat were good predictors of the PEV. The same authors (Sweley et al. 2013) further indicated that the most relevant quality traits for popcorn are expansion volume and "eatability" factors, including unpopped kernels, hull dispersion, and the color, texture, and flavor of popped flakes.

In order to maintain the optimal moisture content for popping expansion, the moisture must be maintained at around 13.5%. Thus, damaging the pericarp during harvest or processing can greatly affect the quality and popping expansion volume (PEV) of the end product. PEV is a measure of a known amount of popcorn in grams, relative to the expansion volume (mL) after popping preferably using the standard popping equipment known as metric weight volume tester or MWVT (Serna-Saldivar 2008).

The typical expansion volumes of popcorn lie between 36 and 55 cm^3/g (Sweley et al. 2011, 2012, 2013). Other quality performance traits include kernel size measured in kernels per 10 grams: 52–67 = large, 68–75 = medium and 76–105 = small (Table 9.1). The test weight of popcorn is significantly higher compared to field corn and ranges between 81 and 84.8 kg/hL (65–68 lb/bu). Kernels are also classified by their shape, where some kernels are round (pearl) and others are pointed (rice). Color varies in both yellow and white popcorn ranging from bright colors to dull; these too are important characteristics depending on the end use of the product.

Popcorn is merchandized in three kernel types: white, small yellow and large yellow. Color differences in popcorn kernels result from genetic differences in the pericarp, aleurone and endosperm. Kernels of white popcorn are usually rice shaped while small and large yellow popcorns are pearl shaped. The three different types meet different needs within the industry. The range of 1000 kernel weight for the large, medium and small popcorn kernels are 149–152, 133–147 and 95–131 g, respectively (Serna-Saldivar 2008). The role of the pericarp in the popping process has been addressed and explained by Hoseney et al. (1983). The integrity and soundness of the pericarp are critically important to optimize popping volume because it acts as the lid of a pressure cooker. The pericarp thickness has been shown to influence the rate of dehydration of shelled popcorn.

From the nutritional viewpoint, dry popped corn is considered a whole grain food and a relevant source of calories provided by the gelatinized starch. The second major natural ingredient is a highly digestible protein that lacks sufficient amounts of both lysine and tryptophan to sustain optimum infant growth. This whole grain product also contains significant amounts of bran rich in dietary fiber and minerals and germ high in oil, protein, simple sugars, vitamins, and minerals. The detailed nutritional composition of various popcorn products is detailed in Chapter 20.

9.4 POSTHARVEST MANAGEMENT OF POPCORN

Figure 9.1 details the critical postharvest management practices of popcorn. In order to produce the best quality popcorn, farmers should harvest kernels when they reach full maturity. The harvesting operation is critical and is aimed to minimize kernel damage. Most hybrids produced nowadays are harvested by combines despite the fact that there is likely to be more mechanical damage and therefore a reduction in PEV. Popcorn quality is optimally preserved if harvested at the correct moisture (16–17% moisture) by a properly adjusted combine. The shelling losses are high and there is more probability of physical damage when the kernel moisture is above 18%. Likewise, kernels are also more susceptible to impact damage from the mechanical operations of harvesting when harvested at moistures below 14%. The combine should be adjusted in order to minimize kernel damage. The cylinder speed, concave clearance should be set to provide a balance between shelling efficiency machine losses and physical damage to the popcorn. The harvested kernels are cleaned and dried to lower the moisture to around 13 to 13.5%. Kernels must be dehydrated slowly to prevent the formation of stress cracks and fissures but fast enough to prevent mold development (Metzger et al. 1989, Serna-Saldivar 2008, 2010). Kernels should not be dried below 11% moisture otherwise the popping expansion will not be realized.

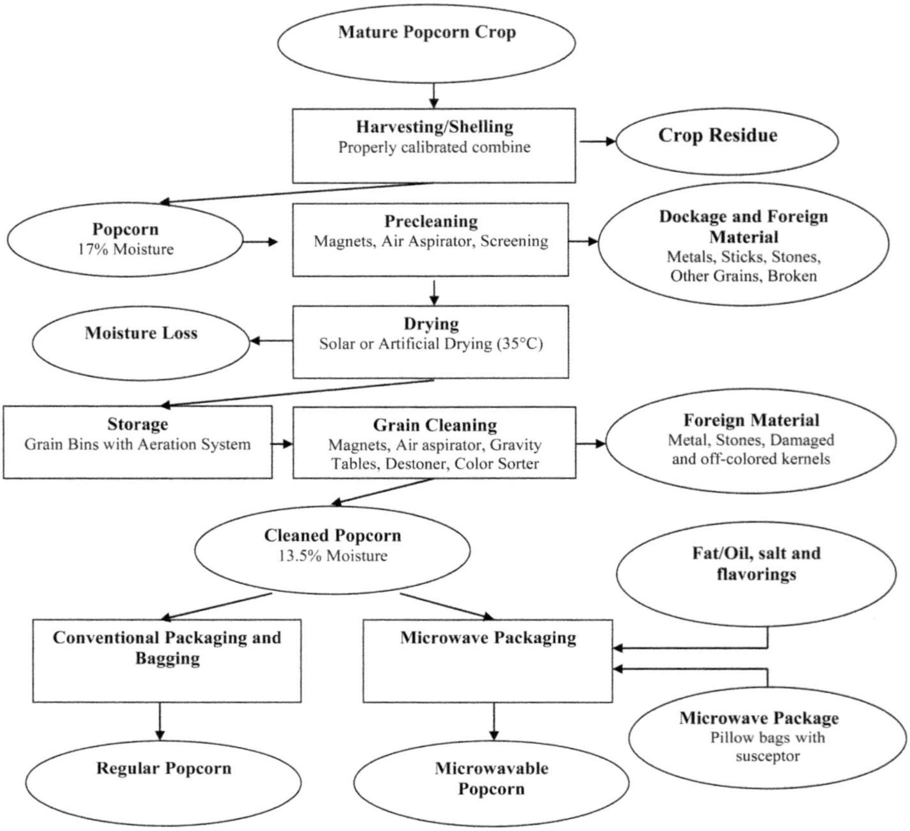

FIGURE 9.1 Postharvesting practices for the production of cleaned and packaged popcorn (adapted from Serna-Saldivar 2008, 2010).

Popcorn must be free of microbial contamination, insects and rodent damage and should have the best popping expansion (volume of popped flakes produced from a given weight of kernels). The ideal moisture content to store popcorn is 13.5–14.5%. At this moisture, the grain usually has an A_w in the range of 0.7 to 0.75. Aeration or ventilation systems similar to those used for other types of cereals should be provided to prevent moisture migration and help maintain kernel quality throughout storage. Before storing all bins should be cleaned thoroughly so as to get rid of insects, dust and other foreign material. Insecticides approved by regulatory agencies should be applied onto contact surfaces and sometimes applied directly to a lot of grain. The storage facilities should also be rodent proof so to minimize contamination with feces, urine and hair (Serna-Saldivar 2008, 2010).

Structural damage to the popcorn kernel due to mechanical harvesting, shelling and during handling drastically reduces PEV and increases the percentage of unpopped grains (Hoseney et al. 1983). The effect of mechanical damage on popcorn seems to be two fold. First, the damaged site acts as a major pathway for the escape of water vapor from the endosperm during the heating of the kernels. Second, the damage additionally weakens the pericarp (Singh et al. 1997). Popcorn kernels purposely damaged with a razor knife to remove the tip cap or cut the sides or germ had lower PEV rates than undamaged counterparts did. The expansion volume of damaged kernels increased by 52 to 86% when the size of the kernel increased from less than 4.36 mm to more than 5.95 mm. Damaging the tip cap and germ caused less loss of expansion than did other type of injures. Damaged kernels had faster popping times than did undamaged counterparts. Popping times increased with larger kernel sizes for all types of damages (Singh et al. 1997).

9.5 POPPING MECHANISM

Popping occurs at about 180 to 190°C, which is equivalent to a steam pressure of 930 kPas or 9.3 bar (135 psi) inside the kernel. The water in the grain is superheated and at the instant of popping converts to steam, which provides the driving force for expanding the endosperm after the kernel ruptures (Hoseney et al. 1983, Byrd and Perona 2005). When the pericarp ruptures, the superheated water rapidly expands in less than 1/15 sec causing the molten starch to expand resulting in the typical popcorn flakes (Byrd and Perona 2005). The pericarp and outer layers of the kernel participate directly in the popping action by serving as a pressure vessel enclosing the endosperm. Hybrids or varieties with high proportions of translucent, flinty or vitreous endosperm attain higher PEV (Pordesimo et al. 1990, 1991).

Byrd and Perona (2005) determined the rate of popcorn expansion in oil and in heated air and found that the activation energy during popping followed a first-order integrated rate law as follows $\ln(N) = -kt + \ln(No)$ where No = the initial number of popcorn kernels, N = number of kernels remaining unpopped at time t. The activation energy for the first-order decrease in the number of unpopped kernels vs. time was 39.9 kcal/mol between 180 and 210°C and 12.9 kcal/mol at temperatures ranging from 210 to 250°C. The data is consistent with the rate of heat transfer into the kernel known as Newton's law of cooling. For an individual kernel, the probability of popping was directly proportional to the difference between the internal vapor pressure and the kernel's critical pressure. The minimum popping temperatures predicted by the model were 181°C and 187°C for oil and air-popped flakes, respectively.

The moisture content of the popcorn greatly affects the expansion rate and volume. According to several authors, the optimum moisture ranges from 11 to 14% with an optimum value of around 13.5% (Song and Eckhoff 1994a). Gokemen (2004) studied the effects of kernel moisture content and popping methods on popping quality of five popcorn genotypes. The popcorn kernels were equilibrated to seven moisture levels ranging from 8% to 20% and were popped by microwave oven, hot-air popper, cooking pan with and without oil and salt. Genotype, moisture content, and popping method affected the expansion volume, flake size and percentage of unpopped kernels. The highest popping quality estimated by PEV, flake size and the percentage of unpopped kernels were obtained from grains with a moisture content of 14%. The microwave oven gave the highest flake size and percentage of unpopped kernels whereas the lowest expansion volume and flake size were obtained from the cooking pan popping with oil and salt. As moisture content deviates from 14%, the PEV and flake size declined significantly. In addition, hybrids with medium kernel size had higher values for the expansion volume and flake size.

Allred-Coyle et al. (2000) studied the effects of kernel moisture content and size, type of hybrid and microwave wattage on the expansion volume of microwave popcorn. The highest expansion volume and the smallest percentage unpopped grains were achieved in medium-size kernels containing 11% moisture. Likewise, Ertas et al. (2009) studied the performance of three popcorn genotypes differing in kernel size and moisture contents (10, 12 or 14%) popped conventionally or in a microwave oven. The optimum moisture levels for the highest expansion volume varied between 12 and 14% for the different genotypes.

Waldrat et al. (1970) identified relevant compounds like pyrazines, furans, pyrroles, carbonyls and substituted phenols, which contribute to the flavor and aroma of popcorn. Thereafter, Buttery et al. (1997) employing capillary GC–MS assays studied flavor and aroma volatiles isolated from microwave oven-produced popcorn. They reported the presence of the previously reported compounds and some additional like dimethyl sulfide, dimethyl di- and trisulfides, 3-methylindole (skatole), α- and β-ionones, 2-methyl-3-hydroxypyran-4-one (maltol), 2,3-dihydro-3,5-dihydroxy-6-methyl-4*H*-pyran-4-one, 5-methyl-4-hydroxy-3(2*H*)-furanone (norfuraneol) and geranyl acetone. Hydrogen sulfide was identified as the major volatile compound emitted during popping.

Borras et al. (2006) determined the relationship between the amount and type of lipids, starch composition, storage proteins and kernel structure on popcorn expansion on seven commercial

Argentinean popcorn samples. A significant negative correlation was observed between oleic acid and popping volume and a positive correlation was observed between linoleic acid and popping volume. α-Zein proteins and glutelins significantly correlated with popcorn expansion volume with correlation coefficients $R^2 = 0.963$ and 0.744, respectively.

9.6 POPPING PROCEDURES

There are basically two types of popping methods: wet and dry (Figure 9.2) and two types of popcorn configurations: mushroom or butterfly. Most popcorn is wet-popped in conventional pans or in microwavable bags because they commonly yield better organoleptic properties. However, dry popping is ideally suited for production of reduced calories or low-fat products. Table 9.2 summarizes the advantages and disadvantages of these popping techniques.

The popped corn with a spherical shape is named ball or mushroom and this particular configuration is preferred in the confection industry because it is less susceptible to breakage, more resistant to handling and more efficiently coated with flavorings and confectionary syrups. The butterfly type has a higher PEV or lower apparent bulk density, more tender and with better mouth feel. These flakes are preferred for on premises popping where it is sold by volume. The popping expansion of mushroom type is lower than butterfly because the individual flakes fit closer together. For production of ball popcorn the kernels are heated approximately 20°C higher compared to the

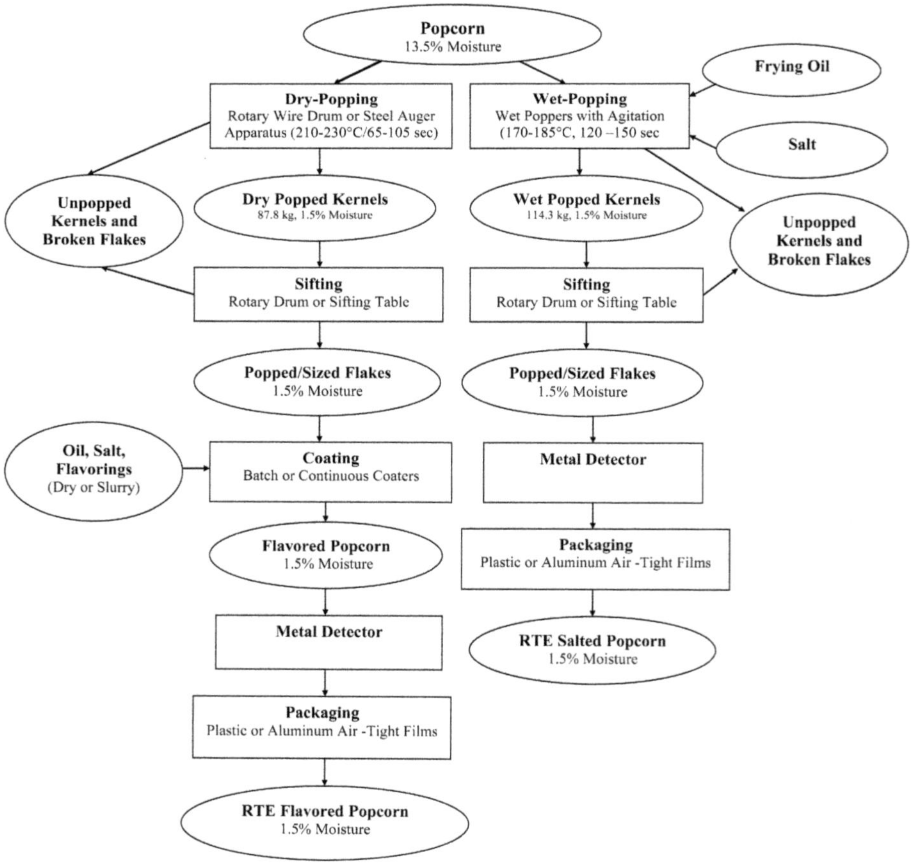

FIGURE 9.2 Diagram of the production of dry or wet-popped flavored popcorn (Serna-Saldivar 2008).

TABLE 9.2
Advantages and Disadvantages of the Wet and Dry Popping Procedures

Wet Popping		Dry Popping	
Advantages	**Disadvantages**	**Advantages**	**Disadvantages**
Animation of the popping	Requires about 25% more oil than dry popping.	Requires about 25% less oil than wet popping.	More expensive equipment.
Better aroma and flavor	The heat of popping is detrimental to oil quality.	Low labor cost especially when automatic equipments are used.	Less cleaning problems compared with wet popping.
Adapted to popping at point of sale	Cleaning problems on the kettles.	Ideally suited for production of flavored popcorn, caramel popcorn, popcorn balls, and molded figurines.	More fire hazard when hand-fed batch types are used.
Relatively inexpensive equipment	Loss of popping volume if the corn:oil ratio is higher than 4:1.	Better distribution of the oil because it is generally sprayed.	
Small space process requirement	The resulting popcorn is higher in energy or calorie density.	Popcorn usually lower in calories is ideally suited for low reduction programs.	
Minimum fire hazard			
Simplicity of operation with minimum labor skill			

Source: Snack Food Association (1992).

butterfly counterpart (temperatures of 235 and 215°C, respectively) (Snack Food Association 1992, Serna-Saldivar 2008, 2010).

Sweley et al. (2011) identified and characterized different polymorphisms of popped flakes from a yellow popcorn hybrid and then determined the impact of flake shape on composition and sensory characteristics. Kernels were popped using a microwave oven and visually sorted into three different polymorphisms depending on whether the appendages were expanded unilaterally, bilaterally or multilaterally. After popping, 9.0, 71.2 and 12.3% of kernels expanded unilaterally, bilaterally and multilaterally, respectively, while 7.6% of kernels did not explode. Expansion volumes for unilaterally, bilaterally and multilaterally expanded polymorphisms were 28.6, 43.0 and 53.5 cm^3/g, respectively. Interestingly the unilateral popcorn flakes retained the most fat and sodium, while the multilaterally expanded flakes contained the highest levels of protein, total carbohydrate, and pyrazines that impart aroma and flavor. Sensory evaluation tests revealed significant differences among polymorphisms for flavor and texture, with the unilaterally expanded polymorphism receiving the highest overall acceptability.

9.6.1 Dry Popping

Dry popping is considered the least popular method but is the one that yields the most nutritionally accepted product. Industrial dry poppers come in two basic types the older method being a rotary wire drum turned over an open flame. The flame heats the kernels and causes them to burst. A unique helical screen arrangement causes the popped maize to be removed from the popper as soon as it pops. The modern method consists of a steel auger wrapped with perforated steel and heated air is forced through the holes. These industrial equipment consist of a continuously fed, fluidized bed oven or drum that is also commonly used to expand third-generation snacks (Chapter 12) and roasted peanuts (Chapter 13) with the advantage of not using oil. The main advantage of this continuous popper is the strict temperature and residence time controls and the recirculation of about 90% of the air used in the process. The basic design of the continuous dry poppers consists of a horizontal auger of about 40.6 cm (16 inches) with nine complete flights wrapped around the shaft. The auger is wrapped with a perforated steel sheet with 0.24 mm holes (3/32 inches). About one-third of the steel is open holes and the rest is metal. The combined auger and perforated steel wrap revolves as one unit and is housed in an oven, which is held at a temperature of about 221°C (430°F). A fan located at the bottom of the oven blows hot air up through the bottom of the perforated rotating drum. Raw popcorn kernels are introduced into one end of the cylinder, and the heated forced air flowing through the perforations fluidize the material lying on the bottom of the drum enhancing the heat transfer. The revolving action of the drum propels the product to the discharge end of the popper. The equipment is highly efficient because it is furnished with adjustable controls of feed rate and product residence time (from only 15 seconds to as long as 5 min). The volume of airflow that fluidizes the product is controlled by the rpm of the blower. Popcorn is usually heated for about 80 seconds (65 to 105 seconds) at 215 to 230°C (Cretors 2001). Dry poppers are designed for large-scale operations and are capable of popping from 70 to 450 kg/hr. When kernels are dry popped they usually lose 12–13% moisture and yield 8 to 10% of scrap (Cretors 2001). After dry popping, the popcorn flakes are separated from unpopped kernels and fines in a rotary sifter with 1.3 cm by 1.3 cm stainless steel wire mesh drum with openings of 1.6 to 2.8 cm^2 (Snack Food Association 1992). The clean and sized popped flakes are then flavored in a coater. The coating process of popcorn is similar to that used for other snack foods. Oil, salt, flavorings or sugar-based flavorings are directly applied or sprayed onto the product. The industry uses batch or continuous coaters (Chapter 3). Batch coating for salty products is typically inclined enrobing pans that turn on an axis inclined 30°. The base product is placed in the pan and the oil coating and seasonings introduced as the pan rotates. The coating is either poured or pumped. The pan is allowed to work until the coating is fully spread and evenly distributed on the product. The most common way to apply coatings is with a continuous machine consisting of a horizontal stainless steel cylinder that rotates on its axis. The speed of rotation and the angle of the cylinder are adjusted to control the product flow rate or residence time. The coater is equipped with either a dry applicator or a pump system that sprays the coating in the first part of the inclined machine (Chapter 3). Dry popping requires about 25% less oil than wet popping. Oil is added to dry popped corn and is often the vehicle for addition of other flavors and colors. Coconut oil is preferred for popping and coating popcorn. Popped corn rapidly absorbs moisture and loses texture at moisture levels above 3% (Snack Food Association 1992).

For production of cheese-flavored popcorn, dry powdered cheese can be applied to the popped corn or sprayed. The disadvantage of the dry powder is that comes off on the consumer's hands. The most common method is to spray cheese slurry onto the popcorn. The cheese flavoring is produced by mixing oil, preferably coconut, with powdered cheese and other ingredients (Chapter 8). In order to apply the coating, the mixture should be heated to around 50°C to melt the cheese and avoid plugging the spray system. Cheese-flavored popcorn generally contains 15% cheese, 28% oil and 57% popcorn (Snack Food Association 1992).

Large quantities of popcorn are popped, flavored and packaged as snacks. Regular ball popcorn is treated with butter-flavored vegetable oil as the base for flavors and colors. Flavored coatings are usually applied by spraying the popcorn as it moves on a conveyor or pouring a free-flowing flavored solution on the popcorn held on a revolving coating pan (Snack Food Association 1992).

9.6.2 Wet Popping

Wet popping is the preferred way to produce popcorn. Industrial wet poppers work much like a dry pan where oil, popcorn and salt are placed into a heated pan. An agitator on the bottom constantly stirs the mixture allowing a better and even heat distribution among the kernels. When the temperature of the popcorn raises high enough kernels gradually burst. At the moment of expansion the oil is absorbed and the salt sticks to the flakes (Cretors 2001, Serna-Saldivar 2008).

The typical industrial wet-popping line consists of one or more banks of oil poppers positioned over a conveyor belt. Normally, four to eight poppers are set in battery and operators add measured amounts of popcorn, oil and salt to the kettles. Operating on a typical 3-min cycle, the operators will dump, empty and refill the apparatus every 30 seconds. The cooking oil may be added to the kettle volumetrically or by a metering pump. In some equipment, the pumps are immersed in the pails themselves and are equipped with thermostatically controlled heating elements to melt frying shortening or oils that are solid at room temperature (Cretors 2001).

These poppers are either heated by gas or electricity being the first and the most preferred in the industry. The most common wet poppers are supplied by C. Cretors & Co in steel construction. Wet poppers are designed to handle charges of 900 to 2,300 g of popcorn with 2.5 min cycles and have hourly capacities of 18 to 45 kg. The rule of thumb for wet popping is to add the raw popcorn to the kettle before the oil and to close the lid immediately. The most frequent popcorn:oil ratio is 3:1 (Figure 9.2). The popping cycle varies from 2.5 to 3 min. The batch is assumed to be finished when five seconds last between pops. The conveyor belt transports the burst popcorn to a sifter, which removes unpopped kernels and other fines from the finished product (Figure 9.2). The sifting operation is similar to the one described for dry popping and should be aimed toward the removal of unpopped kernels and broken flakes. After sifting, the popcorn should be immediately packaged to avoid moisture absorption and loss of texture (Snack Food Association 1992).

The vegetable oil used for wet popping plays an important role in the final sensory properties of the product. The oil for wet popping should be more stable than the oils used for coating dry popped flakes. The oil for wet popping should have a high smoke point (260°C); that is the reason coconut or peanut oil is the most frequently used. Peanut oil imparts better flavor but is less stable and leaves a gummy residue on the surface of the popper. This polymerized residue is very hard to clean (Snack Food Association 1992, Serna-Saldivar 2008).

9.6.3 Microwave Popping

Microwaveable popcorn has increased in popularity for its convenience for in-home popping and snacking. There are three key elements for the production of microwave popcorn: quality of raw kernels, the blend of oil flavorings, which accompany the popcorn and the design of the microwavable bag. Most bags contain about 100 g of the mixture of popcorn and the oil cake that yields about 90 g of burst-flavored flakes. Generally, 75 g popcorn is mixed with 25.5 g melted partially hydrogenated vegetable oil and 1.5 g salt. Other flavorings (butter, margarine, cheese) and even colorants can be included in the cake. The maximum PEV of popcorn occurs when the unpopped kernels have an A_w of 0.5 to 0.65. At higher A_ws the grain is prone to microbial growth and when the A_w is lower than 0.4 a significant lower popping volume is obtained (Labuza and Contreras-Medellin 1981).

The mixture is placed in pillow-style microwave bags with a 15 cm susceptor plate and a heat seal band on the inside of the lip (Chapter 4). The microwave container is placed in sealed cellophane bags in order to protect the product from atmospheric air that can cause moisture changes in the kernels and promote rancidity. For microwave popping, the recommended power output in conventional microwaves is 660 W for 2 min on high power. Generally, popcorn starts bursting 50 seconds after starting the microwave cycle and finishes after 100 seconds. The PEV of microwave popcorn should be in the range of 34.2 to 43.9 cm^3/g, the percentage of unpopped kernels from 2.1 to 6.8 and flake size from 4.6 to 7.1 cm^3.

Mohamed et al. (1993) studied the popping performance of 18 popcorn hybrids using two microwave treatments (with and without salt) compared to the conventional popping procedure. Nearly all hybrids gave significantly higher volumes when popped in a conventional hot oil than those obtained by microwave popping. The addition of salt had a negative effect on popping quality because the salt formed a coating around the grains modifying their thermal and dielectric properties. Pericarp thickness gave the highest correlation with PEV in both microwave and conventional popping while sphericity correlated better with unsalted microwave popping. Lin and Anantheswaran (1988) also studied effects of grain morphology on microwave popping and concluded that within the same popcorn genotype, large kernels produced larger flakes compared to small counterparts. Likewise, Pordesimo et al. (1990, 1991) compared the performance of five varieties in microwave popping and reported that sphericity, kernel size and specific gravity correlated positively to PEV in samples popped in a microwave oven. The greater expansion volume was observed in medium to short broad kernels which had sphericity values greater than 0.7. Within a given variety, expansion volume and flake size were higher for kernels with higher specific gravity.

9.6.4 Confectionary and Caramel-Coated Popcorn Products

There is an array of confectionary or sweetened popcorn snacks available in the market. These products evolved from the American Cracker Jack™ snack produced from more than 150 years. Cracker Jack is the tradename of an old popcorn-based snack flavored with a syrup containing molasses, caramel, colorings and flavorings and commonly blended with roasted peanuts (Chapter 13). As mentioned before, the inventor of this product is attributed to the German migrant Frederick W. Rueckheim who mechandized popcorn products in Chicago, USA. Caramel-coated and other related products are made from mushroom popped kernels covered with a sugar or molasses based caramel candy that is baked in order to form an outer shell normally less than 1 mm thick. Typically a sugar solution or syrup is made and heated until it browns and becomes thick, producing a caramelized candy syrup. This hot slurry is then mixed with mushroom popped corn, baked and cooled down in preparation for packaging. The process creates a sweet flavored and crunchy snack that is blended with roasted peanuts (Chapter 12) and/or tree nuts (Chapter 15) like pecans, almonds or cashews (Figure 9.3). The key steps of the process consist of first producing mushroom-type popcorn free of unpopped kernels and fines that are subsequently coated with slurries produced from vegetable oils, butter, sugar syrups, molasses and other minor additives like flavorings and coloring agents. Most covering formulations contain baking soda, which produces lighter colors and an aerated coating structure that averages only 1 mm of thickness. The slurry is heated to yield a free-flowing solution for effective application onto the popcorn. The hot slurry is poured into a rotating drum for a more effective coating. The revolving drum is designed to lift and blend the popcorn with the sugar coating. After the foamed sugar and popcorn are well coated a small amount of oil/lecithin is sprayed to enhance the separation of the popcorn. The resulting flavored popcorn is cooled down or alternatively baked in order to produce the desired crispiness that characterizes these products (Snack Food Association 1992, Serna-Saldivar 2008).

Sugar-coated popcorn is flavored using batch or automated coating drums. According to Cretors (2001) the coaters used to deliver sweetened slurries differ from other coaters because the amount of coverage is significantly larger and in some instances the coater also cooks the caramel-coated popcorn. The automated batch cooker and coaters generally have a batch size of 40 kg of sugar coating. The efficiency and production rate of the coater may be increased by premixing and heating the slurry ingredients in a steam-jacketed kettle. Once the coating is fully cooked, the popcorn is added to the kettle and the internal revolving system effectively blends the products. When the corn is completely coated, a small amount of oil/lecithin mixture is sprayed on the product to facilitate its separation into individual pieces as it cools. When the process is complete, the contents of the coater drum are discharged with the help of pneumatic or hydraulic cylinders. Then, the hot caramel-coated popcorn is transported to a continuous cooler where it is cooled and agitated to enhance the separation into individual pieces.

The other alternative design for larger operations is the use of a fully automatic continuous system consisting of kettles that holds all sugar ingredients at a high temperature, a pump that delivers the sweetened hot (148°C) premix onto the popcorn in a stainless steel trough with a steam-heated auger. As same as the batch system, a small amount of oil/lecithin mixture is sprayed into the coater

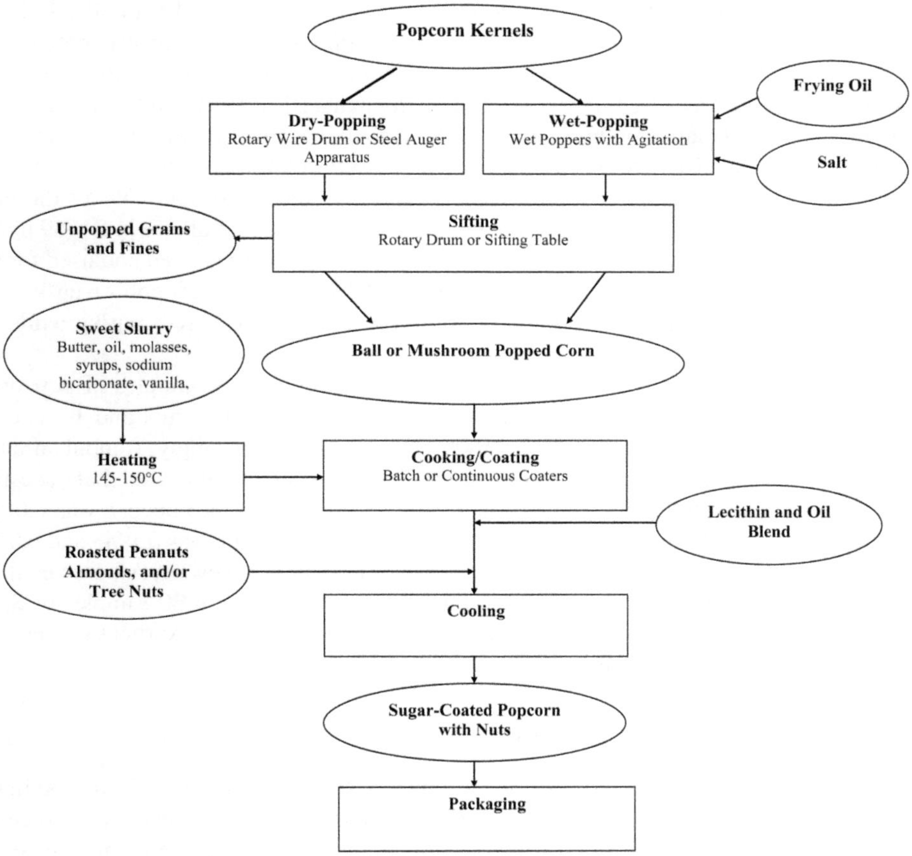

FIGURE 9.3 Diagram of the production of sugar-coated and flavored popcorn blended with nuts.

to promote separation of the product into individual particles. Then the caramel-coated popcorn is cooled down in preparation for packaging (Cretors 2001).

9.7 FOOD SAFETY OF POPCORN

Popcorn is considered a safe snack because is commonly thermal treated at temperatures higher than 200°C. Nevertheless, the industry still worries about the presence of significant levels of mycotoxins (Chapter 20) and cross-contamination with pathogenic bacteria after popping and before packaging. Undoubtedly, the risk of mycotoxins is more worrisome because they are heat resistant and can cause illnesses at very low concentrations. Popcorn is known to possibly contain significant amounts of *Aspergillus*-produced aflatoxins and ochratoxins A (OTA) and *Fusarium*-produced fumonisins, deoxynivalenol (DON) zeralenone and T2 toxins. Among mycotoxins, aflatoxins are the most relevant because of their high toxicogenic effects especially in terms of the risk of cancer. Popcorn can develop *Aspergillus flavus* and *Aspergillus parasiticus* in the field and during faulty storage and these fungi can produce aflatoxins B1, B2, G1 and G2 (Chapter 20). According to Dowd and Johnson (2010) the field incidence of mycotoxins in commercial popcorn is related to ear insect (caterpillar) damage. These authors found higher incidence of fumonisins and DON produced by Fusariums compared to aflatoxins produced by *Aspergillus flavus*. The concentrations of fumonisins and DON averaged 1.7 mg/kg (sample max. 2.77 mg/kg) and 1.9 mg/kg (sample max. 2.66 mg/kg), respectively. Interestingly, insect-damaged kernels had 100 to 500 fold or greater levels of fumonisins compared to sound kernels, while DON levels were closer to 10- to 30-fold higher in insect-damaged versus nondamaged counterparts. *Aspergillus flavus* growth was noted on some popcorn ears in 2005, but aflatoxin was only detected in popcorn at 1.3 ppb or μg/kg.

Sekiyama et al. (2005) assessed the occurrence of aflatoxins B1, B2, G1 and G2, OTA and zearalenone in more than 120 maize foods collected in the Brazilian state of Parana. Only three samples were positive to aflatoxin B1 (8 to 59 μg/kg), two (1.7%) to B2 (2.4 μg/kg), one (0.8%) to OTA (64 μg/kg) and one (0.8%) to zeralenone (448 μg/kg). The study indicated that 8.3% of the popcorn samples tested positive and these contained the highest concentration of aflatoxin B1 (59 μg/kg).

Currently, the FDA regulates food and ingredients intended to be processed and used for human consumption so they do not contain levels of aflatoxin B1 higher than 20 ppb or μg/kg. On the other hand, fumonisin tolerance levels for popcorn in the United States are 3 mg/kg while for the European Union 0.8 mg/kg for corn-based snacks.

Andrade et al. (2018) evaluated the occurrence of 11 mycotoxins in samples of popcorn. The samples were mainly contaminated with fumonisins, aflatoxin B1, neosolaniol and T-2 toxin, with mean concentrations of 592, 10.2 and 10.2 μg/kg, respectively. Interestingly, fumonisin and aflatoxin B1 co-occurred in 52% of the popcorn samples. The same authors indicated that the estimated intake for the Brazilian population of total fumonisins and T-2 toxin for popcorn was 0.0051 and 0.00004 μg/kg body weight/day, respectively, and for the 99th percentile was 0.1943 and 0.0014 μg/kg body weight/day, respectively, suggesting a low risk of ingestion. Likewise, Alborch et al. (2012) studied the co-occurrence of aflatoxins, citrinin, OTA and zeralenone in 30 samples of Spaniard popcorn kernels. Aflatoxins and OTA were detected in two and ten popcorn kernel samples whereas citrinin and zeralenone were not detected.

9.8 POPPING OF OTHER CEREALS

Popped sorghum and millets are snacks similar to popcorn consumed widely in India and in several African countries (Taylor and Duodu 2019). All types of sorghums and millets pop to some extent, but the degree of popping can be quite variable, with hard endosperm types popping the best. With sorghum, grain characteristics like small kernel size, high bulk density, medium hardness, a thick pericarp, and highly amylose content seem to be positively associated with good popping quality (Mishra et al. 2015).

Certain cultivars of sorghum (*Sorghum bicolor*) and various types of millets have been traditionally popped in Africa and India for many decades. A special type of hard endosperm sorghum named Shalu is recognized as the best for popping. Murty et al. (1982) identified more than 30 accessions of sorghum with excellent popping quality among nearly 3,700 Indian sorghum genotypes and concluded that the popping ability is highly heritable. These special sorghums yield small kernels that contain white and thick pericarps, corneous endosperms with a lower germ-endosperm ratio compared to regular counterparts. Similarly, Malleshi and Desikachar (1985) found varietal differences in popping quality of pearl (*Pennisetum americanum*), finger (*Eleusine coracana*) and foxtail (*Setaria italica*) millets. The popping of sorghum and millets resulted in a small, porous and crunchy snack with similar features to those of popcorn. During popping, grains are exposed to high temperatures for a short time. The kernels are heated until the internal moisture turns to steam, and the steam and air expand rapidly and pop the outer shell (pericarp) of the kernel (Mishra et al. 2014). Essentially, during popping the kernel acts as a pressurized vessel due to the rapid expansion of its internal moisture and air as it is heated. Due to the thermal gradient, the moisture inside the grains vaporizes and tries to escape through the micropores, which causes the grain to burst as a result of the expansion of the starchy endosperm.

Grain popping in India has traditionally been done using sand at a temperature of about 250°C. However, with the growing demand for millet- and sorghum-based "pops," a number of mechanical popping machines have been developed. The use of microwave ovens for popping sorghum has also been explored and successfully performed (Mishra et al. 2014, Taylor and Duodu 2019).

9.9 GRAIN-PUFFING TECHNOLOGIES

Besides the conventional method of popping, there are other ways to expand whole grains, refined milled coarse fractions and pellets being oven- and gun-puffing the two technological processes more generally used. Oven-puffing is based on a sudden application of heat at atmospheric pressure so that the water in the grain is vaporized expanding the product. On the other hand, gun puffing is based on a sudden transfer of grain containing superheated steam from high pressure to low pressure, thereby allowing the water to suddenly vaporize and cause expansion. Recently, the application of microwave heating to puff grains and pellets has been practiced. Independently of the type of puffing, the degree of expansion is affected by the type of grain, the grain moisture content and its chemical composition and physical conformation. In most instances, the starch is positively related to the expansion rate whereas the dietary fiber and fat contents are negatively associated. Among, methods, gun puffing is considered the best one to achieve higher expansion rates. Puffed grains are valued primarily for their lightness, texture (crispness) and flavor acquired after expansion.

9.9.1 Oven and Microwave Puffing

Oven-puffed cereals are almost exclusively made from white polished rice or maize since they inherently expand when exposed to high temperatures in conduction or impingement ovens. The high starch content of rice and the morphology of the compound starch granules make this grain one of the preferred for the production of oven expanded grains because they are widely used for the production of RTE breakfast cereals and nutritional bars. Oven-puffing is generally achieved in continuous rotary ovens which transfer most of the heat by convection and more recently by impingement ovens that apply forced air to create a more rapid heat transfer and better expansion rate. Resulting oven-puffed grains are usually classified in order to discard fines and broken expanded kernels and in some instances, products flavored in a tumbler with dry or sprayed liquid seasonings (Chapter 3).

Joshi et al. (2014) optimized the processing conditions in a domestic convective microwave oven for the production of puffed rice. Puffing characteristics were studied for rice at 14% moisture applying different preheating temperatures of the glass base plate (180, 200 and 220°C), microwave power level (300, 600 and 900 W), and residence time (10–100 sec). These authors concluded that the microwave puffing of rice was better at higher power level and by preheating grains on the base plate. The same authors further optimized process parameters using response surface methodology, at four parameters with three levels: moisture content (12, 14 and 16%), preheating temperature of the base plate (180, 200 and 220 °C), microwave power level (300, 600 and 900 W), and residence time (50, 60 and 70 sec). The response surface methodology indicated that the optimized conditions were 14% grain moisture, 220°C of the base plate, 900 W and 60 sec residence time.

In a similar study, Swarnakar et al. (2020) investigated how the microwave settings affected the puffing of brown rice. A full factorial design was chosen for the experiment with three levels of microwave power (800, 900 and 1,000 W), and four levels (20, 25, 30 and 35 sec) of heating time. It was observed that microwave power and heating time had significant effects on puffing percentage, expansion ratio, whiteness index and hardness. The optimized conditions were 900 W of microwave power for 35 sec of heating. Puffing percentage, expansion ratio, whiteness index and hardness of microwave puffed brown rice at optimum conditions were 57.44%, 5.34, 58.31 and 11.28 N, respectively.

In terms of pseudocereals (Chapter 13), Deepak et al. (2020) investigated the expansion performance of quinoa by hot-air oven-puffing. The puffing conditions studied were grain conditioning of 2 ml or 18 ml/100 g without or with the presence of salt (1%) at temperatures of 240 to 280°C for 20 to 60 sec. The puffing yield, expansion ratio, flake size and overall acceptability of puffed seeds increased while bulk density decreased with lower moisture level and salt concentration and augmented with higher oven temperature and residence time. The optimum conditions for puffing were 2 ml moisture/100 g grain sample, 0% salt concentration, 253.8°C oven temperature and 60 sec residence time. At these conditions, a fourfold volume expansion was achieved.

Likewise, amaranth (Chapter 13) grains are reported to have good popping rates when exposed to intense dry heat for relatively short periods. According to Solanki et al. (2018) amaranth can be effectively popped with either a traditional or an improved process. The demand for popped amaranth is increasing because expanded grains are quite soft in texture and are RTE or promptly incorporated into new food formulations. Because of their small seed size, different conditions are needed for the popping of amaranth compared to other grains. Solanki et al. (2018) investigated two different processes to pop amaranth grains through the traditional batch popping machine and an improved process consisting of a fluidized bed system. The variables studied were popping volume, expansion or fold and popping rate in the temperature range of 120 to 150°C and time range of 5 to 30 sec. Popping volume, fold and rate ranged from 0.41 to 0.54 cm^3 per 100 grains, 5.5 to 6.7 and 28 to 80%, respectively in the entire range of temperature and time combinations used in the traditional method. In the improved process, the popping volume, fold and rate ranged from 0.42 to 0.48 cm^3 per 100 grains, 5.3 to 6.1 and 44 to 65%, respectively.

9.9.2 Gun Puffing

Gun puffing is a technology widely used by the RTE breakfast cereal industry since the 1960s. The process is defined as the controlled expansion of kernels, dry-milled fractions or extruded pellets (Chapter 12). The necessary condition to achieve puffing or explosion of grains is the application of a strong thermal treatment to increase pressure in a hermetically sealed container or puffing gun programmed to suddenly release the pressure. The moisture content of the grain affects expansion rate because it is instantly converted into water vapor when the pressure is released. The grain explodes due to the quick exit of water vapor when the internal and atmospheric pressures equilibrate (Fast and Caldwell 2000, Serna-Saldivar 2010b). The most common cereals for production of puffed grains are whitepolished rice and durum or hard wheats (Chapter 6). Regular white or parboiled rice with less than 0.5% fat is preferred for gun puffing. Waxy or low amylose rices have comparatively lower expansion volumes. In fact, white polished rice can be puffed without tempering. On the other hand, wheat kernels are normally tempered with water or saline solution before gun puffing in order to optimize expansion and promote the detachment of the pericarp during the severe expansion. After gun puffing, the pericarp is removed by air aspiration and/or screening to yield puffed decorticated grains.

Most commercial puffing guns are batch although there are also continuous. Kernels are placed inside the gun chamber and then the equipment hermetically sealed. The gun usually rotates on its axis to avoid sticking grains to the internal walls. These equipment generally operate at programmed pressures of up to 1,380 kPa (13.8 bars or 200 psi).

The three basic sequential steps of gun puffing are (a) preparation of the raw material for puffing followed by (b) heating or cooking the grains with superheated steam at high pressure in the sealed chamber and (c) sudden release of the pressure (Hsieh and Luh 1991, Serna-Saldivar 2010a). Preparation of raw materials generally consists of cleaning and classification and tempering or conditioning. Once grains are enclosed in the gun, the pressure gradually increases due to heating (200–260°C). The heat evaporates the grain water and causes a notable pressure increase. When the puffing gun reaches the programmed pressure, it suddenly opens releasing the expanded grains. The puffed grains are normally projected and caught in a continuously vented bin. During the microinstant of pressure release and equilibration, the superheated water exits the grain causing expansion because the vapor pressure escapes through the micropores of the grain structure. In addition, the grain loses significant amounts of moisture and the expanded item acquires a different texture, flavor and aroma (Serna-Saldivar 2010a, Mishra et al. 2014). Puffed grains with 6 to 9% moisture are classified in order to remove broken kernels, fines, nonpuffed kernels and pericarp. The classified puffed grains are then dehydrated to drop their moisture to approximately less than 3% and in some instances flavored in coating tumblers. This technology is considered one of the most effective to produce the largest expansion rates. Gun puffing has been also used for making puffed barley, pseudocereals and legumes for the manufacturing of different snacks. Besides the type of grain, the more relevant factors affecting expansion rate are the moisture present in the grain, the moisture involved in the process of steaming within the gun and the time and the temperature at which the grain is preheated (Lai and Cheng 2004).

Hoke et al. (2007) studied the expansion volume or puffing index of naked barley at various operating conditions. The purpose was to optimize the manufacture of expanded naked barley kernels that have exceptional nutritional value for humans. The parameters investigated were the mass and the moisture content of the charge of the puffing barrel, pressure and temperature during the process, the type of barley cultivar, pre-processing of the grain (grinding and polishing) and water addition. A regression equation was established for the calculation of the puffing index as a function of the operating parameters. The puffing index was mainly influenced by the operating temperature and pressure. The puffing index increased with increasing pressure in the range from 0.8 to 1.1 MPa at a constant operating temperature. The best quality expanded naked barley with the optimum sensory properties was obtained at the operating temperature of 550°C and pressure from 0.9 to 1.0 MPa with the moisture content of the charge around 16.5%.

Mariotti et al. (2006) researched changes associated with gun puffing of common wheat, emmer wheat, rye, barley, rice or buckwheat. The results demonstrated that the effect of puffing was strongly influenced by the morphology and composition of the kernel. Puffed rye and rice have a very porous matrix, made up of numerous cavities of different sizes separated by a very thin "wall." On the other hand, exploded wheat, emmer wheat and barley showed a much more compact, homogeneous and less porous structure whereas the puffed buckwheat (Chapter 13) showed a large number of small and regular cavities. Moreover, puffing induced significant changes in the structure and physical properties of the starch and an increased water-holding capacity of both the puffed grains or milled flours. The bulk density of common wheat, emmer wheat, rye, barley, rice and buckwheat diminished 40, 48, 52, 43.5, 74 and 78%, respectively indicating that rice and buckwheat showed the highest expansion rates.

Similarly, Zapana et al. (2020) successfully puffed quinoa for application as RTE breakfast or as an ingredient for snack formulations. In this study, puffed quinoa products with and without starch–chitosan coating were developed by gun, extrusion and microwave puffing at different process conditions (pressure, power, moisture content and energy consumption). Size, bulk density, color, expansion index, water absorption and solubility, microstructure, mechanical and thermal properties, chemical composition and *in vitro* digestibility of organic matter and proteins of popped quinoa were assessed. Optimal process conditions for gun and microwave puffing were 1.31 MPa after 780 sec and 1200 W for 60 sec at 18–20% moisture contents. The authors concluded that gun puffing yielded high-quality popped quinoa seeds with high nutritional value.

REFERENCES

Alborch, L., Bragulat, M.R., Castellá, G., Abarca, M.L., and Cabañes, F.J. 2012. Mycobiota and mycotoxin contamination of maize flours and popcorn kernels for human consumption commercialized in Spain. *Food Microbiology*, 32(1): 97–103.

Allred-Coyle, T.A., Toma, R.B., Reiboldt, W., and Thakur, M. 2000. Effects of moisture content, hybrid variety, kernel size, and microwave wattage on the expansion volume of microwave popcorn. *International Journal of Food Sciences and Nutrition*, 51(5): 389–394.

Andrade, G.C.R.M., Pimpinato, R.F., Francisco, J.G., Monteiro, S.H., Calori-Domingues, M.A., and Tornisielo, V.L. 2018. Evaluation of mycotoxins and their estimated daily intake in popcorn and cornflakes using LC-MS techniques. *LWT-Food Science and Technology*, 95: 240–246.

Borras, F., Seetharaman, K., Yao, N., Robutti, J.L., Percibaldi, N.M., and Eyherabide, G.H. 2006. Relationship between popcorn composition and expansion volume and discrimination of corn types by using zein properties. *Cereal Chemistry Journal*, 83(1): 86–92.

Buttery, R.G., Ling, L.C., and Stern, D.J. 1997. Studies on popcorn aroma and flavor volatiles. *Journal of Agricultural and Food Chemistry*, 45(3): 837–843.
Byrd, J.E., and Perona, M.J. 2005. Kinetics of popping of popcorn. *Cereal Chemistry Journal*, 82(1): 53–59.
Cretors, C. 2001. Popcorn products. In: *Snack Foods Processing*, 1st edition, eds. E.W. Lusas and L.W. Rooney, 385–420. Boca Raton: CRC Press.
Crumbaker, D.E., Johnson, I.J., and Eldredge, J.C. 1949. Inheritance of popping volume and associated characters in crosses between popcorn and dent corn 1. *Agronomy Journal*, 41(5): 207–212.
Deepak, S., Sharmila, T., Maheswari, M., and Shivaswany, M. 2020. Optimization of variables that influence puffing quality of hot air oven puffed quinoa. *International Journal of Pharmaceutical Research*, 12(3): 736–746.
Dofing, S.M., Thomas-Compton, M.A., and Buck, J.S. 1990. Genotype x popping method interaction for expansion volume in popcorn. *Crop Science*, 30(1): 62–65.
Dowd, P.F., and Johnson, E.T. 2010. Field incidence of mycotoxins in commercial popcorn and potential environmental influences. *Mycotoxin Research*, 26(1): 15–22.
Ertas, N., Soylu, S., and Bilgicli, N. 2009. Effects of kernel properties and popping methods on popcorn quality of different corn cultivars. *Journal of Food Process Engineering*, 32(4): 478–496.
Fast, R.B., and Caldwell, E.F. 2000. *Breakfast Cereals and How are They Made*, 2nd edition. Saint Paul: American Association of Cereal Chemists.
Finkel, E. 2020. State of the industry. Snacks 2020. Popcorn Sales Explode [Online]. Available: https://www.snackandbakery.com/state-of-the-industry-snacks.
Gökmen, S. 2004. Effects of moisture content and popping method on popping characteristics of popcorn. *Journal of Food Engineering*, 65(3): 357–362.
Hoke, K., Houška, M., Průchová, J., Gabrovská, D., Vaculová, K., and Paulíčková, I. 2007. Optimisation of puffing naked barley. *Journal of Food Engineering*, 80(4): 1016–1022.
Hoseney, R.C., Zeleznak, K., and Abdelrahman, A. 1983. Mechanism of popcorn popping. *Journal of Cereal Science*, 1(1): 43–52.
Hsieh, F., and Luh, B.S. 1991. Breakfast rice cereals and baby foods. In: *Rice. Volume II: Utilization*, ed. B.S. Luh, 177–194. New York: Van Nostrand Reinhold.
Jele, P., Derera, J., and Siwela, M. 2014. Assessment of popping ability of new tropical popcorn hybrids. *Australian Journal of Crop Science*, 8(6): 831.
Johnson, I.J., and Eldredge, J.C. 1953. Performance of recovered popcorn inbred lines derived from outcrosses to dent corn 1. *Agronomy Journal*, 45(3): 105–110.
Joshi, N.D., Mohapatra, D., Joshi, D.C., and Sutar, R.F. 2014. Puffing characteristics of parboiled milled rice in a domestic convective–microwave oven and process optimization. *Food and Bioprocess Technology*, 7(6): 1678–1688.
Karababa, E. 2006. Physical properties of popcorn kernels. *Journal of Food Engineering*, 72(1): 100–107.
Labuza, T.P., and Contreras-Medellin, R. 1981. Prediction of moisture protection requirements for foods. *Cereal Foods World*, 26: 335–343.
Lai, H.M., and Cheng, H.H. 2004. Properties of pregelatinized rice flour made by hot air or gum puffing. *International Journal of Food Science and Technology*, 39(2): 201–212.
Lin, Y.E., and Anantheswaran, R.C. 1988. Studies in popping of popcorn in a microwave oven. *Journal of Food Science*, 53(6): 1746–1749.
Lu, H.J., Bernardo, R., and Ohm, H.W. 2003. Mapping QTL for popping expansion volume in popcorn with simple sequence repeat markers. *Theoretical and Applied Genetics*, 106(3): 423–427.
Malleshi, N.G., and Desikachar, H.S.R. 1985. Milling, popping and malting characteristics of some minor millets. *Journal of Food Science and Technology*, 22: 400–403.
Mariotti, M., Alamprese, C., Pagani, M.A., and Lucisano, M. 2006. Effect of puffing on ultrastructure and physical characteristics of cereal grains and flours. *Journal of Cereal Science*, 43(1): 47–56.
Metzger, D.D., Hsu, K.H., Ziegler, K.E., and Bern, C.J. 1989. Effect of moisture content on popcorn popping volume for oil and hot air popping. *Cereal Chemistry*, 66(3): 247–248.
Mishra, G., Joshi, D.C., Mohapatra, D., and Babu, V.B. 2015. Varietal influence on the microwave popping characteristics of sorghum. *Journal of Cereal Science*, 65: 9–24.
Mishra, G., Joshi, D.C., and Panda, B.K. 2014. Popping and puffing of cereal grains: A review. *Journal of Grain Processing and Storage*, 1(2): 34–46.
Mohamed, A.A., Ashman, R.B., and Kirelis, A.W. 1993. Pericarp thickness and other kernel physical characteristics relate to microwave popping quality of popcorn. *Journal of Food Science*, 58(2): 342–346.
Murty, D.S., Patil, H., Rasada Rao, K.E., and House, L.R. 1982. A note on screening the Indian sorghum collection for popping quality. *Journal of Food Science and Technology*, 19(2): 79–80.

Pordesimo, L.O., Anantheswaran, R.C., Fleischmann, A.M., Lin, Y.E., and Hanna, M.A. 1990. Physical properties as indicators of popping characteristics of microwave popcorn. *Journal of Food Science*, 55(5): 1352–1355.

Pordesimo, L.O., Anantheswaran, R.C., and Mattern, P.J. 1991. Quantification of horny and floury endosperm in popcorn and their effects on popping performance in a microwave oven. *Journal of Cereal Science*, 14(2): 189–198.

Robbins, W.A., and Ashman, R.B. 1984. Parent-offspring popping expansion correlations in progeny of dent corn x popcorn and flint corn x popcorn crosses'. *Crop Science*, 24(1): 119–121.

Sakin, M.A., Gokmen, S., Yildirim, A., Belen, S., and Kandemir, N. 2005. Effects of cultivar type on yield and quality of popcorn (Zea mays everta). *New Zealand Journal of Crop and Horticultural Science*, 33(1): 17–23.

Scott, P., Pratt, R.C., Hoffman, N., and Montgomery, R. 2019. Specialty corns. In: *Corn: Chemistry and Technology*, 3rd edition, ed. S.O. Serna-Saldivar, 289–303. Cambridge: Elsevier (Woodhead Publishing), AACC-International.

Sekiyama, B.L., Ribeiro, A.B., Machinski, P.A., and Machinski Junior, M. 2005. Aflatoxins, ochratoxin A and zearalenone in maize-based food products. *Brazilian Journal of Microbiology*, 36(3): 289–294.

Serna-Saldivar, S.O. 2008. *Industrial Manufacture of Snack Foods*, 3–23, 105–123. London: Kennedys Publications Ltd.

Serna-Saldivar, S.O. 2010a. Manufacturing of breakfast cereals. In: *Cereal Grains: Properties, Processing and Nutritional Attributes*, 329–354. Boca Raton: CRC Press (Taylor & Francis Group).

Serna-Saldivar, S.O. 2010b. Manufacturing of cereal-based snacks. In: *Cereal Grains: Properties, Processing and Nutritional Attributes*, 355–394. Boca Raton: CRC Press (Taylor & Francis Group).

Singh, V., Barreiro, N.L., McKinstry, J., Buriak, P., and Eckhoff, S.R. 1997. Effect of kernel size, location and type of damage on popping characteristics of popcorn. *Cereal Chemistry Journal*, 74(5): 672–675.

Snack Food Association. 1992. *Section 802. Popcorn Manufacturing. In Corn Quality Assurance Manual*, 59–79. Alexandria: The Snack Food Association.

Solanki, C., Saha, D., Kudos, S.A., and Gupta, R.K. 2018. Effect of popping methods on popping characteristics of amaranth grain. *International Journal of Chemical Studies*, 6(2): 2779–2782.

Song, A., and Eckhoff, S.R. 1994a. Optimum popping moisture content for popcorn kernels of different sizes. *Cereal Chemistry*, 71: 458–460.

Song, A., and Eckhoff, S.R. 1994b. Individual kernel moisture content of preshelled and shelled popcorn and equilibrium isotherms of popcorn of different sizes. *Cereal Chemistry*, 71: 461–463.

Song, A., Eckhoff, S.R., Paulsen, M., and Litchfield, J.B. 1991. Effects of kernel size and genotype on popcorn popping volume and number of unpopped kernels. *Cereal Chemistry*, 71: 464–467.

Swarnakar, A.K., Srivastav, P.P., and Das, S.K. 2020. Puffing characteristics of preconditioned pressure parboiled brown (unpolished) rice in microwave oven and optimization of process parameters. *Journal of Agricultural Engineering*, 57(2): 127–137.

Sweley, J.C., Rose, D.J., and Jackson, D.S. 2011. Composition and sensory evaluation of popcorn flake polymorphisms for a select butterfly-type hybrid. *Cereal Chemistry Journal*, 88(3): 321–327.

Sweley, J.C., Rose, D.J., and Jackson, D.S. 2012. Hybrid and environment effects on popcorn kernel physiochemical properties and their relationship to microwave popping performance. *Journal of Cereal Science*, 55(2): 188–194.

Sweley, J.C., Rose, D.J., and Jackson, D.S. 2013. Quality traits and popping performance considerations for popcorn (*Zea mays* Everta). *Food Reviews International*, 29(2): 157–177.

Taylor, J.R.N., and Duodu, K.G. 2019. Traditional sorghum and millet food and beverage products and their technologies. In: *Sorghum & Millets: Chemistry, Technology and Nutritional Attributes*, 2nd edition, eds. J.R.N. Taylor and K.G. Duodu, 259–292. Cambridge: Elsevier (Woodhead Publishing). AACC-International.

United States Department of Agriculture. Popcorn, ingrained in America's agriculture history. Accessed: http://specialcollections.nal.usda.gov/popcorn-exhibit.

Walradt, J.P., Lindsay, R.C., and Libbey, L.M. 1970. Popcorn flavor: Identification of volatile compounds. *Journal of Agricultural and Food Chemistry*, 18(5): 926–928.

Weathermax, P. 1954. *Indian Corn in Old America*. New York: The MacMillan, Co.

Zapana, F., de Bruijn, J., Vidal, L., Melín, P., González, M.E., Cabrera, G., Williams, P., and Bórquez, R. 2020. Physical, chemical and nutritional characteristics of puffed quinoa. *International Journal of Food Science and Technology*, 55(1): 313–322.

Ziegler, K.E. 2000. Popcorn. In: *Specialty Corns*, ed. A.R. Hallauer, 199–234. Boca Raton: CRC Press.

Ziegler, K.E. 2003. Popcorn. In: *Corn Chemistry and Technology*, 2nd edition, eds. P.E. White and L.A. Johnson. Saint Paul: American Association of Cereal Chemists.

10 Production of Wheat-Based Snacks

Ximena Quintero Fuentes and Sergio O. Serna-Saldivar

CONTENTS

10.1 INTRODUCTION

Wheat (*Triticum aestivum*) has been used as one of the favorite human foods since prehistoric times and nowadays is still widely consumed. According to statistics of the FAO (2021) approximately 765.8 million tons were produced in 2019 and the average human being heavily depends on wheat for the supply of relevant amounts of calories and proteins. In 2013, the daily consumption of

DOI: 10.1201/9781003129066-10

wheat averaged 179 g, which provided 527 kcal and nearly 16 g of protein. Wheat is irreplaceable because its gluten proteins, gliadin and glutelins, yield a viscoelastic dough capable of retaining carbon dioxide or other gases produced by yeast or chemical leavening agents and water vapor produced during fermentation and baking. Most kinds of wheat are produced in temperate areas of China (133.6 million tons), India (103.6 million tons), Russia (74.5 million tons), the United States (52.3 million tons), France (40.6 million tons), Canada (32.3 million tons), Ukraine (28.4 million tons), Pakistan (24.3 million tons) and Australia (17.6 million tons) (FAO 2021). The United States, Canada and Australia have been traditionally the major exporters of this relevant cereal grain.

Common wheat is divided into two major classes: hard and soft. These classes are in most cases roller-milled into refined flours (Chapter 6). The hard wheat flour, with protein content ranging from 10.5 to 12.5%, is mainly used for the production of yeast-leavened bakery items including snacks. The soft wheat flour, with protein content varying from 7.5 to 10%, is chiefly utilized for the production of chemically leavened cookies, cakes and related products. The all-purpose or family flours with protein contents of 10 to 10.5% are generally obtained by blending hard and soft wheats prior to milling and are utilized for the manufacturing of soda crackers or biscuits, wheat flour tortillas, oriental noodles and donuts. Another application of wheat is as raw material for the production of value-added starch and vital gluten (Chapter 6). Durum (*Triticum durum)* wheat is mainly milled into yellow colored semolina for the production of short and long pasta products. The wheat milling procedure and the chemical composition of different sorts of flours are detailed in Chapter 6.

This chapter describes the major uses of wheat for the production of baked snacks around the globe. Emphasis is made on different types of crackers/saltines, pretzels, pita chips and toasted bread items. Dollar sales of crackers and pretzels increased between 5 to 7.0 % to 7.8 billion and 1.3 billion, respectively, based on data for the 52 weeks ending May 17, 2020 (Friedberg 2020, Lorenzi 2020).

10.2 BAKERY INGREDIENTS

Most wheat-based snacks are manufactured following conventional procedures of the bread, cookie and cracker industries and in some instances via extrusion cooking. For the specific case of toasted breads or crisps, the procedure adds an additional step of toasting or dehydration to yield a long shelf life fermented and convenient crispy product well suited to apply spreads or accompany with different sorts of cheeses and processed meats (Serna-Saldivar 2008).

10.2.1 Bread Ingredients

The fundamental ingredients needed for the production of breads are flour, salt, yeast and water. Most yeast-leavened formulations also contain sugars or sweeteners and shortening. In order to improve product quality and fulfill expectations of microbial and textural shelf life, other ingredients and additives such as malt, preservatives, oxidizing agents, emulsifiers, yeast food, nonfat dry milk and vital gluten are commonly utilized. However, for baked snacks it is not necessary to include antimicrobial agents because toasting or dehydration diminishes water activity to safe levels. On the other hand, the products, especially those formulated with high-fat content, will likely need artificial or natural antioxidants to prevent and retard rancidity (Serna-Saldivar 2010).

10.2.1.1 Hard Wheat Flour

Hard wheat flour is the most relevant ingredient in bread formulations because it affects the functionality and characteristics of the finished product. It also affects processability and requirements for other additives. The functionality of wheat flour depends on the quantity and characteristics of the gluten. For bread making, straight-grade or patent hard wheat flours are preferred because they contain higher protein content that yields a strong gluten network. The optimum water absorption, mixing or dough development time and bread characteristics are strongly associated with the protein content as well.

The bread industry has implemented several types of flour quality control measurements. The most common instruments to assess dough rheological properties are the mixolab, farinograph, extensigraph and alveograph. These methods are discussed in Chapter 19. Other important flour components are starch, pentosans, simple carbohydrates and polar lipids (glycolipids, phospholipids). Starch damage due to activation of diastatic enzymes in the field or due to improper storage and/or mechanical damage during milling greatly affects the functionality of the flour. Starch damage is usually assessed with the falling number apparatus or alternatively with the mixolab or viscoamylograph. In general terms, refined bread flours contain 11–13.5% protein so they can yield strong gluten networks, 0.6% pentosans and less than 0.4% ash. The flour when tested with the farinograph absorbs between 60–64% water, has a dough development time of 6–8 min, dough stability of 10–17 min and a mixing tolerance index of only 20 to 35 farinograph units (Serna-Saldivar 2010).

Whole wheat flours are produced by milling whole wheat or by adding different amounts of bran to refined flours (Chapter 6). These flours are gaining popularity due to the production of high dietary fiber foods including many snacks. From a positive viewpoint, whole wheat snacks contain more dietary fiber, fewer calories and more phytochemicals or antioxidants that favor human health. From the negative viewpoint, the phenolics and other related compounds associated with bran affect crumb and crust colors and flavor. These bakery items usually acquire brown colorations and produce denser loaves of bread due to the dilution of gluten proteins and the interference of the bran with the gluten network. Furthermore, whole wheat flours tend to form stickier and slacker doughs compared to counterparts obtained from refined flours and normally require less water and blending to properly develop the gluten network (Serna-Saldivar 2010).

10.2.1.2 Water

Water is fundamental because it solubilizes most of the baking ingredients, triggers the activity of yeast and diastatic enzymes associated with the barley malt and is the vehicle necessary for starch gelatinization and gluten formation. It is standard practice to blend hard and soft wheat flours with 62–66% and 54–58% water, respectively. The distribution of water in the dough is 31% with the gluten proteins, 46% with starch granules and 23% with pentosans or arabinoxylans (Serna-Saldivar 2010).

10.2.1.3 Yeast and Sour Dough Inoculums

Yeast (*Saccharomyces cereviceae*) is a biological fermenting immobile microorganism. The single yeast cells are nucleated and chemosynthetic. The size of the oval-shaped cells varies from 5 to 16 μm long and 4 to 8 μm wide. The quantity of fresh or dry yeast commonly added to bread formulations varies from 3 to 6% and 1 to 2%, respectively. Yeast reproduces asexually by budding and is capable of transforming simple sugars like glucose, fructose, mannose, galactose, sucrose, maltose and maltotriose into ethanol, carbon dioxide and energy. The carbon dioxide trapped by the gluten system produces the leavening force. The other by-products of its metabolism generate many key compounds like organic acids (acetic, butyric, lactic, succinic) and other flavor compounds such as aldehydes and ketones (acetaldehyde, formaldehyde, propionaldehyde, isobutylaldehyde, methylethyl ketone, isovaleraldehyde and 2-methyl butanol) which impart most of the typical bread flavor and aroma. During fermentation, the pH of the dough gradually drops due to the production of organic acids. This is important because the final bread pH is closely associated with its susceptibility to microbial deterioration. The chemical composition of fresh compressed yeast is 70% moisture, 13.5% protein, 12% soluble carbohydrates, 2% ash, 1.1% crude fat and 1.5% cellulose (Serna-Saldivar 2008, 2010). Yeast is typically added to bread formulations as fresh compressed or dehydrated with moisture contents of approximately 70 and 8%, respectively. Commercially, yeast is grown in a media with molasses, essential minerals, vitamins and small quantities of inorganic nitrogen salts in large aerobic reactors with strict controls of temperature, aeration and agitation.

After several days of reproduction, the cells are harvested via centrifugation and filtration and the resulting fresh yeast is pressed to physically remove water and produce fresh compressed yeast. Alternatively, the compressed yeast is first mixed with protective additives and then carefully extruded into small strips in preparation for drying. Drying is performed at approximately 45°C for 3–6 hr to drop the moisture to around 8%. In practical terms, one part of fresh compressed yeast is equivalent to 0.33–0.45 parts of the dry counterpart. Dehydrated yeast has the advantage of a long shelf life especially when vacuum packaged but its main disadvantage is that it takes longer to reactivate after reconstitution with water and also contains higher amounts of glutathione that have gluten reducing or weakening effect. The parameters to facilitate and regulate yeast activity during fermentation and proofing are the addition of yeast food, temperature of 26–30°C and relative humidity of 85% (Serna-Saldivar 2010).

Sourdough breads are predominantly fermented with bacteria that produce a more acidic pH compared to regular breads leavened with yeast. The main bacteria used to inoculate sourdoughs are *Lactobacillus planctarum, L. San francisco, L. fermentum, L. brevis, Leuconostoc mesenteroides* and/or *Streptococcus thermophiles.* These are mainly utilized for the production of variety breads, especially rye dark-colored products from rye (*Secale cereale*). Because of bacteria fermentation, sour breads possess unique organoleptic properties and internal crumb texture. Due to the higher acidic pH, these breads possess longer shelf lives and customers view them as more nutritious. The organic acids produced by the fermenting bacteria like acetic, lactic, hydroxiacetic, formic and pyruvic are the main contributors to the characteristic flavor profile. Acetic acid is considered the major acidic compound. Other fermenting compounds, such as free amino acids, also contribute to the typical flavor profile. Sourdoughs have different rheological properties compared to regular doughs due to their comparatively lower pH (5.2–5.4). As a result, these doughs usually have better oven-spring and yield breads with higher volumes and lower apparent densities (Kulp and Lorenz 2003, Serna-Saldivar 2010).

10.2.1.4 Salt

Salt or sodium chloride is also considered a flavoring agent and counteracts the flavor of sweeteners. One of its major effects is that it strengthens the gluten via the ionic modification of its proteins. Salt also slightly lowers the water activity or A_w and therefore acts as an antimicrobial agent. Most bread formulations are still supplemented with 1.5–2.5% salt (Serna-Saldivar 2010) although there is a new movement toward the reduction of the sodium intake via partial substitution with potassium chloride and addition of flavor enhancers.

10.2.1.5 Sugar and Other Sweeteners

Sugar and other sweeteners impart three basic functionalities in bread systems: flavor, color and shelf life via regulation of yeast activity. The most used sweeteners are table sugar or sucrose, invert syrup, glucose syrup and high fructose corn syrups. These carbohydrates affect flavor in two ways: The residual sweet flavor and more importantly the fermented products generated throughout fermentation. Reducing sugars also affect color development during baking via Maillard and caramelization reactions. The amount of sugar varies according to the bread type, generally French bread and baguettes, Chinese steamed breads do not contain added sugars in the formulation whereas table breads normally supplemented with 4–7% sugar based on flour weight (Serna-Saldivar 2010).

10.2.1.6 Shortening/Lard

Plastic shortenings or lards are used in bread formulations for several purposes. These fats act as a lubricant, lower dough stickiness and thus improve dough machinability. However, their main functionality is to improve crumb texture and decrease staling. The plastic fat forms thin films between the gluten network and other dough components lowering the rate of starch retrogradation, phenomenon responsible for progressive staling. The level generally used of these fats is 3–3.5% in table breads and up to 10% or more in sweet pastries (Serna-Saldivar 2010). Fractionated palm stearin and zero *trans* plastic shortenings (Chapters 3 and 7) with good oxidative stabilities are gaining popularity to produce healthier products and declare green food labels.

10.2.1.7 Malt

Most bread formulations contain barley malt flour. Diastatic malts with amylolytic enzyme activity (α and β amylases) are added to hydrolyze mainly mechanically damaged starch (3–4% of the total starch granules) of flours. The amylases gradually degrade the starch molecules yielding dextrins, maltose and other fermenting sugars. Maltotriose, maltose and glucose are yeast substrate and thus contribute to bread volume and flavor. The malt added to bread formulations varies according to the diastatic activity or falling number of the flour. Generally, malt is added to bread formulations from 0.2 to 1% (Serna-Saldivar 2010). Nondiastatic malt flours are also used as flavoring and coloring agent in the fabrication of bakery items including snack foods. These malts are thermally treated to deactivate their enzymes and promote Maillard and caramelization reactions that generate flavorful and colored compounds. Both nondiastatic malt flours and syrups are commercially available to supplement and enhance a malty flavor and color in snack foods. The dark-colored malt syrups rich in maltose possess about half the sweetness of refined table sugar.

10.2.1.8 Vital Gluten

Vital gluten is mainly composed of the wheat endosperm proteins gliadin and glutelins and is frequently used in whole wheat or composite bread formulations to counteract the negative effects of bran and addition of nongluten forming flours such as rye, oats and soybean. The vital gluten is commonly obtained from the Martin wheat wet milling procedure detailed in Chapter 6. When used, the vital gluten with approximately 70% protein is added to bread formulations from 1 to 3% based on flour weight (Serna-Saldivar 2010).

10.2.1.9 Dough Conditioners

Dough conditioners or emulsifiers are chemical compounds that have hydrophilic and lipophilic groups or ends. They are widely used because the polar and non-polar groups bind with dough components diminishing the rate of staling. Staling is delayed because the polar side of the emulsifier binds to the outer part of the amylose whereas the lipophilic side binds to the inner part of the amylose helix. The most popular emulsifiers are stearoyl-2 sodium lactylate, DATEM, monoglycerides and lecithin. They are usually added in levels varying from 0.01 to 0.3% based on flour weight (Serna-Saldivar 2010).

10.2.1.10 Oxidation Agents

Oxidation agents have little or no effect on yeast activity; however, they affect the dough rheological properties and therefore their gas retaining capacity. Oxidation agents promote the formation of disulfide bridges resulting in stronger gluten networks. Breads treated with oxidative compounds usually have a better crumb texture. The most common and popular oxidative agent is potassium bromate although several countries around the globe prohibit its use because toxicological tests have shown that it produces cancer in laboratory animals. When used, the level added is 20 ppm based on flour weight and is generally accompanied with ascorbic acid (30–100 ppm). Ascorbic acid and bromates act synergistically improving the functionality of low-protein flours. Other oxidizing agents used are azodicarbonamide and cupric sulfate (Serna-Saldivar 2010).

10.2.1.11 Mold Inhibitors

From the microbial viewpoint, the shelf life of bread products depends on the growth of different genus of molds (*Aspergillus*, *Penicillium*, *Rhizopus* and *Fusarium*). Propionate salts are widely used to prolong shelf life by inhibiting mold growth without affecting yeast activity. The organic acids produced during fermentation lower the dough pH acting synergistically with these antimycotic agents. The dehydration of bread for production of toasted products lowers the water activity so these products do not need addition of mold inhibitors (Serna-Saldivar 2010).

10.2.1.12 Yeast Food

Yeast food consists of mineral salts and nitrogenous compounds essential for yeast growth. The mixture usually contains calcium acid phosphate, dicalcium phosphate and ammonium sulfate. These compounds are usually blended with starch or flour so they can be added in higher amounts with low variability (for instance 0.5% based on flour weight) (Serna-Saldivar 2010).

10.2.1.13 Nonfat Dry Milk

Powder milk is added to some bread formulations (1–2.5 % based on flour weight) to improve flavor and crust color. It is worth mentioning that yeast does not have the capacity to break down lactose. Dry milk also affects positively the nutritional value of bread because it complements the amino acid profile of wheat proteins and contains significant amounts of calcium, magnesium and other essential nutrients (Serna-Saldivar 2010).

10.2.2 Ingredients for Cookies and Crackers

Besides the array of ingredients described above, cookies and crackers are normally formulated with soft or all-purpose flours and raised with chemical leavening agents or baking powders.

10.2.2.1 Soft and All-Purpose Wheat Flours

Soft and all-purpose wheat flours are the most functional key ingredients for cookies and crackers, respectively. Soft wheat flours had significantly smaller granulation, protein and ash contents and less mechanical starch damage compared to hard and all-purpose flours. Therefore, soft wheat flours absorb less water and require lower dough mixing to achieve optimum gluten development. The most relevant quality characteristics are cookie spread factor, water absorption (regular and alkaline), pH, dough strength and extensibility (Chapter 19). Most soft wheat flours have lower gluten strength but are more elastic compared to hard wheat flours.

Soft and all-purpose whole wheat flours are produced by milling whole kernels or by supplementing different amounts of bran to refined flours containing 8.5 and 10.5% protein respectively. These flours are employed for the production of high dietary fiber products that are viewed as healthful (Serna-Saldivar 2010).

10.2.2.2 Baking Powders and Chemical Leavening Agents

Most products elaborated with soft wheat flours are chemically leavened with baking powders or chemical agents that release carbon dioxide after hydration. The CO_2 is retained by the dough in miniscule bubbles during both dough mixing and baking affecting crumb volume, texture and mouthfeel (Faridi 1994, Faridi et al. 2000).

A chemical leavening system contains two functional components: a leavening base and an acid. Sodium bicarbonate is the most widely used and is chemically neutralized by the acid. These chemical agents yield gas and the rate of gas production is closely associated with the dough pH. Normally, higher rates of CO_2 production are observed in more acidic doughs. The most common leavening agents are monocalcium phosphate (MCP), dicalcium phosphate (DCP), sodium aluminum sulfate (SAS) and sodium acid pyrophosphate (SAP). These agents are classified into slow, fast and double acting. Slow-acting agents (DCP, SAS and SAP) need elevated temperatures (baking stage) to yield most of the gas, whereas fast-acting agents (MCP) generates most of the gas at ambient temperature (mixing and dough proofing stages). Double-acting agents are produced by blending leavening agents and are the most popular because they release carbon dioxide at both room and baking temperatures.

Most commercial baking powders are manufactured from a blend of chemical leavening agents, acidulants and an inert compound generally consisting of starch. Both the acid and acid salts are the key elements to control the release of carbon dioxide. An important property of chemical leavening agents is the neutralization value that is defined as the parts of sodium bicarbonate that neutralizes 100 parts of the chemical leavening agent under controlled baking conditions. For example, MCP possesses a neutralization value of 80 whereas DCP of only 33. The sodium aluminum salts have values of 100. The acid is relevant because it controls the release of carbon dioxide and affects flavor. The most common acids are tartaric and gluco-delta-lactone. The inert agent minimizes the interaction between leavening agents and the acid. Examples of fast-acting baking powders are: tartaric acid 6%, tartar cream 44.9%, sodium bicarbonate 26.7% and starch 22.4% or MCP 33.4%, sodium bicarbonate 26.7% and starch 39.8%. A regular slow-acting formulation contains 40.4% SAP, 30.6% sodium bicarbonate and 29% starch (Faridi 1994, Faridi et al. 2000). There are many examples of double-acting baking powders. Regular formulations consist of MCP (6.7 to 13.2%), SAS (19 to 21%), sodium bicarbonate (26.7%) and starch (40 to 45%) (Serna-Saldivar 2010).

10.3 PRODUCTION OF TOASTED BREAD

The industrial baking process to obtain toasted bread is practically identical to the production of fresh bread with the additional operation of bread toasting or drying. There are many bread formulations generally produced from refined wheat flour, whole wheat flour or composite flours produced from mixtures of wheat flour with oat (*Avena sativa*), rye or various crushed grains. Sesame (*Sesame indicum*) and poppy (*Papaver somniferum*) seeds are often used as toppings. Typical formulations to obtain yeast-leavened and sourdough breads are depicted in Table 10.1. The nutritional compositions of these toasted/dehydrated breads are detailed in Chapter 20.

10.3.1 Baking Processes

There are basically three industrial procedures to obtain pan breads: straight dough, sponge and continuous. These processes demand different flour requirements, additives, labor and yield breads with somewhat different features (Serna-Saldivar 2010).

10.3.1.1 Straight-Dough Process

Breads obtained by the straight-dough process are normally produced by small bakeries and require the highest labor input. It consists of blending dry ingredients, dough mixing, fermentation, molding, proofing and baking. The straight-dough procedure is regularly used for production of French bread and baguettes that in some instances are dehydrated to produce shelf-stable toasted slices or sticks (Figure 10.1). The straight-dough baking process starts when wheat or composite flours are mixed with the other ingredients and water until a fully developed dough is obtained (Calvel 1987). The correct water addition and mix time are critically important to obtain good quality bread. Generally, most breads require from 60 to 67% water. The dough is manually or mechanically divided into portions of equal weight or size for fermentation. Pieces of dough are placed inside fermentation cabinets with temperature (28–36°C) and relative humidity controls (> 85%). Doughs are fermented for different periods depending on the sort of bread desired. For French bread the first dough fermentation is short because the yeast does not have enough substrate or fermenting sugars to sustain fermentation. For pan bread which contains around 5–6% sugar, the dough is fermented for about 1.5–2 hr before final proofing. Doughs are manually or mechanically punched by hand or mechanically molded in preparation for panning and final proofing. Proofing times also vary according to the type of bread and proofing temperature but most doughs are proofed for 45–90 min or until achieving the desired proof height. The dough is finally placed in an oven set at temperatures of 200–230°C for 20–25 min. Some ovens are equipped with a steam injection to produce better crust texture. During this operation, French breads and baguettes are cut with a knife so the

TABLE 10.1
Bakers Formulations Normally Employed to Manufacture an Array of Yeast-Leavened and Sourdough Breads[a]

	Bakers Formulation, %							
	French Bread		Pan Breads				Sour Breads	
Ingredient	**Regular**	**Whole Wheat**	**White**	**Whole Wheat**	**Oat**	**Rye**	**Pumper-nickel**	**Whole Wheat**
Hard Wheat flour	100		100	50	60	60	24	
Whole Wheat flour		100		50				100
Oat flour					40			
Rye flour						40	76	
Water	60–64	60–65	66	66	65	65	71	60
Sugar			6	5	5	6		6
Honey or fructose syrup				2	2			
Shortening			3.5	3.5	3.5	3.5	1	2
Fresh yeast[b]	5–6	5	5	5	6	5		3
Sourdough							10	10
Salt	2	2.2	1.5	1.5	2	1.5	2–5	2
Nonfat dry milk			0.5–1	2		2		
Vital gluten		1.5		2	2	2		
Diastatic malt	0.5		0.2	0.2	0.2	0.2		0.5
Nondiastatic malt							1	
Yeast food			0.5	0.5	0.5	0.5		
Calcium propionate			0.2	0.2	0.2	0.2		
Emulsifiers or dough conditioners			0.1	0.1	0.2	0.1		
Potassium bromate/ ascorbic acid	0.2/0.1		0.02/0.1	0.02/0.1	0.02/0.1	0.02/0.1		

[a] From Calvel (1987) and Serna-Saldivar (2010).
[b] Dry-yeast can be used instead of compressed fresh yeast at a concentration of 1-2%.

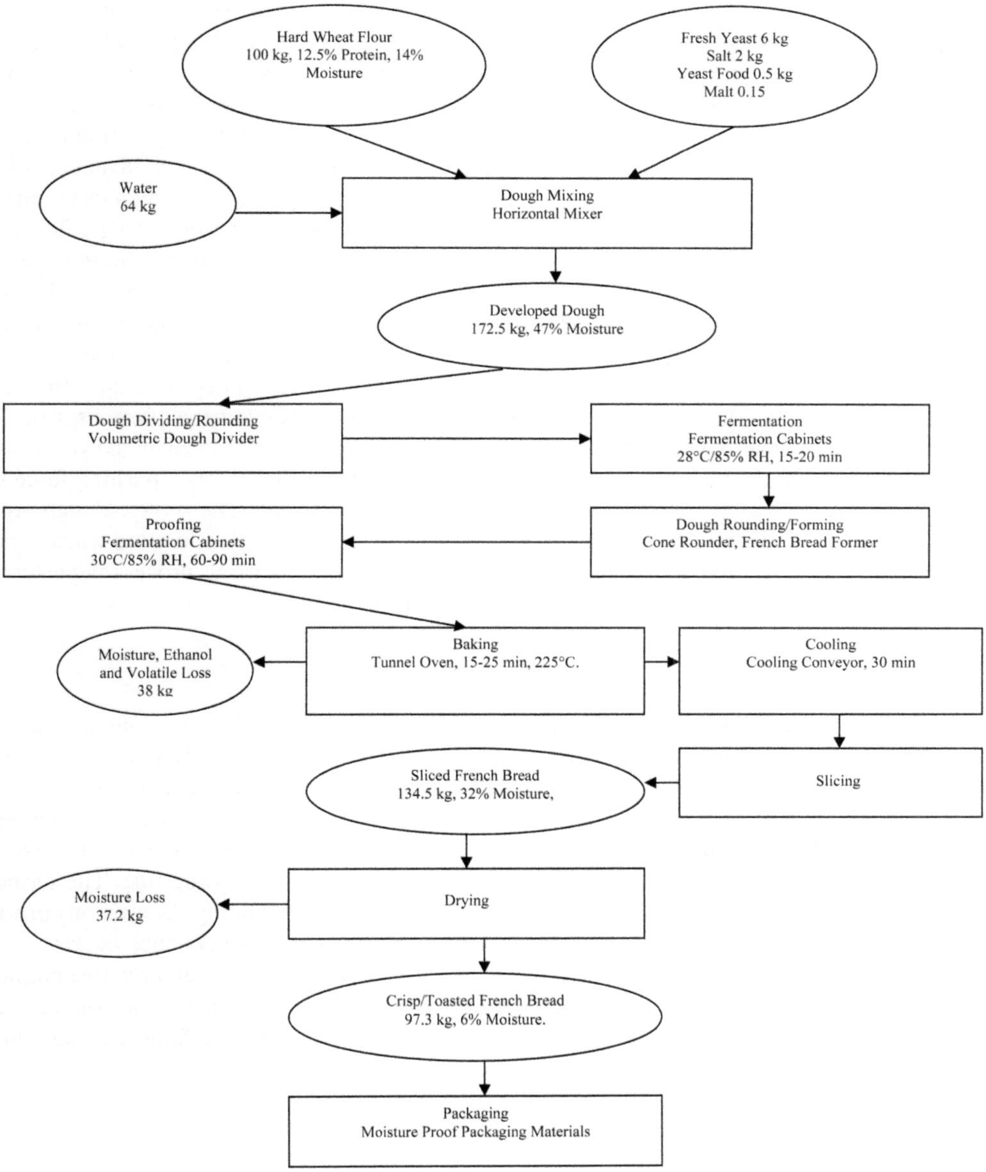

FIGURE 10.1 Flowchart and mass balance for the production of toasted French bread using the straight-dough procedure.

crust can break upon baking. Resulting breads are allowed to cool at room temperature, sliced if required and further toasted to obtain slices with approximately 6% moisture. From 100 kg flour are obtained about 97.5 kg of toasted French bread or sticks with 6% moisture content (Figure 10.1).

10.3.1.2 Sponge-Dough Process

Undoubtedly, the most common commercial baking procedure is the sponge-dough process that is subdivided into regular and liquid sponges. The sponge-dough procedure starts dividing the flour into two lots: the sponge and the dough flour. The sponge dough is obtained after mixing the sponge flour with water, yeast and yeast food. The resulting dough is fermented in troughs at 28–30°C in a fermented room with strict relative humidity controls for 3–5 hr. At the completion of the fermentation, the sponge is lifted in troughs to a horizontal mixer where the dough flour and rest of the ingredients and water are blended until fully developing the dough. The mix time varies according to flour characteristics and type of mixer but it usually takes 8–12 min. The resulting dough is then divided into uniform pieces, fermented, molded and panned in preparation for proofing. After the last fermentation period, the dough is ready for baking. The main advantage of the sponge procedure is that the dough attains its optimum rheological properties after shorter mixing times, the inplant fermentation schedules are shortened, and the bread acquires stronger flavor and better crumb texture. The dough is only fermented for 30–40 minutes before punching degassing in a conical rounder for panning and final proofing. The proof time varies from 50 to 65 min or until achieving the desired proof height. The fermented and molded dough is normally baked in continuous ovens set at temperatures ranging from 200 to 230°C. During this thermal operation, the dough loses water, the ethanol volatizes, the yeast is inactivated, the gluten denatures, the starch gelatinizes, and the bread increases its height and volume due to the phenomena of oven spring. The oven spring is due to the production of CO_2 during the initial stages of baking when the yeast is still alive, the expansion of the trapped CO_2, and the water evaporation that exerts a pushing force that increases height and volume especially during the first 5 min of baking. The sponge-dough procedure described before can be easily adapted for the production of a variety of toasted whole wheat, composite (rye, oats etc.) and sourdough breads (Figure 10.2). The toasting process is performed after baking at lower temperatures in order to almost completely dehydrate the slices of bread and attain the desired color and texture of the toasted breads.

10.3.1.3 Continuous Baking Procedure

Continuous baking processes are highly mechanized saving labor requirements, plant space and processing time. The continuous baking system is based on the utilization of a liquid-acting ferment named brew and liquid ingredients that are pumped into a high-speed mixer. The use of high-speed mixers demands the use of cold-tempered water to compensate heat generated due to friction or mechanical work. In addition, these formulations usually contain more yeast and sweeteners. The aim is to lower fermentation time and to increase the output of the processing line. The fermenting dough is divided, molded, panned and proofed for 90 min before baking. Continuous proffers generally operate at higher temperatures than the other baking systems to enhance the rate of fermentation. Baking is performed in continuous ovens with strict temperature and relative humidity controls. Table bread is baked at 220°C for 20–25 min. The loaves of bread are cooled on racks and sliced before toasting and packaging. This system demands the best quality flour and ingredients and quality controls of raw ingredients and process.

10.3.2 Toasting

Fresh breads are discarded by the consumers when they present mold growth or stale. These sorts of deterioration cause huge losses. Bread is a product that has a limited microbial shelf life because it contains sufficient water or A_w for mold growth. In most instances, bread is deteriorated by amylolytic molds like *Rhizopus, Penicillum, Aspergillus* or *Fusarium*. On the other hand, bakery items rapidly stale due to the phenomenon of retrogradation. The French and related breads rapidly stale because their formulations do not contain shortening and emulsifiers. Toasted or crispbreads have prolonged microbial shelf life because they generally contain approximately 6% water equivalent to an A_w of less than 0.5 and a pH below 5. Under these conditions, it is impossible for microorganisms to prevail. The toasting process enhances staling and the crispy and brittle texture that characterize crispbreads. Fresh breads usually contain from 30 to 34% moisture and they should be carefully dehydrated to produce crispbreads. The rate of product dehydration depends on the oven settings, and product volume or dimensions. Toasting is different from baking because the baked product is dehydrated at a lower temperature. The presence of residual reducing sugars can greatly affect color

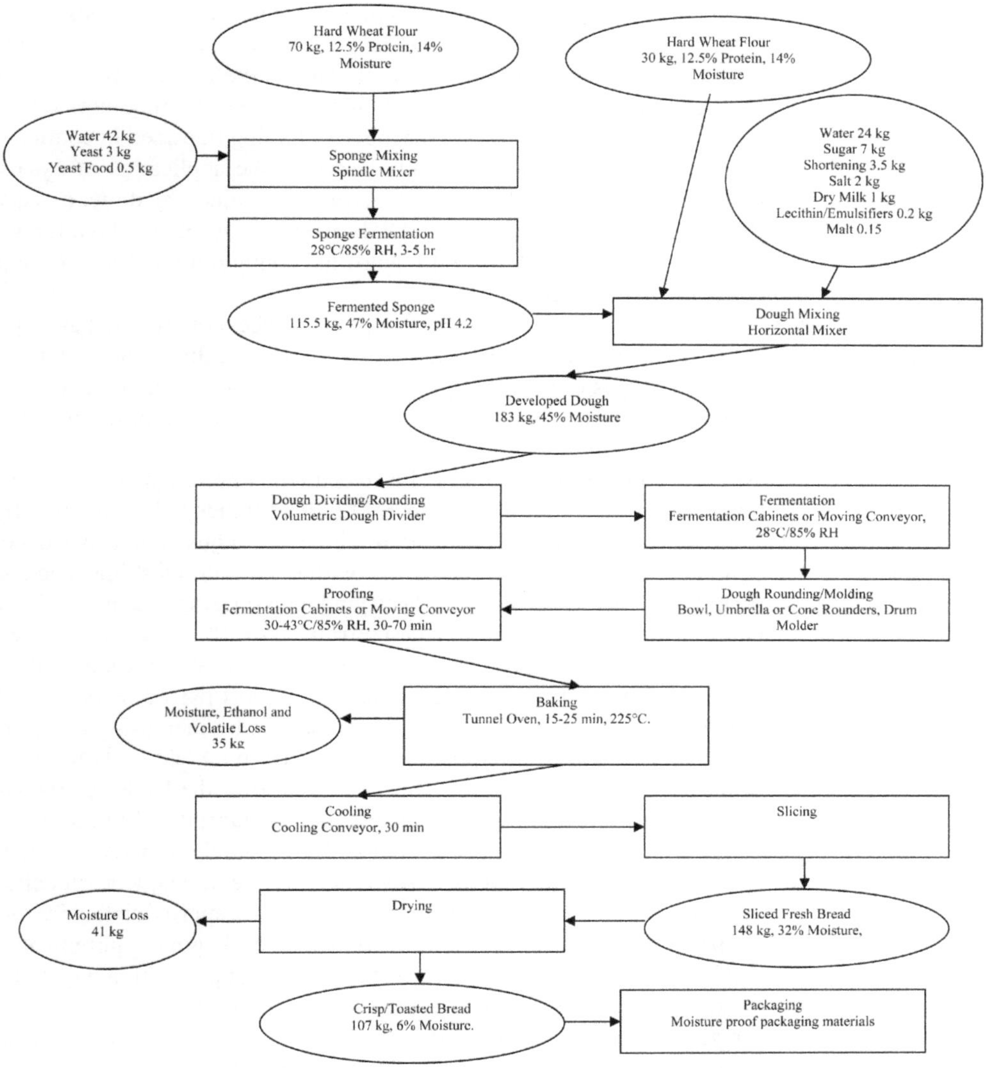

FIGURE 10.2 Flowchart of the manufacturing procedure used for the production of toasted breads by the sponge-dough baking process.

if the oven temperature is not properly controlled. For instance, sliced French breads or baguettes can be dehydrated at higher temperatures than table bread that contains around 6% sugar in the formulation. Toasted breads are usually dehydrated at temperatures ranging from 70 to 150°C. It is common in the industry to produce smaller loaves of sliced bread in order to enhance drying. Crispbreads are allowed to cool and equilibrate at room temperature before packaging with materials that possess an excellent barrier against air moisture and oxygen (Chapter 4). These products should be protected from lipid oxidation or rancidity because they are expected to last longer on the shelf (Serna-Saldivar 2008).

10.4 PRODUCTION OF EXTRUDED CRISPBREADS

The term crispbread is a misnomer because the product is not fermented with yeast and the product texture is greatly influenced by the starch containing raw materials rather than the wheat gluten.

Crispbreads produced with extrusion technology have created a whole new array of snacks that are low in calories and widely accepted especially in Europe. The possibility for a wide variation in the design of the extrudate has been exploited. The European market has the traditional rectangular-shaped crisps but new filled forms such as U shape and hollow tubes are becoming more relevant. In this instance, the extrusion line replaces the large and expensive baking line used in commercial crispbread manufacturing. Meuser and Wiedmann (1989) described the application of extrusion cooking to produce crispbread analogs. The typical crumb texture is obtained by direct expansion caused by vapor water when the extrudate leaves the slit die. Existing pieces of crispbreads cool as the water evaporates and shrink a little and solidify into their final configuration. The extrudate is transversally cut into slices before packaging.

Figure 10.3 depicts a flowchart of the crispbread extrusion process. The more relevant unit operations are the twin-screw extrusion, cutting, baking and toasting (Serna-Saldivar 2008). There are many recipes to produce different types of crispbreads. The most common formulation consists of wheat flour, salt, sugar, milk, egg solids, fat/oils and water. The nutritional composition of crispbread is detailed in Chapter 20.

The process starts when a pneumatic mixer thoroughly blends the dry raw materials that are pneumatically fed via a rotary discharge valve into a hopper. The blend is then fed volumetrically to the conveyor into the inlet of the twin extruder (Figure 10.3). The rate is adjusted to achieve a constant mass flow. The twin-screw extruder is designed for corotating and intermeshing. The barrel has five different zones: feed, liquid metering, venting, plastifying, pressurizing and die. In the feed zone, the moisture content of the incoming blend is increased from 14 to 19%. The liquid metering should be adjusted to deliver the water to achieve the target moisture, otherwise the extrudate will not have the desired texture and quality. The barrel has a deaeration port placed before the plastifying area. The deaeration reduces the degree of filling of the screws by increasing the product density. In the plastifying zone, the screws are equipped with reverse-pitch elements to increase residence time and ensure the air-free plastification. In this particular area, the feedstock increases its temperature to about 100°C. The plastified dough is subsequently transported to the pressure buildup barrel area where the dough raises its temperature to 180°C for only a matter of few seconds. The melted dough immediately expands and releases water vapor and heat when it exits the slit configured die orifices. The extrudate temperature drops to 100°C and approximately 9% water flashes out as steam. The expansion rate is carefully controlled to meet packaging requirements and package weight. The expanded strips of crispbreads are threaded and gripped with a pair of take-off rolls and cross-cut into slices. The cut slices with approximately 9% moisture are then toasted in an electric infrared oven set at 400°C for 10 sec where they will further decrease their moisture to about 6%. Toasting enhances the textural, browning color and organoleptic properties of crispbreads. Resulting crispbreads are finally equilibrated at room temperature in cooling conveyors and immediately packaged. According to Meuser and Wiedmann (1989) the extrusion, barrel heating/cooling and toasting operations demand 35, 14 and 36% of the total energy expenditure. From 100 kg flour are obtained about 91.5 kg of crispbread with 6% moisture (Figure 10.3). The quality of crispbreads is related to the ingredients of the formulation and their composition as it occurs for directly expanded products. Addition of wheat bran to produce lower calorie and high-fiber crispbreads reduces expansion rate and product thickness. There are products containing 65–85% wheat bran in the European market; nevertheless, these high-fiber extrudates are harder and more brittle (Meuser and Wiedmann 1989, Serna-Saldivar 2008).

10.5 PRODUCTION OF PITA CHIPS

Pita breads are circular, leavened two-layered flat breads that have their origin in the Middle East and are also called Arabic bread, balady, shamy, Syrian bread or pocket bread. Pita bread is made with a lean formulation consisting of flour, water, bakers' yeast and salt (Gocmen et al. 2009). Traditional pita breads are made by the sourdough process, but alternatively, they can also be

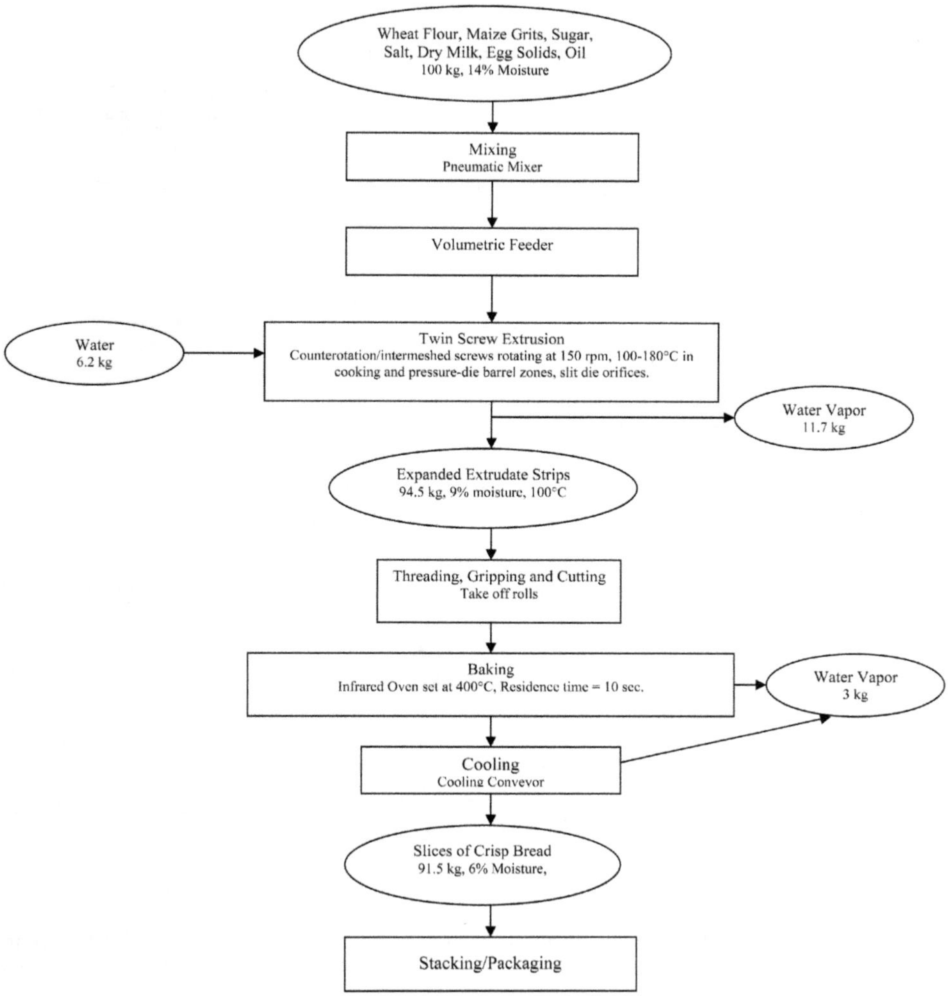

FIGURE 10.3 Flowchart of the extrusion manufacturing process of crispbreads.

produced with yeast (Edwards 2013). In pita bread production, all the ingredients are mixed to form a fully developed dough with a temperature of 24.5–25.5°C and fermented for about 1 hr. The dough is then scaled, rounded by means of a rounding belt, and given an intermediate proof of 15–20 min (Gocmen et al. 2009). The relaxed dough pieces are sheeted and then undergo a final proof of 30 min in a cabinet maintained at a temperature of 30°C and relative humidity of 95% (Gocmen et al. 2009). The pieces are then subjected to a short time, high-temperature cook of 1.5–2 min at a very high baking temperature of 350–400°C. Under these conditions, a crust is formed immediately and the pieces balloon as the air, water, ethanol and carbon dioxide expand. When the bread is removed from the oven it cools and collapses (Edwards 2013).

Pita chips are generally manufactured industrially by splitting and cutting pita bread loaves into chip-sized pieces (Figure 10.4). One approach involves pressing a dough ball between two hot plates to form the pita loaf, baking the loaf and then cutting the loaf into smaller chip sizes. This approach is referred to as a dough ball press method followed by splitting and cutting the bread loaves (Bernard et al. 2014). Another approach is to cut a sheet of dough longitudinally into long strips (Goglanian 2001). These strips are run through an oven, thereby producing a tube-shaped

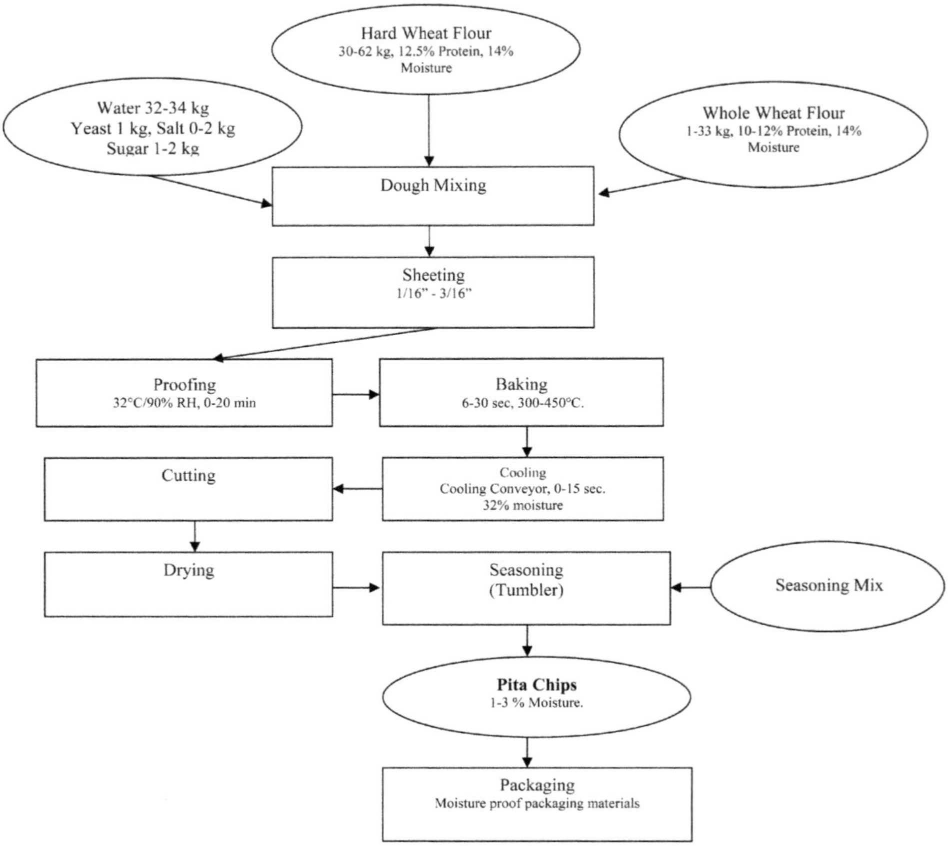

FIGURE 10.4 Flowchart of the manufacturing process for the production of pita chips.

bread product. These sections are cut into chip shapes, thus making chips of both the top and the bottom of the tube.

An improved continuous method for making pita chips involves mixing wheat flour, sugar, salt, yeast and water to form a sheetable dough prior to the sheeting step involving two or more sheeter roller pairs such that the thickness of the sheet is gradually reduced to about 0.2–0.5 cm (Bernard et al. 2014). The sheet is then proofed to allow the dough to rise in a warm, humid environment for a certain time before further processing. A proofer box is a chamber that is humidity and temperature-controlled, for example, at about 50% relative humidity and about 32°C. Proofing relaxes the dough and allows the yeast to work. A softer textured product, for example, typically needs a longer proofing time than a harder textured product. The dough is then sheeted into a continuous dough sheet, which is then cut longitudinally into continuous dough strips (Barnett and Joseph 2015). The dough strips are then formed into continuous bread loaves in a cooking oven with temperatures in the range of about 300° C and about 600°C for 6 to 60 seconds, depending on product thickness and heat intensity. In the oven, the dough strips puff up and form a cavity in the center of each strip (Bernard et al. 2014). This results in tubes of bread. Upon exiting the oven, the bread tubes contain about 32% moisture (Bernard et al. 2014). If a single-layered pita chip is desired, the tube is next subjected to a splitting step. This splitting step separates the flattened tube into a top and bottom strip (Quintero Fuentes et al. 2015). The longitudinal strip is next cut using a cutting roller that forms individual chip-sized pieces. Finally, these individual pieces are dried and seasoned prior to packaging (Quintero Fuentes et al. 2015).

10.6 PRODUCTION OF SALT BISCUITS OR CRACKERS

According to Misra and Tiwari (2014) biscuits are classified into short or soft dough biscuit, hard dough biscuits and fermented or nonfermented crackers. Depending on the method of production, soft dough biscuits are further classified as rotary molded or wire cut. The difference in the viscoelastic properties of soft and hard dough is mainly due to the variation in the level of sugar and fat as well as mixing conditions. Hard dough normally contains a lower level of fat and/or sugar, while short dough contains a higher amount of fat and sugar. Biscuits are also classified on the basis of taste into three categories. Sweet biscuits or cookies, semisweet or tea biscuits and salt biscuits or crackers which are usually manufactured from fermented doughs. Independently of the type of products, biscuits usually contain less than 5% moisture and low A_w, which make them microbial stable.

Salt biscuits or crackers are considered one of the main wheat-based snacks and are mainly used as the base to apply various kinds of spreads. They are rarely consumed alone as it does not occur with most of the other salted snack foods. Dollar sales of crackers increased about 6%–7.8 billion for the 52 weeks ending May 2020 (Friedberg 2020). Sales are high because these products are perceived by the consumers as low-fat and calorie and have versions of high-fiber items. The nutritional composition of various types of crackers including Graham crackers is detailed in Chapter 20.

Manley (2000a,b,c) classified crackers into three major categories: cream, soda and savory crackers. Their main features and manufacturing procedures are summarized in Table 10.2.

Crackers or hard biscuits are considered a hybrid bakery item because it uses bread and cookie manufacturing procedures and the product is leavened with yeast and chemical agents. Saltines are the major cracker category and they are prepared from yeasted doughs that contain little or no sugar. Table 10.3 depicts the typical formulation to prepare these crackers. The sponge procedure is normally followed and when the dough is prepared chemical leavening agents are added. Some industrial processes last more than 20 hr because the sponge phase lasts from 12–24 hr. Crackers are formulated with stronger mixing and higher protein flours than are cookies and are formed and baked with a typical sheeting cookie line. The ideal flour usually contains from 10 to 10.5% protein and is classified as intermediate or all-purpose flour (Manley 2000a, Serna-Saldivar 2010, Misra and Tiwari 2014).

The industrial manufacturing process starts when batches of a mixture of hard and soft wheat or all-purpose flours containing about 10.5% protein are mixed with yeast and water to yield sponges that are fermented at 28°C and 85% relative humidity in proof room (Figure 10.5). The sponge usually contains 60% of the total flour used in the manufacturing process. The Spindle mixer is widely used because it is well suited for mixing and the dough trough also stores the dough during the fermentation steps. The Spindle mixer consists of a removable wheeled dough bowl that can hold from 680 to 1,000 kg that mixes the dough with two or three shafts or spindles that rotate from 15 to 45 rpm (Moreth 1994). Sponges are left to ferment for approximately 20 hr and produced at certain programed time intervals or schedules with the aim to produce enough dough to feed the continuous and automated cracker manufacturing line. Interestingly, natural lactic acid bacteria adhered to the surfaces of the trough and naturally present in previous dough batches develop during the long fermentation and also contribute to the typical cracker flavor and acid development. The initial sponge pH is around 7 and drops to a range of 4.2–4.5 during fermentation. Once the sponge is ripped, it is mixed with the rest of the flour (generally 40%), other ingredients and sodium bicarbonate to produce a dough with a pH of 7–7.4. The resulting dough is placed again in the fermentation room for 2–5 hr fermentation. From this step the process becomes continuous.

Crackers are primarily formed with sheeting cutting forming equipment (Figure 10.5). This is the most popular method of forming cookies. The fermented dough is first passed through a three-roll sheeter and then gradually reduced with two other reduction roll stations. After the first coarse of lamination, the dough is lapped/layered and then gradually sheeted to about 2–3 mm. Layering is essential for the final cracker texture and works required to masticate the product. The sheet of

TABLE 10.2
Classification and Comparison of Main Features and Manufacturing Processes Among the Three Major Types of Crackers

Type of Cracker	Main Features	Manufacturing Process
Cream	Cream crackers are traditional hard products with low moisture content. The term cream is misleading because this cracker does not contain any cream or dairy products in the formulation. These biscuits are unsweetened based on a formulation of wheat flour, yeast, fat and salt (Manley 2000a).	The process starts with a very soft dough, which is usually made from just flour, yeast, salt and water. The wheat flour with about 10.5 to 11% protein is used to produce an sponge that fermented for 12–16 hr to. Then, the ripe sponge is mixed with the rest of the flour, fat, water, salt, sodium bicarbonate, malt extract and other minor ingredients to form a stiff dough that is further fermented for several hrs. Normally, the dough is sheeted and reduced through three or four pairs of rolls to a final thickness suitable for cutting. The dough is usually cut at about 2 mm thickness. The scrap dough, which represents about 10–120% of the total dough weight, is recirculated. Finally the dough is baked into crackers in belt ovens to optimize texture, final moisture content of the finished product and color. During baking, the laminations lift apart irregularly and this is how the blisters may form on the surfaces. Normally the baking temperatures and time are from 250 to 300°C for 3 to 5.5 minutes. The temperature profile is lower at the end of baking in order to prevent checking and breaking (Manley 2000a).
Soda	Soda crackers or saltines derived from cream biscuits and are widely produced in the United States since at least 1840. There are many similarities between soda and cream crackers. However, soda crackers undergo and alkaline reaction after baking which affects flavor, texture and surface finish. Typically, soda crackers are square configurations of about 5 × 5 cm and 4 mm thick with about 2.5% moisture. The formed pieces of dough are usually perforated with nine docker holes in three rows of three in order to prevent excess blistering (Manley 2000b).	The process of soda crackers is very similar to cream crackers. In short, the dough is typically fermented in two distinctive stages with a wet sponge lying for 18 hours followed by a dough stage which is left to ferment for about 3–4 additional hrs. During yeast fermentation, lactic acid bacteria normally grow affecting flavor and pH. When the sponge dough is blended with the rest of the ingredients to produce the final dough soda or sodium bicarbonate is added increasing the pH to levels of 7.2 to 8.0. The dough is allowed to ferment for several hours before sheeting and cutting. Typically soda crackers are baked for 2.5–3 minutes at temperatures around 300°C at the first zone and then at 250°C at the end of the oven. During cooling additional moisture (about 1.5%) is lost to around 2.5% (Manley 2000b).
Savory	Hard crackers that are normally salted, flavored and fat sprayed after baking. These crackers are characterized by very open textures and soft eating mouth feel. Normally they are manufactured as simple units or cream sandwiched with a savory, nonsweet, cream often based on cheese powder (Manley 2000c).	Savory crackers are produced from well-developed doughs modified with sodium metabisulfite and/or proteolytic enzymes aimed to break gluten disulfide bonds. Less commonly, the dough is fermented with yeast followed by sheeting/forming like cream or soda crackers. These crackers are comparatively more delicate in texture and more aerated due to the leavening with ammonium bicarbonate. Various savory crackers are covered on their top with poppy, sesame or celery seeds, and garnished with flaked or coarse salt before baking. Most savory crackers are sprayed with oil while they are still hot immediately after exiting the oven. The oil dressing applied at around 8–18% of the biscuit weight, greatly improves the appearance of the biscuit surface, enhancing the color and overall organoleptic properties (Manley 2000c).

TABLE 10.3
Formulation Commonly Used to Manufacture Soda Crackers by the Sponge Dough Process[a]

	Bakers Formulation, %	
Ingredient	**Sponge**	**Dough**
Intermediate Wheat Flour (10.5% Protein)	60	40
Water	30	6
Fructose syrup		1.4
Shortening		8.2
Yeast[b]	0.25	
Yeast food	0.5	
Salt		1.4
Fermented dough	2.0	
Sodium bicarbonate		1.5
Diammonium phosphate		0.1
Diastatic malt	0.15	
Lecithin		0.2
Amylases	0.02	
Proteases		0.0025

[a] From Serna-Saldivar (2010).
[b] Dry yeast can be used instead of compressed fresh yeast at a concentration of 0.8%.

dough is rotary cut into the typical cracker form and the trim or scrap dough carried back to the dough hopper and reused. The formed pieces of dough could be salted or treated with toppings such as sesame or poppy seeds on a mash conveyor with a dispenser or egg washed prior to baking. Baking is performed on continuous direct gas-fired ovens equipped with moving bands and different baking zones. Most of the heat is transferred by conduction in these types of ovens. The bake time varies from 2 to 3 minutes at temperatures ranging from 220 to 240°C. During baking of crackers produced from 100 kg flour, approximately 45 kg water is lost (Figure 10.5). The baking conditions affect the sensory properties and color of the resulting crackers. Crackers generally exit the baking oven at moistures of about 2% and are cooled before packaging with materials resistant to the atmospheric moisture. Approximately 103 kg of saltines are obtained from 100 kg flour (Figure 10.5). From the microbial viewpoint, crackers are shelf stable because of their low A_w. The preservation of the characteristic crisp and brittle texture and the prevention of lipid oxidation are fundamental for crackers.

Most snack crackers and saltines are oil sprayed immediately after baking and before cooling. Round-shaped crackers are usually coated with an oil base. The oil is generally applied at a temperature of 65–70°C and from 1% up to 22% or more, based on the weight of the unsprayed crackers. The oil could be applied with a manifold that applies a continuous curtain of oil onto the product, or more commonly with spray nozzles or spinning disks (Figure 10.5). The last oil-delivering machine consists of disks that spin at high speeds (1800 to 3600 rpm) so when the oil drops on the disks it is sprayed into a fine mist (Moreth 1994). Due to its melting properties, the preferred oil for spraying is coconut although mixes of peanut and plastic shortenings are also used (Chapter 7). Sprayed crackers have higher energy density than regular crackers due to the oil coating (Chapter 20).

According to Manley (2000b) savory or snack crackers are a broad group of biscuits that are variously salted, flavored and fat sprayed after baking. These products differing in size and shapes are regarded as savory snacks, nibbles or biscuits for cheese. Normally they are simple biscuits but

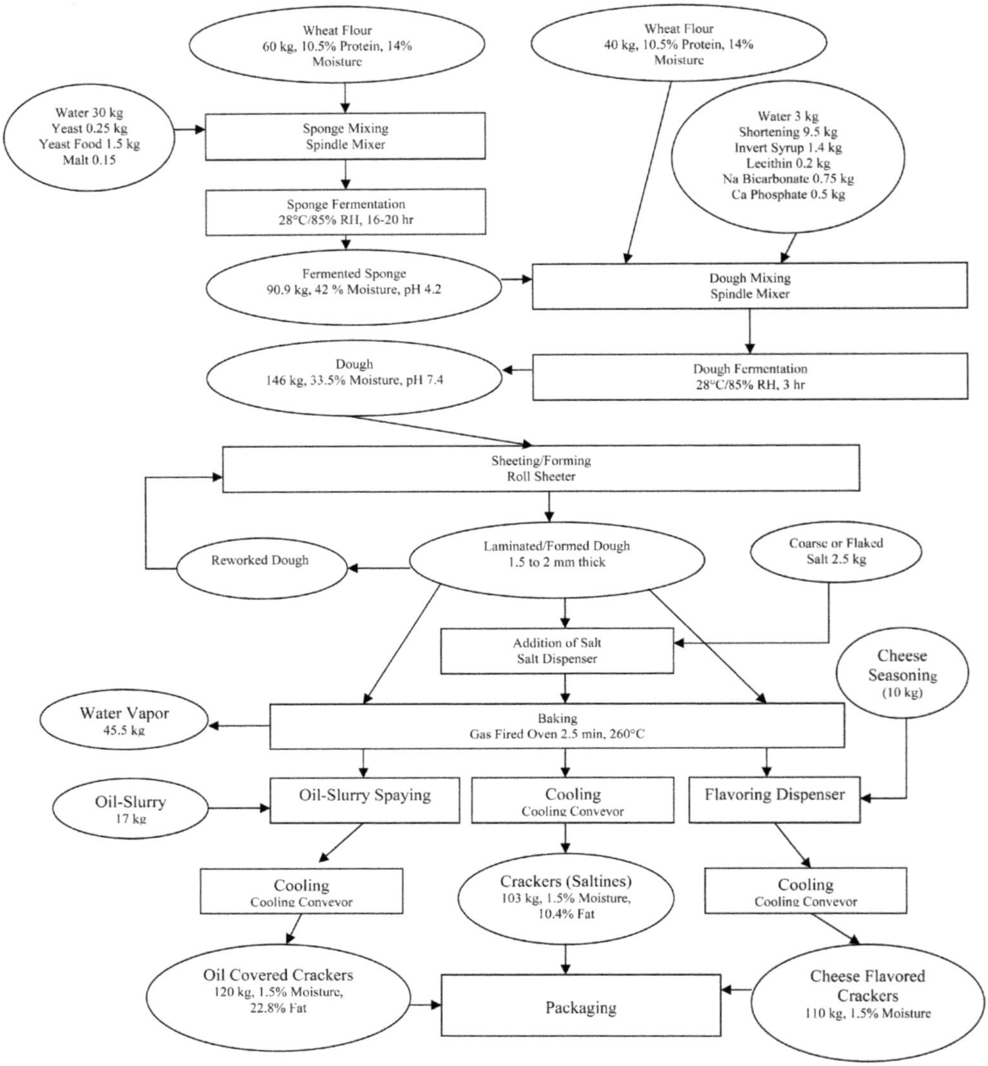

FIGURE 10.5 Flowchart of the manufacturing procedures for the production of several types of crackers.

sometimes they may be cream sandwiched with a savory, nonsweet, cream filling often based on cheese powder. The combination of the fermented flavor complements the cheese flavor so they are ideally suited for social events and for snacking at schools. These crackers are generally colored with paprika to intensify the color. Additional flavorings such as cayenne and sage are added to improve organoleptic properties. The crackers normally have a distinctive orange-red coloring that is easily recognized by the customer. Fresh or dehydrated, aged or regular cheddar cheese is used (Chapters 8 and 17). The preferred method is to incorporate the dry powdered form to achieve a better distribution during the sponge producing stage (Matz and Matz 1978). The key to maintaining a crisp cracker and a soft filling is to deliver a cracker and a filling with the same A_w to avoid moisture migration from the filling to the cracker. Generally, the filling material comprises a shortening together with a seasoning powder containing dehydrated cheese and oil. Additional optional ingredients that may be included are emulsifiers to avoid oil separation and migration into the cracker, salt, corn syrup solids or maltodextrin (Ardisson-Korat et al. 2013).

Filled sandwich crackers containing a creamy, low A_w, and bake stable filling between two crisp oven-baked crackers are manufactured by first mixing with high shear an oil phase, an aqueous phase and a solids phase. Then, top and bottom sheets of cracker dough are prepared, the bottom of which is moved at a predetermined horizontal velocity for depositing continuous or intermittent streams of a filling from a depositor with various openings. The top dough sheet is then placed over the bottom sheet, and the sheets are cut and/or scored in a predetermined pattern to form a composite unbaked dough and filling. Finally, the composite is baked sufficiently to provide a crisp outer crust with textural and microbiological stability (Mihalos et al. 2005).

Filled extruded crackers can be made with a high-pressure high shear extruder and filled at the die with a creamy cheese filling. Raw dry materials are first introduced along with water into a blender or mixer thoroughly combining and moistening the dry ingredients. Blended and moisturized raw materials may be transferred via a bucket elevator into an extruder hopper for transport into an extruder. The extruder forms a continuous tube of extrudate with a filling in the center of the extruded tube. The cofilled extrudate is then conveyed through a pillow cutter to help form or shape the extruded, filled dough tube into individual pieces while sealing the filling inside each piece. Filled dough pieces are then conveyed on one or more conveyor belts to an oven, which bakes the products to provide for a desirable crisp baked texture of the outer dough shell and to yield a shelf-stable, RTE snack product. Baked products can then be transported into a coating tumbler to be seasoned with oil-based slurries (Ardisson-Korat et al. 2013).

10.6.1 Checking/Stress Cracks

In the cracker and bread chip industry, certain groups of products are renowned for breakage at variable times after baking as a result of a hairline cracking phenomenon known as "checking". Cracker checking is usually observed as cracks, which extend across a central portion of the biscuit rendering it very prone to complete breakage on subsequent handling (Bailhache et al. 2019).

Experiments using a range of storage conditions confirm that it is the presence of moisture gradients, in combination with a low environmental RH, which increases the chance of checking (Saleem et al. 2005). Baking causes hardening of the cracker or bread chip surface and seals in moisture. As the products equilibrate after baking, the center moisture diffuses to the surface causing surface expansion while the center contracts. The expansion is caused by the uptake of moisture at the rim, which is initially dry after baking, and the contraction is attributable to loss of moisture from the more moist center (Saleem et al. 2005). Avoiding significant moisture gradients from center to crust at the oven exit will decrease checking after baking and during storage. Thicker products have a higher moisture gradient from the center to the outer crust thus thinner products will be less prone to checking.

Radiofrequency drying can be used as a solution to mitigate checking by evening out the moisture distribution after baking targeting the remaining moisture pockets (Orsat and Vijaya Rghavan 2014). In radiofrequency (RF) heating electromagnetic energy is transferred directly to the heated product. The longer wavelengths of RF compared with microwaves are able to penetrate further into the food products resulting in more even heating and reducing moisture gradients (Altemimi et al. 2019).

10.7 PRODUCTION OF PRETZELS

According to Terry Groff (2001) and Seetharaman (2014) pretzels is one of the oldest snacks and have been produced for at least 900 years although some scholars believe that this snack was first produced about 1,500 years ago. The term pretzels derives from the Latin word pretiola, which means little reward. The typical pretzel configuration with three holes represented the trinity holy of the Christians. Historically it is known that monks gave away pretzels to the underprivileged as a religious symbol. The typically twisted configuration that resembles the Christina prayer pose was a sign of prosperity and luck. Hitherto, pretzels are still popular in the European Community and the United States. Remarkably, pretzels are the third largest selling salted snack in the United States. According to Lorenzi (2020), the total sales exceeded USD 1.3 billion dollars in the calendar year from May 2019 to May 2020. Pretzels are popular because of their unique flavor and energy density. Pretzels contain fewer calories and fat compared to other snacks (see nutrient composition and nutritional value in Chapter 20).

Pretzels are divided into two categories: hard and soft. The first category is considered as a snack whereas the second as a fresh baking item. Hard and soft pretzels' surface is treated with an alkaline solution prior to baking (Figure 10.6). Pretzels are unique because the flavor and color are due to this sodium hydroxide wash. The typical flavor profile is also affected by the products yielded during fermentation, nondiastatic malt and salt (Table 10.4). For soft and hard textured pretzels hard and soft wheat flours are recommended. Wheat flour is mixed with dry ingredients and water (42% water absorption) to form a stiff dough that is mechanically rolled and formed into cylinders that are molded into traditional hard pretzel forms: bow tie, or other forms. Pretzels are fermented with yeast for a short time (30 min), formed and washed for 25–30 sec with a 1% sodium hydroxide solution heated to approximately 93°C (Seetharaman 2014). Pretzels are immediately salted, flavored and baked. The baking is regularly performed in three sequential steps: the first lasting approximately 10 min at high temperature and is aimed toward the formation of the typical dark brownish crust. The other two baking steps are to further dehydrate the product and impart the desired typical texture. A good pretzel should have a dark brownish outer crust and a white inner crumb with tiny and homogenously distributed gas cells or *loci*.

10.7.1 Manufacturing Process

The production of hard and soft pretzels follows the typical steps used in bread baking. The most relevant operations are: dough mixing, fermentation, molding/forming, application of sodium hydroxide, surface salting and baking (Figure 10.6, Serna-Saldivar 2008, Seetharaman 2014). The main difference between pretzels and other fermented baked products is the cooking of the dough with sodium hydroxide. This is considered the key step of the process.

Research has shown that pretzel texture and hardness are positively correlated with wheat flour protein content, dough water absorption, dough development time and stability. Dough mixing is accomplished using the same equipment utilized for bread making. However, the pretzel dough is formed with less water (40–45%) when compared with table bread (62–65%). Therefore, the dough is more difficult to mix, and handle and the mixer requires more energy or potency to blend the dough. Most processing lines use batch mixers although some high capacity industries use continuous mixers that consist of a gravimetric feeder for the flour and a liquid metering device to deliver water and the other ingredients. The dough is mixed by the action of a single shaft with impellers and mixing blades. The dough exiting from the discharge die is cut to predetermined dough pieces and transferred to the forming/molding machines.

Most forming machines used nowadays are extruders because they adapt to fully automated high-volume manufacturing lines. The extruder is fed with the pretzel dough and the screw forces the dough through the die to produce the predetermined pretzel form. The thickness of the pretzel is controlled by the speed of the reciprocating blade or knife working immediately after the die. The extruder has a short barrel and a special screw design (low shear) in order to avoid the modification of the dough rheological properties. Some extruders have the capacity of processing 800 kg/hr (Serna-Saldivar 2008). A twin-roll cutting machine can also form pretzels. A roll that rotates on top of the dough positioned on a moving belt cuts the laminated pretzel dough. Scrap dough is recirculated to the hopper, reworked and reused (Matz and Matz 1978, Serna-Saldivar 2008).

From the forming machine, the cut pretzel is transferred into proof cabinets consisting of automated systems of a continuous conveyor. Regardless of the type of proofing device, the pretzels are rested for 5 to 20 min at temperatures ranging from 28 to 37°C. For this specific product, it is

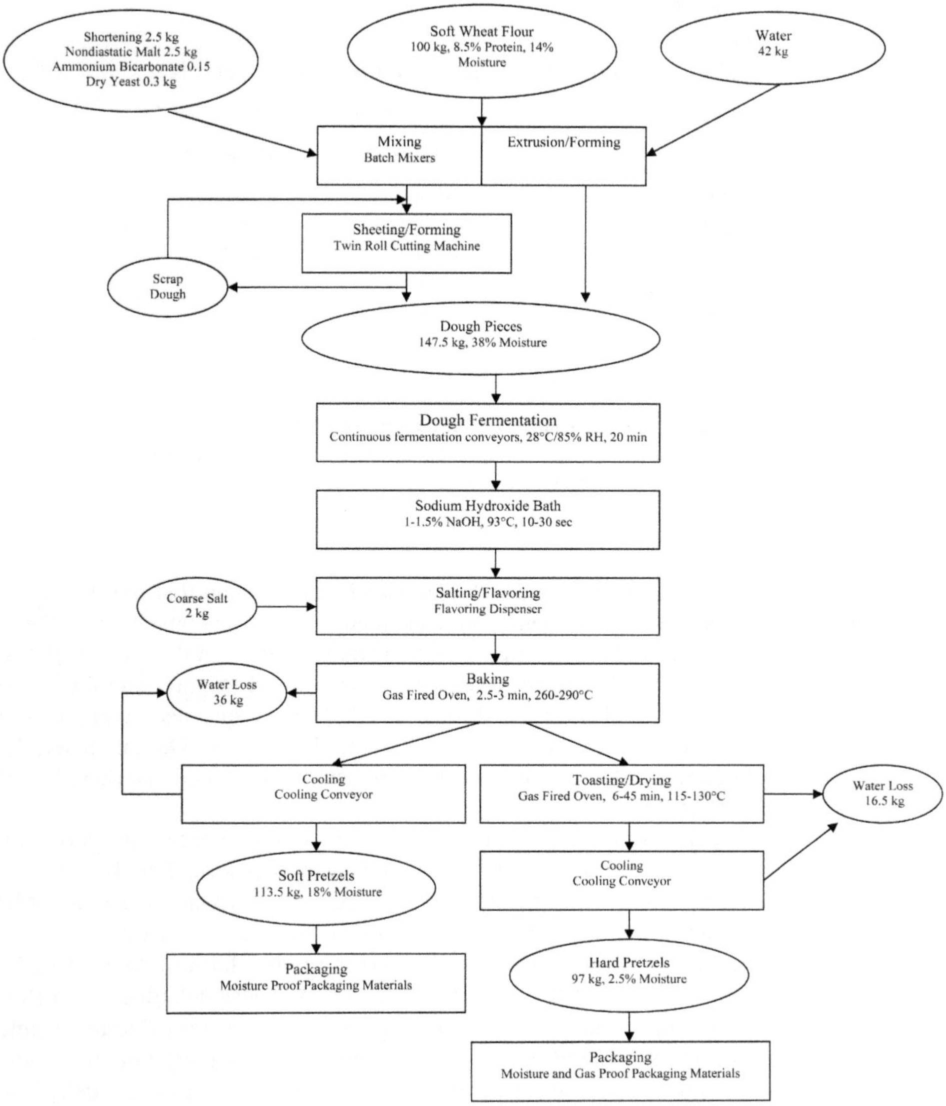

FIGURE 10.6 Flowchart of the manufacturing procedure for production of pretzels.

desirable that the pretzel forms a partially dehydrated crust; thus, proofers do not work with a high relative humidity setting as in other types of fermented bakery products. The aim of the short time proofing is to activate the yeast so to produce carbon dioxide and flavor compounds. The carbon dioxide is necessary to slightly increase the pretzel volume and form the typical crumb internal structure and the generation of fermented flavor compounds, which contribute to the pretzel flavor profile and pH. Rest times should be increased if dry yeast is used in the formulation because it takes longer for the dehydrated yeast cells to activate. The most important and critical operation, which distinguishes pretzel production and characteristics, is the lye (NaOH) cooking. The preformed dough pieces are simply cooked by dipping in a hot sodium hydroxide bath or cooker. Variables to control in this important operation are residence or dwell time, solution temperature and pH or strength of the lye solution. Most pretzels are cooked for 10–15 seconds at 90°C using a 1 to 1.5% sodium hydroxide solution. The aim is to change the pH and gelatinize the surface starch

TABLE 10.4
Formulation Commonly Used to Manufacture Pretzels[a]

Ingredient	Bakers Formulation, %	
	Hard Pretzel	**Soft Pretzel**
Soft wheat flour	100.00	
Hard wheat flour		100.00
Water	46.70	45.00
Vegetable shortening	2.50	
Nondiastatic malt	2.50	2.00
Dry yeast[b]	0.30	0.50
Ammonium bicarbonate	0.15	

[a] Coarse or flaked salt is applied onto the pretzels before baking.
[b] Compressed fresh yeast can be used instead of dry yeast at a concentration of 0.9–1% for hard pretzels and 1.5% for soft pretzels.

located on the crust. Cookers have tube heat exchangers and a belt system that moves forward the pretzels along the cooker. Since most fermented pretzels tend to float, the conveying system forces the pretzels to submerge or the cooker is equipped with a waterfall curtain that applies the hot-lye solution. After the sodium hydroxide treatment, excess lye is allowed to drain and the precooked pretzels are conveyed under a salt dispenser and then baked. The salt dispenser equipment works with a special system to recirculate salt that does not adhere to the product. The salt generally used is coarse or flaked. Soft pretzels or low sodium pretzels are usually nonsalted and therefore bypass this operation (Serna-Saldivar 2008).

Baking is continuously performed in wire mesh, gas-fired and automated ovens that have at least three recognized temperature areas. The dwell time varies from 2.5 to 10 min depending on the type of pretzel being produced. The temperature profile also varies according to the product but in most cases, the first zone is set at 10–15°C higher temperature than the next baking zones. The common oven conditions to manufacture regular hard pretzels are 288, 260 and 260°C for 2.5–3 min. The temperature is adjusted to avoid excess dehydration of the crust and blistering. During this important operation, the typical color develops due to caramelization and browning reaction exacerbated by the alkaline pH of the surface. For the production of hard pretzels, the baked pretzels containing approximately 10% moisture are dehydrated in bands at temperatures ranging from 115 to130°C for 6 to 45 minutes. The final moisture content of hard pretzels should be about 2–3.5%. Over drying yields hard pretzels that are more prone to break during packaging and distribution while under drying yields pretzels without the typical texture. Pretzels are usually only flavored with coarse or flaked salt because they are formulated to be low or no fat. Addition of seasonings postbaking is difficult because the flavorings will not adhere to the smooth crusty surface of the product. To overcome this difficulty, pretzels are first treated with a liquid adhesive blend followed by the application of a powdered seasoning mix. The most common way to deliver the seasonings is on a moving belt (see Chapters 3 and 8). Pretzels are cooled and generally packaged in a form-fill-seal machine (Chapter 4). The packaging material is laminated or coextruded and should be an excellent barrier against air humidity and gases. Hard pretzels are usually packaged with an inert gas to protect the flavor and decrease the rate of product breakage during handling. About 97 kg of hard pretzels are obtained from 100 kg flour (Figure 10.6, Serna-Saldivar 2008). The nutritional composition of pretezels is detailed in Chapter 20.

REFERENCES

Altemimi, A., Naji Aziz, S., Al-Hilphy, A.R.S., Lakhssassi, N., Watson, D.G., and Ibrahim, S.A. 2019. Critical review of radio-frequency (RF) heating applications in food processing. *Food Quality and Safety*, 3(2):81–91.

Ardisson-Korat, A.V., Graham, D.W., Howard, S., Huang, I., Morales-Alvarez, J.C., Narayanan, B., Ozturk, B., Rao, M.V.N., Quintero-Fuentes, X., Siricururatana, P., Smith, R.T., and Yurdakan-Ince, S.G. 2013. *Method and Formulation for Producing Extruded Snack Food Products and Products Obtained Therefrom*. US 2013 0045317A1.

Bailhache, C., Pontoire, B., Ribourg, L., Jonchère, D., Queveau, M., Irled, E., Thie-rye, A., Le-Bail, A., and Le-Bail, P. 2019. A study of cracks in dry cereal products. *Food Hydrocolloids*, 94:528–536.

Barnett, M., and Joseph, P.K. 2015. *Continuous Process and Apparatus for Making a Pita Chip*. US 20150150269A1.

Bernand, D.C., Diermeier,D.J., Elsawy, T.M., Joseph, P.K., Mathew, R., Simbashivan, S., and Williams, R. 2014. *An Improved Continuous Process and Apparatus for Making a Pita Chip*. WO 2014/022701 A2.

Calvel, R. 1987. *O pâo francés e os productos correlatos: Tecnologia e prática da panificaçâo*. Fortaleza: J. Macedo S.A., Comercio, Administrçâo e Participaçôes.

Edwards, W.P. 2013. Flat breads. In: *Science of Bakery Products*, ed. W.P. Edwards. London: Royal Society of Chemistry.

FAO. 2021. Statistical database. Rome, Italy. Electronic page http://apps.fao.org. Accessed May 12, 2021.

Faridi, H. 1994. *The Science of Cookie and Cracker Production*. New York: Chapman and Hall.

Faridi, H., Gaines, C.S., and Strouts, B.L. 2000. Soft wheat products. In: *Handbook of Cereal Science and Technology*, eds. K.J. Lorenz and J.G. Ponte, 575–614. New York: Marcel Dekker, Inc.

Friedberg, J. 2020. State of the industry 2020. Crackers grow through innovation. https://www.snackandbakery.com/authors/1934-joyce-friedberg.

Gocmen, D., Inkaya, A.N., and Aydin, E. 2009. Flat breads. *Bulgarian Journal of Agriculture Science*, 15:298–306.

Goglanian, G. 2001. *Method for Preparing Elongated Pita Bread*. US6291002.

Kulp, K., and Lorenz, K.J. 2003. *Handbook of Dough Fermentation*. New York: Marcel Dekker.

Lorenzi, N. 2020. State of the industry 2020. Pretzel market enjoys growth. https://www.snackandbakery.com/articles/94980-state-of-the-industry-2020-pretzel-market-enjoys-growth.

Manley, D. 2000a. Cream crackers. In: *Technology of Biscuits, Crackers and Cookies*, ed. D. Manley, 1–8. Cambridge: Woodhead Publishing Company.

Manley, D. 2000b. Soda crackers. In: *Technology of Biscuits, Crackers and Cookies*, ed. D. Manley, 1–8. Cambridge: Woodhead Publishing Company.

Manley, D. 2000c. Savoury or snack crackers. In: *Technology of Biscuits, Crackers and Cookies*, ed. D. Manley, 1–8. Cambridge: Woodhead Publishing Company.

Matz, S.A., and Matz, T.D. 1978. *Cookie and Cracker Technology*. 2nd edition. Westport: The AVI Publishing Company, Inc.

Meuser, F., and Wiedmann, W. 1989. Extrusion plant design. In: *Extrusion Cooking*, eds. C. Mercier, P. Linko and J.M. Harper, 91–155. Saint Paul: American Association of Cereal Chemists.

Mihalos, M.N., Antonini, A., Wines, C., Lewis, G.A., Thomas, J.K., Gerard, K., DesRochers, J.L., and Forte, Z.E. 2005. *Process for Preparing Filled Cracker Products*. US 20050249845A1.

Misra, N.N., and Tiwari, B.K. 2014. Biscuits. In: *Bakery Products Science and Technology*, eds. W. Zhou, Y.H. Hui, I. DeLeyn, M.A. Pagani, C.M. Rosell, J. Selman, and N. Therdthai, 585–601. Chichester, West Sussex: Wiley Blackwell.

Moreth, N. 1994. Engineering and processing. In: *Cookie Chemistry and Technology*, ed. K. Kulp. Manhattan: American Institute of Baking.

Orsat, V., and Vijaya Rghavan, G.S. 2014. Radiofrequency processing. In: *Emerging Technologies for Food Processing*. 2nd edition, ed. D.W. Sun, 385–398. London: Academic Press.

Quintero Fuentes, X., Joseph, P.K., and Williams, R. 2015. *Continuous Process for Making a Pita Chip*. US009210941 B2.

Saleem, Q., Wildman, R.D., and Huntley, J.M. 2005. Improved understanding of biscuit checking using speckle interferometry and finite-element modelling techniques. *Physical and Engineering Sciences*, 461(2059):2135–2154.

Serna-Saldivar, S.O. 2008. Manufacturing of wheat based baked snacks. In: *Industrial Manufacture of Snack Foods*, ed. S.O. Serna-Saldivar, 265–296. London: Kennedys Publications Ltd.

Serna-Saldivar, S.O. 2010. Manufacuring of bakery products. In: *Cereal Grains: Properties, Processing and Nutritional Attributes*, 259–328. Boca Raton: CRC Press (Taylor & Francis Group).

Seetharaman, K. 2014. Pretzel production and quality control. In: *Bakery Products Science and Technology*, eds. W. Zhou, Y.H. Hui, I. DeLeyn, M.A. Pagani, C.M. Rosell, J. Selman, and N. Therdthai, 611–618. Chichester, West Sussex: Wiley Blackwell.

Terry Groff, E. 2001. Perfect pretzel production. In: *Snack Foods Processing*, eds. E.W. Lusas and L.W. Rooney, 369–384. Lancaster: Technomic Publishing Co.

11 Alkaline-Cooked Snack Foods

Sergio O. Serna-Saldivar

CONTENTS

DOI: 10.1201/9781003129066-11

11.1 INTRODUCTION

Alkaline or lime-cooking has become one of the most relevant facets of maize preparation for human foods. Nixtamalization is the process of lime-cooking, steeping, and washing mature maize kernels to produce *nixtamal*, which is subsequently stone-ground to form a soft, moist, and malleable dough typically named *masa*. *Masa* is the backbone for the manufacturing of an array of products like table or soft tortillas, other traditional Mexican foods such as thick porridges (*Atole*) and Tamales, and the two major categories of deep-fat-fried snacks: corn chips and tortilla chips. Today *masa* is obtained from two processes: fresh *nixtamal* or dry *masa* flour (DMF).

Snacks produced from lime-cooked maize are the fastest growing market within the salted category. In the United States, the sales of tortilla chips/tostadas in 2019 were estimated on 6 billion dollars with an annual growth of 7.6% (Decker 2020). Today, extruded corn chips and tortilla chips are manufactured throughout the world. The chemistry, technology, manufacturing, and nutritional value of nixtamalized products have been extensively reviewed before (Serna-Saldivar 2008, Serna-Saldivar and Rooney 2015, Serna-Saldivar and Chuck-Hernandez 2019, Escalante-Aburto et al. 2020, Ramirez-Araujo et al. 2019, Serna-Saldivar 2021).

There are fundamentally three types of snacks produced from alkaline-cooked corn: parched corn, also known as corn nuts; corn chips; and tortilla chips (Figure 11.1). Corn nuts are manufactured from special types of maize genotypes like Cuzco or Cacahuacintle that are alkaline-cooked, deep-fat-fried, and seasoned. Extruded corn chips are produced from coarsely ground *masa* that is formed and deep-fat-fried, whereas tortilla chips, taco shells, and *tostadas* are produced from pieces or whole-baked tortillas that are deep-fat-fried (Figure 11.1). Baking of tortillas prior to frying greatly affects the organoleptic properties and nutritional value of these snacks. Tortilla chips have a stronger alkaline-cooked flavor and a crispier texture than corn chips. In addition, tortilla chips are less energy-dense (Chapter 20) because they absorb about 12% units less oil during frying. These snacks are sold as retail products and offered in restaurants, amusement parks, and convenience stores. Lime-cooked snacks are relatively inexpensive and easier to process with greater flexibility and profitability compared to potato chips. One of the most popular snacks in baseball and football stadiums are nachos (a combination of tortilla chips with melted cheddar cheese and jalapeños).

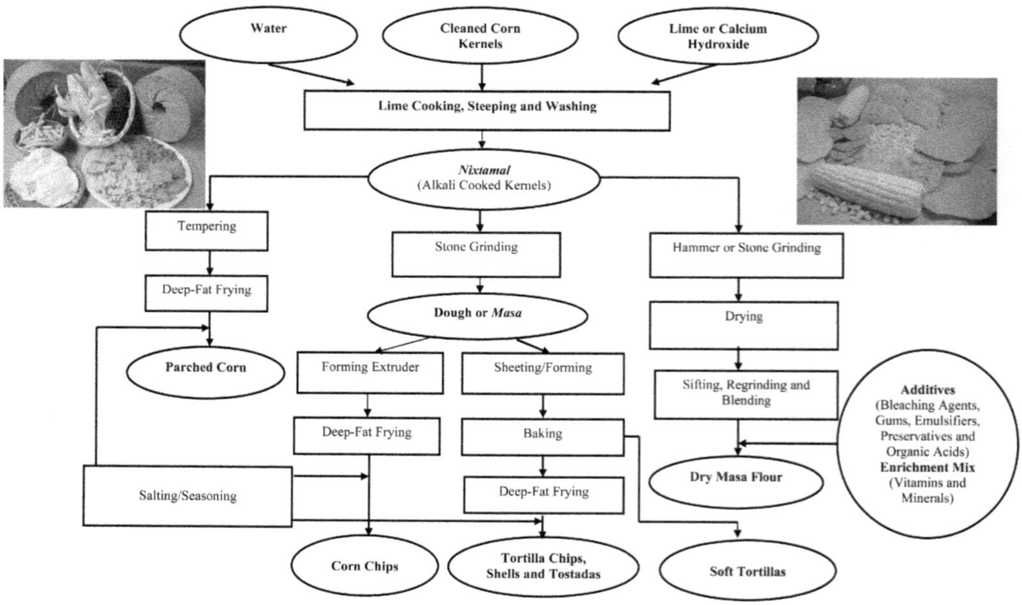

FIGURE 11.1 General flowchart of the industrial processes for the manufacture of lime-cooked products.

Nixtamalized snacks can be alternatively produced from specialty types like blue, quality protein, high-oil, *cacahuacintle*, and nutritionally enhanced. Among these, blue maize keeps increasing in popularity because of their unique flavor, color, anthocyannins, and higher antioxidant capacity (Serna-Saldivar 2021). In addition, the industries have developed novel products containing other grains or flours from sesame, flax, amaranthus, soybean, chickpeas, and common beans (Serna-Saldivar et al. 2015).

This chapter reviews the functionality of ingredients, physicochemical changes incurred during processing, and the current industrial state of nixtamalized processes for production of snacks from regular and specialty maize types.

11.2 HISTORICAL OVERVIEW

The development of Mesoamerican cultures started about 10,000 years ago when the indigenous maize plant became a cultivated crop and chiefly responsible for the initiation of formal agriculture in Mesoamerica. According to Garcia-Lara and Serna-Saldivar (2019), the origin of this crop occurred during the invention of the new world agriculture. The oldest paleoethnobotanical evidence of its domestication is in an archeological site named "Nac Neish," located in northeast of Mexico. It consists of three primitive caves presenting data between 6,000 and 20,000 years BC (Serna-Saldivar 2015a, Garcia-Lara and Serna-Saldivar 2019). In one of the caves, it was found a tomb that contained human remains covered with reed mats (*petlatl*) and an offering consisting of a basket with corncobs. The American Indians created corn from a tiny two-rowed ear of teosintle of about 3 cm long into the first tiny corn ear with its four ranks of paired female spikelets. This transformation took perhaps only a couple of centuries. Thereafter, two different corns classified as pyramidal or cylindrical corns originated the modern corns. The oldest remains of 8- and 12-rowed corns from Mesoamerica are 7,200 and 3,000 years old, respectively. These relevant developments originated different races varying in kernel morphology, color, endosperm texture, and other important features that make these types highly useful for snack processors (white, yellow, blue, and purple corns for corn and tortilla chips, *cacahuacintle* for parched corn, popcorn, etc.). More than 25 corn races have been used to produce high-yielding varieties and hybrids that make corn the most productive crop in the planet with yearly production exceeding 1.1 billion tons in 2019 (FAO 2021).

After the discovery of the Americas, the crop eventually reached and adapted to the Iberia Peninsula and from there distributed to other parts of the globe. The Portuguese conquistadors and merchants were more responsible than the Spaniards for spreading corn around Europe and Africa. However, the Spaniards catalyzed the Asian connection because they traded goods from the Pacific ports of Mexico to the Philippines. This occurred as early as 1565, but the crop became relevant in Asia after the seventeenth century.

The ancient Mesoamerican cultures provided not only the regular maize and popcorn to the whole world but also the nixtamalization process that is key to obtain corn and tortilla chips, which are considered as the second most sold savory snack category in the world. Historically, it is known that the Aztecs produced *totopochtli*, an ancient prototype of free-fat tortilla chips, by toasting tortillas on a hot *comal* or griddle. This shelf-stable product was carried as snack by travelers or used during wars (Serna-Saldivar 2015a).

The first DMF was produced by the Mexican company Molinos Azteca more than 70 years ago. Most DMF are produced following the basic steps of the patent filed by Gonzalez and Rubio in 1965 (Gonzalez and Rubio 1968). Industrial production of pregelatinized and shelf-stable DMF is accomplished by continuous lime-cooking in large reactors, washing, *nixtamal* grinding, drying, sieving, regrinding coarse particles, resieving, classifying, and blending to meet certain particle size requirements. The pregelatinized DMF is formulated into flours with carefully controlled particle size distribution, water absorption, and other characteristics such as pH and supplementation with different additives (Serna-Saldivar et al. 1990, Serna-Saldivar and Rooney 2015).

The dramatic increase in sales of lime-cooked snacks occurred during the past 50 years. Latin Americans inhabiting in the southwestern states of the United States first introduced fresh and toasted tortillas (tostadas and taco shells) and then the industry-designed high-output industrial lines aimed toward the production of fried corn chips and tortilla chips. These snacks have experienced one of the most relevant growth rates in terms of volumes and sales during the past 40 years and nowadays are manufactured practically in all developed countries of the globe (most European countries, Russia, China, Australia, South Korea, Brazil, Argentina, etc.). The first use of industrial equipment to produce corn chips occurred in San Antonio Texas in 1932. Elmer Doolin, founder of the Frito Company, bought rights from a Mexican who wanted to return to his homeland. Dooling started the business in his kitchen and soon the sales boomed and within a year moved to Dallas. When Dooling passed away, he had transformed the US$100 loan from his mother into an industry with yearly sales of more than US$50 million (Serna-Saldivar 2015a). I.J. Filler was granted a trademark for Corn Chips in Texas and the US Patent Office in 1932 and 1935, respectively. That is possibly the reason why Dooling chose the name Fritos™. In terms of high-throughput machinery, one of the major advances was the design and development of a mechanical screw press, developed by Doolin and the Lawson Brothers Machine Shop, that extruded the lime-cooked *masa* into ribbons in preparation for deep fat frying. The machine consisted of a vertical positioned barrel and a mechanically driven piston that pushed the *masa* into the die located at the end of the barrel. When the piston retracted, a new load of *masa* was put into the machine. This press was improved when the piston was driven by compressed air and a 90-degree elbow tube was fitted to the discharge end of the barrel. A wire cutter was also driven by a pneumatic assembly with an air regulator to control speed. These new extruders were capable of producing 200 kg/h. Hydraulics machines widely used nowadays replaced the pneumatic systems. The advantages of these corn chip extruders were increased production and a more consistent chip length. Frito Lay/L.C. Miller Co. first built these improved special extruders in the early 1960s (Burtea 2001).

The more popular produced and consumed tortilla chips developed about 20–30 years later than corn chips. A. West and G. Chesquiere of the newly firm Frito Lay Incorporation got the idea of producing tortilla chips after seeing Mexican style tostadas and US taco shells. The trade name Doritos was selected for these new type of products. The first triangle-shaped Doritos were manufactured at Alex Foods in Anaheim, California. The tests went so well that soon the products were not only produced in California but also in southwestern United States. Doritos™ went to the national market in 1966. Few years later, sales of tortilla chips surpassed sales of corn chips. The market of tortillas chips has steadily increased above the rate of potato chips in such a way that tortilla chips has almost reached the current market and sales of potato chips. Sophisticated and highly automated machinery has been developed in order to produce large quantities of tortilla chips. J.C. Ford Company, Casa Herrera, Electra Food Machinery, Lawrence Equipment, Maddox Metal Works, and Heat and Control are the major suppliers of grinders, formers, and frying equipment and fully automated lines. The larger production lines are capable of producing 1.5 tons of chips/h.

11.3 RAW MATERIALS

Mature maize kernels, food-grade lime, and water are the three fundamental ingredients required for producing *masa* and soft tortillas. For production of snacks, frying oils (Chapter 7), salt, and seasonings (Chapter 8) are usually used. Product quality and processing parameters depend on the kernel characteristics. However, the oil, salt, and seasonings used for the preparation of maize snacks have a profound effect on flavor, mouth feel, and overall acceptability.

11.3.1 Maize Kernels

Maize is the most produced cereal worldwide with a current annual production exceeding 1.1 billion tons (FAO 2021). Nevertheless, most of this output is used for production of livestock and poultry feeds and fuel bioethanol. Among the various direct food human uses, the lime-cooked products have positioned as one of the principal not only in Latin America but also all over the globe. The quality and physical properties of the maize kernels play the most relevant role in manufacturing of lime-cooked snacks because they affect product quality and strongly influences processing parameters (Rooney and Suhendro 2001, Rooney and Serna-Saldivar 2015).

11.3.1.1 Regular Corns

Selected white, yellow, and blue dent open pollinated or hybrids are preferred for the manufacturing of lime-cooked snacks. The major advantages of growing hybrids are improved yields and more uniform kernel type and maturity. Both genetics, environment, and postharvesting practices of grain handling and storage affect the quality composition and physical properties of the kernel. The main criteria used to select maize for the manufacturing of tortillas and related snacks are related to their physical properties because they affect chemical composition, functionality, processing parameters, and end use. Most processors contract nongenetically modified (GMO) food-grade corns that are sold with a small premium, especially in the United States. In addition, kernels should be free of heat, insect, or mold damage and meet the maximum permitted level of aflatoxins. Heat- and insect-damaged kernels yield defective (sticky) doughs and off-colored end-products (Garcia-Lara et al. 2013).

At present, white maize is the most reliable source of non-GMO maize in the United States and other parts of the world. The ideal characteristics are dent and should have an intermediate endosperm texture with bright color coming from white cobs. Kernels are intermediate to large (1,000 kernel weight > 320 g) and preferably should have a rounded crown and a shallow unwrinkled dent. Optimally, they have a test weight of 76–79 kg/hL and a true density of 1.3 g/cm^3 (Table 11.1). A high proportion of hard or vitreous to soft or chalky endosperm is required but flinty is not desirable because they require higher cooking regimes, which demand more energy. The physical grain features are relevant because they are closely related to dry matter losses during cooking, steeping, and washing, which in turn affect tortilla yield, processability, and end-product quality. In addition, the pericarp or bran should be easily removed during the first steps of nixtamalization. A good food-grade maize should tolerate overcooking and abuse during handling and industrial manufacturing. According to Serna-Saldivar et al. (1991), the ease of pericarp removal is heritable and can be selected by the nylon bag-cooking and drying test. Corn chips and tortilla chips are normally produced from white kernels or blends of white and yellow kernels (Serna-Saldivar et al. 1990, Rooney and Serna-Saldivar 2015).

Tortilla color is the result of corn kernel and cob color, the amount of lime used during cooking, the extent of washing, and the final pH. Generally, the higher the pH, the stronger the coloration. Kernels from red cobs can yield off-colored tortillas because the tissue that remains after threshing is high in phenolic pigments; therefore, the most suitable kernels come from white cobs (Rooney and Serna-Saldivar 2015). Subtle differences in kernel color can help to predict the overall quality of alkaline-cooked products. As indicated before, kernels should be clean and bright white or yellow. Orange, reddish, or other off-yellow colors are unacceptable for lime-cooking because they greatly affect the appearance of table tortillas and chips as well as their flavor. For instance, a high xanthophyll pigment in the kernel yields chips with an acrid odor because this pigment degrades during deep fat frying. Likewise, white kernels with a lemon or off-white color produce chips with a dull, dirty, and grayish appearance (Rooney and Suhendro 2001).

Most white corns planted in the United States and Mexico are significantly harder compared to the typical soft textured yellow dent corns and therefore are preferred and grown primarily to supply food industries (Table 11.2). About 6 and 3 million tons of these high-yielding corns are annually harvested in Mexico and the United States, respectively. The US prices for white corn generally are from 40 cents to half a dollar per bushel, higher than for regular yellow corn. White corn must be grown in isolation to avoid cross-pollination from yellow corn. The yields of high-producing white corns in Mexico usually surpass yellow types, but the typical GMO yellow corns planted in the United States have higher yields compared to the white non-GMO types (Rooney and Serna-Saldivar 2015).

TABLE 11.1
Relationships of Corn Specifications to Processing and Quality of Alkaline-Cooked Snacks[a]

Standard	Measurement	Usefulness
Grade USA 1 white or yellow dent corn	Federal Grain Inspection Service Standards	Provides a useful baseline to ensure minimum quality
Moisture content (11–14%)	Electronic moisture meter, NIRA, standard oven method	Related to dry matter and storage quality. Corn with excess moisture yields less tortillas and related snacks and is more susceptible to biotic agents during storage. Dry corn requires special care to rehydrate, breaks during handling, and requires longer alkaline cooking time
Test or volumetric weight (76–79 kg/hL or 56–61 lb/bushel)	Weight of corn in a cup with a given volume	Relates to soundness and condition of the kernel. Higher test weights indicate sound corn with properly matured kernels, whereas lower values kernel damage
Foreign material	Dockage test meter or use of a set of sieves	Determination of foreign or extraneous material that affects yields. A high foreign material makes the corn more susceptible to insects during storage
Kernel size and shape	Subjectively evaluate or use screens to remove odd kernels	Uniformity is highest when kernels from each end of the ear are mechanically removed. This operation increases cost of corn significantly. Small kernels cook faster than larger counterparts
Broken/damaged kernels (5–10% maximum)	Visual inspection/sieving	Cracked, chipped, and broken kernels should be removed during cleaning operations that greatly affect yield
Fissures/stress cracks (<30%)	Candle with a light box	Stress cracks make corn more susceptible to mechanical damage and breakage during handling, possibly affecting cooking and *masa* texture
Dead germ (<0.1%)	Visual observation of the germ – standard FGIS procedure for heat damage	Heat-damaged kernels indicate that kernels had high respiration rates during storage and degraded nutrients due to generation of enzymes These kernels are unfit for processing because they produce sticky doughs and low-quality tortillas and snacks, especially in terms of color and texture
Mycotoxins (aflatoxins <20 ppb; fumonisins <4.0 ppm)	ELISA, HPLC	Aflatoxins less than 20 ppb specify 5 ppb or less to be safe. Fumonisins limit not set, probably 4.0 ppm. Sampling is a major problem

Source: [a]Rooney and Suhendro (2001), Rooney and Serna-Saldivar (2015).

Many nations around the world like Mexico and most European countries have established mandatory labeling of foods containing GMO corn despite the fact that it is generally recognized as safe for human consumption. The major concern is in terms of glyphosate-tolerant GMO kernels that are in most instances detected with the polymerase chain reaction (PCR) assay (Rooney and Serna-Saldivar 2015). Glyphosate is a broad-spectrum systemic herbicide classified as an organophosphorus compound. This herbicide acts as a strong inhibitor of the plant enzyme 5-enolpyruvylshikimate-3-phosphate synthase. This herbicide has created a lot of controversy because the supporters claim that it increases significantly maize production without affecting health (Hammond et al. 2004, Healy et al. 2008), whereas the detractors allege that it harms pollinators like bees and is toxic to humans and a possible carcinogenic (International Agency for Research on Cancer. 2015). This is the reason why the direct use of corn for preparation of human foods do not allow the utilization of GMO genotypes and some countries like Mexico are banning the use of these tolerant glyphosate and other GMO.

TABLE 11.2
Specialty Maize Types with Potential to Produce Nixtamalized Products with Distinctive Functional, Nutritional and Nutraceutical Properties

Type of Specialty Maize	Description and Potential Use for Production of Lime-Cooked Snacks
Food-grade (white and yellow types)	Genotypes tailor-bred for production of lime-cooked snacks. These dent types have improved processing efficiency and possess an intermediate endosperm texture, medium to large kernels, and comparatively higher values of test weight, true density, and thousand-kernel weight. Their pericarp should be easily removed after lime-cooking and steeping. In addition, kernels should come preferably from white cobs because pink- or red-colored cobs yield undesirable off-colors, especially in snacks (Rooney and Serna-Saldivar 2015, Serna-Saldivar and Chuck-Hernandez 2019, Serna-Saldivar 2021)
Quality protein maize (QPM)	QPM has upgraded protein quality because it contains about twice as much lysine and tryptophan compared to regular maize (Serna-Saldivar and Rooney 1994, Serna-Saldivar 2015b, Serna-Saldivar 2021). QPM is the result of breeding opaque-2 maize that was combined with modifier genes to improve hardness and agronomic performance. White and yellow genotypes with intermediate endosperm texture have been successfully transformed into high-protein-quality tortillas and chips (Serna-Saldivar et al. 1992, Serna-Saldivar et al. 2008). The improved protein quality also positively affects the physiological and brain development of laboratory animals and humans (Amaya-Guerra et al. 2004, 2006, Serna-Saldivar 2015b)
Blue/pigmented	Blue and other pigmented (red) maizes possess floury or chalky endosperm and are usually planted in highlands, although new types are adapted to subtropical regions The aleurone layer and bran contain anthocyanins that impart the blue appearance and distinctive antioxidant properties. The amount of calcium hydroxide remaining in the nixtamal greatly affects the color of chips because the anthocyanins change coloration at different pHs. The darkest coloration is obtained at an alkaline pH of 8. Fumaric acid can be added to the dough or *masa* to lower the pH and coloration. The anthocyanins are significantly lost during processing into tortilla chips. Specialty blue tortillas and chips, including organic items, are manufactured and sold with a premium price because consumers perceive products more healthful (De la Parra et al. 2007, Garcia-Lara et al. 2012, Urias-Peraldi et al. 2013, Urias-Lugo et al. 2015,Mora-Rochin et al. 2016, Serna-Saldivar 2021)
Cuzco/*Cacahuacintle*	The Cuzco maize comes from eight-rowed ears that produce the largest known kernels. Cuzco corn grows at high altitudes and produces white kernels with soft endosperm texture and bland flavor. Cuzco maize kernels are mainly used to manufacture parched maize (Cornnuts™) and *Cacahuacintle* kernels for hominy and *pozole* production (Bonifacio-Vazquez et al. 2005, Serna-Saldivar 2008, Serna-Saldivar and Chuck-Hernandez 2019, Serna-Saldivar 2021)
High-oil corn (HOC)	Most HOC types contain at least twice the lipids compared to regular types and therefore are more energy-dense. The major drawback of HOC breeding programs is that new types usually yield less compared to regular counterparts. To compensate this deleterious correlation, the use of heterosis has been employed for the production of white and yellow new parental lines and hybrids (Preciado-Ortiz et al. 2013). The potential use of high-yielding white and yellow HOC for tortillas and chips is promising because new hybrids contain up to 9% lipids that can change the functionality of the *masa* and snacks. The potential of HOC for production of baked (no-fat added) tortilla chips is promising
Waxy or high amylopectin corn	These genotypes contain more than 95% amylopectin and have been experimentally used to manufacture special types of tortilla chips (Quintero-Fuentes et al. 1999). However, kernels are difficult to nixtamalize because they tend to easily overcook and upon stone-grinding yield sticky doughs that are difficult to process. However, the combination of regular and waxy corns before lime-cooking change the characteristics of tortilla chips. The major advantage of waxy supplemented chips is that they acquire different texture with smaller and more air pores that impart higher friability and acceptability (Quintero-Fuentes et al. 1999)

(*Continued*)

TABLE 11.2 (CONTINUED)
Specialty Maize Types with Potential to Produce Nixtamalized Products with Distinctive Functional, Nutritional and Nutraceutical Properties

Type of Specialty Maize	Description and Potential Use for Production of Lime-Cooked Snacks
Provitamin-A enhanced	Recently, plant breeders of CIMMyT developed orange-colored maize genotypes containing high provitamin-A (beta-carotene, beta-cryptoxanthin, and alpha-carotene) levels. Consumption of this maize improves total body vitamin A stores and visual function in marginally vitamin A-deficient infants. Some of these kernels contain 15 μg/g of β-carotene equivalents instead of the 2 μg/g normally observed in regular maize types. Tortillas obtained from regular nixtamalization and extrusion cooking contained 85–100% of the total carotenoids assayed in raw kernels (Lozano-Alejo et al. 2007, Rosales et al. 2016, Prasanna et al. 2020, Serna-Saldivar 2021). The utilization of provitamin A enhanced kernels has not been tested in fried corn or tortilla chips, although the excess yellow pigmentation can negatively affect the color and appearance of fired products
High-zinc corns	High-zinc open-pollinated varieties and hybrids were developed by breeders associated with the CIMMyT and the International Institute of Tropical Agriculture. The breeding effort has achieved the target level of 33 ppm of Zn (dwb) without sacrificing field yields. The main objective of these nutritionally enhanced types is to provide additional levels of Zn that favor child growth and strengthen the immune system. Studies are needed in order to assess the quality of tortilla chips produced with these experimental maize types (Andersson et al. 2017, Prasanna et al. 2020, Serna-Saldivar 2021)

From the chemical composition viewpoint, the kernels should contain 65–70% starch, 8–10% crude protein, 2.5–4.5% crude fat, and 1.5–2.1% crude fiber. Kernels should possess a regular endosperm with 25% amylose and 75% amylopectin and low fat acidity values. The generation of lipases, proteases, and amylases demerit dough machinability and end-product quality, especially in terms of flavor, odor, and color. Heat-damaged kernels should be avoided because they produce sticky doughs or *masas* and off-colored products, especially when fried. The yellow endosperm kernels contain significant amounts of β-carotenes, lutein, zeaxanthin, and cryptoxanthin that impart the yellowish coloration to tortillas and snacks, whereas the white counterparts are practically devoid of these lipophilic pigments (Rooney and Serna-Saldivar 2015, Serna-Saldivar 2021).

11.3.1.2 Specialty Corns

Many specialty corn types have been successfully nixtamalized for the production of soft tortillas and snacks with different chemical, phytochemical, nutritional, and organoleptic properties (Table 11.2). The main specialty types are blue, quality protein maize, cacahuacintle, and high-oil corn (HOC). Historically, the use of blue and Cuzco/*Cacahuacintle* landraces for production of foods by the Aztecs (Serna-Saldivar 2015a) has been documented. Besides, some experimental emerging nutritionally enhanced new types like high vitamin A caratenoid and Zn corns are being promoted throughout the globe (Table 11.2).

Most blue types are still landraces endemic of the highlands of Mexico or Peru. However, recent plant breeding efforts have generated hybrids adapted to tropical and subtropical ecosystems with harder endosperm texture suited for production of tortillas and snacks. Smaller blue corn kernels produce darker blue colorations because they possess relatively higher proportion of aleurone layer and are diluted less by the starchy endosperm (Garcia-Lara et al. 2012, Urias-Peraldi et al. 2013).

The ancient Cuzco or *Cacahuacintle* (from the *Nahualt* words *cacahuatl* = cocoa beans and *centli* = kernel) races characterized for yielding the largest kernels are mainly grown in highlands of Mexico and Peru and were used by Mesoamerican cultures to produce *pozole* and tortillas (Table 11.2).

Most races yield white-colored large kernels with thousand-kernel weight of up to 800 g. According to Bonifacio-Vazquez et al. (2005), yellow endosperm and blue *Cacahacintle* landraces are still planted and used for preparation of tortillas in Mexico. The Cuzco maize was introduced in the United States in 1964 with the aim to adapt and transform it into high-yielding hybrids for special food uses. This was achieved after nine years of dedicated plant breeding effort that resulted in the production of white large kernels suitable for manufacturing alkaline-cooked parched snacks (Serna-Saldivar and Chuck-Hernandez, 2019).

One of the most promising specialty types is the quality protein maize (QPM) because it is non-GMO and contains higher levels of lysine and tryptophan compared to regular maize. White and yellow QPMs have intermediate endosperm texture because they maintained the cysteine-rich gamma-zeins. These genotypes are the most suitable for tortillas and chips. Serna-Saldivar et al. (1992a) compared the processing and composition of regular and QPM genotypes and concluded that tortilla chips produced from intermediate endosperm hybrids were practically identical to regular counterparts (Table 11.2).

Waxy corn containing 95% amylopectin have been used to manufacture special types of tortilla chips. However, grains are difficult to nixtamalize because they tend to easily overcook and upon stone-grinding yield sticky doughs that are difficult to machine. The tortilla baking process also yields sticky products that upon frying produce a distinctive spongy-textured chip with improved fragility, more air pores, and acceptability (Quintero-Fuentes et al. 1999). Therefore, if waxy maize is employed, it is recommended to blend it with regular maize and added it to the cooker after regular corn.

Although HOC are not currently used for production of nixtamalized foods, they have good potential, especially if kernels are white-colored. This is because HOC types contain at least twice the amount of oil compared to regular counterparts and the oil is rich in polyunsaturated fats and liposoluble phytochemicals that favor human health (Table 11.2). The main potential of HOC is for the production of low-fat or fat-free tortilla chips.

Besides the specialty maize types described earlier, scientists associated with CIMMyT have developed nutritionally enhanced genotypes that have been improved through conventional breeding. The two most promising are maize biofortified with provitamin A and high-Zn-enhanced maize (Table 11.2). The high-carotenoid orange maize has been tested in tortillas produced conventionally and via extrusion with adequate carotenoid retention results (Rosales et al. 2016), but unfortunately they have not been tested for production of fried snacks.

11.3.2 Lime or Calcium Hydroxide

Archeological evidence implies that wood ashes were the first source of alkali used by ancient Mesoamerican civilizations to alkali-cooked maize. Later, different forms of lime mined from natural quarries or produced via cremation of shells were used instead of ashes (Serna-Saldivar 2015a). Nowadays, food-grade lime, such as quicklime and hydrated lime, consisting principally of calcium oxide with less than 5% magnesium oxide are commercially employed. The lime should be food-grade to ensure that is free of heavy metals like arsenic and lead. The concentration of lime commonly used during cooking is 1% based on kernel weight with levels varying from 0.8% up to 2%. The lime addition facilitates pericarp removal during cooking and steeping, controls microbial activity, and affects texture, flavor, aroma, color, shelf life, and nutritional value. Long-term storage of lime results in the loss or activity and consequently negatively affects pericarp removal. Lime also affects the amount of solids discharged into the processing plant's sewage system. Higher lime concentrations used during cooking increase the total dry matter losses, which reduces product yields and increase costs of effluent treatments. Since only a small quantity of lime (<20%) is retained by the *nixtamal*, the remainder is discharged into the drainage system (Serna-Saldivar et al. 1990, Serna-Saldivar and Rooney 2015, Serna-Saldivar and Chuck-Hernandez 2019, Serna-Saldivar 2021).

The combination of thermal treatment and lime facilitates pericarp removal during cooking–steeping, controls microbial activity, and affects texture, flavor, aroma, color, shelf life, and the nutritional value of end-products (Serna-Saldivar and Chuck-Hernandez, 2019, Serna-Saldivar and Rooney 2015, Serna-Saldivar 2021). According to animal trials, the calcium absorbed by the *nixtamal* and present in tortillas is highly bioavailable (Serna-Saldivar et al. 1992).

In terms of functionality, the calcium oxide hydrolyzes bonds holding the hemicelluloses in cell walls, thereby facilitating the removal of the pericarp. The degraded pericarp, rich in arabinoxylans (Ayala-Soto et al. 2016), acts as a hydrocolloid and imparts desirable textural properties to the *masa*. This weak alkali also improves niacin bioavailability but negatively affects the phytochemical profile because cooking enhances the leaching of phenolics associated with cell walls, arabinoxylans, B-vitamins, and other important nutrients into the *nejayote* (De la Parra et al. 2007, Serna-Saldivar et al. 2015, Serna-Saldivar 2021). For corn and tortilla chips, the nixtamal is more extensively washed in order to remove the pericarp and avoid darker colorations in end-products.

11.3.3 Water

Water is fundamental because it acts as the main medium to transfer heat to the kernels during cooking and steeping and enhances starch gelatinization and protein denaturation. The pH of the water and the minerals that it carries can affect the color of the final product. The lime is not fully solubilized in the water normally employed for cooking. For the manufacture of fresh *masa*, kernels are generally cooked with three parts water and 1% lime based on the total kernel weight. Kernels absorb about 28–30% water during cooking and 10–15% additional water during the steeping stage (Serna-Saldivar et al. 1990). The amount of water absorbed by the *nixtamal* depends upon the extent of cooking, which varies with the type of product and the desired *masa* texture. In fresh *masa* operations, additional amounts of water (2–4% based on the *nixtamal* weight) are added during the milling operation. The amount is carefully dosified for two main purposes: increase the moisture *masa* content to the optimum level for processing and to cool down the milling stones. Normally, *nixtamal* for production of snacks is milled with lesser quantities of water compared to *masa* intended for table tortillas. The final *masa* moisture is a key critical control point in processing operations because it affects dough machinability and regulates oil uptake during deep fat frying and the subsequent quality of the snacks (McDonough et al. 2001, Serna-Saldivar and Rooney 2015).

11.3.4 Oil

Frying oil is the second major component in all lime-cooked snacks. The selection of frying oil depends on cost, stability or shelf life, flavor, and amount of *trans*- and saturated FA. Soybean, palm, cottonseed, canola, and sunflower oils alone or blended are the most popular (Chapter 7). Premium peanut oil, although more expensive, is often blended with other oils to impart a nutty flavor. Frying oils are usually blended with more stable oils like palm or partially hydrogenated to improve stability and prevent oxidation in final products. Partially hydrogenated oils are losing use because they contain significant amounts of *trans*-fats that are negatively perceived by educated consumers. For soybean oil, the slight hydrogenation process is aimed toward the reduction to about one-half of the original linolenic acid content (from 4% to less than 2%). The shelf life of frying oils depends on the oil composition, use of antioxidants, and manufacturing practices discussed in Chapter 3. Manual or automatic filtering of the frying oil, skimming, and maintenance of the proper oil temperature improve oil performance and shelf life of end-products. Oil uptake mainly depends on the moisture content of the *masa* or unfried tortilla chips and the degree of starch gelatinization, frying temperature–residence time, and sort of frying oil. Generally speaking, the lower oil temperature and longer frying residence time increase the amount of oil retained by the snack food product (Serna-Saldivar et al. 1990, Serna-Saldivar and Rooney 2015).

11.3.5 Salt and Flavorings

Salt and flavorings are added as powders or liquids to extruded corn and tortilla chips to enhance the organoleptic properties and appearance of the finished products. The industrial application and formulation of different types of seasonings are discussed in Chapters 4 and 8, respectively.

11.3.6 Antioxidants and Chelating Agents

Synthetic and natural antioxidants are used in frying oils and finished snack foods to prevent oxidative rancidity and protect them from rancidity that leads to development of off-flavors and odors. They act by adsorbing free radicals and disrupt the first step of the chain reaction of autoxidation named initiation described in Chapter 3. This first reaction is catalyzed by ultraviolet light, heat, and the presence of metals and occurs at faster rates in polyunsaturated FA. The use of antioxidants is generally accompanied with a metal sequestrant or chelating agent. The most common antioxidants used are BHA and BHT and common chelating agents are EDTA, phosphoric, ascorbic, and citric acids. The maximum level permitted of synthetic antioxidants in foods is generally 0.02% based on the oil/fat weight present in the snack food. Synthetic antioxidants are lost due to distillation and thermal degradation during prolonged heating of frying oils. Natural tocopherols, widely used as dietary supplements, are gaining popularity replacing synthetic antioxidants, although they are considerably more expensive. These have a potent antioxidant effects both in snack foods and in the human body and are considered GRAS. Like synthetic antioxidants, citric, phosphoric, and/or ascorbic acids act synergistically with tocopherols. The tocopherol level commonly used is 0.02–0.05% (Serna-Saldivar 2008).

11.4 PRODUCTION OF PARCHED ALKALINE-COOKED CORN PRODUCTS

Alkaline cooked, parched, and fried corn kernels are popular salty snack items commonly produced from large and soft textured kernels belonging to the ancient races Cuzco or *Cacahuacintle*. This category of snacks is classified within the snack nuts and seed category. The ideal corn kernels for this snack food was developed from the Giant Cuzco race from Peru or alternatively produced from white-colored *Cacahuacintle* genotypes. These genotypes that bear kernels in eight-rowed ears yield the largest known corn kernel (Table 11.2). The Peruvian Cuzco corn cultivars grow at high altitudes and yield white kernels with soft endosperm texture and bland flavor. The first Cuzco-derived hybrid adapted to the Unites States was introduced in 1964 after nine years of painstaking breeding work. Several new hybrids have been developed since then. Kernels are harvested when the moisture content of the grain is less than 30%. The kernels are cleaned, sized, and carefully dehydrated to lower the moisture to less than 15% (Serna-Saldivar 2008).

The industrial production of parched corn starts when kernels are alkali-cooked in excess water and washed to remove the bran or pericarp (Figure 11.2). Then, the pregelatinized corn kernels are transferred into steep tanks where they are blanched for a few hours in warm water. After draining the alkaline solution, the cooked kernels are deep-fat-fried to evaporate most of the moisture and develop the characteristic texture, flavor, and color. Most of the fried and parched corn are seasoned before cooling and packaging (Serna-Saldivar 2008, Serna-Saldivar and Chuck-Hernandez 2019). The final product commonly contains about 14–15% oil and provides about 440–450 kcal/100 g (Chapter 20). Approximately, 93 kg of salted/seasoned parched and fried kernels are obtained from 100 kg of Cuzco corn kernels (Figure 11.2).

11.5 PRODUCTION OF LIME-COOKED SNACKS

Most lime-cooked maize products consumed outside Mexico and Central America are in the form of fried snacks. Nixtamalized snacks have gradually gained market within the salty snack food

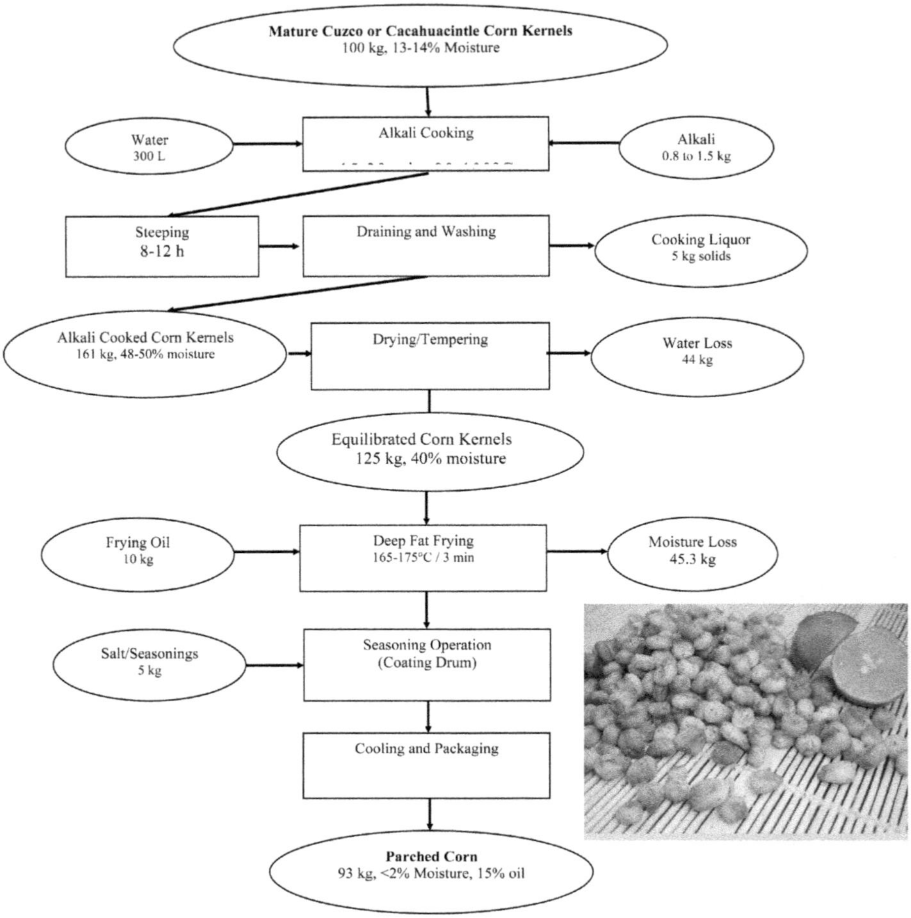

FIGURE 11.2 Flowchart of the industrial process for the manufacture of alkaline-cooked parched corn kernels.

segment during the past 30 years. Today, production volume and sales of both corn and tortilla chips are higher than the traditional leader of the market: potato chips. Baked and baked/fried snacks are gaining popularity because of their lower caloric content and new market global trends.

Tortilla chips and extruded corn chips are usually manufactured from coarsely ground fresh *masa* or alternatively from dry *masa* flour. In contrast to table tortillas, the maize kernels are cooked less extensively and/or quenched immediately after cooking and the *nixtamal* ground into a coarse *masa* with a slightly lower moisture content (approximately 2%). The idea is to gelatinize the starch to a lower extent and produce coarser particles that will allow stem to scape during baking and deep fat frying. For corn chips, the *masa* is extruded and cut into different forms and fried directly, whereas tortilla chips are preformed into thin *masa* pieces in preparation for baking before frying. The additional baking step used for tortilla chips diminishes the moisture content, so these products uptake less oil during frying and acquire a firmer texture and a stronger alkaline flavor and color (Figure 11.1). The baking of table tortillas and tortilla chips also greatly differs. Table tortillas are usually baked to retain more moisture and enhance puffing at the last stages of the thermal process, whereas tortilla chips are baked to remove moisture and the oven temperature regulated to avoid puffing or blistering (Serna-Saldivar et al. 1990, Serna-Saldivar 2008, Serna-Saldivar and Rooney 2015).

The most common tortilla chip configuration is triangular followed by round and strip forms. New configurations exist on the market such as rolled and three-dimensional triangles. *Tostadas*, taco shells, and baskets are tortillas produced from coarse *masa* fried flat, folded, or in a basket, respectively. These products are filled/topped with meat or vegetables and served in restaurants and fast-food establishments. Restaurant-style tortilla chips are produced similarly, but fine *masa* is formed and baked into thin table tortillas. Tortillas are equilibrated for a long time at ambient or refrigeration temperature and normally cut into four pie-shaped pieces before frying (Serna-Saldivar et al. 1990, Serna-Saldivar 2008, Serna-Saldivar and Rooney 2015).

A general flowchart summarizing the typical end uses of *masa* presented in Figure 11.1 includes lime-cooking, steeping, grinding, and baking for production of table tortillas and additional frying for production of both corn and tortilla chips.

11.5.1 Production of Fresh *Masa*

Dough or *masa* is considered the most critical ingredient needed for the production of snack foods. The industrial processes to obtain *masa* are divided into fresh and dehydrated *masa* flour. For production of fresh *masa*, snack food processors need to select and clean the grain, lime-cook the kernels to obtain *nixtamal*, and grind the lime-cooked kernels into a cohesive dough or *masa* containing moisture contents >45%. Figure 11.3 details the typical procedure to obtain coarse *masa*

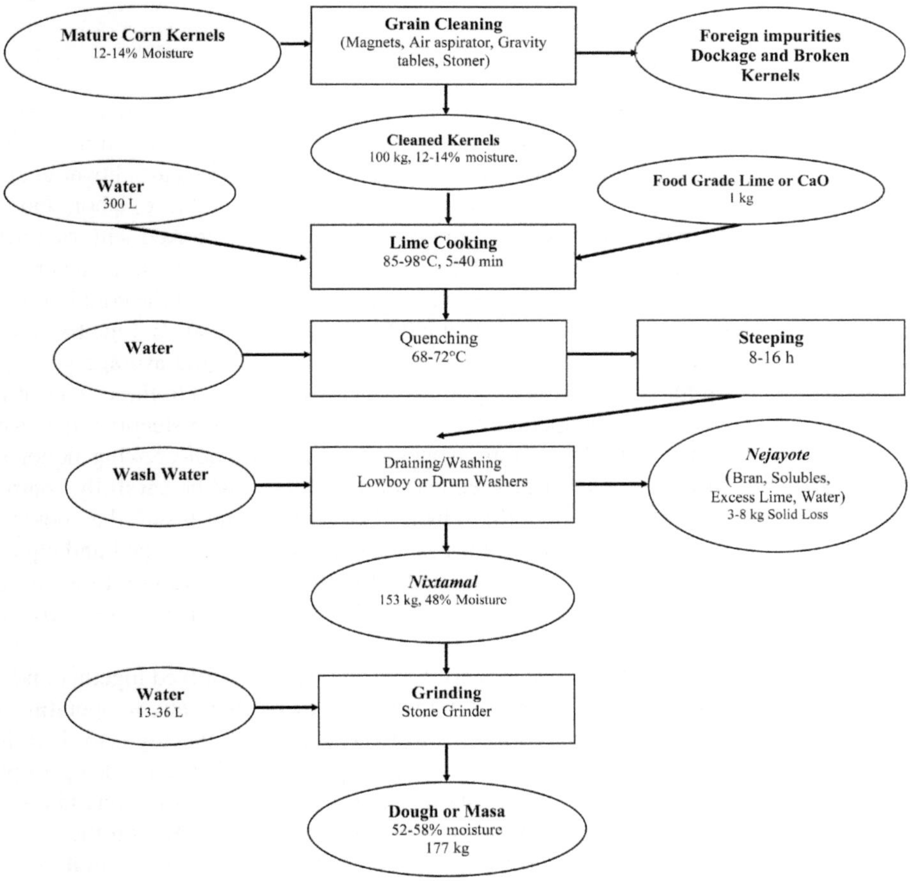

FIGURE 11.3 Flowchart of the production of fresh *masa* needed for the production of lime-cooked snacks.

suitable for the production of both corn chips, tortilla chips, and related products. The key steps are grain cleaning, lime-cooking, and wet milling of the *nixtamal* into the dough.

11.5.1.1 Storage and Grain Cleaning

Good storage conditions must be maintained to retain grain quality. Careful monitoring to avoid insect and rodent growth and mycotoxins, mainly aflatoxins and fumonisins, contamination is necessary. If maize is contaminated with one or several of these biotic agents, the resulting snacks can have sanitary problems such as excess insect fragments, presence of rat hair/feces/urine, or mycotoxins (Chapters 19 and 20). Maize should be maintained below 14% moisture with rotation, ventilation, or drying practices to avoid heat and insect damage and moldy kernels. If maize is to be dried, the temperature should be controlled to avoid stress cracks and fissures. As a result, the lime-cooking properties are significantly improved in low-temperature or field-dried corn. The corn lot is optimally cleaned with air aspiration to remove light contaminants such as vegetable residues, empty or insect-damage kernels, and other light or low-density contaminants followed by sifting with at least two sieves and stoner or gravity tables. Some sifters have an integrated aspiration system to remove the chaff and light contaminants. The last step is passing the cleaned grain through a gravity table or stoner with the aim of removing foreign material with the same size but different density than the clean kernels. These equipment are calibrated and used to remove damaged-unfilled kernels that are generally high in aflatoxins and insect fragments and stones that have lower and higher density than kernels. During these cleaning steps, the grain is usually passed through magnets integrated into the cleaning devices in order to trap ferrous metals that can damage milling and forming equipment.

11.5.1.2 Lime-Cooking and Steeping

Maize is usually batch-cooked in fresh *masa* operations (Figure 11.3) and continuously cooked in dry *masa* operations. In fresh *masa* operations, three basic types of cookers are commercially used: open vats, Hamilton steam jacketed kettles, and vertical closed cooker (Serna-Saldivar et al. 1990). The open vat is the oldest and more intensive in terms of energy expenditure, labor, and control. The Hamilton steam cooker is widely used and the maize is indirectly cooked with steam and the heat transfer ensured by mechanical agitation. The vertical cooker employs direct steam injection to heat and agitate the maize and lime solution in a large tank. Regardless of the cooking procedure and type of cooker, the grain is normally mixed with 2.5–3 parts water and 1% lime based on grain weight. Maize is considered to cook at temperatures above 68–70°C (the average gelatinization temperature of maize starch). The cooking program is generally divided into three steps: rise time, hold at the programmed maximum temperature, and temperature decrease-steeping. In some snack food operations, the last step is performed after quenching the *nixtamal*-cooking liquor to drop the temperature to 65–70°C. Graphs of time–temperature profiles must be carefully controlled to achieve uniform cooking and quality of end-products. The temperature inside the cooker is usually recorded from the time heating begins until the maize mixture has returned and equilibrated to room temperature. The integration of the area under the >68°C temperature–time curve is the best to predict optimum cooking schedules, extent of cooking, and produce uniform *nixtamal* for optimum processing.

During the first cooking step, the maize kernels, lime, and water are mixed together and heated. Most processors gradually increase the temperature near or to boiling. The temperature is then hold for a predetermined time and the heat source turned off. Steeping usually lasts 8–16 h and is important because during this period the *nixtamal* keeps uptaking moisture needed for optimum. The *nixtamal* is either steeped in the cooking tank or pumped to steeping or holding tanks.

The extent of lime-cooking depends on the type of product to obtain and on the grain physical properties (test weight, 1,000 kernel weight, endosperm texture), the interaction of temperature, time and lime concentration, the size of the cooking vessel, and the frequency of agitation. Optimum cooking and steeping are determined subjectively by evaluating the extent of pericarp removal, kernel softening, water uptake, and overall appearance of the *nixtamal*. The *nixtamal* final moisture content is an excellent index of the extent of cooking because it is closely correlated with the degree of starch gelatinization. Generally, *nixtamal* for table tortillas is cooked and steeped to increase its moisture content to about 50%, whereas *nixtamal* for tortilla chips to approximately 46–48%. *Nixtamal* made for snack chips is cooked to a lesser extent either by decreasing cooking time or by quenching as explained above. The main aim of quenching is to quickly drop the temperature, avoid overcooking, and absorb excess water, especially in large cooking vessels.

11.5.1.3 Nejayote Draining and Nixtamal Washing

After steeping, the *nixtamal* and steeping liquor are pumped to or dropped by gravity into mechanical washers. The steep liquor or *nejayote* is first drained and separated from the *nixtamal*, which is subsequently washed with pressurized water to remove the loosen pericarp and excess lime (Figure 11.3). Washing in commercial processes is done using two major equipment: the drum washer or the "lowboy" system (Serna-Saldivar et al. 1990). A drum washer consists of a conveyor that transports the *nixtamal* into a rotating perforated cylinder with internal flights and water sprayers located within the drum. On the other hand, the lowboy system consists of a receptacle equipped with internal screens and sprayers in which the cleaned *nixtamal* is continuously removed from the bottom of the receptacle by an inclined belt conveyor. The washing operation is aimed toward the removal of excess lime and chemically hydrolyzed bran or pericarp enhancing color of snacks.

The drained cooking liquor and the wash waters are normally channeled to the drain or to an effluent treatment plant. The high alkaline pH plus the soluble and insoluble maize solids make residual water high in biological oxygen demand (BOD) or chemical oxygen demand (COD). Therefore, the recovery of the insoluble solids lost during cooking, steeping, and washing and the treatment of the resulting effluent are critically important, especially for those plants located in or nearby cities. Common practice is to screen the liquor to remove insoluble solids and then further reduce the level of suspended solids in sedimentation lagoons (Serna-Saldivar 2008, Serna-Saldivar and Rooney 2015).

11.5.1.4 Stone-Grinding

Grinding is a key operation because it promotes plastic and cohesive properties in the *masa*. The degree of starch gelatinization plus the milling action interacts to form a *masa* with optimum properties for further processing.

In fresh *masa* operations, the clean *nixtamal* is normally wet-milled using a system of two matched carved lava or synthetic stones: one stationary and the other that rotates at 500–700 rpm (Serna-Saldivar et al. 1990) or a grinder equipped with steel plates (Mehta 2001). Volcanic and synthetic (aluminum oxide) stones are the most commonly used by the industry. Synthetic stones last longer and require less maintenance; however, they cannot be dressed in the manufacturing plant (Serna-Saldivar et al. 1990). According to Mehta (2001), natural or lava stones should be redressed every three to four weeks based on two shifts per day. These grinders can process up to 1,800 kg *nixtamal*/h. The typical stone is circular shaped and has an inner central orifice. Stones are generally 10 cm thick and with different diameters for diverse milling capacities. Commercial grinders use stones 25–46 cm in diameter. The internal face is carved radially or spirally. The grooves become progressively shallower as they approach the perimeter of the stone. The number, design, and depth of the grooves vary with the intended product. Stones carved for the manufacturing of snacks have deeper grooves to produce coarser *masas*, whereas stones for production of fine *masas* for table tortillas have shallower and more grooves. The grinding starts when a screw conveyor at the base of the hopper forces the *nixtamal* through a center opening and into the gap between the stones, where attrition and shearing occurs. The material travels outward from the center to the perimeter of the stones where it is gradually cut, kneaded, and mashed. Particle size reduction is the result of several interacting factors: (1) maize type, (2) the degree of *nixtamal* cooking, (3) the design of the stone carving (groove size and depth), (4) the gap or pressure between the grinding stones, and (5) the amount of water used to assist the milling operation. As a result of grinding, the *masa* increases its temperature to 38–50°C.

Water is added during milling for two main reasons: to counteract the heat produced by friction of the stones, which reduces the *masa* temperature, and to increase the *masa* moisture content to the targeted level. For a grinder with a capacity of 600 kg/h, about 0.6–1.8 L of water per minute is added.

Mehta (2001) describes grinders equipped with steel plates that can run at higher rpm and process up to 2,700 kg/h and yield *masas* with lower temperature and reduced stickiness.

In some processes, especially taco shell *masa* production, the *nixtamal* is cut into particles with an Urschell mill. These particles must be subjected to additional attrition milling to obtain the optimum texture. The attrition process breaks down gelatinized starch granules and forms a continuous phase holding the *masa* particles together (McDonough et al. 2001).

11.5.2 Industrial Production of Dry *Masa* Flour

Production of snack foods from pregelatinized or instant DMF has increased significantly during the past 25 years, especially in developed countries around the globe. The industry produces a wide array of flours that vary in color, particle size distribution, pH, content of preservatives, level of additives such as dough conditioners, texture improvers (gums, emulsifiers, etc.), and coloring agents. The major difference among flours is in terms of the intended terminal use. DMF are commonly classified into two broad categories: DMF for table tortillas and for snack foods. Flours for table tortillas usually contain more additives and finer granulations compared to counterparts tailor-made for snack food processors.

Manufacturing snacks from DMF facilitates the process because processors do not have to concern about selection and storage of the corn, the elaborated lime-cooking/steeping, and grinding steps that require expertise, large amounts of water, and energy expenditure. Consequently, the use of DMF greatly reduces processing time, floor space, labor, capital investment for processing equipment, and wastewater treatment. Disadvantages include the comparatively higher price of DMF and the subtle differences in flavor and texture in favor of fresh *masa* products.

11.5.2.1 Industrial Manufacturing of DMF

The industrial production of DMF is commonly achieved following similar steps of fresh *masa* with the obvious addition of drying steps and dry milling procedures to achieve the desired particle size distribution in final flours (Figure 11.4).

The physical properties and color of the corn kernels greatly affect processing parameters and quality of end flours. The mature corn kernels are checked in terms of important parameters such as test weight, endosperm texture, thousand kernel weight, starch damage, insect damage, and aflatoxins. These physical grain properties are carefully assessed because they are closely related to extent of cooking. Before processing, the kernels are thoroughly cleaned in order to remove chaff, extraneous materials, broken kernels, and other impurities.

Most DMF are processed using highly mechanized continuous procedures. The calcium hydroxide commonly added varies from 0.8% to 1% based on the kernel weight. The lime is mixed beforehand with the water, which is generally added with equal parts in relationship with the corn. This operation is in most instances performed in large reactors equipped with a screw conveyor fitted with steam jets. The kernels cook in the different segments of the reactor while moving along the conveyor. Generally, the grain is cooked for 20–30 min at temperatures 75–95°C (Serna-Saldivar et al. 1990, Serna-Saldivar and Chuck-Hernandez 2019). One major difference between production of fresh *masa* and DMF is that the *nixtamal* is not extensively steeped. Normally the *nixtamal* with approximately 36% moisture is immediately separated from the wastewater or *nejayote* in continuous draining and washing equipment such as drum or lowboy systems. The cleaned *nixtamal* is then stored in a bin that feeds specialized hammer mills that continuously grind the wet *nixtamal*

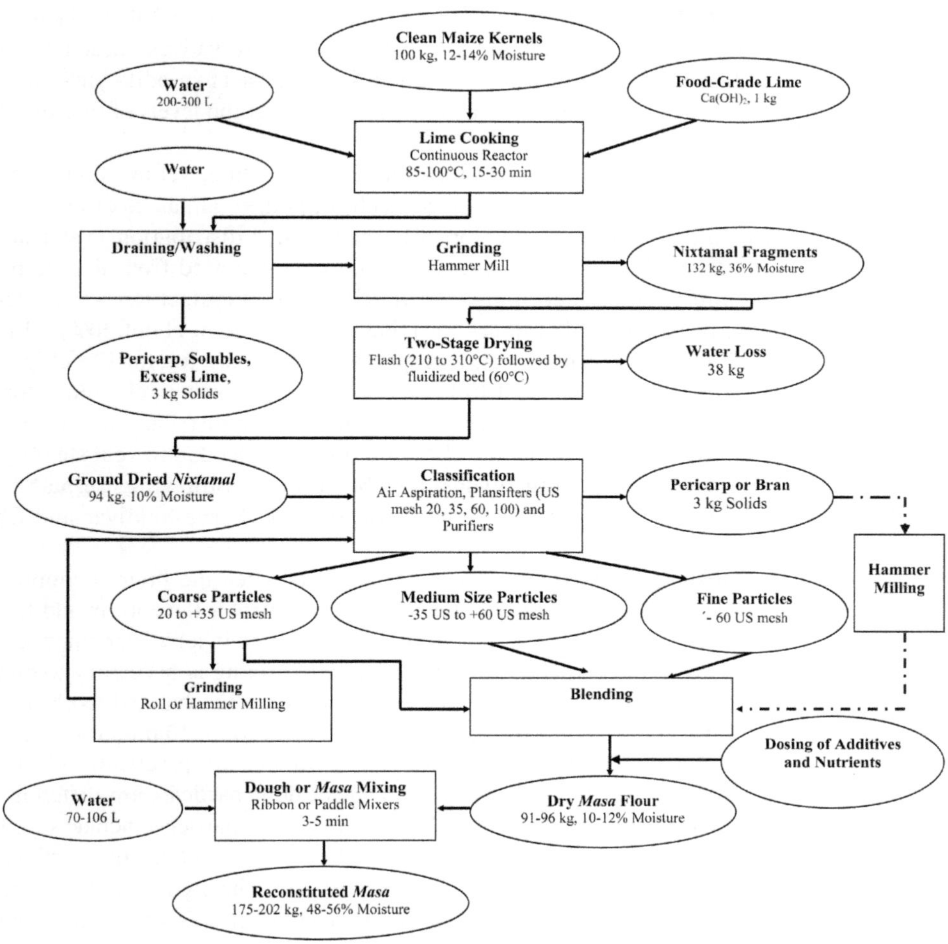

FIGURE 11.4 Flowchart of the milling process for the production of dry *masa* flour and its reconstitution into fresh masa.

to yield fragmented coarse pieces. Alternatively, stone or plate-attrition mills can be employed. The key unit operation for production of DMF is the dehydration of the fragmented particles because it plays a critical role in the final quality of the flour. This operation is usually performed in large tunnels or drying towers in which warm air flows counter-currently to the wet *nixtamal* pieces. The first dehydration step is flash and usually performed at 230–310°C for only few seconds with the goal of removing most of the free water. During this short time, the moisture content of the *nixtamal* pieces drops from 36% to about 20–22%. Then, particles are continuously and gradually dehydrated at lower temperature (60°C) to avoid over-gelatinization and to reduce the moisture to approximately 10% (Serna-Saldivar et al. 1990, Serna-Saldivar 2008, Serna-Saldivar and Rooney 2015, Serna-Saldivar and Chuck Hernandez 2019). The final stages of DMF production consist in classification of dry particles, regrinding particles that do not meet the granulation requirement, and reformulation. The dried particles are air-aspirated and sized or sifted in plansifters and purifiers. The aim of air aspiration is to remove leftover pericarp pieces released during grinding. In some operations, the bran is finely ground and reincorporated into the DMF to yield a whole-grain product. Various flour streams with different granulations are selected and combined to produce DMF with optimum granulation for different applications. The common yield of DMF with 10% moisture is 92–93% based on the original cleaned corn weight (Serna-Saldivar et al. 1990, Serna-Saldivar

2008, Serna-Saldivar and Rooney 2015, Serna-Saldivar and Chuck-Hernandez 2019). The final step of the DMF process is supplementation with gums, organic acids, preservatives, bleaching agents, and a package of selected vitamins and minerals. These additives are used to modify yields and end-product properties, enhance microbial and textural shelf life, and meet the levels of micronutrients specified by regulatory agencies.

The different types of DMF are tailor-made for production of white, yellow, and blue table tortillas with different shelf lives, corn chips, and tortilla chips. Some companies offer more than 25 different DMF formulations that vary in color, pH, particle size distribution, water absorption, and starch gelatinization. Table tortillas are commonly manufactured from fine flours that have more gelatinized starch, whereas corn and tortilla chips are manufactured from medium and coarse flours, respectively. Generally, flours for table tortilla contain about 36%, 29%, and 34% of −100, +100, and +60 US mesh particles, respectively, whereas for tortilla chips 12%, 11%, 23%, and 54% of −100, +100, +60, and +40 US mesh particles, respectively. Flour for corn chips had an intermediate particle size distribution. The particle size distribution is coarser for tortilla chips, taco shells, tostadas, and corn chips because pores are needed to vent steam generated during baking and frying and avoid the undesirable blistering or pillowing (Serna-Saldivar et al. 1990, Almeida-Dominguez et al. 1996, Serna-Saldivar 2008, Serna-Saldivar and Chuck-Hernandez 2019).

Production of nixtamalized snacks from dry *masa* flour starts when the flour is simply rehydrated in a mixer (Figure 11.4). The most important parameter to control is the correct addition of a predetermined amount of water (1–1.2 L/kg flour). Mixing is performed in a blender equipped with a sigma blade that rotates at slow speed (15–25 rpm). The gentle mixing is necessary to enhance hydration and avoid the production of sticky doughs due to excessive mechanical work or shear. Mixing usually last 3–10 min. The rehydrated *masa* should be rested for 5–10 min after mixing to allow water equilibration. After mixing, *masas* from fresh operations are practically identical to rehydrated *masa* from nixtamalized flours. However, dry *masa* flour particles are different from those of fresh *masa* (Gomez et al. 1989, Gomez et al. 1987). Fresh *masa* particles include significant quantities of free starch granules with low protein content, while dry *masa* flour has particles that contain starch and protein similar to levels in the original endosperm. The highly hydrated nature of *nixtamal* (48% moisture) for fresh *masa* and the more extensive stone-grinding procedure liberates the starch from the protein matrix affecting the texture of the *masa*. In the dry *masa* flour process, the particles are obtained by hammer milling from a lower moisture *nixtamal* (36% moisture) and the final particle size is also contributed by further dry milling.

11.6 PRODUCTION OF EXTRUDED CORN CHIPS

Extruded corn chips are made from coarsely ground *masa* either from fresh *masa* (Figure 11.3) or DMF (Figure 11.4) that is extruded and cut into *masa* chips with different configurations, deep-fat-fried, and salted/seasoned (Figure 11.5). The critical operations are the extrusion of the *masa* into different configurations and the following deep fat frying step. The *masa* extruder consists of a horizontal hydraulic piston that forces the dough with at least 52% moisture to pass through a 90° forming die or head. Extrusion starts automatically after *masa* is manually loaded and the cylinder is locked into position. Easily interchangeable dies for forming custom shapes are available. A rotary cutoff assembly cuts dough into uniform pieces. When the cylinder is empty, the piston retracts for refilling. Equipment that can form up to 900 kg/h exist on the market (www.heatandcontrol.com). The extruded *masa* pieces vary in forms and sizes, the most common strips being approximately 1.2 cm wide and 4.0 cm long. Scoops and ribbons are also widely manufactured. The high moisture extruded pieces lost most of the water during the first 10 s frying at temperatures 175–195°C. The frying time also varies, but commonly averages 1.5 min. During deep fat frying, the *masa* exchanges water for oil and the product that exists in the fryer has moisture and oil contents of about 1.5% and 36%, respectively.

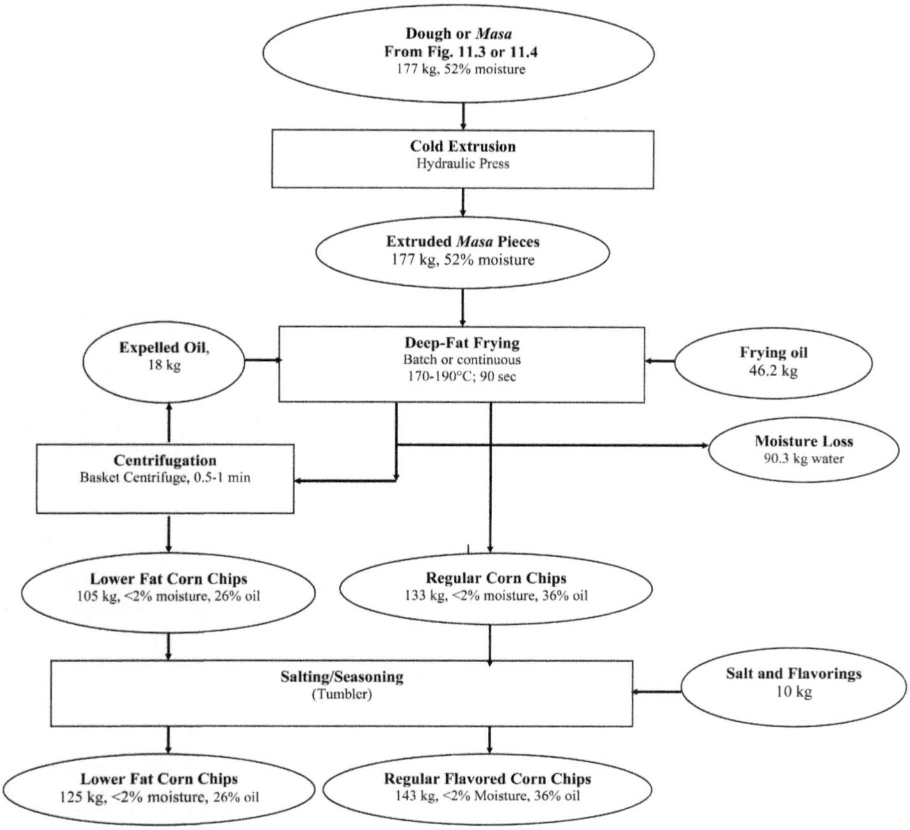

FIGURE 11.5 Flowchart for the process of regular and partially defatted extruded corn chips.

Resulting chips are normally seasoned with 2% salt and various amounts of cheese, citric acid, or chili flavorings (Chapters 3 and 8). In Mexico, red-colored corn chips are demanded by many consumers because of unique flavor and color. For this application, the *masa* is blended with natural ingredients like paprika and/or with FD&C approved colorants before extrusion. One hundred kilograms of cleaned corn kernels with 13% moisture yield approximately 134 kg of salted/seasoned corn chips containing around 36% oil (Figure 11.5).

Partially defatted corn chips are also manufactured in Mexico. The hot fried chips with approximately 36% oil are immediately placed in perforated wire-type baskets for 30–60 s centrifugation. The centripetal force removes part of the free oil without damaging the preformed structure of the chip and decreases the oil content to approximately 26–28%. The removed oil is generally collected and returned to the fryer.

The nutritional composition and caloric density of extruded corn chips is detailed in Chapter 20.

11.7 PRODUCTION OF TORTILLA CHIPS

Tortilla chips are normally produced from coarsely ground *masa* either from fresh *masa* (Figure 11.3) or reconstituted DMF (Figure 11.4) that is laminated and cut into different configurations (triangles, small circles, strips) in preparation for the subsequent steps of baking, cooling, deep fat frying, and seasoning (Figure 11.6). The restaurant-style tortilla chips are normally manufactured from fine *masa* that is sheeted into thin tortillas that are cut into four wedges before frying.

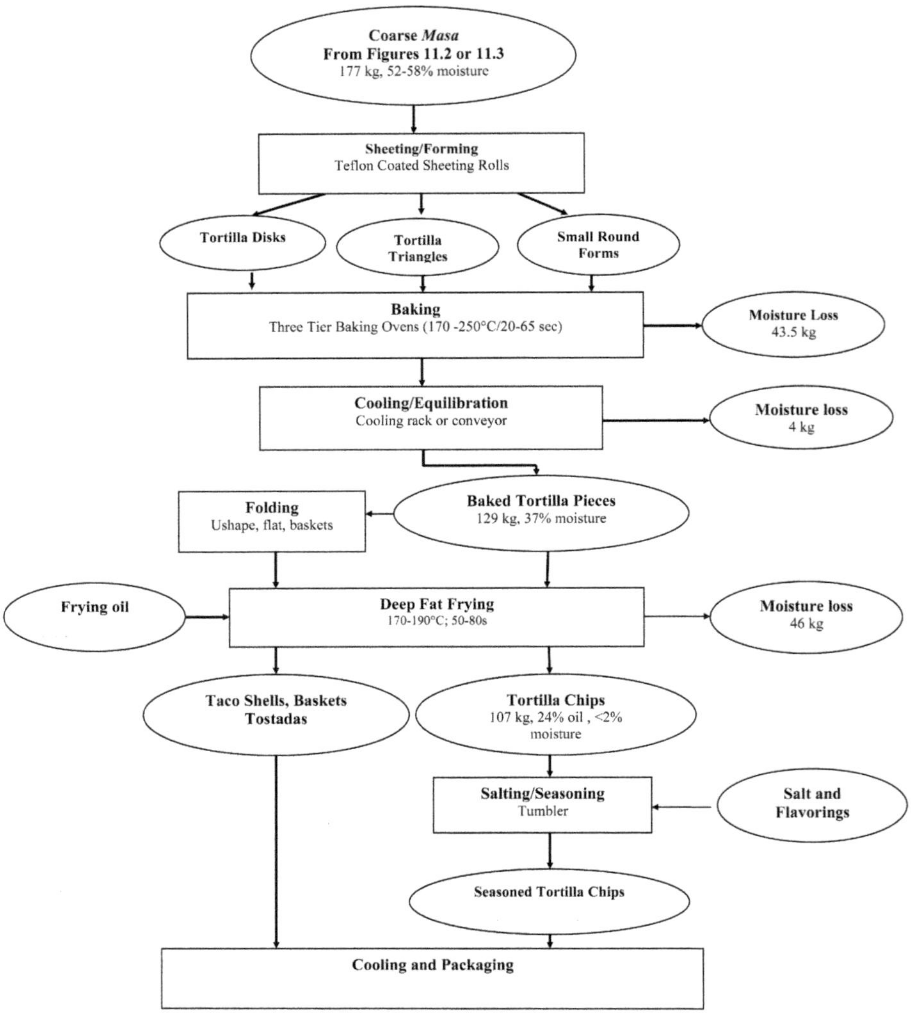

FIGURE 11.6 Flowchart for the production of regular tortilla chips and related products.

11.7.1 Regular Tortilla Chips

11.7.1.1 Kneading and Sheet-Forming Operations

The dough or *masa* is kneaded and mixed into a plastic and cohesive dough by mixers or extruders. This operation is generally performed with a *masa* feeder. In smaller operations, the *masa* is mixed and hand-formed into large loaves, which are manually placed into the hopper of forming equipment.

Tortilla chips are usually formed by two rotating smooth Teflon-coated rolls that automatically press the *masa* into a thin sheet. The gap between the rolls determines product thickness and weight. The resulting sheet of *masa* is cut into large or small disks, triangles, pie-shaped pieces, strips, or other configurations by an attachment located underneath the front roll. Different interchangeable cutters are used for production of various configurations. Sheeting heads that cut up to eight rows of products exist (Serna-Saldivar et al. 1990). A set of cutting wires also helps discharge the *masa* disks onto the conveying belt that feeds the oven. The most common current method takes advantage of the speed differential of the two rolls. As the *masa* passes through the gap between the rolls, it leaves the slow roll and sticks to the faster roll. These sheeters are equipped with a hinged rear roll support that ensures precise gap control and a pneumatically controlled containment saddles to prevent *masa* leaks (www.heatandcontrol.com). The sheet of *masa* is cut and scrap dough recycled similarly as it occurs in the traditional sheeter (Serna-Saldivar and Rooney 2015, Serna-Saldivar and Chuck-Hernandez 2019).

11.7.1.2 Baking

The newly formed *masa* configurations are usually baked into tortilla pieces destined for frying in a three-tier gas-fired oven at temperatures 175–250°C for 20–60 s. These ovens are built with atmospheric gas combustion or proportional mix burners. The baking belts are either of the slat type or woven wire. For snacks, woven wire is preferred. One side of the preformed tortilla is baked twice as long as the other side (Serna-Saldivar et al. 1990). For production of tortilla chips, the burners of the third or final tier are regulated at a lower temperature to avoid pillowing or blistering. The capacity of commercial ovens vary, the larger are capable of baking more than a ton per hour. The best ovens are equipped with fuel-efficient burners and high insulation to lower heat loss and save energy. Food processing manufacturers are also building single-pass ovens tailored for *tostadas* and taco shells. These ovens are longer than conventional triple pass ovens (http://heatandcontrol.com). During baking, about 10–14% moisture is lost from the *masa* to yield baked tortilla pieces containing 42–46% moisture. During this critical operation, the starch further gelatinizes, protein denatures, flavor and color develop due to Maillard reactions, and the microbial counts are eliminated. Texture wise, the tortilla develops its basic structure because the gelatinized starch acts as cohesive glue.

11.7.1.3 Cooling Equilibration

The baked tortilla pieces are cooled for 3–20 min through a series of open tiers that discharge directly into the fryer. The industry supplies models with one, three, and five tiers (http://heatandcontrol.com). Cooling is necessary to decrease the temperature of the tortilla pieces to about 25–30°C and enhance starch retrogradation that significantly improves the texture of the fried products. Some cooling conveyors or racks are supplied with fans to speed up the process. Equilibration time permits some of the moisture inside the chip to migrate to the outside and evaporate, effectively lowering the oil uptake during frying and the formation of blisters and oil pockets. The rate of moisture loss will depend on the temperature, relative humidity, and surrounding air velocity. Commonly, 2–4% moisture is further lost during cooling and equilibration. In some commercial operations, the cooled tortilla chips are left equilibrating for several hours or even overnight. For instance, for many restaurant-style tortilla chips, the baked tortillas are equilibrated for 12 h or more at refrigeration temperatures before frying. The equilibrium of baked chips prior to frying critically affects the oil content, color, texture, and appearance of the tortilla chips. During frying, cell structure and pore size of the chip are affected because the water escapes in a different fashion when the baked tortilla pieces are adequately equilibrated. The surface pores of the chip affect light refraction, which relates to appearance and color. A proper equilibrated baked tortilla piece will have fewer defects such as pillowing and better texture after frying (McDonough et al. 2001).

11.7.1.4 Frying

Frying of tortilla chips is characterized by starch gelatinization and formation of a yellow crust and the characteristic flavor as a result of Maillard reactions. In tortilla chips, most of the starch gelatinization occurs during baking and throughout the first 10–15 s of frying (Moreira et al. 1995).

Most tortilla chips are deep-fat-fried in continuous commercial fryers (Chapter 3) with an hourly production output of up to 1.4 ton (Serna-Saldivar et al. 1990). According to Moreira et al. (1995), the thermal conductivity values for raw and fried tortilla chips decreased to 0.23 W/m/°C and 0.09 W/m/°C, respectively, whereas the specific heat varied from 3.36 to 2.19 kJ/kg/°C indicating that Cp decreased as moisture content diminishes and oil content increases. During the frying of tortilla chips, heat is transferred from the hot oil onto the product surface by convection and from the surface into the center of the chip by conduction. The liquid water moves from the inside to the evaporation zone, leaving the product surface as vapor. Some of this vapor may remain trapped within the intracellular pores due to restrictive diffusion. The vapor will expand distorting the pore walls and contributing to the internal porosity (Moreira et al. 1995, 1999). When unfried tortilla chips are placed into the hot oil, the moisture is gradually removed and the evaporated water partially replaced by the frying oil. In other words, there is a mass exchange. Most of the water is released and oil absorbed during the first 10 s of frying. McDonough et al. (1993) evaluated the structural changes of tortilla chips using scanning electron microscopy and concluded that the frying oil immediately coats and clings to the chip surface and diffuses into the item as soon as it contacts the hot frying oil in the fryer.

For the production of lap-folded or conically shaped tortilla shells, tostadas, baskets, and related products, tortillas from coarse *masa* are either bent or formed into the desired configuration using special forming devices and then fried (Serna-Saldivar et al. 1990, Serna-Saldivar and Chuck-Hernandez 2019). The frying oil temperature and product residence time depend on the type of product. Baked tortilla chip pieces from yellow corns are normally fried at lower temperatures compared to counterparts produced from white kernels (Moreira et al. 1999) because carotenoids degrade at high temperatures into β-ionones, affecting both the color and flavor of the fried product (McDonough et al. 2001).

11.7.1.5 Seasoning

Salt and flavorings are applied immediately after frying in rotating tumblers or drums equipped with liquid sprayers and/or auger or electrostatic powder dispensers (Chapter 3). The hot chips are typically conveyed into a cylindrical rotating cylinder equipped with internal paddles to prevent bridging and improve product spread where a seasoning applicator evenly distributes flavorings onto the product (Serna-Saldivar et al. 1990) or to a belt type of seasoning depositor, which dispenses dry seasonings onto the chips (http://heatandcontrol.com). Levels of salt and seasoning usually added are 1–1.5% and up to 8%, respectively. Approximately 109 kg of salted/seasoned regular tortilla chips with 24% oil are obtained from 100 kg of maize with initial moisture of 13% (Figure 11.6). For more information about seasoning mixes and coating technologies, the reader can refer to Chapters 3 and 8. After flavoring, the chips are allowed to cool down and immediately packaged in moisture proof or aluminized bags to avoid rancidity and loss of crispiness (Chapter 4).

11.7.2 Specialty Types of Tortilla Chips

11.7.2.1 Low-Fat and Fat-Free Tortilla Chips

Low-fat snacks are the driving force of the snack food industry nowadays. Tortilla chips are not the exemption; in fact, sales of light and reduced fat tortilla chips are steadily increasing. There are basically two types of low-fat tortilla chips: fried and baked or toasted (Figure 11.7). The first category usually contains one-third less oil compared to regular counterparts and the second do not contain any added frying oil. The nutrient composition of these sorts of tortilla chips are detailed in Chapter 20. Tortilla chips with no added oil have different taste and texture and therefore are more challenging for food product developers.

For production of reduced oil snacks, moisture is removed from the baked tortilla chip by further baking or toasting (180°C for about 3 min) or flash frying followed by baking to yield items with less than 2% moisture. In the Mexican food market, there are several brands of low-fat fried tostadas that are produced from sun-dried tortillas (Figure 11.7). These popular products have about half of the oil content (12–14% oil) compared to regular tortilla chips. During solar drying, moisture content of baked tortillas is reduced from 34% to about 7%. The equilibrium moisture content is reached after 120 min sun-drying. The advantage of solar drying is that the moisture is lost slowly minimizing the cell wall internal stress, allowing baked tortillas to shrink down to an almost solid core minimizing

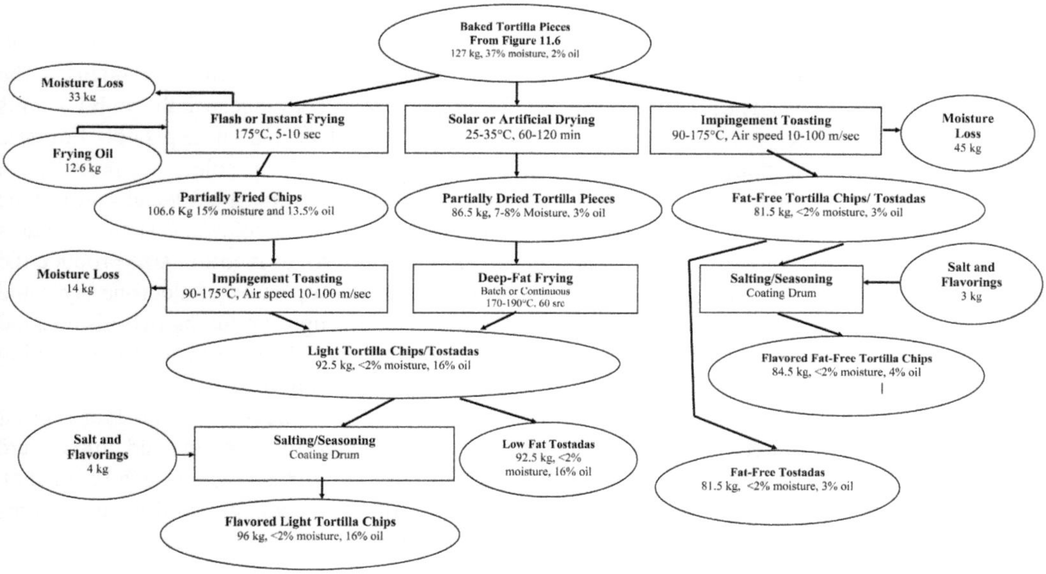

FIGURE 11.7 Flowchart for the production of low-fat tortilla chips and related products.

cracks. Sensory tests indicated that sun-dried tortilla chips had better scores than commercial baked chips (Moreira et al. 1999).

For production of low-fat snacks, hot-air impingement ovens are finding more applications and use because the heat transfer in these ovens are higher and air temperatures lower than in commercial ovens. The temperature, humidity, and air velocity are adjustable above and below the product and the product residence time adjusted by controlling the belt speed. Typically, the temperature and jet air velocity of these ovens are 100–350°C and 10–100 m/s, respectively. The enhanced heat transfer is mainly due to the arrangements of the nozzle(s) and the air velocity. Most of the heat is transferred by convection. Moreira et al. (1999) and Lujan-Acosta and Moreira (1997) described the engineering and heat transfer coefficient in impingement drying and developed a process to produce low-fat tortilla chips using this technology. The key difference between the regular and low-fat manufacturing processes is the air impingement drying after tortilla baking and prior to deep fat frying. These authors subjected 1 mm × 50.8 mm baked tortilla disks with moisture content of around 50% to drying at temperatures 93–177°C under the impinging hot air to further reduce the moisture to 14%. The dehydration time varied from 6 to 10 min, depending on the temperature setting. Immediately after drying, baked and dehydrated chips were deep-fat-fried for only 20 s at 200°C. The resulting chips contained approximately 14% oil and pieces shrunk between 8% and 10%, but increased their thickness in approximately 10%. Other potential alternative is the use of superheated steam impingement drying. In comparison with normal dry air, the superheated steam gelatinizes more the starch and yield softer products after deep fat frying. Approximately 81.5 kg of fat-free tortilla chips with less than 3% oil or 92 kg of light tortilla chips with 13% oil are obtained from 100 kg of maize or 127 kg of baked tortilla pieces containing 37% moisture (Figure 11.7).

11.7.2.2 Blue-Colored Tortilla Chips

Blue maize products, such as tortillas and chips, have recently become popular even though blue maize considerably costs more than regular maize. Blue-colored tortilla chips are marketed in most convenience stores as regular or organic items. In Mexico, the lead manufacturer of DMF merchandises blue maize nixtamalized flours.

Blue maize (Table 11.2) is a floury or soft endosperm type that generally grows in long ears (8–12 rows) derived from landraces cultivated by the ancient Inca, Aztec, and Hopi cultures (Betran et al. 2000, Serna-Saldivar 2015a). The soft textured kernels of blue maize must be lime-cooked for a short time if good quality snacks are desired. Tortilla chips made from blue maize are produced following procedures described before with the adjustment of the lime-cooking time to compensate the softness of the endosperm. The amount of lime or pH greatly affects color because the anthocyanins change coloration at different pHs. The darkest coloration is obtained at an alkaline pH of 8. Fumaric acid can be added to the dough to lower the pH and the coloration. Violet colors are prevalent at lower pH, while higher pH gives a greenish color. The anthocyanins impart a special and unique flavor to blue maize chips. These compounds are mainly located in the monolayered aleurone that is located between the pericarp and starchy endosperm. Kernels of small to medium size produce the darkest blue color chips because the smaller kernels have a higher proportion of aleurone layer and are diluted less by the starchy endosperm (Betran et al. 2000). New high-producing hybrids adapted to semitropical conditions with intermediate endosperm texture have been bred for lime-cooking purposes (Urias-Peraldi et al. 2013, Serna-Saldivar et al. 2015, Serna-Saldivar 2021).

The blue maize contains higher levels of anthocyanins and antioxidant activity than regular counterparts (Garcia-Lara et al. 2012). When blue maize is processed into tortillas and chips, there are considerable losses of anthocyanins and antioxidant activity into the *nejayote* (De la Parra et al. 2007, Mora-Rochin et al. 2016). However, the fried blue tortilla chips contain significant amounts of these antioxidants that protect humans against oxidative stress.

11.7.2.3 High-Quality Protein Maize Tortilla Chips

QPM (Table 11.2) is a high lysine-tryptophan cultivar selected from a mutant maize named opaque-2. The mutant genes that significantly increase the levels of these essential amino acids were first identified by Mertz et al. (1964). Unfortunately, the *opaque*-2 maize had a chalky soft, opaque endosperm and many undesirable agronomic characteristics such as low field grain yield and high susceptibility to insect and plant diseases. After years of continued efforts, scientists from CIMMYT gradually transformed the opaque-2 maize into an array of yellow and white QPM adapted mainly to subtropical regions. QPM was the result of many years of recurrent selection where the high lysine-tryptophan potential of opaque-2 was combined with modifier genes that significantly improved the physical properties of the kernel, especially in terms of endosperm texture and the agronomic performance of the crop. QPM varieties with white or yellow endosperm and with similar yield potential as commercial counterparts have been released in many countries around the globe.

Serna-Saldivar et al. (1992) compared the properties and processing of regular and QPM for production of tortillas and chips and found that the new intermediate endosperm hybrids produced in the United States were practically identical to regular counterparts. In addition, high-producing white and yellow QPM hybrids were developed in Mexico with excellent results when transformed into tortillas and chips (Serna-Saldivar et al. 2008). As expected, the utilization of QPM hybrids resulted in the production of tortillas and snacks with higher levels of essential amino acids, better protein quality, and physiological and brain development of laboratory animals and humans (Amaya-Guerra et al. 2004, 2006, Serna-Saldivar 2015b).

11.7.2.4 Tortilla Chips from Waxy Corn

Waxy maize is a mutant cultivar that contains starch granules with more than 95% amylopectin and has potential to produce special types of tortilla chips. The main disadvantages of waxy maize is cost and the difficulty to nixtamalize it. Waxy maize is very simple to overcook and upon stone-grinding yields sticky doughs that are more problematic to machine. Therefore, waxy maize kernels require a significantly lower cooking time and special grinding conditions. In addition, baking produces sticky raw tortilla chips which upon frying yields a spongy-textured chip with improved fragility and texture compared to chips made with normal maize. Quintero-Fuentes et al. (1999) determined that the production of tortilla chips with waxy starch or flours (rice, maize, sorghum) yielded snacks with different acceptability and texture The chips were more porous (smaller air pores were formed during baking) and friable and additionally had potential for production of lower fat snacks (Table 11.2).

TABLE 11.3
Novel Nixtamalization Technologies Developed for the Production of Fresh *Masa*, Dry *Masa* Flour and Baking of Tortillas

Type of Emerging Technology	Description
	Lime-cooking technologies
Thermo extrusion	This is the most promising alternative technology to produce both fresh *masa* or dry *masa* flour (DMF). Fresh *masa* is prepared via thermoplastic extrusion in short processing times, with low energy consumption and without generation of wastewater (*nejayote*). The preferred method extrudes a blend of corn flour, with relatively lower amounts of lime and water, which are heated into a homogeneous dough that is cooled down immediately after extrusion in a special holding chamber. The patent of Figueroa-Cardenas et al. (1996) claims that the total processing time is reduced to 1.5–7 min and yields fresh *masa* suitable for table tortillas or chips. The extrusion process for production of lime-cooked dough should be carefully controlled, especially in terms of minimizing starch damage. The extrusion process is more versatile in terms of mixing other potential ingredients and retaining relevant nutrients and phytochemicals (Mora-Rochin et al. 2010, Enriquez-Castro et al. 2020)
	This patent protects the production of DMF preferably from degermed ground maize and the separation of the horny endosperm fractions, which are tempered to 32–42% moisture in preparation for cooking to rupture the endosperm particles and enhance partial gelatinization of the starch granules. After cooking, the endosperm particles are dehydrated to yield DMF for table tortillas, taco shells, or corn chips (Lange and Wolt 1997)
Low-shear laminar transport	This technology was conceptualized to replace extrusion which produces high stress and shear. This system is based on low-shear extrusion by the inclusion of a transport system consisting of a laminar cooker without the typical die. The process consist of preconditioning ground maize that is mixed with lime and water in preparation for pumping into the cooking section and gradually transported in a laminar flow fashion, eliminating the internal stress and shear force. The water absorption and solubility indexes of DMF are similar to the ones produced by traditional nixtamalization. The main advantages are the reduction of processing time, energy expenditures, and null emission of *nejayot*e (Mercado-Pedraza 2014, Ortega-Moody et al. 2014, Escalante-Aburto et al. 2020)
Infrared cooking	Infrared heating involves the exposure of a material to electromagnetic radiation at a wavelength of 1.8–3.4 μm, enhancing the vibration of water frequencies of 60,000–150,000 MHz, and internal heating. This technology has been successfully utilized to obtain tortillas and snacks. The infrared nixtamalization retained more lipids and tryptophan and lose less solids compared to the conventional process. Tortillas elaborated with this innovative process yielded tortillas with improved nutritional and textural characteristics with a significant reduction of aflatoxins. However, this technology did not diminish water use and processing time. An important issue of NIH technology is the reduction of aflatoxin contents in tortillas, which can be promoted as an excellent benefit in terms of public health (Mendez-Albores et al. 2014, Escalante-Aburto et al. 2020)
Ohmic heating	Ohmic heating consists of applying an electrothermal technology for efficient and rapid cooking. The application of an electrical current generates internal heat, so products are cooked more uniformly diminishing the energy expenditure and processing time. In terms of lime-cooking, several relevant variables should be carefully controlled like the current and voltage applied, water-to-ground maize ratio, and the particle size of meals. The preferred manufacturing technology consists of first milling of raw maize to a 0.88-mm particle size followed by the addition of 0.3% lime and enough water to achieve 50% moisture, and finally subjecting the blend to continuous ohmic cooker. Alternatively, the precooked *masa* could be dried and ground into instant DMF. This novel technology presents the main advantages of generating nil *nejayote* and products with higher contents of total dietary fiber and phenolic compounds like ferulic acid and flavonoids (Gaytan-Martinez et al. 2012, Varghese et al. 2014, Ramírez-Jiménez et al. 2019)

(*Continued*)

TABLE 11.3 (CONTINUED)
Novel Nixtamalization Technologies Developed for the Production of Fresh *Masa*, Dry *Masa* Flour and Baking of Tortillas

Type of Emerging Technology	Description
Microwave cooking	Whole-maize kernels or ground maize are mixed with water and calcium hydroxide and heated for 4–20 min in a microwave oven applying up to 2,450 MHz and the resulting lime-cooked kernels subsequently dried and ground into DMF. The reconstituted *masa* showed good consistency and tortillas showed comparable color, flavor, and rollability to control tortillas. The microwave-assisted nixtamalization offers the advantage of short processing times without the generation of *nejayote*. Microwave nixtamalization produced tortillas with acceptable physicochemical, textural, quality, organoleptic, and starch-pasting properties. This technology caused 68–84% decrease in aflatoxin content, although the acidification as occurs in the stomach caused some reformation (Martinez-Bustos et al. 2000, Perez-Flores et al. 2011, Mendez-Albores et al. 2012)
Ultrasound/ cavitation	For ultrasonic cooking, kernels are placed in a stainless steel transducer tank set at 80°C for cooking with an ultrasound frequency of 25 kHz because the cavitation action is much more vigorous at this frequency. In addition, the power output of the transducer was set at 550 W. Ultrasound nixtamalization has been successfully used to produce *pozole* (alkali-cooked maize similar to hominy used for a popular traditional soup). The new technology significantly reduced cooking time for the production of comparable quality of *pozole*. Ultrasound-assisted lime-cooking has been effective in shortening cooking times in *pozole* elaboration and to control the extent of gelatinization in nixtamalization of corn flours (Mendez-Albores et al. 2014, Robles-Ozuna et al. 2016)
Enzyme-assisted	This patented technology of Jackson and Sahai (2002) focus on the minimization of the amount of wastewater and processing time. Corn kernels are first tempered to 25–80%, then an alkali (calcium, sodium, or potassium hydroxides) is added to increase pH levels which favor enzyme activity, and then digested with an alkaline protease which hydrolyze the protein contained in the pericarp. The enzyme treatment weakens cell walls and eases the subsequent milling step. Normally, hard kernels are cooked at 90ºC/15 min before the enzymatic treatment to obtain the desired softness and facilitate the milling, whereas soft textured kernels can be processed without cooking. Alternatively, the *masa* could be dehydrated (80–150ºC) and reground to obtain DMF, which could be further processed into fried snacks
Tortilla baking technologies	
Infrared baking	The use of infrared baking of *masa* disks for soft tortillas is performed in reduced time by applying infrared radiation waves simultaneously on both sides of the product. The infrared radiation waves are preferably applied by means of an optical arrangement formed by two parallel arrays of infrared emitters, one on each side of the food product (Gonzalez-Hernandez et al. 1996)
Radio wave baking	De la Luz Martinez et al. (1997) patented a process for cooking corn tortillas by applying very-low and low-frequency radio waves simultaneously on both sides of the tortilla disk. This reduced cooking forms layers that retain a high degree of moisture within the body of the tortilla without excessive dehydration. The radio frequency waves are preferably applied by an antenna formed by two parallel arrays of electrodes, one on each side of the tortilla; the arrays of electrodes have alternate electrodes of opposite polarity to constitute dipoles which irradiate the radio frequency waves approximately parallel to both sides of the tortilla

11.7.2.5 Tortilla Chips from High-Oil Corns

Most high-oil corns contain at least twice the amount of lipids assayed in regular types. Their cultivars have been bred because the oil provides 2.25 the energy value of the starch and the oil preserved the high levels of linoleic acid. The potential use of high-yielding white and yellow HOC for tortillas and chips is promising because many new hybrids adapted to temperate and subtropical regions contain up to 9% oil. Vázquez-Carrillo et al. (2015) investigated the performance of tortillas produced with HOC and concluded that tortillas contained higher oil (up to 4.5%) than counterparts made with landraces (2.9–3.0%). Likewise, Preciado-Ortiz et al. (2018) indicated that the HOM kernels contained more linoleic acid than normal maize and this omega-6 fatty acid decreased significantly in tortillas. More research is needed in order to determine yields and quality of baked and fried tortilla chips and the effects of the higher oil, phospholipids, phytosterols, and other liposoluble compounds in their functional, organoleptic, and nutraceutical profiles.

11.7.2.6 Tortilla Chips Supplemented with Other Grains

There are special types of tortilla chips supplemented with other grains like sesame, flax, amaranthus, soybeans, common beans (Chapter 13), and fresh or dehydrated vegetables such as cactus (*Opuntia* spp.) and spinach (Chapter 16). These enriched products are rich sources of proteins, polyunsaturated fats, dietary fiber, and/or natural antioxidants that are known to favor human health (Serna-Saldivar et al. 2015). The properties of these grains are discussed in Chapter 13.

11.8 EMERGING NIXTAMALIZATION TECHNOLOGIES

Table 11.3 summarizes the principles of unconventional methods for producing *masa*, DMF, for baking dough pieces into tortillas. These methods attempt to produce fresh *masa* or DMF continuously, faster, and more efficiently in terms of labor, energy, and plant space. The other relevant advantage is that most of these novel technologies claim the generation of nil wastewaters. Despite all these alternatives, the industrial sector has not embraced these technologies because intermediate and end-products do not have similar functionality and organoleptic properties (Escalante-Aburto et al. 2020, Serna-Saldivar 2021).

One of the most promising technologies are conventional extrusion-cooking and low-shear laminar flow that have been successfully employed for the production of both fresh *masa* and DMF (Table 11.3). The key of these processes is to avoid excessive starch gelatinization and extrudate puffing by controlling shear (Mercado-Pedraza et al. 2014, Escalante-Aburto et al. 2020, Serna-Saldivar 2021) through screw configuration and speed, particle size distribution, tempering moisture content, heat input, and die design. For DMF, whole-corn meal or refined corn grits or meals are first lime-mixed and conditioned to reach high-moisture contents. The tempered blend is continuously fed into the extruder and in some instances cooled down with a special attachment placed after the die with the main objective to avoid over-gelatinization and prevent excessive starch damage (Serna-Saldivar and Rooney 2015, Serna-Saldivar and Chuck-Hernandez 2019, Enriquez-Castro et al. 2020, Serna-Saldivar 2021). The use of extrusion to produce fresh *masa* should make special emphasis on the particle size of the feedstock and the tempering moisture so as to continuously obtain *masa* with the desired texture, degree of starch gelatinization, and moisture.

For production of dry instant flours, the extrudates can be dehydrated to reduce their moisture to 10–12% at temperatures of about 65°C. The dried extrudate is further milled into flour preferably with roller mills because they control better the granulation (Serna-Saldivar et al. 1990, Serna-Saldivar 2021)

One of the major advantages of extrusion is that products retain higher amounts of important antioxidants and phytochemicals that favor human health (Mora-Rochin et al. 2010, 2016).

Other emerging methods such as ohmic heating, microwaving, ultrasound/cavitation, and infrared-cooking have been proposed and patented for both lime-cooking and tortilla baking. The principles and advantages of these technologies are summarized in Table 11.3.

REFERENCES

Almeida-Dominguez, H.D., Cepeda, M., and Rooney, L.W. 1996. Properties of commercial nixtamalized corn flours. *Cereal Foods World*, 41: 624–630.

Amaya-Guerra, C., Alanis-Guzman, M.G., and Serna-Saldivar, S.O. 2004. Effects of soybean fortification on protein quality of tortilla based diet from regular and quality protein maize. *Plant Foods for Human Nutrition*, 59(2): 45–50.

Amaya-Guerra, C., Serna-Saldivar, S.O., and Alanis-Guzman, M.G. 2006. Soybean fortification and enrichment of regular and quality protein maize tortillas affects brain development and maze performance of rats. *British Journal of Nutrition*, 96(1): 161–168.

Andersson, M.S., Saltzman, A., Virk, P., and Pfeiffer, W.H. 2017. Progress update: Crop development of biofortified staple food crops under HarvestPlus. *African Journal of Food, Agriculture, Nutrition and Development*, 17(2): 11905–11935.

Ayala-Soto, F.E., Serna-Saldivar, S.O., and Welti-Chanes, J. 2016. Effect of processing time, temperature and alkali concentration on yield extraction, structure and gelling properties of corn fiber arabinoxylans. *Food Hydrocolloids*, 60: 21–28.

Betran, J., Bockholt, A.J., and Rooney, L.W. 2000. Blue corn. In: *Specialty Corns*, ed. A.R. Hallauer, 293–301. Boca Raton: CRC Press.

Bonifacio-Vázquez, E.I., Moreno-Salinas, Y., Ramos-Rodríguez, A., and Carrillo-Ocampo, A.C. 2005. Calidad pozolera en colectas de maíz cacahuacintle. *Revista Fitotecnia Mexicana*, 28(3): 253–260.

Burtea, O. 2001. Snack food from formers and high shear extruders. In: *Snack Foods Processing*, 1st edition, eds. E.W. Lusas and L.W. Rooney, 73–114. Boca Raton: CRC Press.

De la Luz Martinez, J., Gonzalez-Hernandez, J., Sanchez-Sinencio, F., Martinez-Bustos, F., Figueroa-Cardenas, J.D.D., and Ruiz-Torres, M. 1997. *Method for Cooking Tortillas Using Very Low and Low Frequency Radio Waves, USPO 559371.*

De la Parra, C., Serna-Saldivar, S.O., and Liu, R.H. 2007. Effect of processing on the phytochemical profile and antioxidant activity of corn for production of masa, tortillas and tortilla chips. *Journal of Agricultural and Food Chemistry*, 55(10): 4177–4183.

Decker, K.J. 2020. State of the industry 2020: Tortilla chips go well beyond basic. https://www.snackandbakery.com/topics/2645-tortilla-chips.

Enriquez-Castro, C.M., Torres-Chávez, P.I., Ramirez-Wong, B., Quintero-Ramos, A., Ledesma-Osuna, A.I., Lopez-Cervantes, J., and Gerardo-Rodríguez, J.E. 2020. Physicochemical, rheological, and morphological characteristics of products from traditional and extrusion nixtamalization processes and their relation to starch. *International Journal of Food Science*, 2020: Article ID 5927670.

Escalante-Aburto, A., Mariscal-Moreno, R.M., Santiago-Ramos, D., and Ponce-García, N. 2020. An update of different nixtamalization technologies, and its effects on chemical composition and nutritional value of corn tortillas. *Food Reviews International*, 36(5): 456–498.

FAO. 2021. FAOSTAT. Rome, Italy. http//faostat.fao.org/.

Figueroa-Cardenas, J.D.D., Gonzalez-Hernandez, J., Martinez, J.D.L.L., Martinez-Bustos, F., Ruiz-Torres, M., and Sanchez-Sinencio, F. 1996. *Method for the Preparation of Instant Fresh Corn Dough or Masa, USPO 5532013 A.*

Garcia-Lara, S., Chuck-Hernandez, C., and Serna-Saldivar, S.O. 2013. Detrimental effects on chemical and functional properties of corn tortillas manufactured with larger grain borer (*Prostephanus truncates* Horn.) damaged kernels. *Journal of Stored Products*, 55: 99–105.

Garcia-Lara, S., Gutierrez-Uribe, J.A., and Serna-Saldivar, S.O. 2012. Nutraceutical properties of blue maize. In: *Maize: Cultivation, Uses and Health Benefits*, ed. J.C. Jimenez-Lopez, 113–135. New York: Nova Publishers.

Garcia-Lara, S., and Serna-Saldivar, S.O. 2019. Corn history and culture. In: *Corn: Chemistry and Technology*, 3rd edition, ed. S.O. Serna-Saldivar, 1–19. Cambridge: Elsevier (Woodhead Publishing), AACC-International.

Gaytan-Martínez, M., Figueroa, J.D.C., Vazquez-Landaverde, P.A., Morales-Sanchez, E., Martinez-Flores, H.E., and Reyes-Vega, M.L. 2012. Physicochemical, functional, and chemical characterization of nixtamalized corn flour obtained by ohmic heating and traditional process. *CyTA-Journal of Food*, 10(3): 182–195.

Gomez, M.H., McDonough, C.M., Rooney, L.W., and Waniska, R.D. 1989. Changes in corn and sorghum during nixtamalization and tortilla baking. *Journal of Food Science*, 54(2): 330–336.

Gomez, M.H., Rooney, L.W., Waniska, R.D., and Pflugfelder, R.L. 1987. Dry corn masa flours for tortilla and snack food production. *Cereal Foods World*, 32: 372–377.

Gonzalez-Hernandez, J., De La Luz-Martinez, J., Sanchez-Sinencio, F., Martinez-Bustos, F., Figueroa-Cardenas, J.D.D., and Ruiz-Torres, M., 1996. Method of cooking corn dough tortillas using infrared radiation, USPO 5567459.

Gonzalez, R.M., and Rubio, M.J. 1968. Process for producing tortilla flour. US Patent 3,369,908A.

Hammond, B., Dudek, R., Lemen, J., and Nemeth, M. 2004. Results of a 13 week safety assurance study with rats fed grain from glyphosate tolerant corn. *Food and Chemical Toxicology*, 42(6): 1003–1014.

Healy, C., Hammond, B., and Kirkpatrick, J. 2008. Results of a 13-week safety assurance study with rats fed grain from corn rootworm-protected, glyphosate-tolerant MON 88017 corn. *Food and Chemical Toxicology*, 46(7): 2517–2524.

International Agency for Research on Cancer. 2015. IARC monographs volume 112: Evaluation of five organophosphate insecticides and herbicides. International Agency for Research on Cancer and World Health Organization. http://www.iarc.fr/en/media-centre/iarcnews/pdf/MonographVolume112.pdf and http://monographs.iarc.fr/ENG/Monographs/vol112/mono112-02.pdf.

Jackson, D.S., and Sahai, D. 2002. Enzymatic process for nixtamalization of cereal grains. USPO 6,428,828 B1, Aug 6, 2002.

Lange, J.H., and Wolt, M.J. 1997. *Method for Producing a Masa Flour.* WO 1997029647 A1.

Lozano-Alejo, N., Carrillo, G.V., Pixley, K., and Palacios-Rojas, N. 2007. Physical properties and carotenoid content of maize kernels and its nixtamalized snacks. *Innovative Food Science and Emerging Technologies*, 8(3): 385–389.

Lujan Acosta, F.J., and Moreira, R.G. 1997. Reduction of oil in tortilla chips using impingement drying. *LWT – Food Science and Technology*, 30(8): 834–840.

Martinez-Bustos, F., García, M.N., Chang, Y.K., Sanchez-Sinencio, F., and Figueroa, C.J.D. 2000. Characteristics of nixtamalised maize flours produced with the use of microwave heating during alkaline cooking. *Journal of the Science of Food and Agriculture*, 80(6): 651–656.

McDonough, C.M., Gomez, M.H., Lee, J.K., Waniska, R.D., and Rooney, L.W. 1993. Environmental scanning electron microscopy evaluation of tortilla chip microstructure during deep-fat frying. *Journal of Food Science*, 58(1): 199–203.

McDonough, C.M., Gomez, M.H., Rooney, L.W., and Serna-Saldivar, S.O. 2001. Alkaline cooked corn products. In: *Snack Foods Processing*, 1st edition, eds. E.W. Lusas and L.W. Rooney, 73–114. Boca Raton: CRC Press.

Mehta, S.P. 2001. Tortilla chip processing. In: *Snack Foods Processing*, 1st edition, eds. E.W. Lusas and L.W. Rooney, 261–280. Boca Raton: CRC Press.

Mendez-Albores, A., Martinez-Morquecho, R.A., Moreno-Martinez, E., and Vazquez-Duran, A. 2012. Technological properties of maize tortillas produced by microwave nixtamalization with variable alkalinity. *African Journal of Biotechnology*, 11(85): 15178–15187.

Mendez-Albores, A., Zamora-Rodriguez, D., Arambula-Villa, G., Vazquez-Duran, A., and Moreno-Martínez, E. 2014. Impact of different alkaline-heating processes on technological and nutritional properties of maize tortillas. *Journal of Food and Nutrition Research*, 53(1): 60–70.

Mercado-Pedraza, E.B., Morales-Sanchez, E., Reyes-Vega, M., Gaytan-Martinez, M., and Ortega-Moody, J.A. 2014. Effects of a low-shear transport system on the physicochemical characteristics of nixtamal corn flour. *Journal of Food Processing and Preservation*, 38(3): 1330–1338.

Mertz, E.T., Bates, L.S., and Nelson, O.E. 1964. Mutant gene that changes protein composition and increases lysine content of maize endosperm. *Science*, 145(3629): 279–280.

Mora-Rochin, S., Gaxiola-Cuevas, N., Gutierrez-Uribe, J.A., Milan-Carrillo, J., Milan-Noris, E.M., Reyes-Moreno, C., Serna-Saldivar, S.O., and Cuevas-Rodriguez, E.O. 2016. Effect of traditional nixtamalization on anthocyanin content and profile in Mexican blue maize (*Zea mays* L.) landraces. *LWT – Food Science and Technology*, 68: 563–569.

Mora-Rochin, S., Gutiérrez-Uribe, J.A., Serna-Saldivar, S.O., Sánchez-Peña, P., Reyes Moreno, C., and Milán-Carrillo, J. 2010. Phenolic content and antioxidant activity of tortillas produced from pigmented maize processed by conventional nixtamalization or extrusion cooking. *Journal of Cereal Science*, 52(3): 502–508.

Moreira, R.G., Castell-Perez, E.M., and Barrufet, M.A. 1999. *Deep-Fat Frying Fundamentals and Applications*. Gaithersburg: Aspen Publishers.

Moreira, R.G., Palau, J.E., and Sun, X. 1995. Deep fat frying of tortilla chips: An engineering approach. *Food Technology*, 49(2): 146–152.

Ortega-Moody, J.A., Morales-Sánchez, E., Mercado-Pedraza, E.V., Ríos-Moreno, J.G., and Trejo-Perea, M. 2014. Design and construction of low shear laminar transport system for the production of nixtamal corn dough. *DYNA*, 81(185): 48–55.

Perez-Flores, G.C., Moreno-Martinez, E., and Mendez-Albores, A. 2011. Effect of microwave heating during alkaline-cooking of aflatoxin-contaminated maize. *Journal of Food Science*, 76(2): 48–52.

Prasanna, B.M., Palacios-Rojas, N., Hossain, F., Muthusamy, V., Menkir, A., Dhliwayo, T., Ndhlela, T., San Vicente, F., Nair, S.K., Vivek, B.S., Zhang, X., Olsen, M., and Fab, X. 2020. Molecular breeding for nutritionally enriched maize: status and prospects. *Frontiers in Genetics*, 10: 1392.

Preciado-Ortiz, R.E., García-Lara, S., Ortiz-Islas, S., Ortega-Corona, A., and Serna-Saldivar, S.O. 2013. Response of recurrent selection on yield, kernel oil content and fatty acid composition of subtropical maize populations. *Field Crops Research*, 142: 26–35.

Preciado-Ortíz, R.E., Vázquez-Carrillo, M.G., de Dios Figueroa-Cárdenas, J., Guzmán-Maldonado, S.H., Santiago-Ramos, D., and Topete-Betancourt, A. 2018. Fatty acids and starch properties of high-oil maize hybrids during nixtamalization and tortilla-making process. *Journal of Cereal Science*, 83: 171–179.

Quintero-Fuentes, X., McDonough, C.M., Rooney, L.W., and Almeida-Dominguez, H. 1999. Functionality of rice and sorghum flours in baked tortilla and corn chips. *Cereal Chemistry Journal*, 76(5): 705–710.

Ramirez-Araujo, H., Gaytan-Martinez, M., and Reyes-Vega, M.L. 2019. Alternative technologies to the traditional nixtamalization process: Review. *Trends in Food Science and Technology*, 85: 34–43.

Robles-Ozuna, L.E., Ochoa-Martínez, L.A., Morales-Castro, J., Gallegos-Infante, J.A., Quintero-Ramos, A., and Madera-Santana, T.J. 2016. Effect of nixtamalization conditions ultrasound assisted on some physicochemical, structural and quality characteristics in maize used for pozole. *CyTA-Journal of Food*, 14(2): 324–332.

Rooney, L.W., and Serna-Saldivar, S.O. 2015. Food grade corn quality for lime cooked tortillas and snacks. In *Tortillas: Wheat flour and corn products*, eds. L.W. Rooney and S.O. Serna-Saldivar, 227–246. Saint Paul: American Association of Cereal Chemists.

Rooney, L.W., and Suhendro, E.L. 2001. Food quality of corn. In: *Snack Foods Processing*, 1st edition, eds. E.W. Lusas and L.W. Rooney, 37–72. Boca Raton: CRC Press.

Rosales, A., Agama-Acevedo, E., Bello-Perez, L.A., Gutierrez-Dorado, R., and Palacios-Rojas, N. 2016. Effect of traditional and extrusion nixtamalization on carotenoid retention in tortillas made from provitamin A biofortified maize (*Zea mays* L.). *Journal of Agricultural and Food Chemistry*, 64(44): 8289–8295.

Serna-Saldivar, S.O. 2008. Snacks from alkaline-cooked maize products. In: *Industrial Manufacture of Snack Foods*, ed. S.O. Serna-Saldivar, 183–233. London: Kennedys Publications Ltd.

Serna-Saldivar, S.O. 2015a. History of corn and wheat tortillas. In *Tortillas: Wheat Flour and Corn Products*, eds. L.W. Rooney and S.O. Serna-Saldivar, 1–28. Saint Paul: American Association of Cereal Chemists.

Serna-Saldivar, S.O. 2015b. Nutrition and fortification of corn and wheat tortillas. In *Tortillas: Wheat Flour and Corn Products*, eds. L.W. Rooney and S.O. Serna-Saldivar, 29–63. Saint Paul:American Association of Cereal Chemists.

Serna-Saldivar, S.O. 2021. Understanding the functionality and manufacturing of nixtamalized maize products. *Journal of Cereal Science*, 99: 103205.

Serna-Saldivar, S.O., Almeida-Dominguez, H.D., Rooney, L.W., Gómez, M.H., and Bockholt, A.J. 1991. Method to evaluate ease of pericarp removal on lime-cooked corn kernels. *Crop Science*, 31(3): 842–844.

Serna-Saldivar, S.O., and Chuck-Hernandez, C.E. 2019. Food uses of lime-cooked corn with emphasis in tortillas and snacks. In: *Corn: Chemistry and Technology*, ed. S.O. Serna-Saldivar, 469–500. Cambridge: Elsevier (Woodhead Publishing), AACC-International.

Serna-Saldivar, S.O., Gomez, M.H., Islas-Rubio, A.R., Bockholt, A.J., and Rooney, L.W. 1992. The alkaline processing properties of quality protein maize. In: *Quality Protein Maize*, ed. E.T. Mertz, 273–293. Saint Paul: American Association of Cereal Chemists.

Serna-Saldivar, S.O., Gomez, M.H., and Rooney, L.W. 1990. The technology, chemistry and nutritional value of alkaline cooked corn products. In: *Advances of Cereal Science and Technology*, Volume X, ed. Y. Pomeranz, 243–307. Saint Paul: American Association of Cereal Chemists.

Serna-Saldivar, S.O., Gutierrez Uribe, J., and García Lara, S. 2015. Phytochemical profiles and nutraceutical properties of corn and wheat tortillas. In *Tortillas: Wheat Flour and Corn Products*, eds. L.W. Rooney and S.O. Serna-Saldivar, 65–93. Saint Paul: American Association of Cereal Chemists.

Serna-Saldivar, S.O., Herrera-Macias, P., Amaya-Guerra, C.A., Melesio-Cuellar, J.L., Preciado-Ortiz, R.E., Terron-Ibarra, A.D., and Vazquez Carrillo, G. 2008. Evaluation of the lime-cooking and tortilla making properties of quality protein maize hybrids grown in Mexico. *Plant Foods for Human Nutrition*, 63(3): 119–125.

Serna-Saldivar, S.O., and Rooney, L.W. 1994. Quality protein maize processing and perspectives for industrial utilization. In: *Quality Protein Maize 1964–1994*, eds. B.A. Larkins and E.T. Mertz, 69–120. Sete Lagoas: EMBRAPA/CNPMS.

Serna-Saldivar, S.O., and Rooney, L.W. 2015. Industrial production of maize tortillas and snacks. In *Tortillas: Wheat Flour and Corn Products*, eds. L.W. Rooney and S.O. Serna-Saldivar, 247–281. Saint Paul: American Association of Cereal Chemists.

Urias-Lugo, D., Heredia, B.J., Serna-Saldivar, S.O., Muy-Rangel, D., and Valdez-Torres, B. 2015. Total phenolics, total anthocyanins and antioxidant capacity of native and elite blue maize hybrids (*Zea mays* L.). *CYTA- Journal of Food*, 13(3): 336–339.

Urias-Peraldi, M., Gutiérrez-Uribe, J.A., Preciado-Ortiz, R.E., Cruz-Morales, A.F., Serna-Saldivar, S.O., and Garcia-Lara, S. 2013. Nutraceutical profiles of improved blue maize (*Zea mays*) hybrids for subtropical regions. *Field Crops Research*, 141: 69–76.

Varghese, K.S., Pandey, M.C., Radhakrishna, K., and Bawa, A.S. 2014. Technology, applications and modelling of ohmic heating: A review. *Journal of Food Science and Technology*, 51(10): 2304–2317.

Vazquez-Carrillo, M.G., Santiago-Ramos, D., Gaytán-Martínez, M., Morales-Sánchez, E., and de Jesús Guerrero-Herrera, M. 2015. High oil content maize: Physical, thermal and rheological properties of grain, masa, and tortillas. *LWT – Food Science and Technology*, 60(1): 156–161.

12 Production of Snacks by Extrusion Cooking

Erick Heredia-Olea, Sergio O. Serna-Saldivar, and Esther Perez-Carrillo

CONTENTS

12.1 INTRODUCTION

Extrusion cooking is considered as one of the most relevant food technologies invented in the 20th century (Panak-Balentic et al. 2017). Initially, it was used to produce cold-extruded pastas, but it rapidly evolved to manufacture a variety of products, from ready-to-eat (RTE) snacks and breakfast cereals to center-filled or coextruded items. This technology offers high versatility and has many other advantages from the processing and energy use. For this reason, extrusion has increasingly been used for producing many food items for more than 70 years.

Extruders are used for a broad range of food applications, including agglomeration, cooking, dehydration, expansion, gelatinization, protein denaturation or texturization, homogenization, pasteurization/sterilization, and shaping. In terms of snack processing, the industry employs two main processes, second-generation and third- generation, aimed to obtain direct-extruded puffs and pellets or half-products. These extrusion systems are categorized as high temperature and short time and are capable of generating a wide variety of snacks with different textures, colors, flavors, and aromas (Serna-Saldivar 2008, Masatcioglu et al. 2014). Nowadays, extruded snacks are considered as the category with the greatest versatility and growth potential. While its dominance in the snack market is unquestionable, this technology still offers flexibility and area of innovation in matters of machinery, raw material development, and applications. For example, extrusion can also be implemented to produce analogs of traditionally manufactured snacks, while coextrusion offers the advantage of creating new and innovative products with fillings.

DOI: 10.1201/9781003129066-12

Food extrusion is the act of forcing edible materials through an orifice or die. Inherent in the extruder design is a continuous flow of material from the feed section of the screw to the discharge end. These concerted actions performed continuously yield an array of snacks with different physical and chemical properties, configurations, and nutritional values. One of the main advantages of cooking and forming extruders is that these processing lines save site space, offers more versatility, adapts to highly productive continuous factories, and saves labor and energy when compared with expenditures of traditional processes. The whole process can effectively mix different raw materials in the preconditioner or inside the extruder barrel. The feedstock is continuously exposed to shear and heat (cooking) and finally shaped and cut into uniform extrudates or pieces (Lohani and Muthukumarappan, 2017).

Recent statistics indicate that the global extruded snacks market size was valued at US$51.6 billion in 2019 and is expected to witness significant growth in the years to come (Grand View Research Report Summary 2020). In the United States, this category sold about US$10 billion (Parker 2020). The global extruded snacks size is expected to reach US$74.5 billion by 2027, expanding at a CAGR of 4.7% over the forecast period.

This chapter aims to describe conventional and innovative extrusion technologies commonly used for production of second- and third-generation snacks, as well as coextruded snack foods.

12.2 EXTRUDERS

Extruders can be classified based on the number of screws (single or twin screw) and the type of thermal expenditure (cold or thermoplastic extruder). Twin-screw extruders have a much broader operational range and can produce practically any product that a single extruder manufactures; however, the reverse is not necessarily true. Thermoplastic extruders are subdivided into barrel heated or polytropic, and adiabatic or autogenous. The latter are also named low-cost extruders (LCE) because during the process, heat is generated by friction and therefore does not use any exogenous source of heat (Mościcki 2011).

Inside the snack food industry, the single-screw heated extruder is the most common processing technique. In this setting, the barrel acts as heat exchanger, transferring energy from the barrel directly to the sample as it moves along the extruder. The barrel has hollow sleeves, where circulating steam or a heating medium, such as hot oil or direct electrical heating can be applied (Yacu 2020).

Regardless of the type of extruder, the typical equipment consists of a feeding hopper, barrel, screw(s), die, and cutter (Figure 12.1). The barrel encases the screws, which are rotated by a motor. Some extruders are equipped with a preconditioner, especially when the feedstock needs to be tempered to high moisture prior to extrusion, such as the case for production of high-quality pellets or half-products.

12.2.1 Feeding System

The extrusion operation requires a consistent and uniform feedstock delivery system to produce continuously good quality extrudates. This system generally consists of a hopper, which holds the raw material and is commonly located above the feeding device. The material is supplied through the extruder mouth directly to the first barrel section, also known as the conveying segment, while the feeding rate is monitored by a metering system to maintain a uniform stream. In a basic setting, the screw feeder is equipped with a variable speed drive, and the mass flow inlet must be manually correlated to the screw speed and density of the feedstock. More recent models are adapted with auto-adjustable scales to sense changes in the extrudate density, and through a feedback loop, adjust the screw speed to obtain a uniform product. This allows constant uniform feed to the barrel and provides the system with a tool to sense formulation deviations or changes through the process. In an industrial setting, this translates to fewer personal, shorter line turnover and higher production

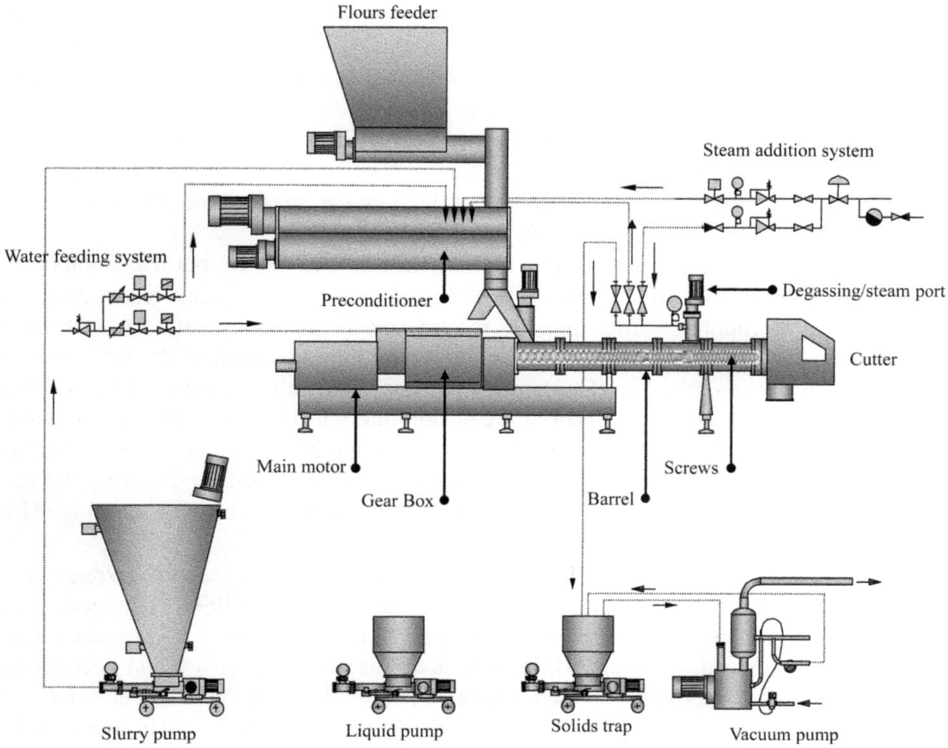

FIGURE 12.1 Typical extruder components (courtesy of Buhler Group).

rate. All these advantages are achieved at a relatively lower cost based on the initial investment (Richards and Harrington 2020).

In more complex systems, the equipment can have more than one feeder of different sizes in order to supply different raw materials. The mixed feedstock can be delivered directly to the barrel. In this type of setting, a preconditioning step is recommended to mix or treat (addition of water or steam) the ingredients before they reach the barrel. This allows the feedstock blend to experience a longer cooking and/or modification time as it moves along the barrel (Mościcki 2011, Yacu 2020).

Screw feeders come in single- and twin-screw types, closed or open helix screws, and with or without agitators (Moore 1994). Some extruders are also equipped with liquid intake devices that are fed by slurry-jacketed tanks. Positive displacement metering pumps are normally used to deliver liquids at constant flow rates into the preconditioning devices or directly into the first part of the extruder barrel.

12.2.2 Barrel

The barrel or stator is a stationary pipe like retainer in which the screw is encased. The screw rotates on its own axis, conveying the feedstock along the stator. Some barrels are built in segments, where each piece can have a cooling or heating sleeve. The heating sleeve normally uses steam, hot water, or electrical coils, while cold water or forced cool air can be used for cooling purposes. This allows tuning the temperature of each zone independently to the needs of the process and the final product (Harper 1989, Pichmony and Ganjyal 2020)

The extruder relies on pressure and shear to achieve the axial movement of the feed along the barrel and to reach the die. This increase in pressure will also be critical to increase the temperature inside the system and transform the feedstock into the final product. To aid the generation of the required pressure, the inside surface of the barrel can show a grooved design. The most common patterns are longitudinal or helical grooves, and this internal design is selected based on the desired product as well as the extruder equipment employed. In the case of the single-screw system, these grooves enhance the movement of the feedstock as it travels through the barrel, reducing slippage and increasing shear stress. Sometimes in single screw-extrusion systems, the diameter of the screw increases as it approaches the die, creating a volume restriction and increasing the pressure, drag, and shear on the product as well as the temperature, helping the mixing, kneading, and cooking of the dough (Morales-Alvarez 2020, Yacu 2020).

On the other hand, barrels in twin-screw systems tend to be smooth, but exceptions can exist for certain types of processes. In this assembly, both screws move in accordance, corotating (move in same direction) or counterrotating (moving in opposite directions) moving the material along the whole barrel. The pressure and shear are generated by the movement of the feedstock against itself and the walls of the screw-barrel system as it travels between the screws. These types of systems can be further categorized as intermeshing or non-intermeshing, according to the interaction between the screw flight of each piece or the degree of overlapping between the flights. These types of interactions between the two screws and the barrel are responsible for conveying the material and increasing the area of contact, as well as the expected cooking degree in this setting (Morales-Alvarez 2020).

In thermoplastic extruders, the barrel is broadly divided into three consecutive zones: the feeding, transition, and the rheological or cooking zone (Jaluria and Wang 1999). The feeding zone is primarily designed to blend and transport the material. In the transition or compression zone, the feed is compacted and further mixed, while there is an increase in pressure and heat-generation. Finally, in the cooking zone, the feedstock is expected to melt due to the heat and elevated pressure in second-generation snack production or compress the feedstock to produce dense pellets for production of third-generation snacks (Rauwendaal 2014, Yacu 2020).

12.2.3 Screw(s)

The screw configuration is the heart or core of the extruder. Its main function is to move the feedstock along the barrel and subject it to the shear controlled by its design and the temperature and pressure. The screw(s) are encased inside the barrel and rotates on its own axis by the extruder drive, consisting of a motor, a speed reduction system, a torque transfer system, and a bearing support mechanism (Rokey 2000).

The extruder screw can be manufactured from a solid metal shaft, as a whole piece, or as modular parts. In the latter, the screw elements can be arranged along a splinted metal shaft, each element with specific pattern (Table 12.1) to create a specific configuration. The modular design enables easy reconfiguration, opening the possibility of manufacturing numerous types of products. The final part of the screw is usually conical, especially when the extruder operates at high die pressure (Rauwendaal 2014, Yacu 2020).

The screw rotation can be clockwise or counterclockwise. For twin-screw extruders, rotation can be opposite or counterrotating or corotating. The most common is the corotating configuration because it can be operated at low speeds. The most common twin-screw designs employ the self-wiping features by having a high degree of intermeshing. Intermeshing screws are designed to show a small distance between the axis of each screw (small interaxis), normally smaller than the sum of the screw's diameters, while in non-intermeshing, the screws are further separated, with spaces higher than the sum of both screw's diameter (Ganjyal 2020; Morales-Alvarez 2020).

The screw configuration is the main factor that dictates the type of application. The flight is defined as the helical conveying surface of the screw that moves the feedstock forward and the pitch is the angle of the flight relative to the axis. Some screws have interrupted or cut flights in which studs or bolts are inserted through the extruder barrel wall into the empty flight section to induce additional shear. The main characteristic of the screw is the length-to-diameter ratio, commonly abbreviated *L/D*, and the

TABLE 12.1
Different Screw Elements and Their Functionality

Type	*L/D* Ratio	Description/Function	Image
Forward conveying (right-hand)	<1	Narrow pitch element. Compresses the dough more than a wide pitch element. Its channels are more filled increasing the local pressure	
	1	Two flights. Typical element symmetry. A right-hand screw element pushes the dough toward the die	
	>1	A wide-pitch screw conveys better than a narrow-pitch screw; its channels are less filled; the downstream pressure is greater. Usually it is located at least the first barrel length, and probably longer, to move the feed material away from the feed throat, so more material can come into the extruder barrel	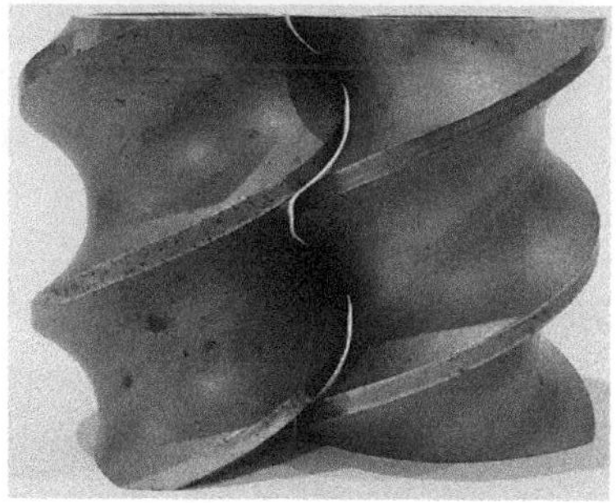

(Continued)

TABLE 12.1 (CONTINUED)
Different Screw Elements and Their Functionality

Type	*L/D* Ratio	Description/Function	Image
	0.5	These elements are used to create offsets together with spacers to increase the volume inside the barrel and the degree-of-fill increasing pressure and shear stress. They are also used in combination with the reverse elements to increase the dough cooking	
Reverse conveying (left-hand)	0.5	Displaces the flux inside the barrel backward, creating a hydraulic seal increasing the pressure, kneading, friction, and temperature. Pushes dough away from the die, increasing the degree-of-fill in the elements behind it, increasing the mean residence time, broadening the residence time distribution, increasing the SME, and lowering die pressure	
Mixing (right-hand)	<1	Good mixers. Have low conveying and pressure buildup capacity	

(*Continued*)

TABLE 12.1 (CONTINUED)
Different Screw Elements and Their Functionality

Type	*L/D* Ratio	Description/Function	Image
Mixing (left-hand)	<1	Good mixers. Can increase pressures and input a high shear rate	
Kneading/mixing (right-hand)	>1	Generates high shear stress, kneading, high pressure. Increases contact between the barrel and the dough, has low conveying effect. Effective for gelatinization or texturization	
Spacer ring	-	Placed between elements with offset, generates space for dough displacement, and prevents elements edges from hitting each other	

(*Continued*)

TABLE 12.1 (CONTINUED)
Different Screw Elements and Their Functionality

Type	*L/D* Ratio	Description/Function	Image
Screw tip	–	Holds the screw element tight preventing the entry of melted product inside the space between the shaft and the elements	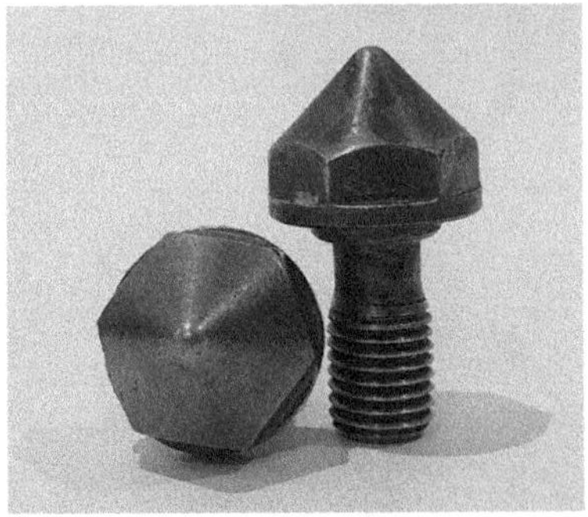

Photographs are courtesy of Buhler Group.

compression ratio. *L*/*D* is calculated by measuring the distance from the internal rear edge to the discharge end of the barrel divided by the diameter of the bore. There are many screw *L*/*D* ratios ranging from 1:1 to 25:1. The short *L*/*D* extruders are commonly used for second-generation snacks, whereas the long *L*/*D* extruders are used for production of pellets. On the other hand, the compression ratio is defined as the volume of the full flight of the screw at the feed opening divided by the volume of the last full flight before the die plate. Typical compression ratios range from 1:1 to 5:1. The root is the solid part of the screw in which the flight is wound. The screws normally contain shear-locks or steam-locks consisting of a ring that locks together individual sections to reduce blowback of steam to cooler sections of the barrel (Riaz 2000). Both single and twin screws can have several parallel flights along their lengths, the most common being single, double, triple, and quadruple. One rotation of the screw should deliver material at a distance in the axial direction equivalent to the pitch (Brent 1999).

12.2.4 Die Plate

Located at the end of the barrel, the die plate has two key functions. This component generates backpressure and increases residence time, allowing shear forces to further alter the rheological characteristics of the material at the melting zone. This segment is also responsible for shaping the final product. The die assembly has a defined design made of drilled orifices or holes that can change in shape and number of dies based on the extrudate, adding complexity to a simple piece. For the correct selection of the design and the number of die orifices required for a given process, the manufacturer must know the expected and desired characteristics of the extrudate, which are affected by the degree of dough cooking and melting occurring inside a barrel–screw system (Guy 2001, Pichmony and Ganjyal 2020).

12.2.5 Cutter

The final section of the extruder is an independent assembly normally formed by a metal blade(s) rotating at controlled speed, which cut extrudates into uniform pieces. This setting is known as "face cutter" since the knife(ves) slices the extrudates immediately as they exit the die orifice. Another less common option uses a belt to transport the product for further treatments before cutting. Although not as popular as the "face cutter" approach, the process and the final characteristics of the product will dictate the most suitable (Guy 2001, Pichmony and Ganjyal 2020).

12.3 HIGH-SHEAR EXTRUDED PRODUCTS

Second-generation baked and fried puffed snacks are typically manufactured with high shear or friction extruders (Figure 12.2). These extruders are called dry extruders because raw materials are not conditioned or, if conditioned, are tempered to low-moisture contents (normally less than 19%). Most extruders used to produce puffs are single-screw and short-barreled with *L*/*D* of 4 or less. The major difference between baked and fried collets centers on the type of extrusion system. The baked expanded products are produced using a high-shear extruder and extrudates are baked or dehydrated to impart the desired texture and crispness. The liquid or powdered seasonings are applied prior to packaging and have an impact in the overall oil content. A specific type of fried expanded collet is obtained using a special type of extruder, which produces denser and irregular-shaped extrudates. The latter product can be fried and seasoned, or simply seasoned with an oil-based product (Pichmony and Ganjyal 2020).

12.3.1 Production of Direct Expanded-Baked Puffs

In order to produce good quality extrudates or collets, the raw material should be highly refined and properly conditioned. In most instances, refined grits conditioned to moisture contents from 14%

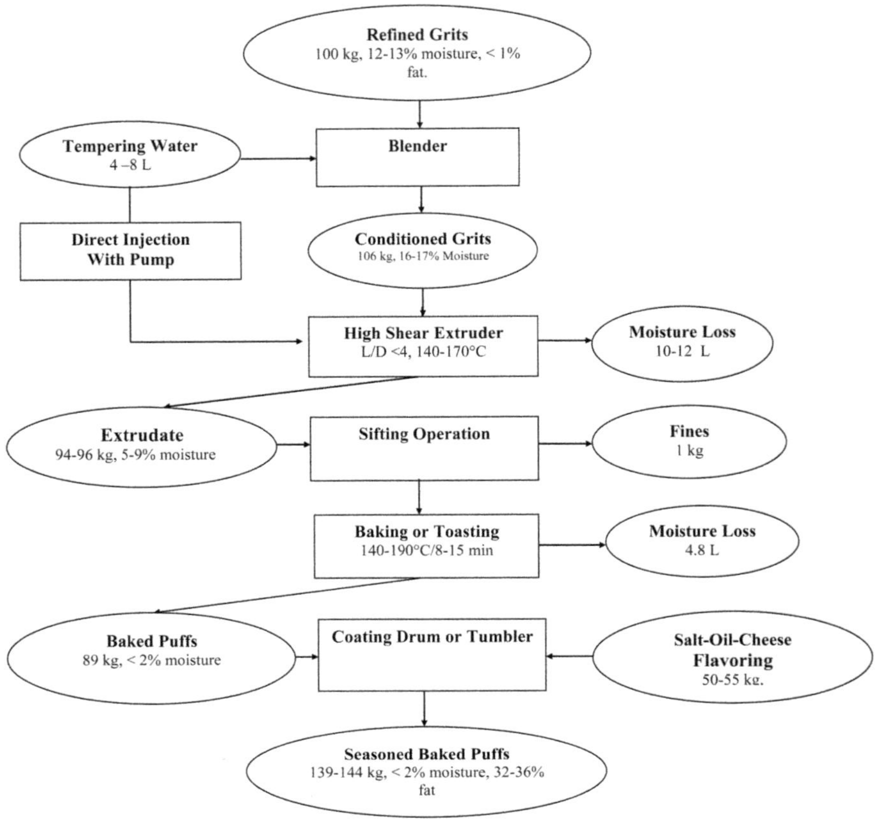

FIGURE 12.2 Flowchart of the extrusion technology for the production of expanded and seasoned baked snacks.

to 19% produced the best-quality products (Fletcher et al. 1985). The properties of these materials are covered in Chapter 6. High shear extruders can be adiabatic or thermoplastic. The adiabatic are less expensive and the heat is generated by the conversion of mechanical energy to thermal energy. The helical screw design and die restriction play important roles in the expansion rate. To enhance shear and slippage, the internal wall of the barrel normally contains helical grooves that become shallower toward the end of the barrel. The screw(s) is the main part of the extruder, this piece oversees material transport, and its design greatly affects shear rate. The main screw characteristics are distance between flights, depth of the flight, angle of the flight, and the type of flight (full radius or square cut). Higher shear is produced when the distance between and depth of the flights are shorter. High shear extruders are designed to produce the maximum shear and compression at the exit end of the screw to increase the pressure and enhance expansion rate (Riaz 2004). The higher the pressure differential between the end of the screw and the atmospheric pressure, the higher the expansion index. Adiabatic extruders usually contain flow plates before the extruder die. These pressure plates have numerous holes in various sizes. They increase backpressure, minimize product surging, and therefore guarantee the uniformity of the pressure before reaching the die. The die is located after the pressure plate(s) and consists of a plate with the orifice(es) that will form the particular forms of extrudates. The expansion rate is calculated by measuring the average collet diameter divided by the diameter of the die orifice (Hsieh et al. 1990). For instance, a puff with an average diameter of 14 mm, extruded through a 2-mm orifice, had an expansion rate of 7. The multi-orifice dies are commonly used because they yield higher outputs. These plate dies are interchangeable, and manufacturers offer more than 50 designs. The most common designs are balls, rings, and curls. Special dies that produce three-dimensional shapes have become popular in recent years. These machines are easy to operate and maintain and a single equipment is designed to extrude 100–1,000 kg/h (Mościcki, 2011).

A flowchart with the typical yield of baked maize extruded is depicted in Figure 12.2. The quality of the refined grits or feed material, conditioning, and extruder conditions are the main and key elements for obtaining the optimum collets. Refined maize grits, with an average particle size of 40–60 US mesh (Chapter 6) and 12–13% moisture, are normally conditioned to approximately 16–17% before extrusion (Garber et al. 1997). Refined grits are introduced into the inlet of the barrel or screw feeding zone. In this particular place, the distance between flights is longer and deeper than in the other zones. The pitch or flight angle favors the movement of the feedstock. The flow channel of the screw is typically filled with grits. In this area, the incoming material is slightly compressed, and trapped air expelled. In some extruders, the tempering water is injected into the first part of the feeding zone where the temperature is generally 60–80°C. Thereafter, the conditioned grits enter the kneading or transition zone that is characterized to contain a higher barrel temperature (90–120°C) and a screw design that favors compression by lowering the distance between flights and the flight height, so the flow channels achieve a higher degree of fill. The starch-rich extrudate starts to lose its granular structure and begins to gelatinize due to the temperature and pressure. The combination of heat and pressure converts the grits into a plastic dough. When the extrudate or dough-like material reaches the last zone, called cooking, the dough melts and becomes a viscous fluid because the barrel is adjusted to the maximum temperature (140–180°C) and the screw produces the maximum shear, compression, and pressure (Riaz 2004). In this section, the screw flights are closer, shallower, and in some instances interrupted. In short, the starchy materials subjected to direct extrusion cooking suffer the phenomena of melting and gelatinization. Melting is a physical process, whereas gelatinization a chemical process. When subjected to heating, starch granules with high water content gelatinize, whereas those with low water melt. At the very instant, that the viscous fluid exits the die orifice, it will instantaneously equilibrate with the external atmospheric pressure causing the expansion and setting the collet structure. During this instant, most of the moisture vaporizes, so the extrudate drops its moisture to about 5–8%. The puff or extrudate is cut and sized to the proper length with a rotating knife. The collet expansion rate is generally between 6 and 8. For instance, if the puff was extruded through a 2-mm die orifice, it will have a diameter of 12–16 mm and a bulk density of 50–65 g/L. The puffs are conveyed into a rotary dryer or oven (Figure 12.2) to drop their moisture to around 1–2%. The driers, set to approximately 150°C, dehydrate the low-density collets in only 4–6 min. The dehydration rate is fast due to the low density and porosity of the collets. The high temperature does not affect color because the feedstock does not contain significant quantities of reducing sugars that promote upon heating Maillard reactions and caramelization (Hsieh et al. 1990). The baked expanded collets are finally flavored in a rotary drum or tumbler (Chapter 3) generally with a salt-oil-cheese slurry that is sprayed onto the extrudates. The amount of coating applied ranges from 34% to 42% based on the final puff weight (Serna-Saldivar 2008). The product is finally cooled and immediately packaged to avoid loss of texture or crispiness. A 100 kg of refined corn grits normally yield 136 kg of cheese-flavored puffs, depending on the amount of covering (Figure 12.2). The puffs generally contain 27–36% oil and about 2% salt. The nutritional composition of baked puffs and other extruded snacks are thoroughly covered in Chapter 20.

The most relevant processing controls are the proper conditioning of the grits, the feedstock flow rate, the barrel temperature at the different zones, and the cutter speed. Tempering of grits or feedstock plays an important role in the final characteristics of the direct expanded collets. If the collet density is too low, a higher tempering may be required, whereas when the opposite occurs, tempering should be reduced. The quality of the grits plays an important role in setting the extrusion parameters and greatly affects end-product quality (Fletcher et al. 1985). The factors of major control are the particle size distribution because it affects the movement of the feedstock through the extruder and decreases one of the most common problems in direct extrusion: surging (Garber et al. 1997). Surging refers to the buildup of grits in the inlet or barrel mouth. The feedstock should be highly refined with a low fiber, ash, and oil contents (Chapter 3). The increase of these values beyond the specifications results in lower expansion rates. The particle size of raw ingredients is important when extruding puffed snacks (Serna-Saldivar et al. 2008). Huber and Rokey (1990) reported that a soft texture resulted from a fine granulation and a coarser meal led to a harder and denser collet. Mohamed (1990) found that extruding maize grits varying in size from 402 to 618 μm increased pressure with increased particle size but expansion and hardness decreased.

In addition, the starch should be undamaged or native. Cereals that experienced defective storage or are heat- or mold-damaged, yield grits with higher starch damage. These grits usually had less expansion rates and therefore yield low-quality puffs (Garber et al. 1997).

Maize grits are the preferred raw material for production of direct extruded snacks. Other refined cereals fractions with good potential are refined rice and sorghum grits. Moore (1994) reports the manufacture of potato stick products made from a blend of 65% potato granules and 35% refined maize meal. This blend is extruded similarly to other direct-expanded products to obtain puffs with the unique potato flavor.

Direct extruded collets can be alternatively produced supplementing small quantities of lime to the formulation. Martinez-Bustos et al. (1998) and Chang et al. (1999) successfully produced a hybrid direct-extruded collets using extrusion and calcium hydroxide to impart the typical tortilla flavor (Chapter 11) and increase the nutritional value via calcium concentration. The authors studied the effects of calcium hydroxide, tempering moisture, and barrel temperature on properties of expanded puffs obtained from a single-screw extruder. As expected, the highest radial expansion and the lowest density and breaking force were achieved for the treatment treated without calcium hydroxide. The best expansion rates were obtained when maize meal was tempered to 16% moisture and the extruder barrel was set at 130°C. Calcium hydroxide decreased expansion rate and increased density, but allowed the production of nixtamalized extrudates with unique tortilla flavor.

Likewise, Zazueta-Morales et al. (2001, 2002) successfully produced nixtamalized blue maize extrudates. These authors concluded that it was possible to elaborate appropriate extruded puffs from partially degermed blue maize meal when calcium was added between 0.02% and 0.078% and the extruder operated at 117–180 rpm. Lime concentration significantly affected the expansion index, bulk density, and penetration force. Extruded puffs with lime concentration of 0.05% or less showed good expansion characteristics and had similar appearance to the control. The development of blue maize expanded products have nutritional advantages compared to regular counterparts because blue maize is rich in anthocyanins and other natural antioxidants with positive effects on human health (Chapter 21).

12.3.2 Production of Fried Expanded Puffs

Fried expanded curls are still widely accepted among snack consumers because they possess a different flavor and configuration. These snacks have been produced for more than 75 years and are manufactured using a special type of extruder that generates most of the heat via friction. Technically, this extruder falls into the medium shear category and works similarly as the direct expanded high shear extruders. The major difference is that the fry-type snack extruder does not extrude the product through a typical die. Instead, it uses a set of plates: one rotating and the other stationary. These plates shape and expel the extrudate that is internally cut by rotor fingers (Moore 1994, Burtea 2001).

Tempered or untempered maize grits, rice grits potato flours, and regular or modified starches are introduced and transported into the barrel section of the stator and the shear applied by the screw movement increases both the pressure and internal temperature (>120°C), so the starch gelatinizes, melts, and plasticizes. The rotor fingers cut the extruding material, which are then immediately extruded radially between the two plates. The main way of controlling this extruder is by adjusting the gap between the rotors. According to Moore (1994), three things occur in this special die: the

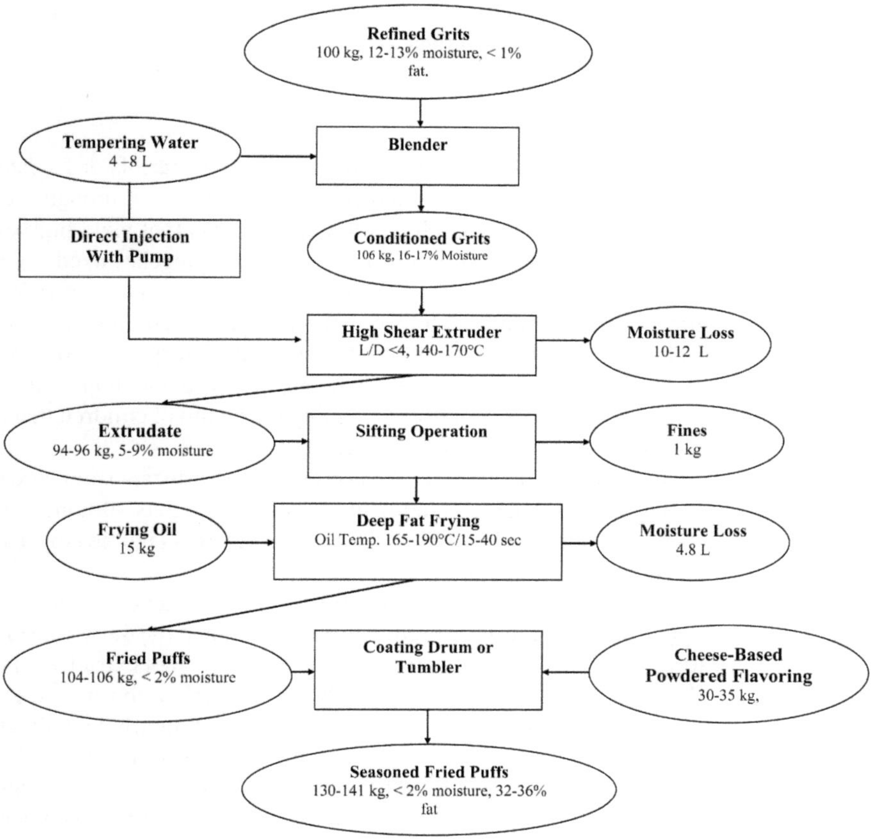

FIGURE 12.3 Flowchart of the extrusion technology for the production of seasoned fried puffs.

maize meal or grits are subjected to high shear and pressure which generates most of the heat to cook the raw material, a rapid pressure loss causes the superheated water to turn into steam which expands the collets and the flow of the raw material between the plates twists and forms the collets. Without this special die assembly, the curls would not have the characteristic shape and texture. As the material exits from between the plates, the pressure drops from 48 atm to atmospheric and the water superheats and vaporizes expanding the extrudate (Burtea 2001). However, lower expansion rates are obtained in contrast with direct extruded baked products. The higher density curls make these extrudates suitable for a short frying schedule (Figure 12.3). The extrudates are generally fried to reduce the moisture level from around 8% to 1.5% to grant texture and stability, and the fried curls pick up from 20–25% oil (Moore 1994). The extremely porous surface of extruded snacks enhances oil absorption and adherence. After frying, the collets are seasoned using a two-stage seasoning system, where an oil-based slurry is first applied followed by a dry seasoning that normally is cheddar cheese flavored. Usually, a snack seasoning containing 10–15% salt (base-weight) is dosed onto the extrudates (for additional information, refer to Chapters 3 and 8). This brings the total fat content to a level of around 30–35% (Chapter 20). Fry-type extrudates are easily recognized because of their higher density, especially due to their irregular rough-shaped forms in varying lengths. One disadvantage is that the extruder produces more fines than products extruded through die orifices. The main control of the fry-type curls process is the quality of the maize grits in terms of their oil, total starch, and starch damage contents. In addition, the particle size distribution of the grits, tempering of the raw material, and the gap between the stator-rotor plates is controlled. Generally, a coarser granulation will produce a curl with a crunchier texture and a slightly larger cell size. A finer granulation would give a softer texture but may also cause surging and operational problems (Moore 1994, Burtea 2001, Serna-Saldivar 2008).

12.3.3 Extruded Whole Grains

As explained before, the refined fractions of cereals are the main raw materials or feedstocks currently used for production of extruded snacks. Refined fractions of cereals, thoroughly discussed in Chapter 6, are preferred due to their high-starch and low-oil contents that yield high expansion index, crunchy texture, and low density. In addition, refined native, pregelatinized, or modified starches from maize, wheat, potato, and tapioca are used to increase the expansion index of both extruded second- and third-generation snacks. However, the new and more educated consumers are demanding snacks with lower calories and higher protein and dietary fiber to favor health. According to Hegazy et al. (2017), snacks are increasingly being produced with non-starchy sources to avoid the high carbohydrate intake that is negatively affecting the health of children that normally consume about 25% of their total energy intake from snacks.

This new trend has demanded the utilization of whole-grain flours or meals from cereals, pseudocereals, legume pulses (Berrios 2016), dehydrated fruits or vegetables, nuts, almonds, hazelnuts, and by-products rich in fiber and proteins like soybean Okara and defatted legume (soybean, chickpeas, peas) meals and concentrates (Grasso 2020).

Extrusion of whole-cereal grains is gaining momentum with the emergence of consumers concerned about diet and weight loss. The new dietary guidelines recommend the daily consumption of three or more 1-ounce equivalents of whole-grain products. The production of extruded whole-grain products is a complicated task because the fiber, particularly the insoluble rich sources, and the fat associated with the germ, reduce expansion. The type and composition of the whole grain also affect the degree of radial expansion. Additional issues associated with extrusion of this type of feedstock include rancidity, caused by unstable lipids, and the development of bitter flavors (Eastman and Lee 2005). According to Slavin et al. (2001), extrusion does not change the total dietary fiber content, but it significantly affects the ratio of soluble to insoluble fractions. Some studies showed that this ratio was higher in extruded products than in control samples.

Whole-grain ingredients commonly used in extrusion include wheat, brown rice, maize, and oats. Barley, sorghum, and millets are also used but to a lesser extent. Non-cereals commonly used include flax, grain amaranth, and quinoa, discussed in Chapter 13. Whole wheat yields denser extrudates and a nutty flavor, whereas brown rice has an acceptable expansion but extrudates are very prone to rancidity. Extruded products containing whole-maize meal are acceptable but have reduced expansion when compared with degermed fractions. Oats is one of the most popular whole grains used in extrusion and bakery products because it contains a favorable ratio of soluble to insoluble dietary fiber. It is well-known that the highly prebiotic beta-glucans reduce the risk of cardiovascular disease (Chapter 21). However, oats has a low expansion rate and is the cereal most susceptible to rancidity, so the extrusion of this cereal is generally accompanied by the supplementation of natural or synthetic antioxidants in order to prolong the shelf life of whole-oats-based snacks. Like oats, barley has a high beta-glucan content and is lower in fat and susceptibility to rancidity. When whole-barley flour is used in extrusion, very poor radial expansion is obtained. In order to improve expansion rate, the whole-barley flour is blended before extrusion with starch or refined flours (Grasso 2020).

Whole sorghum has been successfully direct-expanded to produce flavored snacks. The advantage is that the sorghum does not need any milling procedure and the resulting snacks have more dietary fiber and keep nutraceuticals associated with the pericarp, germ, testa, and aleurone layer. White whole sorghum has potential because it yields gluten-free extrudates. Therefore, it is currently used in some snacks marketed to consumers who have celiac disease or gluten intolerance.

Berrios (2016) recently summarized the extrusion processing of main commercial legume pulses making emphasis on dry beans, dry peas, and chickpeas (Chapter 13). The effects of direct extrusion of these seeds effectively diminish antinutritional factors and greatly improve protein digestibility. The use of whole-grain meals rich in dietary fiber, resistant starch, and prebiotics also favors the phytochemical profile rich in antioxidants. Among legume pulses, chickpeas have recently gained attention because it does not contain allergenic compounds like soybeans and peanuts and has a mild or neutral flavor, especially due to its starch content that makes it more suitable for direct extrusion processes. Meng et al. (2010) optimized the extrusion conditions of a formulation containing 60% chickpea, 23% potato flour, and 15% whey protein concentrate in terms of tempering moisture content, screw speed, and barrel temperature. The extrudates with high expansion ratio and low bulk density and hardness were obtained at low feed moisture, high screw speed, and medium to high barrel temperature. The resulting chickpea-based puffs had an adequate nutritional profile, especially in terms of protein, dietary fiber, resistant starch, and micronutrients.

The combination of cereals and legumes is highly effective to improve both the quantity and quality of protein, the dietary fiber, and the micronutrient and phytochemical profiles (Lopez-Barrios et al. 2018). Thermoplastic extrusion normally improves the starch and protein digestibility and in some cases modifies the ratio of insoluble to soluble dietary fiber because it can transform insoluble fiber into soluble with more prebiotic effects (Espinosa-Ramírez et al. 2021).

In addition to these raw materials, sprouted or germinated grains have been recently used for the production of novel extruded snacks. The germination generates intrinsic enzymes that hydrolyze macromolecules and increase the bioavailability of important nutrients like phosphorus. Lopez-Barrios et al. (2018) produced direct-expanded extrudates with cotyledons of sprouted black beans and found that extrudates contained significant amounts of bioactive peptides with antioxidant activity. This technology has been successfully applied to other legume pulses like chickpeas. Hegazy et al. (2017) extruded corn snacks supplemented with germinated chickpea and tomato powder and found that the novel snack contained higher antioxidant properties. Other researchers produced snacks with high antioxidant activity using a combination of brown rice and germinated pumpkin flour. This snack exerted high antioxidant properties due to the use of whole-grain materials and the disruption of fiber components during germination and thermoplastic extrusion. The optimum extrusion conditions were low tempering moisture (10–13%), 350 rpm screw speed, and 140°C in the last zone of the barrel (Nakhon et al. 2018).

The problem for the use of raw materials rich in protein or fiber is the reduction in the expansion index of the extrudates. The regular extrusion had negative effects on important nutrients contained in the raw materials because they are degraded or transformed into lower molecular weight moieties. However, the high-molecular-weight components like starch, protein, and fiber are disrupted by the high shear forces inside the extruder (Lohani and Muthukumarappan 2017). This synergistic effect of high temperature and pressure degrades large structures into low-molecular-weight dextrins, peptides, and lignans, which have improved water solubility, digestibility, and nutraceutical properties. Viacava et al. (2020) extruded dehydrated carrot strips and concluded that the process enhanced the water solubility and the degradation of fiber components increasing the assayable concentrations of free phenolic compounds. The same authors suggest that the puffs were easily extruded with a starchy feedstock (maize, rice, or wheat flout) at high levels (>25%) to obtain nutraceutical snacks with improved functionality and less net energy

Soybean is the most popular ingredient to add in order to increase protein content and quality of the product. Whole and defatted meals and protein concentrate and isolates can be used (Chapter 13). The advantage of whole meal is that it contains high fiber, minerals, lecithin, nutraceutical compounds, and PUFA, but it greatly reduces expansion rate and the resulting extrudates are prone to rancidity. Defatted soybean meal has approximately 50% protein, retains most of the nutraceuticals, and has the advantage of being practically free of lipids. The problem is that it significantly changes the color and the flavor of the extrudates. Concentrates and isolates contain 65% and 90% protein, respectively, and therefore are normally used in lesser amounts; therefore, the functionality is not compromised as much. These proteins have a bland flavor and produce the best extrudates in terms of color, texture, expansion rate, and flavor. The disadvantage is that these products are practically fiber-free and do not contain significant amounts of phytochemicals with known nutraceutical effects.

Amaranth is an ancient grain widely used by Mesoamerican cultures and its use has increased recently because of its high nutritional value. The amaranth protein has high levels of lysine and is well-balanced and the grain also has significant levels of starch. However, extrudates from amaranth have low radial expansion and the finished products have a particular flavor that is not liked by some consumers.

12.4 EXTRUSION OF PELLETS OR HALF-PRODUCTS

Third-generation snacks are also referred to as half-products. Following extrusion cooking and forming, the resulting pellets are carefully dehydrated to yield a shelf-stable product that could be stored for long periods. Pellets are called half-products because they need additional processing to reach the consumer. For snacks, the key additional process is frying and less frequently conventional or microwave baking. During these heat operations, pellets usually puff or expand and loose moisture, so a crisp snack is produced. Different types of raw materials and combinations of ingredients can be incorporated into half-products. Most are formulated with wheat, potato, rice, maize and oat flours or meals, starches refined from maize, potato, wheat and tapioca, oils and fats, baking powder, salt, hydrocolloids, and emulsifiers like lecithin. These raw materials described in Chapters 6 and 14 are rich in starch needed to maximize the expansion of the final snack product. Normally, these formulations contain more than 70% of total starch. The half-products are normally fried into a puff snack that is usually flavored with powdered seasonings or oil-based liquid slurries (Huber and Rokey 1990, Huber 2001, Serna-Saldivar 2008, Moscicki and Zuilichem 2011, Ghandi et al. 2016, Panak-Balentic et al. 2018).

There are two major sorts of indirect expanded products: pellets and products that are sheeted and cut into ribbons or other configurations. In this last category, fabricated potato chips are included. Pellets are cooked in an extruder and forced through a die at temperatures below 100°C. The lower temperature prevents expansion and favors the production of a high dense pellet that is further dried to yield a shelf-stable intermediate product. Pellets, in contrast to the directly expanded puffs, offer the advantage of being distributed with high weight and low volume (Panak et al. 2018). Thus, they can be available at any time and place for further processing into the finished products. The basis for the wide variety of these snacks is their shape imparted by the die design and configuration. Most common shapes are flat (for production of pork skin analogs), tube, shell, ring, screw, and wheels weighing from 250 to 750 mg each piece (Meuser and Wiedmann 1989). For production of laminated products, the dense extrudate from the forming extruder is sheeted and cut into square, rectangular, or similar configurations before the critical dehydration step. Within this category are the fabricated chips that are gaining popularity worldwide. The raw materials are cooked similarly to pellets, but the extrudate is not dried for storage. Instead, they are conveyed to the deep fat fryer or high-convection ovens to remove moisture and impart texture and flavor (Moore 1994, Aguilar-Palazuelos et al. 2006, Serna-Saldivar 2008, Moscicki 2011, Ghandi et al. 2016, Panak-Balentic et al. 2018).

12.4.1 PRODUCTION OF PELLETS

A typical flow diagram production of third-generation snacks is presented in Figure 12.4. Most pellets are manufactured from mixtures of starch-rich raw materials that include dehydrated potato products and/or potato starch, wheat flour, and refined maize grits or meals (Chapter 6). Optimally, the starch content of the blends range between 80% and 90% of the dry matter. The starch is the most important chemical component because it imparts most of the functionality during extrusion and greatly affects the product expansion rate during frying. The plastification attained during the thermoplastic extrusion is stopped by the dehydration process and the gelatinized starch plays an important role during deep fat frying or baking expansion (Panak et al. 2018). Key operations for

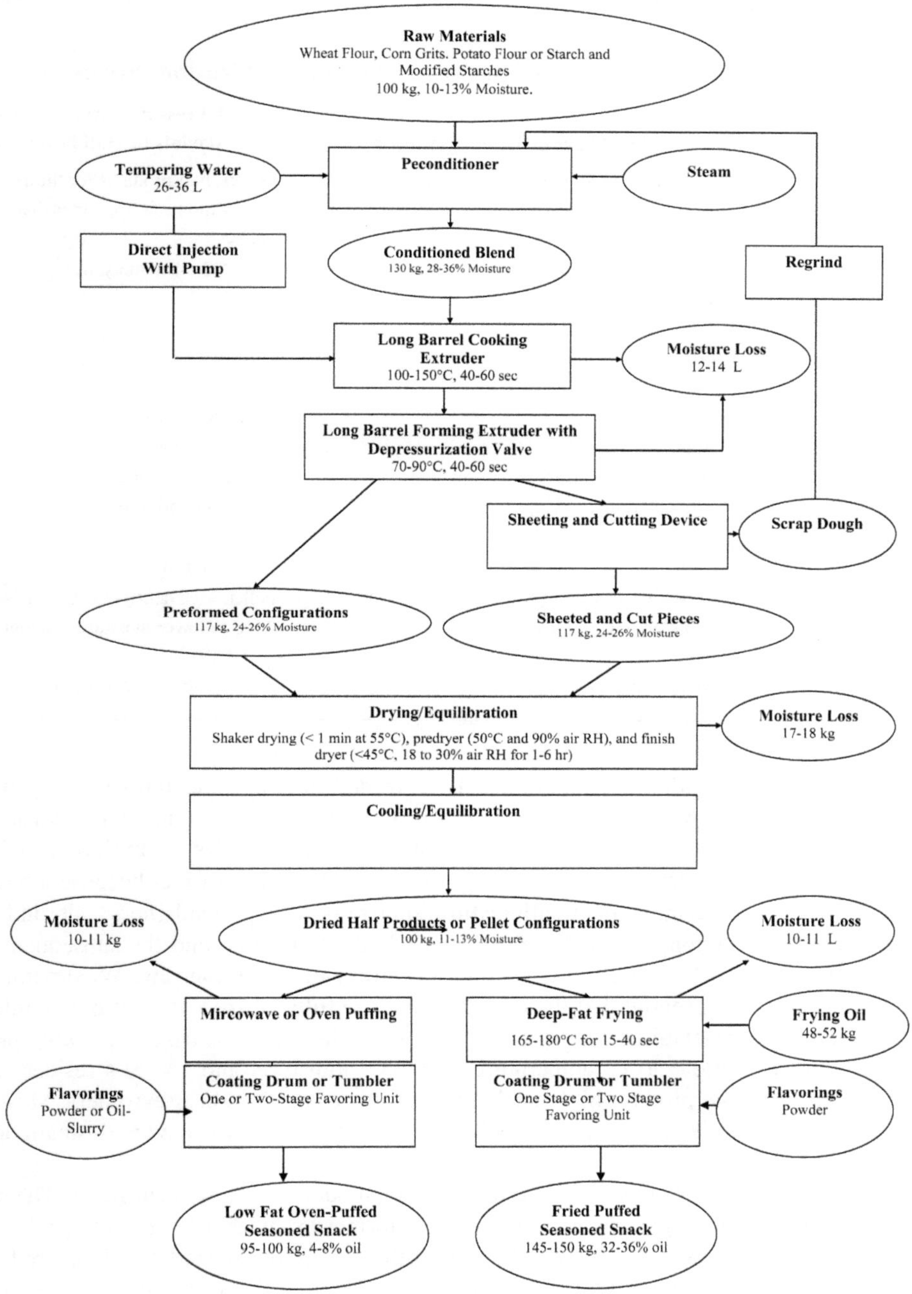

FIGURE 12.4 Flowchart of the production of extruded pellets and third-generation savory snack foods.

production of third-generation snacks are extrusion, pellet drying, and frying (Figure 12.4). The first step of the manufacturing process is the mixing of raw materials. The amount of water and mixing rate are important quality control parameters because the aim is to homogeneously distribute the water into the blend of raw materials. This operation is also named tempering.

In contrast to direct extruded products, pellets are usually tempered to moistures from 25–35% (Table 12.2). The higher tempering enhances starch gelatinization and decreases the expansion of

TABLE 12.2
Comparison between High-Shear and High-Pressure Forming Extrusion Processes

	High Shear or Direct Extrusion	High Pressure Forming Extrusion (Pellets or Half Products)
Product	Direct expanded, puffs or second-generation snacks, coextruded pillows, texturized vegetable protein	Pellets, half products for third-generation snacks, gluten-free pasta, flakes
Feed moisture	Low (12–20% moisture)	High (28–48% moisture)
Extrudate moisture before further processing	4–8%	20–25%
Maximum barrel temperature	110–180°C	20–100°C
Temperature gradient	Increasing	Homogeneous or decreasing
Shear rate and energy input	High	Relatively low
Screw design	Produces high shear	Produces low shear
Screw rpm	150–400 rpm	Less than 150 rpm
Screw *L/D* ratio	<4:1	6–30:1
Pressure at die	High	Low medium
Product or extrudate	Expanded, low density with 5–10% moisture	Pellet, high density, >20% moisture
Drying	150°C to lower moisture to <2%	70°C to lower moisture to about 10–12%
Flavorings	After extrusion	Before and after extrusion

the extrudate. Third-generation snacks are usually extruded in two sequential steps or with only one special extruder that performs cooking and forming in the same equipment. Extruders for these applications usually have a long *L/D* ratio and a vent to depressurize the extrudate before die forming (Table 12.2). The first extruder is commonly named cooking extruder because it heats and gelatinize the starchy materials mixed with texture improvers such as emulsifiers, salt, and flavorings. The inlet barrel section is cooled with water to facilitate feeding and the subsequent barrel zones heated to achieve the desired level of starch gelatinization (Meuser and Wiedmann 1989). In the cooking step, it is essential that the raw materials are fully cooked or gelatinized unless the formulation contains pregelatinized or modified starches. The typical extrusion cooking processing conditions include 100–150°C temperatures, 25–35% moisture, and 15–20 s residence time. The screw design does not produce high mechanical energy as in high shear extruders. Most of the energy required to gelatinize the starch is supplied by the barrel usually heated with steam or water vapor (Table 12.2).

The most common setups are configured to provide a moderate to low shear input. The cooked dough existing in the cooking extruder is fed into the former extruder, which works under totally different conditions. This extruder cools and densifies the cooked plasticized dough before forming the pellets. To achieve this, the screw is designed to provide low shear and pressure. Most of the pressure in the forming extruder is produced by the die restriction (Panak et al. 2018). The forming zone moves forward and cools down the extrudate to a temperature of 90–110°C. The viscoelastic plastic material is continuously shaped by the die into the desired pellet configuration. Some forming extruders are equipped with a pressure release device or vent located immediately before the die or forming zone. This prevents the expansion of the extrudate and enhances the formation of a high dense pellet. The cooked, formed pellet exits the extruder with moisture contents 20–25%. The extrudate retains the shape of the die orifice because there is no expansion taking place. In most instances, the material existing the die is cut into thin pieces by a cutter blade, which is riding around the external die face. In some cases, the die is a thin slot, which makes a continuous ribbon that is cut into uniform pieces with a rotary cutter (Moore 1994). According to Meuser and Wiedmann (1989), approximately 3–4% moisture is lost in extruders equipped with vents at the discharge end. These extruders are relatively easy to operate and maintain and are designed to produce hourly up to 25 tons (Serna-Saldivar 2008).

Many third-generation snacks are produced with the same formulations and extrusion technologies described above with the only difference that the forming extruder is equipped with a dough sheeter and cutter. Production of laminated, perforated, and spatial forms are normally made from wide sheets of pre-rolled dough on special devices which set their shape and uniform thickness. The extruded dough should be free from air bubbles and have adequate flexibility and rheological properties for further processing and product performance after frying or baking operations. The formation of dough with suitable rheological properties for extrusion and sheeting and the dehydration process used are associated with the puffing quality of these half-products. In terms of formulation, this requires good selection of ingredients rich in starch and the water conditioning to attain optimum dough consistency. In terms of processing, the dough should be properly cooked to achieve the desired level of starch gelatinization and consistency. Finally, the dehydration should be controlled to produce half-products with 10–12% moisture without any stress cracks and fissures (Guy 2001, Huber 2001, Serna-Saldivar 2008, Moscicki 2011).

Pellets or cut dough configurations are immediately dehydrated to lower their moisture content to about 10–12% (Figure 12.5). The drying process is critical because water should be carefully removed from the pellet by evaporation in a manner that the quality characteristics of the fried pellets are controlled. Optimally, drying should be performed in two sequential steps (pre-drying and drying) (Meuser and Wiedmann 1989, Serna-Saldivar 2008). The short-time pre-drying step is normally performed at 55°C in a dryer equipped with a shaker. The main aim of this step is to evaporate the surface water to prevent sticking of pellets or dough configurations. During this particular step, the moisture content can drop 2–4%. Then, the moist half products are dehydrated in large belt dryers, which have strict controls of air temperature–relative humidity and air speed. The goal is to reduce the moisture to 10–12% without producing stress cracks, cracking, and blisters. Other alternative is the use of drum driers set at temperatures around 45–90°C. The residence time required to lower the moisture to 12% or less is 1–8 hours, depending on the pellet size, density, and initial moisture content. It is recommended to equilibrate or temper the dried pellets for at least one day before further processing (Serna-Saldivar 2008, Moscicki and Zuilichem 2011) because the frying or baking of recent dried pellets yield products with less expansion rate because moisture is more and less concentrated in the geometric center and surface of the pellet configuration, respectively. However, most industrially produced pellets or half-products are stored for long periods before further processing or merchandized as half-products. Snack pellets are normally expanded in deep fat fryers, impingement, or microwave ovens. Frying is usually performed at an oil temperature of 170–175°C for 10–40 s, depending on the pellet size. The expanded snacks exit the fryer onto a mesh belt for oil draining. During the deep fat frying operation, heat is transferred from the hot oil onto the product surface by convection and from the surface into the center of the pellet by conduction. The water moves from the inside to the evaporation zone, leaving the product surface as vapor. Some of this vapor remains trapped within the intracellular pores due to restrictive diffusion. The vapor will expand distorting the pore walls and contributing to the internal porosity. Most of the water is released and oil absorbed during the first 10-s frying. The result is production of an expanded product having suitable cell structure and desirable texture or mouthfeel. The bulk density of these expanded products fluctuates from 40 to 100 g/L (Moore 1994). During this critical operation, the extrudate drops its moisture to less than 2% and develops the characteristic texture, flavor, and color due to Maillard reactions.

Regular and microwave baking is the other common alternative for expanding half-products. Puffing at environmental pressure requires the use of high baking temperatures such as provided by air impingement ovens or similar equipment that provide high rate of heat transfer. The development of products for microwaving with similar microwavable bags like the ones used for popcorn

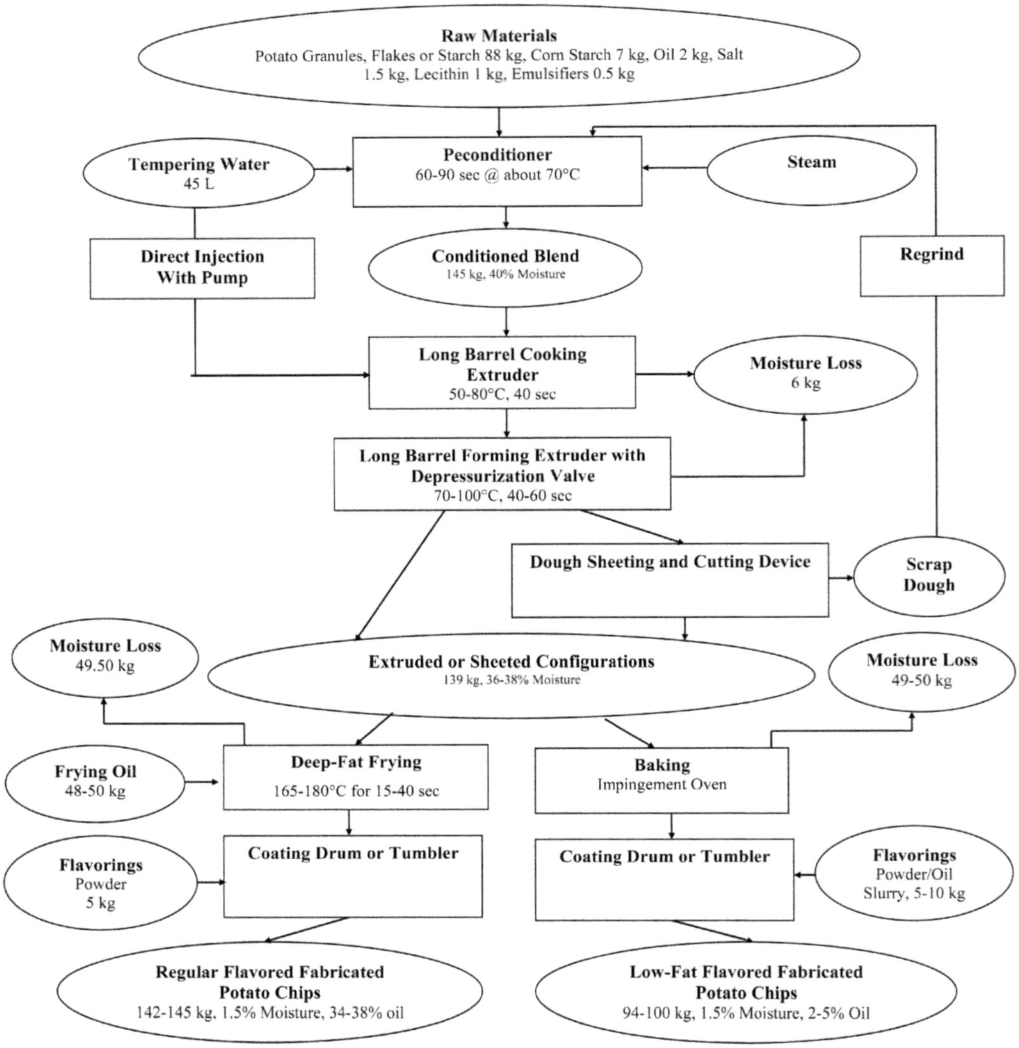

FIGURE 12.5 Flowchart of extrusion processes for production of regular and low-fat fabricated potato chip.

are gaining acceptance, especially for the puffing of pellets and production of low-fat convenient snacks (Serna-Saldivar 2008). Cooking, heating, and expanding foods with microwaves is fast, convenient, and inexpensive technique. Pellets are placed in an electromagnetic field where heat is produced by friction of the water molecules contained in the food matrix. In some cases, the packaging with susceptor helps to develop and concentrate heat when subjected to microwaving (Chapter 4). The mechanism of microwave expansion is based on the transition of amorphous matrix to the rubbery and flow states occurring during heating. Moisture generates the superheated steam, which accumulates at nuclei in the glassy matrix, creating a locally high pressure and expansion. When moisture is lost from the matrix and microwave heating is ceased, the puffed snack cools down and reverts to the glassy state and the final structure. As for the expansion by hot air or during deep fat frying, the shape and internal air cell structure is affected by the degree of gelatinization of the pellets. Optimal expansion by microwave heating is achieved when the starch is approximately half-gelatinized and the moisture content in the pellets is about 10% (Panak-Balentic et al. 2018).

According to Meuser and Wiedmann (1989), most of the energy consumed during the manufacturing process of deep fat fried snack pellets is in the extrusion, drying, and deep fat frying or baking operations. The extrusion and drying steps use about 70% of the electrical consumption, whereas the thermal energy expenditure used to fry 1 kg pellets is 244 kcal.

Third-generation snacks are normally flavored with powdered seasonings generally applied with a belt-type flavor depositor or in a revolving tumbler (Chapter 3). The first equipment has an adjustable dispenser to promote uniform application of dry seasonings and a wire mesh belt that moves and sifts the flavorings to eliminate lumps. Salt and flavorings can also be applied in single- or two-stage rotating tumblers or drums equipped with liquid sprayers or auger or electrostatic powder dispensers. After seasoning, fried products are cooled and immediately packaged in moisture-proof bags to avoid rancidity and loss of crispiness (Chapter 4).

12.5 EXTRUSION OF FABRICATED CHIPS

Fabricated chips are gaining popularity because the public perceives these products as low fat. The extrusion technology for potato chip fabrication started in the 1960s but was widely used until the 1990s. Sales of fabricated potato chips keep increasing in the world, mainly in the United States.

Fabricated chips are produced with a similar extrusion line as pellets (Figure 12.5). The major differences are that the raw materials are generally conditioned to higher moisture contents and cooked less extensively to achieve less-starch gelatinization and that the extrudate is not dried prior to frying. In other words, this technology does not produce half-products. The partial starch gelatinization is required to control the oil uptake during frying. Other critical factor to control is the extrudate thickness. The extrudates are generally forced through dies with opening of less than 1 mm. Some processes use sheeting rolls to further decrease the extrudate thickness. After cutting, the chips are directly channeled to the deep fat fryer to remove their moisture to less than 2% and impart the characteristic flavor, color, and texture. The hot fabricated chips are then seasoned in a tumbler or drum (Chapter 3). Low-fat fabricated chips are obtained by baking in impingement ovens. Like fried fabricated chips, the product is baked to drop its moisture content to less than 2% and the temperature, air speed, and other settings calibrated to obtain good tasting and crispy products. The main advantage of fabricated chips are the flexibility and uniformity of the end-product which allows manufacturers to offer a wide variety of fabricated products with different nutritional attributes and the packaging of uniform pieces (Serna-Saldivar 2008, Moscicki and Zuilichem 2011, Ghandi et al. 2016).

12.6 COEXTRUSION

Coextruded or co-injected snacks are manufactured since the 1980s by inserting or pumping a filling into a void in an expanded item. The coextrusion technology allows manufacturers to make products with two textures, two colors, and/or two flavors. The two dissimilar materials can come from two different extruders, or one extruder and a pump, as described in Figure 12.6. The key is the design of the die assembly, which should be capable of co-injecting a cream, paste, or filling at the die in direct expanded extrudates. This technology requires injecting the filling into a piston extruder die operating at low pressure compared to the pressure normally used for the extrusion of the cereal-based outer shell (Gandhi et al. 2016, Morales-Alvarez and Rao 2016).

Moore (1994) classifies coextruded cereal-based products into three categories: tubes with cereal-based fillings, tubes with fat-based fillings, and tubes with water-based fillings. Most dual textured snacks fall into the category of tube/fat-based filling, cheese-filled being the most common savory snack. Figure 12.7 depicts the key steps for the production of cream-filled or coextruded snacks containing an outer shell based on extruded refined cereal base formulation. The use of fat-based fillings will minimize the problem of water migration from the inner fill to the outer crust and prevent loss of crispiness because both layers will have similar water activities (A_w of

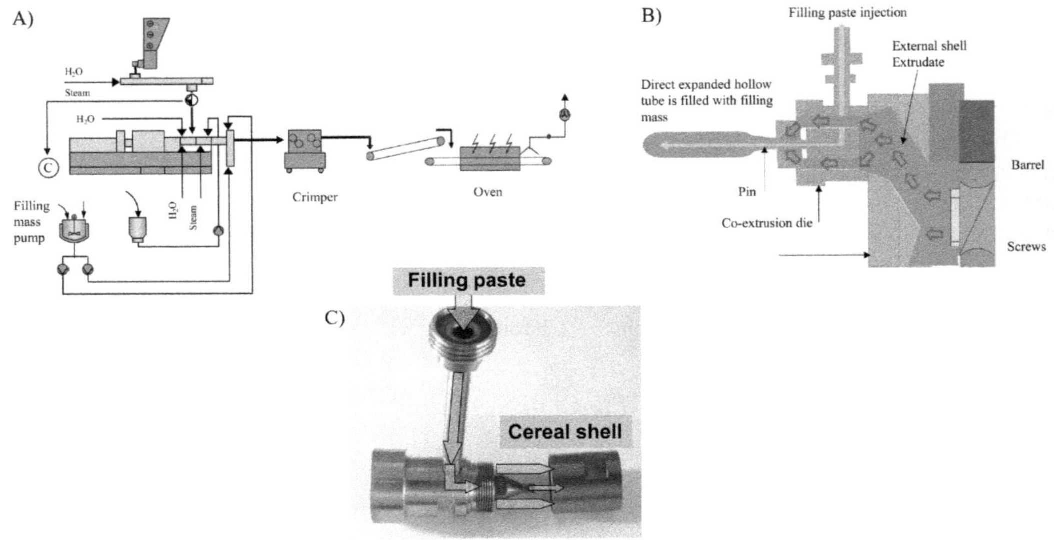

FIGURE 12.6 Drawing of the design of a coextrusion system: (a) coextrusion line; (b) coextrusion die elements; (c) nozzle (courtesy of Buhler Group).

approximately 0.65). In addition to coextruded tubes with an internal filling, a cereal-based outer shape in a U-shape is produced. The U-shape shell is first filled and then rolled to cover the filling. The rheological matching between the outer dough/shell and filling promotes complete filling and minimize undesirable voids (Seker 2016). Twin-screw extruders are the most preferred for these applications (Harper 1989, Morales-Alvarez and Rao 2016), although single short-barreled extruders are also used. Many variations of coextrusion may be made with dies and post-extrusion equipment such as rollers, stampers, and belt-type cutters (Huber 2001). Few truly coextruded snacks have found success in the market due to their lower shelf life and the loss of texture due to migration of moisture from the filling to the outer extruded shell. The engineering and design of the die for coextrusion is of paramount importance. The extruded portion of the snack flows directly through the die parallel to the direction of flow through the barrel. The filling is normally pumped into the die perpendicular to the flow of the outer shell extrudate and into the center of the extrudate flow region (Figure 12.6). The amount of filling pumped controls the coextruded product dimensions and affects the filling-to-shell ratio (Huber 2001). The technologies and additives are constantly improving, so these innovative products have a bright future because they can be manufactured with low processing and labor requirements.

12.7 SUPERCRITICAL EXTRUSION TECHNOLOGY

Supercritical carbon dioxide (CO_2) has been successfully applied in extrusion for the production of novel and nutritionally improved snacks. The main advantage is that the supercritical CO_2 is a nontoxic, nonflammable, and inexpensive material. This extrusion process known as SCFX found its application primarily in the processing and manufacturing of plastics, but recently it has been applied to produce innovative food products and snacks (Panak-Balentic et al. 2017).

As explained before, the typical extrusion processes combine elevated temperatures, shear, and pressure. These conditions negatively affect raw material, degrading beneficial molecules for the consumer, such as carotenoids, flavonoids, anthocyanins, tocopherols, vitamins, and the functionality of proteins and aromatic compounds (Masatcioglu et al. 2014). Another negative effect is production of harmful molecules, such as hydroxymethylfurfural caused by the interaction between proteins and sugars through Maillard reactions. These molecules decrease the overall protein,

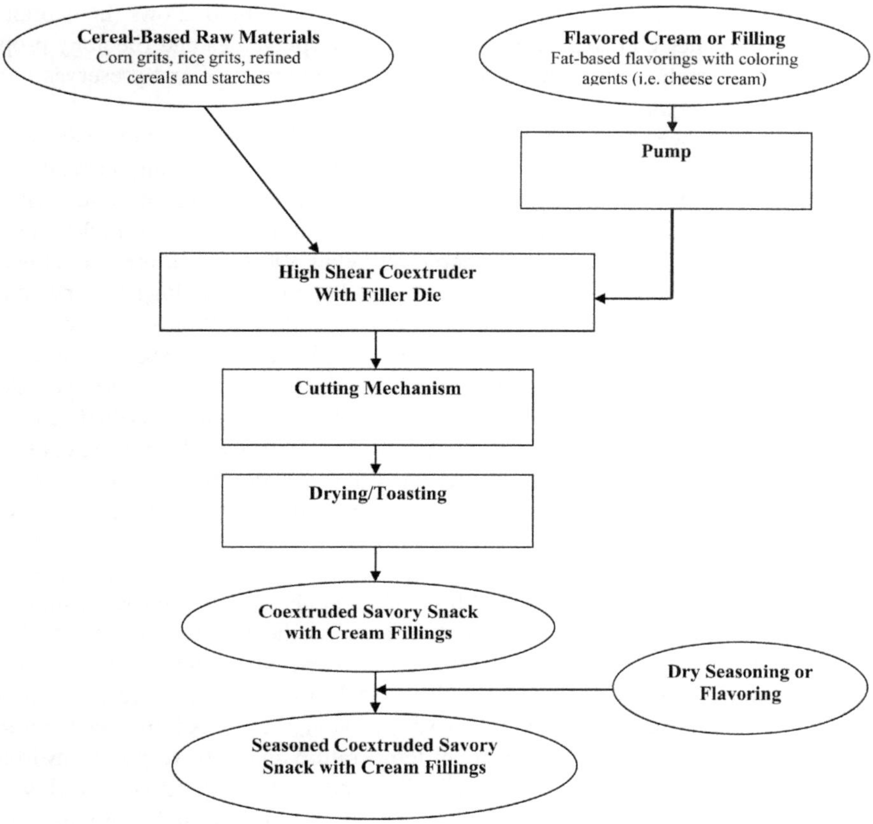

FIGURE 12.7 General diagram of the coextrusion process for production of cream-filled cereal-based snacks.

increase the bitterness, and at elevated concentrations can lead to cancer. Therefore, the reduction of the extrusion processing temperatures is considered advantageous to diminish the inactivation of thermolabile nutrients, while avoiding the generation of harmful compounds. To achieve the temperature reduction is necessary to decrease the amount of mechanical energy (screws speed) and/or reduce the barrel hitting. Even if we can ward off nutrient's degradation, the low temperature reduces the energy of the superheated vapor inside the barrel and the volume of direct expanded snacks. Then, the low-temperature extrusion has a challenge, to increase the inner cell structure of the snacks.

The liquid carbon dioxide is stored in a cool pressurized vessel to prevent its vaporization. The pressure is controlled by a compressor that holds the pressure above the critical pressure. The CO_2 is passed through a heat exchanger and directly injected into the extruder barrel in both sub- or supercritical form.

The addition of supercritical fluids can overcome the problem of expansion at low temperatures (<100 °C). Supercritical fluids are substances at pressure and temperature that are above critical temperature and critical pressure (Felder et al. 2016). The properties of the gases above the critical point are similar to the liquid phase, like density, but the viscosity is more like a low-pressure gas. Supercritical fluid can be easily incorporated into the extruder barrel, the SCF is solubilized with the high pressure inside the barrel, and is added when the dough inside the extruder is melted. The supercritical fluid is normally added in a final segment of the extruder barrel before the endplate or die. Supercritical fluids act as a blowing agent; when it is precipitously decompressed at the end of the die plate, it immediately turns into gas increasing its own volume and also the volume of the extrudate (Masatcioglu et al. 2014, Seker 2016). The supercritical fluid allows the manipulation in expansion of the extrudates cell structure, changing the morphologies and rheology properties of the dough inside the barrel; reduces the shear stress during extrusion, and preserves most of the original color and flavor associated with raw materials.

Different gases have been tested but CO_2 has proven to be the most practical and versatile (Panak-Balentic et al. 2017). CO_2 has a critical temperature of 31.05°C and critical pressure of 72.9 atm (7386 KPa) (Felder et al. 2016) with the main advantages that it is accessible, cheap, nonharmful, safe, and nonexplosive. The supercritical CO_2 can be mixed in a single phase with the dough at high pressure inside the barrel at temperatures below 100°C. For this reason, long barreled extruders are preferred for this application due to the need of a relatively long cooling zone. In this process, the internal pressure is increased gradually above supercritical pressure and reverse screw elements normally used before the point of supercritical CO_2 injection to prevent the backward flow. A reverse screw elements is also placed at the die to blend the CO_2 with the feedstock (Seker 2016). Another property is the control of the internal structures of the collets as described before. The gas cells size varies depending on the CO_2 mass flow and the post-extrusion drying and cooling, changing the final density of the collet (Panak-Balentic et al. 2017). According to Sokhey et al. (1996), the addition of whey powder, corn fiber, and resistant starch as texturizing agents increase the strength of extrudates and prevent their collapse.

On the other hand, the snacks produced with supercritical CO_2 develop a smooth external surface within symmetrical internal structures (Masatcioglu et al. 2014). Compared with the conventional thermoplastic extrusion, the supercritical CO_2 presents the advantages of lower degradation of thermolabile molecules as B1 vitamin (thiamine) in corn extrudates and vitamins A and C in rice puffs. SCFX also minimizes losses of amino acids in snacks fortified with different protein sources (Balentic et al. 2017). For products enriched with whey protein, the SCFX improved the final texture and generated emulsions and gels stable at room temperature and with very high hydrosolubility (Mustapha et al. 2012). SCFX increases the solubility index of the dried collets, this effect has been proven in different products like maize, rice, pre-gelatinized rice, and sorghum, (Lohani and Muthukumarappan 2017, Panak-Balentic et al. 2017). Furthermore, SCFX with CO_2 has been successfully used to improve the starch properties since this fluid acts as a plasticizer. Masatcioglu et al. (2014) assayed higher concentrations of resistant starch type-3 after extruding a blend of maize flour and hull-less barley flour at low temperature (80–130°C) with CO_2-SCFX. On the other hand, Wang and Ryu (2013) extruded maize grits with maize fiber with supercritical CO_2 at low temperature (<105°C) and concluded that extrudates were more rigid and had better superficial and inner texture compared with traditionally extruded counterparts. Likewise, Panak-Balentic et al. (2018) took advantage of low-temperature SCFX of maize grits enriched with defatted cakes of hazelnut, pumpkin, camelina, and hemp to produce high-protein third-generation snacks designed to puff during microwaving. Results indicated that third-generation snacks had good expansion index, structure, and overall acceptability. Despite all the aforementioned benefits of SCFX and the goodness of supercritical CO_2, there are very few commercial snacks produced with this technology.

REFERENCES

Aguilar-Palazuelos, E., Zazueta-Morales, J. de J., and Martínez-Bustos, F. 2006. Preparation of high-quality protein based extruded pellets expanded by microwave oven. *Cereal Chemistry Journal*, 83(4):363–369.

Berrios, J.de J. 2016. Extrusion processing of main legume pulses. In: *Advances in Food Extrusion Technologies*, eds. M. Maskan and A. Altan, 209–236. Boca Raton: CRC Press.

Brent, J.L. 1999. Principles of extrusion cooking technology: Single and twin screw extruders. In: *Advances in Extrusion Technology Aquaculture/Animal Feeds and Food*, eds. Y.K. Chang and S.S. Wang, 73–84. Lancaster:Technomic Publishing Co.

Burtea, O. 2001. Snack foods from formers and high shear extruders. In: *Snack Foods Processing*, eds. E. Lusas and L.W. Rooney, 281–314. Lancaster: Technomic Publishing Co.

Chang, Y.K., Martinez Bustos, F., Matinez Flores, H.E., Bannwart, A.C., Guedes, P.A., and Gaiotti, E.R. 1999. Some functional properties of an expanded snack and pregelatinized flours made from corn meal treated with calcium hydroxide by extrusion process. In Advances in extrusion technology: Aquaculture/animal feeds and foods. *Proceedings of the International Symposium on Animal and Aquaculture Feedstuffs by Extrusion Technology and the International Seminar on Advanced Extrusion Technology in Food Applications*, eds. Y. Kil, S.S. Wang, Y. Cahng, and S. Shaw, 311–320. Lancaster: Technomic Publishing Co.

Eastman, J., and Lee, G. 2005. Whole grains in extruded products. *Cereal Foods World*, 50(4):168–172.

Espinosa-Ramírez, J., Rodríguez, A., De la Rosa-Millán, J., Heredia-Olea, E., Perez-Carrillo, E., and Serna-Saldivar, S.O. 2021. Shear-induced enhancement of technofunctional properties of whole grain flours through extrusion. *Food Hydrocolloids*, 111:1–13.

Felder, M.R., Rousseau, R.W., and Bullard, L.G. 2016. *Elementary Principles of Chemical Processes*, 4th edition, 228–236, 269. Hoboken: John Wiley & Sons Inc.

Fletcher, S.I., Richmond, P., and Smith, A.C. 1985. An experimental study of twin crew extrusion cooking of maize grits. *Journal of Food Engineering*, 4(4):291–312.

Gandhi, N., Kaur, K., Kaur, A., Singh, B., and Singh, P. 2016. Technological innovations in third generation snack foods. Progress. *Research Intternational Journal*, 11:4948–4959.

Ganjyal, G.M. 2020. *Extrusion Cooking. Cereal Grains Processing*, 2nd edition, 1–564. Duxford: Woodhead Publishing and AACC International Press.

Garber, B.W., Hsieh, F., and Huff, H.E. 1997. Influence of particle size on the twin-screw extrusion of corn meal. *Cereal Chemistry Journal*, 74(5):565–661.

Grand View Research Report Summary. 2020. *Extruded Snacks Market Size, Share & Trends Analysis Report by Product (Potato, Corn, Rice, Tapioca, Mixed Grains), by Distribution Channel, by Region, and Segment Forecasts, 2020–2027.* Market Analysis Report.

Grasso, S. 2020. Extruded snacks from industrial by-products: A review. *Trends in Food Science and Technology*, 99:284–294.

Guy, R. 2001. Snack foods. In: *Extrusion Cooking Technologies and Applications*, ed. R. Guy, 161–181. Boca Raton: CRC Press.

Harper, J.M. 1989. Food extruders and their applications. In: *Extrusion Cooking*, ed C. Mercier, P. Linko and J.M. Harper, 110–143. Saint Paul: American Association of Cereal Chemists.

Hegazy, H.S., El-Bedawey, A.E.A., Rahma, E.H., and Gaafar, A.M. 2017. Effect of extrusion process on nutritional, functional properties and antioxidant activity of germinated chickpea incorporated corn extrudates. *American Journal of Food Science and Nutrition Research*, 4(1):59–66.

Hsieh, F., Peng, I.C., and Huff, H.E. 1990. Effects of salt, sugar and screw speed on processing and product variables of corn meal extruded with a twin-screw extruder. *Journal of Food Science*, 55(1):224–227.

Huber, G.F. 2001. Snack foods from cooking extruders. In: *Snack Foods Processing*, eds. E. Lusas and L.W. Rooney, 315–367. Lancaster: Technomic Publishing Co.

Huber, G.F., and Rokey, G.J. 1990. Extruded snacks. In: *Snack Food*, ed. Booth, R.G., 107–138. New York: AVI Publishing Co.

Jaluria, P., and Wang, S.S. 1999. Unified approach to extrusion studies. In: *Advances in Extrusion Technology Aquaculture/Animal Feeds and Food*, eds. Y.K. Chang and S.S. Wang. 85–96. Lancaster:Technomic Publishing Co.

Lohani, U.C., and Muthukumarappan, K. 2017. Process optimization for antioxidant enriched sorghum flour and apple pomace based extrudates using liquid CO_2 assisted extrusion. *LWT-Food Science and Technology*, 86:544–554.

Lopez-Barrios, L., Heredia-Olea, E., Guajardo-Flores, D., Perez-Carrillo, E., and Gutierrez-Uribe, J.A. 2018. Bioactive peptides by in vitro digestion of germinated bean cotyledons extrudates. *Journal of Food Research*, 7(1):76–85.

Martinez Bustos, F., Chang, Y.K., Bannwart, A.C., Rodriguez, M.E., Guedes, P.A., and Gaiotti, E.R. 1998. Effects of calcium hydroxide and processing conditions on corn meal extrudates. *Cereal Chemistry Journal*, 75(6):796–801.

Masatcioglu, M.T., Yalcin, E., Hwan, P.J., Ryu, G.H., Celik, S., and Koksel, H. 2014. Hull-less barley flour supplemented corn extrudates produced by conventional extrusion and CO_2 injection process. *Innovative Food Science and Emerging Technologies*, 26:302–309.

Meng, X., Threinen, D., Hansen, M., and Driedger, D. 2010. Effects of extrusion conditions on system parameters and physical properties of a chickpea flour-based snack. *Food Research International*, 43(2):650–658.

Meuser, F., and Wiedmann, W. 1989. Extrusion plant design. In: *Extrusion Cooking*, eds. C. Mercier, P. Linko and J.M. Harper. Saint Paul: American Association of Cereal Chemists.

Mohamed, S. 1990. Factors affecting extrusion characteristics of expanded starch-based products. *Journal of Food Processing and Preservation*, 14(6):437–452.

Moore, G. 1994. Snack food extrusion. In: *The Technology of Extrusion Cooking*, ed. N.D. Frame, 110–114. Glasgow: Blackie Academic & Professional.

Morales-Alvarez, J.C. 2020. Engineering aspects of extrusion: Extrusion processing as a multiple-input and multiple-output system. In: *Extrusion Cooking: Cereal Grains Processing*, ed. G.M. Ganjyal, 1–27. Cambridge: Woodhead Publishing.

Morales-Alvarez, J.C., and Rao, M. 2016. Industrial applications of extrusion for development of snack products including co-injection and pellet technologies. In: *Advances in Food Extrusion Technologies*, eds. M. Maskan and A. Altan, 257–274. Boca Raton: CRC Press.

Moscicki, L. 2011. *Snack Pellets. Extrusion-Cooking Techniques. Applications, Theory and Sustainability*, ed. L. Mosccki, 81–90. Weinheim: Wiley VCH.

Mośćicki, L., and van Zuilichem, D.J. 2011. Extrusion-cooking and related technique. In: *Extrusion-Cooking Techniques: Applications, Theory and Sustainability*, ed. L. Mośćicki, 1–24. Weinheim: WILEY-VCH Verlag & Co. KGaA.

Mustapha, N.A., Rattarattanamongkol, K., and Rizvi, S.S.H. 2012. The effects of supercritical fluid extrusion process on surface hydrophobicity of whey protein concentrate and its relation to storage and heat stability of concentrated emulsions. *Food Research International*, 48(2):470–477.

Nakhon, P.P.S., Jangchud, K., Jangchud, A., and Charunuch, C. 2018. Optimization of pumpkin and feed moisture content to produce healthy pumpkin-germinated brown rice extruded snacks. *Agriculture and Natural Resources*, 52(6):550–556.

Panak-Balentic, J., Ackar, D., Jozinovic, A., Babic, J., Milicevic, B., Jokic, S., Pajin, B., and Subaric, D. 2017. Application of supercritical carbon dioxide extrusion in food processing technology. *Hemijska Industrija*, 71(2):127–134.

Panak-Balentic, J., Babic, J., Juzinovic, A., Ackar, D., Milicevic, B., Muhamedbegovic, B., and Subaric, D. 2018. Production of third generation snacks. *Croatian Journal of Food Science and Technology*, 10(1):98–105.

Panak-Balentic, J., Juzinovic, A., Ackar, D., Babic, J., Milicevic, B., Bensic, M., Jokic, S., Saric, D., and Subaric, D. 2018. Nutritionally improved third generation snacks produced by supercritical CO_2 extrusion I. Physical and sensory properties. *Journal of Food Process Engineering*, 42(2):1–9.

Parker, L. 2020. State of the industry 2020: Puffed and extruded snacks grow through innovation. Snack Food & Wholesale Bakery. SFWB. https://www.snackandbakery.com/articles/94981-state-of-the-industry-2020-puffed-and-extruded-snacks-grow-through-innovation.

Pichmony, E., and Ganjyal, G.M. 2020. Basics of extrusion processing. In: *Extrusion Cooking: Cereal Grains Processing*, ed. G.M. Ganjyal, 1–27. Cambridge: Woodhead Publishing.

Rauwendaal, C. 2014. *Polymer Extrusion*. 11–48, 509–652, 697–762. Cincinnati: Hanser Publishers.

Riaz, M.N. 2000. Introduction to extruders and their principles. In: *Extruders in Food Applications*, ed. M.N. Riaz. Lancaster: Technomic Publishing Co.

Riaz, M.N. 2004. Snack foods, processing. In: *Encyclopedia of Grain Science*, 1st edition, eds. C. Wrigley, C. Walker and H. Corke, 98–108. Oxford: Elsevier.

Richards, J., and Harrington, J. 2020. Extrusion plant design. In: *Extrusion Cooking: Cereal Grains Processing*, ed. G.M. Ganjyal, 391–413. Cambridge: Woodhead Publishing.

Rokey, G.J. 2000. Single screw extruders. In: *Extruders in Food Applications*, ed. M.N. Riaz. Lancaster: Technomic Publishing Co.

Seker, M. 2016. Extrusion of snacks, breakfast cereals and confectionaries. In: *Advances in Food Extrusion Technologies*, eds. M. Maskan and A. Altan, 169–208. Boca Raton: CRC Press.

Serna-Saldivar, S.O. 2008. Extruded snack foods. In: *Industrial Manufacture of Snack Foods*, ed. S.O. Serna-Saldivar, 235–264. London: Kennedys Publications Ltd.

Slavin, J., Jacobs, D., and Marquart, L. 2001. Grain processing and nutrition. *Critical Reviews in Biotechnology*, 21(1):1.

Sokhey, A.S., Rizvi, S.S.H., and Mulvaney, S.J. 1996. Application of supercritical fluid extrusion to cereal processing. *Cereal Foods World*, 41:29–34.

Viacava, F., Santana-Galvez, J., Heredia-Olea, E., Perez-Carrillo, E., Nair, V., Cisneros-Zevallos, L., and Jacobo-Velázquez, D.A. 2020. Sequential application of postharvest wounding stress and extrusion as an innovative tool to increase the concentration of free and bound phenolics in carrots. *Food Chemistry*, 307(1):1–9.

Wang, Y.Y., and Ryu, G.H. 2013. Physical properties of extruded corn grits with corn fibre by CO_2 injection extrusion. *Journal of Food Engineering*, 116(1):29–34.

Yacu, W. 2020. Extruder screw, barrel, and die assembly: General design principles and operation. In: *Extrusion Cooking: Cereal Grains Processing*, ed. G.M. Ganjyal, 73–117. Cambridge: Woodhead Publishing.

Zazueta-Morales, J.J., Martinez-Bustos, F., Jacobo Valenzuela, N., Ordorica-Falomir, C., and Paredes Lopez, O. 2001. Effect of the addition of calcium hydroxide on some characteristics of extruded products from blue maize (*Zea mays* L) using response surface methodology. *Journal of the Science of Food and Agriculture*, 81(14):1379–1386.

Zazueta-Morales, J.J., Martinez-Bustos, F., Jacobo Valenzuela, N., Ordorica-Falomir, C., and Paredes Lopez, O. 2002. Effect of calcium hydroxide and screw speed on physicochemical characteristics of extruded blue maize. *Journal of Food Science*, 67(9):3350–3358.

13 Snacks Based in Legumes, Pseudocereals, and Other Seeds

Sergio O. Serna-Saldivar

CONTENTS

DOI: 10.1201/9781003129066-13

13.1 INTRODUCTION

Grain legumes and pseudocereals play a relevant role in the world food supply. These grains are subjected to several primary processes, such as dehulling, cooking, roasting, splitting, and grinding, before used in different food preparations. Roasting or puffing of legumes by subjecting them to high temperatures for a short time has been practiced in Asia, the Middle East, and South America for many years. These seeds are an excellent source of protein, complex carbohydrates, and dietary fiber (DF). More importantly, legume proteins, rich in globulins, lysine, and tryptophan, complement the essential amino acid profile of cereal grains improving the overall protein quality. Both the complex carbohydrates and DF are known to lower serum cholesterol and postprandial glycemia in hypercholesterolemic and diabetic patients (Guillon and Champ 2002, Serna-Saldivar and Sanchez-Hernandez 2020). Furthermore, these grains are rich sources of oligosaccharides or flatus producing sugars raffinose, stachyose, verbascose, and manninotriose, which are currently regarded as relevant prebiotics. These undigestible sugars are readily fermented in the large intestine benefitting the growth of bifidogenic bacteria, which generate short-chain volatile fatty acids (FA) associated with many health benefits (Serna-Saldivar and Sanchez-Hernandez 2020, Tosh and Yada 2010).

After cereals, pulses and legumes are the main suppliers of protein for humankind. In most countries, conventional diets include at least one pulse as complement of cereals. The most utilized oil-bearing legumes are soybeans and peanuts, whereas the most common starchy pulses are common beans, cowpeas, lentils, peas, and chickpeas. Regrettably, most of the 336.7 million tons of soybeans harvested in 2019 worldwide are still regularly used for production of feeds for domestic animals (FAO 2021). On the other hand, nearly 88.4 million tons of pulses produced worldwide in 2019 are generally directly consumed by humans and their contribution to the diet is especially relevant in developing countries because they are cheaper compared to animal foods. Among pulses, beans are the most relevant because they constitute about one-third of the total production.

Yellow and green peas are widely consumed in Asia, common beans in Latin America, cowpeas in Africa, chickpeas in India, and lentils in the Middle East. However, all these grains are gaining popularity in developed countries due to their nutritional value and positive health implications. Unlike cereals grains that are regularly consumed as refined milled fractions (Chapter 6), legumes and pulses are typically consumed as whole grains. Legumes and pulses contain in their raw forms different concentrations of antinutritional factors, trypsin inhibitor, urease, and lectins or hemagglutinins, so to safely consume these seeds, they need thermal treatments or undergo germination/sprouting or fermentation.

Pseudocereals are also gaining interest as ingredients that favor health. Pseudocereals are non-cereal plants that generate grains that are commonly processed into flours that usually supplement cereal-based foods, including snacks. The major types are amaranth, quinoa, buckwheat, and chia. These grains are gaining acceptance because they are viewed as highly nutritious and with health-promoting properties.

This chapter reviews the main nutritional and phytochemical characteristics of legumes, pseudocereals, and some oilseeds commonly used for the production of healthy snacks and their processing into various types like roasted grains and extruded products.

13.2 PROPERTIES AND CHEMICAL VALUE OF LEGUME SEEDS (PULSES)

After cereal grains, legumes are considered as the second most relevant types of grains for humankind. These seeds are broadly categorized into two groups: legume oilseeds and pulses. The main oilseeds are soybeans and peanuts, which contain 19% and 47% lipids rich in polyunsaturated FA. On the other hand, the pulses are legumes that contain at least twice as much protein compared to cereals and additionally contain significant amounts of starch and relatively lower amounts of lipids (<6%). In terms of production of snacks, undoubtedly, peanuts are the most relevant followed by soybean proteins that are commonly used to increase protein content of extruded products and nutritional bars. The major types of pulses are common beans, chickpeas, cowpeas, peas, and lentils. Today, these grains are gaining acceptance because they are viewed as highly nutritious and with health-promoting properties.

All these seeds belong to the Leguminosae family, which produce dicotyledon fruits in a pod. Seeds are constituted by three major anatomical parts: testae, two cotyledons, and the hilum. The testae usually comprises less than 10% of the seed weight, whereas the hilum less than 2%. Therefore, seeds are mainly constituted by the cotyledons that act as reserve tissue rich in proteins, lipids, and complex carbohydrates. The testae or hull that protects seeds is rich in DF and phytochemicals that are known as important antioxidants that protect the human system from oxidative stress and chronic diseases.

Table 13.1 depicts the nutritional profile of the main legume seeds, including the main types of soybean proteins (defatted flour, concentrate, and isolate) widely used in the preparation of snack bars and other products.

13.2.1 Soybeans (*Glycine max*)

Among grain legumes, soybeans are the most grown in the globe. According to the FAO (2021), the word production exceeded 333.6 million tons with Brazil, the United States, Argentina, and China producing 34%, 29%, 16.5%, and 4.7% of the total production, respectively. Among seeds, soybeans contain the highest protein (35–37%) and contains 17–20% polyunsaturated oil rich in linoleic acid (Chapter 7). The rest is constituted by carbohydrates, DF, and minerals. Among legumes, soybeans are considered as one of the best sources of important phytochemicals like phenolics, flavonoids, isoflavones, and other hydrophilic and lipophilic antioxidants (Xu and Chang 2008). For these reasons, soybean is recognized as a healthy food because it is an excellent source of essential nutrients at a relatively lower cost compared to animal products. Although soybean has been used for many centuries for direct production of human foods in Asia, there is a limitation to its use in other parts of the globe. The main drawback of raw soybean is that it has a beany and astringent flavor and antinutritional factors that can be controlled via thermal processing (Bagheri and Kashaninejad 2018, Serna-Saldivar and Sanchez-Hernandez 2020).

In terms of direct food uses, soybeans are generally channeled to the oil-crushing industry where oil and defatted protein meals are obtained as major products. The crude soybean oil is refined and in some instances modified and widely used as frying oil (Chapter 7), whereas the defatted soybean meal containing approximately 50% protein (Table 13.1) is used as feedstock for production of texturized vegetable proteins, concentrates (65–70% protein), and isolates (>88% protein). These concentrated forms of soybean protein are used for the production of different sorts of snacks, chiefly protein bars. The protein is rich in all essential amino acids and has a high digestibility rate. The addition of about 10% of defatted soybean flour, 6% concentrate, or 5% isolate to cereal-based snacks increases protein by about 20% and greatly improves protein quality because soybeans are rich in lysine and tryptophan. This is the main reason why soybean proteins are increasingly being used for the manufacturing of protein and granola bars.

13.2.2 Peanuts (*Arachis hypogea*)

Peanuts presumably originated in Brazil or Peru where they have been utilized as snack for almost 3,000 years. In the United States, peanuts started to be consumed after the Civil War (Woodroof 1983).

The peanut crop started to expand in the United States at the beginning of the 20th century where specialized equipment for its cultivation, harvesting, shelling, and processing was developed. Salted, roasted, and processed peanuts are one of the most popular snacks because they have a unique nutty flavor, smooth texture, and an exceptional aroma. Peanuts are one of the most nutritious snack foods because they provide high amounts of energy, protein, vitamins, and minerals

TABLE 13.1
Comparison of the Nutritional Profile of Major Legume Seeds and Some of Their Extracted Protein Products Commonly Used to Prepare Snacks

	Soybean				Peanuts			Pulses				
Per 100 g	Whole	Defatted Meal	Concentrate	Isolate	Spanish	Valencia	Virginia	Pinto Beans	Chickpeas	Cowpeas	Pigeon Peas	Lentils
Moisture, g	8.54	7.25	5.8	4.98	6.39	4.26	6.91	11.33	7.68	11.04	10.59	8.26
Calories, kcal	446	327	328	335	570	570	563	347	378	34	343	352
Protein, g	36.49	51.46	63.63	88.32	26.15	25.09	25.19	21.42	20.47	23.85	21.7	24.63
Lipids, g	19.94	1.22	0.46	3.39	49.60	47.58	48.75	1.23	6.04	2.07	1.49	1.06
Ash, g	4.87	6.15	4.7	3.58	2.03	2.17	2.61	3.46	2.85	3.39	3.45	2.71
Carbohydrates, g	30.16	33.92	25.4	0	15.83	20.41	16.54	62.55	62.95	59.64	62.48	63.35
Dietary fiber, g	9.3	17.5	5.5	0	9.5	8.7	8.5	15.5	12.2	10.7	15	10.7
Calcium, mg	277	338	363	178	106	62	89	113	57	85	130	35
Iron, mg	15.7	7.34	10.78	14.5	3.91	2.1	2.55	5.07	4.31	9.95	5.23	6.51
Magnesium, mg	280	313	140	39	188	184	171	176	79	333	183	47
Potassium, mg	1797	2480	450	81	744	332	690	1393	718	1375	1392	677
Sodium, mg	2	2	900	1005	22	1	10	12	24	58	17	6
Zinc, mg	4.89	4.44	4.4	4.03	2.12	3.34	4.43	2.28	2.76	6.1	2.76	3.27
Thiamin, mg	0.87	0.54	0.32	0.18	0.68	0.64	0.65	0.71	0.48	0.68	0.64	0.87
Riboflavin, mg	0.87	0.3	0.14	0.1	0.14	0.30	0.13	0.21	0.21	0.17	0.19	0.21
Niacin, mg	1.62	3.43	0.72	1.33	15.93	12.88	12.38	1.17	1.54	2.8	2.97	2.61
Pyridoxin, mg	0.38	0.62	0.13	0.1	0.35	0.34	0.35	0.47	0.54	0.36	0.28	0.54
Folate, mcg	375	305	340	176	240	246	239	525	557	639	456	479
Type of fat, g												
Saturated	2.88	0.14	0.05	0.42	7.10	7.33	6.43	0.24	0.6	0.54	0.33	0.15
Monounsaturated	4.4	0.21	0.08	0.65	22.32	21.41	25.58	0.23	1.38	0.17	0.01	0.19
Polyunsaturated	11.26	0.53	0.2	1.65	17.24	16.50	14.88	0.41	2.73	0.89	0.81	0.53

Source: USDA (2021).

(see Table 13.1 and also refer to Chapter 20). Peanuts contain more protein than meat and a good balance of essential amino acids. The sulfur-containing amino acids limits protein quality. As other legumes, peanuts contain important amounts of nutraceutical compounds such as flavonoids, phytosterols, natural antioxidants, and an excellent balance of saturated to polyunsaturated FA. It is the only commercial oilseed that contains significant quantities of omega-6 arachidonic acid (20:4).

Among legumes, peanuts are undoubtedly the most relevant seed used for production of snacks. From the processing viewpoint, peanuts can be channeled to the snack food or the oil-crushing industries. The first obtains a wide array of snacks and the second a premium oil used for various purposes, including fabrication of blended oils for frying (Chapter 7).

According to the FAO (2021), the world production of groundnuts with shells was 49.75 million tons with China, India, and the United States producing 35.2%, 13.5%, and 5.1% of the total 2019 world production, respectively. Unlike soybeans, peanuts are regularly directly consumed by humans, and one of its major uses is in the form of snacks. Four major types of peanuts are recognized by the global markets: Spanish, Valencia, Virginia, and runner. Table 13.2 describes the main differences among the different types and the features of the pods and shelled kernels. The Spanish and Valencia crops mature earlier compared to the Virginia and runner crops. The runner type is the most widely planted and produced in the United States and is used for roasting and as peanut butter. Table 13.1 depicts the slight differences in the nutritional composition among these types. The Virginia and Valencia kinds yield thick-shelled pods and are favored by snack food processors. The pods of the most popular Virginia type bear one or two large kernels weighing 0.5–1 g each, whereas the Valencia counterparts up to four kernels weighing each from 0.36 to 0.5 g. The Spanish and runner varieties yield thin-shelled pods with one or two elliptical colored seeds with 242–440 kernels per 100 g (Serna-Saldivar 2008). Among commercial legumes, peanuts are known to contain the highest oil content (47–49%, Table 13.1). The average protein and total DF contents are about 26% and 9%, respectively.

The timing of the peanut harvest is critical since it can greatly affect quality. If peanuts are harvested too early, many of the pods will be immature with shriveled kernels. If harvesting is too late, many of the pods may be lost or contaminated with molds, including the aflatoxin-producing *Aspergillus* (Chapter 20). Additionally, the maturity of the harvested kernels is important to end-users because it greatly affects flavor, texture, and color. Peanuts are usually picked at an average kernel moisture content of 18–30% and dehydrated to 5–10%. For marketing, the desired moisture content of shelled peanuts is about 7–8%. If the peanuts are dried too quickly, they will tend to break producing splits; and if they are dehydrated too slow, they will have more chances of mold and aflatoxin contamination. The shell of the pod comprises 20–30% of the whole weight and normally the total sound mature kernels yield 68–79% of the pod weight (Table 13.2).

13.2.3 Chickpeas (*Cicer arietinum*)

Chickpea, also known as garbanzo, is one of the earliest cultivated pulses, and recently it has gained relevance because of its nutritional value, phytochemicals, and health-promoting benefits (Jukanti et al. 2012, Milan-Noris et al. 2018, 2019a). Guardado-Felix et al. (2019) and Milan-Noris et al. (2019b) recently reviewed the health effects of proteins/bioactive peptides and starch and undigested carbohydrates associated with chickpea-based foods. Among legumes, it is considered as nonallergenic. The seeds have been consumed for more than 7,500 years, especially in the Middle East and Turkey. The worldwide production of chickpeas in 2019 was estimated in 14.2 million tons with India as the major producer with nearly 10 million tons of production. Mexico is the main producer in America with a production of 0.2 million tons in 2019 (FAO 2021).

Chickpeas are divided into Kabuli or Desi types, the first being the most popular nowadays because it yields larger seeds with lighter colorations. The smaller seeded Desi types are usually green, red, or even black-pigmented and have bright future because of their phytochemical profile richer in antioxidants and DF. Chickpeas are usually boiled and eaten in salads and stews or ground

TABLE 13.2
Main Features of the Plant, Pods and Kernel Belonging to the Four Major Types of Peanuts

	Type of Peanut			
	Spanish	**Valencia**	**Virginia**	**Runner**
Plant maturity	Matures within 90–120 days. Typically bloom and produce pods on the upright main stem	Matures within 90–110 days. Typically bloom and produce pods on the upright main stem	Mature within 130–170 days	Mature within 130–165 days. Generally planted in warmer weather and yield in poor drained soils
Pods	Thin-shelled pods with one or two and infrequently three round to elliptical flesh colored kernels	Pods are cylindrical, slender, and contain two to four kernels	Thick-shelled pods are large, plump, and usually contain two kernels	Thin-shelled pods with normally two medium-sized kernels
Type of kernel	Mostly ellipsoidal. Kernels of older varieties are normally smaller compared to newer types (medium size). The testae or skin is smooth and has a delicate texture. Kernel color ranges from pale pinkish buff when fresh to light brown	Mostly spherical. Kernels are small to medium, with blocky, flattened ends. Color ranges from bright red when fresh to darker red during storage	Kernels are medium to large and generally elongated. Most kernels have a distinct taper at each end and a tendency to be pointed at the sprout end. The skin has about the same texture as runners. Color ranges from light pinkish-tan when fresh to reddish-brown during storage	The somewhat elongated kernels are medium sized and very uniform in shape. Many kernels have a blunt and flattened ends. The skin is usually a little thicker and less smooth than the Spanish type. The color ranges from pinkish-brown when fresh to reddish brown
Total sound mature kernels, %	62–71.5	68–79	67–74.5	71–78
Number of kernels/100 g	242–440	198–273	100–115	138–182
Main use	These peanuts have a reddish skin around the nut and is often used in confectionary, cookies and peanut butter. The relatively smaller kernels contain the highest oil content suited for the extraction of peanut oil	Roasting in the shell, salted peanuts. They are generally perceived sweeter than the other types	Production of large salted peanuts	Runner peanuts are mostly used in the production of roasted peanuts, Japanese-style crusted peanuts, and peanut butter

Source: Serna-Saldivar (2008) and USDA (2020).

into flour for production of batters and bakery products. Hummus, which is the Arabic word for chickpeas, is prepared by cooking and grinding into a paste that is further mixed with sesame paste or other pastes.

13.2.4 Dry Beans, Common (*Phaseolus vulgaris*)

The worldwide production of dry beans in 2019 was estimated to be 28.9 million tons, with India, Myanmar, and Brazil as the major producers (FAO 2021). Dry beans come in a wide variety of market classes that differ in color from white to cream: pinto, kidney, navy, and black. Beans normally contain approximately 20% protein rich in globulins (Table 13.1). The protein lacks sulfur-containing amino acids and is less digestible compared to other legumes and pulses. However, the thermal processing of whole beans improve digestibility and inactivates antinutritional factors. Beans are considered good sources of slow-release carbohydrates (Martín-Cabrejas et al. 2004), oligosaccharides, and both insoluble and soluble DF (Ramirez-Jiménez et al. 2014, Serna-Saldivar and Sanchez-Hernandez 2020).

13.2.5 Cowpeas (*Vigna unguiculata*)

Cowpea is a pulse consumed as a high-quality protein source in many parts of the globe. Its high protein and carbohydrate contents with a relatively low fat complement the amino acid pattern to cereal grains. Cowpea has gained popularity among consumers and researchers as a result of its exerted health beneficial properties, including antidiabetic, anticancer, antihyperlipidemic, anti-inflammatory, and antihypertensive properties. Among the mechanisms that have been proposed in the prevention of chronic diseases, the most proven are attributed to the presence of compounds such as soluble and insoluble dietary fiber, phytochemicals, and bioactive peptides (Gutierrez Uribe et al. 2011, Jayathilake et al. 2018).

The worldwide production of dry cowpeas in 2019 was 8.9 million tons. Nigeria and Niger harvested about 67% of the global production (FAO 2021). Cowpeas have similar food uses as common beans. The three major chemical constituents of raw cowpeas are total carbohydrates, protein, and DF (Table 13.1). Starch constitutes the major form of total carbohydrates with values ranging from 37% to 44% of the total seed weight. Recently, Rengadu et al. (2020) indicated that cowpeas contain 18.72–19.67% amylose and 9.42–13.74% prebiotic resistant starch. According to Benítez et al. (2013), raw cowpeas contained 34.1 (dwb) of total DF of which 31% and 3.1% were insoluble and soluble fractions, respectively.

13.2.6 Peas (*Pisum sativum*)

The pea is an annual and cool-season crop grown mainly in Canada, Russia, and China. The global production of this relevant pulse was 14.1 million tons in 2019 (FAO 2021) with Canada producing about 30% of the total world output. The commercial production of green and yellow pea flours and their protein concentrates and isolates are increasing mainly in Canada and the United States in order to substitute other vegetable proteins that are viewed as allergenic. The chemical composition of peas summarized in Table 13.1 shows that dry seeds contain more than 62% carbohydrates and 21% protein. Starch is the major constituent of the total carbohydrates accounting for approximately 49% of the total weight. Peas are also rich in resistant starch with values varying from 6.2% to 12.9% and from 31.2% to 33.4% in products subjected to low- and high-pressure cooking. As a result, the predicted glycemic indices were in the range of 42–84 (Skrabanja et al. 1999). Peas are also rich in oligosaccharides raffinose, stachyose, and verbascose with values of 0.9%, 2.3%, and 1.4%, respectively (Alonso et al. 2000). The limiting amino acids of the pea proteins are sulfur-containing methionine and cystine. De Almeida Costa et al. (2006) determined that raw peas contain 22.5% and 1.9% insoluble and soluble DF (dwb), respectively.

13.2.7 Lentils (*Lens culinaris*)

The lentil is a dietary staple throughout the regions of India, Sri Lanka, Pakistan, and neighboring regions. As a food crop, the majority of world production comes from Canada and India, which respectively produced 2.1 and 1.2 million tons of the total global production of 6.7 million tons in 2019 (FAO 2021). This particular pulse can be eaten whole or dehulled after soaking, cooking, sprouting, frying, and baking. Dehulling significantly reduces contents of DF and phytochemicals (Wang and Daun 2004). Wang et al. (2009) assayed a set of eight different lentils that were cooked or dehulled. The total DF values for the raw and dehulled samples were 14.1% and 7.8% (dwb), respectively. The major fiber fraction was insoluble with values for the same samples of 12.2% and 6.8% (dwb).

13.3 PROPERTIES AND NUTRITIONAL VALUE OF PSEUDOCEREALS AND OTHER GRAINS

Pseudocereals are non-cereal and legume plants which generate grains that are not commonly produced massively and have gained interest because of their nutritional and health-promoting properties, including those that are considered gluten-free and recommended for celiacs. Recently, many snacks have been developed containing these special grains or flours. The major types are amaranth, quinoa, buckwheat, and chia. Besides pseudocereals, other specialty grains are also used by snack processors to prepare healthy foods. The most relevant are sesame and flaxseed or linseed. Table 13.3 depicts the nutritional profile of these pseudocereals and selected seeds used for the preparation of snacks, mainly nutritional bars.

13.3.1 Buckwheat (*Fagopyrum esculentum*)

Buckwheat is a pseudocereal first cultivated in Southeast Asia, likely the Yunnan region of China. From Asia, the crop migrated to Eastern Europe where it has been utilized for at least 6,000 years. Buckwheat contains several compounds that may have potential in reducing risk of certain chronic diseases (Mazza and Oomah 2005).

The plant yields triangle-shaped dicotyledonous achenes containing about 13% protein rich in essential amino acids and 10% total DF (Table 13.3). According to the FAO (2021), the global production in year 2019 was estimated to be 1.61 million tons with Russia producing approximately half of this output (0.78 million tons). Buckwheat has potential in reducing certain types of diseases (Mazza and Oamah 2005). Bonafaccia et al. (2003) reported 27.4% total DF in buckwheat kernels and that the insoluble and soluble fractions of groats are similar to oats and wheat. Milling of buckwheat is aimed toward the maximum yield of groats, which contain 55% starch, 12% protein, 4% lipid, 2% soluble carbohydrates, 7% DF, 2% ash, and 18% other components (organic acids, phenolic compounds, tannins, phosphorylated sugars, nucleotides and nucleic acids, unknown compounds). Starch is concentrated in the central endosperm, whereas the protein, lipids, soluble carbohydrates, and minerals in the embryo. Commercial light-colored flours obtained from the inner endosperm contains 75% starch, 6% protein, 1% lipid, 1% soluble carbohydrates, and 3% total DF, whereas the fiber-rich bran contains 18% starch, 36% protein, 11% lipid, 6% soluble carbohydrates, 15% DF, and 7% ash (Serna-Saldivar and Sanchez-Hernandez 2020).

13.3.2 Amaranth (*Amaranthus caudatus, hypochondriacs, cruentus*)

Amaranth is an ancient crop of the Americas, and in the pre-Columbian times it was considered as a relevant staple food. After the Spanish conquest, amaranth fell into disuse, and in the Americas, it survived only in small pockets of cultivation in scattered mountain areas of Mexico and the Andes. Three species are recognized being the pre-Columbian, with *Caudatus* being the most relevant.

TABLE 13.3
Comparison of the Nutritional Profile of Main Pseudocereals and Oilseeds Used for the Production of Snacks

	Pseudocereals				Other Seeds		
Per 100 g	**Amaranth**	**Chia**	**Quinoa**	**Buckwheat**	**Sesame**	**Flax**	**Squash/Pumpkin**
Moisture, g	11.29	6.96	13.28	9.75	3.75	6.96	2.03
Calories, kcal	371	534	368	343	631	534	584
Protein, g	13.56	18.29	14.12	13.25	20.45	18.29	29.84
Lipids, g	7.02	42.16	6.07	3.4	61.21	42.16	49.05
Ash, g	2.88	ND	2.38	2.1	ND	ND	ND
Carbohydrates, g	65.25	28.88	64.16	71.5	11.73	28.88	14.7
Dietary fiber, g	6.7	27.3	7	10	11.6	27.3	6.5
Calcium, mg	159	255	47	18	60	255	52
Iron, mg	7.61	5.73	4.57	2.2	6.36	5.73	8.1
Magnesium, mg	248	392	197	231	345	392	550
Potassium, mg	508	813	563	460	370	813	788
Sodium, mg	4	30	5	1	47	30	18
Zinc, mg	2.87	4.34	3.1	2.4	6.73	4.34	7.64
Thiamin, mg	0.12	1.64	0.36	0.1	0.7	1.64	0.07
Riboflavin, mg	0.2	0.16	0.32	0.43	0.09	0.16	0.15
Niacin, mg	0.92	3.08	1.52	7.02	5.8	3.08	4.43
Pyridoxin, mg	0.59	0.47	0.49	0.21	0.4	0.47	0.10
Folate, mcg	82	87	184	30	115	87	57
Type of fat, g							
Saturated	1.46	3.66	0.71	0.74	9.05	3.66	8.54
Monounsaturated	1.69	7.53	1.61	1.04	23.9	7.53	15.73
Polyunsaturated	2.78	28.7	3.29	1.04	25.49	28.73	19.86

Source: USDA (2021).

Today, grain amaranth is a minor crop of increasing importance in Asia, especially in Nepal and mountainous areas of North and South India. Amaranths are noted for high tolerance to arid conditions and poor soils, where cereals cannot grow. *A. caudatus* is native to Peru and is known to thrive at altitudes above 2,500 m (Serna-Saldivar and Sanchez-Hernandez 2020).

This pseudocereal is highly nutritional with a superior amount of proteins when compared to true cereals. It is a reasonably well-balanced food with functional properties that provide health benefits by decreasing plasma cholesterol, stimulating the immune system, exerting antitumor activity, reducing blood glucose levels, and improving conditions of hypertension and anemia. In addition, it has been reported to possess antiallergic and high-antioxidant activities. Amaranth contains protein contents ranging from 12.5% to 17.6% and about 6.5% total dietary fiber (Table 13.3). In addition, it contains adequate levels of lysine (5.58 g/100 g protein) and methionine (1.58 g/100 g total protein). Amaranth also contains high levels of important vitamins and minerals (Table 13.3) (Caselato-Sousa and Amaya-Farfan 2012).

Repo Carrasco Valencia et al. (2009) evaluated two varieties of Andean *A. caudatus* as sources of DF and some bioactive compounds. The impact of low-cost extrusion on the content of these components was studied for technological applications. The total DF in raw grains ranged from 13.8% to 16.4% varieties and their contents decreased during extrusion. Most of the DF was insoluble and values significantly diminished after extrusion. The soluble DF ranged from 1.46% to 2.45%.

Pedersen et al. (1990) investigated the energy and fiber of raw and processed (popping, toasting, and flaking) amaranth grains with growing rats. The pale seeds contained about 8% of the total DF, whereas the black counterparts about twice as much. The soluble DF fraction made up 33–44% of the DF fraction in the pale-seeded varieties, but only 18% in the black-colored seeds. The monomer sugar composition of the DF was very similar in all products. However, the black seeds were very high in lignin. Interestingly, the DF of the black-seeded products were more resistant to digestion compared to the pale-colored counterparts. Processing of pale seeds had only minor effects on total DF but increased the proportion of soluble DF.

The major use of amaranth in snacks is for the formulation of nutritional bars or processed into expanded products via extrusion cooking. In bars, the amaranth seeds can be supplemented popped, toasted, or as flour.

13.3.3 Quinoa (*Chenopodium quinoa*)

Quinoa is gaining popularity throughout the globe, although the world production is still relatively low (161,415 tons) and only currently planted in the Andean regions of Peru, Bolivia, and Ecuador (FAO 2021). Today, the cultivation of quinoa has been extended to other countries, such as Australia, Canada, China, and the United Kingdom. The plant belongs to the same family of Amaranthus (Amaranthaceae) and grains possess a high nutritional value (Table 13.3) with positive health implications. This pseudocereal is valued because of its nutritional value and bioavailability in regard to proteins, carbohydrates, minerals, FA, and tocopherols. Pereira et al. (2019) made a detailed evaluation of the nutritional value and chemical composition of several quinoa grains varying in color (black, red, and white). The results demonstrated an excellent composition, namely, in the presence of many compounds of interest such as proteins, carbohydrates, unsaturated FA, and tocopherols. The grains contained 75.3–77% carbohydrates of which only 2.9–3.1% were sugars (sucrose and glucose), 14.4–15.6% protein, and 6–6.8% fat.

Repo Carrasco Valencia and Serna (2011) assayed 12–14.4% (dwb) of insoluble DF and 1.4–1.6% (dwb) of soluble DF in four different quinoas. The grain, which also contains some starch, is traditionally employed for the preparation of soups, or ground into flours for the preparation of traditional breads, cookies, and even fermented beverages such as *chicha*. Recently, quinoa has been employed for the formulation of nutrition bars. Quinoa has gained popularity in other parts of the globe because it is gluten-free and viewed as wholesome and highly nutritious. Quinoa also contains a high amount of health-beneficial phytochemicals, including saponins, phytosterols, phytoecdysteroids with positive effects on metabolic, cardiovascular, and gastrointestinal health (Navruz-Varli and Sanlier 2016).

13.3.4 Chia (*Salvia hispanica*)

Chia is indigenous grain of Mesoamerica which recently has gained importance and market due to its oil composition rich in omega-3 (α-linolenic acid), protein quality, DF, and an array of antioxidants that prevent cardiovascular diseases, hypertension, diabetes, and inflammatory and nervous system disorders. Some of most relevant applications of the seeds include their use as a nutritional supplement and as an ingredient in cereal bars, biscuits, pasta, bread, snacks, and yogurt. The biannually cultivated plant is part of the mint or Lamiaceae family and yields mottle-colored seeds with brown, gray, dark red, and white, which are oval, smooth, and shiny. The chemical composition of this pseudocereal indicates that it contains more than 42% of highly polyunsaturated lipids and 18% protein (Table 13.3). The seeds contain around 27% of the total DF, so the intake of 100 g of whole chia fulfills the daily recommended requirement for adults of 25–30 g. The defatted flour contains about 40% DF. Most fiber is comprised by insoluble and only 5–10% is soluble DF (Muñoz et al. 2013).

13.3.5 Sesame Seed (*Sesame indicum*)

Sesame is an annual and drought-resistant crop of the Pedaliaceae family used in several Asian and African countries since ancient times. The plant yields dehiscent capsules that enclose small kernels with thousand kernel weights of only 2–4 g. The seed color varies from white, gray, brown, chocolate, and black. India, Myanmar, China, Sudan, Uganda, and Nigeria in descending order harvest most of total world production estimated at 6.55 million tons in 2019 (FAO 2021). It is considered an oilseed because sesame seeds containing 45–61% oil (Table 13.3) are mechanically or solvent extracted for the production of a premium oil rich in oleic acid (Chapter 7). In addition, the bland-flavored seeds contain about 20% protein rich in methionine and tryptophan and 12% DF (Table 13.3). The seeds contain relevant phytochemicals like sesamin, sesamol, sesamolin, and tocopherols with antioxidant, anti-inflammatory, antihypertensive, and anticarcinogenic activities (Anilakumar et al. 2010). Most of the fiber in the whole sesame seed is found in the testae, which constitutes about 13.6% of the seed weight. According to Elleuch et al. (2007), the insoluble and soluble DF fractions of whole seeds averaged 14.0% and 5.4%, respectively.

In the Western Hemisphere and Europe, sesame is normally consumed as whole, dehulled (decorticated), or as defatted meal (Elleuch et al. 2007, 2012). Sesame is also widely used as topping in hamburger buns, salads, and for the production of nutritional bars. In Turkey and neighboring countries, roasted sesame seeds are processed into a regular or sweetened paste or sesame butter named *tehineh.*

Sesame seeds are commonly decorticated before processing because the husk provides undesirable darker color and bitter flavors. In addition, dehulling eliminates most oxalates associated with the hulls and reduces the DF content (Moharram et al. (1990). The two most common approaches to decorticate whole seeds are mechanical or chemical. The conventional methods for the preparation of dehulled seeds involve cleaning, mechanical dehulling, drying, whitening, toasting, and color sorting (Figure 13.1). The other alternative is the chemical decortication method that involves similar operations used in the conventional milling process. Highly automated sesame dehulling plants base their technologies in removing the hulls by the sequential steps of lye (potassium hydroxide) –soaking, mechanical dehulling, separation, washing of decorticated seeds, and drying. The key step is the alkaline soaking in a heated lye solution aimed to soften hulls via the chemical disruption of fiber components. The chemical process saves processing time because the soaking operation lasts less than 45 min instead of 6–10 h employed with water or saline-water soaking. Resulting alkali-treated seeds are dehulled in special abrasive machines that feed separators that obtain two

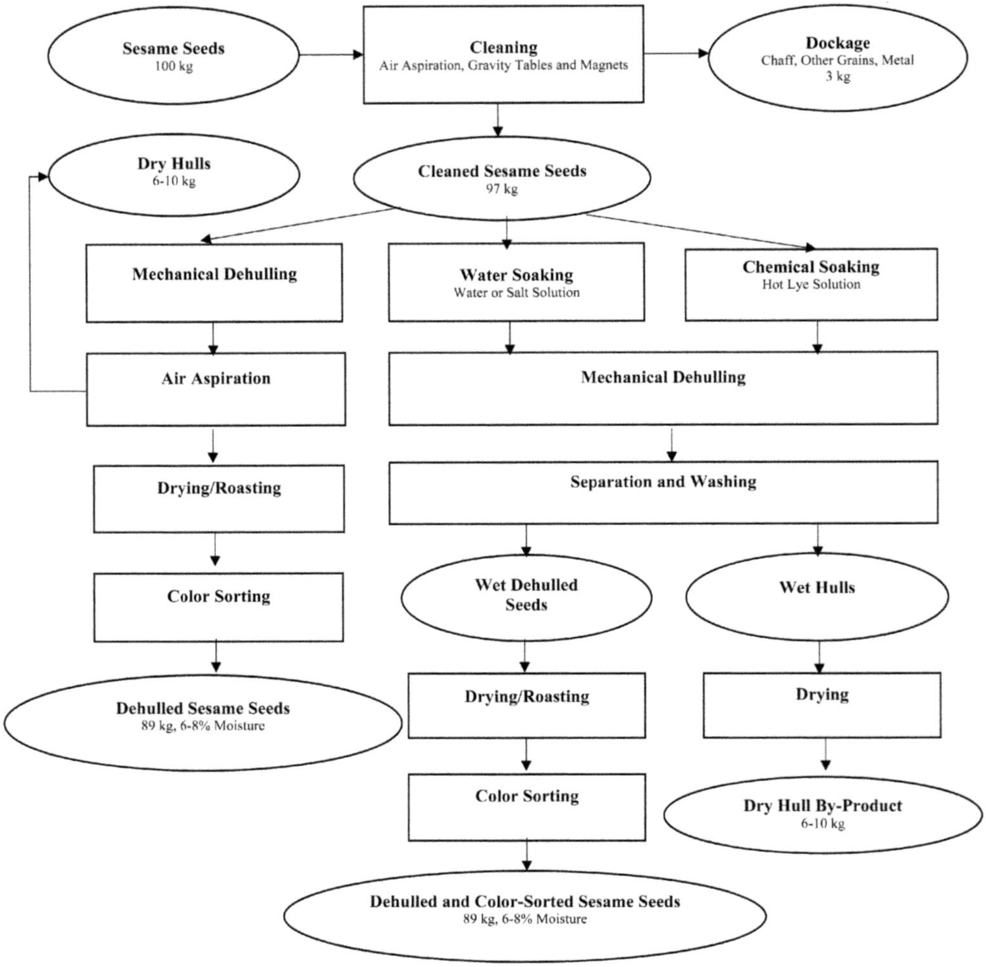

FIGURE 13.1 Mechanical or chemical dehulling process of sesame seeds.

streams: decorticated kernels (large particles) and disrupted hulls (small particles). Then, the dehulled seeds are washed to remove residual lye and dehydrated to reduce the moisture content to less than 10% (Figure 13.1). Many industries apply the technology of electronic color sorting to discard off-colored seeds.

According to Moharram et al. (1990), the hulls of lye-treated sesame seeds treated with concentrated brine yielded kernels with better appearance and color than those obtained by sieving. The same authors indicated that dehulling losses accounted for 5–8%. Carbonell-Barrachina et al. (2009) proposed other chemical dehulling methods with sodium hypochlorite (NaClO) and hydrogen peroxide (H_2O_2) in order to produce decorticated seeds with lighter color values. Apparently, the NaClO treatment was able to remove the hulls without causing damage to the cotyledons.

13.3.6 Flaxseed or Linseed (*Linum usitatissimum*)

The most relevant linseed-growing countries are Russia, Canada, Kazakhstan, China, and the United States, which harvested most of the world production estimated in 3.1 million tons in 2019 (FAO 2021). Flaxseed has been used for food and textile fiber for over 5,000 years and during the past decades, it has received particular attention because of the overwhelming evidence of its tremendous potential in disease prevention, particularly cardiovascular, osteoporosis, rheumatoid arthritis, cancer (breast, colon, and prostate cancer), and constipation (Singh et al. 2011, Kajla et al. 2015).

Flaxseed contains a seed coat or testae rich in fiber and phytochemicals, including lignans, a thin endosperm, two embryos, and an embryonic axis. Embryos form 55% of the total weight of hand-dissected flaxseed, whereas the seed coat and the endosperm account for 36% of the total weight, and the embryo axis is 4% (Singh et al. 2011, Kajla et al. 2015).

Flaxseed or linseed is the commercial seed that contains the highest amounts of omega-3 linolenic acid, although seeds are well-balanced in terms of proteins, carbohydrates, minerals, and vitamins. The grains contain more than 40% oil rich in polyunsaturated FA and more than 18% protein (Table 13.3). Although the proteins of flaxseeds are limited by lysine, they are characterized by a high coefficient of digestibility and biological value (Matinchick et al. 2012). According to Singh et al. (2011), the total DF of whole flaxseed ranges from 36.7% to 46.8% and according to Martinchik et al. (2012) and the USDA (2021), from 27% to 30.8% (Table 13.3). The proportion of insoluble to soluble fiber fractions varies between 80:20 and 60:40 (Martinchik et al. 2012). The insoluble fraction is rich in cellulose and lignin, whereas the soluble DF is rich in pentosans (Serna-Saldivar and Sanchez-Hernandez 2020).

The main advantage and disadvantage of this oilseed is that it contains high amounts of omega-3 linolenic acid, but this FA is very susceptible to oxidation, especially when the seed is ground to produce full-fat flours.

13.3.7 Squash of Pumpkin Seeds (*Cucurbita* spp.)

Pumpkins are native of Mesoamerica and are extensively grown worldwide as a short seasoned horticultural fruit crop, especially in temperate and subtropical regions. According to the FAO (2021), the world production of pumpkin and squash fruits in year 2019 was 22.9 million tons. Of this total production, China, Ukraine, Russia, Spain, and Mexico harvested 8.37, 1.34, 1.20, 0.70, and 0.68 million tons, respectively. The fruit pulp is mainly utilized for the production of numerous desserts and foods. On the other hand, the pumpkin seeds are one of the food processing by-products most underutilized because they are excellent source of protein and polyunsaturated oil (Jittanit 2011). In many countries, pumpkin seeds are consumed as roasted and salted snack (Yoshida et al. 2006, Sacilik 2007, Jittanit 2011, Devi et al. 2018). The seed coats represent about 20% of the total seed weight. The most relevant species is *Cucurbita pepo* followed by *C. maxima*. The edible seed contains nearly 30% protein and up to 50% oil rich in polyunsaturated FA (Table 13.3). The squash seeds are also considered as good source of micronutrients and are valued as a healthy food because of their phenolics, sterols, and phytoestrogens (secoisolariciresinol and lariciresinol) that prevent hyperlipidemia and osteoporosis of menopausal women and reduction of hormone-dependent cancer tumors. The pumpkin seeds are especially rich in tocopherols that act as free radical scavengers, antiaging, and antioxidants (Fu et al. 2006, Saavedra et al. 2015, Lestari and Meiyanto 2018).

13.4 PROCESSING BY ROASTING

Roasting is a key process for the production of grain and tree nut–based snacks because it enhances their flavor, crispiness or texture, appearance, and overall acceptability of end-products. In addition, this thermal treatment inactivates antinutritional factors (Chapter 20), microorganisms, and reduces the beany flavor and odor of legumes, particularly soybeans, due to the inactivation of enzymes such as lipoxygenase and lipases. Furthermore, roasting dehydrates kernels lowering their A_w, gelatinizes the starch, denatures proteins, and removes unwanted volatile compounds. Commonly, kernels are roasted by hot air, hot pan, or oven, although novel methods such as infrared and microwave roasting are being increasingly adopted. Usually, the conventional methods are time-consuming and demand more energy and therefore are less efficient (Bagheri and Kashaninejad 2018, Bagheri 2020).

13.4.1 Peanuts

13.4.1.1 Selection

The proper selection of peanuts is considered the most relevant issue when processed into snacks. Processors should screen lots of incoming peanuts in order to ensure that they are free of mold damage and mycotoxins, chiefly aflatoxins produced by *Aspergillus flavus*. The maximum tolerance or level of aflatoxins for food-grade peanuts is less than 20 ppb. Aflatoxins are highly regulated by regulatory agencies throughout the world because they are potent carcinogens. The toxicology and negative health implications related to aflatoxins are thoroughly discussed in Chapter 20. Cleaning can effectively reduce the level of moldy seeds and consequently the amount of mycotoxins. The separation of damaged seeds is usually done by size, density, and visual appearance. Normally, moldy kernels have lower density and darker colorations than healthy counterparts. The typical cleaning procedure consists of air aspiration to remove light foreign material, sorting in gravity tables, and electronic color sorting machines (Serna-Saldivar 2008, USDA 2020).

13.4.1.2 Storage

Storage is considered as one of the most relevant postharvest practices since peanuts are highly susceptible to mold attack and rancidity. In-shell peanuts have good storage stability providing that they contain less than 10% moisture. The storage of unshelled peanuts at 20°C extends quality for six months, whereas shelled kernels to about four months. The shell which constitutes 20–30% of the pod weight protects kernels from insects and delay rancidity. Shelled kernels should preferably be stored at refrigeration temperatures with low relative humidity. Under these conditions, the storage life of peanuts is about two years. A low relative humidity is required because peanuts absorb moisture from a high relative humidity air and the higher moisture enhance deterioration and negatively affects sensory properties. Isotherms indicate that peanuts equilibrate to about 7% moisture when exposed to air with 70% relative humidity (Serna-Saldivar 2008).

13.4.1.3 In-Shell Roasted and Salted Peanuts

Figure 13.2 depicts the sequential steps commonly utilized to produce in-shell roasted and salted peanuts. The preparation and processing of in-shell roasted peanuts starts with a thoroughly cleaning operation usually consisting of two consecutive steps. The first consisting of removing foreign material (stems, stones, empty pods, chaff) by air aspiration, sifters, and magnets, whereas the second consisting of wet cleaning by washing the pods with coarse wet sand that removes stains and discoloration by abrasion. The sand is screened out and reused while the in-shelled peanuts dried and in some instances powdered with talc or kaolin (Figure 13.2). Production of in-shell salted peanuts is normally produced after pressure-cooking in-shell peanuts immersed in a brine (36% salt) solution. After pressure-cooking, the brine is drained and peanuts washed with water. After excess water drains off, the wet peanuts are centrifuged to further removed water in preparation for roasting in a rotary dryer set at temperatures 135–175°C up to 90 min to lower the moisture content to about 6% (Woodroof 1983, Serna-Saldivar 2008).

13.4.1.4 Shelling

The pods of the peanuts are commonly mechanically removed in a drum sheller. The equipment consists of curved grates forming the lower half of the drum and a revolving beater. The beaters crush the pods against the crests of the grates freeing the kernels. The clearance is adjusted to avoid harming the kernels when pods are mechanically compacted. Peanuts and broken pods pass through the openings of the gates and the fiber-rich and low-density pods effectively removed by air-aspiration. The stream of peanuts is subsequently passed over a screening device to separate

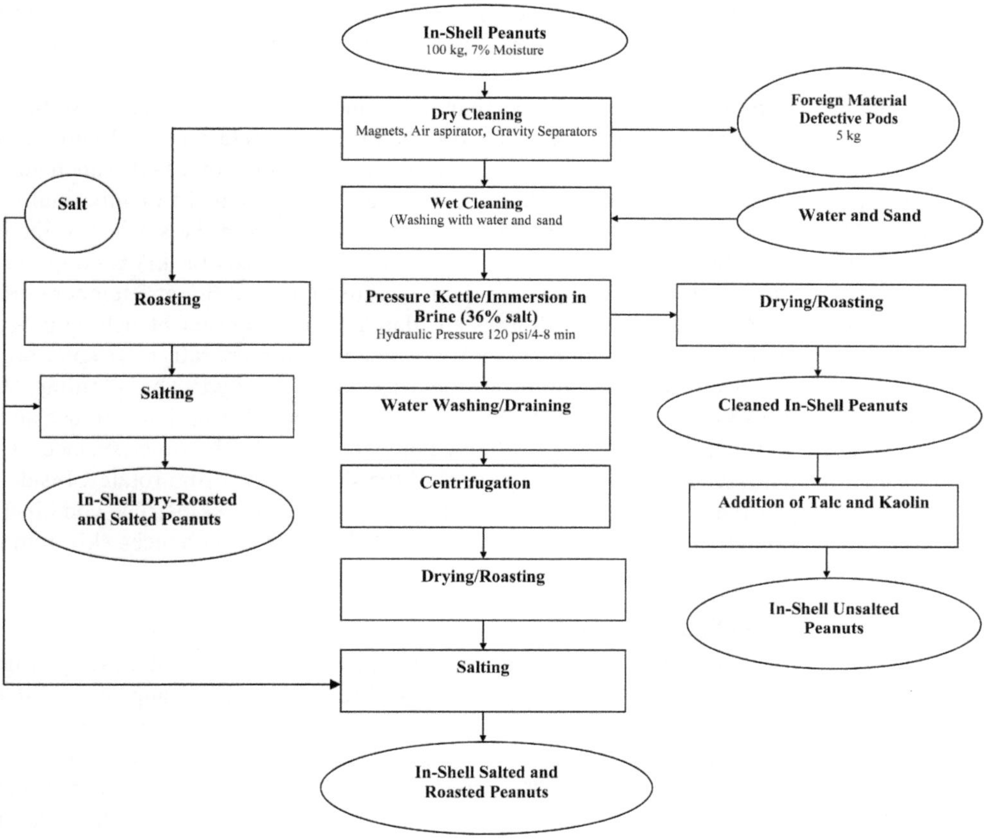

FIGURE 13.2 Production of salted and unsalted in-shell peanuts.

foreign material, small kernels, unshelled peanuts, and splits or broken kernels. The cleaned and sized kernels are then sorted by color by hand or with an electronic eye sorter (Woodroof 1983, Serna-Saldivar 2008).

13.4.1.5 Blanching

In the peanut snack settings, blanching is the operation that removes the red-colored testae or skin and hypocotyls or hearts, which represents about 8% of the total weight. Blanching or testae removal is considered one of the most relevant operations for the production of roasted peanuts and peanut butter because it affects color, flavor, and overall acceptability. Blanching also cleans the seed of dust, mold, filth, and other foreign material. The testae contains considerable amounts of phenolic compounds and tannins and therefore when left in the product, it imparts an astringent off-flavor. There are several ways or methods to blanch peanuts: dry, water, spin, alkali, and hydrogen peroxide blanching. Dry blanching is accomplished by heating the shelled peanuts to 138°C for about 25 min to crack and loosen the skins or testae. The heated peanuts are then cooled and passed through either brushes or ribbed rubber belting to rub off the skins. Screening is used to separate the hilum from the cotyledons (splits). The roller blancher consists of two neoprene-coated counterrotating rolls operating with a speed differential. The skins are loosened while passing through the rolls and removed by vibration and air-aspiration (Woodroof 1983, Serna-Saldivar 2008). The removed testae are air-aspirated, whereas the hearts segregated by sieving. During this operation, there is a loss of about 12% in weight. For certain applications, it is desired to obtain whole skin-free nuts in which case the hearts are retained. Whole-nut blanchers are equipped with resilient rubber rolls adjusted to minimize splitting, whereas split nut blanchers employ brushes or rubber elements (Woodroof 1983).

The wet-blanching process is more expensive than dry-blanching process. This sort of process has the advantage that peanuts do not require heating and therefore retain natural antioxidants and nutraceuticals; however, its main disadvantage is that it requires drying. Wet blanchers first are designed to pass peanuts on conveyors through stationary blades that slit longitudinally the peanut skins. Then peanuts are sprayed with hot water to further loosen the slashed testae that are mainly removed via abrasion under oscillating canvas-covered pads on knobbed conveyor belts. The resulting nuts are dehydrated to lower the moisture to around 6% at a temperature of 50°C for approximately 6 h. The spin blanching is an adaptation of the water-blanching process, but instead of applying the hot water spray, kernels are steamed in preparation for spinning on revolving spindles as the peanuts move, single file, down a grooved conveyor. The spinning action unwraps the peanut skins yielding seeds free of skins or testae. The other option is to use an air-impact horizontal drum or cylinder in which the peanuts are rotated. The inner surface of the drum has an abrasive surface that aids in the removal of the skins as the drum rotates. Inside the drum are special air jets that blow the peanuts with the aim to loosen the skins. The combination of air impact and the friction with the drum inner face of the cylinder enhances skin removal (Woodroof 1983, Serna-Saldivar 2008).

13.4.1.6 Roasting of Shelled Peanuts

There are two common processes to obtain roasted-salted peanuts: dry- and oil-roasting (Figure 13.3). The snack industry has focused to market dry-roasted salted peanuts to appeal the calorie-conscious consumers. Regardless of the process, selected and mycotoxin-free peanuts are blanched by any of the processes described before. For oil-roasting, blanching is performed before frying; whereas for dry-roasting, blanching is performed after the thermal treatment. The final part of the process consists of feeding the roasted peanuts onto cross-feeding belts that remove the skins. The testae are usually removed by air-aspiration (Woodroof 1983, Serna-Saldivar 2008).

13.4.1.7 Low-Fat Peanuts

Low-fat peanuts are increasing in popularity due to concerns about metabolic syndrome and related chronic diseases associated with high caloric intake. In addition, the removal of the fat yields peanuts with longer shelf life from the oil rancidity viewpoint. The process consists of subjecting shelled peanuts to solvent extraction followed by desolventization. Solvent extraction removes almost any desired amount of oil but results in loss of other flavorings. Mechanical removal of a portion of the oil results in better flavor retention but destroys the original peanut configuration (Woodroof 1983, Serna-Saldivar 2008). According to Willich and Feuge (1957), hexane extraction of whole blanched peanuts for 60 h removes approximately 75% of the oil. The partially defatted peanuts are whiter in color, crispier, denser, more hygroscopic, and contain at least twice as much protein than regular counterparts. Defatted peanuts are regularly salted by dipping in saturated brine at room temperature for 5–10 min to gain up to 20% brine by weight followed by oven dehydration to reduce the moisture to about 5%. De-oiled peanuts require better packaging in order to retain their texture because they are comparatively more hygroscopic compared to regular counterparts (Woodroof 1983, Serna-Saldivar 2008). A novel process developed by the USDA Southern Regional Research Laboratory first mechanically compresses raw kernels to remove or expel approximately 80% of the oil followed by immersion in hot water. This last operation returns most of the resulting partially defatted peanuts to their original oval shape (Serna-Saldivar 2008).

13.4.1.8 Japanese-Style Peanuts

The Japanese or crusted peanuts are covered with a hard crat consisting of a mixture of syrup and wheat flour flavored with soybean sauce and other flavorings. The manufacturing process starts

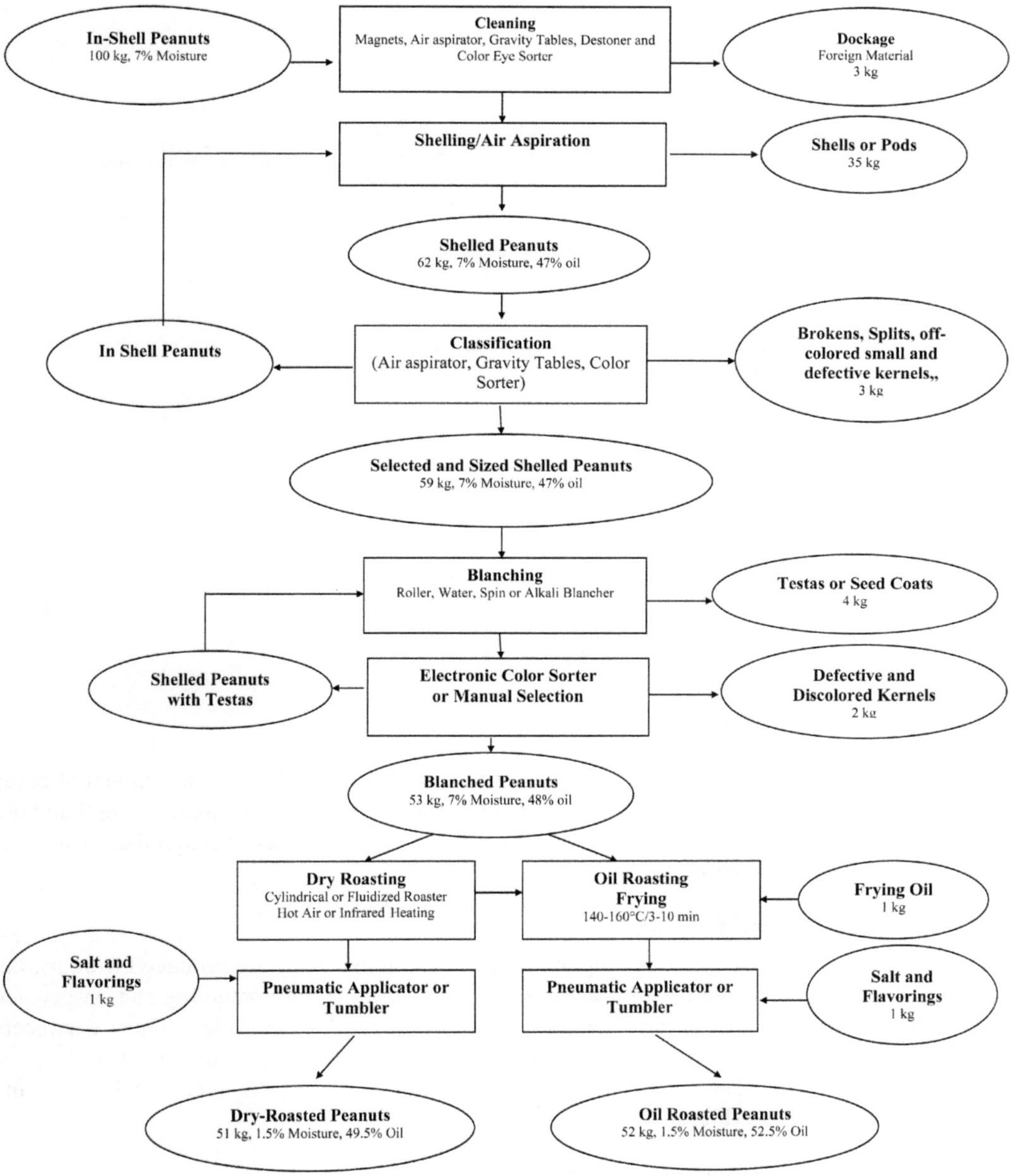

FIGURE 13.3 Production of dry or oil-roasted peanuts.

when the peanuts containing skins or testae are placed on automatic machines or coating tumblers where a mixture of a syrup (water, sugar, glucose, and maltodextrin) containing antioxidants is added to enhance the sticking of the wheat flour mix which contains modified or waxy starches, baking powder, and maltodextrin. The addition of both major raw materials is sequentially performed in order to achieve the desired amount of coating (45–55%) (Figure 13.4). The total operation lasts from several minutes in automated dusting equipment working with centripetal force to up to 45 min when hand-fed into tumblers. Finally, the Japanese peanuts are in some instances dehydrated and allowed to equilibrate at room temperature from several hours up to one day and dry-roasted in special ovens working at temperatures 170–190°C. The hot-roasted peanuts are finally flavored with a soybean sauce liquid base mixed with other flavorings in a one-stage tumbler, cooled to ambient

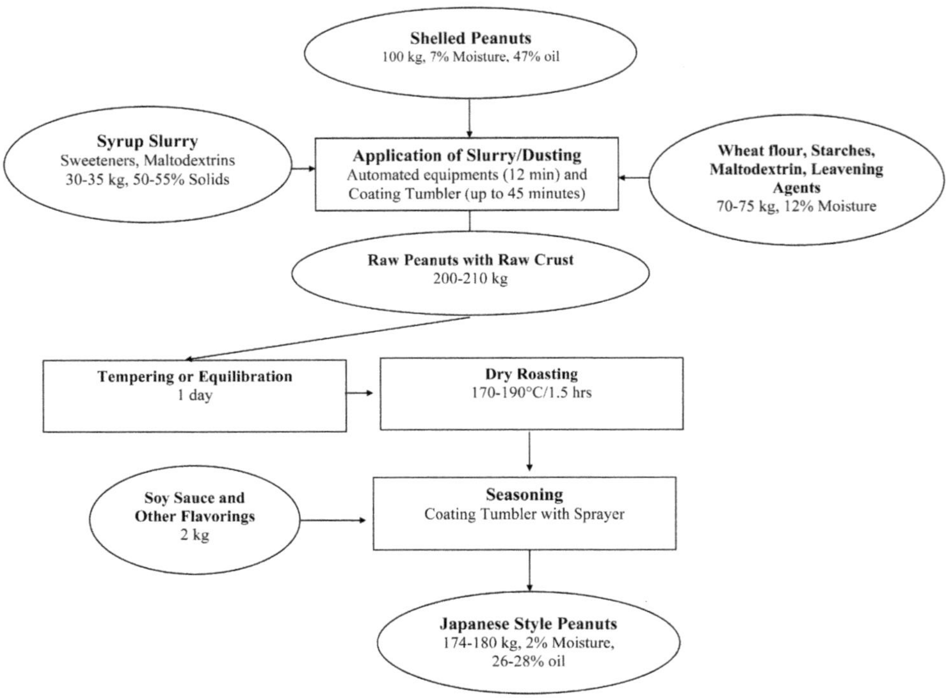

FIGURE 13.4 Production of Japanese-style peanuts.

temperature, and immediately packed to preserve their characteristic texture. The amount of covering of Japanese-style peanuts averages 42%, although some products might contain more than 50%. The thickness and internal crumb texture of the coating can be modified through the formulation and usage of chemical-leavening agents.

13.4.1.9 Honey-Roasted Peanuts

Honey-roasted peanuts have gained acceptability as snack items. They are produced from roasted peanuts that are further treated with a layer of powdery formulation containing salt, sweeteners (sucrose, honey, high-fructose syrup, molasses, or combination of these), spices, flavor enhancers, and adhesive materials (maltodextrins and gums). The honey-roasted peanuts are then deep-fat-fried at temperatures 140–190°C (Serna-Saldivar 2008). During frying, the sugars melt and form a network around the kernel (Figure 13.5).

13.4.1.10 Emerging Roasting Processes of Peanuts

Several novel methods have been recently proposed for roasting peanuts. The most promising is the use of infrared or a combination of infrared and hot air (Bagheri et al. 2016). The infrared is in between the range of microwaves and visible light with wavelengths ranging from 0.75 to 1,000 μm and energies from about 30 THz up to about 430 THz. The application of higher temperature, power, and time decreased moisture, soluble solids, pH, and compressive energy-hardness (texture) and increased browning, total phenolics, and total acceptance of peanuts. The highest and lowest values of specific energy consumed in the roasting process was observed at 120°C, 130 W, 30 min and at 100°C, 200 W, 10 min, respectively. The combination of infrared and hot air produced high-quality roasted peanuts compared to counterparts obtained by conventional methods and more importantly demanded less energy and processing time. Likewise, the utilization of pulsed infrared roasting. Kumar et al. (2009) concluded that the novel roasting method significantly reduced processing time by 33% and 60% compared to sand and drum-roasting, respectively. Roasting at 178–188°C

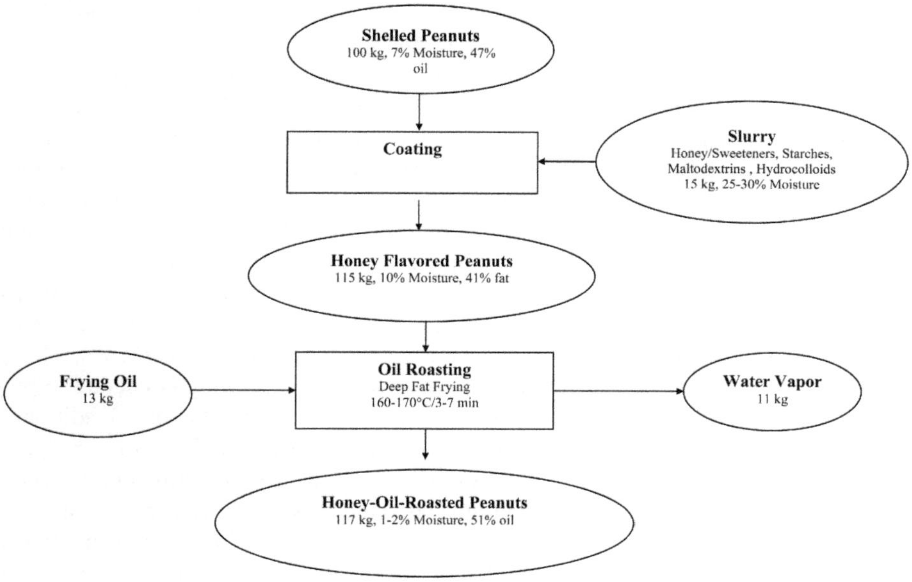

FIGURE 13.5 Production of honey oil-roasted peanut.

for 6.8–9.2 min produced the best roasted peanuts in terms of color, free FA, and texture, whereas processing at 180°C for 8 min produced the best organoleptic properties.

Jiao et al. (2015) investigated the physicochemical properties, volatile compounds, shelf life, and sensory quality of roasted peanuts processed with hot-air-assisted radio frequency and concluded that this novel technique produced excellent quality roasted peanuts. Interestingly, these authors identified 69 volatile compounds in the roasted peanuts, in which 3 new volatiles were generated after about 13 weeks of storage. During the storage, the relative concentrations of favorable volatile compounds, mainly pyrazines and furans, decreased, whereas the relative concentrations of some undesirable flavor compounds increased. The roasted peanuts had 31 weeks of shelf life based on the industrial standard and peroxide values.

13.4.2 Roasting of Other Seeds

The other legumes and pseudocereals are roasted following similar principles specified above. Boge et al. (2009) studied the effects of oil- or dry-roasting on the composition of soybean nuts obtained from different cultivars and concluded that oil-roasting significantly increased the lipid content and caloric density due to oil uptake during frying. Bagheri and Kashaninejad (2018) developed new roasting methods for soybeans. In this research, the kinetics and modeling of soybean roasting using combined infrared–hot air system and consumption of energy were investigated. The results showed that the effect of hot air temperature, infrared power, and combined air temperature and infrared potency on the roasting rate of soybean was statistically significant and the roasting process occurred within the falling rate period. The comparison of roasting methods showed that the minimum and maximum energies were consumed in the infrared process (0.0905 kWh) and hot air roasting (1.752 kWh) processes, respectively. Thus, the infrared heating could be considered ideal for roasting of soybeans with more acceptance due to the formation of lower amounts of bitter or astringent compounds.

Chickpeas have been traditionally used for preparation of roasted grains. Roasting yields a product named *leblebi*, which is considered a popular traditional snack in Turkey, the Mediterranean region, and the Middle East. There are two different types of leblebi, dehulled leblebi (Sari

Leblebi and Girit Leblebi) and whole-grain leblebi (Beyaz Leblebi and Saklz Leblebi). This popular snack food normally contains 23–29% protein and 45–53% starch (Coşkuner and Karababa 2004). The processing stages for all different kinds of leblebi include grain cleaning, soaking, tempering, boiling in excess water, equilibration, roasting, and dehulling. At the end of the production process, the chickpeas swell and soften, and after roasting the cooked grains acquired the desired crispness and texture. The preferred grains for leblebi production are Kabuli varieties, which yield large-seeded (8–9 mm in diameter and 300–500 g of 1,000 kernel weight), light-colored, round, and smooth-surface seeds. Furthermore, the grains must have a thick seed coat and the hull should be easily removed during processing. After cleaning the chickpeas with sifters or gravity tables, the seeds are soaked in preparation for roasting. Grains are soaked overnight in excess water containing different inorganic salts. Thereafter, the chickpeas are heated for 5–8 min at around 100°C and allowed to rest in a sack for 12–18 h to two days. Then, chickpeas are spread on a wooden or concrete area for slow cooling. Cooking is needed to produce a tender edible product, develop aroma, and to inactivate antinutritional factors. Traditionally, dry or soaked seeds are cooked in boiling water for an extended period of 1–2 h in an open pan or for 10–15 min in pressure cookers. There are three types of heating-roasting equipment that are commonly used in leblebi processing: the cylindrical drum roaster, the roaster and husk remover, and the roaster and speckler. The most common drum toaster is used for tempering and roasting of chickpeas and other snacks such as peanuts, pumpkin seeds, sunflower seeds, pistachios, and hazelnuts. The roaster and hull remover equipment consist of a heating pan made of copper or iron generally 50 cm in diameter. The inside surface of the heating pan is roughened and nicked to enhance the roasting area and thus facilitate the removal of the hulls. The temperature of the heating pan ranges from 80°C to 130°C. The roaster and speckler consists of a stainless steel heating pan of 50 cm diameter equipped with a motor that provides a rotary movement of 60 rpm and a paddle used for mixing. The operational temperature of this roasting pan varies from 100°C to 130°C. The roasting time fluctuates according to the temperature but generally lasts about 10 min. Interestingly, several types of leblebi (Girit and Sakiz) are obtained with the use of fine granulated sand during the roasting stages. The sand is removed and recovered from the product by sifting. After leblebi processing, the volume of chickpeas increases but density and kernel weight decreases. At the later stage of roasting, steam exits and the starchy matrix is dehydrated. These operations yield the desired porous-like internal structure. Inclusion of a large number of air spaces also causes the leblebi to have an opaque and chalky appearance due to the scattering of light (Coşkuner and Karababa 2004).

Roasted sesame seeds are widely used in the East Asian and Middle Eastern countries. The roasting process renders a unique nutty flavor to the seeds. Roasting is the key step in processing sesame because it influences color, texture, and flavor of roasted seeds affecting the overall palatability. El-Adawy and Mansour (2000) studied several roasting methods (steam, vacuum, hot plate, and hot air roasting) and concluded that the conventional hot air at 130 °C for 1 h yielded the best roasted seeds. Likewise, Rizki et al. (2015) studied the effects of roasting temperature and time on natural antioxidants (phenolics and flavonoids) and showed that roasted seeds can be considered as a good source of these natural antioxidants. The authors indicated that the optimum roasting time and temperature to obtain the most antioxidants and total phenolics and flavonoids were 90 min and 150°C, respectively.

Yoshida et al. (2006) treated whole pumpkin seeds of two different cultivars to microwave roasting for 6–30 min at a frequency of 2,450 MHz. After the thermal treatment, kernels were separated from whole seeds and assessed in terms of triacylglycerols and tocopherols. Results indicated that only minor increases in chemical changes of the oils, such as peroxide, carbonyl, *p*-anisidine, and thiobarbituric values occurred after a prolonged roasting period. Compared to the original level, more than 85% tocopherols remained after 20 min of roasting. With a few exceptions, microwave roasting for 12 min caused no significant loss of tocopherols and polyunsaturated FA.

13.5 SNACKS OBTAINED BY EXTRUSION

Obatolu Veronica et al. (2006) fortified corn puffs with defatted soybean meals replacing 10%, 20%, or 30% of total weight and analyzed the expansion ratio, density, color, texture, nutritional–chemical composition, and organoleptic characteristics. Results showed that the protein increased from 9.3% in the control corn puffs to 19.3% in those fortified with 30% defatted meal. The decrease in expansion ratio of extrudates containing more than 20% soybean was not perceived by panelists as less acceptable. The fortified puffs containing 20% defatted soybean contained an improved essential amino acid composition and were perceived slightly less acceptable compared to the control puffs.

Likewise, Omwamba and Mahungu (2014) developed a protein-rich RTE-extruded snack from a blend of rice, sorghum, and soybean flour and concluded that the thermoplastic extrusion cooking carried out at barrel temperature of 110–150°C, screw speed of 350–450 rpm, and tempering moisture of 12–14% improved both protein and starch digestibilities. The response surface analysis indicated that the expansion rate and bulk density were highly dependent on feed moisture and barrel temperature. The expansion ratio and bulk density ranged from 2.0 to 2.6 and 0.79 to 0.95 g/mL, respectively. The authors concluded that 100 g of the extruded product supplied 45% of the recommended daily allowance for protein in children aged up to 12 years.

Meng et al. (2010) studied the effects of extrusion conditions on physical properties of chickpea flour–based expanded snacks. Response surface methodology was used to study the effects of feed moisture content (16–18%), screw speed (250–320 rpm), and barrel temperature (150–170°C) on extruder system parameters (product temperature, die pressure, motor torque, and specific mechanical energy or SME) and physical properties (expansion, bulk density, hardness) of puffs. Desirable products, characterized by high expansion ratio and low bulk density and hardness, were obtained at low feed moisture, high screw speed, and medium to high barrel temperatures. This study clearly demonstrated that chickpeas can be used to produce nutritious snacks with desirable expansion and texture attributes.

Batistuti et al. (1991) designed experiments through response surface methodology to optimize the production of extruded snacks from defatted chickpea flours. The independent variables were process temperature (123–137°C) and feedstock tempering moisture (13–27%) in the extrusion of defatted chickpea flour. The expansion ratio increased steadily with a decrease in tempering moisture. These authors demonstrated that it was feasible to produce 100% chickpea extrudates and that the most acceptable puffs were rated by panelists higher compared to a commercial corn puff.

The use of common beans for production of snacks is scarce. However, several investigators have designed direct extruded puffs with combinations of refined fractions of cereals and beans flours. For instance, Anton et al. (2009) extruded blends of navy and red bean flours with corn starch at levels of 15%, 30%, or 45%. As expected, the extrudates containing higher levels of beans were denser and harder. However, puffs produced with 30% of the bean flours produced extrudates with crispness values comparable to corn starch alone. At this level, the protein increased 12-fold, whereas total phenols and antioxidant activities, assayed with the DPPH and ORAC assays, also increased. Red bean fortification yielded extrudates with higher levels of phenols and antioxidant properties. However, total phenols, DPPH, and ORAC values of the navy and red bean extrudates diminished by 10%, 17%, and 10%, and by 70%, 62%, and 17% after extrusion, respectively. As expected, the trypsin inhibitor levels were totally inactivated after these thermal treatments.

A similar study conducted by Perez-Navarrete et al. (2006) in which different combinations of maize and lima bean (*Phaseolus lunatus*) flours were thermoplastically extruded indicated that extrudates had higher *in vitro* protein digestibility. On the other hand, the available lysine and resistant starch values were highest in the 50:50 blend. This particular extrudate contained 17.6% and 7.94% total and soluble DF, respectively. According to these authors, extrusion of blends was feasible with combinations containing less than 50% beans.

Balandran-Quintana et al. (1998) studied the functional and nutritional properties of extruded whole pinto bean meals tempered to 18%, 20%, or 22% moisture and extruded at 140, 160, or 180°C at 150, 200, or 250 rpm. The extrusion temperature and tempering moisture influenced expansion index, bulk density, water absorption index, and *in vitro* protein digestibility. Trypsin inhibitors were inactivated completely for all conditions. The best product was produced with 22% tempering moisture extruded at 160°C.

Smith and Hardacre (2011) developed an extruded expanded snack product from finely ground faba beans. Like the previous study, the physical properties of extrudates were altered by varying the screw speed, the particle size of the flour, and presence or absence of the fiber-rich hulls. The optimum extruded faba bean product had similar properties and a high level of consumer acceptance compared to the commercially available extruded corn snacks. Importantly, the faba bean snack contained three and four times the protein and fiber compared to the control corn puff.

In terms of direct extruded snacks, Falcone and Phillips (1988) studied the effects of feed composition, feed moisture, and barrel temperature on the physical and rheological properties of snack-like products prepared from cowpea and decorticated sorghum flours. Decorticated sorghum flours and cowpeas and blends (33–67%) were tempered to 20.5%, 23.0%, and 25% moisture and thermoplastic extruded at 175, 190, or 205°C. Both the extrusion temperature and tempering moisture affected important properties related to expansion rate and texture. From the sorghum: cowpea ratio viewpoint, as cowpeas increased in the blends, the expansion ratio decreased. The experimental extrudates produced from composite feedstocks were slightly tougher compared to commercial puffs, although not drastically different from them. Likewise, Asare et al. (2004) researched the properties of extruded rice–cowpea–groundnut blends processed via thermoplastic extrusion. The combined effect of cowpea (0–20%), groundnut (0–10%), and feed moisture (14–48%) levels was varied. The puffs with the best bulk density values were obtained with lower tempering moisture contents. Increasing the legume addition affected color in finished products. Results suggested that the optimal process variables for the production of a puffed snacks with enhanced nutritional value and adequate expansion rate and internal structure were feed moisture contents of 14–20% and maximum additions of 20% cowpeas and 10% groundnuts.

The effects of extrusion cooking on the nutritional properties of peas indicated that extrudates had better *in vitro* protein and starch digestibilities and reduced levels of antinutritional factors, such as protease inhibitors and lectins. The concentrations of the limiting amino acids methionine and cystine further reduced by the thermal extrusion process. The weight gains and protein efficiency ratios of laboratory rats fed extruded pea diets supplemented with amino acids were significantly higher than those achieved by rats fed supplemented raw pea diets (Alonso et al. 2000).

Philipp et al. (2017) studied the properties and sensory attributes of extruded rice snacks fortified with up to 45% pea protein isolate. Results indicated that the extruder screw speed strongly affected the expansion rate and textural attributes of extrudates. Panelists preferred extrudates with high expansion rate and rated with good scores those fortified with 13% pea protein isolate.

Koksel and Masatcioglu (2018) investigated the physical properties of yellow pea puffs using the novel technique of nitrogen-assisted extrusion cooking (Chapter 12). Extrudates were prepared at three feed moisture contents (14%, 16%, or 18%). Nitrogen gas was injected into the extruder barrel (150 °C at the die) at five different pressures (100–500 kPa). High correlations were found between extrudate expansion and hardness ($r = -0.899$), between bulk density and crunchiness ($r = -0.874$), and between nitrogen gas injection pressure and crispness ($r = 0.822$). The expansion of puffs was mainly affected by feed moisture content and nitrogen pressure. This research showed the high potential of physical blowing agent–assisted extrusion for the production of high-quality yellow pea extrudates in terms of physical, textural, and microstructural properties.

Chan et al. (2019) studied the effects of different blowing agents (N_2 and CO_2) on physical properties of extruded puffed snacks made from red lentil flour. Microscopy studies indicated that the extrudate cross sections had higher number of cells due to gas injection. Some textural parameters, such as crunchiness, were positively affected by gas use. The most significant change in overall extrudate color and lightness was observed for red lentil processed with N_2 gas injection. The authors concluded that the use of these blowing agents during extrusion has great potential in manipulating the extrudate expansion rate, microstructure, texture, and color. Particularly, the N_2 gas was well-suited for the extrusion of red lentil flour.

Wojtowicz et al. (2013) studied the influence of buckwheat addition on physical properties, texture, color, and sensory characteristics of extruded corn snacks supplemented with different amounts of buckwheat flour (10–50% of the corn grits). The expansion ratio, fracturability, hardness, and sensory notes lowered with increasing levels of buckwheat flour. The results showed that buckwheat was useful and valuable for direct-extruded snacks. The best acceptable enriched extrudates were obtained with less than 30% substitution. Li et al. (2019) confirmed that buckwheat is an adequate feedstock for production of highly nutritious snacks providing the proper selection of the extrusion parameters. Brennan et al. (2012) assessed the glycemic indexes of extruded cereals enriched with buckwheat and amaranth flours and concluded that enriched products had a significantly lower glycemic index compared to the control.

Diaz et al. (2013) extruded 80% corn and 20% amaranth or quinoa blends in a corotating twin-screw extruder varying the tempering moisture content from 15% to 19%, the screw speed from 200 to 500 rpm, and temperature at the end of the barrel or die from 150°C to 170°C. The extrudates containing amaranth or quinoa showed adequate expansion and less lipid oxidation suggesting that the natural antioxidants improved stability of whole extrudates after exposure to high relative humidity.

Zapana et al. (2020) indicated that quinoa can be successfully puffed via extrusion. In this study, puffed quinoa products with and without starch–chitosan coating were developed by extrusion. The optimum condition for the production of high-quality puffs were screw speed of 500 rpm and 180 s. The application of a coating layer consisting of starch, chitosan, and sorbitol improved physicochemical characteristics after extrusion puffing due to higher capacity to retain water vapor within the extrudates.

Nikmaram et al. (2015) investigated the production of high-quality expanded corn extrudates supplemented with up to 30% sesame processed in a twin-screw extruder at different screw speeds (140–220 rpm) and barrel temperatures (120–160°C). Results showed that increasing the sesame level reduced the expansion rate and yielded extrudates with more density, harder texture, and darker appearance. These properties remarkably improved when the extruder operated at higher temperatures. According to sensory assessment, sesame snacks could be used up to the level of 15% to produce high-quality snack extrudates with acceptable sensory properties and enhanced nutritional value.

Wu et al. (2007) studied the effects of flaxseed meal (0%, 5%, 10%, and 15%) substitution of corn meal and extrusión processing variables, including tempering moisture content (16–20%) and screw speed (200–400 rpm), on properties of extrudates. Results showed that increasing the levels of flaxseed or tempering moisture and decreasing screw speed diminished expansion rate and yielded extrudates that were harder and darker (more reddish and less yellow). The lignan content in the raw flaxseed was 2.27 mg/g. These important phytochemicals were retained in higher extents when more flaxseed meal substituted the corn meal and when the extruder operated at higher screw speeds. About 25–52% of the lignans were lost after extrusión. However, the extrusion at higher tempering moistures had an opposite effect.

Min et al. (2015) also optimized the extrusion conditions for production of corn snacks enriched with up to 50% flaxseed meal. Four extrusion conditions were varied: temperature (100–160°C), screw speed (60–180 rpm), feedstock rate (18.35–54.95 kg/h), and tempering moisture content (14–30% w/w). The extrudates were analyzed in terms of expansion index, bulk density, texture (hardness and brittleness), and *in vitro* protein digestibility. The optimum extrusion conditions were obtained at feed composition with flaxseed content of 20%, tempering moisture of 17.37–22.43%, extrusion temperature of 134.3–156.1°C, screw speed of 114–165.7 rpm, and feeding speed of 34.38–45.95 kg/h.

Tobias-Espinosa et al. (2019) recently investigated the effects of addition of different levels of amaranth and flaxseed in the development of extruded puffs. The enriched extrudates contained higher protein (>12%), soluble and insoluble DF, and polyunsaturated FA. The extruded products with low levels of maize grits had lower expansion rate but higher levels of fiber and protein. Results indicated that good quality puffs were obtained with 8.6–9.3% flaxseed meal or 18.7–22.9% amaranth.

Bombo-Trevisan and Gomes-Arêas (2012) tested different combinations of refined corn grits (>1% fat) and partially defatted flaxseed flour (9% fat) were extruded for production of expanded puffs. The feed moisture, process temperature, and flaxseed flour addition were varied and extrudates mainly assessed in terms of expansion rate. The optimum corn–flaxseed snack contained approximately sevenfold more total dietary fiber and twice the protein compared to the control corn snack. The enriched puffs were well-accepted indicating the potential of this new product for commercialization.

In other similar studies (Ganorkar and Jain 2015, Ganorkar et al. 2016), response surface methodologies were employed to develop extrudates formulated with flaxseed, rice, and corn meals. The flaxseed product substituted part of the rice flour and maintained the same amount of corn meal. Ganorkar and Jain (2015) investigated the substitution from 15% to 25% of roasted flaxseed flour in the blend that was conditioned from 12% to 16% moisture and extruded at temperatures 120–140°C and screw speeds 300–330 rpm. The supplementation with higher levels of flaxseed flour decreased expansion rate and overall acceptability of extrudates, whereas the tempering moisture was positively related to expansion rate. The extruder barrel temperature was negatively related to expansion rate and overall acceptance. The feedstock containing 15% flaxseed and extruded at 16% moisture, 120°C, and 330 rpm yielded the best enriched extrudates with 3.08 of expansion rate and 0.106 g/cm^3 of bulk density. The same research group (Ganorkar et al. 2016) conducted similar experiments with defatted flaxseed flour, which substituted 7.5–20% corn and rice flours. Again, the level of defatted flaxseed flour was found to be the most significant independent variable affecting the characteristics of extrudates followed by extruder barrel temperature and then screw rpm. As the level of defatted flaxseed meal increased, the expansion rate and overall acceptability decreased. The best enriched extrudates were obtained with 10% substitution and extruding a feedstock tempered to 20% moisture at screw speed and barrel temperature of 312 rpm and 125°C.

13.6 OTHER SNACKS

Maetens et al. (2017) studied the histochemical and nutritional properties of a snack chip developed from germinated soybeans. The moisture, protein, and fat contents of soybeans germinated for 1, 3, or 5 days were 2.4–5.3, 42.7–43.3, and 21.4-25.5%, respectively, whereas lipoxygenase-1 and lipoxygenase-3 activity and trypsin inhibitor reductions were 2.8–17.2%, 16.2–26.1%, and 16.6–31.1%, respectively. The flour obtained from five-day sprouted grains showed the highest reduction in both lipoxygenases activities and trypsin inhibitor. The snack was produced from doughs prepared by mixing dried germinated soybean flour with table salt (1.2 g/100 g), baking soda, and boiling water in a flour-to-water ratio of 2.5:1. The dough was sheeted and cut to yield 2 × 2 cm pieces and then chips were baked at 190°C for 8–10 min. The prototype snack chip made from germinated soybeans showed excellent nutritional properties and low glycemic index.

Furthermore, chickpeas have been successfully used in the preparation of a variety of fried snacks (Annapure et al. 1998). Recently, chickpea flours in combination with refined starch-rich cereals grits or meals have been used to produce nutritious snacks with desirable expansion and texture properties (Bhattacharya and Prakash 1994, Meng et al. 2010), although the incorporation of chickpeas decreased product expansion and increased density and breaking strength. Debnath et al. (2003) demonstrated that the oil content of deep-fat-fried chickpea snacks could be modulated by pre-drying processes. The pre-fry drying and deep fat frying of chickpea ribbons for 40 min and 45 s, respectively, yielded snacks with 54% less oil content without compromising the sensory quality.

Akubor (2004) determined the physical, sensory, and nutritional properties of the typical Nigerian wheat-based snack foods (cake, chin-chin, and puff-puff) fortified with cowpea flour. Chin-chin is categorized as a baked or fried crispy snack produced from a wheat-based stiff dough supplemented with salt, sweeteners, eggs, margarine, or fats. The properly developed dough is sheeted to about 0.6 cm thick and then cut into small squares before frying. All the snack foods showed higher protein content when compared with the control snack counterpart made with 100% wheat. As expected, the protein content increased with the increasing levels of cowpea flour in the blend. Importantly, the physical properties of the enriched snacks were not negatively affected by supplementation with cowpeas. Similarly, at all levels of cowpea supplementation, the sensory characteristics of the snack foods did not differ significantly.

Nomanyi et al. (2020) formulated and analyzed the physical, nutritional, and sensory attributes of baobab-based ready-to-eat sorghum and cowpea blend snack bars. Popped sorghum, baked cowpeas, and baobab pulp powder were blended in five different formulations (45:55:0; 50:45:5; 55:35:5; 55:35:10; 60:25:15; 65:15:20). Protein in the formulations ranged between 11.38% and 21.3%, whereas fat and crude fiber contents ranged between 2% and 3.3% and from 1.6% to 2.76%, respectively. Sensory evaluation tests indicated that color, taste, texture, aroma, appearance, and overall acceptability were acceptable, although bars with high concentrations of baobab were the least preferred.

The enrichment with cowpeas of an indigenous Nigerian fried corn snack named Ipekere or Ipakere was researched by Shakpo and Osundahunsi (2016) with the main aim of improving the nutrient content of the snack, especially in terms of protein and essential minerals. This snack is made from a stiff paste from coarsely milled maize meal, spiced with onion, pepper, and salt that is deep-fat-fried generally with palm oil. Flour blends were produced from corn and cowpea flours in the following ratios 100:0 (control), 90:10, 80:20, and 70:30. Stiff pastes of the blends seasoned with onion and pepper were divided into two parts: the first part was molded and deep-fried in palm oil, while the second part was oven-dried at 150°C for 2 h. Resulting snacks contained proteins from 5.86% to 12.50%, total ash 2.32% to 4.32%; fat content 8.89% to 25.04%; moisture content 8.90% to 17.47%; and fiber 7.32% to 18.76%. As expected, it was observed that fat content was affected by oil-frying but not by cowpea substitution. Results indicated that potassium, sodium, zinc, and iron contents increased in blends containing more cowpeas. Sensory evaluation showed that there were significant differences ($p < 0.05$) among the snacks produced in terms of aroma, appearance, texture (crispness), taste and overall acceptability. Oven-dried samples were scored lower than oil-fried in aroma, taste, and overall acceptability, but were preferred in terms of appearance and crispness. The overall result showed that 20% cowpea substitution was the most adequate to produce an acceptable Ipekere with enhanced nutritional value.

Kerr et al. (2001) successfully made chips with a combination of different levels of cowpea flour milled through 0.5, 1.0, and 2.0 mm screens and starch. Color values L^*, a^*, and b^* indicated that chips with higher cowpea protein/cornstarch ratios were darker and browner. The authors concluded that both milling conditions and ratio of cowpea flour to starch influenced textural parameters, color, and quality of chips.

Nomanyi et al. (2020) formulated and analyzed the physical, nutritional, and sensory attributes of baobab-based ready-to-eat sorghum and cowpea-blend snack bars. Popped sorghum, baked cowpeas, and baobab pulp powder were blended in five different formulations (45:55:0; 50:45:5; 55:35:5; 55:35:10; 60:25:15; 65:15:20). Protein in the formulations ranged between 11.38% to 21.3%, whereas fat and crude fiber contents ranged between 2% and 3.3% and between 1.6% and 2.76%, respectively. Sensory evaluation tests indicated that color, taste, texture, aroma, appearance, and overall acceptability were acceptable, although bars with high concentrations of baobab were the least preferred.

Zapana et al. (2020) indicated that quinoa can be gun or microwave puffed and used as an ingredient in snack formulations. Gun-puffing consists of the sudden release of accumulated superheated vapor by changing elevated internal pressure to atmospheric pressure that results in an expanded grain matrix with porous structure. The optimum process conditions for gun puffing were maximum 1.31 MPa after 780 s, whereas for microwave puffing 1,200 W for 60 s in grains previously tempered to 18–20% moisture. Gun and puffing yielded high-quality popped quinoa. The application of a coating layer consisting of starch, chitosan, and sorbitol improved physicochemical characteristics after gun puffing due to higher capacity to retain water vapor within the grains.

Coorey et al. (2012) studied how the incorporation of 5%, 10%, 12%, or 15% ground chia affected the nutritive profile and consumer acceptance of baked rice-potato chips. There were no significant differences in appearance, color, flavor, and overall preference between a commercial chip sample and the 5% chia chips. The chemical analysis indicated that all supplemented chips were excellent sources of omega-3. A 50 g portion of the 5% chia chip would provide almost half of the Recommended Daily Intake of omega-3 FA for healthy women and just over a quarter for healthy men. The levels of dietary fiber, protein, calcium, and antioxidants were also higher in 5% chia chip samples compared to the control. In terms of optimal consumer acceptance and nutritional benefit, the 5% chip has the highest rating and is as acceptable as a commercially available product. This investigation showed that gluten-free chips containing 5% chia, 3% DF, and 15 mg/100 g of omega-3 and higher antioxidant activity yielded chips with good consumer acceptance.

Giacomino et al. (2013) researched the utilization of extruded flaxseed meal incorporated in cereal-based bars to demonstrate its nutritious usefulness. The authors concluded that the extrusion process did not affect amino acid content and that the flaxseed protein was limited by lysine with a chemical score of 83%. The formulated cereal bars contained high protein, DF, and a more favorable ratio of ω6:ω3.

Khouryieh and Aramouni (2013) evaluated the effect of flaxseed flour addition on the physical and sensory characteristics of oat-based snack bars. Four formulations of the flaxseed snack bars were prepared by partially replacing oats with flaxseed flour added at levels of up to 18%. There were no significant differences in water activity, moisture, and firmness values between the flaxseed bars and the control. Bars supplemented with 12% flax had similar sensory attributes and overall acceptability compared to the control. These results indicated that flaxseed flour incorporation up to 12% substantially enhanced the nutritional qualities of the oat-based snack bars, especially in terms of DF and omega-3 FA.

REFERENCES

Akubor, P.I. 2004. Protein contents, physical and sensory properties of Nigerian snack foods (cake, chin-chin and puff-puff) prepared from cowpea–wheat flour blends. *International Journal of Food Science and Technology*, 39(4):419–424.

Alonso, R., Grant, G., Dewey, P., and Marzo, F. 2000. Nutritional assessment in vitro and in vivo of raw and extruded peas (*Pisum sativum* L.). *Journal of Agricultural and Food Chemistry*, 48(6):2286–2290.

Anilakumar, K.R., Pal, A., Khanum, F., and Bawa, A.S. 2010. Nutritional, medicinal and industrial uses of sesame (*Sesamum indicum* L.) seeds-an overview. *Agriculturae Conspectus Scientificus*, 75(4):159–168.

Annapure, U.S., Singhal, R.S., and Kulkarni, P.R. 1998. Studies on deep-fat fried snacks from some cereals and legumes. *Journal of the Science of Food and Agriculture*, 76(3):377–382.

Anton, A.A., Fulcher, R.G., and Arntfield, S.D. 2009. Physical and nutritional impact of fortification of corn starch-based extruded snacks with common bean (*Phaseolus vulgaris* L.) flour: Effects of bean addition and extrusion cooking. *Food Chemistry*, 113(4):989–996.

Asare, E.K., Sefa-Dedeh, S., Sakyi-Dawson, E., and Afoakwa, E.O. 2004. Application of response surface methodology for studying the product characteristics of extruded rice–cowpea–groundnut blends. *International Journal of Food Sciences and Nutrition*, 55(5):431–439.

Bagheri, H. 2020. Application of infrared heating for roasting nuts. *Journal of Food Quality*, Article ID 8813047:1745–4557.

Bagheri, H., and Kashaninejad, M. 2018. Modeling of soybean snack roasting by infrared heating using artificial neural network (ANN). *Journal of Food Technology and Nutrition*, 15(4):19–30.

Bagheri, H., Kashaninejad, M., Ziaiifar, A.M., and Aalami, M. 2016. Novel hybridized infrared-hot air method for roasting of peanut kernels. *Innovative Food Science and Emerging Technologies*, 37:106–114.

Balandran-Quintana, R.R., Barbosa-Canovas, G.V., Zazueta-Morales, J.J., Anzaldúa-Morales, A., and Quintero-Ramos, A. 1998. Functional and nutritional properties of extruded whole pinto bean meal (*Phaseolus vulgaris* L.). *Journal of Food Science*, 63(1):113–116.

Batistuti, J.P., Barros, R.M.C., and Areas, J.A.G. 1991. Optimization of extrusion cooking process for chickpea (*Cicer arietinum* L.) defatted flour by response surface methodology. *Journal of Food Science*, 56(6):1695–1698.

Benitez, V., Cantera, S., Aguilera, Y., Molla, E., Esteban, R.M., Díaz, M.F., and Martín-Cabrejas, M.A. 2013. Impact of germination on starch, dietary fiber and physicochemical properties in non-conventional legumes. *Food Research International*, 50(1):64–69.

Bhattacharya, S., and Prakash, M. 1994. Extrusion of blends of rice and chick pea flours: A response surface analysis. *Journal of Food Engineering*, 21(3):315–330.

Boge, E.L., Boylston, T.D., and Wilson, L.A. 2009. Effect of cultivar and roasting method on composition of roasted soybeans. *Journal of the Science of Food and Agriculture*, 89(5):821–826.

Bombo-Trevisan, A.J., and Gomes-Arêas, J.A. 2012. Development of corn and flaxseed snacks with high-fibre content using response surface methodology (RSM). *International Journal of Food Sciences and Nutrition*, 63(3):362–367.

Bonafaccia, G., Marocchini, M., and Kreft, I. 2003. Composition and technological properties of the flour and bran from common and tartary buckwheat. *Food Chemistry*, 80(1):9–15.

Brennan, M.A., Menard, C., Roudaut, G., and Brennan, C.S. 2012. Amaranth, millet and buckwheat flours affect the physical properties of extruded breakfast cereals and modulates their potential glycaemic impact. *Starch-Stärke*, 64(5):392–398.

Carbonell-Barrachina, Á.A., Lluch, M.Á., Pérez-Munera, I., Hernando, I., and Castillo, S. 2009. Effects of chemical dehulling of sesame on color and microstructure. *Food Science and Technology International*, 15(3):229–234.

Caselato-Sousa, V.M., and Amaya-Farfán, J. 2012. State of knowledge on amaranth grain: A comprehensive review. *Journal of Food Science*, 77(4): R93–R104.

Chan, E., Masatcioglu, T.M., and Koksel, F. 2019. Effects of different blowing agents on physical properties of extruded puffed snacks made from yellow pea and red lentil flours. *Journal of Food Process Engineering*, 42(3):e12989.

Coorey, R., Grant, A., and Jayasena, V. 2012. Effect of chia flour incorporation on the nutritive quality and consumer acceptance of chips. *Journal of Food Research*, 1(4):85–95.

Coşkuner, Y., and Karababa, E. 2004. Leblebi: A roasted chickpea product as a traditional Turkish snack food. *Food Reviews International*, 20(3):257–274.

de Almeida Costa, G.E., da Silva Queiroz-Monici, K., Reis, S.M.P.M., and de Oliveira, A.C. 2006. Chemical composition, dietary fibre and resistant starch contents of raw and cooked pea, common bean, chickpea and lentil legumes. *Food Chemistry*, 94(3):327–330.

Debnath, S., Bhat, K.K., and Rastogi, N.K. 2003. Effect of pre-drying on kinetics of moisture loss and oil uptake during deep fat frying of chickpea flour-based snack food. *LWT – Food Science and Technology*, 36(1):91–98.

Devi, N.M., Prasad, R.V., and Sagarika, N. 2018. A review on health benefits and nutritional composition of pumpkin seed. *International Journal of Chemical Studies*, 6(3):1154–1157.

Diaz, J.M.R., Kirjoranta, S., Tenitz, S., Penttilä, P.A., Serimaa, R., Lampi, A.M., and Jouppila, K. 2013. Use of amaranth, quinoa and kañiwa in extruded corn-based snacks. *Journal of Cereal Science*, 58(1):59–67.

El-Adawy, T.A., and Mansour, E.H. 2000. Nutritional and physicochemical evaluations of tahina (sesame butter) prepared from heat-treated sesame seeds. *Journal of the Science of Food and Agriculture*, 80(14):2005–2011.

Elleuch, M., Bedigian, D., Besbes, S., Blecker, C., and Attia, H. 2012. Dietary fibre characteristics and antioxidant activity of sesame seed coats (testae). *International Journal of Food Properties*, 15(1):25–37.

Elleuch, M., Besbes, S., Roiseux, O., Blecker, C., and Attia, H. 2007. Quality characteristics of sesame seeds and by-products. *Food Chemistry*, 103(2):641–650.

Falcone, R.G., and Phillips, R.D. 1988. Effects of feed composition, feed moisture, and barrel temperature on the physical and rheological properties of snack-like products prepared from cowpea and sorghum flours by extrusion. *Journal of Food Science*, 53(5):1464–1469.

FAO. 2021. Food and Agriculture Organization of the United Nations, Rome. Electronic page http://www.fao.org/faostat/.

Fu, C.L., Shi, H., and Li, Q.H. 2006. A review on pharmacological activities and utilization technologies of pumpkin. *Plant Foods for Human Nutrition*, 61(2):70–77.

Ganorkar, P.M., and Jain, R.K. 2015. Development of flaxseed fortified rice–corn flour blend based extruded product by response surface methodology. *Journal of Food Science and Technology*, 52(8):5075–5083.

Ganorkar, P.M., Patel, J.M., Shah, V., and Rangrej, V.V. 2016. Defatted flaxseed meal incorporated corn-rice flour blend based extruded product by response surface methodology. *Journal of Food Science and Technology*, 53(4):1867–1877.

Giacomino, S., Peñas, E., Ferreyra, V., Pellegrino, N., Fournier, M., Apro, N., Olivera-Carrion, M., and Frías, J. 2013. Extruded flaxseed meal enhances the nutritional quality of cereal-based products. *Plant Foods for Human Nutrition*, 68(2):131–136.

Guardado-Felix, D., Milán-Noris, A.K., Serrano-Sandoval, S.N., and Serna-Saldivar, S.O. 2019. Chickpea proteins and peptides in health. In: *Handbook of Chickpeas: Nutritional Value, Health Benefits and Management*, eds. A.T. Lund and N.D. Schultz. New York: Nova Science Publishers.

Guillon, F., and Champ, M.-J. 2002. Carbohydrate fractions of legumes: Uses in human nutrition and potential for health. *British Journal of Nutrition*, 88(supplement 3):293–306.

Gutiérrez-Uribe, J.A., Romo Lopez, I., and Serna-Saldivar, S.O. 2011. Phenolic composition and inhibition of mammary cancer cells by extracts of cowpeas (*Vigna unguiculata*) and its anatomical parts. *Journal of Functional Foods*, 3(4):290–297.

Jayathilake, C., Visvanathan, R., Deen, A., Bangamuwage, R., Jayawardana, B.C., Nammi, S., and Liyanage, R. 2018. Cowpea: An overview on its nutritional facts and health benefits. *Journal of the Science of Food and Agriculture*, 98(13):4793–4806.

Jiao, S., Zhu, D., Deng, Y., and Zhao, Y. 2015. Effects of hot air-assisted radio frequency heating on quality and shelf-life of roasted peanuts. *Food and Bioprocess Technology*, 9(2):308–319.

Jittanit, W. 2011. Kinetics and temperature dependent moisture diffusivities of pumpkin seeds during drying. *Kasetsart Journal (Natural Science)*, 45(1):147–158.

Jukanti, A.K., Gaur, P.M., Gowda, C.L.L., and Chibbar, R.N. 2012. Nutritional quality and health benefits of chickpea (*Cicer arietinum* L.): A review. *British Journal of Nutrition*, 108(S1): S11–S26.

Kajla, P., Sharma, A., and Sood, D.R. 2015. Flaxseed-a potential functional food source. *Journal of Food Science and Technology*, 52(4):1857–1871.

Kerr, W.L., Ward, C.D.W., McWatters, K.H., and Resurrection, A.V.A. 2001. Milling and particle size of cowpea flour and snack chip quality. *Food Research International*, 34(1):39–45.

Khouryieh, H., and Aramouni, F. 2013. Effect of flaxseed flour incorporation on the physical properties and consumer acceptability of cereal bars. *Food Science and Technology International*, 19(6):549–556.

Koksel, F., and Masatcioglu, M.T. 2018. Physical properties of puffed yellow pea snacks produced by nitrogen gas assisted extrusion cooking. *LWT-Food Science and Technology*, 93:592–598.

Kumar, S., Debnath, S., and Hebbar, U.H. 2009. Pulsed infrared roasting of groundnuts and its quality. *International Journal of Food Engineering*, 5(4):21 https://doi.org/10.2202/1556-3758.1158.

Lestari, B., and Meiyanto, E. 2018. A review: The emerging nutraceutical potential of pumpkin seeds. *Indonesian Journal of Cancer Chemoprevention*, 9(2):92–101.

Li, H., Xiong, Z., and Li, X. 2019. Optimization of the extrusion process for the development of extruded snacks using peanut, buckwheat, and rice blend. *Journal of Food Processing and Preservation*, 43(12):e14264.

Maetens, E., Hettiarachchy, N., Dewettinck, K., Horax, R., Moens, K., and Moseley, D.O. 2017. Physicochemical and nutritional properties of a healthy snack chip developed from germinated soybeans. *LWT- Food Science and Technology*, 84:505–510.

Martín Cabrejas, M.A., Sanfiz, B., Vidal, A., Molla, E., Esteban, R., and López Andreu, F.J. 2004. Effect of fermentation and autoclaving on dietary fiber fractions and antinutritional factors of beans (*Phaseolus vulgaris* L.). *Journal of Agricultural and Food Chemistry*, 52(2):261–266.

Martinchik, A.N., Baturin, A.K., Zubtsov, V.V., and Molofeev, V. 2012. Nutritional value and functional properties of flaxseed. *Voprosy Pitaniia*, 81(3):4.

Mazza, G., and Oomah, B.D. 2005. Buckwheat as a food and feed. In: *Specialty Grains for Food and Feed*, eds. A.A. Elsayed and P. Wood, 375–393. St. Paul: AACC-International.

Meng, X., Threinen, D., Hansen, M., and Driedger, D. 2010. Effects of extrusion conditions on system parameters and physical properties of a chickpea flour-based snack. *Food Research International*, 43(2):650–658.

Milan-Noris, A.K., De la Rosa-Millan, J., and Serna-Saldivar, S.O. 2019a. Comparative analysis of techno-functional properties, starch digestion and protein quality of pigmented chickpea flours. *International Journal of Food Science and Technology*, 54(6):2288–2299.

Milan-Noris, A.K., Gutierrez-Uribe, J.A., Santacruz, A., Serna-Saldivar, S.O., and Martinez-Villaluenga, C. 2018. Peptides and isoflavones in gastrointestinal digests contribute to the anti-inflammatory potential of cooked or germinated desi and Kabuli chickpea. *Food Chemistry*, 268:66–76.

Milan-Noris, A.K., Montoya-Rodriguez, A., De la Rosa-Millan, J., and Serna-Saldivar, S.O. 2019b. Health effects of starch and undigested carbohydrates associated to chickpea based foods. In: *Handbook of*

Chickpeas: Nutritional Value, Health Benefits and Management, eds. A.T. Lund and N.D. Schultz. New York: Nova Science Publishers.

Min, W., Yi, L., Lijun, W., Dong, L., and Zhihuai, M. 2015. Effects of extrusion parameters on physicochemical properties of flaxseed snack and process optimization. *International Journal of Agricultural and Biological Engineering*, 8(5):121–131.

Moharram, Y.G., Osman, H.O., and Abou-EI-Khier, Y.I. 1990. Wet decortication of sesame seeds by new methods. *Food and Nutrition Bulletin*, 12(1):1–6.

Momanyi, D., Owino, W., and Makokha, A. 2020. Formulation, nutritional and sensory evaluation of baobab based ready-to-eat sorghum and cowpea blend snack bars. *Scientific African*, 7:e00215.

Muñoz, L.A., Cobos, A., Diaz, O., and Aguilera, J.M. 2013. Chia seed (Salvia hispanica): An ancient grain and a new functional food. *Food Reviews International*, 29(4):394–408.

Navruz-Varli, S., and Sanlier, N. 2016. Nutritional and health benefits of quinoa (*Chenopodium quinoa* Willd.). *Journal of Cereal Science*, 69:371–376.

Nikmaram, N., Garavand, F., Elhamirad, A., Beiraghi-Toosi, S., and Goli-Movahhed, G. 2015. Production of high quality expanded corn extrudates containing sesame seed using response surface methodology. *Quality Assurance and Safety of Crops and Foods*, 7(5):713–720.

Obatolu Veronica, A., Omueti Olusola, O., and Adebowale, E.A. 2006. Qualities of extruded puffed snacks from maize/soybean mixture. *Journal of Food Process Engineering*, 29(2):149–161.

Omwamba, M., and Mahungu, S.M. 2014. Development of a protein-rich ready-to-eat extruded snack from a composite blend of rice, sorghum and soybean flour. *Food and Nutrition Sciences*, 5(14): Article ID: 48537.

Pedersen, B., Knudsen, K.E., and Eggum, B.O. 1990. The nutritive value of amaranth grain (*Amaranthus caudatus*). 3. Energy and fibre of raw and processed grain. *Plant Foods for Human Nutrition*, 40(1):61–71.

Pereira, E., Encina-Zelada, C., Barros, L., Gonzales-Barron, U., Cadavez, V., and Ferreira, I.C.F.R. 2019. Chemical and nutritional characterization of *Chenopodium quinoa* Willd (quinoa) grains: A good alternative to nutritious food. *Food Chemistry*, 280:110–114.

Perez-Navarrete, C., González, R., Chel-Guerrero, L., and Betancur-Ancona, D. 2006. Effect of extrusion on nutritional quality of maize and lima bean flour blends. *Journal of the Science of Food and Agriculture*, 86(14):2477–2484.

Philipp, C., Buckow, R., Silcock, P., and Oey, I. 2017. Instrumental and sensory properties of pea protein-fortified extruded rice snacks. *Food Research International*, 102:658–665.

Ramírez Jiménez, A., Reynoso Camacho, R., Mendoza Díaz, S., and Loarca Piña, G. 2014. Functional and technological potential of dehydrated *Phaseolus vulgaris* L. flours. *Food Chemistry*, 161:254–260.

Rengadu, D., Gerrano, A.S., and Mellem, J.J. 2020. Physicochemical and structural characterization of resistant starch isolated from *Vigna unguiculata*. *International Journal of Biological Macromolecules*, 147:268–275.

Repo Carrasco Valencia, R.A.M., and Serna, L.A. 2011. Quinoa (Chenopodium quinoa, Willd.) as a source of dietary fiber and other functional components. *Food Science and Technology*, 31:225–230.

Repo Carrasco Valencia, R.A.M., Peña, J., Kallio, H., and Salminen, S. 2009. Dietary fiber and other functional components in two varieties of crude and extruded kiwicha (*Amaranthus caudatus*). *Journal of Cereal Science*, 49(2):219–224.

Rizki, H., Kzaiber, F., Elharfi, M., Ennahli, S., and Hanine, H. 2015. Effects of roasting temperature and time on the physicochemical properties of sesame (*Sesamum indicum* L) seeds. *International Journal of Innovation and Applied Studies*, 11(1):148.

Saavedra, M.J., Aires, A., Dias, C., Almeida, J.A., De Vasconcelos, M.C.B.M., Santos, P., and Rosa, E.A. 2015. Evaluation of the potential of squash pumpkin by-products (seeds and shell) as sources of antioxidant and bioactive compounds. *Journal of Food Science and Technology*, 52(2):1008–1015.

Sacilik, K. 2007. Effect of drying methods on thin layer drying characteristics of hull-less seed pumpkin (*Cucurbita pepo* L.). *Journal of Food Engineering*, 79(1):23–30.

Serna-Saldivar, S.O. 2008. Peanuts, almonds and tree nuts. In: *Industrial Manufacture of Snack Foods*, ed. S.O. Serna-Saldivar, 125–148. London: Kennedys Publications Ltd.

Serna-Saldivar, S.O., and Sanchez-Hernandez, D. 2020. Dietary fiber in cereals, legumes, Pseudoceros and other seeds. In: *Science and Technology of Fibers in Food Systems*, eds. J. Welti-Chanes, S.O. Serna-Saldivar, O. Campanella, and V. Tejada Ortigoza. New York: Springer.

Shakpo, I.O., and Osundahunsi, O.F. 2016. Effect of some processing methods on the proximate, mineral, microbiological and sensory qualities of cowpea enriched maize snack (Ipekere Agbado). *Research Journal of Food Science and Nutrition*, 1(1):1–9.

Singh, K.K., Mridula, D., Rehal, J., and Barnwal, P. 2011. Flaxseed: A potential source of food, feed and fiber. *Critical Reviews in Food Science and Nutrition*, 51(3):210–222.

Skrabanja, V., Liljeberg, H.G., Hedley, C.L., Kreft, I., and Björck, I.M. 1999. Influence of genotype and processing on the in vitro rate of starch hydrolysis and resistant starch formation in peas (*Pisum sativum* L.). *Journal of Agricultural and Food Chemistry*, 47(5):2033–2039.

Smith, J., and Hardacre, A. 2011. Development of an extruded snack product from the legume *Vicia faba* minor. *Procedia Food Science*, 1:1573–1580.

Tobias-Espinoza, J.L., Amaya-Guerra, C.A., Quintero-Ramos, A., Perez-Carrillo, E., Núñez-González, M.A., Martínez-Bustos, F., Melendez-Pizarro, C.O., Baez-Gonzalez, J.G., and Ortega-Gutiérrez, J.A. 2019. Effects of the addition of flaxseed and amaranth on the physicochemical and functional properties of instant-extruded products. *Foods*, 8(6):183.

Tosh, S.M., and Yada, S. 2010. Dietary fibres in pulse seeds and fractions: Characterization, functional attributes, and applications. *Food Research International*, 43(2):450–460.

USDA. 2020. *Milled Peanuts. Inspection Instructions.* Marketing and Regulatory Programs, Agricultural Marketing Service, Specialty Crops Program, Specialty Crops Inspection Division, 1–153. Washington, DC: USDA.

USDA. 2021. Agricultural Research Service, Nutrient Data Laboratory. USDA national nutrient database for standard reference, Release 28. http://www.ars.usda.gov.

Wang, N., and Daun, J.K. 2004. The chemical composition and nutritive value of Canadian pulses. *Canadian Grain Commission Report*, 19–29.

Wang, N., Hatcher, D.W., Toews, R., and Gawalko, E.J. 2009. Influence of cooking and dehulling on nutritional composition of several varieties of lentils (*Lens culinaris*). *LWT – Food Science and Technology*, 42(4):842–848.

Willich, R.K., and Feuge, R.O. 1957. De-oiling of peanuts to yield a potentially useful food product. *Food Technology*, 11:332–336.

Wojtowicz, A., Kolasa, A., and Moscicki, L. 2013. Influence of buckwheat addition on physical properties, texture and sensory characteristics of extruded corn snacks. *Polish Journal of Food and Nutrition Sciences*, 63(4):239–244.

Woodroof, J.G. 1983. *Peanuts: Production, Processing and Products*, 3rd edition. Westport: AVI Publishing Co.

Wu, W., Huff, H.E., and Hsieh, F. 2007. Processing and properties of extruded flaxseed-corn puff. *Journal of Food Processing and Preservation*, 31(2):211–226.

Xu, B., and Chang, S.K. 2008. Total phenolics, phenolic acids, isoflavones, and anthocyanins and antioxidant properties of yellow and black soybeans as affected by thermal processing. *Journal of Agricultural and Food Chemistry*, 56(16):7165–7175.

Yoshida, H., Tomiyama, Y., Hirakawa, Y., and Mizushina, Y. 2006. Microwave roasting effects on the oxidative stability of oils and molecular species of triacylglycerols in the kernels of pumpkin (*Cucurbita* spp.) seeds. *Journal of Food Composition and Analysis*, 19(4):330–339.

Zapana, F., de Bruijn, J., Vidal, L., Melín, P., González, M.E., Cabrera, G., Williams, P., and Borquez, R. 2020. Physical, chemical and nutritional characteristics of puffed quinoa. *International Journal of Food Science and Technology*, 55(1):313–322.

14 Snacks from Potatoes

Carmen Hernandez-Brenes, Perla A. Ramos-Parra, and Sergio O. Serna-Saldivar

CONTENTS

DOI: 10.1201/9781003129066-14

14.1 INTRODUCTION

Potatoes (*Solanum tuberosum*) are among the most important crops globally due to their significant contribution to human health and high consumption volume. They are good sources of carbohydrates, vitamins, and minerals (USDA 2021). China is the main worldwide potato producer, followed by India, the Russian Federation, and the United States (FAOSTAT 2018). In the United States, the potato production was worth about 4,250 million dollars in 2019 (USDA 2020), of which 9%, 6%, and 25% of total potatoes were used in the manufacture of chips and shoestrings, dehydrated products (starch and flour), and frozen French fries, respectively. Potato "chips" or, as called in the United Kingdom, potato "crisps" are a trendy snack worldwide. In the United States, potato chips are the number one salty snack food with annual sales of around 69,273.1 million dollars accounting for 18.8% of the total dollars spent on global snack foods in 2020.

In 1853, chef George Crum invented prototypes of potato chips in a New York hotel named Saratoga Springs, where potato chips were made the house's specialty (Mottur 1989). However, the first formal potato chip industries did not start to flourish until the early decades of the twentieth century. In the United States, this market based on sales is dominated by large corporations such as Frito-Lay Inc., Kellog Co, Private Label, Cape Cod Potato Chip, and Utz Quality Foods, among others (Statista 2020). Nevertheless, many small businesses selling "home-style" products still prevail in the market.

Normally, traditional potato chips are produced by deep fat frying thin slices of raw potato, followed by seasoning, but there are many variations to this conventional process. For instance, unit operations may be introduced to produce potato chips with different shapes or crispier products with lower fat content.

Other than chips, potato-based snacks are also commercialized, for instance, products made from expanded pellets produced by extrusion. These types of snacks were discussed in greater detail in Chapter 12. There are, however, imitations of potato chips, which are called fabricated chips, which are becoming every day with more commercial relevance. These snacks are made from a potato dough of dehydrated or mashed potatoes as the main ingredients. Potato dough is mechanically formed into slices, which are then fried. The original potato chip deep fat frying method and the production process to obtain dough-based fried potato snack products will be described in more detail in the following sections.

This chapter reviews the structure of the raw potato and the optimum properties of the tuber for the production of regular potato chips and potato flour, flakes, and starch commonly used for the manufacturing of fabricated potato chips or as raw materials for the production of other snack foods.

14.2 STRUCTURE OF THE RAW POTATO TUBER

The potato (*Solanum tuberosum*) is an annual plant that is generally propagated asexually. The tuber starts forming post-anthesis from the tip of the rhizome. The tuber consists of the skin containing the outermost layer called the periderm and cortex. The inner part consists of the vascular ring, vascular parenchyma, and pith. The parenchyma is the storage tissue rich in starch granules. The tuber also contains many buds located in different places on the surface of the tuber. The tuber's main chemical component is starch stored in large elliptical-shaped granules, which is constituted by about 75% amylopectin and 25% amylose (Figure 14.1).

Table 14.1 depicts typical chemical compositions of the fresh potato and raw materials derived from this tuber. Potatoes have a high moisture and starch content, which together represent about

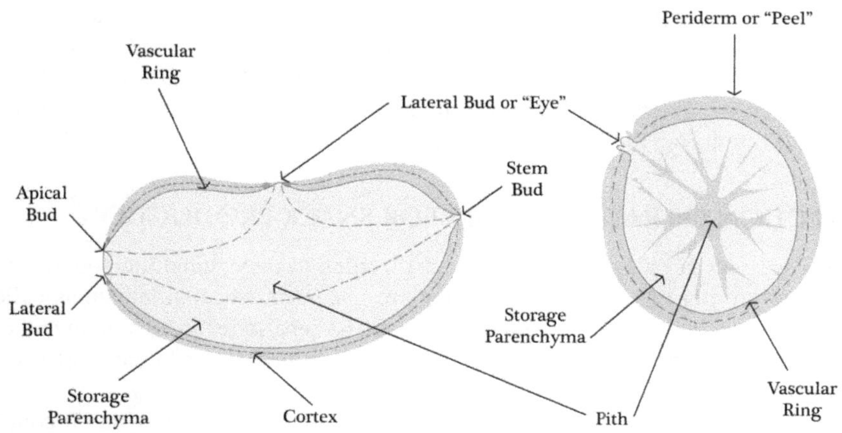

FIGURE 14.1 Structure of the raw potato tuber.

TABLE 14.1
Average Composition of Raw Potato (*Solanum tuberosum*) Tubers and Shelf-Stable Potato Flour and Flakes[a]

Nutrient per 100 g	Raw Potatoes[b]	Red Raw Potato	Potato Flour	Potato Flakes
Water, g	83.29	80.96	6.52	6.58
Energy, kcal	58	70	357	354
Protein, g	2.57	1.89	6.9	8.34
Tryptophan, g	0.021	0.019	0.115	0.988
Lysine, g	0.107	0.098	0.413	0.456
Aspartic acid, g	0.48	0.442	1.173	1.864
Methionine, g	0.032	0.029	0.107	0.456
Lipids, g	0.10	0.14	0.34	0.41
Saturated, g	0.026	0.035	0.09	0.169
Monounsaturated, g	0.002	0.003	0.008	0.009
Polyunsaturated, g	0.044	0.059	0.15	0.143
Carbohydrates, g	12.44	15.9	83.1	81.17
Dietary fiber, g	2.5	1.7	5.9	6.6
Minerals, mg				
Calcium	30	10	65	27
Iron	3.24	0.73	1.38	1.21
Sodium	10	19	55	77
Vitamins				
Thiamine, mg	0.021	0.081	0.228	0.988
Riboflavin, mg	0.038	0.031	0.051	0.11
Niacin, mg	1.033	1.149	3.507	6.264
Folate total, mcg	17	18	25	46
Vitamin A, IU	0	0	0	1
Vitamin C, mg	11.4	8.6	3.8	81

[a] Obtained from USDA-Agricultural Research Service (2019). [b]FoodData Central ID: 170032, raw potato; 170029, red raw potato; 168446, potato flour; 170445, potato flakes dry form.

94.6% of the whole tuber weight. Because of their high moisture content, potatoes are more susceptible to respiration and microbial deterioration, so they should be stored at low temperatures. The potatoes' optimum storage conditions should be 8–10°C with high relative humidity (95%) in a well-ventilated facility. The periderm plays an important role in preventing microbial deterioration and mainly develops during the first storage stage.

14.3 SELECTION OF RAW POTATOES FOR SNACK PRODUCTION

Potato chip processors or chippers find it very challenging to understand the potato tuber. Unlike carrots, potatoes are not considered roots. They are underground modified stems called tubers, making them very actively growing tissues that respond to light, moisture, and many other environmental conditions. The chemical composition of potatoes at harvest and their response to storage conditions are the main determinants of their suitability for snack processing.

Moreover, potato chips' characteristic crispiness depends on the potato starch content, which comprises about 65–80% of its dry weight (Storey and Davies 1992). A direct correlation between starch and dry matter content in potatoes is well documented in the literature. Thus, measurements of specific gravity and solid contents can be considered indirect estimations of starch content (Chapter 19) and are commonly used by industry to determine the processing application of a particular potato cultivar or a particular batch of tubers. Since the potato crop is very complex, the dry matter alone cannot determine a particular processing application, but the average tuber size can improve the decision. As a result, the optimum potatoes for chip manufacture should have a size from medium to large and specific gravities above 1.085, equivalent to approximately 22% of dry matter content. Chippers frequently use large potatoes to simplify peeling and obtain higher yields, yet small to medium potatoes may be used for small bags of chips of individual serving portions.

Potato tubers of relatively small size (100–200 g) and low specific gravity (1.070–1.080) tend to slough when boiled, maintaining their cell integrity. They are generally used to manufacture dehydrated potato flakes or mashed potato dough to manufacture fabricated potato chips or dough-based fried potato snacks.

14.3.1 Raw Potato Factors That Affect Potato Chip Yield

14.3.1.1 Potato Cultivars

The cultivar is the most important factor directly associated with the potato starch content, and consequently, it affects the potato chip yield. A second relevant factor is the production location since the crop can be greatly affected by environmental factors (Gould and Plimpton 1985). Worldwide there are more than 4,500 varieties of potatoes (Dourado et al. 2019). However, the favorite potatoes of consumers in the United States include the Russet (33%), Red (19%), White (15%), and Yukon (10%), among others (Statista 2020). Many cultivars are used to manufacture potato chips globally; popular varieties include Russet Burbank, Ranger Russet, and Atlantic, among others. Since the cultivar is the key to obtaining a successful product, large snack producers work closely with potato breeders. Their goal is to improve the specific gravity, reducing sugar content and storability of their raw potatoes. They also frequently protect their own cultivars by patenting them or keeping them as industrial secrets. Several tests are usually performed to determine if a specific cultivar is suited for potato chip processing. These include slicing, frying, moisture, yield, oil absorption, and defects (blisters, hash, and bulking), among others, which are described in the quality control section.

14.3.1.2 Relationship between Specific Gravity and Oil Absorption

The specific gravity of a particular potato batch directly relates to the yield of recovery of potato chips. High specific gravity tubers produce potato chips with lower oil content and higher production yields (Table 14.2).

TABLE 14.2
Influence of Specific Gravity on Fried Potato Chip Yield and Oil Content[a]

Specific Gravity	Yield of Fried Potato Chips (%)	Oil Content of Fried Potato Chips (%)
1.060	28.44	47.04
1.065	29.22	45.71
1.070	30.00	44.38
1.075	30.78	43.05
1.080	31.56	41.72
1.085	32.33	40.39
1.090	33.11	39.06
1.095	33.89	37.73
1.100	34.67	37.40
1.105	35.45	35.07
1.110	36.23	33.74

[a] Reproduced from Gould and Plimpton (1985).

In commercial operations, tubers with specific gravities above 1.085 produce chips with oil contents in the range of 35–40% and with acceptable processing yields for the finished product (28–36%). Although potatoes' specific gravity is frequently used to evaluate raw potato batch acceptability and possible performance for potato chip production, the specific gravities of raw potato can vary greatly within a single batch. The range of variation in the tubers translates into large variability in the potato chips. Since the chipping industry wants to deliver a consistent product, this is one of the most common concerns in its quality control. For instance, in a sample of 150 tubers from a commercial potato chip cultivar (Snowden), the specific gravities ranged from 1.065 to 1.095. A broader dispersion in smaller tubers' specific gravity was observed for the same sample compared to the range of specific gravities for medium-sized potatoes (Hernández-Brenes 2000). Similar observations were reported by Meredith (1988).

Extra-large tubers tend to have low-starch content due to their higher moisture content. To decrease the finished product variability, potato chip processors tend to select raw potatoes with lower standard deviation for specific gravity and starch. The specific gravity distribution for a certain chipping cultivar during the growing season is very consistent from one season to the next. As a result, this opens the potato chip processors' opportunity to increase potato chip products' consistency. Therefore, specific gravity distribution documentation can help determine the best harvesting times to minimize the defective product percentages (product falling outside the desired performance). For example, an immature potato lot, which may contain a wide range of starch content (large standard deviation and frequency distribution with multiple modes), would be more likely to produce a large percentage of defective (soft/oily) chips and only a small percentage with the desired quality. On the other hand, a batch of mature tubers with a single-mode for starch and dry matter would result in many chips with the desired quality specifications.

14.3.2 Impact of Raw Potato Composition on Potato Chip Color and Flavor

14.3.2.1 Reducing Sugars

The Maillard reaction during frying produces the desirable flavor and yellow color of potato chips, although some end-products can generate off-flavors that can reduce sensory acceptability (Shallenberger et al. 1959, Maga 1994). Glucose and fructose are the main reducing sugars in potatoes, so they are used as quality biomarkers for raw potatoes. Upon harvest, reducing sugars should be below an upper limit range of 2,875–3,300 ppm fresh weight (equivalent to 0.29–0.33% w/w); above these levels, potato slices will fry too dark for acceptable quality (Lisinska and Leszczynsky 1990, Harris 1992). Consequently, potatoes with more than 0.33–0.41% of reducing sugars (equivalent to 2% dry weight) are rejected for fried products, such as potato chips (Riaz 2004, Pedreshi et al. 2008, Álvarez-Torre and Canet-Parreño 2016). It is interesting to note that Maga (1973), using a blindfolded sensory panel, demonstrated that panelists preferred dark chips' odor and flavor. Panelists indicated that dark-colored chips had a more characteristic potato chip flavor. Nevertheless, consumers may be biased by the dark color of chips and rank them as less acceptable than the lighter ones.

14.3.2.2 Long-Term Storage and Reconditioning

Even though the potatoes reducing sugar content increases during storage, it can be high right after harvest. When fresh potatoes are not available regularly during late fall and winter, the potato chip and French fry industries in the United States and Europe use stored potatoes instead. Before long-term storage, the chipping industry usually cures potatoes after harvest at 15°C and 95% RH for one to two weeks. Then the tubers are stored for ten months at 10°C and 95% RH. Before processing, long-termed stored potatoes need to be reconditioned at 15–20°C for a maximum of six weeks. Reconditioning lowers reducing sugars but generally only reverts the levels formed during the long-term storage (Isherwood 1976). For this reason, reducing sugar levels at harvest need to be below 0.3% w/w. It has been documented that approximately 80% of the reducing sugars are converted into starch during reconditioning, and the remaining 20% levels are used in respiration.

It is worth mentioning that potatoes, which are stored at colder temperatures (<10°C), result in higher contents of reducing sugars due to starch degradation, which is also known as the "cold-sweetening" phenomenon (Coffin et al. 1987, Amrein et al. 2003, Datir et al. 2020) In economic terms, this phenomenon impacts both the chip manufacture and the potato producer (Isherwood 1976, Coffin et al. 1987, Pathirana et al. 2008). Storage of potato tubers at 3°C in air resulted in an accumulation of sucrose and hexoses and a significant decrease in the acceptability of potato chip color within two weeks. Storage of counterparts in modified atmospheres containing 1,000 ppm of ethylene and low oxygen in the air delayed the decline in color by six to eight weeks. It was then demonstrated that a low-oxygen atmosphere reduces the extent of sucrose accumulation and the rate of sucrose to hexose conversion (Parkin and Schwobe 1990).

14.3.2.3 Amino Acids

Although reducing sugar content is usually a limiting factor for the flavor and color development of potato chips (Marquez and Añon 1986), they are not the only potato components involved in Maillard reactions. Free amino acids are also involved in the Maillard reaction, and asparagine is the main one in most potato cultivars (Shepherd et al. 2010, Halford et al. 2012). Potatoes with high free amino acid content produce browner and bitter potato chips due to pyrazine formation during frying (Roe et al. 1991, Brierley et al. 1996, Hernández-Brenes 2000). During long-term storage, the free amino acid content of tubers can increase significantly due to protein breakdown (Fitzpatrick and Porter 1966, Brierley et al. 1996, Yang et al. 1999).

While the reducing sugars are affected by the reconditioning step, amino acids are not, so this process is not a viable option to reverse the effects of amino acids during long-term storage. By selecting cultivars with low free amino acid levels or reduced protease activity, potato chip producers can control free amino acid content and prevent release during long-term storage. Strict controls on nitrogen fertilization and selection/application of sprout inhibitors are also recommended.

14.3.2.4 Amino Acids and Acrylamide Formation during Frying

Acrylamide is a toxic compound that is classified as a "probable carcinogen to humans." It is generated during the manufacture of potato snacks as an end-product of the reaction between free amino acids and reducing sugars (Schouten et al. 2020) (Chapter 20). The European Commission Regulation (2017/2158) requires determining acrylamide contents in foods, such as fried potatoes (European Commission 2017). As a result, the food industry is, at present, being challenged with the development of strategies to reduce acrylamide levels in potato snacks.

Potential solutions to reduce acrylamide formation in potato snacks, including selecting specific cultivars and storage conditions, have been proposed (Amrein et al. 2003). In 2004, the US Department of Agriculture (USDA) approved the commercial planting of genetically modified potato crops to reduce acrylamide contents in French fries and potato chips. Other solutions have focused on the manufacturing process, such as pretreatments, with various ingredients applied to potato slices before frying (Genovese et al. 2019, Schouten et al. 2020). Some other strategies have focused on modifying processing unit operations and conditions (Alan-Sanders et al. 2003, Lawrance-Barry et al. 2011), further discussed in this chapter's processing section.

14.3.2.5 Other Potato Components

Other potato components such as phenolic and ascorbic acids have been reported to affect potato chip color (Rodriguez-Saona and Wrolstad 1997). Ascorbic acid can react with amino acids during frying. The autoxidation of phenolic compounds, which is favored by high temperatures and alkaline pH of the product, may also produce nonenzymatic browning (Cilliers and Singleton 1989).

14.3.3 Impact of Raw Potato Composition on Potato Chip Texture

The desirable texture of potato chips is generated during frying, mainly induced by starch gelatinization and water loss, which increase oil uptake and crispness. Raw potatoes' specific gravity is the most relevant factor to consider to attain the characteristic potato chip texture. Specific gravity is an indirect measurement of starch content and is directly related to oil uptake, as previously stated. Excessive oil uptake (>40%) results in defective soft/oily potato chips. However, raw potato cell wall composition, particularly in relationship with pectic substances, affects oil absorption and potato chip texture. Normally, long-term stored potatoes have elevated water-soluble pectic substances, which produces a softer texture and more oil absorption during frying (Sterling and Bettelheim 1955, Hernández-Brenes 2000). Infections by phytopathogenic fungi and bacteria could cause an increase in solubility characteristics of pectic substances throughout storage. Microorganisms secrete pectic enzymes such as endo-pectate lyase and endo-polygalacturonase that weaken the cell wall-middle lamella and cause cell separation (Van den Broek et al. 1997).

14.4 CONVENTIONAL POTATO CHIP MANUFACTURING PROCESS

The actual manufacturing process consists basically of frying thin slices of peeled potato in oil. The desirable crispy texture of potato chips is caused by a rapid moisture content reduction from 80% to 0.5–1.5% in about 2 min. The basic flow diagram for the original process contains nine processing operations: peeling, trimming, slicing, washing, draining, frying, inspection, seasoning, and packaging (Figure 14.2). Additional unit operations are sometimes incorporated in commercial processing facilities to reach a specific goal; for example, preheating treatments can improve the texture of stored tubers, or other steps like baking, microwave cooking-dehydration, and de-oiling are added to lower oil content and enhance texture. The following sections describe in detail the different unit operations for the manufacturing of fried potato chips.

14.4.1 Peeling

Potatoes are moved using hoppers and conveyors to remove any foreign material and be cleaned in flumes and brushed to eliminate adhered dirt before peeling. The most commonly used peeling methods are abrasive rolls, brushes, both with continuous rinsing and, in some facilities, lye or steam. Abrasive peelers consist of rotating cylinders or drums coated with carborundum, an abrasive material frequently used. By this peeling method, the tubers' weight is reduced by approximately 10–15%, which is mainly influenced by the potato variety and tuber maturity. When potatoes are newly harvested, their skin is fragile and can be removed easily, even by touch. Yet, after these potatoes are allowed to be set before

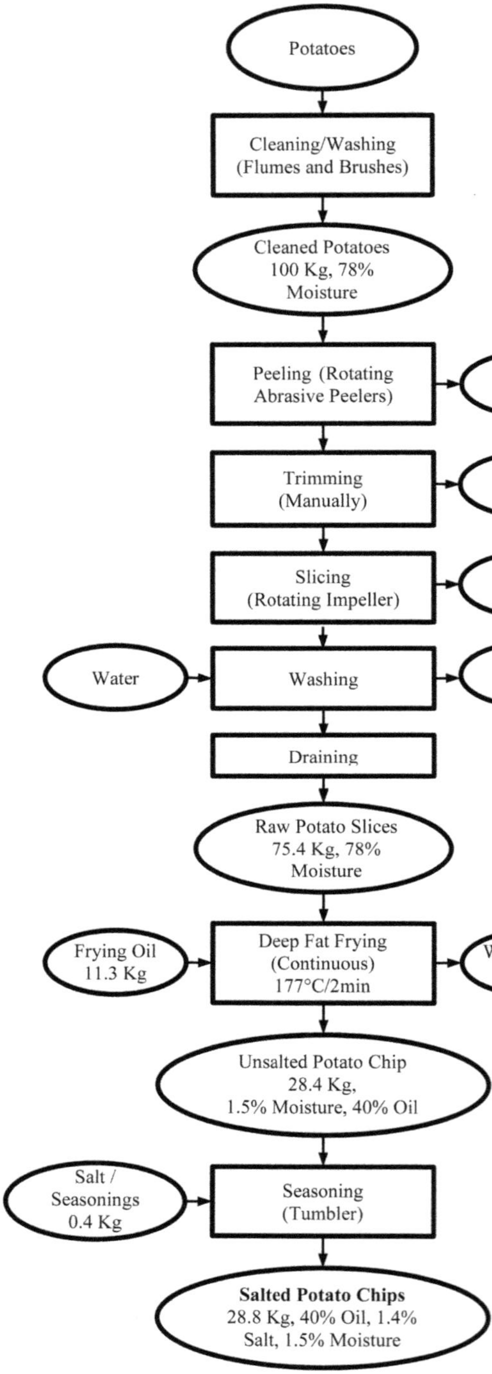

FIGURE 14.2 Flow diagram of the conventional deep fat frying process for regular chips.

long-term storage, their skin becomes thicker and produces higher losses during peeling. Steam is also commercially used to peel potatoes even though it has the disadvantage of partially cooking the surface of the potato, which affects the finished product's appearance.

14.4.2 Trimming

After peeling, defective tubers should be removed in conveyors, and damaged portions should be trimmed manually. The trimming losses vary significantly depending on the potato lot's condition, yet some authors have reported losses from 0.3% to 20% (Matz 1993, Mottur 1989). Postharvest diseases, pests, and partial sprouting during storage usually cause stored and mature potatoes to be more defective. When selecting a particular cultivar for potato chip processing, it is recommended to include tests to estimate yields after peeling and trimming. These simple tests are commonly performed during cultivar evaluation. They may also be used to adjust the degree of peeling and trimming necessary during regular production for a particular incoming potato lot.

14.4.3 Slicing

The slice thickness is essential to obtain the typical crispy potato chips created by the rapid moisture loss and crust formation during frying. The thickness for straight-cut slices usually varies in the range of 1.00–1.60 mm. Potato slices above this range are used to manufacture "kettle chips" and other thicker potato products (Patent US10548341B2), such as 2.54–3.05 mm for crinkle cut chips. The slices with a thickness of 1.85–1.91 mm tend to absorb slightly less oil and present a more rigid texture, although they require adjustments during frying to reduce the moisture content to <1.5% at the end of this process (Bernard 1985). Continuous or batch slicers can be used for this process. In the United States, it is common to see processing scale slicers from Urschel Laboratories Inc. (Valparaiso, Indiana), a company that specializes in slicing applications. Even though the equipment designs vary and are continually being modified, peeled potatoes are generally dropped onto a horizontally rotating impeller which moves them constantly through vertical blades fixed in a stationary ring. Disposable knives are used in newer equipment because they do not require sharpening and are inexpensive to replace.

On the other hand, older equipment requires constant adjusting and sharpening. In addition to blade sharpness, losses during the slicing operation are affected by the raw tuber's texture. Older or dehydrated potatoes have less firmness (lower cell-wall turgor pressure), resulting in flaccid slices that stick to the equipment and produce a large percentage of defects.

14.4.4 Slice Washing

When the unwashed slices are placed directly into the fryer, they tend to stick together, holding moisture and oil between them, for cell wall contents like starch and internal water are liberated during slicing. Therefore, it is crucial to wash potato slices to avoid this issue. When the product is not correctly cooked, potato chips will cluster together and have a soft texture, which may cause mold growth inside the package. Clustering can be avoided by washing the slices in mesh tumblers by immersion, high-pressure water sprayers, or both. Heating units can be included in washers to control the temperature of the water. The washing step produces the second most significant dry matter loss after peeling, which is estimated to be 10–20% of the initial potato weight, causing a significant economic impact.

14.4.5 Preheating

An optional unit operation for potato chip production is preheating, which consists of rinsing potato slices in warm water at a temperature below potato starch gelatinization (55–60°C). This operation is applied to lower reducing sugars and amino acids or to improve the color and texture of stored potatoes. The main reducing sugars in potatoes, glucose, and fructose are reduced below 0.25–0.3% because the warm water increases their solubility. Likewise, asparagine levels are also lowered to reduce acrylamide formation during the frying, up to 60% compared to unblanched potatoes (Haase et al. 2003, Pedreschi et al. 2007). Consequently, these changes prevent browning defects in potato chips.

The mechanism for texture enhancement implicates stimulating the endogenous potato enzyme pectin methylesterase (PME), which activates at 50–70°C. PME reduces the pectin degree of esterification and creates binding sites for cationic metals, strengthening the cell wall-middle lamella complex. Preheating is also thought to cause the release of starch-bound calcium in potatoes, making it available for pectin binding (Bartolome and Hoff 1972). Moreover, slice preheating treatments can also be combined with cell wall strengthening agents that decrease surface blisters, reduce fat content, and improve texture (Schroeder et al. 1981, Pedreschi et al. 2007, Trejo-Copado 2009). For instance, calcium salts (up to 0.04 M of Ca^{+2}) or acacia gum solution (at least about 3% by weight), as claimed by Patents 4,272,554 and US20090169710A1, respectively, can be added in the warm water to improve the preheating unit operation.

Preheating treatments can be beneficial to improve the texture of soft long-term stored tubers and may be applied to whole potatoes or slices. Furthermore, this treatment with whole tubers is more challenging to control since the variations in size increase the time to achieve internal temperatures of 55–60°C (Hernández-Brenes 2000). Nevertheless, it produces less dry matter losses and higher slicing yields.

14.4.6 Slice Draining

The excess washing water must be eliminated to reduce oil uptake during frying. Some washing equipment is designed as two-stage processes. The first stage is to wash the product in rotating drums equipped with water sprayers. The second stage is to remove excess water through continuous motion, which may include air injection. Other draining procedures include a vibrating mesh conveyor belt with vacuum suction applied from below and a compressed air curtain on top to facilitate water removal: a step also called "air knife sweeping." In addition, centrifugal devices that are very effective may include hot air injection to dry the slice surface and reduce frying time and oil uptake.

14.4.7 Frying

After the previous processing operations, raw potato slices that are either peeled, unpeeled, washed, and/or unwashed are ready for deep fat frying. The loss of moisture from 78% to 0.5–1.5% in approximately 2 min is essential to produce a crispy potato chip (Figure 14.2). This crispiness can be obtained if the drop in the temperature when the cold product enters the fryer can be preserved below 5°C. The frying mediums are frequently fats and oils from both animal and vegetable sources, and the average initial temperature for frying oil should be at least 177°C. The frying medium can be hydrogenated or nonhydrogenated. It may include one or a mixture of oils, such as corn, olive, soybean, palm, cottonseed, canola, rapeseed, sunflower, and peanut oils, among others (Chapter 7). The snack industry has been desiring to eliminate *trans*-fats from their products as a health-related marketing strategy due to changes in food labeling that includes the declaration of *trans*-fatty acid content (Chapter 20). Moreover, synthetic triglycerides may also be used as frying media like low-calorie or zero-calorie fat substitutes. For instance, sucrose polyester Olean® from Procter & Gamble Co. can be used to eliminate the fat content of the finished product, or it can also be used blended with other oils to reduce fat content (Chapter 7).

The current batch fryers use a modern and automatic system to produce kettle-style potato chips, characterized by generating a hard, crunchy texture and making 163 kg/h of the finished chip. Batch deep-fat fryers typically exhibit a U-shape temperature profile starting above 160–177°C and dropping very fast to 150–167°C, for when the product enters the fryer, it remains at a lower temperature until the oil can recover its initial temperature by the end of the frying cycle. The deficiency of batch operations to control temperature presents a temperature profile that is ideal for producing home-style potato chips. Unfortunately, controlling the temperature is critical for the characteristic crispiness due to temperature drop (ΔT). In batch operations, the ΔT can be affected by the amount of product that enters the fryer, introduced with buckets without weighting it appropriately in small businesses.

Large potato chip companies use a batch fryer with a system that continuously compensates the carried oil level into the finished product. On the other hand, the rudimentary batch frying equipment makes it difficult to conserve the oil levels as it absorbs it continually during this process. All these variations can affect the consistency of the final product, especially in moisture and oil content. In the quality control section (Chapter 19), these two factors are described as critical for assessing potato chip quality and are monitored regularly.

Large companies with more outstanding production tend to use continuous fryers that can have the capacity to produce 2,948 kg/h potato chips. The product is placed in small net cages transported through the oil tank by a conveyor belt, and the oil is heated in a tube-type heat exchanger before being pumped into the fryer. Paddles slow down the product to achieve the correct residence time while it moves through the tank. Large continuous fryers require various pieces of equipment, which include a tank containing the frying oil. This equipment has an exhausting hood under the oil tank to eliminate vapors and a heat exchanger with a thermostat. The best quality fryers contain an oil filtering system for recycling and, therefore, extend the fried potatoes' shelf life. For example, a potato chip fryer, manufactured by Heat and Control Inc., has an excellent quality system since the complete oil volume is passed through a filter to remove particles every minute. After the product is fried, it is transported into a mesh conveyor to drain the excess oil. Potato chip de-oiling can also be included as optional unit operations, which are described in the following section.

Furthermore, the temperature and quality of the oil affect the efficiency of this process. Frying oil quality is measured with the levels of free fatty acids to indicate the degree of triglyceride hydrolysis and total polar component (TPC) to determine nonvolatile compounds generated during frying (Asokapandian et al. 2019) (Chapter 3). Free fatty acid levels should not be above 1% since it may increase lipid oxidation reactions in the final product, breaking down to low-molecular-weight acids, aldehydes, and ketones responsible for rancid aromas and off-flavors in potato chips. At the same time, legislation in some Asian and European countries does not allow vegetable oil with TPC higher than 25%, which is produced during the reuse of frying oil (Takeoka et al. 1997, FSSAI 2011). Moreover, aldehyde hexanal can also be used as a measurement of rancidity in a potato chip. Chapter 19 describes quality control methods for potato chips and related products.

14.4.8 Low-Fat Potato Chips Process

The demand for healthy foods has recently become a worldwide trend. Thus, the snack industry is continually challenged to produce alternatives to consumers, as "low and light" potato chip products. Low-fat potato chips can be produced with many methods. Nevertheless, the oil content of potato chips depends on the raw potato's initial solid content (Table 14.2). Covering potato chips with aqueous solutions of high-amylose starches, gelatins, pectins, or even treating the surfaces with alpha-amylases are known methods that can reduce fat content. As previously mentioned, oils and synthetic fats can be combined to fry raw potato slices and produce a low-fat final product (20%). Some methods depict modifications of the conventional potato chip process to reduce fat content. These methods may include operation units like reduced pressure fried, kettle frying, baking, microwave vacuum drying (Lemos-Monteiro et al. 2020), and/or high-temperature air-drying as pretreatment before frying. Other potato chip processing facilities have adopted automatic centrifuge after frying to reduce oil content to 22% or less in fried potatoes, considering their specific gravity (Table 14.2). Following this process, the final steps may include seasoning and packaging.

14.4.8.1 De-Oiling/Drying

Potato chip producers frequently use additional unit operations to reduce oil content like the ones described in Figure 14.3. There are several patents for potato chip de-oiling (Peters 1966, Bernard 1985, Neel et al. 1990, Andersson and Martensson 2006, Andersoon and Grigonis 2008). Some of these can be used freely since the time limits for exclusive industrial rights already expired, but others require licenses since the assignees still own them.

The optional de-oiling steps consist of deep-fat frying at low temperatures (150–167°C) and drying. Frying potato chips at these temperatures reduces the moisture content to 10% or lower. After frying, hot chips are dried with heated air, superheated steam, or inert gas to reduce the moisture content to approximately 2%. US Patent 4,537,786 states that higher oil removal can be achieved when the hot air temperatures are within the range of 143–154°C at a pressure of 7.1 mm of water. For the hot air blast, potato chips should be ideally placed in a porous conveyor belt where the hot air flows from above with a suction underneath at a negative pressure of 40.6 mm of water (Bernard 1985). This unit operation makes it possible to lower oil content from 40% to 25% or less than the traditional method for the composition and processing conditions established in Figure 14.3.

Regardless, a patent assigned to Frito-Lay Inc. declares to reduce oil content even further to 18–24% (Neel et al. 1990), which consists of applying superheated steam at 149–166°C and impinging velocities of 549–640 m/min under a reduced oxygen environment (<14% O_2) to partially fry slices reaching moisture levels of 5–8%. Then, the potato chips are dehydrated until the moisture level is around 0.8–2% as an additional process.

On the other hand, a patent assigned to Kraft Foods R&D Inc. claims the production of potato chips with fat and moisture content of less than 19% and 5% by weight, respectively (Andersson and Martensson 2006). First, the invention included a blanching step with an aqueous solution and 0.05–2% divalent cations by weight at a temperature of 40–100°C. Then, the potato slices were entirely or partially submerged in a frying medium, either hot oil or fat, not higher than 195°C. Later, the oil-removing step is with a stream of superheated steam, saturated steam, and hot gas such as air or nitrogen at a temperature higher than 110°C with ambient pressure. The commercial low-fat oil-stripping system contains high-volume fans to circulate the stripping atmosphere and steam to reduce atmosphere oxygen. The steam itself or a thermal fluid heat exchanger raises the temperature of the atmosphere inside. As the product is conveyed through the oil-stripper, the atmosphere flows through the product pack (2.5–15 cm bed depth) to remove surface oil, and the stripped oil is recycled back to the fryer.

14.4.8.2 Vacuum Frying

Additionally, the vacuum frying process, discussed in Chapter 3, performed at around 3.115 kPa and 144°C, represents an alternative method to obtain a potato chip with lower oil content and attractive sensory attributes, especially in terms of color and texture (Garayo and Moreira 2002).

14.4.8.3 Low Surface Oil Potato Chip

Lemos-Monteiro et al. (2020) and the international patent WO2012104213A9 of Bailey et al. (212), assigned to Frito Lay Inc., claimed the novelty of a product with low surface oil potato chip produced by microwave cooking-dehydration steps. According to the authors, the process generates a product with equal or superior organoleptic properties (flavor and microstructure) and shelf life compared to a conventional fried potato chip (Figure 14.4). Another method for producing low oil content fried potatoes claims a first vacuum drying step with the integration of microwave vacuum frying and de-oiling under vacuum (Zhang et al. 2015).

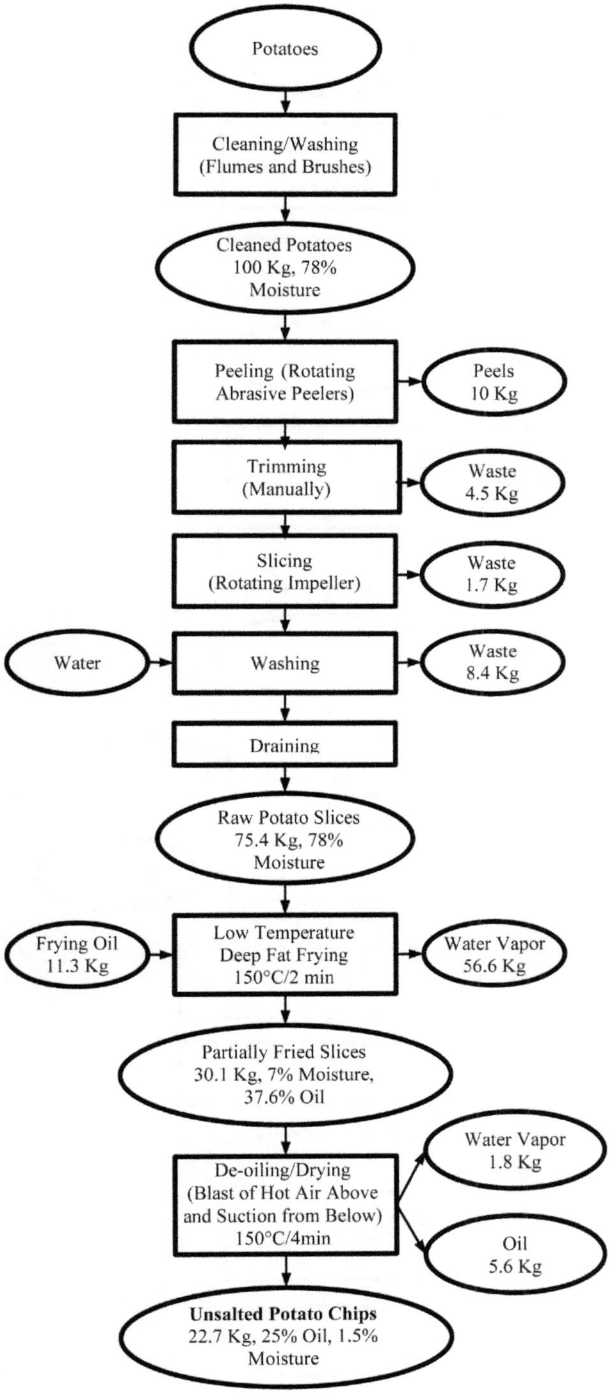

FIGURE 14.3 Flow diagram of the conventional process for lower oil potato chips.

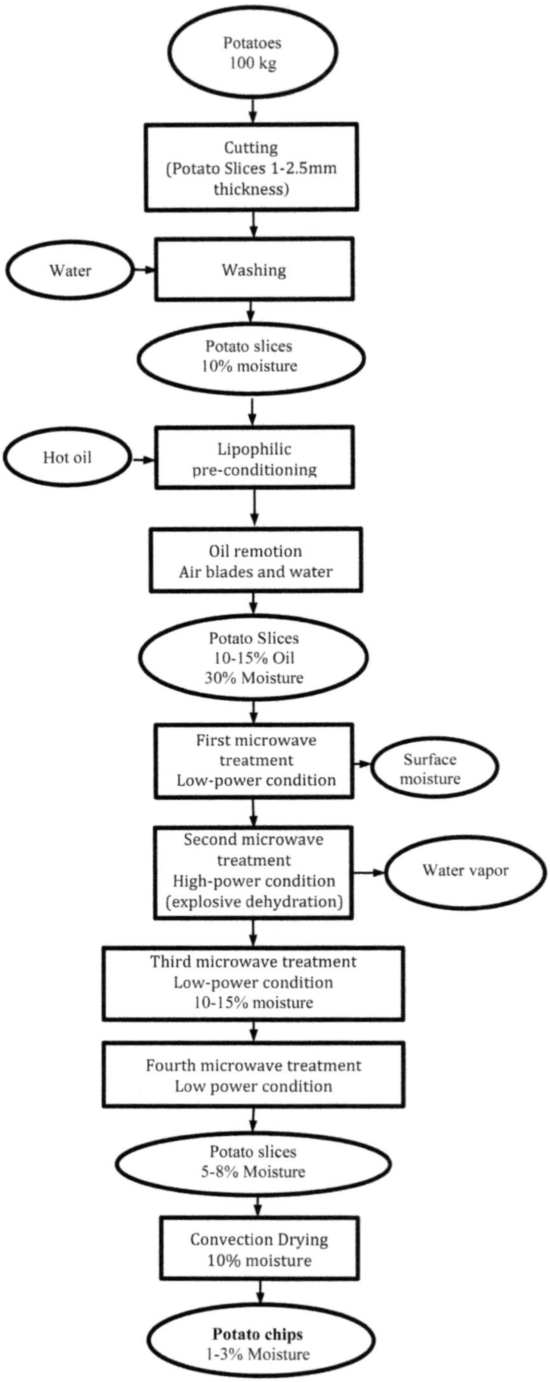

FIGURE 14.4 Flow diagram of the low-fat potato chip process using microwave treatments (adapted from WO 2012/104213A9 International Patent).

14.4.8.4 Kettle-Cooked

The Kettle® potato chip manufactured by Kettle Foods Limited, UK, has an oil content of 11.8 wt.%, which is baked rather than fried, so it has a lower oil content than conventional fried potato chips (30–35 wt.%). Typically, kettle-style potato chips imply batch processes, but the US Patent WO2016048338A1, assigned to Frito-Lay North America Inc., claims a continuous fryer with no flume to manufacture kettle-style potato chips without mimicking the traditional U-shaped temperature-time profile used in the conventional process.

14.4.9 Inspection

Operators visually inspect potato chips manufactured on a small scale to remove defective ones manually. On the other hand, processing facilities with large production use machine vision equipment or color sorters, which need to be calibrated to identify the most frequent defects in potato chips to remove them from the line. These defects may be potato chips with excessive browner color or dark spots, which may be produced when a batch of tubers have very high reducing sugar content, or green areas that can be a health hazard because of the presence of toxic glycoalkaloids solanine and chalconine. Even though glycoalkaloids are colorless, it turns green when the tubers are exposed to sunlight due to chlorophyll synthesis. Growing and postharvest disorders are also considered defects in potato chip production. An example of this is the hollow heart, which is observed as black spots in the chips.

Air pockets, blisters, and pillows are not classified as common defects, but they can reduce consumer acceptance or cause the product to break. Blisters or ruptures of tissue are formed by the pressure that is built within the slices. When the amount of steam produced inside is more significant than the permeability of the overlying layers of cells, pillows will most likely be created. Gould and Plimpton (1985) recommended recording the number and size of blisters for each potato lot after its frying test.

14.4.10 Seasoning

Before the product is packed, they are usually salted or seasoned. Seasoning is frequently done while the potato chips are hot since it allows the salt and seasoning to penetrate the chip's surface. Ordinarily, this process's commercial equipment comprises salt and/or seasoning being sprinkled over the chips in a conveyor belt or tumbled with the chips in a rotating drum (Chapter 3). Frequently, very fine crystals of salt are added in a proportion of 1–2% of the final product weight. Seasonings include spice mixtures combined with other ingredients. There are many flavored potato chips, such as barbecue, cheese, sour cream and onion, and vinegar, among others. Flavors are usually added in a proportion of 5–10% of the final product weight.

14.4.11 Packaging

Finally, potato chips are packaged after seasoning. The first versions of potato chips packaged using cellophane or wax paper bags have a shelf life of one to three weeks. Many new packages have been developed throughout the years, giving this product more extended shelf life (Chapter 4).

Packages for potato chips are currently made of complex laminates of synthetic polymeric materials such as metalized polypropylene manufactured by the aspersion of aluminum to a polypropylene ply, which is the material most frequently used. Each material has a specific function. For example, polypropylene provides mechanical strength, good sealing, and partial moisture barrier properties. On the other hand, metallization generates low-oxygen permeability, additional moisture barrier properties, and protection against light, which catalyzes oxidative reactions (Rojas de Gante 1997). These package materials prolong the shelf life of the potato chips to approximately two to three months, but it varies depending on the different factors, as packaging characteristics and frying oil quality (Chapter 7).

The potato chip package should have an excellent barrier of moisture water vapor that allows less than 3% moisture absorbance; for potato chips with a 4% increase in moisture, being 1.5% the moisture in which they are packaged, usually become soft and stale. Small bags of chips require a better moisture barrier, which makes them more expensive. For instance, a small bag of 21 g needs more packaging material with a moisture vapor transmission four times higher than the one required for a large bag of 213 g (Hirsch 1991).

Oxygen plays an essential role in potato chip packaging since UV light reacts with unsaturated fats producing peroxides, precursors to rancidity. For this reason, potato chip packages must have very low oxygen permeability. When the package has good oxygen barrier properties, a common way to increase shelf life for potato chips is to reduce oxygen headspace to less than 1% by introducing nitrogen gas to the package (Robertson 2012). Moreover, this may prevent the breakage of potato chips during transportation and handling as the package is inflated by the injected inner gas (Chapter 4).

In the last decades, there has been a trend toward using bioplastics as food packaging material instead of plastics, such as high-density polyethylene, low-density polyethylene (LDPE), polyethylene terephthalate, and polystyrene (PS). For example, large companies such as Pepsico's Frito-Lay have developed applications of polylactic acid (PLA) bags for packaging potato chips, which is a biopolymer widely used for the food industry (Jabeen et al. 2015).

14.5 RAW MATERIALS DERIVED FROM POTATOES

Besides the use of the raw potato tuber, the snack food industry demands large amounts of shelf-stable flours, flakes, and native and modified starches. These raw materials are commonly used for the production of second-generation potato chips and sticks and third-generation extruded snacks (Chapter 12). The main advantage of these products is convenience, shelf life, and functionality, especially water absorption and retention. Also, processors do not worry about the procurement of raw potato tubers and storage. There are two major categories of shelf-stable potato products: flour/flakes and refined starch obtained after wet milling. The potato starch is highly valued due to its unique viscoamylograph properties, especially in terms of paste viscosity.

14.5.1 Manufacturing of Potato Flour and Flakes

Raw and pre-gelatinized potato flours and flakes are frequently used to manufacture potato-based snacks or improve the properties of other snack items. These products are mainly comprised of starch followed by protein. Table 14.1 compares the typical composition of raw potato with potato flour and flakes.

The fundamental manufacturing process of potato flours applies the same principles of old traditional procedures. It consists of washing raw potatoes, followed by peeling, cooking, and mashing before drying, generally on a single-drum drier equipped with rolls. The resulting dehydrated potato is then ground into flours with the desired particle size (Willard and Hix 1987). There are two categories of potato flours: raw or low-cooked and fully cooked or gelatinized flours. Both types are processed following similar procedures, with the exception that gelatinized flours are normally pressure-cooked before dehydration (Figure 14.5). Pressure cooking enhances starch gelatinization, so flours readily absorb water and rehydrate. Gelatinized flours are regularly used in the formulation to produce half-products or pellets (Chapter 12). The raw or less cooked counterparts are preferred for direct expansion processes because they enhance radial expansion. About 16–17% flour is obtained from raw potatoes, which normally contain only 20% solids (Serna-Saldivar 2008).

Yadav et al. (2006) researched the influence of drying conditions on functional properties of potato flour assessing, solubility, swelling, consistency, viscoamylograph pasting behavior, and sedimentation volume of flours obtained with two different drying procedures. The functional properties of flours dehydrated in drum drying were more degraded or gelatinized compared to

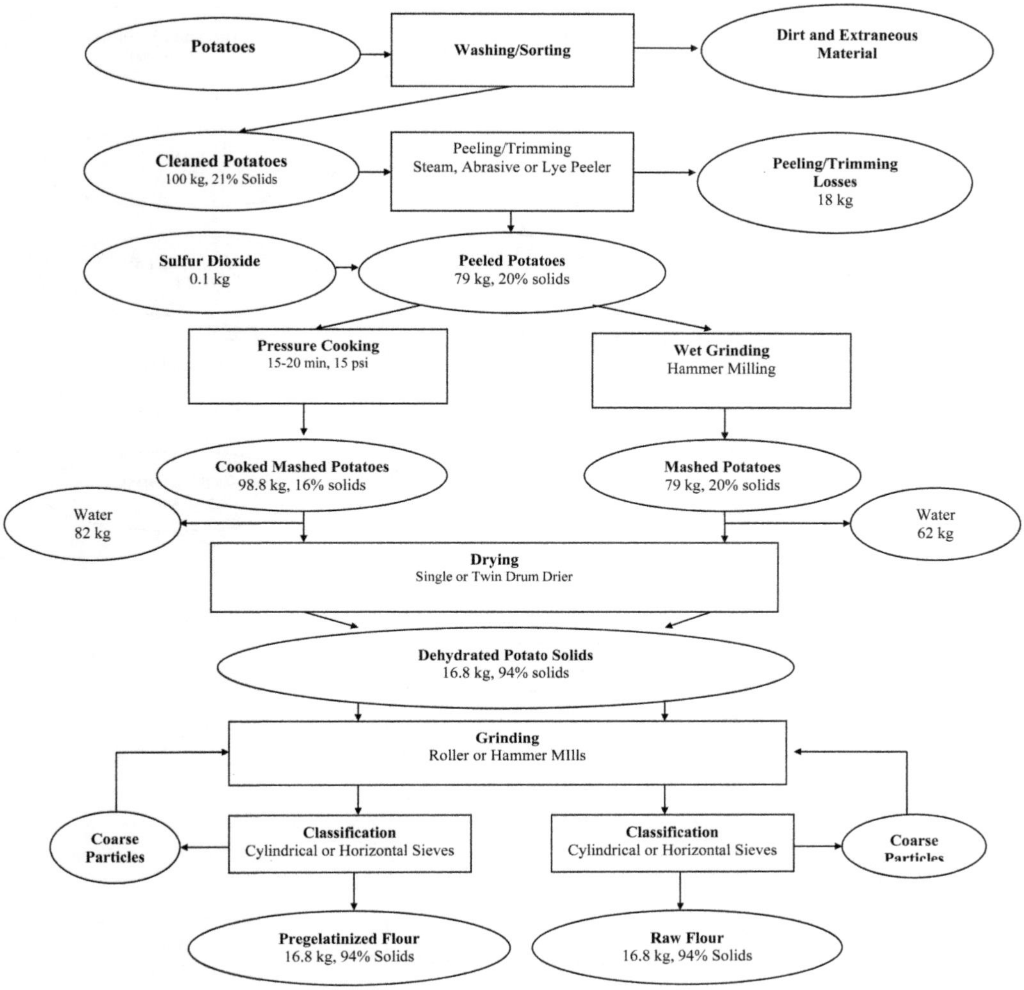

FIGURE 14.5 Diagram of the dry-milling processes for the production of raw or pre-gelatinized potato flours (adapted from Serna-Saldivar 2008).

counterparts dehydrated with hot air. Interestingly, the soluble amylose of drum-dried flours almost doubled as a result of this dehydration process. Moreover, the crystallinity index of drum-dried flour was much lower, indicating the loss of granular surface and internal arrangement of the starch granules.

On the other hand, the typical process to obtain potato flakes consists of the sequential steps of peeling, slicing, cooking, cooling, mashing, drying, and grinding to achieve the desired flake size (Figure 14.6, Serna-Saldivar 2008). The major step is cooking commonly performed with steam to gelatinize the starch. The cooked potato mash is treated with sodium bisulfite to improve color or retard nonenzymatic browning and with other approved additives like emulsifiers (monoglycerides, sodium stearoyl-2 lactylate), synthetic antioxidants (BHA and BHT), and chelating agents (citric acid) to improve texture and shelf life. Sodium acid pyrophosphate is also utilized in order to prevent discoloration. Potato flakes are similar to gelatinized potato flours, although they differ in their particle size and are usually mixed with these food additives. From the functional viewpoint, the potato flakes readily rehydrate when exposed to water and form cohesive doughs. Flakes used for fabricated potato chips are normally ground to a finer size than conventional counterparts. The

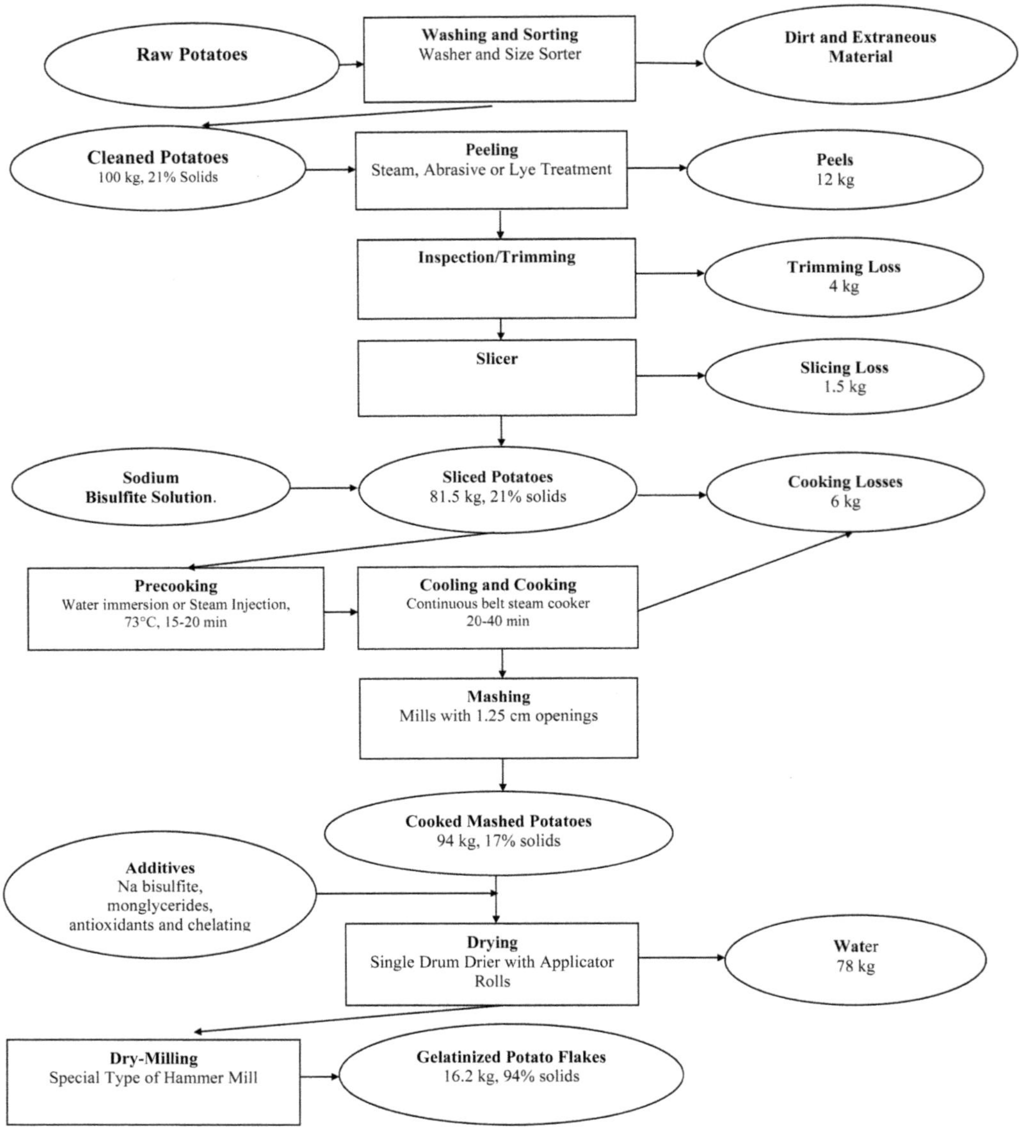

FIGURE 14.6 Diagram of the process for the production of pre-gelatinized potato flakes (adapted from Serna-Saldivar 2008).

viscoamylograph peak or maximum paste viscosity is relevant to raw flours and pre-gelatinized flours/flakes because it closely relates to functionality and water absorption (Serna-Saldivar 2008).

Martinez-Serna Villagran et al. (2000) patented the dehydration of potato flakes for use in dough compositions used to make fabricated products. The dehydrated flakes containing 16–27% amylose yielded more cohesive, nonadhesive, and machinable doughs with water absorption indexes ranging from 6.7 to about 9.5 g water/g dry flakes.

Reyniers et al. (2018) indicated that the main advantage of pre-gelatinized potato flakes is that they readily develop viscosity upon hydration, even with cold water. The same authors researched the use of cellulases to degrade fibrous-rich cell walls with the aim of improving functional starch properties. The enzyme-treated flakes enhanced peak viscosity values up to 75%.

Recently, Zalpouri et al. (2020) compared physicochemical parameters of dried potato flakes obtained by refractive and convective drying techniques. The processes were optimized using response surface methodology varying drying temperature and blanching times. The optimized dried potato flakes obtained with the connective drying with high protein and low soluble sugars were obtained at a temperature of 53.6°C and a blanching time of 4.98 min. For refractive drying, the optimized values for drying temperature and blanching time were 71.2°C and 5 min, respectively. The authors finally concluded that flakes obtained by refraction-based drying had higher nutritive and visible quality attributes and, therefore, better potential for the food industry.

14.5.2 Wet Milling for the Production of Refined Potato Starch

Refined potato starch has unique viscoamylograph properties and is highly demanded by the food industry, especially for the production of third-generation snacks and fabricated potato chips and sticks. The process to obtain starch starts by selecting the appropriate potatoes, which composition varies according to variety, maturation, area of growth, cultural practices, and storage conditions. The best tubers should contain as much starch as possible. Thus, in modern potato starch factories in Europe, only special varieties and types are used. Culled potatoes are not recommended because they are more difficult to process because of their comparatively lower solid content (glassy potatoes) and relatively small starch granules (Grommers and van der Krogt 2009).

Most fresh potatoes contain 18–22% solids, of which 13.5–16.5% are comprised of starch. According to Grommers and van der Krogt (2009), the solids of potatoes contain an average of 76.6% starch with values ranging from 65% to 85%.

The wet-milling potato starch process starts when dirt and foreign impurities are removed, and then tubers are washed in a trough with paddles or tumblers (Figure 14.7). After cleaning, the potatoes are milled with either rotary saw blade (European technology) or special vertical hammer mills enclosed in a screen that is generally employed in the United States. The macerated potatoes are impacted against the screen, and the pulp is forced through the holes. In the first stage of grinding, the screen has 4.8–6.4 mm orifices. Sulfur dioxide is normally added to inhibit the action of oxidative enzymes such as polyphenol oxidase that might discolor the starch. The resulting macerated potato pulp is further screened in rotary sieves positioned in a battery to separate first the coarse fiber (US mesh 80–120) from the finer material (US mesh 120–150). During this operation, most of the starch granules are liberated from the pulp that is further ground in hammer mills equipped with 1.5-mm screens to finish releasing the longitudinal-shaped starch granules. The starch-rich stock from the screens and mills is diluted with water and concentrated and purified in continuous centrifuges operating in series. The resulting semi purified starch slurry is diluted with water and reconcentrated in hydro cyclones/batch centrifuges that further remove impurities. In this last operation, any remaining impurities such as fine fiber are concentrated on the top of the formed cake and scraped off with a blade. The final clean starch is washed again in the presence of sulfur dioxide, dewatered using a vacuum filter to about 60–65% solids, and flash dried at a temperature of about 48°C (Figure 14.7). Industrial grade potato starch is generally dried to moisture below 17% and will contain about 0.35% minerals and traces of protein and sugars (Mitch 1984, Serna-Saldivar 2008). The yield of the refined potato starch from raw tubers is about 12–13% with 14% moisture.

Bergthaller (2004) describes native potato starch as more refined in comparison to competing cereal counterparts (Chapter 6). The high level of purity, especially in terms of lipids (<0.2%), grants potato starch a neutral taste and makes it less prone to develop off-flavors under prolonged storage conditions. The protein level of refined potato starch (0.05–0.2%) is also lower compared to other commercial starches.

14.5.2.1 Properties of Native and Modified Potato Starches

The potato starch possesses unique functional properties, especially in terms of paste viscosity. Among commercial starches, the oval-shaped potato starch granules produce the largest sizes that

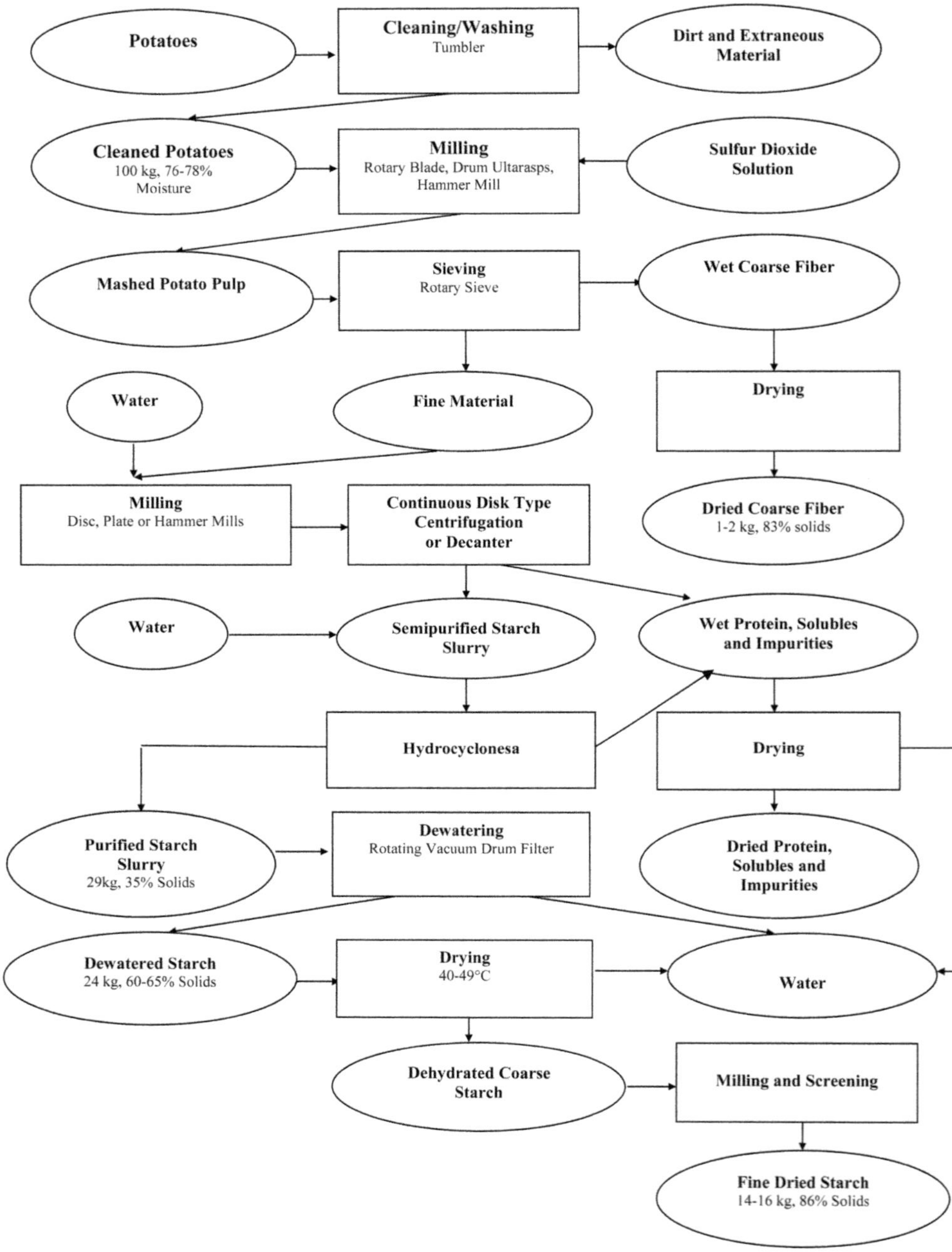

FIGURE 14.7 Diagram of the wet milling process for the production of potato starch (adapted from Serna-Saldivar 2008 and Grommers and van der Krogt 2009).

range from 5 to 100 μm. The regular starch granules are comprised of 21–25% linear amylose (AMY), and 75–79% branched amylopectin (AMP). The molecular size distribution of AMY is 0.2–3.9 × 10^6, whereas the degree of polymerization ranges between 840 and 21,800. For potato AMP, its molecular weight is 60.9 × 10^6 with a degree of polymerization subdivided into three groups containing 16,100, 5,500, and 2,100 units. These starch macromolecules are organized in amorphous and semicrystalline granular rings, commonly known as growth rings, which are 100–400 nm thick. The physical state of the AMY component is amorphous, and therefore it is found in the amorphous parts of the potato starch granules, although there is some cross-linking between AMY and AMP (Bertoft and Blennow 2016, Grommers and van der Krogt 2009).

Concerning functionality in food systems, potato starch is well-known for providing high viscosity and forming pastes and gels of long consistency. This behavior of pastes and gels is combined with high clarity (Mitch 1984, Bergthaller 2004). However, the prolonged heat treatment of potato starch disrupts granules increasing shear-thinning, so regular potato starch is considered very heat sensitive.

The potato starch has distinctive properties, especially in terms of the swelling power of the granules that yield high peak viscosity values and a low glass transition temperature and paste clarity. According to Singh et al. (2016), the DSC thermal properties of potato starch are onset temperature T_o = 59.7–60.7°C, peak temperature T_p = 63.3–64.6°C, final temperature T_c = 67.3–70.3°C, and enthalpy of gelatinization ΔH gel = 12.6–17.9 J/g.

More than 500 modifications of potato starch are currently known. However, three categories are commercially relevant: substituted, converted, and cross-linked. Substituted potato starches are esters and ethers forms aimed to prevent the formation of ordered structures in pastes and retard retrogradation. The main commercial esters are starch phosphate and acetate, whereas the major ethers are hydroxyalkyl, carboxymethyl, and cationic starches. Converted starches, also known as thin-boiling, are produced by the degradation of the starch chains into small segments. As a result, they can be cooked at higher concentrations compared to native starches. There are four subgroups of commercially converted starches: dextrins, acid-modified, oxidized, and enzyme-hydrolyzed. Dextrins are mainly used for the formulation of seasonings (Chapter 8), widely utilized to produce savory snacks, whereas acid-modified starch for the production of candies (Grommers and van der Krogt 2009).

According to Singh et al. (2016), the reagents used to oxidize potato starch include periodate, chromic acid, permanganate, nitrogen dioxide, and sodium hypochlorite. The main factors controlling the degree of oxidation are reagent concentration, pH, and temperature. During oxidation, hydroxyl groups at the C-2, C-3, and C-6 positions are first oxidized to carbonyl and then to carboxyl groups. Oxidized and bleached starches are used in foods as coating and sealing agents, as emulsifiers or dough conditioners for bread-making, and as agents to improve the adhesion of batter and breading mixes in fried foods.

Commercial cross-linked starches are used to reduce calories in snacks and as a new source of dietary fiber with prebiotic properties (Singh et al. 2016). The high level of cross-linking diminishes their digestion in the gastrointestinal tract, and the undigested fractions, when they reach the hindgut, they stimulate the growth of probiotic bacteria.

14.6 MANUFACTURING PROCESSES FOR FABRICATED POTATO SNACKS

Even though fabricated and traditional potato chips tend to have the same fat content, around 38–40% (Table 14.3), depending on the conditions, consumers consider that fabricated chips appear healthier. Due to this, Pringles®' stackable chips from Procter and Gamble Co. became more popular in Europe in the late 1990s. Chips produced from dough are made from potato flakes, flour, starches, granules, or mixtures. Baked Lays® chips baked with dehydrated potato, not fried, have 65% less fat than conventional potato chips (Bradshaw and Ramsay 2016). Frito-Lay snacks, part of PepsiCo, manufactures the product, including Baked Ruffles with 9.5% market share of the leading potato chip brands based on sales in the United States in 2017, similar to the market share from Pringles® (9.6%) by Procter and Gamble Co. (Statista 2017).

Pringles' success might be because some consumers favor tastier snacks in place of the healthier ones. Although low-fat baked potato snacks prevail, several healthy products have been removed from the market since they were perceived as tasteless. The products are usually aimed to be bought by women, in some countries, for instance surveys indicate that American and British men

TABLE 14.3
The Oil Content of Commercial Potato Chips*

Potato Chip	Oil Content wt.%	% Oil as Free Surface Oil
Low-fat potato chips process	15	0.68
Baked Kettle® potato chip	11.8	2.11
Red Sky® potato chip	22	2.51
Lights® potato chip	25	1.39
Conventional® potato chip	33	2.42

* Table modified from the WO 2012104213A9 International Patent (Bailey et al. 2012).

frequently avoid them. Many new potato snacks have been developed over the years and effectively achieved dough-based products with less fat and better flavor based on the Pringles® strategy. From 1997 to 2003, the compound annual growth rate (CARG) in mass terms (kg of product) for the global potato chip industry was 3.7% (Datamonitor 2004, Kirkman 2007). In contrast, a CARG of 4.40% during the period 2021–2026 has been forecasted, with North America and Pacific Asia being the largest and the fastest-growing market, respectively (Mordor Intelligence 2021).

There are several patents about the manufacturing processes of dough-based fried potato snacks. Some of these reduce potato chips' fat content and improve their texture by adding ingredients, and others are founded on shape, processing equipment, and packaging, among other aspects. The general manufacturing process consists of creating a dough that can then be formed into sheets from which snack pieces are cut, and then they are fried with the time and temperature required to reach a crispy texture (Villagran 2004). These snacks usually have 18–40% oil content, depending on the ingredients and unit operations used. A flow diagram is demonstrated in Figure 14.8 with the process to manufacture low-fat dough-based potato snacks with 18–25% oil content.

14.6.1 Ingredient Functionality and Dough Formation

Potato flour, starch-based flour, hydrolyzed starch, water, and optional materials (e.g., emulsifiers, flavorings, salt, seasonings, and calcium carbonate) are mixed to form the dough, the principal ingredient being by weight potato flour. Dehydrated potato flakes and granules, dried potato products, and mashed potatoes with intact cells (US Patent 6,808,734 by Villagran et al. 2004) are regularly used to make potato flour.

Nevertheless, starch-based flours may be integrated into the dough formulation like maize, wheat, barley, peanut, tapioca, oat, rice, and soybean flours. These additional flours are incorporated with potato flour to obtain products with distinctive compositions and flavors. For instance, the addition of oat flour to the formulation raises soluble fiber content. The production of starch-based flours must comprise a starch pre-gelatinization step (at least 80% conversion), and then they are dried and milled.

Hydrolyzed starches are oligosaccharides or dextrins prepared by acid or enzymatic hydrolysis regularly from maize starch. On some occasions, maltodextrins and corn syrup solids are included in the formulation as hydrolyzed starches with dextrose equivalents (DE values) in the range of 10–20. These ingredients are fundamental for the dough structure by reducing the amount of work necessary to form it. For "light" snacks, hydrolyzed starches are included at levels of 5–12% to create a sheetable dough. This dough contains less water content and expands adequately during frying, and, consequently, the final product has a lower oil content. The dough elasticity reduces as their percentage in the formulation decreases, creating harder and more brittle snacks (Villagran et al. 2004).

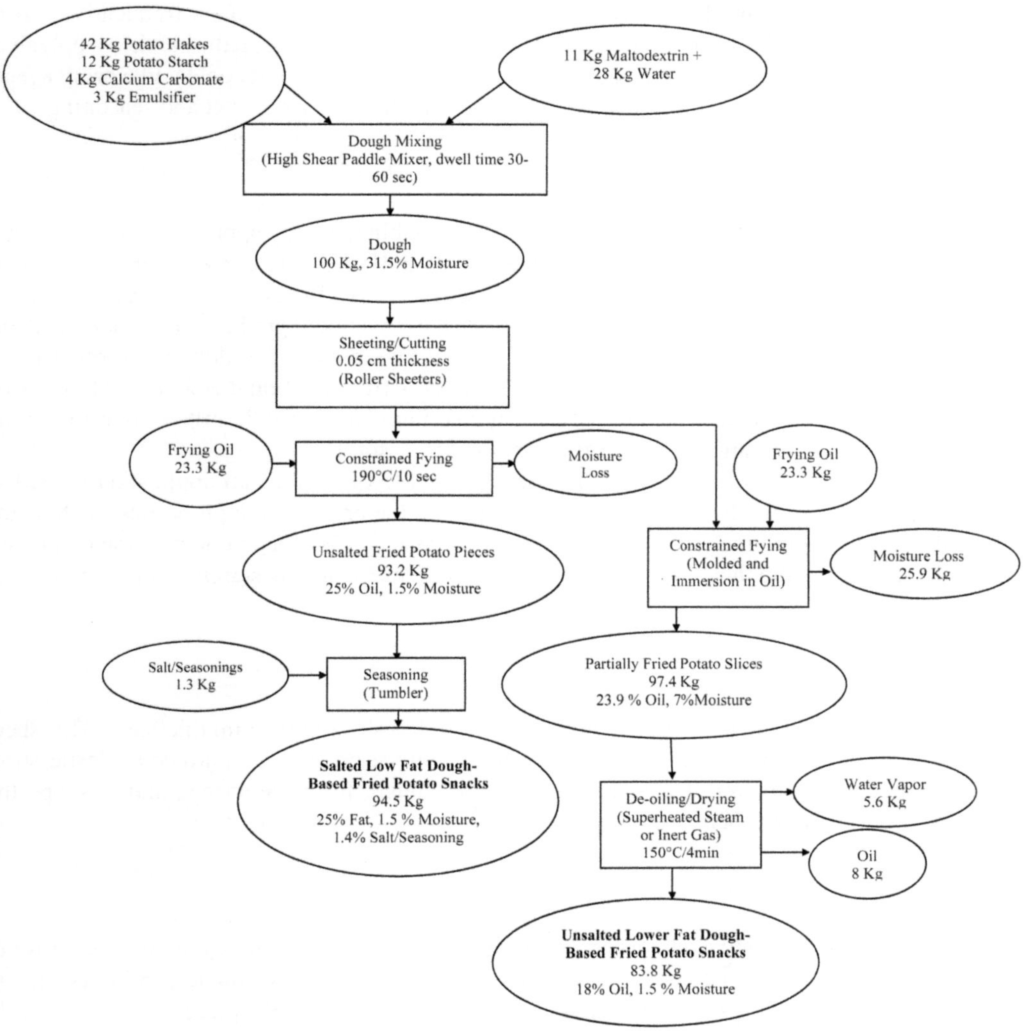

FIGURE 14.8 Flow diagram for manufacturing low-fat fabricated potato snacks according to US Patent 5,464,642 (adapted from Villagran et al. 1995).

Water used for the dough formation ranges from 40% to 50%, but it may be reduced to 23–32%, depending on the additives. The function of water in the dough is to allow cohesive sheet formation, acting as a plasticizer. The water from dried ingredients like potato flakes and hydrolyzed starches (3–8%) is not considered in the added water.

The dough should have a relatively low moisture content, approximately 40–50%, to reduce oil levels and improve the snack's texture. Calcium carbonate and hydrolyzed starches are ingredients used to reduce dough moisture levels and still develop cohesive sheets. Calcium carbonate use at 4–6% causes the dough to expand during frying without increasing fat uptake secondary to such expansion. It does not promote either excessive blistering or delamination, stated in US Patent 5,464,642 (Villagran et al. 1995).

The original formulations of fabricated potato chips had high moisture content because they did not use functional additives like calcium carbonate, so they also had higher fat levels, about 38–40%. The highly moistened dough must be overworked to reach the appropriate expanded texture in the final product. Emulsifiers act as sheeting agents in the potato snack dough, decreasing stickiness in the dough and puffing and blistering in the final product. Some emulsifiers frequently used to manufacture fabricated chips are mono- and diglycerides of stearic acids, palmitic acids, polyglycerol esters, polyethoxysorbitan esters, and sucrose esters of long-chain fatty acids. Flavoring agents could also be included in dough formulation to improve the flavor, such as salt and seasonings.

14.6.2 Dough Formation

For the dough preparation, the ingredients are regularly combined by any appropriate method. First, the dry ingredients are mixed with water to achieve a 40–50% moisture content. These dry ingredients are potato flakes, supplemental starch sources, and optional materials, such as emulsifiers, calcium carbonate, flavorings, salt, and seasonings. However, some ingredients like maltodextrins are combined with water before being added to the dry ingredients. The dough is formed using conventional mixers. Batch operations generally use mixers from the brand Hobart® (Hobart Co., Troy, Ohio), while continuous operations typically use high shear paddle mixers like the brand Turbulizer® (Bepex International LLC, Minneapolis, Minnesota).

The US Patent 558,730B1 assigned to Procter & Gamble Co. claims an application related to dough composition and method for selectively controlling the organoleptic properties, such as the texture, flavor, and mouth melt of the fabricated potato snacks. The dough composition comprises a potato-based flour component, one or more polysaccharides (including starch, gum, or mixtures), and water.

14.6.3 Sheeting

The next step is to laminate the potato dough into a very thin sheet of 0.5 mm thickness. This sheet is formed using a pair of counter-rotating cylindrical rollers, creating a continuous elastic sheet without air pockets. However, extruders may mix the dough, form the sheets, and/or shape the dough into pieces (Chapter 12).

14.6.4 Cutting

The sheet is typically cut into oval-shaped pieces during this operation unit by another set of rotating cutting rolls. They are sometimes cut into shapes like circles, squares, bowties, and stars, among others. Any proper stamping or cutting equipment can be used to form the pieces.

14.6.5 Constrained Frying

The next step is to position each potato piece between two molds of the chosen shape, with openings that allow the frying medium to enter and water to evaporate. The first patent that reported this apparatus was released in 1971 assigned to Procter and Gamble Co. in 1971 (Liepa 1971) (US Patent 3,626,466). This equipment consisted of a conveying mechanism with many pairs of molds attached to it to carry potato chips through the frying medium. According to their invention, the constrained potato slices were fried at 180–190°C to reach moisture contents between 0.5% and 2%.

In 2003, US Patent 6,875,458 was assigned to Frito-Lay North America Inc. (Dove et al. 2005) described a mold-form fryer with a top conveyor with no bottom mold or conveyor. The apparatus is described as a forming fryer with a top conveyor placed above a frying pan with oil, transporting snack pieces at a constant velocity. The raw potato snack pieces are introduced into the frying unit operation by a bottom entrance conveyor belt. Once in oil within the fryer, the snacks meet with molding surfaces on the top conveyor that transports them while frying. At the exit portion of the fryer, a bottom exit conveyor receives the cooked snack pieces from the top conveyor. Authors claimed that the molding surfaces could produce uniformly shaped snack pieces that are either convexly or concavely shaped when viewed upon the bottom exit conveyor.

There are some dough-based products with irregular shapes in the market, even though the leading dough dough-based potato snacks are saddle-shaped. In a non-constrained frying operation, the potato pieces can be immersed in the oil by the moving belt or basket of a continuous fryer, or they may also be produced by batch frying.

14.6.6 De-Oiling/Drying

Additional unit operations to reduce final oil content, previously described in other sections, also apply to fabricated potato snacks. For instance, potato pieces with initial moisture contents of 30–40% are fried to moisture contents of 7–10% or less, then exposed to overheated steam or hot air to reduce their oil content, and also subjected to drying further to reduce the moisture levels to 1.5–2%. These optional processing operations include a combination of frying and baking steps, which are also called impingement ovens. Snacks produced as described in Figure 14.8 apply these optional de-oiling/drying processing steps; therefore, finished potato chips may contain approximately 18% instead of 25% oil content. A different process to produce potato chips with reduced calories was described in European Patent EP1597977A1, in baking individually the manufactured potato chips (Shimaoka 2005).

14.6.7 Seasoning

The process of seasoning may vary depending on if it is non constrained (irregular shaped) or constrained frying (equally shaped). For irregularly shaped snacks, this step consists of the dosification of flavorings onto the product, which is then introduced into a rotating drum, similarly to regular potato chips. However, constrained fried snacks are sprayed with oils and flavors to minimize breakage. Natural or flavored oils can be used, including olive, avocado, safflower, sesame, walnut, or peanut pure or combined with aromatics of butter, herbs, lemon, chili pepper, garlic, onion, barbecue, cheddar cheese, sour cream, vinegar, etc. Fat contents of the fried snack may increase due to the incorporation of oils; for instance, the original 18–40% fat may increase to a level as high as 44%. In most chips, salt contributed 1–2% of end-product potato chip weight, whereas the flavoring levels could vary from 5% to 10% of the final product weight (Serna-Saldivar 2016).

14.6.8 Packaging

The package design is a crucial issue for the preshaped snacks (e.g., saddle-shaped) since they are stacked side-by-side and need protection when the product is dropped or exposed to lateral shock forces. Various patents have been granted protecting the packaging features for this type of product (Johnson 1979, Gruenbacher et al. 1999). The product Pringles® from the Procter & Gamble Co. (Cincinnati, Ohio) has been packaged traditionally as a single stack of snacks in a cylindrical fiberboard laminated with a re-closable plastic lid located at the top end of the body. During packaging, the cylindrical fiberboard cans are formed, one of the end lids is placed, and then the constrained-fried snacks are placed inside under nitrogen. The cylindrical containers are selected because they can maintain their shape without deformation during pressure changes; for instance, at high altitudes, pressures may be in the range of 0.14–0.28 kg/cm^2, which are approximately 66% less than at sea level (Gruenbacher et al. 1999).

The newer preshaped product Lay's Stax® from Frito-Lay Inc. (Plano, Texas) is packaged in plastic bottles made by blow molding and protected as a shape trademark. Blow-molded containers can be made with multiple layers to incorporate different types of materials that possess barrier properties (e.g., oxygen barrier); biodegradability of the material is also a relevant concern for consumers. Plastic containers offer the advantage of color and also can be recycled. Packaging manufacturers hold intellectual property related to design for functionalities such as pressure and vacuum resistance. Saddle-shaped snacks, for instance, are placed in U-shaped bottom packages designed to cushion them to prevent breakage.

14.7 ACKNOWLEDGMENTS

The authors would like to thank Irma Catalina De Anda-Lobo, B.Eng., for her technical insights and suggestions for presenting material on the different unit operations for potato snack processing. The authors also want to acknowledge the assistance of Carmen M. Roesch-Hernandez in the preparation of the figures of this chapter.

REFERENCES

Alan-Sanders, R., Cammiade, D.G., Gerald, R.S., Keeney, J.H., Marko, S., Martin-Villagran, M.D., Tak Lin, P.Y., and Vincent-Zyzak, D. 2003. Method for reducing acrylamide in foods, foods having reduced levels of acrylamide. Patent No. CN100438773C. Assigned to Procter & Gamble Co.

Álvarez-Torres, M.D., and Canet-Parreño, W. 2016. Thermal processing of potatoes. In: *Advances in Potato Chemistry and Technology*, eds. J. Singh and L. Kaur, 403–457. New York: Academic Press.

Amrein, T.M., Bachmann, S., Noti, A., Biedermann, M., Barbosa, M.F., Biedermann-Brem, S., Grob, K., Keiser, A., Realini, P., Ezscher, F., and Amadó, R. 2003. Potential of acrylamide formation, sugars, and free asparagine in potatoes: A comparison of cultivars and farming systems. *Journal of Agricultural and Food Chemistry*, 51(18):5556–5560.

Andersson, A., and Grigonis, D. 2008. Low fat potato chips and method for production. European Patent No. EP2130442A. Assigned to Estrella Maarud Holding AS.

Andersson, P.A., and Maartensson, T.S. 2006. Low fat potato chips and method for their production. European Patent No. EP1900289A1. Assigned to Kraft Food R&D, Inc., Munich Alemania.

Asokapandian, S., Swamy, G.J., and Hajjul, H. 2019. Deep fat frying of foods: A critical review on process and product parameters. *Critical Reviews in Food Science and Nutrition*, 60(20):3400–3413.

Bailey, R.A., Bartlett, G., Clark, N.J., Dobson, L.A., Herbert, O., Newberry, B.R., Roberts, D.J., Spurr, M.A.J., Tomlinson, P.F., and Warburg, B.L. 2012. Low surface oil potato chip and manufacture thereof. International Patent No. WO 2012104213A9. Assigned to Frito-Lay.

Bartolome, L.G., and Hoff, J.E. 1972. Firming of potatoes: Biochemical effects of preheating. *Journal of Agricultural and Food Chemistry*, 20(2):266–270.

Bergthaller, W. 2004. Developments in potato starches. In: *Starch in Food: Structure, Function and Applications*, ed. A.C. Eliasson, 241–257. Cambridge: Woodhead Publishing.

Bernard, D.C. 1985. Method for preparing low oil fried potato chips. United States Patent No. 4,537,786. Assigned to Frito-Lay Inc., Dallas, TX.

Bertoft, E., and Blennow, A. 2016. Structure of potato starch. In: *Advances in Potato Chemistry and Technology*, eds. J. Singh and L. Kapur, 57–75. London: Academic Press.

Bradshaw, J.E., and Ramsay, G. 2016. Potato origin and production. In: *Advances in Potato Chemistry and Technology*, eds. J. Singh and L. Kaur, 1–32. London: Academic Press.

Brierley, E.R., Bonner, P.L.R., and Cobb, A.H. 1996. Factors influencing the free amino acid content of potato (*Solanum tuberosum L*) tubers during prolonged storage. *Journal of the Science of Food and Agriculture*, 70(4):515–525.

Cilliers, J.J.L., and Singleton, V.L. 1989. Nonenzymic autoxidative phenolic browning reactions in a caffeic acid model system. *Journal of Agricultural and Food Chemistry*, 37(4):890–896.

Coffin, R.H., Yada, R.Y., Oarkin, K.L., Grodzinski, B., and Stanley, D.W. 1987. Effect of low temperature storage on sugar concentrations and chip color of certain processing potato cultivars and selections. *Journal of Food Science*, 52(3):639–645.

Datamonitor. 2004. *Potato Chips: Global Industry Guide*. New York: Datamonitor.

Datir, S.S., Yousf, S., Sharma, S., Kochle, M., Ravikumar, A., and Chugh, J. 2020. Cold storage reveals distinct metabolic perturbations in processing and non-processing cultivars of potato (*Solanum tuberosum* L.). *Scientific Reports*, 10(1):6268.

Dourado, C., Pinto, C., Barba, F.J., Lorenzo, J.M., Delgadillo, I., and Saraiva, J.A. 2019. Innovative non-thermal technologies affecting potato tuber and fried potato quality. *Trends in Food Science and Technology*, 88:274–289.

Dove, K.M., Henson, W.D., Joseph, P.K., and Tatsch, D.J. 2005. Single mold form fryer with enhanced product control. United States Patent No. 6,875,458. Assigned to Frito-Lay North America, Inc., Plano, TX.

European Commission. 2017. Commission Regulation (EU) 2017/2158: Establishing mitigation measures and benchmark levels for the reduction of the presence of acrylamide in food. *Official Journal of the European Union*, 304:24–44.

FAOSTAT. 2018. Potatoes production indices. https://www.fao.org/faostat/en/#data/QI. Accessed September, 2020.

Fitzpatrick, T.J., and Porter, W.L. 1966. Changes in the sugars and amino acids in chips made from fresh, stored and reconditioned potatoes. *American Potato Journal*, 43(7):238–248.

FSSAI. 2011. Handling and disposal of used cooking oil. In: June 2018. Food Safety and Standards (Licensing and Registration of Food Businesses) Regulation.

Garayo, J., and Moreira, R. 2002. Vacuum frying of potato chips. *Journal of Food Engineering*, 55(2):181–191.

Genovese, J., Tappi, S., Luo, W., Tylewicz, U., Marzocchi, S., Marziali, S., Romani, S., Ragni, L., and Rocculi, P. 2019. Important factors to consider for acrylamide mitigation in potato crisps using pulsed electric fields. *Innovative Food Science and Emerging Technologies*, 55:18–26.

Gould, W.A., and Plimpton, S.L. 1985. Quality evaluation of potato cultivars for processing. *Research Bulletin*, Ohio-Agricultural-Research and Development Center. No. 1172.

Grommers, H.E., and van der Krogt, D.A. 2009. Potato starch: Production, modifications and uses. In: *Starch Chemistry and Technology*, 3rd edition, eds. J. Bemiller and R. Whistler, 511–539. New York: Academic Press.

Gruenbacher, D.P., Zeik, D.B., and Kinne, D.J. 1999. Container for multiple side-by-side stacks of fragile articles. United States Patent No. 5,921,429. Assigned to the Procter & Gamble Company, Cincinnati, OH.

Haase, N.U., Matthäus, B., and Vosmann, K. 2003. Minimierungsansätze zur Acrylamid-Bildung in pflanzlichen Lebensmitteln-aufgezeigt am Beispiel von Kartoffelchips. *Deutsche Lebensmittel-Rundschau*, 99:87–90.

Halford, N.G., Muttucumaru, N., Powers, S.J., Gillatt, P.N., Hartley, L., Elmore, J.S., and Mottram, D.S. 2012. Concentrations of free amino acids and sugars in nine potato varieties: Effects of storage and relationship with acrylamide formation. *Journal of Agricultural and Food Chemistry*, 60(48):12044–12055.

Harris, P.M. 1992. Introduction. In: *The Potato Crop, the Scientific Basis for Improvement*, ed. P.M. Harris. London: Chapman & Hall.

Hernández-Brenes, C. 2000. *Factors Contributing to the Color, Flavor and Texture of Potato Chips*. University of Arkansas. Doctoral Dissertation.

Hirsch, A. 1991. What materials are best suited for snack food packaging? In: *Flexible Food Packaging: Questions and Answers*, ed. A. Hirsh, 147–156. Boston, MA: Springer.

Isherwood, F.A. 1976. Mechanism of starch-sugar interconversion in *Solanum tuberosum* (potatoes). *Phytochemistry*, 15(1):33–41.

Jabeen, N., Majid, I., and Gulzar, A.N. 2015. Bioplastic and food packaging: A review. *Cogent Food and Agriculture*, 1(1):1–6.

Johnson, C.L. 1979. Package liner and fragile snack chip combination. United States Patent No. 4,164,588. Assigned to the Procter & Gamble Company, Cincinnati, OH.

Kirkman, M.A. 2007. Global markets for processed potato products. In: *Potato Biology and Biotechnology, Advances and Perspectives*, eds. D. Vreugdenhil, J. Bradshaw, C. Gebhardt, F. Govers, M. Taylor, D. MacKerron, and H. Ross, 27–44. Oxford: Elsevier Science.

Lawrance-Barry, D., Jeffrey-Burnham, C., Desai, P.M., Joseph, P.K., Masson, J.R., Rao, V.N.M., Saunders, R.W., Stalder, J.W., and Topor, M.G. 2011. Method for reducing acrylamide formation in thermally processed foods. Patent No. EP2277391A1. Assigned to Frito-Lay North-America, Plano, TX.

Lemos-Monteiro, R., Oliveira de Moraes, J., Domingos, J.D., Mattar-Carciofi, B.A., and Borges-Laurindo, J. 2020. Evolution of the physicochemical properties of oil-free sweet potato chips during microwave vacuum drying. *Innovative Food Science and Emerging Technologies*, 63:102317.

Liepa, A.L. 1971. Molding device for preparing chip-type products. U.S. Patent No. 3,626,466. Assigned to the Procter and Gamble Co., Cincinnati, OH.

Lisiniska, G., and Leszczynski, W. 1990. Manufacture of potato chips and French fries. In: *Potato Science and Technology*, eds. G. Lisinska and W. Leszczynski, 165–232. London and New York: Elsevier Applied Science.

Maga, J.A. 1973. Influence of freshness and color con potato chip sensory preferences. *Journal of Food Science*, 38(7):1251–1252.

Maga, J.A. 1994. Potato flavor. *Food Reviews International*, 10(1):1–48.

Marquez, G., and Añon, M.C. 1986. Influence of reducing sugars and amino acids in the color development of fried potatoes. *Journal of Food Science*, 51(1):157–160.

Martines-Serna Villagran, M.D., Beverly, D.J., and Williamson, L. 2000. Dehydrated potato flakes. US Patent No. 7,998,522. Patent and Trademark Office, Washington, DC.
Matz, S.A. 1993. *Snack Food Technology*, 3rd edition, 286–287. Mc Allen: Pan-Tech International.
Meredith, P. 1988. The development and properties of the tuber of the potato (*Solanum tuberosum*). Part I. Variability of tubers. *Starch/Stärke*, 10:369–374.
Mitch, E.L. 1984. Potato starch: Production and uses. In *Starch: Chemistry and Technology*, eds. R.L. Whistler, J.N. BeMiller, and E. Pachall, 479–490. Orlando, FL: Academic Press.
Mordor Intelligence. 2021. Potato chips market – growth, trends, covid-19 impact, and forecasts (2021–2026). https://www.mordorintelligence.com/industry-reports/potato-chips-market. Accessed December, 2021.
Mottur, G.P. 1989. A scientific look at potato chips-the original savory snack. *Cereal Foods World*, 34:620–626.
Neel, D.V., and Reed, R.B. 1990. Process for preparing low oil potato chips. U.S. Patent No. 4,933,199. Assigned to Frito-Lay Inc., Dallas, TX.
Parkin, K.L., and Schwobe, M.A. 1990. Effects of low temperature and modified atmosphere on sugar accumulation and chip color in potatoes (*Solanum tuberosum*). *Journal of Food Science*, 55(5):1341–1344.
Pathirana, R., Harris, J.C., and McKenzie, M.J. 2008. A comparison of microtubers and field-grown tubers of potato (*Solanum tuberosum* L.) for hexoses, sucrose and their ratios following postharvest cold storage. *Postharvest Biology and Technology*, 49(1):180–184.
Pedreschi, F., Bustos, O., Mery, D., Moyano, P., Kaack, K., and Granby, K. 2007. Color kinetics and acrylamide formation in NaCl soaked potato chips. *Journal of Food Engineering*, 79(3):989–997.
Pedreshi, F., Mery, D., and Marique, T. 2008. Quality evaluation and control of potato chips and French fries. In: *Computer Vision Technology of Food Quality Evaluation*, ed. D. Sun. Amsterdam: Academic Press.
Peters, A.B. 1966. Improvements in or relating to frying potato crisps and sticks. British Patent No. GB1034035. Original Assigned to Meredith and Drew Ltd.
Reyniers, S., De Brier, N., Matthijs, S., Brijs, K., and Delcour, J.A. 2018. Impact of physical and enzymatic cell wall opening on the release of pre-gelatinized starch and viscosity forming potential of potato flakes. *Carbohydrate Polymers*, 194:401–410.
Riaz, M. 2004. Snack foods, processing. In: *Encyclopedia of Grain Science*, eds. C.W. Wrigley, H. Corke and C. Walker. Oxford: Academic Press.
Robertson, G.L. 2012. *Food Packaging – Principles and Practice*, 3rd edition. Boca Raton, FL: CRC Press.
Rodriguez-Saona, L.E., and Wrolstad, R.E. 1997. Influence of potato composition on chip color quality. *American Potato Journal*, 74(2):87–106.
Roe, M.A., and Faulks, R.M. 1991. Color development in a model system during frying: Role of individual amino acids and sugars. *Journal of Food Science*, 56(6):1711–1713.
Rojas de Gante, C. 1997. Trends in food packaging. In: *Food Engineering 2000*, eds. P. Fito, E. Ortega-Rodriguez, and G.V. Barbosa-Canovas, 347–362. New York: Chapman & Hall.
Schouten, M.A., Genovese, J., Tappi, S., Di Francesco, A., Baraldi, E., Cortese, M., Caprioli, G., Angeloni, S., Vittori, S., Rocculi, P., and Romani, S. 2020. Effect of innovative pre-treatments on the mitigation of acrylamide formation in potato chips. *Innovative Food Science and Emerging Technologies*, 64:102397.
Schroeder, O.E., Wohlman, A., and Topor, M.G. 1981. Process for preparing blister-inhibited potato chips. U.S. Patent No. 4,272,554. Assigned to Frito-Lay, Inc., Dallas, TX.
Serna-Saldivar, S.O. 2008. Manufacturing of cereal based dry milled fractions, potato flour, dry masa flour and starches. In: *Industrial Manufacture of Snack Foods*, ed. S.O. Serna-Saldivar, 25–103. London: Kennedys Publications Ltd.
Serna-Saldivar, S.O. 2016. Snack foods: Types and composition. In: *Encyclopedia of Food and Health*, eds. B. Caballero, P.M. Finglas and F. Toldrá, 13–18. Oxford: Elsevier Science.
Shallenberger, R.S., Smith, O., and Treadaway, R.H. 1959. Role of sugars in the browning reaction in potato chips. *Journal of Agricultural and Food Chemistry*, 7(4):274–277.
Shepherd, L.V.T., Bradshaw, J.E., Dale, M.F.B., McNicol, J.W., Pont, S.D.A., Mottram, D.S., and Davies, H.V. 2010. Variation in acrylamide producing potential in potato: Segregation of the trait in a breeding population. *Food Chemistry*, 123(3):568–573.
Shimaoka, K., and Uwaya, K. 2005. Baked potato chips and their method and system of manufacture. European Patent No. EP1597977A1. Applicants: Central Bussan Co.
Singh, J., Colussi, R., McCarthy, O.J., and Kaur, L. 2016. Potato starch and its modification. In: *Advances in Potato Chemistry and Technology*, eds. J. Singh and L.Kapur, 195–248. London: Academic Press.
Statista. 2017. Market share of the leading potato chip brands in the United States in 2017, based on dollar sales. https://www.statista.com/statistics/262117/sales-share-of-the-leading-10-potato-chip-brands-in-the-us/. Accessed June, 2021.

Statista. 2020. Leading vendors of potato chips in the United States in 2020, based on sales. https://www.statista.com/statistics/188220/leading-potato-chips-vendors-in-the-united-states-in-2011/. Accessed September, 2020.

Sterling, C., and Betteliieim, F.A. 1955. Factors associated with potato texture. 3. Physical attributes and general conclusions. *Food Research*, 20:130–137.

Storey, R.M.J., and Davies, H.V. 1992. Tuber quality. In: *The Potato Crop*, ed. P. Harris, 507–569. London: Chapman & Hall.

Takeoka, G.R., Full, G.H., and Dao, L.T. 1997. Effect of heating on the characteristics and chemical composition of selected frying oils and fats. *Journal of Agricultural and Food Chemistry*, 45(8):3244–3249.

Trejo-Copado, L.F. 2009. Reduced fat potato chip. Patent No. US20090169710A1. Assigned to Sabritas S de R L de CV, D.F., México.

USDA-National Agricultural Statistics Service. 2020. Potatoes 2019 summary. https://www.nass.usda.gov/Statistics_by_State/Washington/Publications/Potatoes/index.php. Accessed September, 2020.

USDA-Agricultural Research Service. 2021. USDA national nutrient database for standard reference. https://fdc.nal.usda.gov. Accessed March, 2021.

Van den Broek, L.A.M., Den Aantrekker, E.D., Voragen, A.G.J., Beldman, G., and Vincken, J.P. 1997. Pectinlyase is a key enzyme in the maceration of potato tuber. *Journal of the Science of Food and Agriculture*, 75(2):167–172.

Villagran, M.D. 2004. Potato dough. U.S. Patent, 6,808,734. Assigned to the Procter & Gamble Co., Cincinnati, OH.

Villagran, M.D., Toman, L.J., Byars, K.D., Dawes, N.C., and Zimmerman, S.P. 1995. Process for making reduced-fat fried snacks with lighter, more expanded snack structures. U.S. Patent 5,464,642. Assigned to the Procter & Gamble Company, Cincinnati, OH.

Willard, M.J., and Hix, V.M. 1987. Potato flour. In: *Potato Processing*, 4th edition, eds. W. Talburt and O. Smith, 665–681. New York: Van Nostrand Reinhold Co.

Yadav, A.R., Guha, M., Tharanathan, R.N., and Ramteke, R.S. 2006. Influence of drying conditions on functional properties of potato flour. *European Food Research and Technology*, 223(4):553–560.

Yang, J., Powers, J.R., Boylston, T.D., and Weller, K.M. 1999. Sugars and free amino acids in stored Russet Burbank potatoes treated with CIPC and alternative sprout inhibitors. *Journal of Food Science*, 64(4):592–596.

Zalpouri, R., Kaur, P., Kaur, A., and Sidhu, G.K. 2020. Comparative analysis of optimized physiochemical parameters of dried potato flakes obtained by refractive and convective drying techniques. *Journal of Food Processing and Preservation*, 45(1):e15077.

Zhang, Y., Quan, X., Shen, S.S., and Wang, Y. 2015. Method for producing low oil content potato chips by combined process of vacuum drying and follow-up microwave vacuum frying. China Patent No. CN104642824A. Applicants: Gen Mills China Invest Co. Ltd.; Univ Jiangnan.

15 Almonds and Tree Nuts

Johanan Espinosa-Ramírez and Sergio O. Serna-Saldivar

CONTENTS

15.1 INTRODUCTION

Almonds, cashew, walnuts, hazelnuts, pecans, macadamias, pistachio, and pine nuts are categorized as edible seed kernels of trees. These tree nuts have been consumed for years and are considered as healthy and nutritious foods (Payne 2005). There are several products derived from almonds and tree nuts; however, their consumption as snacks is one of the main uses extended worldwide. For the production of snack foods, nuts can be sold individually or blended with other snack items such as roasted peanuts (Chapter 13) and dehydrated fruits (Chapter 16). According to the US Food and Drug Administration, mixed nuts is the snack food consisting of a mixture of four or more shelled tree nuts with or without one or more optional shelled nut ingredients. Each of the nut ingredients should be present in a quantity more than 2% and less than 80% by weight of the finished snack food. These snacks are considered first-generation or minimum processed since kernels are only roasted or heat-treated and sometimes fried prior to flavoring and packaging.

DOI: 10.1201/9781003129066-15

TABLE 15.1
Almond and Tree Nuts Kernel Production and Main Producers

Tree Nut	Kernel Basis Production (ton)[a]	World Estimated Consumption (ton)[b]	Top Producers in 2019/2020 (Kernel Basis)
Almond	1,368,703	1,304,051	USA (77%), Australia (8%), Spain (6%), and Iran (4.3%)
Cashew	786,266	720,170	Western Africa (47%), India (22%), and Eastern Africa (9%)
Walnuts	965,402	887,731	China (46%), USA (27%), Chile (7%)
Hazelnuts	528,068	520,028	Turkey (75%), Italy (7%), Azerbaijan, USA, and Chile (3% each)
Pecans	139,739	141,492	Mexico (47%), USA (43%), and South Africa (8%)
Macadamias	60,057	58,821	South Africa (29%), Australia (22%), and Kenya (12%)
Pistachios	655,200	678,984	USA (51%), Iran (31%), Turkey (11%)
Pine nuts	17,220	22,745	China (34%), Russia (20%), Korea (17%)
Sum of almond and tree nuts	4,520,655	4,334,022	

Source: International Nut & Dried Fruit (2020).

[a] Reported in kernel basis, except for inshell pistachios.

[b] Based on the estimated percentage of the population consuming the specific product.

Almonds and tree nuts are widely consumed around the globe. In 2018, the worldwide consumption of almonds and tree nuts was more than 4 million tons. Almonds and walnuts are the most consumed followed by cashews, pistachios, and hazelnuts (Table 15.1). The regions with higher consumptions were Europe (26%), North America (23%), Asia (20%), and the Middle East (12%), whereas Latin America, Africa, and Oceania represented the regions with relatively lower intakes (10, 7, and 2%, respectively). Interestingly, increasing consumption trends are experienced each year in all regions, especially in high- and middle-income economies. Therefore, the global production has also increased significantly to satisfy this demand. Although the production of walnuts and almonds has been higher during the past years, other tree nut crops – cashews, pecans, and macadamias – have also shown a significant rise in production (International Nut & Dried Fruit 2020).

The high increment in the consumption of almonds and tree nuts observed worldwide may be associated with the health benefits linked to their consumption. In fact, almonds and tree nuts are considered as one of the most nutritious snacks since they contain high concentrations of oil rich in unsaturated fatty acids, adequate concentrations of protein and dietary fiber (DF), and a variety of vitamins and minerals (Chapter 20). In addition, they are loaded with bioactive phytochemicals such as phenolic compounds, carotenoids, phytosterols, and tocopherols (Chapter 21). These phytochemicals have been associated with antioxidant, anti-inflammatory, anti-proliferative, antiviral, chemopreventive, and hypocholesterolemic actions (Bolling et al. 2011). As a result, the consumption of nuts has been related with health benefits such as reduction in the incidence of coronary heart disease, cholesterol-lowering effects, and beneficial influence on blood lipid profile, blood pressure, visceral adiposity, and metabolic syndrome, among others (Ros 2010). These natural snacks have also gained attention as a mechanism to prevent and control weight gain attributed to satiety and prebiotic effects of their DF and phytochemicals in terms of modulating beneficial gut microbiota (Amr and Abeer 2011, Sugizaki and Naves 2018). Consumption of these snacks is expected to grow significantly due to the higher number of scientific studies that confirm their beneficial health effects.

The aim of this chapter is to summarize the main features and operations for processing almonds and tree nuts into snack foods. The chemical composition and phytochemicals associated with these nutritious kernels are also thoroughly covered.

15.2 STATISTICS OF TREE NUT PRODUCTION AND UTILIZATION

In general, commercial tree nuts are available in two major categories: (1) in-the-shell, also called inshell, and (2) shelled. Inshell tree nuts are marketed with the intact shell after collection and removal of the husks, which are normally removed mechanically or manually following drying to yield a more stable product and enhance the separation of the shell from the kernel. Resulting nuts are finally cleaned by bleaching and polishing. The preference of buying inshell tree nuts depends on the country and season. For instance, within Germanic traditions, it is popular to buy inshell nuts during the fall/winter holidays for cracking by hand or with nutcrackers. In other European regions, nuts are marketed inshell and hand-cracked to ensure freshness and produce bakery items. In other countries such as the United States, marketing of inshell nuts is limited (Payne 2005). It is recommended to perform a disinfestation step before prolonged storage to avoid pests that compromise the quality of the nut crop. Chemical fumigation may be an option, although it is forbidden in many countries. Alternative physical methods such as radio-frequency heating have proven effective to disinfest some nut crops such as pistachios, walnuts, and almonds (Ling et al. 2016).

Shelled nuts are the most common forms used in food processing. Nuts destined for the shelled market are selected from varieties with greater proportions of meat to shell. The removal of shell is performed by mechanical cracking after drying the inshell nut to a desirable level, which differs according to the nut type. In some regions of the world, traditional cracking methods using hammer or anvil are still used. However, equipment that are more sophisticated have been developed, including pressure crackers and grips. The former is designed to pass nuts between two hard rubber rollers that split the shell, whereas the latter gently squeezes nuts to remove shells. Independent of the method used, it must guarantee the complete removal of the shell without damaging the interior meat or kernel. After cracking, shelled nuts are sorted by size and packaged. If nuts do not meet the minimum screen size specifications, pieces are marketed as "nut meal" or "fluff," which are highly appreciated and demanded by the baking and other food industries. Natural shelled nuts, defined as those kernels left with the skin intact, are available in several forms: whole, sliced, chopped, or diced and are used in various processes and formulations to obtain value-added products (Payne 2005). These processes are summarized and described in Table 15.2.

For snack production, whole nuts are preferred and these can be procured in their natural, blanched, or roasted forms or optionally seasoned with sweeteners, salt, or savory coats. Roasting is a process used to modify the organoleptic profile of nuts in terms of flavor, crispiness, and color (Bagheri 2020). This process is performed at 125–200°C for 5–30 min. The roasting time must be optimized depending on the type of tree nut in order to attain the desired organoleptic features and to avoid the formation of undesirable metabolites such as furan. This carcinogenic compound is formed during roasting due to the oxidative degradation of polyunsaturated fatty acids (PUFA), which are abundant in these food products (Chapter 20). Cha and Lee (2020) recently evaluated the effects of different roasting times (5–25 min) and temperatures (150–200°C) in the production of furan in several types of nuts and found that as time and temperature increased, linolenic and linoleic acids decomposed into higher concentrations of furans. The lowest furan contents were observed when roasting was performed at the lowest tested temperatures (150–160°C) and processing time (<10 min). Other authors recommended to perform roasting under mild conditions such as 125°C and 15 min to avoid formation of toxic compounds associated with this thermal treatment (Duduzile Buthelezi et al. 2019). Traditionally, the roasting of nuts is performed with hot air, on hot pans or ovens. These methods are simple and low-cost but are time-consuming and have high expenditure of thermal energy. Moreover, conventional methods may yield uneven products that are over- and under-roasted on the surface and center, respectively. Recently, infrared (IR) heating

TABLE 15.2
Processes Commonly Employed to Add Value of Tree Nuts

Process	Description
Blanching	This process involves the complete removal of the skin of nut kernels, leaving a clean ivory-colored meat. Generally, blanched nuts are more pliable and resistant to brittle, and can be slivered, sliced, or diced. Pieces are ideal for grinding into pastes and fillings, they do not impart color or bitter and astringent taste as a result of the skin removal. Soaking in hot water is the conventional method used for blanching almonds. Alternatively, nuts may be blanched using hot lye followed by an acid rinse or using glycerin and a sodium carbonate solution. Typically, this operation is not performed on pecans or walnuts
Dicing	Performed to achieve individual product specifications or buyer requirements, where sizing is important. Dicing may be performed on blanched, natural, or roasted nuts. Finely diced nuts are used as toppings and incorporated into fillings, batters, doughs, and snack bars. Diced nuts are used as a garnish, to add body and texture or to add flavor in different products
Coating	Coating of nuts with sweet or hot/spice seasonings to add secondary flavors to nuts. The common way to coat nuts uses a confectionery kettle made of copper. Sugar slurry is heated in the kettle, which rotates to yield even coating, and nuts are added when the adequate temperature is reached. Prior to coating, raw nuts need to be pretreated to avoid staling and sogginess of coated kernels. Pretreatment includes drying and/or treatment with modified protein–starch that contains an antioxidant. Coated nuts are often vacuum packed to reduce staling and lipid oxidation
Roasting	Process that creates a crisp texture, yields a golden color, and improves the flavor of raw nuts by changing the "green" taste for a full-bodied flavor. Roasting can be performed by drying or frying. Dry roasting is the heating of nuts with radiant microwave heat and in oil roasting, raw nuts are immersed in boiling oil with low melting point. Roasting decreases the moisture content to around 1%, contributing to the crisp texture. Treatment of roasted nuts with an edible coating may be performed to help extend shelf life
Salting	During this process, roasted nuts are covered with an oil dressing containing an antioxidant and coated with a highly soluble, fine flake salt. The salt is sealed on the nut by the use of an oil spray. Salted nuts are popular as snacks and are commonly used in cocktail mixes. Package may be performed in tins under vacuum conditions, foil pouches, and glass and plastic containers.
Nut paste	Sweetened mixture of pulverized nuts produced with natural or blanched nuts. Are spreadable but with grainy texture
Nut butter	Puree of nuts with a smooth consistency that may be produced with raw or roasted nuts. Are good replacers of animal fats and are used in bakery and confectionery. May be blended with chocolate
Nut syrup	Syrup produced by the addition of small and medium nutmeats to corn syrup or honey. Are used as toppings and fillings in desserts and pastry

Source: Payne (2005).

has been recognized as an alternative to improve the roasting of nuts due to its low energy cost and high heat efficiency. IR is energy in the range of electromagnetic spectrum between microwaves and visible light. IR has frequencies in the range of 30–430 THz and wavelengths from 0.75 to 1,000 μm. During IR heating, energy is transferred in the form of electromagnetic waves, leading to a quicker heat diffusion and shorter treatment times in comparison to convection and conduction mechanisms. Moreover, IR radiation penetrates and transfers thermal energy throughout the food material, making the heating more homogeneous. However, the main limitations of IR heating are low penetration power and fracturing of materials when they are exposed to prolonged treatments. To overcome the restrictions in the penetrating power of IR radiation, this technology may be combined with other thermal methods such as hot air. The combination of these methods leads to a more efficient roasting in terms of energy requirement and processing times than using them individually. When using IR heating to roast nuts, the optimization of process conditions is essential to avoid nut fractures caused by extensive treatments. Tree nuts that have been roasted using IR heating include pistachios, almonds, and hazelnuts (Bagheri 2020).

The high fat content of all tree nuts is a relevant factor to consider during handling and processing, especially after the shell is cracked and removed because lipids become more prone to oxidation. Therefore, to extend and guarantee the nut quality, it is recommended to store shelled nuts at low moisture, low temperature, and controlled relative humidity (55–65%). These conditions will help to maintain their texture and prevent microbial spoilage and rancidity (Payne 2005).

15.3 ALMONDS

Almond (*Prunus communis*) is a valuable crop that originated in the Mediterranean and has been cultivated since ancient times in temperate and subtropical regions of the globe (Yada et al. 2011). Among tree nuts, almonds are the most consumed (Table 15.1) because they have been recognized for a long time as a rich source of nutrients and currently their consumption is increasing even more due to their perception as healthy snacks. In 2019, the world production of inshell almonds was 3.497 million tons harvested from 2.126 million Ha with a production yield of 1.644 tons/Ha (FAO 2021). The leading world producer of almonds is the United States where the production concentrates in the state of California. Other important producers are Australia, Spain, and Iran (Table 15.1).

The almond tree is well adapted to deep, fertile, well-drained, and loamy soils. The optimum elevation for almond cultivation is 750 m above the sea level (Desphande et al. 1995). Almonds are propagated asexually by grafting and must be cross-pollinated by bees to bear fruits. It takes up to four years for trees to begin bearing fruit yielding the highest productions at approximately seven to eight years. Most commercial orchards harvest almonds mechanically. To do so, the distance between almond trees must be around 8 m apart to allow space for the harvesting equipment consisting of shakers and sweepers (Payne 2005).

The almond fruit is a drupe that resembles peach until it ripens. At maturity, the leathery outer covering, commonly named hull, splits and curls outward, discharging the pit. Therefore, although almonds are commonly labeled as nuts or dry fruits, they are seeds which are enclosed in a hard fruit covering and therefore are not true nuts (Petruzzello 2020).

Almonds are divided into two major categories: bitter and sweet. The most cultivated and used for snacks and direct food use is the sweet type. On the other hand, bitter almonds are commonly used for the manufacturing of flavoring extracts and cosmetics (Payne 2005). They contain high levels of cyanogenic glycosides, specifically amygdalin (Yada et al. 2011). Almonds are also classified as hard- or soft-shelled. In soft-shell almonds, the chipping and breaking is minimum during its removal, and therefore this type is preferred when the consistent appearance of the kernel is important. The hard-shell almonds do not blanch well and are used in mixes due to their rich flavor. Examples of soft-shell varieties are Nonpareil and NePlus, whereas a popular hard-shell variety is Mission. Besides the shell hardness, these varieties also differ in size, shape, and color (Payne 2005).

15.3.1 Almonds Processing and Storage

After harvesting, almonds are hulled and dehydrated to drop moisture contents lower than 8% (Payne 2005). Then, they are mechanically cracked, sized, and processed into the desired final presentation (Table 15.2). For most snack applications, almonds are blanched and skinned followed by roasting or frying before packaging in airtight containers. The typical blanching process consists of placing kernels in hot water (80–85°C) for approximately 3 min and then skinned manually or mechanically. Roasting is performed at 150–200°C for 5–25 min (Cha and Lee 2020). Roasting times are optimized to achieve desired organoleptic characteristics but also to avoid excessive formation of furan, which is a recognized carcinogenic compound (Chapter 21). A thermal treatment is mandatory to assure the inactivation of pathogenic bacteria that can cross-contaminate the product during harvesting. This treatment could be fulfilled by roasting but when raw almonds are desired, they should be pasteurized. In this sense, recent studies have evaluated the use of radio frequency for the pasteurization of inshell almonds (Li et al. 2017), and this novel method proved to be effective for the reduction of microorganisms in raw inshell almonds as well as to maintain the product quality. Yang et al. (2010) used IR heating to dry-roast and pasteurize almonds. These authors found that after sequential IR and hot air roasting at 130, 140, or 150°C, bacterial reductions from 4.1-, 5.8-, and 7.0-log were obtained, respectively. Also, a reduction of up to 62% in the treatment time was achieved, in comparison to hot-air roasting treatments. Moreover, the sensory quality of medium roasted almonds did not show significant differences when processed with IR radiation and hot air, compared to those treated only with hot air. Therefore, authors concluded that sequential IR and hot-air roasting would be an efficient method to produce dry-roasted pasteurized almonds. The authors further suggested that for industrial purposes, the IR heaters should be installed upstream to the regular almond roasting process with hot air, which could heat the almonds to the desired roasting temperature rapidly. The combined process helps to reduce the processing time for roasting and pasteurization.

Coating especially with sweet seasonings and salting of raw or roasted almonds are common practices to enhance the snack flavor profile. Snacks produced with almond flour have been successfully developed. Hashemi et al. (2017) produced high-quality extruded snacks from corn and partially defatted almond powder with upgraded protein and DF contents.

The storage conditions for almonds is of high relevance since it affects the shelf life of the product, which under adequate conditions may reach up to three years without loss of quality. Once shelled and processed, almonds should be stored at low temperatures (3–5°C), under dry conditions (65–70% relative humidity), away from strong odors and sunlight. The adequate selection of packaging materials (Chapter 4) to block oxygen, ambient moisture, and light is highly relevant to extend the shelf life of almonds (Payne 2005).

15.3.2 Chemical Composition of Almonds

The chemical and nutrient composition of raw almonds (Table 15.3) clearly indicates that its main chemical components are lipids with an average content around 50% followed by proteins, which represent approximately 20% of the kernel weight. The lipid content varies significantly according to cultivar, growing region, and climatic conditions, and therefore by the year of harvest. In this sense, lipid values in almonds may range from 40 to 67% (Yada et al. 2011). The high levels of lipids in kernels are responsible for their high caloric load. However, the lipid profile is appreciated from the nutritional perspective and has been the focus of several studies related to health benefits (Ros 2010). The fatty fraction in almonds is mainly stored in the cotyledons as an intracellular oil droplet and is constituted primarily of mono and PUFA (Yada et al. 2011). Specifically, the major fatty acids found are oleic 18:1 and linoleic 18:2, which represent around 87% of the total lipids. These fatty acids are followed by palmitic 16:0, stearic 18:0, and palmitoleic 16:1, which represent 6.2, 1.4 and 0.5% of the total oil, respectively (USDA 2021).

Protein is the second more predominant component (Table 15.3) and its levels may differ, mainly depending on the variety. The range of protein values varies from 14 to 26%. The major storage protein in almonds is a globulin called amandin. Around 30% of the proteins are constituted by essential amino acids, where methionine, lysine, and threonine are reported as the limiting acids (Yada et al. 2011). However, one of the major limitations of almonds is that their proteins may trigger allergenic reactions in susceptible people. Recently, Zhang and Jin (2020) reviewed the molecular properties of almond allergens. The authors reviewed the allergens designated by the World Health Organization, which included Pru du 3, a lipid transfer protein; Pru du 4, a profiling; Pru du 5, a ribosomal protein; Pru du 6, a legumin commonly named pruning; and Pru du 9, an antimicrobial protein and other five proteins classified as possible allergens. Among the latter, authors identified almond vicilin, γ-conglutinin, 2S albumin, Pru du 1, and Pru du 2.

The predominant carbohydrates in almonds are non-starch polysaccharides associated with the DF fraction. In fact, their DF content is the highest compared to the tree nuts group (Table 15.3). The

TABLE 15.3
Nutrient Composition of Different Types of Nut Products

Component	Almonds	Cashews	Walnuts	Hazelnuts	Pecans	Macadamias	Pistachios	Pine Nuts
Moisture, %	4.41	5.20	4.07	5.31	3.52	1.36	4.37	2.28
Energy, kcal/100 g	579	553	654	628	691	718	560	673
Protein, %	21.15	18.22	15.23	14.95	9.17	7.91	20.16	13.69
Lysine, %	0.568	0.930	0.424	0.420	0.287	0.018	1.138	0.540
Methionine, %	0.157	0.360	0.236	0.220	0.183	0.023	0.360	0.259
Total lipids, %	49.93	43.85	65.21	60.75	71.97	75.77	45.32	68.37
Saturated fat, %	3.80	7.78	6.13	4.46	6.18	12.06	5.91	4.90
Monounsaturated, %	31.55	23.80	8.93	45.65	40.80	58.88	23.26	18.76
Polyunsaturated, %	12.33	7.84	47.17	7.92	21.61	1.50	14.38	34.07
Carbohydrate by difference, %	21.55	30.19	13.71	16.70	13.86	13.82	27.17	13.08
Starch	0.72	23.49	0.06	0.48	0.46	1.05	1.67	1.43
Dietary fiber, %	12.5	3.3	6.7	9.7	9.6	8.6	10.6	3.7
Total sugars, %	4.35	5.91	2.61	4.34	3.97	4.57	7.66	3.59
Calcium, mg/100 g	269	37	98	114	70	85	105	16
Iron, mg/100 g	3.71	6.68	2.91	4.70	2.53	3.69	3.92	5.53
Magnesium, mg/100 g	270	292	158	163	121	130	121	251
Phosphorus, mg/100 g	481	593	346	290	277	188	490	575
Potassium, mg/100 g	733	660	441	680	410	368	1025	597
Sodium, mg/100 g	1	12	2	0	0	5	1	2
Zinc, mg/100 g	3.12	5.78	3.09	2.45	4.53	1.30	2.20	6.45
Vitamin C mg/100 g	0	0.5	1.3	6.3	1.1	1.2	5.6	0.8
Thiamin, mg/100 g	0.205	0.423	0.341	0.640	0.660	1.195	0.870	0.364
Riboflavin, mg/100 g	1.138	0.058	0.150	0.113	0.130	0.162	0.16	0.227
Niacin, mg/100 g	3.618	1.062	1.125	1.800	1.167	2.473	1.300	4.387
Pantothenic acid, mg/100 g	0.471	0.864	0.570	0.918	0.863	0.758	0.52	0.313
Vitamin B6 mg/100 g	0.137	0.417	0.537	0.563	0.210	0.275	1.7	0.094
Folate, µg/100 g	44	25	98	113	22	11	51	34
Vitamin E (α-tocopherol), mg/100 g	25.63	0.90	0.70	15.03	1.4	0.54	2.86	9.33

Source: USDA (2021).

DF in almonds is mainly constituted by arabinose-rich polysaccharides, including pectic substances (Yada et al. 2011). Among sugars, sucrose is the predominant molecule, representing 90% of the total sugars (USDA 2021). Almonds are almost free of starch molecules and this complex carbohydrate represents less than 1% of the kernel weight.

The micronutrient composition of essential minerals and vitamins associated with almonds is depicted in Table 15.3. The mineral contents are affected by the soil, water supply, especially during fruit development and ripening, and also by the plant species, and therefore may be different among varieties (Yada et al. 2011). The predominant minerals in almonds are potassium and phosphorus. They also contain relevant amounts of calcium, being the highest when compared to tree nuts discussed herein (Table 15.3). Almonds also contain an adequate content of vitamins, especially niacin, riboflavin, and vitamin E. In fact, almonds contain the highest vitamin E concentration among the different tree nuts.

The phytochemicals associated with almonds (Table 15.4) clearly indicate that these kernels contain flavonones and flavonols with average contents of 0.38 and 7.93 mg/100 g, respectively (Bolling et al. 2011). The predominant flavonone in raw almonds is naringenin and among the main flavonols are kaempferol-3-*O*-glucoside, kaempferol-3-*O*-rutinoside, and quercetin-3-*O*-rutinoside (Chang et al. 2016). As for other nutrients, the phenolic profile and thus the antioxidant properties of almonds are affected by the type of variety. In this sense, the polyphenols may range from 4 to 11 mg/100 g, total phenols from 58 to 159 mg gallic acid equivalents (GAE)/100 g, and the ferric-reducing antioxidant power (FRAP) from 368 to 891 μmol Trolox equivalents (TE)/100 g (Bolling et al. 2011). The predominant phenolic acids in natural almonds are chlorogenic acid (1.61 mg/g dw) and 5-hydroxybenzoic acid (1.26 mg/g dw) (Chang et al. 2016). Almonds also contain significant concentrations of hypocholesterolemic phytosterols, especially β-sitosterol (Table 15.4).

Consumption of almonds has been related to potential health benefits such as controlled levels of lipids in blood, cardioprotective effects, and potential prebiotic properties. These benefits are likely related to the presence of bioactive compounds in almonds, including DF, phytosterols, and phenolic compounds (Yada et al. 2011). However, intensive processing of almonds such as blanching or roasting at high temperatures may decrease levels of phytochemicals such as phenolic compounds and, therefore, the associated antioxidant capacity (Bolling et al. 2011).

15.4 CASHEWS

Cashew (*Anacardium accidentale*) is among the three most important nuts in the international market and commonly mingled with almonds and walnuts (International Nut & Dried Fruit 2020). It is estimated that cashew is the second most consumed tree nut only after almond (Table 15.1). In 2019, the world production of inshell cashew was 3.960 million tons with a harvested area of 7.091 million Ha and a production yield of 0.558 tons/Ha (FAO 2021). The cashew tree is believed to be native to Brazil, but nowadays the world regions with higher production are Western Africa, India, and Eastern Africa (Table 15.1). In Western Africa, Côte d'Ivoire is the main producer (FAO 2021). Since the processing of inshell cashew nuts is difficult and expensive due to the specific characteristics of the shell, African cashews are mostly processed in India and Vietnam (Rico et al. 2016).

The nut production around the world is mainly from wild-grown and small farms. Harvest is laborious and commercial production on large plantations is limited. It takes three years for trees to start bearing fruits and seven years to reach maximum production. The cashew nuts are allowed to ripen on the tree and let them fall to the floor where they are collected. The tree yield is from 45 to 70 kg of nuts and fruit, representing approximately 9 kg of unhulled nuts and only 3 kg of raw, dried kernels (Payne 2005). This comparatively lower yield is the reason why the harvested area and the tons of production of inshell cashews are significantly higher in comparison to almond values (Section 15.2). However, recent breeding efforts to develop trees with higher yields have increased yields by about 40% (Payne 2005, FAO 2021).

TABLE 15.4
Phytochemicals in Almonds and Tree Nuts

Phytochemical Compound	Almonds	Cashew	Walnuts	Hazelnuts	Pecans	Macadamia	Pistachio	Pine Nuts
Total phenols (mg GAE/100 g)	261	242	1602	447	1588	233	703	206
Flavonoids (mg/100 g)	25.01	1.12	0.54	13.21	2713.49	137.9	136.45	0.03
Anthocyanins (mg/100 g)	2.46	0.00	2.71	6.71	18.02	0.00	6.06	0.00
Flavan-3-ols (mg/100 g)	4.47	1.98	0.00	5.25	15.99	0.00	6.85	0.49
Total isoflavones (mg/100 g)	0.01	0.01	0.03	0.03	0.00	NR	3.63	NR
Total proanthocyanidins (mg/100 g)	184.0	8.7	67.2	500.7	494.0	10.0	252.7	1.0
Carotenoids (μg/100 g)	2	31	21	106	55	NR	22832	26
Phytates (mg/100 g)	2542	698	2070	1285	852	471	1563	200
Sphingolipids (mg/100 g oil)	0.63	NR	0.68	0.32	0.55	NR	0.82	0.57
Total sterol content (mg/100 g)	192	154	198	132	234	116	189	191
β-Sitosterol (mg/100 g)	130	113	87	102	117	108	198	132
Stigmasterol (mg/100 g)	4	0	0	1	3	0	5	0
Campsterol (mg/100 g)	5	9	5	7	6	8	10	20
Sum of tocopherols[a] (mg/100 g)	26.57	6.60	23.57	15.36	26.70	0.54	24.07	20.83
β-tocopherol (mg/100 g)	0.23	0.03	0.15	0.33	0.39	0	0	0
γ-tocopherol (mg/100 g)	0.64	5.31	20.83	0	24.44	0	20.41	11.15
δ-tocopherol (mg/100 g)	0.07	0.36	1.89	0	0.47	0	0.8	0

Source: Bolling et al. (2011) and USDA (2021). NR: not reported.

[a] Include vitamin E (α-tocopherol) presented in Table 15.3.

Cashew trees yield a fruit known as cashew apple, which is a swollen peduncle with sweet flavor and aroma that is consumed in beverages and jams. The cashew nut grows externally to the fruit (at the end of the peduncle) and inside a kidney-shaped hard shell which is composed of two layers that form a honeycomb structure where in-between layers lies an oily substance known as phenolic cashew nut shell liquid. Inside the hard shell, and protected from this oily substance by the second layer of the shell, is the semilunar-shaped kernel or cashew nut (Mendes et al. 2019). The kernel is the chief product and its processing aims to extract intact kernels from the nut. Kernels weigh approximately 1.93 g and have an average length of 2.37 cm (Serna-Saldivar 2008).

15.4.1 Cashews Processing and Storage

Processing of cashew nuts involves the recovery of the kernel or edible portion by mechanical means. Immediately after harvesting, the nuts are sun-dried until the kernel shrinks from the shell. Then, fruits are cleaned, conditioned, roasted, shelled, dried, peeled, graded, and packaged. The main cleaning operation is subjecting the cashew nuts to sieving to remove dirt, stones, and foreign material. Inshell raw cashews must be steam cooked or drum roasted to remove a toxic nut shell liquid known as cashew nutshell liquid (CNSL) composed of anacardic acid and cardol. This liquid is the reason why cashews are not sold inshell since it causes severe skin irritation. In order to remove CNSL, cashews are traditionally conditioned by soaking nuts in water followed by roasting over a sieve in an open-air fire. The resulting smoke generated by the liquid that runs out from the nuts is toxic. Modern removal of CNSL include roasting in hot oil bath treatments or by solvent extraction techniques (Payne 2005). An alternative process which is extended in India is the steam cooking. In this process, the well-dried raw nuts are subjected to steam boiling with 4.5–5 kg/cm^2 steam pressure for about 30 min and then cooled (Mohod et al. 2010). After roasting or steam cooking, cashews are carefully decorticated, which is one of the key steps. This operation commonly consists of manual cracking the fruit to separate the kernels by hand and foot operated shelling machines. The resulting kernels are dehydrated to reduce the moisture to about 4% and facilitate the removal of the adhered testa. The dried kernels are conditioned in moist chambers for approximately 24 h before the removal of the testa. Finally, the cashew nuts are graded by size and packaged (Serna-Saldivar 2008).

Cashews are mainly designated for the snack market, where they are salted and consumed as whole kernels and halves (Payne 2005). Cashews are generally preserved in cold storage or vacuum-packaged tins with the headspace refilled with carbon dioxide. The use of this inert gas delays rancidity and preserve freshness. The optimum storage conditions are at a temperature of 0–2°C and a relative humidity of 60–75% (Serna-Saldivar 2008).

15.4.2 Cashews Chemical Composition

Cashews are considered highly nutritious because they provide calories from carbohydrates, proteins, and lipids. The chemical and nutrient composition of raw cashew is depicted in Table 15.3. Although the nutritional composition of cashews may vary depending on the production region, lipids are consistently the main components comprising more than 40% of the total weight and being the reason for the high caloric load of this food. The major fatty acids are oleic 18:1 and linoleic 18:2, followed by palmitic 16:0 and stearic 18:0 acids, which account for approximately 54, 18, 8.9, and 7.3% of the total lipids, respectively (USDA 2021).

Unlike almonds and other tree nuts, cashews contain significant amounts of starch, which comprises approximately 20% of the total weight (Table 15.3). Other carbohydrates contained in cashews include soluble sugars (~6%), especially sucrose which represents up to 98% of total sugars, and DF in a lower proportion (~3%).

Cashews are among the nuts that contain higher protein content (Table 15.3) with values up to 23%, depending on the variety (Rico et al. 2016). More importantly, its protein quality has been highlighted due to the high lysine content, which is among the highest compared to other tree nuts

(Table 15.3). Yet, depending on the cashew variety, the concentration of this amino acid could be still below the requirement for 1–2-year infants, though it accomplishes the requirements of adults (WHO/FAO/ONU 2007). Still, cashew proteins are known to trigger allergenic reactions. There are three proteins identified as the major allergens in cashew nut: Ana o 1 and Ana o 2 classified as globulins and the main storage proteins in nuts, and the albumin Ana o 3, rich in arginine, glutamine, asparagine, and cysteine, similar to other prolamin proteins that are recognized to trigger allergic reactions in other grains. As seen with other tree nuts, a trace amount of cashew protein (1 mg) is enough to induce an adverse immunological response. Therefore, adequate labeling of products containing cashews is necessary since there is a worldwide increasing prevalence of allergic reactions associated with increased consumption of this tree nut, especially in industrialized countries (Mendes et al. 2019).

Cashew also contains adequate levels of hydrosoluble and liposoluble vitamins and minerals (Table 15.3). Compared to the rest of nuts, cashew is especially rich in iron, magnesium, phosphorus, and zinc, but it also contains the highest sodium content. As for other components, the sodium content differs depending on the variety ranging from 10.5 to 20.5 mg/100 g (Rico et al. 2016).

The content of phytochemical compounds in cashews is summarized in Table 15.4. The main phytochemicals found in cashews are phytates, phenolic compounds, and sterols; however, their contents are lower when compared with other tree nuts. The main phenolic acid is syringic acid (0.61 mg/g defatted meal) and the main flavan-3-ol is catechin (11.7 mg/ g defatted meal), although epicatechin and epigallocatechin are also present (Chang et al. 2016). β-Sitosterol is the predominant sterol representing up to 85% of total sterols followed by δ-5-avenasterol and campesterol representing ~7.5 and ~6%, respectively (Rico et al. 2016). The consumption of cashews has been related to several health benefits, which are likely associated with these phytochemicals. Besides, cashew nuts have been used in traditional medicine to treat diseases such as diabetes, diarrhea, inflammation, and skin infections due to their documented antibacterial and antifungal activities (Mendes et al. 2019). However, the selection of the cashew shelling method which involves drying and steaming or oil roasting is critical to avoid losses of phytochemical compounds (Bolling et al. 2011).

15.5 WALNUTS

The Persian walnut (*Juglans regia* L.) is the most important member of its genus as a commercial nut producer. The ancient Greeks knew the walnut and its common name suggests Persia (today Iran) as its native homeland, but its original distribution was larger and extended from the Caucasus to Turkistan. From Europe, the Persian walnut reached North America with the early settlers during the 16th and 17th centuries. The term English walnut is also used because these nuts were transported in English ships (Serna-Saldivar 2008). Walnut is one of the three most important nuts in terms of worldwide production. In 2019, the inshell walnut world production reached 4.498 million tons with a harvested area of 1.305 million Ha and a production yield of 3.446 tons/Ha (FAO 2021). The current main world producers are China and the United States (Table 15.1).

Walnut plantations do best in areas with good, well-drained, and nonsaline loamy soils. This crop is wind-pollinated and orchards perform the best in areas with little spring or fall frosts and with adequate winter rains and chilling. Walnut orchards have a normal productive life of 50–100 years. The trees start to bear fruit from four to ten years after plantation and the best production is achieved in 10–20-years-old trees. The fruit of walnut trees consists of a husk, which encloses the woody nut. Walnuts are commonly handpicked after mechanically shaking the trees (Serna-Saldivar 2008).

15.5.1 Walnut Processing and Storage

After harvesting, the kernel moisture content may be up to 35%. Therefore, husks should be immediately removed, nuts dried to reduce the moisture to about 4%, and then kept under dry storage

(Serna-Saldivar 2008). The low moisture and water activity aids to preserve the quality of the walnut and avoids deterioration during prolonged storage (Payne 2005).

Walnuts are marketed as both inshell and shelled nuts. To make them more appealing for inshell commercialization, walnuts are washed and sometimes bleached to improve the outer surface appearance. Inshell consumption is popular during the Christmas holidays. Shelled walnuts are produced by specialized equipment to mechanically crack the shell and air separators to remove shell and fiber fractions. Once shelled, storage conditions must be designed to prevent deterioration. Ideal conditions involve low temperatures (0–3.5°C) and relative humidity (55–65%), avoiding strong odors and selecting adequate materials that serve as vapor barriers (Payne 2005).

For snacking, walnuts may be found natural, roasted, blanched, or coated with honey or sugar (Payne 2005). Walnuts are roasted for about 30 min at temperatures around 120°C to achieve the most desirable kernel color (Ghafoor et al. 2020).

15.5.2 Walnut Chemical Composition

The chemical and nutritional composition of walnuts is presented in Table 15.3. Walnuts have a high caloric density (~650 kcal/100 g) due to their lipid content, which averages 65% of the total weight. However, in a study performed in walnuts cultivars produced in Portugal, the lipid contents reached 82% (Pereira et al. 2008). The lipid fraction is appreciated from the nutritional viewpoint since more than 70% of the fat is composed of PUFA. Linoleic (18:2 ω-6) and linolenic (18:3 ω-3) are the two main fatty acids comprising 58 and 14% of total lipids, respectively. These fatty acids are followed by 13.5% oleic (18:1 ω-9) and 7% palmitic (16:0) acids (USDA 2021).

The second and third major components in walnuts are proteins and DF, respectively (Table 15.3). The DF content of walnuts is intermediate and the total sugar content, mainly comprised of sucrose, is the lowest compared to almonds and other tree nuts (USDA 2021). Walnuts contain a variety of essential minerals and vitamins (Table 15.3) and particularly their folate concentration is comparatively higher compared to other tree nuts.

The main phytochemicals found in walnuts (Table 15.4) indicate that they contain one of the highest total phenolic contents among tree nuts. In fact, unlike other tree nuts, hydrolysable tannins have been reported in walnuts with concentrations of 23.56 mg/100 g (Bolling et al. 2011). The main phenolic acid in raw walnut is gallic acid (37.2 mg/100 g), while among flavan-3-ols glansreginin A (484 mg/kg) and procyanidin trimer (278 mg/kg) predominate (Chang et al. 2016). The walnuts are also rich in phytates and sterols where β-sitosterol predominates. In addition, it contains significant amounts of γ-tocopherol, which represents 88% of these compounds (Table 15.4). Reports have found that walnuts have low levels of auto-oxidation, even though they have high contents of unsaturated fatty acids and relatively low amounts of α-tocopherol. The low susceptibility to rancidity is mainly attributed to the high content of antioxidants such as phenolic acids and tannins (Hayes et al. 2016). Due to the high load in phytochemicals, walnut extracts have also proved to have antibacterial and antimycotic properties (Pereira et al. 2008). Roasted walnuts presented higher antioxidant activity compared to the raw counterpart (Hayes et al. 2016). A slight degradation of tannins (reported as catechin equivalents) has been reported after roasting at high temperatures (~200°C) (Bolling et al. 2011). Likely, the hydrolysable tannins may be causing the enhanced antioxidant activity observed in roasted walnuts.

There has been considerable interest in the health benefits of walnuts especially because of their high levels of PUFA, DF, polyphenols, tocopherols, and other nutraceuticals that reduce the risk of cardiovascular disease, diabetes, and other disorders (Hayes et al. 2016). In a study reported by Bamberger et al. (2017), authors found that the daily consumption of 43 g of walnuts decreased the fasting cholesterol, non-HDL cholesterol, LDL cholesterol, triglycerides, and apoB in plasma of healthy patients.

15.6 HAZELNUTS

Hazelnuts or filberts (*Corylus avellana* or *Corylus maxima*) are distributed throughout the temperate parts of the northern hemisphere. In 2019, the world production of inshell hazelnut was 1.125 million tons with a harvested area of 1 million Ha and a production yield of 1.125 tons/Ha (FAO 2021). Hazelnuts are more common in Europe than in America. Turkey is the main world producer with more than 70% of the total output (Table 15.1).

Hazelnut orchards start producing fruits after four to six years and reach full production when they are 10–12 years old. The normal productive life of the trees is about 40 years (Serna-Saldivar 2008). According to FAO (2021), the hazelnut orchard yields in the world are lower compared to almonds (Section 15.2) and walnuts (Section 15.3). The major problem with hazelnut orchards is the biennial bearing (Serna-Saldivar 2008). Hazelnut fruits are constituted of a husk that encloses a woody and round nut. To harvest hazelnuts, mechanical shakers and manual methods are used (Payne 2005). Nuts should be picked up from the tree within the first days after falling; otherwise they will be more prone to deterioration. The fruits should be immediately dehusked and then stored in a cool, dry shed or storage facility with adequate air circulation (Serna-Saldivar 2008).

15.6.1 Hazelnut Processing and Storage

Freshly harvested nuts are first washed in preparation for disinfection. Then, the inshell nuts must be dehydrated, which is commonly performed with hot air at temperatures lower than 50°C. The aim of this step is to reduce the moisture from 15–25% to approximately 10% (Wang et al. 2020). Recently, Wang et al. (2020) evaluated the hot-air-assisted continuous radio frequency (HARF) heating to improve the drying efficiency of inshell hazelnuts. This technology led to a reduction in the drying time from 7 h with the traditional hot-air method to 22 min. Moreover, the HARF dried nuts retained higher contents of bioactive compounds and led to lower lipid oxidation compared to hazelnuts conventionally dried with hot air. However, their study led to more shell cracking during HARF drying, which indicated that drying conditions need to be properly optimized to avoid this type of potential defect.

As for other tree nuts, shelled hazelnuts should be stored under cold storage preferably in tightly sealed containers and away from strong odor materials. These conditions may lead to a shelf life of one to two years for hazelnuts stored at refrigeration or frozen conditions (Payne 2005).

Roasting hazelnuts may be performed after immersing the hazelnut halves in lemon juice, salt, and starch (Shafiei et al. 2020). Roasting could be done at temperatures ranging from 120 to 180°C for about 5–60 min. Both time and temperature affect the textural properties of roasted hazelnuts. Therefore, time–temperature combinations should be optimized to determine the better roasting schedule for hazelnuts in terms of texture (Demir and Cronin 2005). Although roasting may positively modify the texture and flavor of filberts, it has negative effects on the nutritional composition and shelf life. Roasting reduces significantly the levels of riboflavin and thiamine (up to 30 and 50%, respectively) and may also lead to the reduction of amino acids, especially with increasing roasting degrees (Özdemir et al. 2001). Interestingly, according to the results found by Özdemir et al. (2001), the use of a two-stage roaster where two different temperatures can be applied (the second step with a lower temperature) could help to reduce the losses of these nutrients. Roasting is also known to promote oxidative reactions, which leads to a shorter shelf life due to the decrease of antioxidant compounds that are naturally present in the kernel (Shafiei et al. 2020). Belviso et al. (2017) compared two roasting systems (hot air and IR) at different time–temperature combinations and found that a better oxidative stability was obtained in hazelnuts roasted with the hot-air system at low temperature (120°C) for 40 min. Based on their results, these authors recommend storage times of six months at 4°C to maintain a high antioxidant capacity. The combination of microwave and IR heating for roasting hazelnuts was tested by Uysal et al. (2009). Results indicated that hazelnuts roasted at optimum conditions with the proposed microwave-IR methodology had comparable quality in terms of color, texture, and moisture, compared to hazelnuts roasted with a conventional roasting performed in an electrical oven. Moreover, the roasting time was significantly reduced from 20 to 2.5 min when the proposed microwave-IR methodology was applied.

Hazelnuts are a popular snack and are commonly included in snack mixes as whole, sliced, or chopped natural, blanched, or roasted forms. Flour, oil, and hazelnut butter are also other products commonly found on the market and specially used in dairy, bakery, and confectionery products (Köksal et al. 2006).

15.6.2 Hazelnut Chemical Composition

The main component of hazelnuts is the lipid fraction which may reach values up to 68% (Köksal et al. 2006). The high content of fat and relatively low moisture makes hazelnut a high caloric product (620 kcal/100 g, Table 15.3). However, as seen with other tree nuts, hazelnuts have a favorable fatty acid profile. Hazelnuts are rich in monounsaturated oleic acid, which accounts for 75% of total lipids. Other relevant fatty acids in hazelnuts are linoleic 18:2 and palmitic 16:1, which represent 13 and 5% of total lipids, respectively.

The carbohydrate fraction represents around 16% of the total weight and is mainly formed by DF and sugars (Table 15.3). The main sugar present in hazelnuts is sucrose, which comprises up to 97% of the total soluble sugars (USDA 2021).

The protein contents in hazelnuts represent approximately 15% of the kernel weight (Table 15.3) and it varies depending on the hazelnut variety. For instance, Köksal et al. (2006) evaluated Turkish varieties and assayed protein contents up to 21%. These authors also found that hazelnuts are rich in the essential amino acid leucine (from 924 to 1,271 mg/100 g of kernel), but deficient in methionine (162 ± 19 mg/100 g of kernel) and lysine (454 ± 48 mg/100 g of kernel).

The vitamin and mineral contents of hazelnuts depicted in Table 15.3 indicate that in general, hazelnuts contain intermediate contents compared to almonds and other tree nuts. Although the USDA (2021) does not report sodium in hazelnuts, they may contain slight quantities depending on the variety (from 2.0 to 3.8 mg/100 g) (Köksal et al. 2006). In terms of vitamins, filberts contain the highest concentrations of vitamin C, pantothenic acid, vitamin B6, and folates among tree nuts. They also contain an adequate content of the antioxidant vitamin E (Table 15.3).

Filberts also contain an array of relevant phytochemical compounds, which are summarized in Table 15.4. Hazelnuts and pecans contain the highest concentrations of proanthocyanidins when compared to the other tree nuts. The main phenolic acids are protocatechuic acid (2.54–7.83 mg/100 g) and gallic acid (1.70–4.11 mg/100 g). Condensed and hydrolysable tannins have also been reported in hazelnuts (Chang et al. 2016). After roasting hazelnuts, the total phenolic content is not significantly affected. This occurs because individual forms as protocatechulic acid, phloretin-2-*O*-glucoside, catechin, and epicatechin contents decrease while other compounds such as gallic acid increase and compensate the losses of other compounds. The increment in gallic acid could be likely associated with the degradation of hydrolysable tannins during roasting (Bolling et al. 2011).

15.7 PECANS

Pecan (*Carya illinoensis*) is a deciduous tree of the Juglandaceae family that is growing in popularity due to the nutritional composition and positive health implications of its delicious and edible nuts. Pecan is closely related to walnuts (*Juglans regia*), hickory (*Carya ovata*, *Carya texana*, and *Carya alba*), and the Chinese wing nut (*Peterocarya stenopters*). This valuable nut crop is indigenous to the territories that neighbor northern Mexico and southern United States (Andersen 2019). Major disadvantages of pecan culture are the length of time before pecans come into significant production (five to ten years) and that it commonly bears fruits heavily and lightly in alternate years. Undoubtedly, plant hormones play a role as a plant signal in alternate bearing

(Thompson and Conner 2012, Andersen 2019). The alternate bearing has been recognized as a key component limiting the expansion of the pecan industry by producers, shellers, and processors. The cultivar has a major impact on alternate bearing. Besides, several cultural practices have been developed to reduce alternate bearing. These include management of the tree canopy for light interception, mineral nutrition, groundcover vegetation, water drainage, pest control, and cropload (Smith 2012)

The pecan nuts have been used as food for thousands of years before the discovery of America in 1492. Members of the Spanish expedition described them in their journals and documented that it was a subsistence crop for at least two months every other year for native Indians. Native Americans and fur traders disseminated pecans from the Mississippi Valley eastward (Andersen 2019) and soon became an important trade item. The first recorded shipment to England was documented in 1761. The pecan name was adopted from the Algonquin Indian tribe language, meaning nuts were so hard that needed a stone for cracking. Mexico and the United States are the main producers of pecans in the world (Table 15.1). Other countries that produce this tree nut are South Africa, Australia, and Brazil, which accounted jointly for 10% of the production in 2019 (International Nut & Dried Fruit 2020).

The perennial pecan trees grow in regions with long warm growing seasons. The length of the growing season is the period of the last frost in the spring and the first killing frost or freeze in the fall. The trees begin to produce fruits after at least five years and in some instances take to fruit about ten years. The tree can last for more than a century. Pecan trees grow on a wide variety of soil types, although the most productive ones grow preferably on alluvial depth soils. Trees require abundant soil moisture for normal growth and fruiting, but they do not withstand excessive moisture nor poor soil aeration that accompanies poor drainage. A 10–20-year-old commercial pecan tree yields from 20–50 kg/year equivalent to 1.3–3.5 tons/Ha (Alasalvar and Shahidi, 2008, Thompson and Conner 2012, Smith 2012, Rabago-Panduro et al. 2020).

Pecan trees yield dehiscent fruits constituted by a shuck or husk and a seed. The seed is divided into hard shell, middle septum, and the edible kernel (Vendrame and Wetzstein, 2005, Janick and Paull 2008). The pecan nut kernels represent less than 30% of the total mass of the whole nut (Idowu et al. 2017). During fruit development, the kernel stores oil, protein, polysaccharides, and an array of lipophilic and hydrophilic phytochemicals with positive human health implications (Alasalvar and Shahidi, 2008, Rabago-Panduro et al. 2020).

The most suitable varieties for shelling and direct consumption are the thin-shell pecans also called "paper shelled nuts." These varieties yield nuts with a higher kernel-to-shell ratio and generally produce larger nuts. The indigenous pecan nuts weigh approximately 2 g each, have an average length and width of 2.25 cm and 1.38 cm, respectively, and yield around 50% edible kernels. In contrast, current paper-shelled varieties produce nuts with an average weight of 7.7–9.3 g and yield up to 60% kernels. The round- and lenticular-shaped pecans have an average length and width dimensions of 3.3 × 2.4 and 6.1 × 2.5 cm, respectively (Serna-Saldivar 2008). According to Smith (2012), the

most popular commercial varieties of pecans are Burkett, Cherokee, Cheyenne, Mahan, Mohawk, Pawnee, Sioux, Wichita, and Western. Recently, Rabago-Panduro et al. (2020) evaluated the detailed composition of dry Western pecan kernels.

15.7.1 Pecan Processing and Storage

According to Byford (2005), the optimum maturation point for pecan harvesting is when the shuck splits into four wedges and contains a moisture content averaging 10%. Immediately after harvesting, the pecan nuts are normally washed, cleaned, and dehydrated to reduce their moisture to approximately 3%. The nuts are sized, graded, and sold with shells or alternatively shelled or cracked with the aim of obtaining halves of the kernels. The kernels normally constitute 56–61% of the total nut weight. Smith (2012) performed a 5-year study with Pawnee pecans that yielded 18–29 kg/tree among years. The nut weights varied from 7 to 10 g, whereas the kernel weights from 5.4 to

6 g. Therefore, the %kernel yields varied from 59 to 61%. Interestingly, the pecan trees with a low alternate-bearing index produced the greatest 5-year cumulative yield (Smith 2012)

The percentage of shelled kernels marketed has steadily increased through the years. The shelling process consists of first storing the harvested nuts since the harvesting season generally lasts only a couple of months. The common method is to store at 0°C until they are needed for shelling. Pecans are cleaned in rotating drums, which selectively remove dockage or foreign material. Pecans can also be sized in cylindrical drums to separate nuts into different diameter sizes. This is to facilitate efficient cracking since the cracking machines are adjusted according to the nut size. Sizing also helps to produce half kernels with similar sizes after shelling. The nuts are soaked in excess chlorinated water or subjected to live steam in preparation to cracking. The treatment increases the water content of the shell, sterilizes the nuts, softens the shell, and after cracking yields kernels that are more pliable and less brittle or susceptible to breakage. The moisture content of the resulting kernels increases from 3–4% to around 9–10%. The most important operation is cracking, and this is accomplished with a machine that breaks every single nut. Oblong pecans crack better than round ones since they fall more readily in place for end to end in between the plungers. After cracking, partial separation of kernels and shells is accomplished by a series of shaker-vibrating screens with different size opening and different speeds to disengage kernels from shells. Air jets are used to remove small shell pieces and corky materials. After separation, kernels are passed over vibrating screens with openings of progressively larger sizes in order to grade kernels into halves and different size pieces. Shellers strive to optimize the yield of halves because they are usually sold at higher prices than broken pieces. The final step is drying the sized kernels to return to the original moisture content of 3–4%. Air-forced dryers are commonly used to drop the moisture content of the kernels (Serna-Saldivar 2008).

After shelling, pecans may be roasted by heating at approximately 180°C for 15 min with hot air or infrared radiation. The heat treatment improves flavor and aroma but decreases shelf life because it accelerates rancidity and destroys intrinsic natural antioxidants. Synthetic antioxidants such as BHA and BHT (0.02% based on fat content) can be added to prolong shelf life (Serna-Saldivar 2008).

The storage of pecans or shelled kernels should be strictly controlled and aimed toward the preservation of the original characteristics and freshness. Rancidity is the main factor to control and is strongly influenced by storage temperature. Pecan kernels can be stored for one year at 0°C or for two years at –15°C in odor-free environments. The relative humidity of 65% is the most effective in maintaining the ideal moisture content of pecan kernels at 3–4%. If the relative humidity is too high, kernels will absorb moisture, lose texture, and become susceptible to molds. If the relative humidity is too low, kernels will lose weight and become brittle and more prone to rancidity (Serna-Saldivar 2008).

15.7.2 Pecans Chemical Composition

The chemical and nutrient composition of pecan kernels is summarized in Table 15.3. The major constituent of the kernel is the lipidic fraction (around 72% of the total kernel weight). The most abundant class of lipids are triglycerides, rich in mono- and polyunsaturated fatty acids. Their lipid content is the reason why pecan kernels are considered high caloric (716 kcal/100 g dwb). Oleic and linoleic acids are the two main fatty acids comprising about 62.5–74.9% and 17.8–27.5% of the total lipids, respectively (Oro et al. 2008, Pinheiro do Prado et al. 2013). Pecans contain less than 8% of saturated fatty acids palmitic (16:0) and stearic (18:0) (Oro et al. 2008, Pinheiro do Prado et al. 2013, USDA 2021). However, the fatty acid composition differs according to variety and maturation. Rudolph et al. (1992) concluded that more matured kernels contained higher and lower concentrations of oleic and linoleic acids, respectively. Furthermore, Flores-Córdova et al. (2016) compared the fatty acid composition of kernels shelled from the Wichita and Western varieties and found that oleic and linoleic acids were present in higher concentrations in Wichita (45% oleic) and Western (47% linoleic) kernels, respectively. Among polar lipids, pecans contain significant amounts of several types of phospholipids, monoglycerides, and diglycerides.

The second and third major components of the pecan kernels are carbohydrates, especially DF and sugars, and proteins (Table 15.3). As seen with other tree nuts, pecan proteins trigger allergenic reactions in susceptible patients. The major allergens recognized in pecans are the 2S albumin or Car i 1, 11S legumin, or Car i 4 and the pecan vicilin which is a copper-binding protein (Zhang et al. 2016).

Pecans also contain a variety of essential minerals and vitamins (Table 15.3) and phytochemicals (Table 15.4). Scientific data suggest that both lipidic and hydrophilic phytochemicals associated with pecans may reduce the risk of chronic diseases associated with oxidative stress (Ros 2010). Pecans have the highest content of sterols and tocopherols among tree nuts (Table 15.4). β-Sitosterol is the major phytosterol associated with pecans and approximately 90 and 5% of the total tocopherols of pecans are constituted by γ- and α-tocopherols, respectively. The presence of these compounds in pecans has gained relevance due to their antioxidant capacity that retard lipid rancidity and their nutraceutical and health benefits such as hypocholesterolemic effect. Among hydrophilic phytochemicals, pecans outstand for their high content of total phenols, flavonoids, anthocyanins, flavan-3-ols, and proanthocyanidins compared to the rest of the tree nuts reviewed (Table 15.4). Polyphenols represent more than 97% of the total antioxidant capacity of pecans (175.2 μmol trolox equivalents [TE]/100 g [wb]) (Wu et al. 2004, USDA 2021). Procyanidins and prodelphinidins are among the most abundant polyphenols in pecan kernels (Bakkalbaşi et al. 2008, Smeriglio et al. 2017). Ellagic acid is also a predominant polyphenol in pecan kernels and is found in significant amounts (301 ± 7 mg/100 g) (Abe et al. 2010). Although not edible, pecan hard shells are considered as a potential source of both simple phenolics and polyphenols (Flores-Córdova et al. 2016). Chlorogenic, gallic, and phydroxybenzoic acids have been shown as the principal simple phenolics of pecan shells, as well as epigallocatechin, epigallocatechin-3-gallate, and procyanidin with high antioxidant activity (Engler Ribeiro et al. 2017). However, thermal processes such as frying pecans may reduce their phytochemical content. In a study performed by Yang et al. (2015), authors found that frying pecans at low temperatures (200°C) and short times (10 min) slightly increased the content of phenolic compounds, tocopherols, and phytosterols due to a matrix breakdown that released these compounds. In the same way, the antioxidant capacity also increased. However, with higher processing times and temperatures, their concentration decreased significantly. Therefore, special attention should be paid while selecting adequate conditions during heat treatments to avoid loss of phytochemicals.

15.8 MACADAMIAS

The macadamia (*Macadamia integrifolia*) or Queensland nut is indigenous to Australia. South Africa, Australia, and Kenya are the major producers of this particular nut (Table 15.1). This crop is one of the tree nuts least produced. Nevertheless, its production has shown an increasing trend in the past ten years with a linear increment of 3,459 tons per year (International Nut & Dried Fruit 2020).

The evergreen tree of 10–15 m high is a member of the Proteaceae family. Macadamia orchards start producing fruits after 5–12 years and reach full production when they reach 10–15 years of age. The normal productive life of the trees is 25–30 years (Serna-Saldivar 2008). Macadamia trees flourish in subtropical climates with rich soils and high annual rainfalls but require good drainage. Although trees are self-pollinated, to increase yields for commercial production, it is recommended to establish more cultivars within the orchard. Space among trees should be considered to allow mechanical harvesting (Payne 2005).

When ripe, the nuts are encased in a fibrous green husk or pericarp with high moisture which makes them prone to quick deterioration (Phatanayindee et al. 2012). Under the fleshy green husk, macadamias consist of an edible nut kernel encased in a hard shell (Payne 2005). The round-shaped kernel weighs about 3 g and has a diameter of approximately 2 cm (Serna-Saldivar 2008).

There are two categories of macadamia nuts used for commercial production: the smooth shell (*M. integrifolia*) and the rough shell (*M. tetraphylla*). The majority of production corresponds to

M. integrifolia, which is preferred due to its flavor and finished product quality. *M. tetraphylla* is easier to crack but tends to have less oil and more sugar. These two types of macadamias must not be mixed when processing since *M. tetraphylla* kernels tend to brown more rapidly due to its 2–4% higher content of sugar (Payne 2005).

15.8.1 Macadamias Processing and Storage

Nuts should be picked up from under the tree within the first couple of days after falling; otherwise, the fruit will deteriorate and become rancid. They should be immediately dehusked and dried. Freshly husked macadamias contain up to 20% moisture, which must be reduced to less than 4% if good quality is desired (Serna-Saldivar 2008). When moisture is lowered to values around 1.5%, the kernel shrinks away from the inside of the shell helping to recover a higher proportion of whole kernels (Phatanayindee et al. 2012).

Drying is usually done in forced air dryers preferably at temperatures below 40°C. The drying methods affect the quality of macadamias, especially due to their high content of unsaturated fatty acids. Therefore, this process should be carefully selected to avoid the decomposition of fatty acids. Phatanayindee et al. (2012) evaluated a combination of heat pump drying with tunnel drying, which was compared to hot-air drying. Authors found that the hybrid process led to a reduced processing time, lower concentrations of degraded fatty acids and undesirable volatiles, and higher stability of macadamias during storage.

After cracking, nuts yield about 35–45% edible kernels. The nuts are selected according to their specific gravity. The less-dense nuts containing more lipids are graded by flotation in water (Serna-Saldivar 2008). The storage stability of raw macadamias depends on moisture and temperature. The moisture content should be up to 1.4% in raw kernels and when moisture is higher than 2%, macadamias would have poor storage stability. As for other nuts, macadamias must be stored under low oxygen, cold storage, and low relative humidity and away from strong odors as these are readily absorbed while in cold storage. Even though macadamias contain high fat content, their shelf life is comparable with other nuts when stored under adequate conditions (Payne 2005).

Macadamias may be merchandized as raw or roasted forms. Roasting may be performed with hot air or in oil. This operation is primarily performed to improve the organoleptic characteristics especially in terms of color and flavor; however, it may also inactivate undesirable enzymes and improve the storage stability (Duduzile Buthelezi et al. 2019). Air-roasting is performed at temperatures around 125°C for approximately 15–25 min. When roasting is performed in oil, coconut oil is commonly used at temperatures around 120°C. Some problems associated with this method are the degradation of oil yielding peroxidant contamination and loss of endogenous oil to the frying medium. However, this method permits an even color in the oil-roasted macadamias (Wallace and Walton 2011). After cooling, oil-roasted kernels are treated with adhesive oils and sprinkled with salt (1.5% w/w) (Payne 2005). To improve oxidative stability, roasted kernels could be treated with antioxidants before packaging (Desphande et al. 1995, Serna-Saldivar 2008).

Roasted or fried macadamias must be packed immediately after processing to avoid moisture absorption. A vacuum package is preferred and film pouches, tins, and glass jars may be alternatively used. This package promotes long-term shelf life, especially when stored in cool and dry conditions (Payne 2005). Interestingly, Duduzile Buthelezi et al. (2019) found that roasting macadamias improved their quality in terms of lower peroxide values, higher concentrations of phenolic compounds, and therefore higher antioxidant activity when compared to raw counterparts. The enhancement in the antioxidant properties of macadamias led to higher stability during storage.

15.8.2 Macadamias Chemical Composition

The chemical and nutrient composition of macadamia nuts is summarized in Table 15.3. The main chemical component in this nut is fat (75%) leading to a product with more than 700 kcal/100 g.

Macadamias contain the highest average fat content, caloric load, and proportion of monounsaturated fatty acids (78% of total lipids) compared to the rest of the nuts reviewed in this chapter. Oleic (18:1) and palmitoleic (16:1) acids are the two main fatty acids representing 58 and 17% of total lipids.

The second main component is carbohydrates, especially DF and sugars (Table 15.3). The third component in macadamias is proteins with an average content of 8%. This value is comparatively lower than almonds and other tree nuts. Macadamias also contain proteins that trigger allergic reactions. Anaphylaxis in macadamia-allergic patients and IgE reactivity to a 17-kDa protein in both raw and roasted macadamia extracts have been reported (Geiselhart et al. 2018).

Macadamias contain low contents of phosphorous, potassium, and zinc when compared to other tree nuts, but it has the highest thiamin content (Table 15.3). The phytochemicals found in macadamias are summarized in Table 15.4. Although a variety of phytochemicals are found in macadamias, their contents are generally among the lowest values among tree nuts. For instance, the only tocopherol reported in macadamias are α-tocopherols, which are found in lower amounts compared to other nuts reviewed herein. Similarly, the sterol content in macadamias is the lowest among tree nuts.

Duduzile Buthelezi et al. (2019) found that roasting macadamias under mild conditions (125°C for 15 min) led to a higher concentration of flavonoids, total phenolic compounds, as well as improved antioxidant activity compared to the raw macadamias. Needless to say, this treatment also improved the sensory quality of roasted kernels. However, special attention should be taken to avoid the formation of toxic compounds during roasting operations.

15.9 PISTACHIOS

Pistachios (*Pistacio vera* L.) are indigenous to semi-desert areas and mountains of south-central Asia (Serna-Saldivar 2008). Pistachio nuts are different in each country, depending on the varieties raised and the growing conditions. For instance, pistachios in Iran are large, have light yellow kernels, and the meat is drier than Sicilian, Syrian, and Turkish varieties, which have the advantage of producing more intense flavors. On the other hand, Italian and Afghanistan pistachios have a distinctive green color (Payne 2005). In 2019, the world production of inshell pistachios was 0.911 million tons with a harvested area of 1.034 million Ha and a production yield of 0.881 tons/Ha (FAO 2021). The leading producers are the United States, Iran, and Turkey (Table 15.1). According to the International Nut & Dried Fruit (2020), it is estimated that pistachio is the fourth most consumed nut after almonds, walnuts, and cashew.

The pistachio is a dry-climate crop very resistant to droughts, and trees are known for their longevity, which reaches up to 200 years. Trees tend to be biennial and commonly start producing after seven years and yield maximum production when they are 15–20 years old. One tree can yield up to 40–45 kg when they reach full production. Collection in some parts of the world remains manual, but can also be mechanically harvested when nuts are ripe and start to split (Payne 2005).

Fruits in pistachios grow in clusters, similar to grapes. Although they are called nuts, botanically, pistachios are drupes consisting of an ivory shell that shields an edible meat portion called seed (Payne 2005). Unlike other nuts, pistachios have a unique feature since there is a natural shell splitting before the harvest of pistachios. This favors their marketing as inshell nuts since it would be easily removed by the consumer. The weight of 100 nuts is around 119 g with an average size of 21 mm and kernel percentage of 51% (Pourian et al. 2019).

15.9.1 Pistachios Processing and Storage

Pistachio fruits have a reddish hull when they are at their physiological maturity and are ready to harvest (Pourian et al. 2019). Pistachios are mechanically shaken from the tree or by hand and fall directly onto catching structures. The hull must be removed as soon as possible since it has high moisture (40–50%) that eases contamination with microorganisms once they fall to the orchard floor. In addition, dehulling prevents shell staining (Payne 2005, Tavakolipour 2013). After dehulling, the pistachio kernels contain average moisture of up to 47% (Tavakolipour 2013). This moisture is reduced to about 3–4% by exposing pistachios to hot air at temperatures of about 37°C (Serna-Saldivar 2008).

Inshell pistachios are available in the natural ivory coloration or in red-colored shells produced after the application of edible dyes (Serna-Saldivar 2008). They are classified or graded into different sizes (extra large, large, medium, and small), followed by roasting, salting, and packaging. Although pistachios are mainly found in inshell form, shelled pistachios are also marketed. When shelled with rollers, pistachios are first classified into whole, whole and broken, large pieces, and small pieces and in some instances sorted by color. Shelled pistachios are available dry or roasted and with or without salt, depending on the final application (Payne 2005).

The traditional Iranian method to roast pistachios consists in soaking pistachios in salty water (~20% NaCl) followed by a hot-air drying process at 135°C for 20 min. This method was compared by Hojjati et al. (2015) with a microwave roasting method. To do so, pistachios were also soaked in 20% NaCl, then dried at 60°C, and finally roasted in a microwave oven. The conditions to improve the concentrations of volatile compounds and aroma profile in pistachios were 640 W for 4 min. Microwave roasting represents an adequate alternative to the traditional method since it is rapid and economical. Moreover, the microwaved pistachios contained higher amounts of unsaturated fatty acids.

Pistachios are very stable in storage if they are properly dried. Their shelf life can be extended for up to two years when nuts are stored at 0°C (Serna-Saldivar 2008). However, roasted pistachios start to deteriorate after 12 months. For shelled pistachios, the moisture to extend shelf life should not exceed 5% and their storage is recommended at 5°C (Payne 2005). The selection of the proper packaging material may also help to improve the shelf life. In this sense, nylon proved to be the best material to package pistachios when compared to metal and cellophane, giving even better results when packaging was performed under vacuum conditions (Tavakolipour 2013).

15.9.2 Pistachios Chemical Composition

The main component in pistachios is fat with an average content of 45% (Table 15.3). However, higher contents of fat (55–60%) have also been reported in Persian pistachio kernels (Tavakolipour 2013). Their fat content and caloric load are among the lowest values within nuts and are similar to contents assayed in almonds and cashews. About 50 and 32% of the lipid fraction are composed of monounsaturated and polyunsaturated fatty acids, respectively. Oleic (18:1) and linoleic (18:2) are the predominant fatty acids in pistachios accounting for 50 and 31% of total lipids, respectively.

Pistachios contain a relatively higher protein content compared to other tree nuts. However, their content may vary depending on the variety ranging from 13.7 to 23.6% (Pourian et al. 2019). Similar to cashew, pistachios also contain high contents of lysine and methionine, which gives them good protein quality. In fact, unlike cashew, the lysine content in pistachio meets the WHO/FAO/ONU (2007) requirement for preschool children.

The DF in pistachios is around ~10% of the kernel weight (Table 15.3). Total sugars are also present in significant quantities with an average content of 7.7% being sucrose, the main soluble sugar component (USDA 2021). As for most of the tree nuts, except cashew, the starch content in pistachios is lower than 2%.

The mineral and vitamin contents depicted in Table 15.3 clearly indicate that potassium is the mineral present in higher quantities. These nuts also contain high contents of phosphorous, magnesium, and calcium. The main vitamin found in pistachios is vitamin C, and they also contain adequate quantities of B-vitamins pyridoxine, niacin, and thiamin.

The phytochemicals found in pistachios are summarized in Table 15.4. These nuts contain high contents of phenolic compounds, although they are not as high as in walnuts and pecans. Pistachios contain the highest contents of isoflavones, carotenoids, phytates, and sphingolipids among tree nuts. In fact, it is the only nut that contains a significant amount of carotenoids and related xantophylls (Table 15.4). The main phenolic acids in raw pistachios are chlorogenic (4.96 mg/100 g) and protocatechuic (3.14 mg/100 g) acids (Chang et al. 2016). Other phenolics found in pistachio seeds are gallic acid, catechin, eriodictyol-7-*O*-glucoside, naringenin-7-*O*-neohesperidoside, quercetin-3-*O*-rutinoside, eriodictyol, genistein-7-*O*-glucoside, genistein, daidzein, and apigenin. Pistachio skins contain a higher content of total phenolic compounds and better antioxidant capacity compared to the seeds. Skins also contain phenolic compounds that are not present in the seed. Among these compounds are epicatechin, quercetin, naringenin, luteolin, kaempferol, cyanidin-3-*O*-galactoside, and cyanidin-3-*O*-glucoside (Tomaino et al. 2010). The significant amounts of lutein (xanthophyll) and anthocyanin are responsible for their unique green and purple colorations. Lutein and zeaxanthin are the primary carotenoids present in pistachios. The lutein + zeaxanthin content in pistachios represents around 1,160 µg/100 g of dry-roasted pistachios (Dreher 2012). Pistachios also contain stilbenes (803 µg/100 g) and flavonols (1.46 mg/100 g) which are not commonly found in other tree nuts (Bolling et al. 2011). Pistachios contain adequate quantities of tocopherols, γ-tocopherol being the predominant form representing around 85% of total tocopherols. On the other hand, their sterol content is low compared to other tree nuts, but consistent with this group of nuts, β-sitosterol is the predominant form (Table 15.4). However, postharvest processing may lead to decreased phytochemical contents. For instance, sun drying may reduce up to 80% of total phenol content, flavonoids, and stilbenes, while bleaching reduced the cyanidin-glucoside and cyanidin-galactoside as well as the antioxidant capacity of pistachios (Bolling et al. 2011). Rodríguez-Bencomo et al. (2015) found increased total phenolic contents and antioxidant activity after roasting pistachios at 160°C for 20–22 min. However, not all the phenolic compounds showed this trend. Among the phenolic compounds that increased after roasting were protocatechuic acid, catechin, and rutin. This increment is explained by the releasing of bound phenolic acids as a result of the thermal processing due to the breakdown of cell wall constituents (Rodríguez-Bencomo et al. 2015). However, as seen with other nuts, extended treatments may lead to the degradation of these compounds. The storage time and temperatures also significantly decreased the lutein content in pistachios (Bolling et al. 2011).

The outstanding phytochemical profile of pistachios is known to exert many health benefits, including lowering blood lipids and cholesterol and preventing oxidative and inflammatory stresses related to chronic diseases and cancer. Several studies also suggest that consumption of pistachios may help in appetite management and weight control (Dreher 2012) and their inclusion within a healthy diet may beneficially affect cardiovascular disease risk factors (Gebauer et al. 2008).

15.10 PINE NUTS

Pine nuts are the edible seeds of certain pine species and are harvested in natural stands and plantations in many regions of the world. Pine nuts were harvested for food even prior to the time of ancient Rome and Greece, and have been a staple in the diet of several Native American tribes in North America and indigenous peoples of Siberia and the Russian Far East. Today pine nuts are harvested in many regions of the northern hemisphere and marketed domestically and internationally as gourmet products (Sharashkin and Gold 2004). Pine nuts are considered a delicacy in many of the world's cultures. In addition to being eaten raw or roasted, they are key ingredients in a variety of dishes, including breads, candies, cookies, sauces, and cakes as well as toppings in salads and meat dishes. China, Russia, and Korea are the main world producers of pine nuts (Table 15.1). Although this is the least produced and consumed tree nut (Table 15.1), they are especially popular in the Middle East countries, China, Mongolia, Japan, Italy, and Brazil (Ciesla 1998).

Approximately 29 species of *Pinus* produce edible seeds or pine nuts. Most of the edible nut-bearing pines are haploxylon (soft) pines, although several species are of the diploxylon (hard) pine group. In North America, the main pinyon pine nut species are the Colorado (*Pinus edulis*), single-leaf (*Pinus monophylla*), and Mexican (*Pinus cembroides*) and other pinyon pines indigenous to

Mexico or southwestern United States are important regional foods but are not widely exported. In Asia, the Korean (*Pinus koraiensis*), Russian (*Pinus sibirica*), and chilgoza (*Pinus gerardiana*) pines are commonly harvested and utilized. The most widespread pine nut species grown in Europe is the stone pine (*Pinus pinea*), which has been cultivated for more than 5,000 years. China is the world's largest producer and exporter of seeds of *P. koraiensis* and *P. sibirica* (Ciesla 1998). *Pinus* genus also comprises species that are not classified as edible and could be used to adulterate edible species. It is believed that this adulteration could be related to consumer complaints following consumption of commercial pine nuts, reporting that they experienced taste disturbance lasting for up to two weeks after consumption (Destaillats et al. 2011).

15.10.1 Pine Nuts Processing and Storage

One of the main limitations of the cultivation of these pines is the time it takes to develop full matured kernels (approximately 1.5 years) in the cones (Ciesla 1998). The pine kernels are covered with a hard casing that is botanically known as a seed coat. The edible nutrients are found in the embryo or nut. The shell represents around 72.5% of the pine nut total weight, while the meat or kernel represents 27.3%. The skin of pine nuts is removed and represents around 0.2% of the total weight (Nergiz and Dönmez 2004). Needless to say, the hard shell must be removed before the pine nut can be eaten. Unshelled pine nuts have a long shelf life if kept under dry and cold (–5 to 2°C) conditions. The storage of the pine nuts at room temperature significantly decreases their shelf life because they quickly develop off-flavors due to oxidative rancidity. That is the main reason nuts are often frozen to preserve their flavor and freshness (Ciesla 1998).

15.10.2 Pine Nuts Chemical Composition

Pine nuts have an exceptional nutritional value. These edible nuts compare favorably with pecans, and walnuts in fat, protein, and carbohydrate contents (Table 15.3). Zuleta et al. (2018) analyzed chemical–nutritional composition of pine nuts harvested in six countries. The major components of these nuts were lipids (34.7–44.3%), protein (32.1–36.6%), and DF (10.9–14.8%). Interestingly, the pine nuts were practically devoid of carbohydrates, which make them a good choice in the prevention of diabetes, metabolic syndrome, and other chronic diseases.

According to the USDA (2021), a 100 g serving of pine nuts provides around 673 kcal, 90% of which is provided by fats or lipids. In *Pinus pinea*, oleic and linoleic acids were the major unsaturated fatty acids, while palmitic, stearic, and lignoceric acids were the main saturated ones (Nergiz and Dönmez 2004). According to the USDA (2021), the linoleic 18:2 and oleic 18:1 acids represent 48% and 26% of total lipids, respectively. *Pinus* spp. seed lipids also contain very specific polyunsaturated fatty acids known as Δ5-olefinic acids, which are useful as taxonomic markers to identify their botanical origin. Therefore, pine nuts from different varieties may be identified by their distinctive fatty acid profile (Destaillats et al. 2011). The establishing of methodologies to identify kernels is important to control processing procedures, protect producers, and avoid unfair competition (Evaristo et al. 2010).

The pine nuts provide an average of 14% protein (Table 15.3). However, the protein content of some pine nuts such as *P. pinea* or *P. sabiniana* exceeds all other commercial nuts reaching values up to 34% protein with a high protein quality (Ciesla 1998). As other tree nuts, allergic reactions after the consumption of pine nuts have also been reported. The allergen has been recognized as a protein of 17 kDa that is sensitive to reducing agents, mechanism by which it loses its IgE-binding properties (García-Menaya et al. 2000).

The carbohydrate fraction in pine nuts is mainly composed of DF, sugars, and a small amount of starch (Table 15.3). Potassium, phosphorus, and magnesium are the predominant minerals found in pine nuts, although zinc, iron, and manganese are also detected in significant amounts (Nergiz and Dönmez 2004, USDA 2021). In terms of vitamins, the nuts are considered an adequate source of tocopherols (vitamin E) and B- vitamins, including thiamin or B1, riboflavin or B2, and niacin or B3 (Table 15.3).

Pine nuts also contain an array of phytochemicals summarized in Table 15.4. Ellagic (5.8–35.4 μg/g) and gallic (2.9–26.0 μg/g) acids are the predominant phenolic acids in raw pine nuts, while the flavan-3-ol gallocatechin (9.2–65 μg/g) is also present in significant amounts (Chang et al. 2016). According to the studies performed with rats fed with high fat and cholesterol diet, those supplemented with pine nuts proved to exert hypolipidemic and hypocholesterolemic effects (Amr and Abeer 2011). The overall beneficial effects of pine nut consumption are related to both the adequate fatty acid profile and bioactive phytochemicals which act synergistically.

REFERENCES

Abe, L.T., Lajolo, F.M., and Genovese, M.I. 2010. Comparison of phenol content and antioxidant capacity of nuts. *Food Science and Technology*, 30:254–259.

Alasalvar, C., and Shahidi, F. 2008. *Tree Nuts: Composition, Phytochemicals, and Health Effects*. Boca Raton: CRC Press.

Amr, A.R., and Abeer, E.E.K. 2011. Hypolipideimic and hypocholestermic effect of pine nuts in rats fed high fat, cholesterol-diet. *World Applied Sciences Journal*, 15(12):1667–1677.

Andersen, P.C. 2019. Pecan tree. Document HS982. Horticultural Sciences Department, UF/IFAS Extension. University of Florida. https://edis.ifas.ufl.edu

Bagheri, H. 2020. Application of infrared heating for roasting nuts. *Journal of Food Quality*, 2020:1–10. doi: 10.1155/2020/8813047

Bakkalbaşi, E., Menteş, Ö., and Artik, N. 2008. Food ellagitannins–occurrence, effects of processing and storage. *Critical Reviews in Food Science and Nutrition*, 49(3):283–298. doi: 10.1080/10408390802064404

Bamberger, C., Rossmeier, A., Lechner, K., Wu, L., Waldmann, E., Stark, R.G., Altenhofer, J., Henze, K., and Parhofer, K.G. 2017. A walnut-enriched diet reduces lipids in healthy Caucasian subjects, independent of recommended macronutrient replacement and time point of consumption: A prospective, randomized, controlled trial. *Nutrients*, 9(10):1097. doi: 10.3390/nu9101097

Belviso, S., Dal Bello, B., Giacosa, S., Bertolino, M., Ghirardello, D., Giordano, M., Rolle, L., Gerbi, V., and Zeppa, G. 2017. Chemical, mechanical and sensory monitoring of hot air- and infrared-roasted hazelnuts (*Corylus avellana* L.) during nine months of storage. *Food Chemistry*, 217:398–408. doi: 10.1016/j.foodchem.2016.08.103

Bolling, B.W., Chen, C.Y.O., McKay, D.L., and Blumberg, J.B. 2011. Tree nut phytochemicals: Composition, antioxidant capacity, bioactivity, impact factors. A systematic review of almonds, Brazils, cashews, hazelnuts, macadamias, pecans, pine nuts, pistachios and walnuts. *Nutrition Research Reviews*, 24(2):244–275. doi: 10.1017/S095442241100014X

Byford, R. 2005. Growth and development of pecan nuts. Guide H-618. Cooperative Extension Service, College of Agriculture and Home Economics. Las Cruces: New Mexico State University.

Cha, C.Y., and Lee, K.G. 2020. Effect of roasting conditions on the formation and kinetics of furan in various nuts. *Food Chemistry*, 331:127338. doi: 10.1016/j.foodchem.2020.127338

Chang, S.K., Alasalvar, C., Bolling, B.W., and Shahidi, F. 2016. Nuts and their co-products: The impact of processing (roasting) on phenolics, bioavailability, and health benefits: A comprehensive review. *Journal of Functional Foods*, 26:88–122. doi: 10.1016/j.jff.2016.06.029

Ciesla, W.M. 1998. *Non-Wood Forest Products from Conifers*. Rome: Food and Agriculture Organization of the United Nations

Demir, A.D., and Cronin, K. 2005. Modelling the kinetics of textural changes in hazelnuts during roasting. *Simulation Modelling Practice and Theory*, 13(2):97–107. doi: 10.1016/j.simpat.2003.11.007

Deshpande, S., Sathe, S., and Kadm, S. 1995. Other nuts. In: *Handbook of Fruit Science and Technology: Production, Composition, Storage and Processing*, eds. D. Salunkhe, and S. Kadam. New York: Marcel Dekker Inc.

Destaillats, F., Cruz-Hernandez, C., Giuffrida, F., Dionisi, F., Mostin, M., and Verstegen, G. 2011. Identification of the botanical origin of commercial pine nuts responsible for dysgeusia by gas-liquid chromatography analysis of fatty acid profile. *Journal of Toxicology*, 2011:316789. doi: 10.1155/2011/316789

Dreher, M.L. 2012. Pistachio nuts: Composition and potential health benefits. *Nutrition Reviews*, 70(4):234–240. doi: 10.1111/j.1753-4887.2011.00467.x

Duduzile Buthelezi, N.M., Samukelo Magwaza, L., and Zeray Tesfay, S. 2019. Postharvest pre-storage processing improves antioxidants, nutritional and sensory quality of macadamia nuts. *Scientia Horticulturae*, 251:197–208. doi: 10.1016/j.scienta.2019.03.026

Engler-Ribeiro, P.C., de Britto Policarpi, P., Dal Bo, A., Barbetta, P.A., and Block, J.M. 2017. Impact of pecan nut shell aqueous extract on the oxidative properties of margarines during storage. *Journal of the Science of Food and Agriculture*, 97(9):3005–3012. doi: 10.1002/j

Evaristo, I., Batista, D., Correia, I., Correia, P., and Costa, R. 2010. Chemical profiling of Portuguese *Pinus pinea* L. nuts. *Journal of the Science of Food and Agriculture*, 90(6):1041–1049. doi: 10.1002/jsfa.3914

FAO. 2021. Food and Agriculture Organization of the United Nations. FAOSTAT Database. http://www.fao.org/faostat/en/

Flores-Córdova, M.A., Sánchez-Chávez, E., Chávez-Mendoza, C., García-Hernández, J.L., and Preciado-Rangel, P. 2016. Bioactive compounds and phytonutrients in edible part and nutshell of pecan (*Carya illinoinensis*). *Cogent Food and Agriculture*, 2(1):1–12. doi: 10.1080/23311932.2016.1262936

García-Menaya, J.M., Gonzalo-Garijo, M.A., Moneo, I., Fernández, B., García-González, F., and Moreno, F. 2000. A 17-kDa allergen detected in pine nuts. *Allergy*, 55(3):291–293. doi: 10.1034/j.1398-9995.2000.00291.x

Gebauer, S.K., West, S.G., Kay, C.D., Alaupovic, P., Bagshaw, D., and Kris-Etherton, P.M. 2008. Effects of pistachios on cardiovascular disease risk factors and potential mechanisms of action: A dose-response study. *American Journal of Clinical Nutrition*, 88(3):651–659. doi: 10.1093/ajcn/88.3.651

Geiselhart, S., Hoffmann-Sommergruber, K., and Bublin, M. 2018. Tree nut allergens. *Molecular Immunology*, 100:71–81. doi: 10.1016/j.molimm.2018.03.011

Ghafoor, K., Al Juhaimi, F., Geçgel, Ü., Babiker, E.E., and Özcan, M.M. 2020. Influence of roasting on oil content, bioactive components of different walnut kernel. *Journal of Oleo Science*, 69(5):423–428. doi: 10.5650/jos.ess19205

Hashemi, N., Mortazavi, S.A., Milani, E., and Tabatabai Yazdi, F. 2017. Microstructural and textural properties of puffed snack prepared from partially deffated almond powder and corn flour. *Journal of Food Processing and Preservation*, 41(5):1–12. doi: 10.1111/jfpp.13210

Hayes, D., Angove, M.J., Tucci, J., and Dennis, C. 2016. Walnuts (*Juglans regia*) chemical composition and research in human health. *Critical Reviews in Food Science and Nutrition*, 56(8):1231–1241. doi: 10.1080/10408398.2012.760516

Hojjati, M., Noguera-Artiaga, L., Wojdyło, A., and Carbonell-Barrachina, Á.A. 2015. Effects of microwave roasting on physicochemical properties of pistachios (*Pistaciavera* L.). *Food Science and Biotechnology*, 24(6):1995–2001. doi: 10.1007/s10068-015-0263-0

Idowu, O.J., Sanogo, S., and Brewer, C.E. 2017. Short term impacts of pecan waste by-products on soil quality in texturally different arid soils. *Communications in Soil Science and Plant Analysis*, 48(15):1781–1791.

International Nut & Dried Fruit Council. 2020. *Nutfruit*. Reus: International Nut and Dried Fruit Council.

Janick, J., and Paull, R.E. 2008. *The Encyclopedia of Fruit and Nuts*. Cambridge: CABI Publishing.

Köksal, A.I., Artik, N., Şimşek, A., and Güneş, N. 2006. Nutrient composition of hazelnut (*Corylus avellana* L.) varieties cultivated in Turkey. *Food Chemistry*, 99(3):509–515. doi: 10.1016/j.foodchem.2005.08.013

Li, R., Kou, X., Cheng, T., Zheng, A., and Wang, S. 2017. Verification of radio frequency pasteurization process for in-shell almonds. *Journal of Food Engineering*, 192:103–110. doi: 10.1016/j.jfoodeng.2016.08.002

Ling, B., Hou, L., Li, R., and Wang, S. 2016. Storage stability of pistachios as influenced by radio frequency treatments for postharvest disinfestations. *Innovative Food Science and Emerging Technologies*, 33:357–364. doi: 10.1016/j.ifset.2015.10.013

Mendes, C., Costa, J., Vicente, A.A., Oliveira, M.B.P.P., and Mafra, I. 2019. Cashew nut allergy: Clinical relevance and allergen characterisation. *Clinical Reviews in Allergy and Immunology*, 57(1):1–22. doi: 10.1007/s12016-016-8580-5

Mohod, A., Jain, S., and Powar, A.G. 2010. Pollution sources and standards of cashew nut processing. *American Journal of Environmental Sciences*, 6(4):324–328.

Nergiz, C., and Dönmez, I. 2004. Chemical composition and nutritive value of *Pinus pinea* L. seeds. *Food Chemistry*, 86(3):365–368. doi: 10.1016/j.foodchem.2003.09.009

Oro, T., Ogliari, P.J., Amboni, R.D.M.C., Barrera-Arellano, D., and Block, J.M. 2008. Evaluación de la calidad durante el almacenamiento de nueces Pecán [*Carya illinoinensis* (Wangenh.) C. Koch] acondicionadas en diferentes envases. *Grasas y Aceites*, 59(2):132–138.

Özdemir, M., Açkurt, F., Yildiz, M., Biringen, G., Gürcan, T., and Löker, M. 2001. Effect of roasting on some nutrients of hazelnuts (*Corylus Avellena* L.). *Food Chemistry*, 73(2):185–190. doi: 10.1016/S0308-8146(00)00260-0

Payne, T.J. 2005. Nuts. In: *Processing Fruits: Science and Technology*, eds. D.M. Barret, L. Somogyi, and H. Ramaswamy, 765–810. Boca Raton: CRC Press. doi: 10.1201/9781420040074.ch20

Pereira, J.A., Oliveira, I., Sousa, A., Ferreira, I.C.F.R., Bento, A., and Estevinho, L. 2008. Bioactive properties and chemical composition of six walnut (*Juglans regia* L.) cultivars. *Food and Chemical Toxicology*, 46(6):2103–2111. doi: 10.1016/j.fct.2008.02.002

Petruzzello, M. 2020. Almond. *Encyclopædia Britannica* (April). https://www.britannica.com/plant/almond

Phatanayindee, S., Borompichaichartkul, C., Srzednicki, G., Craske, J., and Wootton, M. 2012. Changes of chemical and physical quality attributes of macadamia nuts during hybrid drying and processing. *Drying Technology*, 30(16):1870–1880. doi: 10.1080/07373937.2012.703275

Pinheiro do Prado, A.C., Manion, B.A., Seetharaman, K., Deschamps, F.C., Barrera, D., and Block, J.M. 2013. Relationship between antioxidant properties and chemical composition of the oil and the shell of pecan nuts [*Caryaillinoinensis* (Wangenh) C. Koch]. *Industrial Crops and Products*, 45:64–73.

Pourian, M.A., Bakhshi, D., Hokmabadi, H., and Aalami, A. 2019. Pomological attributes and chemical composition of cultivars and wild genotypes of pistachios (*Pistacia vera* L.) in Iran. *Journal of Nuts*, 10:163–173. doi: 10.22034/jon.2019.1870225.1060

Rabago-Panduro, L.M., Martin Belloso, O., Welti-Chanes, J., and Morales de la Peña, M. 2020. Changes in bioactive compounds content and antioxidant capacity of pecan nuts [*Carya illinoinensis* (Wangenh. K. Koch)] during storage. *Revista mexicana de ingeniería química*, 19(3):1439–1452.

Rico, R., Bulló, M., and Salas-Salvadó, J. 2016. Nutritional composition of raw fresh cashew (*Anacardium occidentale* L.) kernels from different origin. *Food Science and Nutrition*, 4(2):329–338. doi: 10.1002/fsn3.294

Rodríguez-Bencomo, J.J., Kelebek, H., Sonmezdag, A.S., Rodríguez-Alcalá, L.M., Fontecha, J., and Selli, S. 2015. Characterization of the aroma-active, phenolic, and lipid profiles of the pistachio (*Pistacia vera* L.) nut as affected by the single and double roasting process. *Journal of Agricultural and Food Chemistry*, 63(35):7830–7839. doi: 10.1021/acs.jafc.5b02576

Ros, E. 2010. Health benefits of nut consumption. *Nutrients*, 2(7):652–682. doi: 10.3390/nu2070652

Rudolph, C., Odell, G., Hinrichs, H., Hopfer, D., and Kays, S. 1992. Genetic, environmental, and maturity effects on pecan kernel lipid, fatty acid, tocopherol, and protein composition. *Journal of Food Quality*, 15(4):263–278.

Serna-Saldivar, S.O. 2008. Peanuts, almonds and tree nuts. In: *Industrial Manufacture of Snack Foods*, ed. S.O. Serna-Saldivar, 125–148. London: Kennedys Publications Ltd.

Shafiei, G., Ghorbani, M., Hosseini, H., Sadeghi Mahoonak, A., Maghsoudlou, Y., and Jafari, S.M. 2020. Estimation of oxidative indices in the raw and roasted hazelnuts by accelerated shelf-life testing. *Journal of Food Science and Technology*, 57(7):2433–2442. doi: 10.1007/s13197-020-04278-9

Sharashkin, L., and Gold, M. 2004. Pine nuts: Species, products, markets, and potential for U.S. production. In: *Northern Nut Growers Association 95th Annual Report.* Columbia: Proceeding for the 95th Annual Meeting.

Smeriglio, A., Barreca, D., Bellocco, E., and Trombetta, D. 2017. Proanthocyanidins and hydrolysable tannins: Occurrence, dietary intake and pharmacological effects. *British Journal of Pharmacology*, 174(11):1244–1262. doi: 10.1111/bph.13630

Smith, M.W. 2012. Fruit production characteristics of "Pawnee" pecan. *Horticulturalscience*, 47(4):489–496. doi: 10.21273/HORTSCI.47.4.489

Sugizaki, C.S.A., and Naves, M.M.V. 2018. Potential prebiotic properties of nuts and edible seeds and their relationship to obesity. *Nutrients*, 10(11):1645. doi: 10.3390/nu10111645

Tavakolipour, H. 2013. Postharvest operations of pistachio nuts. *Journal of Food Science and Technology*, 52(2):1124–1130. doi: 10.1007/s13197-013-1096-6

Thompson, T.E., and Conner, P.J. 2012. Pecan. In: *Fruit Breeding*, eds. M.L. Badenes and D.H. Byrne, 771–802. New York: Springer Science & Business Media.

Tomaino, A., Martorana, M., Arcoraci, T., Monteleone, D., Giovinazzo, C., and Saija, A. 2010. Antioxidant activity and phenolic profile of pistachio (*Pistacia vera* L., variety Bronte) seeds and skins. *Biochimie*, 92(9):1115–1122. doi: 10.1016/j.biochi.2010.03.027

USDA. 2021. USDA national nutrient database for standard reference. http://www.usda.gov

Uysal, N., Sumnu, G., and Sahin, S. 2009. Optimization of microwave-infrared roasting of hazelnut. *Journal of Food Engineering*, 90(2):255–261. doi: 10.1016/j.jfoodeng.2008.06.029

Vendrame, W., and Wetzstein, H. 2005. *Carya illinoinensis* pecan. In *Biotechnology of Fruit and Nut Crops*, ed. R.E. Litz, 298–306. Cambridge: CABI Publishing. doi: 10.1079/9780851996622.0298

Wallace, H.M., and Walton, D.A. 2011. Macadamia (*Macadamia integrifolia*, *Macadamia tetraphylla* and hybrids). In: *Postharvest Biology and Technology of Tropical and Subtropical Fruits. Volume 3: Cocona to Mango.* Cambridge: Woodhead Publishing Limited. doi: 10.1533/9780857092885.450

Wang, W., Wang, W., Wang, Y., Yang, R., Tang, J., and Zhao, Y. 2020. Hot-air assisted continuous radio frequency heating for improving drying efficiency and retaining quality of inshell hazelnuts (*Corylus avellana* L. cv. Barcelona). *Journal of Food Engineering*, 279:109956. doi: 10.1016/j.jfoodeng.2020.109956

WHO/FAO/ONU. 2007. *Protein and Amino Acid Requirements in Human Nutrition.* Geneva: FAO/WHO/UNU Expert Consultation.

Wu, X., Beecher, G.R., Holden, J.M., Haytowitz, D.B., Gebhardt, S.E., and Prior, R.L. 2004. Lipophilic and hydrophilic antioxidant capacities of common foods in the United States. *Journal of Agricultural and Food Chemistry*, 52(12):4026–4037. doi: 10.1021/jf049696w

Yada, S., Lapsley, K., and Huang, G. 2011. A review of composition studies of cultivated almonds: Macronutrients and micronutrients. *Journal of Food Composition and Analysis*, 24(4–5):469–480. doi: 10.1016/j.jfca.2011.01.007

Yang, J., Bingol, G., Pan, Z., Brandl, M.T., McHugh, T.H., and Wang, H. 2010. Infrared heating for dry-roasting and pasteurization of almonds. *Journal of Food Engineering*, 101(3):273–280. doi: 10.1016/j.jfoodeng.2010.07.007

Yang, J., Zhou, F., Xiong, L., Mao, S., Hu, Y., and Lu, B. 2015. Comparison of phenolic compounds, tocopherols, phytosterols and antioxidant potential in Zhejiang pecan [*Carya cathayensis*] at different stir-frying steps. *LWT: Food Science and Technology*, 62(1):541–548. doi: 10.1016/j.lwt.2014.09.049

Zhang, Y., and Jin, T. 2020. Almond allergens: Update and perspective on identification and characterization. *Journal of the Science of Food and Agriculture*, 100(13):4657–4663. doi: 10.1002/jsfa.10417

Zhang, Y., Lee, B., Du, W.X., Lyu, S.C., Nadeau, K.C., Grauke, L.J., Zhang, Y., Wang, S., Fan, Y., Yi, J., and McHugh, T.H. 2016. Identification and characterization of a new pecan [*Carya illinoinensis* (Wangenh.) K. Koch] allergen, Car i 2. *Journal of Agricultural and Food Chemistry*, 64(20):4146–4151. doi: 10.1021/acs.jafc.6b00884

Zuleta, A., Weisstaub, A., Giacomino, S., Dyner, L., Loewe, V., Del Río, R., and Lutz, M. 2018. An ancient crop revisited: Chemical composition of Mediterranean pine nuts grown in six countries. *Italian Journal of Food Science*, 30(1):170–183.

16 Snacks from Fruits and Vegetables

Jorge Welti-Chanes, Zamantha Escobedo-Avellaneda, and Maria Soledad Tapia

CONTENTS

16.1 INTRODUCTION

Healthy eating means consuming a variety of foods that deliver the nutrients and phytochemicals needed to maintain health in terms of preventing malnutrition in all its forms as well as a range of noncommunicable diseases (NCDs) and conditions. Furthermore, the consumption of healthy foods should also meet sensory attributes related to pleasure. These nutrients include protein, carbohydrates, fat, vitamins, and minerals. However, the increased production and consumption of processed foods, rapid urbanization, and changing lifestyles have led to a shift in dietary patterns. People are now consuming more foods high in energy, fats, free sugars, and salt/sodium, and many people do not eat enough servings of fruits, vegetables, and other high-dietary fiber (DF) foods such as whole grains (WHO 2018).

The changes in dietary patterns have been associated with an increased demand for convenience foods and have contributed to the demise of traditional family meals in favor of consumption of foods and drinks between main meals. This inter-meal intake of food is often described as "snacking." Even if there is no scientific consensus of what constitutes a snack, either as an eating occasion or as a snack food, most food-based dietary guidelines and nutritional research frequently refer to snacking habits as desirable or undesirable snack food choices (Potter et al. 2018).

Snacking is practiced by all age groups, and is associated with concepts such as "junk" food or empty calories, loss of appetite control, obesity, and overconsumption (Sadler et al. 2019). The consumption of snacks has often been identified as an important contributor to the rapid increase in overweight individuals (Nederkoorn et al. 2010), with children and adolescents being of special concern, since their snacking habits have been related to caries (Johansson et al. 2010), obesity (Alm

DOI: 10.1201/9781003129066-16

et al. 2008), and ill-eating habits (Drenowatz et al. 2018). Some studies exclude beverages (Wouterts et al. 2018), whereas others advocate that the term "snack" refers to all types of foods (including fruit) and drinks consumed outside the context of main meals (i.e., breakfast, lunch, and dinner) (de Graaf 2006).

In its practical advice on maintaining a healthy diet, the World Health Organization (WHO 2020) explicitly refers to the well-known recommendation of eating at least 400 g, or five servings, of fruit and vegetables per day, since it reduces the risk of NCDs, and helps to ensure an adequate daily intake of DF. Also, among recommendations to increase fruit and vegetable intake are (a) always including vegetables in meals, (b) eating fresh fruit and raw vegetables as snacks, (c) eating fresh fruit and vegetables that are in season, and (d) eating a variety of fruit and vegetables (WHO 2020). Most children do not meet the recommendations of WHO for an intake of at least 400 g daily of fruits and vegetables, so it is very important to identify the factors that affect consumption of these foods by this important population group. The environment unconsciously affects the selection of our food, and therefore it has been proposed that some strategic changes in the environment that have to do with food offers and choices could be potentially useful strategies to guide consumers, especially children, toward the selection of more nutritious options, without thereby affecting their freedom of selection.

Some studies have determined that offering children a variety of vegetables can be a simple and effective strategy to induce them to consume not only more vegetables, but also healthier meals, and this would be valid, not only at home but also in cafeterias and school canteens (Alaimo et al. 2013). This is an opportunity area for developing novel and healthy snacks from fruits and vegetables.

Some examples of the so-called fruit and vegetable "healthy snacks" are dried fruit and vegetables, fruit and vegetable bars, cereal products with fruit pieces in the form of bars or cakes without glucose–fructose syrup. Freeze-dried fruit and vegetable bars are gaining more and more popularity; fruit is rich in simple sugars, so from the point of view of a snack ingredient, it is worth reaching for vegetables that contain less of sugars, and additionally are richer in fiber, vitamins, and some minerals. As compared to fruit, vegetables are less attractive in terms of taste. It seems advisable to make them more attractive by adding spices and giving them an unconventional form, which may be a freeze-dried bar (Ciurzyska et al. 2019).

As Jack et al. (1997) stated back in England, "there may be scope to increase fruit consumption, in particular, by promoting it as a snack." According to Mordor Intelligence (2020), the fruit snack market was projected to grow with a CAGR (Compound Annual Growth Rate) of 8.17% during the forecast period (2019–2024).

The fruit and vegetable snack market is estimated to witness a substantial growth owing to an increase in the demand for refreshments that are rich in nutritional content, along with the rising consumer awareness, primarily related to the consumption of healthy food. Rising consumer preference for processed foods that are rich in nutritional content, as compared to other regular processed food, is expected to aid the fruit market's growth. Aspects such as ingredient origin, artisanal production, ethnicity, plant-based composition, sustainability, delivery of useful nutrition, and convenience in modern life are relevant factors in composing this context (Innova Market Insights 2020). This applies to vegetables too. Moreover, vegans and vegetarians eat a large number of fruits and vegetables, but flexitarians who can be persuaded to eat more vegetarian meals or less meat overall will likely increase the consumption of fruits and vegetables to get the nutrients that would be present in a serving of meat. This is another window of opportunity for snacks (Grand View Research 2020). However, the outbreak of the COVID-19 pandemic in 2020 disrupted some of the existing trends in the fresh fruit and vegetables sector and it has included the snack sector. The retail sales increased, especially for fruit and vegetables that are high in vitamin C. Consumers tried to minimize their retail visits and preferred to go for affordable fruit and vegetables with a good shelf life (CBI 2020). Again, this is expanding the opportunities for the snack sector.

To date, with the scientific and technological advances in food processing, several approaches can be used for design and processing of fruits and vegetables snack items. An attempt to revise them will be done in this chapter, as well as the trends identified in this promising and challenging field, since many of the health benefits of fruit and vegetables can be transferred into related snacks. The goal of producing snacks from fruit and vegetables is an auspicious area that should be accompanied by proper selection of raw materials and processing technologies.

16.2 CLASSIFICATION OF SNACKS FROM FRUITS AND VEGETABLES

Snacks from fruits and vegetables can be classified into three large groups: dehydrated, intermediate-moisture (IM), and minimally processed products. The main characteristics distinguishing a dehydrated from an IM snack are water activity (a_w) and moisture level. In dehydrated products, a_w values range between 0.4 and 0.6 with moisture content around 10%, while in IM fruits values range between 0.65 and 0.90, with moistures between 10% and 50%. In both types of snacks, the reduction of water activity is an important factor in order to control reactions that compromise stability and safety. Nevertheless, in order to guarantee stability of IM products, other conservation factors need to be applied in the so-called hurdle technology that allows a milder process of water removal compared to dried fruits and vegetables (Escobedo-Avellaneda et al. 2012). On the other hand, minimally processed fruits and vegetables (MPFV) usually go through a very slight reduction of water activity, maintaining values close to that of fresh products (0.95–0.98) with moisture values around 90–95%. Their stability is also achieved through the application of hurdle technology (Alzamora et al. 1998, 2016).

Most low- and intermediate-moisture snacks from fruits and vegetables are obtained from the fruit cut into pieces or slices, or from the concentrated and dehydrated pulp to obtain bars or rolled products. However, in recent years, the consumption of extruded snacks obtained in mixtures of cereal starches and fruits and vegetables (dried pulp or skin) has increased. These products are an excellent source of fiber and other functional or nutritional properties that the fruits and vegetables give (Potter et al. 2018). This topic of novel extruded products is discussed in Chapter 12.

16.2.1 Dehydrated Fruits and Vegetables

Dried or dehydrated fruits and vegetables are products obtained after the removal of water mainly by evaporation (air drying in a tunnel or cabinet, packed and fluidized bed, vacuum drying [VD], solar drying [SD]) so as to obtain crunchy or rubbery structured products. Alternatively, the sublimation of water by the process called freeze-drying (FD) produces dried products with a porous structure and much better capacity of rehydration (García-Amezquita et al. 2016, Caballero-Cerón et al. 2018)

WHO classifies traditional dried fruits as "fruit," and like fresh counterparts, their sugar content is not defined as "free sugars" (WHO 2015, Swan et al. 2018). Sadler et al. (2019) categorized dried fruits in four groups: traditional dried fruit, sugar-infused dried fruit, candied dried fruit, and processed fruit snacks. Depending on the selected production/processing method, some fruits may be classified in more than one category.

Dried fruits and vegetables can be supplemented with sugars and other ingredients or prepared from macerated or pureed pastes that are further dried to obtain leathers, pieces, and packaged snacks.

Food choices that constitute "healthy snacking" are highly valued by ample sectors of consumers, especially those interested in appetite control. Observational studies suggest that increased fruit and vegetable consumption can contribute to weight maintenance and facilitate weight loss when substituted for other energy dense foods. This may be more so when considering highly satiating meals or snacks that are high in fiber, protein, or volume and low in energy (Sadler et al. 2019).

Traditional dried fruits and vegetables, even if not high in protein, could be considered as useful strategies to improve appetite control on account of their high palatability, sweetness, high fiber content, low to moderate GI, and chewiness which delivers oral and sensory satiety (Sadler et al.

2019). Also, other health benefits which are associated with an adequate intake of whole fruits and vegetables could be transferred to dried products, as shown in a systematic review of observational studies conducted by Mossine et al. (2020) with respect to cancer risk. Some current available data provides evidence that consumption of dried fruits may be associated with a lower cancer incidence or mortality in populations, as it is the case that a higher intake of raisins prevents cancers of the gastrointestinal system.

Likewise, Dreher (2018) revised the beneficial health effects of consuming adequate levels of whole fruits and vegetables regarding the bioactive fiber prebiotic effects and role in improved weight control, wellness, and healthy aging and concluded that benefits could also be transferred to dried counterparts. Patel et al. (2013) showed that *ad libitum* consumption of raisins and grapes compared to potato chips and chocolate chip cookies (about 200 kcal vs. 450 kcal) had potential as after-school snacks to achieve lower snack intake prior to dinner in children 8–11 years old. Likewise, adults consuming five dried prunes (40 g) as a snack reduced lunch meal energy intake by 100 kcal compared to that after an isocaloric low-fiber and energy-dense snack (Farajian et al. 2010). Both fresh and dried fruits and vegetables help to reduce hunger, increase meal satisfaction, and/or decrease energy intake when consumed as snacks or with meals compared to non-fruit control foods.

The contribution of dried fruit and vegetables to "five a day" fruit and vegetable intake has been discussed as well as their promotion as part of the diet (Sadler et al. 2019) taking into consideration that dried products do not need to be eaten in place of fresh counterparts and are considered nutritive and healthy. If substitution is necessary, dried vegetable products are far better a choice in place of juices and smoothies. The intake of dried products should be encouraged because it could orientate consumers toward a healthier diet and achieve the recommended 5-a-day fruit and vegetable servings. The consumption of confectionery items is reduced when both fresh and dried fruit and vegetables are included in the diet.

Dried fruits and vegetables have similar nutrients compared to their equivalent fresh products, except for the lower vitamin C, which is commonly lost during processing. Nevertheless, the dried products contain higher amounts of bioactive compounds such as polyphenols and related phytochemicals that in turn exert higher antioxidant activity. Dried fruits and vegetables are particularly rich in DF, sugars, vitamins, and minerals and are practically devoid of lipids. Moreover, some dried fruits like prunes and apricots contain high levels of sorbitol, which confers laxative properties and increases stool weight and facilitates intestinal transit. Dried fruits and vegetables are popular, convenient, have a long shelf life, well-accepted, being viable as snacks for appetite control in interventions in susceptible individuals, taking care not to exceed the portion size which should always be smaller than that of the fresh product (Sadler et al. 2019). All these considerations can be extended to IM and HM fruits and vegetables.

16.2.2 Intermediate-Moisture Fruits and Vegetables

IMFV are usually obtained by osmotic dehydration (OD) with the aim to reduce a_w by the addition of solutes, usually sugars, salts, or maltodextrins. These products are usually obtained after water removal by an osmotic process followed by subsequent drying or other preservation methods (Vergara et al. 1997, Valdez-Fragoso et al. 2002). Figure 16.1 depicts a flow diagram of the production of osmotic dehydrated products.

In some cases, fruit and vegetable purees can be concentrated and/or dehydrated to intermediate- or low-moisture bars, leathers, or laminated products (Parn et al. 2015, Maity et al. 2016, Ciurzyska et al. 2019). Figure 16.2 shows examples of these types of products from apple pulp and peel, as well as from currant pulp. One of the main characteristics of laminated fruit snacks is their stickiness, a property that needs to be adjusted depending on the final product. Valenzuela and Aguilera (2015) showed that in lamination technology, stickiness of apple leathers can be diminished using maltodextrin to avoid problems during processing and packaging, while glucose is the ingredient of choice if an increase of stickiness is needed.

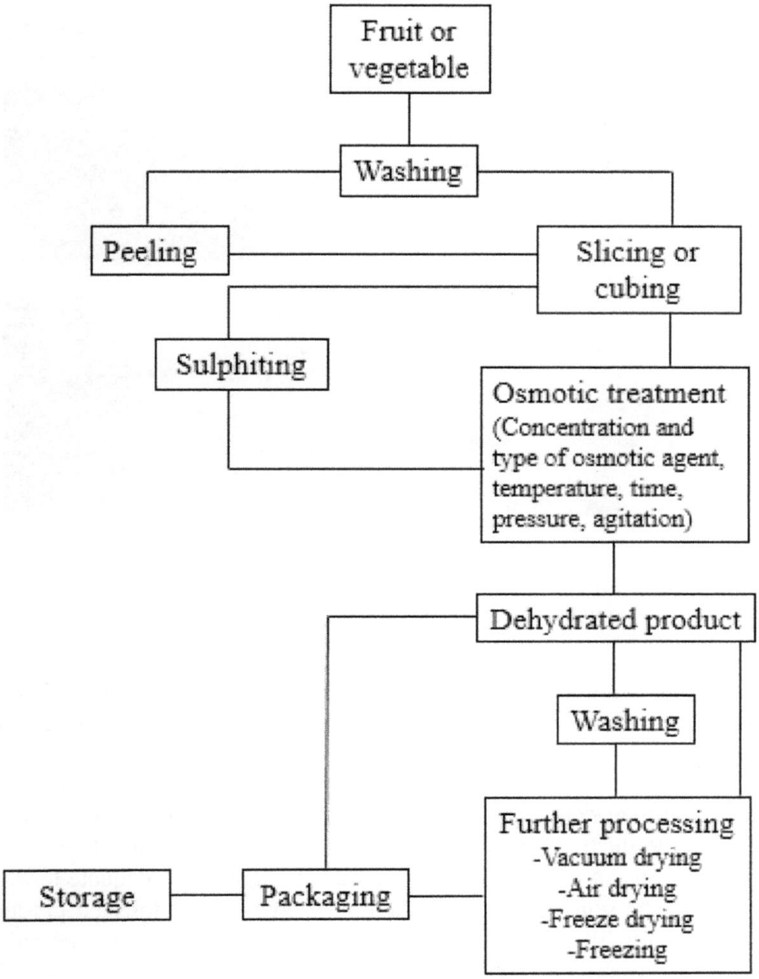

FIGURE 16.1 Flow diagram of osmotic dehydration process (adapted from Yadav and Singh 2012).

Relevant improvements of traditional IM products have been made in recent years by applying the combined action of reducing a_w along with other conservation factors. The combined methods approach enlarges the range of options for development of novel products with enhanced quality, texture, palatability, sensory properties, and nutritional quality. About 25 years ago, Leistner (1985) introduced the concept of "hurdles" to illustrate the fact that in most foods, an intelligent combination of preservation parameters (obstacles or "hurdles") enhances microbial safety and stability. Traditional examples of such factors are acidity (low pH), limited water available for microbial growth (a_w reduction), addition of preservatives, light heat treatments (blanching), and refrigeration. The modern food industry designs new processes based on scientific concepts especially in terms of various other "hurdles" and processes. Vacuum impregnation (VI) is one of such processes. This technology takes advantage of the porous structure of fruits to improve the osmotic process and the quality of final products (Mújica-Paz et al. 2003). Some of the "hurdles" successfully investigated as novel antimicrobials are chitosan, organic acids, bacteriocins, essential oils, modified atmosphere-packaging, usage of edible films as external barriers, and enzymes such as lactoperoxidase. Likewise, new decontamination technologies have been studied such as inductive heating, microwave and radio frequency, pulsed electric field, ohmic heating, high hydrostatic pressure,

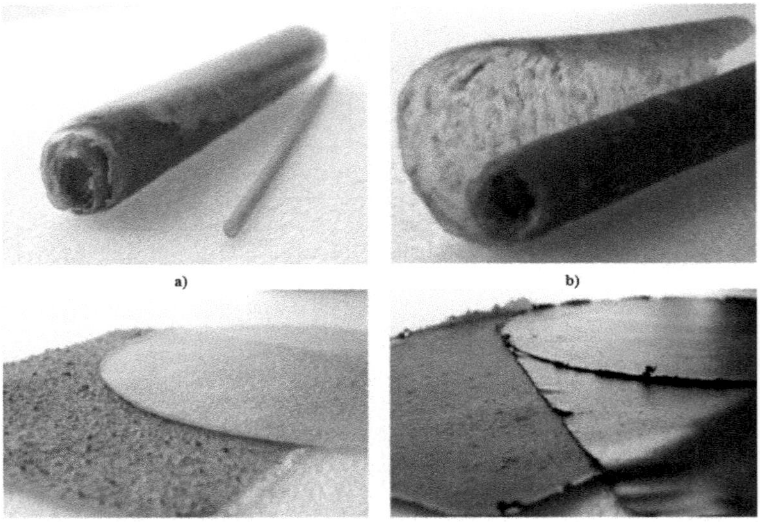

FIGURE 16.2 Fruit leather with (a and b) apple pulp with apple peel, (c) apple pulp with dried apple pomace, and (d) apple and black currant pulp with dried apple pomace (Ciurzyska et al. 2019).

pulsed light, oscillating magnetic field, high-voltage arc discharge, ultrasound (US), X-rays, ultraviolet light, electrolyzed water, and vacuum impregnation (Barbosa-Cánovas et al. 2005, Tapia and Welti-Chanes 2012).

IM fruits are part of traditional eating patterns in populations all over the world. They have been very popular and their consumption level remains significant as of today. However, in order to obtain shelf-stable products at room temperature, high concentrations of solutes are necessary to reduce a_w to safe levels, and not all kind of vegetable products are adequate for this technology. This generally affects the sensory properties and caloric density of finished products. Similarly, pH is relevant, and although fruits are a good example of foods that accept pH reduction without significantly affecting flavor, the pH of IM fruits can be lowered to levels around pH 5.0 in order to not compromise palatability.

Among the issues that have prevented IM fruit snacks from continuing to develop using traditional approaches are consumer health concerns associated with the high levels of humectants (sugars) and preservatives. Additionally, consumers are looking for fresh and natural products without preservatives. The food industry has responded to these demands with so-called minimally processed fruits and vegetables, which have become a very widespread industry.

16.2.3 Minimally Processed Fruits and Vegetables

Minimum processing includes practices such as washing, peeling, cutting, packaging, among others, after which the fruit or vegetable snack is generally kept in refrigerated storage and/or under controlled or modified atmospheres. The stability of these products vary according to the type of fruit or vegetable, processing method, and storage conditions. However, the stability of the product without refrigeration would be ideal, being an issue that should be approached with research and development. The principle used by Leistner (1985) for shelf-stable high-moisture meats ($a_w > 0.90$), where only mild heat treatment is used and the product still exhibits a long shelf life at room temperature can be applied to other foods. High-moisture fruit and vegetable snacks are a good option for the application of this principle to develop safe and palatable minimally processed items that are stable at room temperature.

If the objective is to obtain snacks with characteristics similar to fresh fruits and vegetables, dehydration cannot be solely used in the overall processing scheme. Besides partial dehydration, the further reduction of a_w by adding humectants is commonly employed to keep the product in a high-moisture state. To compensate for the high moisture that remains in the product (in terms of stability), a mild heat treatment (blanching) can be applied without affecting the sensory and nutritional properties. Moreover, the pH could be further diminished and preservatives supplemented to reduce the risk of microbial deterioration without impairing taste.

Alzamora et al. (2016) performed pioneering work to obtain high-moisture peaches and pineapples stable at room temperature. During the 1990s, the use of this approach led to major developments in innovative technologies to obtain "high-moisture fruit products" (HMFP) that could be stored for three to eight months without refrigeration. Examples of some of these processes, and a schematic diagram for the preparation of fruits, in this case, mango and papaya, are presented in Table 16.1 and Figure 16.3, respectively. These products considered snacks which are stable at room temperature are manufactured by the combined methods approach. The principle of these methods is to slightly modify individual stresses to control microbial growth and avoid the severity of techniques based on the use of a single preservation factor (Barbosa-Cánovas et al. 2003).

16.3 TECHNOLOGIES BASED ON WATER ACTIVITY REDUCTION TO OBTAIN LOW- AND INTERMEDIATE-MOISTURE CONTENT SNACKS FROM FRUIT AND VEGETABLES

Fruit-based snacks can be produced/preserved with methods such as convective drying (CD), vacuum drying, freeze-drying, microwave drying (MD), OD, VI, among others. These technologies are focused and based in a_w control. VD, FD, and MD are mainly used to manufacture low a_w snacks, whereas OD and VI to produce intermediate a_w counterparts (Figure 16.4).

The market of dried fruits and vegetables has grown rapidly in many countries, since they are widely used as a basic ingredient in the confectionary, bakery, and cereal industries (Zhang et al. 2006). Drying is an antique preservation technique used to increase food security and shelf life. It is a simultaneous heat and mass transfer process based on the removal of water at values below 10% (Figure 16.5). Drying is aimed to reduce a_w to levels that avoid microbial growth and do not significantly affect biochemical stability. Dry foods decrease transportation, package, and storage costs. In drying, water is removed by using a heating source such as hot air, superheated steam, or hot surfaces. For most drying techniques, the heat used to vaporize water can be transported by convection, conduction, or radiation (Orsat et al. 2006). According to the physical conditions used to transfer heat and to remove water, drying can be classified into CD, VD, FD, and MD. In the first, the heat is applied directly to the food by a gas (air, superheat vapor, or combustions gases) and the water vapor eliminated by the same gas, whereas in VD heat is transferred as CD or indirectly by contact with a hot surface or by radiation removing water faster due to the low pressure. Finally, in FD the water is removed by sublimation, whereas in MD the heat is generated volumetrically in situ by placing the food in an electromagnetic field (Mujumdar 2007, Geankoplis 2016).

VD is commonly combined with CD, MD, or FD to improve quality. VD is performed at subatmospheric pressure to allow the drying at low temperature obtaining products of better quality compared to counterparts dried at atmospheric pressure, especially in terms of porosity, color, shrinkage, and nutritional value. It is particularly preferred for foods with temperature- and oxygen-sensitive components. The vapor pressure gradient provides the driven force for mass transfer during VD.

The selection of the drying technique depends on the type of product, its final quality, and energy consumption (Sagar and Kumar 2010, Karam et al. 2016). Table 16.2 compares the different drying methods mentioned herein. As an example of the ease of moisture removal and the final quality of the product, Figure 16.6a and b shows the effect of different drying methods on the rate of a_w change

TABLE 16.1
Some New Technologies for Preserving Fruits by Combined Methods to Obtain High-Moisture Shelf Stable Snacks

Fruit	"Hurdle"	Storage Temperature (°C)	Shelf Life (Months)
Passion fruit, *Passiflora edulis flavicarpa* (Pulp)	pH: 2.7–3.0 a_w: 0.98 (sucrose) KS: 1,500 ppm 4-hexilresorcinol: 200 ppm Ascorbic acid: 2,500 ppm	30	60 days
Peach, *Prunus persica* (halves)	a_w: 0.97 (sucrose) pH: 3.7 KS: 1,000 ppm $NaHSO_3$: 150 ppm	35 23	3 8
Peach, *Prunus persica* (halves)	a_w: 0.94 (glucose) pH: 3.5 KS: 1,000 ppm $NaHSO_3$: 150 ppm	20 30	4 4
Mango, *Mangifera indica* L. (puree)	a_w: 0.93 (sucrose) pH: 3.6 SB: 1,480 $NaHSO_3$: 150 ppm	30	6
Mango, *Mangifera indica* L. (slices)	a_w: 0.93 (sucrose) pH: 3.3 SB: 1,400 ppm $NaHSO_3$: 150 ppm	35	6
Mango, *Mangifera indica* L. (slices)	a_w: 0.97 (sucrose) pH: 3.0 KS: 1,000 ppm $NaHSO_3$: 150 ppm	35	4–5
Papaya, *Carica papaya* (slices)	a_w: 0.97 (sucrose) pH: 3.8 KS: 1,000 ppm $NaHSO_3$: 150 ppm	35 25	4–5
Pineapple, *Ananas comosus* (slices)	a_w: 0.97 (glucose) pH: 3.1 KS: 1,000 ppm $NaHSO_3$: 150 ppm	27	4
Banana, *Musa paradisíaca* (puree)	a_w: 0.92 (sucrose) pH: 4.2 KS: 1,500 ppm SO_2: 300 ppm	26	3
Naseberry fruit, sapodilla, *Achras zapota* (slices)	a_w: 0.92–0.94 (sucrose) pH: 4.0–4.3 $NaHSO_3$: 150 ppm	26	3

Source: Adapted from Alzamora et al. (1993).
KS: potassium sorbate.

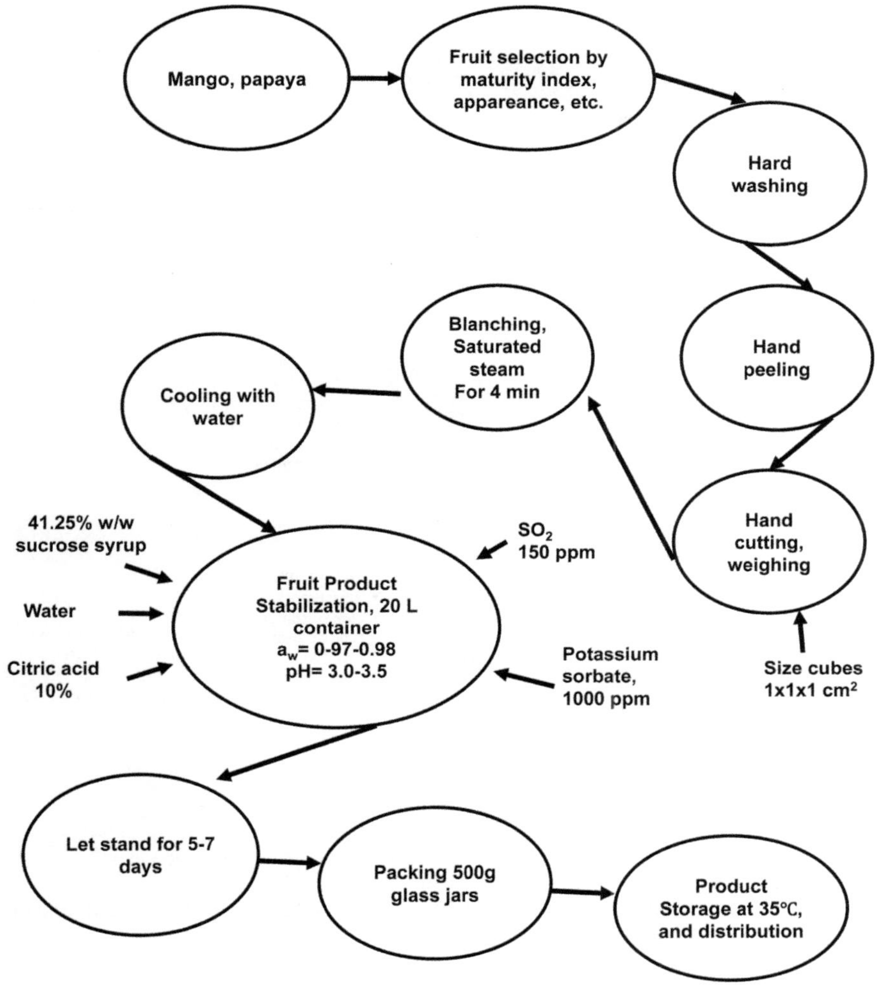

FIGURE 16.3 Schematic diagram for the preparation of high-moisture mango and papaya of extended shelf life by combined methods (adapted from Barbosa-Cánovas et al. 2003).

and total color change of laminated guava snacks. The utilization of MD or FD processes in comparison with the use of hot-air drying diminished significantly the drying time and improved color retention in the laminated guava snacks obtained after freeze-drying (Patel et al. 2016).

16.3.1 Convective Drying

Drying is a complex process that accounts for about 10–25% of the total energy required to produce foods (Strumillo and Adamiec 1996, Mujumdar and Passos 2000). Sun drying is a simple and common method used to reduce water content in fruits and vegetables; nevertheless, the changes in sensory and nutritional characteristic as well as the hygiene problems have motivated the development of more sophisticated and faster methods. CD, also known as conventional drying or hot-air drying (HAD), is an economic technique widely adopted in the food industry. In CD, the hot air transfers heat to the food surface by conduction and then the liquid moves from the interior of the food to the surface and eliminated by air convection (Ahmed 2011, Karam et al. 2016). Several types of dryers are commercially used to remove water from fruits and vegetables. Continuous processes as tunnel,

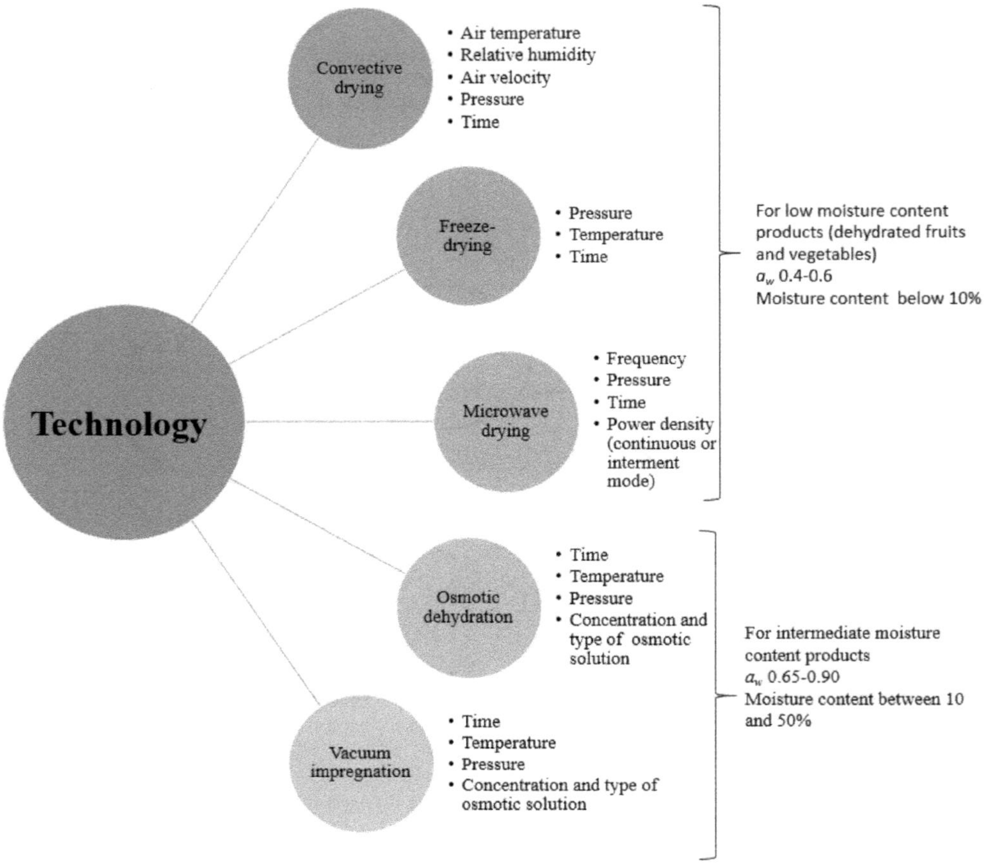

FIGURE 16.4 Main methods based on water activity control used to produce low- and intermediate-moisture content snacks derived from fruits and vegetables.

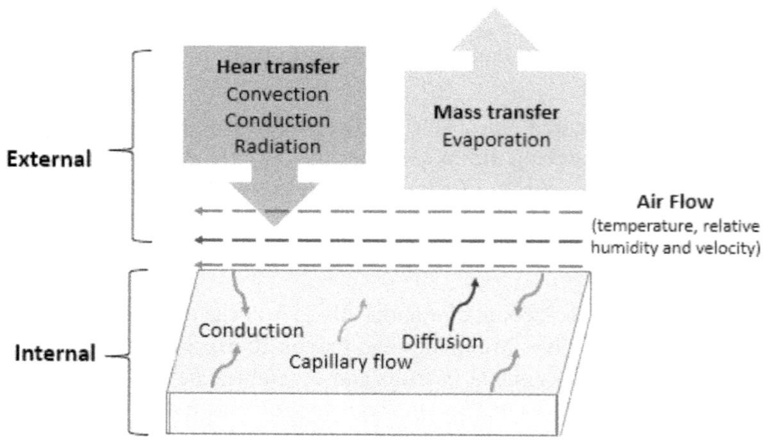

FIGURE 16.5 Conceptual representation of thermal drying of a solid food.

TABLE 16.2
Drying Conditions of Fruits in Tunnel Dryer and Characteristic of the Finished Product

	Drying Conditions			Finished Product	
Fruit	**Load (kg/m²)**	**Temperature (°C)**	**Time (h)**	**Moisture (%)**	**Yield (%)**
Plum	15	40–50	6	18–20	25–35
		75–80	14		
Apple (rings)	10	55–75	5–6	20	10–12
Apricot (halves)	10	60–70	10–15	15–20	10–15
Cherry	10	55–70	6–8	12–15	25
Pear (halves and	15	65–70	15–22	18–20	10–15
quarters)	15	60–70	10–15	15–20	10–15

Source: Adapted from http://www.fao.org/3/V5030E/V5030E0j.htm.

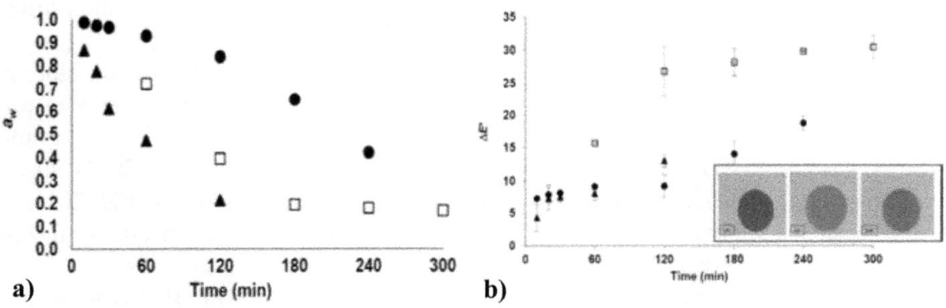

FIGURE 16.6 Effect of different drying methods (convective drying = circle, freeze drying = square, and microwave drying = triangle) on (a) water activity and (b) total color change (ΔE^*) of laminated guava snacks (Patel et al. 2016).

belt trough, fluidized bed and foam-mat drying are preferably used for vegetables. Factors related to the food to be dried like shape, size, composition, and others and features of end-dried product dictate operating conditions and the selection of a particular dryer (http://www.fao.org/3/v5030e/V5030E0b.htm) and the related operation costs. Table 16.2 summarizes some of the tunnel dryer conditions for the processing of fruits and the features of finished products.

Because of the complexity of drying, many theories have been proposed to describe the internal mechanism that dictate moisture movement. For capillary and porous materials such as fruits and vegetables, some of the proposed mechanism are the liquid diffusion driven by concentration gradients, liquid transport attributed to capillary forces, and vapor diffusion due to shrinkage and partial vapor pressure differences. Furthermore, the moisture movement could be ruled by liquid or vapor transport due to the gradient in total pressure caused by external pressure and temperature, evaporation and condensation effects due to temperature gradients, and liquid transport due to gravity (Gavrila et al. 2008).

The drying process is normally characterized by two stages: the constant rate and the falling rate periods. In the first one, water moves at a constant rate from the interior of the product to the surface and therefore there is always a thin layer of water on the food surface that is gradually removed by vaporization. The falling rate stage appears when the a_w of the product is reduced and the removal of water becomes difficult and therefore the drying rate diminishes constantly. During this falling rate period, the process requires more time and energy (Geankoplis 2016). The drying rate and process time depend on air properties such as its relative humidity, temperature, and flow rate, as well as on internal factors related with the food such as its composition, water content, geometry, thickness, structure, and physical state (Khraisheh et al. 1997, Mulet et al. 2000, Ratti 2001). All these factors contribute to the mass and heat transfer rates closely related to the drying rate and time. The application of lower temperature improves product quality but increases the drying time and operation costs. The use of vacuum during convective drying reduces the water boiling point and therefore nutrients are less damaged. Ultrasound and pulsed electric field (PEF) have been proposed as pretreatments before CD to enhance mass transfer due to the microstructural changes originated. Therefore, these technologies can decrease drying time and quality loss.

Power ultrasound is a process characterized by the use of high energy and intensity that is applied at low frequencies (20–100 kHz). The cavitation phenomena generated during the US application increases the mass transfer reducing the time required to dry foods. In HAD processes, US can be applied before, simultaneously, or intermittently. The simultaneous and intermittent processes are often carried out using the airborne ultrasound, which consists of using air as sound energy transmitting medium (Jambrak et al. 2008, Witrowa-Rajchert et al. 2014, Onwude et al. 2017).

The main advantages of combining US and HAD are (1) reduction of drying time and energy; (2) increase of drying rate and mass and heat transfer rates; (3) effectivity at low air velocity, temperature, and US power; and (4) improved product quality. However, there are still some limitations to the application of this technology at an industrial level (Onwude et al. 2017).

The direct application of US in carrot was better than in airborne US due to the higher penetration of the acoustic energy, which resulted in faster water removal (Gallego-Juarez et al. 2007). Compared with HAD, 13–57% reduction in drying time has been observed for the simultaneous application of US and HAD for apples (Sabarez et al. 2012, Rodríguez et al. 2014), strawberries (Gamboa-Santos et al. 2014), carrots (Cárcel et al. 2011), and potatoes (Ozuna et al. 2011). In addition, the simultaneous application of US and HAD is reported to yield higher quality dry products (Rodríguez et al. 2014, Kowalski et al. 2017).

PEF consists of the application of pulses of high voltage and short duration from microsecond to millisecond in foods placed between two electrodes (San Martin et al. 2007). The electric field generated due to the voltage applied can cause electroporation resulting in increased mass transfer. The application of PEF before HAD in fruits and vegetables has shown to reduce drying time and improve drying rate. The use of PEF and HAD in apples (Wiktor et al., 2013), potatoes (Lebovka et al. 2007), and carrots (Wiktor et al. 2016) significantly decreased drying time. The higher mass transfer rates and lower drying times obtained when combining PEF and HAD have the potential of reducing the total energy required. The energy expenditure for water evaporation during PEF-assisted HAD depends on temperature and pressure and has been reported to be 2.5–2.7 MJ/kg, whereas for HAD 4–6 MJ/kg (Raso and Barbosa-Cánovas 2003). Some of the advantages of combining PEF and HAD are the uniform rise of product temperature, adequate for heat-sensitive products dried at moderate temperatures, and better or more attractive color scores. Some limitations are the complex equipment design and installation, the suitability and electrical properties of fruits and vegetables, and the electrical damage to the product (Kowalski et al. 2017).

CD-treated products can be stored for about one year, although some nutritional, physicochemical, and sensory characteristics of the fruits and vegetables can be lost during processing as affected by both drying conditions and the heat sensitivity of the intrinsic food components (Karam et al. 2016). The loss of color and nutritional value, the changes of texture and flavor, as well as the volumetric alterations like shrinkage and collapse are some of the quality problems observed in CD products subjected to high temperatures and long drying times (Mayor et al. 2005, Orsat et al. 2006). It has been observed that HAD of fruits and vegetables at temperatures 40–70°C causes losses of functional compounds. For instance, significant losses of vitamin C were found after HAD of apple (Joshi et al. 2011) and apricot (García-Martínez et al. 2013). Likewise, the reduction of carotenoids was observed in apricot (Ihns et al. 2011), carrot (Zhao et al. 2014), and sweet potato (Vimala et al. 2011), whereas total anthocyanins diminished in blueberries (Sablani et al. 2011) and mulberries (Chottamom et al. 2012). Decrement of total phenolics and antioxidant activities was found for mango (Chong et al. 2013) and mulberry (Chottamom et al. 2012).

16.3.2 Freeze-Drying

Based on the sublimation of water, freeze-drying is recognized for its capacity to generate dried foods of high quality in terms of color, shape, aroma, and nutritional value. These advantages are due to the low temperature and the vacuum conditions used during the process. FD products rehydrate four to six times faster than HAD counterparts because they maintain their porous structure. Moreover, freeze-dried products have minimal changes in color. Nevertheless, the main limitation of this technology is the cost that is estimated to be four to eight times higher than for HAD, so it is preferred for high-value products (Ratti 2001). During FD, frozen products are placed in a hermetic chamber that is maintained at subatmospheric pressure by using a vacuum pump. Sublimation of water is achieved at temperature and pressure below 0°C and 4.58 mmHg, respectively. The vacuum, together with the heat supplied in the freeze-drying chamber, leads to an immediate extraction of moisture from the frozen material. The driving force for water removal is the water vapor pressure difference between the ice interface and the drying chamber. During sublimation, about 90% of the water is removed and water vapor is eliminated as it is generated. The sublimation rate depends on the resistance of the food to heat transfer toward the sublimation front, and of mass or steam from it toward the chamber. Once the frozen water is removed by sublimation, more bound water can be desorbed by heating the product at temperatures of 20–60°C (Clementz and Delmoro 2011).

A number of freeze-dried fruits and vegetables have been dried and evaluated in terms of quality. The freeze-drying of strawberry at 20°C retained a better quality than the one dried at 60°C because the former presented more shrinkage (Hammami and Rene 1997). Carrots (Regier et al. 2005), starfruit, papaya, muskmelon (Shofian et al. 2011), strawberries (Abonyi et al. 2002), and sweet potatoes (Yang et al. 2010) showed a high retention of carotene after FD. Likewise, high ascorbic acid retention was observed for strawberries (Abonyi et al. 2002), carrots (Lin et al. 1998), mango, starfruit, watermelon, and muskmelon subjected to FD (Shofian et al. 2011). An example of a flowchart to obtain freeze-dried snacks is presented in Figure 16.7.

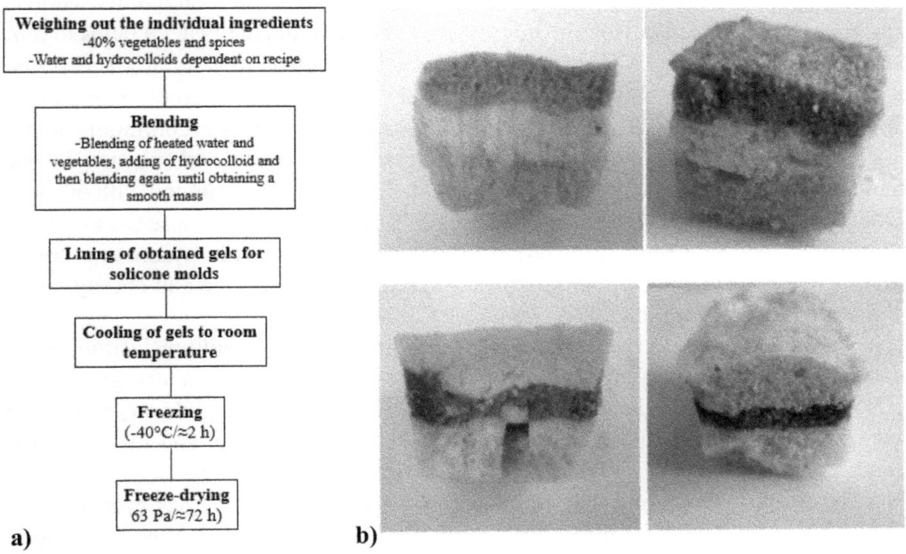

FIGURE 16.7 (a) Flowchart for obtaining freeze-dried snacks and (b) freeze-dried vegetable snacks from carrot, potato, cauliflower, broccoli, pepper, green bean corn, and other ingredients (Ciurzyska et al. 2020).

These researchers showed that the production of freeze-dried vegetable snacks based on frozen vegetables not used during the conventional production of frozen foods is possible and seems to be promising. All snack variants have been characterized by favorable physical properties (high porosity, low shrinkage, appropriate hardness, low water activity, and attractive appearance for consumer). The type of hydrocolloid used has had the greatest impact on the results. Snacks with sodium alginate have had more fine pores, which results in higher porosity and hardness as compared to snacks based on hydrocolloid mixtures, which probably consist of a large number of smaller pores that yielded lower porosity. Any other differences have not been noticed in terms of energy consumption during freeze-drying or blending of vegetable bars. The costs of freeze-drying account for more than 22% up to almost 35% of the electricity costs (Ciurzyska, et al. 2020).

16.3.3 Microwave Drying

MD consists of the application of MW as source of thermal energy inside an oven, often at a frequency from 60 Hz to 2,450 MHz. MD involves generation of heat throughout the food (volumetric heating) by molecular excitation (Vadivambal and Jayas 2009, Motevali et al. 2011). Ionic conduction causes the acceleration of ions toward opposite polarity by electric fields originating the collision with other molecules increasing kinetic energy and temperature. The increment in product temperature is related with the power density applied, which can be supplied in continuous or interment mode (Orsat et al. 2006). Lower power density and internment mode generate lower temperatures. Dielectric constant and the loss factor determine the interaction of MW energy with foods and they are related with the frequency level (Orsat et al. 2006). The former indicates the amount of energy absorbed by the food, while the loss factor describes the dissipation of electric energy (Mullin 1995). These properties also depend on the temperature, moisture content, composition, and particle density of the food, while, thickness, geometry, and dielectric properties of the food matrix affect the microwave heat transfer (Meda and Raghavan 2004).

Some advantages of MD are fast volumetric heating, higher drying rate, better quality and lower processing time, usage of energy consumption, and operational cost. Some drawbacks are the high initial investment, nonuniform heating, partial loss of aroma, changes in sensory attributes including texture, and dependency on sample size and shape (Karam et al. 2016, Zhang et al. 2006).

MW can be applied alone or combined with CD or VD to increase food quality while reducing processing time. The air used during the microwave convective drying (MCD) helps to remove the vaporized water during drying. MW-assisted processes have shown to yield products of high quality. The lower processing time makes MD a moderately low-energy-consumption process (Orsat et al. 2006, Tulasidas et al. 1995). When combined with HAD, MW energy is commonly applied at the beginning of the HAD process to rapidly heat the interior of the product and enhance the removal of water from the surface (Karam et al. 2016, Zhang et al. 2006). When combined with other technologies, MD is more advantageous when applied in the falling rate period because the absorbed energy rises the temperature in the center of the product increasing mass transfer. It has been shown that the application of MW energy during the falling period reduces drying time and energy consumption while improving quality (Maskan 2001, Orsat et al. 2006).

The use of vacuum in MD reduced the boiling point of water resulting in improved quality. Microwave-assisted vacuum drying (MVD) implies the application of MW energy under vacuum. This technology generated products of similar or superior texture, color, flavor, and acceptability compared with FD products, and of lower apparent density, shrinkage, and higher porosity than HAD (Karam et al. 2016).

The drying of many fruits and vegetables with MD, MCD, or MVD has been investigated to evaluate the effects on drying kinetics and quality attributes. The MD of kiwifruit slices caused a high shrinkage and low rehydration capacity (Maskan 2001). In carrots, MD generated products of adequate color and rehydration properties (Wang and Xi 2005). Grapes dried with this technology presented better color and less damage compared with the HAD products (Tulasidas et al. 1995). MVD has been used for drying many vegetable products obtaining good quality. For example, MVD generated apple slices of better quality in terms of vitamin C content, texture, and flavor compared to CD. In addition, the energy consumption decreased about 32% (Qinghua et al. 2008). Carrot slices treated by VMD showed higher rehydration capacity, β-carotene, and vitamin C contents compared to counterparts treated by CD and they presented similar quality to the ones obtained with FD (Tein et al. 1998). In a similar study conducted with carrots, a better retention of carotenoids was found for MVD than for FD and CD (Regier et al. 2005). Products such as cranberry (Sunjka et al. 2004), banana (Drouzas and Schubert 1996, Mousa and Farid 2002), potato (Bondaruk et al. 2007), kiwifruit, apple, and pear (Kiranoudis et al. 1997) have been successfully dried with this technology obtaining good quality. Regarding MCD, a higher ascorbic acid loss was observed in apricot treated by MD (100 W) using HAD as pretreatment than for the one treated only with MD at 100 W (García-Martínez et al 2013). The concentration of carotenoids decreased in higher extent in apricots treated with MD (100 W) pretreated with HAD (García-Martínez et al 2013) and in carrots (140–175 W, 900 W, 2,450 MHz) (Arikan et al. 2012, Zhao et al. 2014).

16.3.4 Osmotic Dehydration

OD is a complex dynamic mass transfer process used to decreasea a_w of whole or pieces of fruits and vegetables when immersed in concentrated hypertonic aqueous solutions consisting of sucrose, fructose, sodium chloride, lactose, maltodextrin, glucose, glycerol, sorbitol, and corn syrups (Escobedo-Avellaneda et al. 2018). Osmosis is known as a partial dehydration process in which about 30–50% of water is eliminated mainly by diffusion and in which gain of solutes could happen depending on the process conditions (Ahmed 2011). About 50% of the weight of fruits and vegetables is lost during OD at room temperature obtaining IM content products (Sagar and Kumar 2010). OD increased shelf life, improves sensorial and nutritional properties, and reduce the negative effect of thermal treatment on flavor and color (Khan 2012). Some of the drying conditions used for some osmotically dehydrated products are summarized in Table 16.3.

The use of vacuum (50–100 mbar) during OD favors the expansion and elimination of the gas occluded into the food pores, increasing the available mass transfer area. The empty spaces are then occupied by the osmotic solution (Sagar and Kumar 2010). The difference between the osmotic

TABLE 16.3
Osmotic Dehydration Conditions to Obtain Vegetable Snacks

Product	Type of Cut	Treatment Conditions
Banana	5-mm slices	2 h, 80% sugar 2,000 ppm SO_2 70°C
Carrot	10 × 10 × 2 mm dices 5-mm slices	4 h, 60% sugar, 10% salt 4,000 ppm SO_2
Green mango	8-mm slices	2 h, 25% salt 8,000 ppm SO_2
Ripe mango	8-mm slices	2 h, 60% sugar 8,000 ppm SO_2
Papaya	8 × 8 mm slices	4 h, 80% sugar 2,000 ppm SO_2 70°C
Strawberry	Whole	4 h, 80% sugar 4,000 ppm SO_2

Source: Adapted from http://www.fao.org/3/V5030E/V5030E0j.htm

pressure of the solution and the food or the chemical potential gradient between the solution and the intracellular fluid act as the driving force for water removal. Water is removed by osmo-diffusion and capillary flow phenomena (Escobedo-Avellaneda et al. 2018). Factors such as type of osmotic solution, concentration, temperature, sample:solution ratio, agitation, sample size and shape, and pretreatments influence the osmotic process (Bolin et al. 1983, Lerici et al. 1985).

OD is also used as a pretreatment before drying by MD, HAD, FD, or VD with the aim to improve the quality of final products and decrease energy consumption (Beaudry et al. 2004, Rastogi and Raghavararo 1997). The application of OD in strawberry before MD generated products of similar quality to those obtained with FD in terms of dehydration capacity and sensory characteristics (Venkatachalapathy and Raghavan 1999). OD has also been applied as a pretreatment before explosion puffing drying (EPD). EPD mix the advantages of CD and FD and overcomes the drawbacks of low-temperature vacuum frying, allowing obtaining natural fruit and vegetable chips, fruit and vegetable powder, or ingredients of high quality and texture used to generate healthy food and snacks. There is great interest of products obtained by this method around the world (Xiaoping et al. 2018). A schematic representation of an explosion puffing drier and of a product obtained by this technology is presented in Figures 16.8 and 16.9, respectively.

Prior to OD, products can be pretreated with the methods mentioned before in order to increase mass transfer rate and efficiency. Pretreatments such as PEF and US have been used before OD to increase mass transfer (Sagar and Kumar 2010). Likewise, the OD of apple slices, mango, and carrot was improved in terms of mass transfer rate due to the application of PEF (Rastogi et al. 1999, Taiwo et al. 2003, Lai and Sharma 2005). The application of US on apple cubes before OD increased water loss and solute gain (Simal et al. 1998).

16.3.5 Vacuum Impregnation

VI is a process used to obtain minimally processed fruits with IM content. It is also used as a strategy to improve the nutritional quality of fruits and vegetables and as a pretreatment of other preservation processes, including drying (Fito et al. 1996, Tapia de Daza et al. 1996). In VI, solid foods are immersed in a concentrated solution and then the pressure is reduced to accelerate the incorporation of solids (Fito 1994, Mújica-Paz et al. 2003). The expansion of the gas occluded in the food pores causes deformation of the solid matrix and when pressure equals the pressure of the system, the

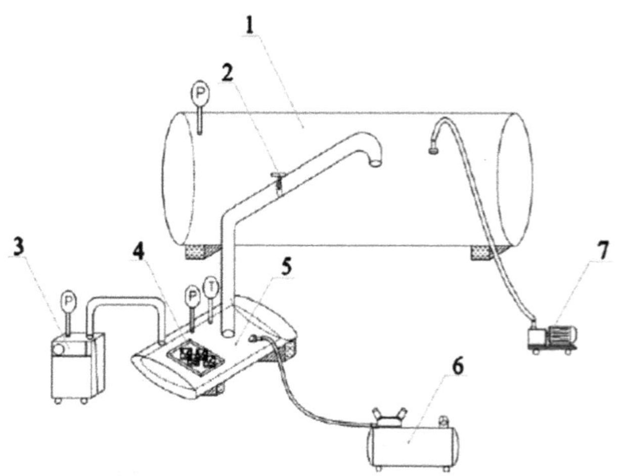

FIGURE 16.8 Diagram of the explosion puffing drier and main components of the equipment: 1 = vacuum chamber, 2 = decompression valve, 3 = steam generator, 4 = samples, 5 = puffing chamber, 6 = air compressor, 7 = vacuum pump (Xiaoping et al. 2018).

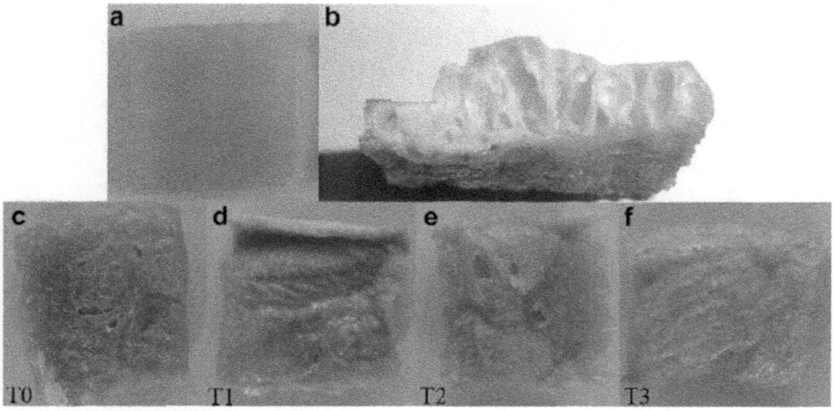

FIGURE 16.9 Effect of osmotic pretreatment on quality of mango chips obtained by explosion puffing drying: (a) fresh sample, (b) cross-sectional view of sample T0, (c) no osmotic drying (OD) pretreatment, (d) 30 min of OD pretreatment, (e) 60 min of OD pretreatment, and (f) 120 min of OD pretreatment (Zou et al. 2013).

external fluid is absorbed in the pores due to the capillary effect. Food impregnation can also be performed at atmospheric pressure, but the quality of the food is normally better at subatmospheric conditions and in addition the process is faster (Fito and Chiralt 1995).

Food impregnation allows the incorporation of different ingredients into the porous structure of the food such as sugars to obtain candied fruits, vitamins, minerals, and functional compounds to improve the functionality of fruits and vegetables. This technology has been successfully used to incorporate probiotic microorganisms to obtain healthy products (Fito and Chiralt 1995, Salvatori et al. 1998, Fito et al. 2001, Zhao and Xie 2004). During impregnation, the food behaves as a semipermeable membrane in which water is lost by diffusion and capillary flow as it occurs in OD. The principle of the technology is that solutes are impregnated into the porous membrane by diffusion and hydrodynamic flows. Diffusion is driven by concentration gradients, while capillary flow is attributed to the pressure difference between the system and external pressure (Shi and Le Maguer 2002, Chiralt and Talens 2005, Khin et al. 2005). Porosity, tissue structure, geometry, size, product ripeness, type of solutes and concentration, sample:solution ratio, pressure, temperature, time, and agitation affect impregnation and food quality (Roa et al. 2001, Shi and Le Maguer 2002, Chiralt and Talens 2005, Occhino et al. 2011).

Fruits such as pineapple cylinders (Lombard et al. 2008), strawberry (Watanabe et al. 2011), apple, mango, and melon slices (Mújica-Paz et al. 2003) and papaya slabs (Moreno et al. 2004) have been impregnated with sucrose solutions, whereas counterparts such as mango cylinders (Ostos et al. 2012) and grapefruit (Moraga et al. 2009) with minerals such as calcium. $CaCl_2$ solutions and/or $CaCl_2$ and pectin methylesterase have been used to improve the texture/firmness of zucchini (Occhino et al. 2011), apples, strawberries, raspberries, (Degraeve et al. 2003), and apple pieces (Guillemin et al. 2008). Apple, banana, and potato cubes were successfully impregnated with phenolic compounds at atmospheric pressure (Rózek et al. 2009). On the other hand, products such as apples (Ursachi et al. 2009) and whole potatoes (Kikuchi et al. 2011) have been impregnated with ascorbic acid. Interestingly, apple cylinders were impregnated with *Lactobacillus casei* (spp. rhamnosus) and then dried showing viable cells greater than 10^6 CFU/g in the dehydrated and stored product (Betoret et al. 2003).

16.3.6 Vacuum Frying of Fruit and Vegetable Snacks

Fried fruits and vegetable snacks are preferably produced today using nonhydrogenated oil, and oils that do not contain saturated fats are free of trans-fatty acids. Some of these products

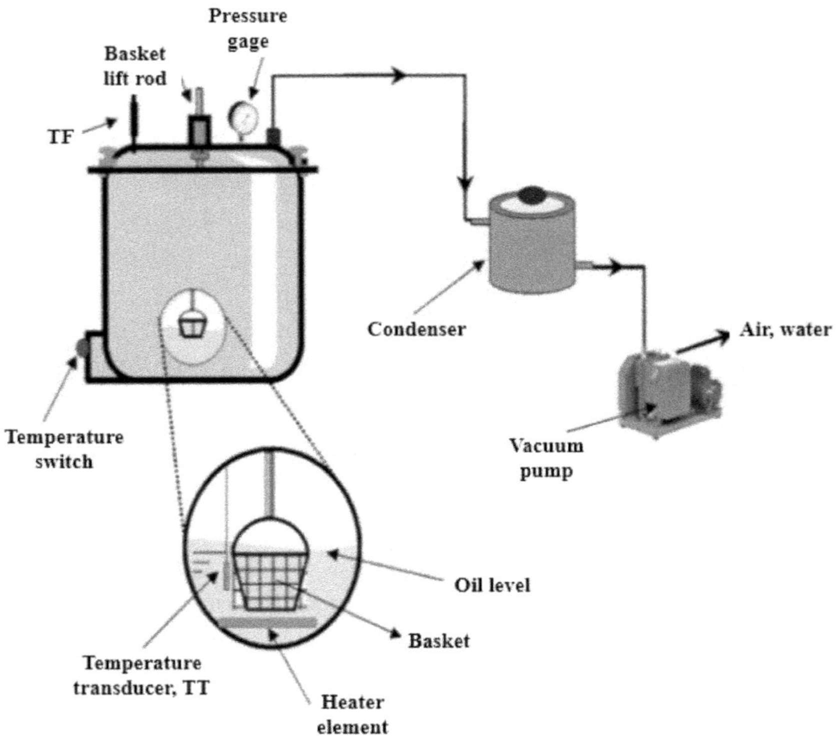

FIGURE 16.10 Setup of a vacuum-frying equipment to obtain high-quality fruit and vegetable-based snacks (adapted from Da Silva et al. 2008).

(sweet-potato chips, apple chips, potato chips-blue) are fried under vacuum yielding less oil absorption (taste less greasy) with higher retention of their natural color and flavors. Vacuum frying is an efficient method of reducing the oil content of fried snacks, maintaining product nutritional quality, and reducing oil deterioration. This technology is used to produce fruits and vegetables with the necessary degree of dehydration without excessive darkening or scorching of the product. In vacuum-frying operations, food is heated at low pressure (<80 kP) causing a reduction in the boiling points of the oil and the moisture in the foods (Shyu et al. 1998). A detailed description of the vacuum-frying process to obtain snacks is described in Chapter 3. The typical setup of a vacuum-frying equipment is shown in Figure 16.10 where the pressure and temperature control elements are the most relevant parts of the equipment (Da Silva et al. 2008). French fries processed in a vacuum fryer can achieve the necessary degree of dehydration without excessive darkening or scorching of the product. Garayo and Moreira (2002) showed that vacuum fryers could produce potato chips with lower oil content (30% less) and the same texture and color characteristics of those fried in conventional (atmospheric) fryers. The frying oil showed lower amounts of free fatty acid (FFA) and a slightly lighter color after 30 batches of frying. Yamsaengsung and Rungsee (2008) found that vacuum fried potato chips and guava slices had lower oil content and more natural colorations than those fried in conventional fryers. Granda et al. (2004) demonstrated that vacuum frying could also produce potato chips with 97% less acrylamide content, a potential carcinogenic found in fried snacks, discussed in Chapter 20, than in the traditionally fried chips. There are many highly nutritious vegetables and fruits that could be vacuum-fried: faba beans, broccoli, cauliflower, carrots, pineapple, mango, etc. This technology improves the nutrition and health by producing products that taste good, keep most of their nutrition values, have lower fat content than the conventionally

FIGURE 16.11 Color differences between products fried under vacuum and traditional frying processes: (a) blue potato fried in the vacuum fried, (b) blue potato fried in the traditional fryer, (c) sweet potato fried in the vacuum fryer, (d) sweet potato fried in the traditional fryer, (e) mango fried in the vacuum fryer, (f) mango fried in the traditional fryer, (g) green bean fried in the vacuum fryer, and (h) green bean fried in the traditional fryer (adapted from Da Silva et al. 2008).

fried snacks, are safer with little or no acrylamide formation, and keep longer. The length of time and the method of frying are important factors that affect phytochemical/nutraceuticals stability (Shirsat and Thomas 1998). In general, vacuum-fried snacks retain more of their natural colors and flavors due to the less oxidation and lower frying temperature. Oil content of vacuum-fried sweet-potato chips was significantly lower than traditional fried products. Blue potato and mango chips showed no significant differences in oil content for each frying method. Anthocyanin and total carotenoids content were significantly high for the products fried in the vacuum fryer than those in the traditional fryer. The texture characteristics of the products for both frying methods were not significantly different. Sensory panelists overwhelmingly preferred the vacuum-fried products for color, texture, taste, and overall quality. The differences in the use of vacuum in the vegetable frying process are clearly observed in Figure 16.11 for potato, sweet potato, mango, and green beans (Da Silva et al. 2008). Vacuum frying is frequently used after applying OD, partial drying, and/or freezing processes to obtain crispy vegetable snacks as reflected in the flow diagram of Figure 16.12 (del Mercado et al. 2020).

16.4 CONCLUSIONS

Several approaches can be used for design and processing of fruit/vegetable-based snacks with the scientific and technological advances in food processing. After revising various processing techniques mainly based on the water removal principle, it is clear that snacks from fruit and vegetables obtained by these operations is an auspicious area that should be accompanied by proper selection of the processing technology. The new trends in the consumption of healthy snacks open a great area of opportunity for the development of novel fruit- and vegetable-based products. The redesign of existing products with more efficient drying processes and the reduction of sugars or other types of solutes in IM products are critically important, especially to maintain the nutritional value of these snacks. In addition, the application of the concepts of hurdle technology and the development of minimally processed snacks are an alternative with a great future and relevance for their application in fruits and vegetables. The outbreak of the COVID-19 pandemic in 2020 has increased the interest of consumers in fruit and vegetables that contain high amounts of vitamin C, which boosts the immune system. The market opportunities for novel and affordable snacks with prolonged shelf life based on fruit and vegetables is promising.

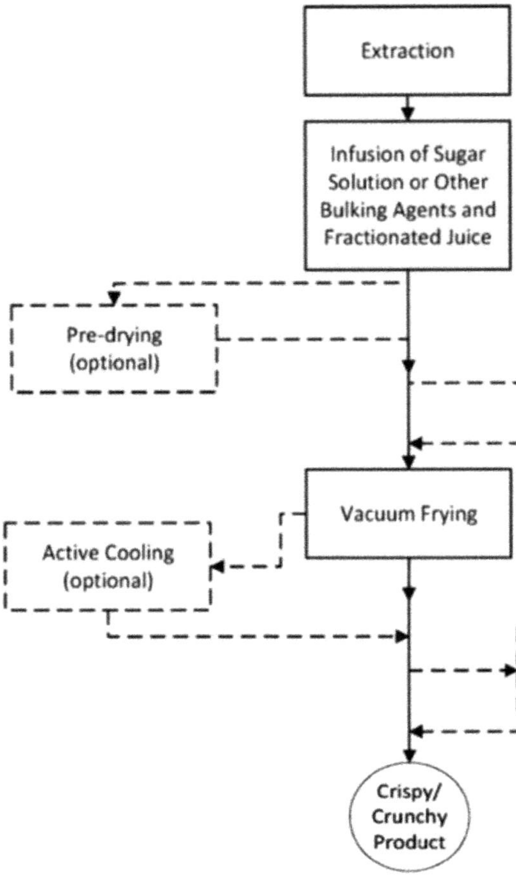

FIGURE 16.12 Flowchart for obtaining crispy/crunchy snacks using osmotic dehydration and pretreatment (del Mercado et al. 2020).

REFERENCES

Abonyi, B.I., Feng, H., Tang, J., Edwards, C.G., Chew, B.P., Mattinson, D.S., and Fellman, J.K. 2002. Quality retention in strawberry and carrot purees dried with refractance window TM system. *Journal of Food Science*, 67(3):1051–1056.

Ahmed, J. 2011. Drying of vegetables: Principles and dryer design. In: *Handbook of Vegetables and Vegetable Processing*, eds. N.K. Sinha, Y.H. Hui, E. Ozgul Evranuz, M. Siddiq, and J. Ahmed, 279–298. Hoboken: Wiley-Blackwell Publishing.

Alaimo, K., Oleksyk, S.C., Drzal, N.B., Golzynski, D.L., Lucarelli, J.F., Wen, Y., and Velie, E.M. 2013. Effects of changes in lunch-time competitive foods, nutrition practices, and nutrition policies on low-income middle-school children's diets. *Child Obesity*, 9(6):509–523.

Alm, A., Fåhraeus, C., Wendt, L.-K., Koch, G., Andersson-Gäre, B., and Birkhed, D. 2008. Body adiposity status in teenagers and snacking habits in early childhood in relation to approximal caries at 15 years of age. *International Journal of Pediatric Dentistry*, 18(3):189–196.

Alzamora, S.M., López-Malo, A., Tapia, M.S., and Welti-Chanes, J. 2016. Minimally processed foods. In: *The Encyclopedia of Food and Health*, eds. B. Caballero, P. Finglas, and F. Toldrá, 767–771. Oxford: Academic Press.

Alzamora, S.M., Tapia, M.S., and Welti-Chanes, J. 1998. New strategies for minimal processing of foods. The role of multitarget preservation. *Food Science and Technology International*, 4(5):353–361.

Alzamora, S.M., Tapia de Daza, M.S., Argaiz, A., and Welti, J. 1993. Application of combined methods technology in minimally processed fruits. *Food Research International*, 26(2):125–130.

Arikan, M.F., Ayhan, Z., Soysal, Y., and Esturk, O. 2012. Drying characteristics and quality parameters of microwave-dried grated carrots. *Food and Bioprocess Technology*, 5(8):3217–3229.

Barbosa-Cánovas, G., Fernández-Molina, J.J., Alzamora, S., Tapia, M.S., López-Malo, A., and Welti-Chanes, J. 2003. *Handling and Preservation of Fruits and Vegetables by Combined Methods at Rural and Village Levels. Technical Manual.* Rome: Food and Agricultural Organization of the United Nations.

Barbosa-Cánovas, G.V., Tapia, M.S., and Cano, M.P. 2005. *Novel Food Processing Technologies.* New York: Marcel Dekker Inc., CRC Press.

Beaudry, C., Raghavan, G.S.V., Ratti, C., and Rennie, T.J. 2004. Effect of four drying methods on the quality of osmotically dehydrated cranberries. *Drying Technology*, 22(3):521–539.

Betoret, N., Puente, L., Díaz, M.J., Pagán, M.J., García, M.J., Gras, M.L., Martínez-Monzo, J., and Fito, P. 2003. Development of probiotic-enriched dried fruits by vacuum impregnation. *Journal of Food Engineering*, 56(2–3):273–277.

Bolin, H.R., Huxsoll, C.C., Jackson, R., and Ng, K.C. 1983. Effect of osmotic agents and concentration on fruit quality. *Journal of Food Science*, 48(1):202–205.

Bondaruk, J., Markowski, M., and Blaszczak, W. 2007. Effect of drying conditions on the quality of vacuum-microwave dried potato cubes. *Journal of Food Engineering*, 81(2):306–312.

Caballero-Cerón, C., Serment-Moreno, V., Velazquez, G., Torres, J.A., and Welti-Chanes, J. 2018. Hygroscopic properties and glass transition of dehydrated mango, apple and banana. *Journal of Food Science and Technology*, 55(2):540–549.

Cárcel, J.A., Garcia-Perez, J.V., Riera, E., and Mulet, A. 2011. Improvement of convective drying of carrot by applying power ultrasound: Influence of mass load density. *Drying Technology*, 29(2):174–182.

CBI. 2020. The Centre for the Promotion of Imports from developing countries. Retrieved on December 30, 2020 from https://www.cbi.eu/market-information/fresh-fruit-vegetables/trends.

Chiralt, A., and Talens, P. 2005. Physical and chemical changes induced by osmotic dehydration in plant tissues. *Journal of Food Engineering*, 67(1–2):167–177.

Chong, C.H., Law, C.L., Figiel, A., Wojdyło, A., and Oziembłowski, M. 2013. Colour, phenolic content and antioxidant capacity of some fruits dehydrated by a combination of different methods. *Food Chemistry*, 141(4):3889–3896.

Chottamom, P., Kongmanee, R., Manklang, C., and Soponronnarit, S. 2012. Effect of osmotic treatment on drying kinetics and antioxidant properties of dried mulberry. *Drying Technology*, 30(1):80–87.

Ciurzyska, A., Ciesluk, P., Barwinska, M., Marczak, W., Ordyniak, A., Lenart, A., and Janowicz, M. 2019. Eating habits and sustainable food production in the development of innovative "healthy" snacks. *Sustainability*, 11(10):2800.

Ciurzyska, A., Marczak, W., Lenart, A., and Janowicz, M. 2020. Production of innovative freeze-dried vegetable snack with hydrocolloids in terms of technological process and carbon footprint calculation. *Food Hydrocolloids*, 108:105993.

Clementz, A., and Delmoro, J. 2011. Snacks frutales. *Invenio*, 14(27):153–163.

Da Silva, P.F., and Moreira, R.G. 2008. Vacuum frying of high-quality fruit and vegetable-based snacks. *LWT: Food Science and Technology*, 41(10):1758–1767.

de Graaf, C. 2006. Effects of snacks on energy intake: An evolutionary perspective. *Appetite*, 47(1):18–23.

Degraeve, P., Saurel, R., and Coutel, Y. 2003. Vacuum impregnation pretreatment with pectinmethylesterase to improve firmness of pasteurized fruits. *Journal of Food Science*, 68(2):716–721.

Del Mercado, B.V., Roy, S., Denson, K., and Blundell, M. 2020. U.S. Patent Application No. 16/762,001.

Dreher, M.L. 2018. Whole fruits and fruit fiber emerging health effects. *Nutrients*, 10(12):1833.

Drenowatz, C., Greier, K., and Klein, P. 2018. Associations between eating habits and food intake in Austrian adolescents. *Annals of Clinical Nutrition*, 2:1010–1018.

Drouzas, A.E., and Schubert, H. 1996. Microwave application in vacuum drying of fruits. *Journal of Food Engineering*, 28(2):203–209.

Escobedo-Avellaneda, Z., García-García, R., Valdez-Fragoso, A., Mújica-Paz, H., and Welti-Chanes, J. 2018. Fruit preservation and design of functional fruit products by vacuum impregnation. In: *Fruit Preservation: Novel and Conventional Technologies*, eds. A. Rosenthal, R. Deliza, J. Welti-Chanes, and G.V. Barbosa-Cánovas, 335–349. New York: Springer.

Escobedo-Avellaneda, Z., Velázquez, G., Torres, J.A., and Welti-Chanes, J. 2012. Inclusion of the variability of model parameters on shelf life estimations for low and intermediate moisture vegetables. *LWT-Food Science and Technology*, 47(29):364–370.

Farajian, P., Katsagani, M., and Zampelas, A. 2010. Short-term effects of a snack including dried prunes on energy intake and satiety in normal-weight individuals. *Eating Behaviors*, 11(3):201–203.

Fito, P. 1994. Modeling of vacuum osmotic dehydration of food. *Journal of Food Engineering*, 22(1–4):313–328.

Fito, P., Andrés, A., Chiralt, A., and Pardo, P. 1996. Coupling of hydrodynamic mechanism and deformation-relaxation phenomena during vacuum treatments in solid porous food liquid systems. *Journal of Food Engineering*, 27(3):229–240.

Fito, P., and Chiralt, A. 1995. An update on vacuum osmotic dehydration. In: *Food Preservation by Moisture Control*, eds. G. Barbosa-Cánovas and J. Welti-Chanes, 351–373. Lancaster: Technomic Publishing.

Fito, P., Chiralt, A., Barat, J., Andrés, A., Martínez-Monzó, J., and Martínez-Navarrete, N. 2001. Vacuum impregnation for development of new dehydrated products. *Journal of Food Engineering*, 49(4):297–302.

Gallego-Juárez, J.A., Riera, E., Fuente Blanco, S., Rodriguez-Corral, G., Acosta-Aparicio, V.M., and Blanco, A. 2007. Application of high-power ultrasound for dehydration of vegetables: Processes and devices. *Drying Technology*, 25(10–12):1893–1901.

Gamboa-Santos, J., Montilla, A., Cárcel, J.A., Villamiel, M., and Garcia-Perez, J.V. 2014. Air-borne ultrasound application in the convective drying of strawberry. *Journal of Food Engineering*, 128:132–139.

Garayo, J., and Moreira, R. 2002. Vacuum frying of potato chips. *Journal of Food Engineering*, 55(2):181–191.

García-Amezquita, L.E., Welti-Chanes, J., Vergara-Balderas, F.T., and Bermúdez-Aguirre, D. 2016. Freeze drying: The basic process. In: *The Encyclopedia of Food and Health vol. 3*, eds. B. Caballero, P. Finglas, and F. Toldrá, 104–109. Oxford: Academic Press.

García-Martínez, E., Igual, M., Martínez-Esparza, E.M., and Martínez-Navarrete, N. 2013. Assessment of the bioactive compounds, color, and mechanical properties of apricots as affected by drying treatment. *Food and Bioprocess Technology*, 6(11):3247–3255.

Gavrila, C., Ghiaus, A.G., and Gruia, I. 2008. Heat and mass transfer in convective drying processes. *Transport Phenomena*, 2:18.

Geankoplis, C.J. 2016. *Procesos de transporte y principios de procesos de separación*. México: Editorial CECSA.

Grand View Research. 2020. *Fruits & Vegetables Market Analysis, 2014–2025 Opportunities beyond COVID-19 Crisis*. San Francisco, CA: Grand View Inc.

Granda, C., Moreira, R.G., and Tichy, S.E. 2004. Reduction of acrylamide formation in potato chips by low-temperature vacuum frying. *Journal of Food Science*, 69(8):E405–E411.

Guillemin, A., Guillon, F., Degraeve, P., Rondeau, C., Devaux, M.C., Huber, F., Badel, E., Saurel, R., and Lahaye, M. 2008. Firming of fruit tissues by vacuum-infusion of pectin methylesterase: Visualisation of enzyme action. *Food Chemistry*, 109(2):368–378.

Hammami, C., and René, F. 1997. Determination of freeze-drying process variables for strawberries. *Journal of Food Engineering*, 32(2):133–154.

Ihns, R., Diamante, L.M., Savage, G.P., and Vanhanen, L. 2011. Effect of temperature on the drying characteristics, colour, antioxidant and beta-carotene contents of two apricot varieties. *International Journal of Food Science and Technology*, 46(2):275–283.

Innova Market Insights. 2020. *Sweets and Snacks Innovation Targets Major Themes of Health, Sustainability and Adventure*. Retrieved on September 30, 2020 from https://www.innovamarketinsights.com/sweets-snacks-innovation-targets-major-themes-of-health-sustainabilityand-adventure/.

Jack, F.R., O'Neill, J., Piacentini, M.G., and Schröder, M.J.A. 1997. Perception of fruit as a snack: A comparison with manufactured snack foods. *Food Quality and Preference*, 8(3):175–182.

Jambrak, A.R., Mason, T.J., Lelas, V., Herceg, Z., and Herceg, I.L. 2008. Effect of ultrasound treatment on solubility and foaming properties of whey protein suspensions. *Journal of Food Engineering*, 86(2):281–287.

Johansson, I., Lif Holgerson, P., Kressin, N.R., Nunn, M.E., and Tanner, A.C. 2010. Snacking habits and caries in young children. *Caries Research*, 44(5):421–430.

Joshi, A.P.K., Rupasinghe, H.P.V., and Khanizadeh, S. 2011. Impact of drying processes on bioactive phenolics, vitamin C and antioxidant capacity of red-fleshed apple slices. *Journal of Food Processing and Preservation*, 35(4):453–457.

Karam, M.C., Petit, J., Zimmer, D., Djantou, E.B., and Scher, J. 2016. Effects of drying and grinding in production of fruit and vegetable powders: A review. *Journal of Food Engineering*, 188:32–49.

Khan, M.R. 2012. Osmotic dehydration technique for fruits preservation-A review. *Pakistan Journal of Food Sciences*, 22(2):71–85.

Khin, M.M., Zhou, W., and Perera, C. 2005. Development in the combined treatment of coating and osmotic dehydration of food: A review. *International Journal of Food Engineering*, 1(1):1–19.

Khraisheh, M.A.M., Cooper, T.J.R., and Magee, T.R.A. 1997. Shrinkage characteristics of potatoes dehydrated under combined microwave and convective air conditions. *Drying Technology*, 15(3–4):1003–1022.

Kikuchi, M., Koaze, H., Sato, T., Kojima, M., Yamamoto, K., Yasuda, K., Mori, M., and Tsuda, S. 2011. Ascorbic acid enrichment of whole potato tuber by vacuum-impregnation. *Food Chemistry*, 127(3):1114–1118.

Kiranoudis, C.T., Tsami, E., and Maroulis, Z.B. 1997. Microwave vacuum drying kinetics of some fruits. *Drying Technology*, 15(10):2421–2440.

Kowalski, S.J., Mierzwa, D., and Stasiak, M. 2017. Ultrasound-assisted convective drying of apples at different process conditions. *Drying Technology*, 35(8):939–947.

Lai, F.C., and Sharma, R.K. 2005. EHD-enhanced drying with multiple needle electrode. *Journal of Electrostatics*, 63(3–4):223–237.

Lebovka, N.I., Shynkaryk, N.V., and Vorobiev, E. 2007. Pulsed electric field enhanced drying of potato tissue. *Journal of Food Engineering*, 78(2):606–613.

Leistner, L. 1985. Hurdle technology applied to meat products of the shelf stable product and intermediate moisture food types. In: *Properties of Water in Foods*, eds. D. Simatos and J.L. Multon, 309–321. Dordrecht: Martinus Nijhoff.

Lerici, C.R., Pinnavaia, G., Rosa, M.D., and Bartolucci, L. 1985. Osmotic dehydration of fruit: Influence of osmotic agents on drying behavior and product quality. *Journal of Food Science*, 50(5):1217–1219.

Lin, T.M., Durance, T.D., and Scaman, C.H. 1998. Characterization of vacuum microwave, air and freeze dried carrot slices. *Food Research International*, 31(2):111–117.

Lombard, G.E., Oliveira, J.C., Fito, P., and Andrés, A. 2008. Osmotic dehydration of pineapple as a pretreatment for further drying. *Journal of Food Engineering*, 85(2):277–284.

Maity, T., Bawa, A.S., and Raju, P.S. 2016. Optimization and quality assessment of ready-to-eat intermediate moisture compressed beetroot bar. *Journal of Food Science and Technology*, 53(8):3233–3243.

Maskan, M. 2001. Drying, shrinkage and rehydration characteristics of kiwifruits during hot air and microwave drying. *Journal of Food Engineering*, 48(2):177–182.

Mayor, L., Silva, M.A., and Sereno, A.M. 2005. Microstructural changes during drying of apple slices. *Drying Technology*, 23(9–11):2261–2276.

Meda, S.V., and Raghavan, G.S.V. 2004. An overview of microwave processing and dielectric properties of agri-food materials. *Biosystems Engineering*, 88(1):1–18.

Moraga, M.J., Moraga, G., Fito, P.J., and Martínez-Navarrete, N. 2009. Effect of vacuum impregnation with calcium lactate on the osmotic dehydration kinetics and quality of osmodehydrated grapefruit. *Journal of Food Engineering*, 90(3):372–379.

Mordor Intelligence. 2020. *Fruit Snack Market- Growth, Trends, and Forecast (2020–2025)*. Retrieved on September 30, 2020 from https://www.mordorintelligence.com/industry-reports/fruit-snackmarket.

Moreno, J., Bugueño, G., Velasco, V., Petzold, G., and Tabilo-Munizaga, G. 2004. Osmotic dehydration and vacuum impregnation on physicochemical properties of Chilean papaya (*Carica candamarcensis*). *Journal of Food Science*, 69(3):102–106.

Mossine, V.V., Mawhinney, T.P., and Giovannucci, E.L. 2020. Dried fruit intake and cancer: A systematic review of observational studies. *Advances in Nutrition*, 11(2):237–250.

Motevali, A., Minaei, S., Khoshtaghaza, M.H., and Amirnejat, H. 2011. Comparison of energy consumption and specific energy requirements of different methods for drying mushroom slices. *Energy*, 36(11):6433–6441.

Mousa, N., and Farid, M. 2002. Microwave vacuum drying of banana slices. *Drying Technology*, 20(10):2055–2066.

Mújica-Paz, H., Valdez-Fragoso, A., López-Malo, E., Palou, P., and Welti-Chanes, J. 2003. Impregnation and osmotic dehydration of some fruits: Effect of the vacuum pressure and syrup concentration. *Journal of Food Engineering*, 57(4):305–314.

Mujumdar, A., and Passos, M.L. 2000. *Developments in Drying*. Bangkok: Kasetsart University Press.

Mujumdar, A.S. 2007. An overview of innovation in industrial drying: Current status and R&D needs. In: *Drying of Porous Materials*, ed. S.J. Kowalski, 3–18. New York: Springer Publishing.

Mulet, A., Garcia-Reverter, J., Bon, J., and Berna, A. 2000. Effect of shape on potato and cauliflower shrinkage during drying. *Drying Technology*, 18(6):1201–1219.

Mullin, J. 1995. Microwave processing. In: *New Methods of Food Preservation*, ed. G.W. Gould, 112–134. Glasgow: Blackie Academic and Professional.

Nederkoorn, C., Houben, K., Hofmann, W., Roefs, A., and Jansen, A. 2010. Control yourself or just eat what you like? Weight gain over a year is predicted by an interactive effect of response inhibition and implicit preference for snack foods. *Health Psychology*, 29(4):389–393.

Occhino, E., Hernando, I., Llorca, E., Neri, L., and Pittia, P. 2011. Effect of vacuum impregnation treatments to improve quality and texture of zucchini (*Cucurbita pepo* L). *Procedia Food Science*, 1:829–835.

Onwude, D.I., Hashim, N., Janius, R., Abdan, K., Chen, G., and Oladejo, A.O. 2017. Non-thermal hybrid drying of fruits and vegetables: A review of current technologies. *Innovative Food Science and Emerging Technologies*, 43:223–238.

Orsat, V., Changrue, V., and Vijaya Raghavan, G.S. 2006. Microwave drying of fruits and vegetables. *Stewart Postharvest Review*, 2(6):1–7.

Ostos, A.S.L., Díaz, M.A.C., and Suarez, M.H. 2012. Evaluation process in different conditions of mango fortification (Tommy Atkins) with calcium by vacuum impregnation. *Revista chilena de nutrición*, 39(2):181–190.

Ozuna, C., Cárcel, J.A., García-Pérez, J.V., and Mulet, A. 2011. Improvement of water transport mechanisms during potato drying by applying ultrasound. *Journal of the Science of Food and Agriculture*, 91(14):2511–2517.

Parn, O.J., Bhat, R., Yeoh, T.K., and Al-Hassan, A.A. 2015. Development of novel fruit bars by utilizing date paste. *Food Bioscience*, 9:20–27.

Patel, B.P., Bellissimo, N., Luhovyy, B., Bennett, L.J., Hurton, E., Anderson, J.E., and Harvey, G. 2013. An after-school snack of raisins lowers cumulative food intake in young children. *Journal of Food Science*, 78(s1):A5–A10.

Patel, P., Ellis, K., Sunkara, R., Shackelford, L., Ogutu, S., Walker, L.T., Herring, J. and Verghese, M. 2016. Development of a functional food product using guavas. *Food and Nutrition Sciences*, 7:927–937.

Potter, M., Vlassopoulos, A., and Lehmann, U. 2018. Snacking recommendations worldwide: A scoping review. *Advances in Nutrition*, 9(2):86–98.

Qinghua, H., Shujun, L., Jiwei, M., Donglin, Z., and Bingnan, Y. 2008. Analysis on energy consumption and product quality of microwave vacuum drying and puffing apple slices. *Transactions of the Chinese Society of Agricultural Machinery*, 39(1):74–77.

Raso, J., and Barbosa-Cánovas, G.V. 2003. Nonthermal preservation of foods using combined processing techniques. *Critical Reviews in Food Science and Nutrition*, 43(3):265–285.

Rastogi, N.K., Eshtiaghi, M.N., and Knorr, D. 1999. Accelerated mass transfer during osmotic dehydration of high intensity electrical field pulse pretreated carrots. *Journal of Food Science*, 64(6):1020–1023.

Rastogi, N.K., and Raghavarao, K. 1997. Water and solute diffusion coefficients of carrot as a function of temperature and concentration during osmotic dehydration. *Journal of Food Engineering*, 34(4):429–440.

Ratti, C. 2001. Hot air and freeze-drying of high-value foods: A review. *Journal of Food Engineering*, 49(4):311–319.

Regier, M., Mayer-Miebach, E., Behsnilian, D., Neff, E., and Schuchmann, H.P. 2005. Influences of drying and storage of lycopene-rich carrots on the carotenoid content. *Drying Technology*, 23(4):989–998.

Roa, V., Tapia, M.S., and Millán, F. 2001. Mass balances in porous foods impregnation. *Journal of Food Science*, 66(9):1332–1336.

Rodríguez, Ó., Santacatalina, J.V., Simal, S., Garcia-Perez, J.V., Femenia, A., and Rosselló, C. 2014. Influence of power ultrasound application on drying kinetics of apple and its antioxidant and microstructural properties. *Journal of Food Engineering*, 129:21–29.

Rózek, A., Achaerandio, I., Güell, C., López, F., and Ferrando, M. 2009. Grape phenolic impregnation by osmotic treatment: Influence of osmotic agent on mass transfer and product characteristics. *Journal of Food Engineering*, 94(1):59–68.

Sabarez, H.T., Gallego-Juarez, J.A., and Riera, E. 2012. Ultrasonic-assisted convective drying of apple slices. *Drying Technology*, 30(9):989–997.

Sablani, S.S., Andrews, P.K., Davies, N.M., Walters, T., Saez, H., and Bastarrachea, L. 2011. Effects of air and freeze drying on phytochemical content of conventional and organic berries. *Drying Technology*, 29(2):205–216.

Sadler, M.J., Gibson, S., Whelan, K., Ha, M.A., Lovegrove, J., and Higgs, J. 2019. Dried fruit and public health – What does the evidence tell us? *International Journal of Food Sciences and Nutrition*, 70(6):675–687.

Sagar, V.R., and Kumar, P.S. 2010. Recent advances in drying and dehydration of fruits and vegetables: A review. *Journal of Food Science and Technology*, 47(1):15–26.

Salvatori, D., Andrés, A., Chiralt, A., and Fito, P. 1998. The response of some properties of fruits to vacuum impregnation. *Journal of Food Process Engineering*, 21(1):59–73.

San Martin, M.F., Sepulveda, D.R., Altunakar, B., Góngora-Nieto, M.M., Swanson, B.G., and Barbosa-Canovas, G.V. 2007. Evaluation of selected mathematical models to predict the inactivation of Listeria innocua by pulsed electric fields. *LWT: Food Science and Technology*, 40(7):1271–1279.

Shi, J., and Le Maguer, M. 2002. Osmotic dehydration of foods: Mass transfer and modeling aspects. *Food Reviews International*, 18(4):305–335.

Shirsat, S.G., and Thomas, P. 1998. Effect of irradiation and cooking methods on ascorbic acid levels of four potato cultivars. *Journal of Food Science and Technology*, 35(6):509–514.

Shofian, N.M., Hamid, A.A., Osman, A., Saari, N., Anwar, F., Pak Dek, M.S., and Hairuddin, M.R. 2011. Effect of freeze-drying on the antioxidant compounds and antioxidant activity of selected tropical fruits. *International Journal of Molecular Sciences*, 12(7):4678–4692.
Shyu, S.L., Hau, L.B., and Hwang, L.S. 1998. Effect of vacuum frying on the oxidative stability of oils. *Journal of the American Oil Chemists' Society*, 75(10):1393–1398.
Simal, S., Benedito, J., Sánchez, E.S., and Rosselló, C. 1998. Use of ultrasound to increase mass transport rates during osmotic dehydration. *Journal of Food Engineering*, 36(3):323–336.
Strumillo, C., and Adamiec, J. 1996. Energy and quality aspects of food drying. *Drying Technology*, 14(2):423–448.
Sunjka, P.S., Rennie, T.J., Beaudry, C., and Raghavan, G.S.V. 2004. Microwave/convective and microwave/vacuum drying of cranberries: A comparative study. *Drying Technology*, 22(5):1217–1231.
Swan, G.E., Powell, N.A., Knowles, B.L., Bush, M.T., and Levy, L.B. 2018. A definition of free sugars for the UK. *Public Health Nutrition*, 21(9):1636–1638.
Taiwo, K.A., Angersbach, A., and Knorr, D. 2003. Effects of pulsed electric field on quality factors and mass transfer during osmotic dehydration of apples. *Journal of Food Process Engineering*, 26(1):31–48.
Tapia, M.S., and Welti-Chanes, J. 2012. Hurdle technology applied in decontamination of whole and fresh-cut produce. In: *Decontamination of Fresh and Minimally Processed Produce*, ed. V.M. Gómez-López, 417–451. New York: Wiley-Blackwell. John Wiley and Sons.
Tapia de Daza, M.S., Alzamora, S.M., and Welti-Chanes, J. 1996. Combination of preservation factors applied to minimal processing of foods. *Critical Reviews in Food Science and Nutrition*, 36(6):629–659.
Tein, M.L., Timothy, D.D., and Christine, H.S. 1998. Characterization of vacuum microwave, air and freeze dried carrot slices. *Journal of Food Research International*, 31(2):111–117.
Tulasidas, T.N., Raghavan, G.S.V., and Mujumdar, A.S. 1995. Microwave drying of grapes in a single mode cavity at 2450 MHz- II. quality and energy aspects. *Drying Technology*, 13(8–9):1973–1992.
Ursachi, C., Segal, R., and Muresan, C. 2009. Vacuum impregnation pretreatment of fresh cut vegetable. *Acta Technica Corviniensis – Bulletin of Engineering*, 20:17–20.
Vadivambal, R., and Jayas, D.S. 2009. Comparison of ultrasound-and microwave-assisted drying of agricultural products–A review. *Stewart Postharvest Review*, 5(5):1–6.
Valdez-Fragoso, A., Mújica-Paz, H., Giroux, F., and Welti-Chanes, J. 2002. Pilot plant for osmotic dehydration of fruits: Design and evaluation. *Journal of Food Process Engineering*, 25(3):189–199.
Valenzuela, C., and Aguilera, J.M. 2015. Effects of different factors on stickiness of apple leathers. *Journal of Food Engineering*, 149:51–60.
Venkatachalapathy, K., and Raghavan, G.S.V. 1999. Combined osmotic and microwave drying of strawberry. *Drying Technology*, 17(4/5):837–853.
Vergara, F., Amézaga, E., Bárcenas, M.E., and Welti, J. 1997. Analysis of the drying processes of osmotically dehydrated apple using the characteristic curve model. *Drying Technology*, 15(3–4):949–963.
Vimala, B., Nambisan, B., and Hariprakash, B. 2011. Retention of carotenoids in orange-fleshed sweet potato during processing. *Journal of Food Science and Technology*, 48(4):520–524.
Wang, J., and Xi, Y.S. 2005. Drying characteristics and drying quality of carrot using a two-stage microwave process. *Journal of Food Engineering*, 68(4):505–511.
Watanabe, Y., Yoshimoto, K., Okada, Y., and Nomura, M. 2011. Effect of impregnation using sucrose solution on stability of anthocyanin in strawberry jam. *LWT – Food Science and Technology*, 44(4):891–895.
Wiktor, A., Iwaniuk, M., Śledź, M., Nowacka, M., Chudoba, T., and Witrowa-Rajchert, D. 2013. Drying kinetics of apple tissue treated by pulsed electric field. *Drying Technology*, 31(1):112–119.
Wiktor, A., Nowacka, M., Dadan, M., Rybak, K., Lojkowski, W., Chudoba, T., and Witrowa-Rajchert, D. 2016. The effect of pulsed electric field on drying kinetics, color, and microstructure of carrot. *Drying Technology*, 34(11):1286–1296.
Witrowa-Rajchert, D., Wiktor, A., Sledz, M., and Nowacka, M. 2014. Selected emerging technologies to enhance the drying process: A review. *Drying Technology*, 32(11):1386–1396.
World Health Organization (WHO). 2015. *Guideline: Sugars Intake for Adults and Children*. Geneva (Switzerland): World Health Organization.
World Health Organization (WHO). 2018. Healthy diet. Fact sheet no. 394. November 2020. https://www.who.int/publications/m/item/healthy-diet-factsheet394.
World Health Organization (WHO). 2020. Healthy diet. Retrieved on April 29, 2020 from https://www.who.int/news-room/fact-sheets/detail/healthy-diet.
Wouters, S., Jacobs, N., Duif, M., Lechner, L., and Thewissen, V. 2018. Affect and between-meal snacking in daily life: The moderating role of gender and age. *Psychology and Health*, 33(4):555–572.

Xiaoping, F., Yajun, W., Zijue, Z., Fangying, W., and Danyang, Y. 2018. A review on explosion puffing technology for fruits and vegetables. *International Journal of Food Engineering*, 4(4):332–336.

Yadav, A.K., and Singh, S.V. 2012. Osmotic dehydration of fruits and vegetables: A review. *Journal of Food Science and Technology*, 51(9):1654–1673.

Yamsaengsung, R., Rungsee, C., and Prasertsit, K. 2008. Simulation of the heat and mass transfer processes during the vacuum frying of potato chips. *Songklanakarin Journal of Science and Technology*, 30(1):109–115.

Yang, J., Chen, J.F., Zhao, Y.Y., and Mao, L.C. 2010. Effects of drying processes on the antioxidant properties in sweet potatoes. *Agricultural Sciences in China*, 9(10):1522–1529.

Zhang, M., Tang, J., Mujumdar, A.S., and Wang, S. 2006. Trends in microwave-related drying of fruits and vegetables. *Trends in Food Science and Technology*, 17(10):524–534.

Zhao, D., An, K., Ding, S., Liu, L., Xu, Z., and Wang, Z. 2014. Two-stage intermittent microwave coupled with hot-air drying of carrot slices: Drying kinetics and physical quality. *Food and Bioprocess Technology*, 7(8):2308–2318.

Zhao, Y., and Xie, J. 2004. Practical applications of vacuum impregnation in fruit and vegetable processing. *Trends in Food Science and Technology*, 15(9):434–451.

Zou, K., Teng, J., Huang, L., Dai, X., and Wei, B. 2013. Effect of osmotic pretreatment on quality of mango chips by explosion puffing drying. *LWT: Food Science and Technology*, 51(1):253–259.

17 Dairy-Based Snacks

Cristina Chuck-Hernandez, Tomás García-Cayuela, and Emilio Méndez-Merino

CONTENTS

DOI: 10.1201/9781003129066-17

17.1 INTRODUCTION

Dairy products as snacks are not a new idea because string cheese and yogurt cups for individual consumption have been commercialized for some time. Ice cream is the second most purchased snack only after cookies and dairy products account for 13% of all between-meal snack choices among 2-year children (Dairy Management 2010, Technomic 2019). On a serving basis, milk, cheese, and prepared foods including milk and cheese (coffee beverages, smoothies, pizza, sandwiches) account for 70% of all dairy servings as snacks, with the other 30% represented by dairy desserts, yogurt, and puddings (Dairy Management 2010). Besides, the dairy industry has followed closely (with new developments, ingredients, and snack processes) the change in consumers' perception of snacks, from the classic definition of "items consumed during nontraditional meal hours" to a broader version of "any item can be a snack and snacks can be consumed at any time of the day." Milk-based products have denoted their capability to maintain a space in the concept of the traditional snack or be used in innovative processes and with novel ingredients.

Now consumers snack nearly three times per day on average, and 42% say they are snacking more often than two years ago. Thirty-nine percent of consumers include more food and beverages as a snack and not only the small portion of portable consumed casually at simple and quick events (Warde and Yates 2017). Besides convenience, the health factor is involved in the consumers' decisions for snacks. Some countries mentioned these products in their dietary guidelines, and in countries like Brazil, Canada, France, and Switzerland, snacks such as cheese, yogurt, or milk are recommended (Hess et al. 2016).

Consumers are looking for different snack purchase locations with the grocery stores as the leading choice (69% of the consumers), closely followed by fast-food burger concepts, coffee, and convenience stores with 62%, 47%, and 44%, respectively. These consumers buy snacks at least once every 90 days with cookies, ice cream, and potato chips/hard pretzels as the top snack alternatives with 74%, 66%, and 62% of the 1,000 consumers surveyed (Technomic 2019). According to the California Milk Advisory Board, the rise in popularity of grab-and-go boxes, comfort foods, healthier, high protein snacks, and beverages different than coffee altogether with the trend of more consumers skipping or replacing one or two meals per day will make possible to see more dairy-derived products in the top ten preferences of the consumers.

This chapter describes classic dairy snacks as cheese-based, yogurt, and ice cream, but also covers new emerging categories and trends highlighting the clean label concept and production of dairy-based foods and beverages as carriers for probiotics, prebiotics, low-sugar, high protein, and phytochemicals. In the following sections, we describe the classic and new categories of dairy snacks consumed nowadays by different population segments. The product history, composition, formulas, processes, and trends are covered and explained.

17.2 SNACKS FROM CHEESES

Recent dairy product innovations have been due to the higher consumer trend of snacking, and cheeses are one of the categories that have received much attention. Health-conscious snacking and smaller portions are more present in recent trends, where cheese is an excellent option because of its great taste and nutrient richness, especially in terms of protein percentage and quality.

TABLE 17.1
Cheese Classification Based on Moisture on Fat-Free Basis (MFBB) and Fat in Dry Solids (FDS)[a]

% Moisture on Fat-Free Basis (MFBB)	Classification I[b]	% Fat in Dry Solids (FDS)	Classification II[c]	Classification III
<41	Extra hard	>60	High fat	1. Cured or ripened (a) mainly surface (b) mainly inferior
49–56	Hard	45–60	Full fat	2. Mold cured or ripened (a) mainly surface (b) mainly interior
54–63	Semihard	25–45	Medium fat	3. Uncured or unripened
61–69	Semisoft	10–25	Low fat	–
>67	Soft	<10	Skim	–

[a] Bylund (2015).
[b] According to MFBB %, classification I is the first phrase used in the cheese designation.
[c] According to FDS %, classification II is the second phrase used in the cheese designation.

Consequently, some cheese snack options are experimenting with new flavors, bit-size portions, convenient and easy-to-open packages, and pairing with dried fruits (Chapter 16), nuts (Chapter 15), or processed meats (Chapter 18), or alternatively incorporating cheese into other healthy snacks like crisps or crackers (Chapter 10) and coextruded items (Chapter 12). According to Dairy Food's Cheese Outlook study, 85% of respondents selected portion-controlled cheese products as an "*in style*" trend for 2018 (Bernier and Labbe 2019).

The many cheese variants can be classified according to their texture, chemical properties as moisture or protein content, flavor, appearance, and type of process (Table 17.1).

The cheese process involves stages similar for most types of these products (Figure 17.1). The first stage is regularly a thermal treatment, followed by the addition of bacteria culture. Then, the enzymatic activity will coagulate the milk to form a firm gel. The coagulum is cut to facilitate whey expelling, and the curds are stirred and heated until the whey separation is finished. At the final stage, the curd is molded and pressed using its weight or a mechanical press.

17.2.1 Dehydrated or Crunchy Cheese Snacks

One innovation trend based on cheeses and snacks is to turn this product into a salty and crunchy alternative, especially for those who want to avoid carbohydrate-loaded chips and crackers. This market segment is looking for snacks made from 100% cheese obtained by different methods and technologies described in the following sections.

17.2.1.1 Sun Drying

Sun drying is one of the oldest preservation methods, with its main advantages being the relatively low cost and the simple operation (Huang and Zhang 2012). The product is placed on a drying tray and covered with a net to prevent dust and insect contamination. The trays are placed directly in the sunlight. The time of drying depends on the product's matrix and can take from seven days to one month (Asgar et al. 2003). Among the main disadvantages of this technology is that conditions are not controlled, and the long drying time can reduce nutritive and sensorial scores especially related to flavor and color (Huang and Zhang 2012).

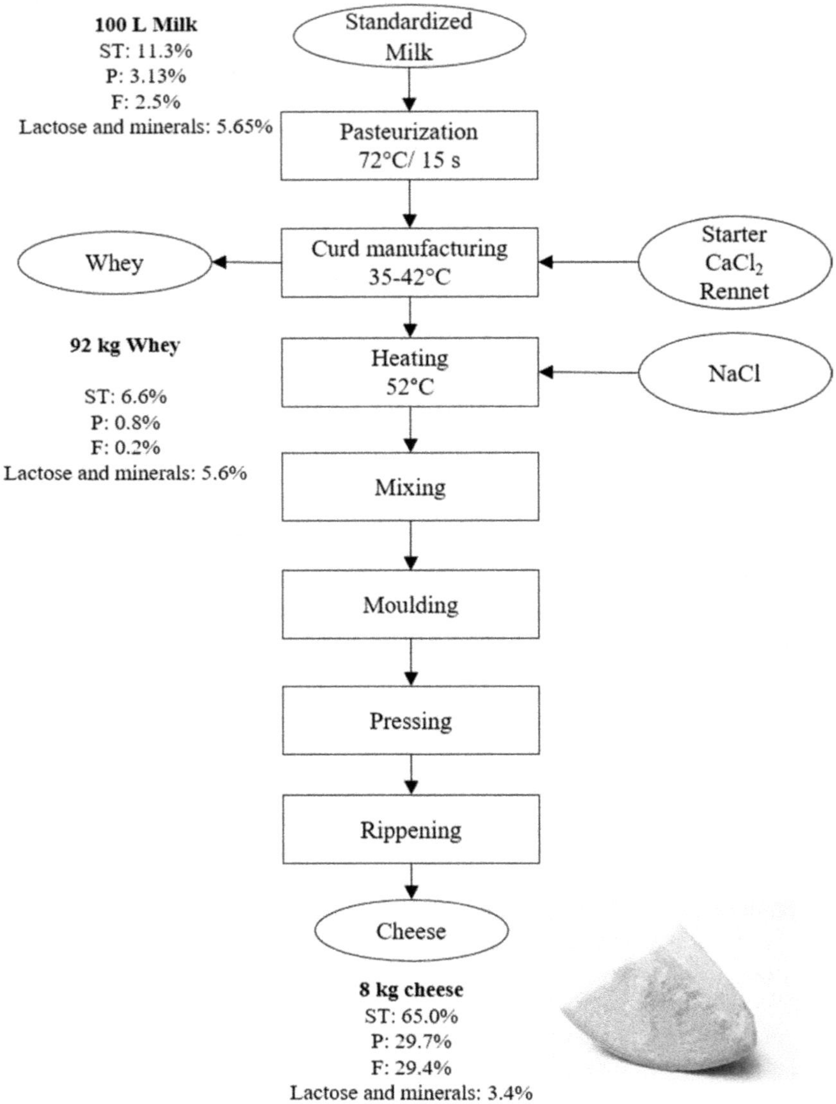

FIGURE 17.1 The general process for parmesan cheese production. ST = total solids, P = protein content and F = fat content.

17.2.1.2 Baking Process

Most savory snacks are obtained after deep-fat frying with the concurrent disadvantage of the significant oil uptake during the operation. As a result of the oil absorption and moisture removal, most snacks contain a high caloric or energy density (Chapter 20). Baking is a healthier alternative to improve the nutritional profile of snack foods. This technology removes moisture without adding oil or fat and gives a toasted flavor to the final product (Tzia et al. 2015). For this purpose, two types of ovens are typically employed and classified according to the type of contact between the food and the dehydration media: (a) direct and (b) indirect ovens. In the direct heating oven, the product is in contact with the gas used as the source of dehydration energy, and of course, the purity of the gas must be considered when planning this process. On the other hand, the indirect heating oven is the equipment where there is a separation between the heating system (steam or combustion gas like natural gas or propane) and the air and thus with the product (Berk 2009).

On the industrial scale, tunnel ovens are the most commonly utilized because temperature, moisture, and air velocity can be controlled. This oven uses an indirect heating system, and usually the air in the tunnel flows tangentially to the product. Recent developments in ovens can achieve the gas flow at higher velocity and perpendicular to the product (Berk 2009).

17.2.1.3 General Dehydration Process for Cheese

The dehydration process (Figure 17.2) starts with cheese, preferably Parmesan, which is already dry and aged. This product is shredded by manual or mechanical means, but the slices should be thin like flat sheets. Thereafter, the sheets are placed in a thin layer or bed and placed in the oven. The temperature and drying time should be 232–246°C and 3–4 min, respectively. Later, the product is removed from the oven and immediately allowed to cool down. The final product is a flat sheet of shelf-stable Parmesan cheese that is crispy. Finally, the product is packed preferably under a controlled atmosphere (Thompson 2006).

There are different products on the market obtained using this process. An example is the tradename "Parm Crisp" from the company "That's How We Roll." The description of the product is a low-calorie snack made with 100% aged parmesan cheese and seasonings. There are different options or flavors such as sour cream, jalapeño, pizza, and other healthier options where cheese is mixed with almonds, cashews, pistachios, and other nuts (Chapter 15).

17.2.1.4 Vacuum Microwave Drying

One of the most promising techniques for drying foods, and in particular, cheese, is vacuum microwave. In this type of drying, the transformation of electromagnetic energy into kinetic energy generates heat within the material.

In general, the mechanism of microwave drying occurs as follows (Rakcejeva et al. 2009): (a) the water absorbs the microwave energy within the food; (b) the absorption of microwaves increases the temperature, and a fraction of the water starts evaporating; (c) internal heating and evaporation of liquid occurs inside the food and generates pressure; and (d) the pressure gradient becomes an additional mechanism to increase moisture transport. A typical temperature limit is 60°C, and since the drying is under vacuum, the pressure is low enough to reduce the boiling point of water, thus enhancing evaporation.

Vacuum-microwave technology has several benefits in the final product, one of the most important being the protection of thermolabile compounds, taste, flavor, and color because of the employment of low temperatures of the process (Ahrens et al. 2006).

In the process of vacuum-microwave drying (Figure 17.3), the cheese is produced and sliced into 4-mm thick laminates. The pieces are then passed through a freezing tunnel at a temperature −40°C

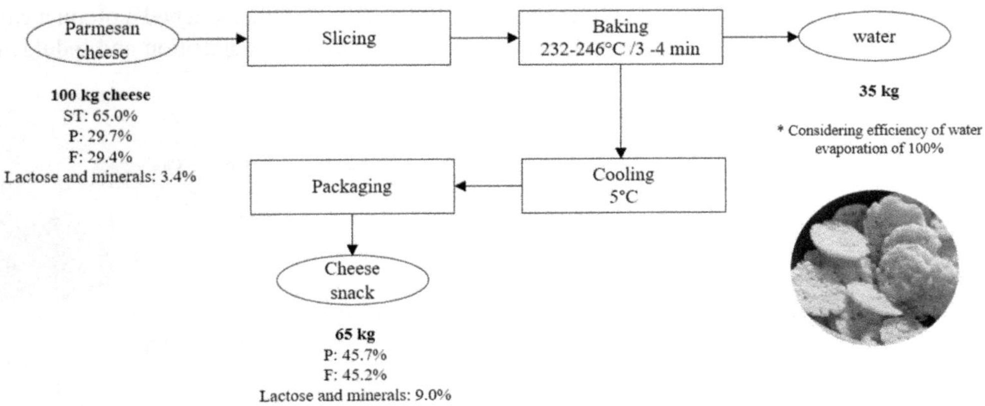

FIGURE 17.2 General process for the production of baked and dehydrated cheese snacks. ST = total solids; P = protein; F = fat.

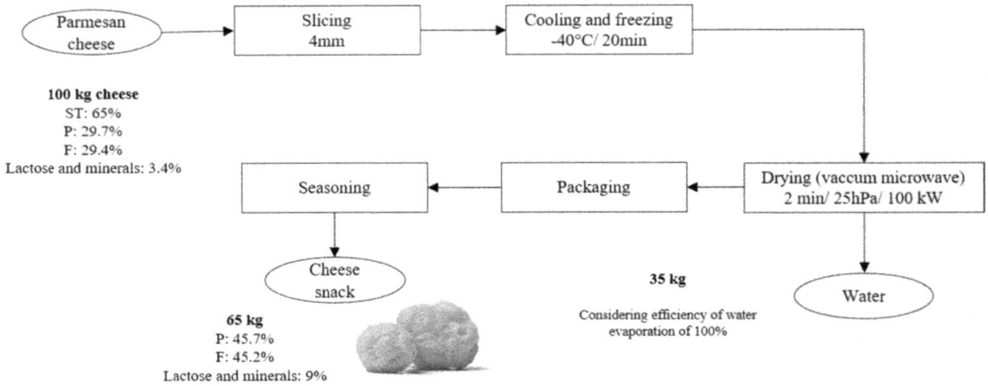

FIGURE 17.3 General diagram of the production of microwave-dried cheese. ST = total solids; P = protein; F = fat.

for 20 min and pre-dried (optional step) at 90°C for 30 min to drop the moisture to approximately 50%. The product is then cooled at 15°C. Once again, the pre-dried (optional step) cheese is frozen at –50°C. Next, the frozen cheese can be seasoned, for example, with paprika and pepper. The spiced cheese passes through a vacuum microwave for 2 min at 25 hPa and 100 kW. Finally, the slices are dried at 40°C for 4 h to 4% moisture. The final product is packed in bags of aluminum triplex (Radas and Pawel 2017).

One example of this type of snack is "Moon cheese" from the company NutraDried Food. The snacks made of cheddar cheese acquire a crunchy texture and because it is dehydrated, it does not require refrigeration. This product has a healthy claim of 10 g of protein, with no sugar and less than 2 g of carbohydrates per 28 g.

17.2.2 Pairing Snack

This type of cheese-based snack does not involve a new cheese process. For example, Sargento created a line called on-the-go cheese "Balance breaks," consisting of a dual compartment with cheese bites on one side and a mix of dried fruits, nuts, or chocolate chunks on the other. This product line earned US$67 million in the first year of launch (2015). More recently, in 2019, Sargento increased this line sales by 28% with a new product more like breakfast, which included quinoa clusters, cocoa clusters, cranberries, seed medley, and raisins.

17.2.3 Rolled-Up Snacks

These snacks consist of a cheese that is wrapped with charcuterie. Sigma launched a product called "Panino" (Figure 17.4), a ready-to-go snack for consumers seeking both nutrition and indulgence.

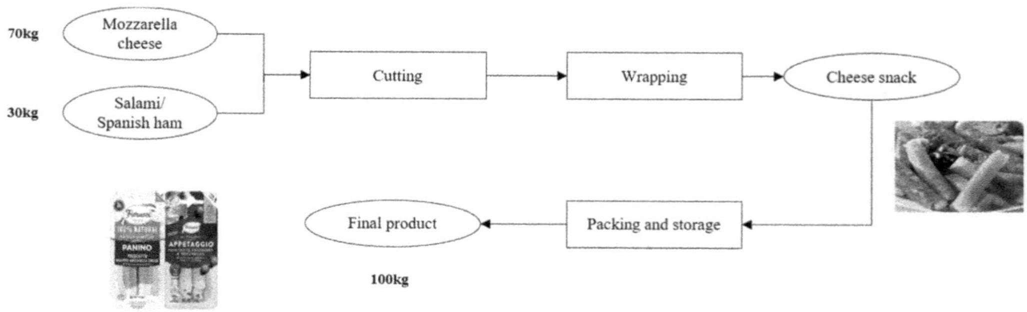

FIGURE 17.4 General process for the production of rolled-up snacks.

They have different combinations such as salami and mozzarella and prosciutto and mozzarella. Moreover, they have some other options with a strip of peach or cranberry mozzarella cheese. These products have healthy claims such as all-natural, no artificial flavors, and 9 g of protein per serving.

17.3 YOGURT AND FERMENTED MILK

It is believed that yogurt and fermented milk dates before cheese production (around 8000–10000 BC) when the first dairy animals were domesticated. This product's popularity had increased in the last decades, associated mainly with research evidence related to its health benefits. In the 1900s, Élie Metchnikoff was the first to propose that lactic acid bacteria from fermented milk exerted an antibacterial effect. This affirmation was based on his observations about the longer life expectancy of the people who consumed yogurt regularly in the Balkan region (Salampessy and Kailasapathy 2011).

The production of cultured milk is one of the earliest forms of food preservation, and the term is used in dairy products where lactose has been converted into lactic acid because of the activity of lactic acid bacteria alone or with the intervention of yeast and molds (Bylund 2015, Shah and Champagne 2016). Many of these products based in milk are produced worldwide and classified according to the dominating microorganism used for its production or based on their metabolites. The three main groups are (a) lactic (mesophilic, thermophilic, and therapeutic), (b) yeast-lactic, and (c) mold-lactic. From the wide variety of fermented milk, some commercially available examples and their type of fermentations are summarized in Table 17.2.

From all products described in Table 17.2, kefir is the most relevant in commercial terms after yogurt. In the last decades, its consumption has risen because of its perceived health image as a snack especially among young consumers. Kefir is a drink with origins in the Caucasian mountains of the former USSR, produced by adding kefir grains to cow's milk. These grains resemble miniature cauliflower florets, and the kefir is fermented with a wide array of symbiotic lactic acid bacteria (*Lactobacillus*, *Lactococcus*, *Leuconostoc*) and yeasts (*Kluyveromyces*, *Candida*, *Saccharomyces*, and *Pichia*). Although lactic acid is the primary fermentative metabolite in kefir, small amounts of ethanol and carbon dioxide are also produced (Shah and Champagne 2016, Aryana and Olson 2017). Probiotics as *Bifidobacterium lactis* HN019, *L. acidophilus* NCFM, and *L. rhamnosus* HN001 have been successfully incorporated into kefir (Kolakowski and Pawlikowska 2012).

On the other hand, Kumys is also a cultured and refreshing fermented milk beverage produced from mare's milk with a mix of yeast and lactic acid bacteria. The final product can contain up to

TABLE 17.2
Examples of Cultured Milk Commercially Available and the Type of Fermentation Used for Their Production[a]

Cultured Milk Examples	Fermentation Type
Ymer	Mesophilic
Yogurt	Thermophilic
Bulgaricus milk	Therapeutic
Gelifus®	Therapeutic
Calpis®	Therapeutic
Yakult®	Therapeutic
Kefir, Kumys	Yeast-lactic
Viili	Mold-lactic

[a] Modified from Shah and Champagne (2016).

2.5% ethanol, carbon dioxide, and lactic acid (1.0%). Another important cultured milk is Viili, a popular product in Finland, manufactured using fermentation with *Lactococcus lactis* ssp. *lactis* biovar *diacetylactis*, *Leuconostoc mesenteroides* ssp. *cremoris*, and the mold *Geotrichum candidum* (Shah and Champagne 2016).

Among all dairy-based products presented in Table 17.2, yogurt is the most popular and best known worldwide. It can be described as fermented or cultured milk where two types of symbiotic microorganisms are used to convert lactose into lactic acid, which acts as a preservative. The increase in acidity (0.17% in milk up to 0.9–1.0% in the final product – at least 0.6% according to CODEX STAN 243-2003-) inhibits the growth of putrefactive bacteria, augmenting the shelf life and viscosity of the product. Yogurt can be sold as a softer gel, as a beverage, or in other presentations according to processes described in Figure 17.5. Yogurt has a distinctive aroma associated with aromatic substances such as acetic acid, diacetyl, and acetaldehyde. The latter derived from the amino acid threonine, whereas diacetyl is produced from pyruvic acid formed during sugar fermentation (Walstra et al. 2006).

Yogurt is, together with cheese-based snacks, one of the highest in sales growth during last years (6% CAGR during 2004–2009 and 4.5% forecasted for 2020–2025). Its global market is forecasted to grow to $26.08 billion during 2019–2023, taking profit from new consumers' eagerness for innovative products. The healthy image associated with this dairy-based food has been one of the main drivers for its growth (Dairy Management 2010, Mordor Intelligence 2020, Schouten 2020). There have been multiple new products or developments in this category, addressing many different

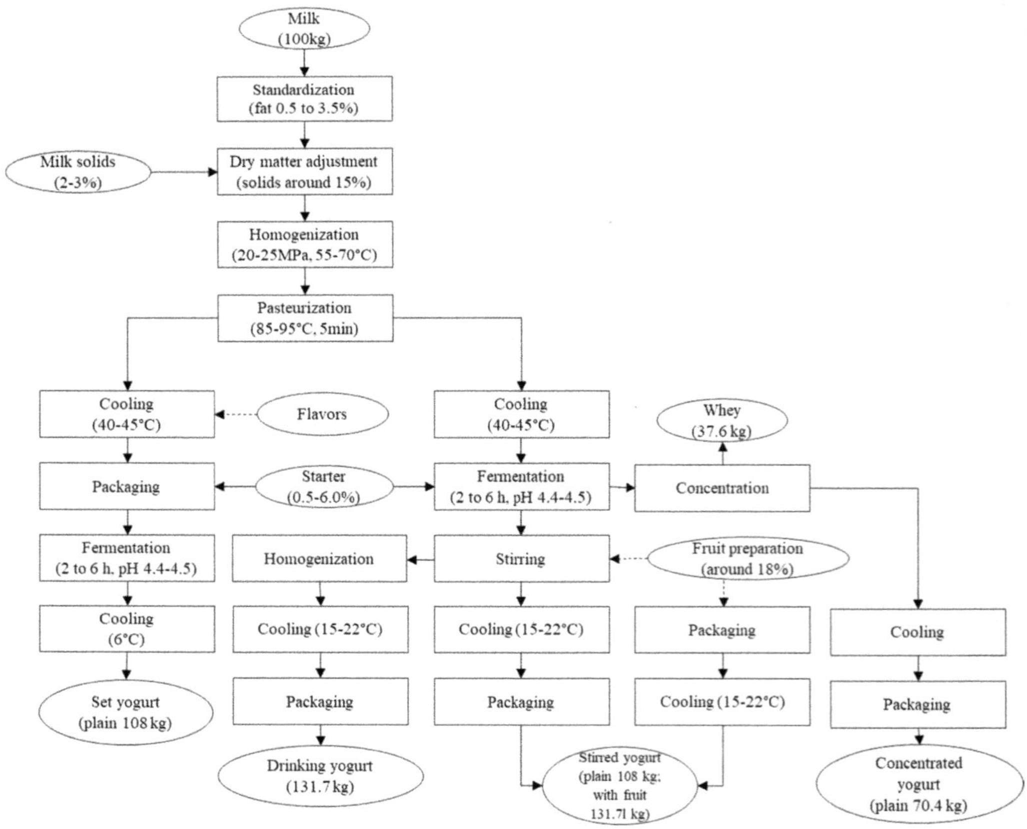

FIGURE 17.5 General flowchart depicting processes for manufacturing set, drinking, stirred, and concentrated yogurts. Adapted from Salampessy and Kailasapathy (2011) and Bylund (2015).

consumer needs in the last decades. Among the most remarkable examples are yogurt oriented to children (see Section 17.3.4), weight management, indulgence, digestive health, probiotics, prebiotics, and high-protein yogurts (Walker et al. 2010, Fisberg and Machado 2015).

Other factors associated with yogurt's popularity is its low price and availability. According to Mordor Intelligence (2020), 45% of the consumers select yogurt as breakfast, and in North America, the consumption is also typical during the afternoon, late morning, or late-night snack. With new developments, drinking yogurt has become a preferable and convenient consumer's alternative because of the healthy image and portability. Besides, yogurt presentation have changed from cups and spoons to drinks, recloseable carton packs, drinking straw, or other alternatives, making it convenient for people on-the-go (Mordor Intelligence 2020).

In yogurt production, the market's major players (Danone S.A., Chobani LLC, General Mills Inc., Nestlé S.A., and China Mengniu Dairy Company Limited) focus on providing innovative and exotic flavors for consumers (see Section 17.3.4) while including functional benefits in each of the categories (Mordor Intelligence 2020, Schouten 2020).

In legal terms, yogurt is defined in the United States as "an acidic-gelled product from culturing cream, milk, partially skim milk, or skim milk alone or in a combination that is fermented by *Streptococcus thermophilus* and *Lactobacillus delbrueckii* subsp. *bulgaricus*" (FDA 2020). In Mexico, this product is defined as the "obtained from fermented milk, standardized or not, using microorganisms *Streptococcus thermophilus* and *Lactobacillus delbrueckii* subsp. *bulgaricus*, resulting in a pH reduction." Alternative cultures as *Lactobacillus* (other than *delbrueckii*) and *Bifidobacterium* are allowed. When alternative cultures are used, the commercial denomination must include the scientific name for the culture together with the word "yogurt" (SCFI/SAGARPA 2018). In Australia, yogurt is defined as "a fermented milk in which lactic acid-producing microorganisms are used to ferment the dairy ingredients, and other additional ingredients and the microorganisms used must remain viable in the final product" (FSANZ 2015). According to the CODEX Standard 243-2003, yogurt is fermented milk produced by the action of cultures of *Streptococcus thermophilus* and *Lactobacillus delbrueckii* subsp. *bulgaricus*, resulting in a reduction of pH with or without coagulation (FAO/WHO 2011). The symbiotic microorganisms that define yogurt must remain active in the final product, with at least 10 million CFU/g. This requirement is mandatory in most of the yogurt standards, including CODEX 243-2003.

Some essential features of the starters used for yogurt fermentation are that they should be thermophilic with an optimum temperature growth range of 40–42°C for *Streptococcus thermophilus* and 43–46°C for *Lactobacillus bulgaricus* and optimally grow at pH of 6.0–6.5 and 5.5–6.0, respectively. As mentioned before, yogurt starters have a symbiotic growth, meaning they have a higher reproduction rate and faster lactic acid production when together than as individual strains. This type of growth occurs because milk contains a small number of peptides and free amino acids required to support lactic acid microorganisms. In yogurt, *L. bulgaricus*, as the highest proteolytic bacteria, will first start fermentation, releasing amino acids and short-chain peptides, and then promotes the growth of *S. thermophilus* that breakdowns lactose very efficiently, yielding formate, carbon dioxide, and acidic compounds that create a friendly environment for further lactobacilli growth (Routray and Mishra 2011, Shah and Champagne 2016).

17.3.1 Yogurt Classification

Yogurt is mainly classified based on fat content, process, or ingredients, as detailed in Table 17.3.

17.3.2 Ingredients

The basic ingredients for yogurt production are milk, starter cultures, sweeteners, fruit, and stabilizers. The most used milk for yogurt manufacture is from cows, but milk from sheep, goats, mares, camels, and even buffalo are also used depending on the geographical region.

TABLE 17.3
Yogurt Classification Based on Different Parameters (Fat Content, Process, or Ingredients)

Parameter	Yogurt Groups
Fat content[a]	*Full fat:* minimum milk fat 3%
	Partially skimmed yogurt: fat content between 0.5% and 3.0%
	No fat: maximum milk fat 0.5%
Process-based[b]	*Set-type yogurt:* fermented and cooled in the package
	Stirred-type: fermented in tanks and cooled before packaging
	Drinking-type: fermented in tanks, cooled, and coagulum broke down to a liquid before packaging
	Frozen-type: fermented in tanks and frozen in a continuous freezer
	Concentrated: also called Greek or strained yogurt, is fermented in tanks, concentrated, and cooled before packaging
Flavored[c]	*Flavored or with fruits:* defined as composite milk products containing a maximum of 50% w/w of nondairy ingredients (nutritive and nonnutritive sweeteners, fruits, and vegetables as well as juices, purees, pulps, preparations, and preserves derived from cereals, honey, chocolate, nuts, coffee, spices, and other natural flavoring foods)
	Plain: defined as yogurt without sweeteners, added sugars, fruits, vegetables, cereals, no flavors, it can have additives as stabilizers and thickeners, allowed according to each national legislation
	Plain with sweeteners: defined as plain yogurt with sugar or any sweetener type and with additives allowed according to each national legislation

[a] From Bylund (2015) and Shah and Champagne (2016).
[b] From Bylund (2015).
[c] FAO/WHO (2011) and SCFI/SAGARPA (2018).

Milk must have excellent bacteriological quality with low bacterial load and without antibiotics, bacteriophages, and residues of cleaning solutions or disinfectants. According to the (FDA 2015), the milk's bacterial load must be less than 100,000 CFU/mL and should not exceed 750,000 somatic cells per milliliter.

The starter cultures used commercially are a mix of *S. thermophilus* and *L. delbrueckii* subsp. *bulgaricus*, but other cultures as *Lactobacillus acidophilus*, *Bifidobacterium*, and *Lactobacillus casei* are also used as either single or mixed cultures. These bacteria develop specific flavors (lactate, acetaldehyde, acetoin, and diacetyl) and improve texture mainly via exopolysaccharide (EPS) production. In addition some are considered as probiotic microorganisms (Vedamuthu 2011).

Other essential ingredients in yogurt manufacturing are sweetening agents and fruits which improve taste, color, and texture. Besides sucrose, the other sweeteners used are saccharin, aspartame, acesulfame-K®, sucralose, stevia, and maltodextrins. Sucrose cannot be used in concentrations over 8% in the yogurt mix because of its significant influence over the media's osmotic pressure. Higher concentrations pose the risk of reducing the starters' development during processing and their counts in the final product (Salampessy and Kailasapathy 2011, Shah and Champagne 2016). Fruits are typically added in pieces, jam, or syrup in percentages from 12% to 18%, with around 50% of their final weight as sugar. It can be added to the pack's bottom, mixed with the yogurt, or in a twin cup with the dairy product (Routray and Mishra 2011, Bylund 2015).

Regarding stabilizers, these are hydrophilic colloids used to bind water, increase final viscosity, and prevent whey separation or syneresis in yogurt. Natural or plain yogurts generally do not require stabilizers because a high-viscosity gel will form naturally. However, fruit-added and stirred types of yogurts are regularly supplemented with hydrocolloids. These most common additives are gelatin, pectin, starch, and agar, supplemented from 0.1% to 0.5% (Tasneem et al. 2014, Rafiq et al. 2020).

17.3.3 Process

Figure 17.5 depicts the primary processing steps for set, drinking, stirred, and concentrated yogurts. Standardization includes adjustments of the milk fat due to normal variations according to sources, breeds, and seasons and complies with commercial requirements. Fat adjustment is performed in a centrifuge or milk separator calibrated to remove different amounts of fat. The most typical final concentration is between 0.5% and 3.5%, although some commercial yogurts are processed with milk containing 0–10% (Bylund 2015). After this stage, the milk is adjusted for solids, usually to reach 15%, using skim milk or nonfat dry milk powders to improve the final product's texture or firmness that also minimizes whey separation in finished products.

The increase of solids can also be reached by evaporating water in an evaporator under vacuum to avoid the thermal damage of milk components The use of ultrafiltration or reverse osmosis membrane filtration is also an alternative to concentrate solids in the milk matrix. When milk powder is added to increase solids, a deaeration step is included using a vacuum vessel to reduce the risk of whey separation, improve viscosity and homogenization, deodorization (removal of volatile off-flavors), and very importantly the significant reduction of fermentation time (Aryana and Olson 2017). In the milk solids adjustment (Figure 17.5), additives as stabilizers and vitamins may also be incorporated.

The next step in the process is homogenization using pressures around 20–25 MPa and temperatures of 55–70°C with the primary objective of preventing creaming during fermentation, assuring a uniform distribution of the milk fat. This stage also improves the final product's stability and consistency, even when the fat content is low (Bylund 2015; Salampessy and Kailasapathy 2011). Once homogenization has been performed, the milk is pasteurized to destroy pathogens and other vegetative cells or inactivate enzymes detrimental to the starter culture. Usually, the yogurt milk matrix's thermal process is higher than for regular milk when fluid products are manufactured. The typical time–temperature profile is 90–95°C for 5 min. The high temperature improves the milk proteins' water-binding capacity, reducing the final product's syneresis. The extensive denaturation of whey proteins allows their interaction with *kappa* caseins stabilizing the final yogurt (Mahomud et al. 2017).

After pasteurization, milk must be cooled before starter addition, and because of this, temperatures should be decreased to 40–45°C, which is optimum for yogurt cultures. Bacteria mix (usually freeze-dried or deep-frozen) is pitched at percentages from 0.5% to 6.0% (depending on the type of yogurt) and also depending on the fermentation system. The fermentation period is 2–6 h until the pH drops to 4.4–4.5. Fruit can be included during packaging (at the bottom of the package) or mixed after fermentation (stirred style). After fermentation, yogurt can be concentrated to produce Greek-style yogurt (strained yogurt), which is defined as a semisolid product with 23–25 g total solids per 100 g, with 8–11 g fat per 100 g (refer to Chapter 21). The concentration can be made using large cloth bags, mechanical separators, ultrafiltration, or product formulation (Ibrahim et al. 1999, Tamime and Robinson 2007).

The product is then cooled to 15–22°C within 30 min after the optimal pH is reached to stop a further increase in acidity. In the stirred yogurt process, cooling occurs in a plate heat exchanger (Bylund 2015), designed to give a gentle mechanical treatment to the product allowing an adequate consistency in the final yogurt.

After or before cooling, depending on the type of process, fruit and flavors can be added using a batch or a continuous mixer. In the latter, a variable speed metering pump is used to feed the ingredients into the yogurt using a continuous fruit-blending unit (Bylund 2015), which is static and designed to guarantee a proper mix in the yogurt. Fruit can be included in a sweet mix (50–55% of sugar) or natural (without sugar). In the first case, pectin as a thickener can be used (≤0.5%). It is important to consider the fruit-based ingredient's proper heat treatment because vegetative microorganisms must be inactivated without affecting the fruit's taste and texture. A scraped surface heat exchanger, tubular heat exchanger, or a tank with scraper units can be used (Salampessy and

Kailasapathy 2011, Bylund 2015). The nutrient composition of the final products is presented in Chapter 21.

17.3.4 Yogurt Trends

New product launches drive the increase in the global yogurt market segment, including offerings as unique flavors, lactose-free, better-for-you options, and dairy-free alternatives (Schouten 2020). Some of the most important trends in yogurt development and commercialization are covered in the following sections.

17.3.4.1 Clean Label

This trend is associated with the consumers' interest to have fewer additives in their food products. Nowadays, there is no general agreement in the food industry regarding what a clean label means. However, the cleanest possible yogurt label is constituted by only two ingredients: milk and starter cultures. This formulation is known as a "classic recipe." Examples of these products are the original plain Greek yogurt in the United States, which has gained popularity in the last decades. These products rely upon the strength of the protein network for its stability and texture. This network is formed during fermentation and endures the mechanical treatment experienced by the product during processing. The flavor is considered one of the most relevant features, and it mainly depends on the milk's quality, the starter culture, and the fermentation process's control. The clean label concept can also include some additives perceived as natural for the consumers, like starch, supplemented to improve and maintain the yogurt's texture (Tetra Pak International 2019a). One example of a clean label yogurt commercialized nowadays is the Organic Valley cooperative with headquarters in La Farge, Wisconsin. This yogurt is made from whole organic milk, 100% grass-fed, and nonhomogenized. Alternatively, this product is manufactured with organic strawberries, wild blueberries, or organic fair-trade vanilla with live cultures, without artificial colors, flavorings, and preservatives, and the milk should be free from antibiotics, synthetic hormones, pesticides, and GMOs.

Another clean label example is Siggi's kids yogurt pouches, launched in 2020 with 50% less sugar and 30% fewer ingredients than other yogurts for children as a target consumer segment. The product is made with pasteurized low-fat milk and pasteurized ultrafiltered skim milk or milk protein concentrate, fruits, cane sugar, natural flavor, fruit pectin, and live active cultures in strawberry banana and blueberry flavors.

17.3.4.2 Ambient Yogurt

This type of product is a short term for "yogurt-based product for ambient distribution," but its definition will vary according to the country. This category has grown because of its convenience for producers and consumers, easy distribution, and the possibility of being consumed using a spoon or even drunk directly from the package. According to Tetra Pak International (2019b), ambient yogurt grew at a compound annual rate of 19% from 2013 to 2019, whereas most other types only 4%. In countries where this product cannot be commercialized as yogurt (because the thermal treatment after fermentation inactivates the yogurt bacteria), some alternative labels are yogurt-based products, post-fermentation heat-treated yogurt, or yogurt-based beverage.

One example of ambient temperature yogurt is Promess brand (commercialized in France), a 200-mL drinking yogurt made with fermented milk (milk and lactic cultures), sugar, modified starch, pectin, gellan gum, locust bean gum, and flavorings (strawberry or banana). According to the company's information, this drinking yogurt has ten months shelf life (300 days) at ambient temperature, contains 3.2% fat, and preservative-free.

17.3.4.3 High Protein, Low Sugar, and Low/Lactose-Free

The demand for high protein yogurts has significantly increased during the past several years and is expected to continue as one of the best and more healthy snack alternatives, especially among people managing their weights or following a high-protein and low-carbohydrate diet (Chapter 21). For instance, the commercial Noosa's yogurt provides 12 g of protein per 5.3 oz cup, whereas the drink based on Greek yogurt of the same brand delivers 13 g of protein per 8.5 fl oz (251 mL). Moreover, these products are made with higher protein and lower sugar than similar foods and produced from rBST-free milk (recombinant Bovine Somatotropin) and available in six flavors: blueberry, mixed berry, peach, strawberry, vanilla bean, and plain.

An example of a low-sugar product is "Activia less sugar & more good." This product was launched in 2019, and according to Danone, the company behind its production, this yogurt is sweetened with honey and fruit, contains 9 g of sugar per serving, 40% less sugar than regular yogurt, and is available in fig-cinnamon, pineapple-passion fruit, pear-ginger, and blueberry-cardamom flavors. The main ingredients are cultured milk, water, fig paste, honey, less than 1% pectin, natural flavors, cinnamon, lemon juice (from concentrate), live cultures (*L. bulgaricus*, *L. lactis*, *S. thermophilus*), and live and active probiotics (*B. lactis*).

Despite the fermentation process, this yogurt's lactose content (3.5–4%) is similar to milk (4.5%). However, lactose in yogurt is digested more efficiently than from other dairy sources, probably for the residual bacterial lactase activity or its capability to be active during the product's gastrointestinal transit (Savaiano 2014, Danone Institute International 2015). According to Dekker et al. (2019), the lactose-free market is the fastest-growing segment in the dairy industry, with a yearly growth of 7.3%. The largest lactose-free dairy category is fluid milk, followed by lactose-free yogurt, which reached 1 billion turnover in the year 2020. The largest and fastest growth area in the world is in Western Europe, followed by Latin America.

An example of this trend is products from Green Valley, a company with a line of lactose-free Greek yogurts that provide 12–16 g of proteins in flavors like plain, vanilla bean, strawberry, and blueberry and presentations nonfat (0%), reduced-fat (2%), and whole milk (4%).

17.3.4.4 New Flavors

There are several examples of yogurts with new and novel flavors. The Chobani line of Greek yogurt, composed of oatmeal, was launched in 2019, providing 4 g of dietary fiber and 10 g of protein. This product is flavored with apple spice, brown sugar, blueberry-maple oatmeal, banana-maple oatmeal, and peach with brown sugar oatmeal.

17.3.4.5 Digestive Health and Enhanced Immunity

The high prevalence of digestive disorders in many consumers increases the demand for probiotic foods, being yogurt one of the best and favorite alternatives to deliver beneficial bacteria. According to Mordor Intelligence (2020), yogurt holds about 38% of the probiotic market share in the United States. A ready-to-drink probiotic yogurt beverage example is from Stonyfield Organic made with fruit and organic low-fat milk. Every 3.1 fl. oz (91.7 mL) bottle contains 60 kcal and the Chobani's probiotic family. This line is aimed to support the immune, digestive, and gut health of adults and kids. Among the probiotic strains that they use is LGG™, a trademark of Christian Hansen. The ingredients are cultured low-fat milk, cane sugar, chicory root fiber, water, fruit pectin, natural flavors, fruit and vegetable juice for color, and locust bean or guar gum.

17.4 DAIRY-BASED BEVERAGES

Beverages have many possibilities to incorporate different flavors, bioactive compounds, and desirable nutrients. In this regard, dairy-based beverages are the most acceptable for consumers in the functional food segment, and thus a wide range of these drinks are found on the market. The global dairy-based beverages market is expected to continue growing in the coming years. However, the great interest that is awakening for nondairy products, especially in the younger population, restricts this growth. Consumers demand healthier dairy products with less saturated fat and cholesterol and significant amounts of functional components such as vitamins, minerals, essential FA, fiber, proteins, and probiotic microorganisms (Nazir et al. 2019).

Dairy-based beverages are often based on whole, semiskimmed, or skimmed milk and/or by-products such as whey and buttermilk. They can be classified into fermented and unfermented dairy beverages (Özer and Kirmaci 2010). The first group includes drinkable yogurt and fermented milk with starter cultures and/or probiotics (see Section 17.3). Simultaneously, flavored and fortified beverages with bioactive compounds, micronutrients, fibers, proteins, etc. are examples of the second group.

17.4.1 Flavored Milk

Flavored milk is a sweetened dairy product with flavors such as chocolate, cacao powder, fruits, coffee, and sometimes food colorings and additional vitamins or minerals. It is important to note that the term "flavored milk drink" is usually used when its fat is below 1–2%. Although most of these beverages are aimed at the children's sector, some developments include adult-friendly flavors such as coffee (Ilao et al. 2017).

The most popular flavored milk/drinks in the world are fruit-flavored milk and chocolate milk. In general, they contain milk with 1–2% fat, sugar (5–10%), food additives (stabilizers, emulsifiers, etc.), and key ingredients such as cocoa powder, fruit pulps, and natural or artificial flavors (banana, strawberry, pineapple, or vanilla, among others) in variable proportions (Azami et al. 2018, Bochnak-Niedźwiecka and Świeca 2020). It is important to consider some issues when preparing flavored milk: adding excessive sugar or fruit syrup could cause problems in the product and distort flavors. The addition of highly acidic fruits can coagulate the milk proteins; thus, its quantity must be controlled or prevented with an acidity regulator's addition. On the other hand, the addition of stabilizers such as pectin, alginate, or carrageenan prevents the phase separation and provides body and viscosity to the product. In contrast, the addition of soy lecithin, mono- and diglycerides of fatty esters as emulsifiers prevents the fat globule breakage and improves the creamy texture (Guneser et al. 2019).

Usually, these flavored milk-based beverages are prepared as follows: (a) the milk is treated in terms of standardization and homogenization; (b) different ingredients (cocoa powder, fruit, flavors, sugar, etc.) are added to warm milk and thoroughly mixed and dissolve; (c) and the resulting blend submitted after that to a stabilization process. Thermal processing such as pasteurization or ultra-high-temperature (UHT) sterilization is the most used treatment to ensure shelf life quality. These beverages are designed to be sold as pasteurized-refrigerated or UHT treated that are stable at room temperature. Nevertheless, various emerging nonthermal technologies, such as low-pressure cold plasma, are also being employed (Mukhopadhyay and Ukuku 2018, Coutinho et al. 2019).

17.4.2 Milk Smoothies and Milkshakes

The milk smoothies and milkshakes are blended beverages composed of fruit and vegetables mixed with water, fruit juice, yogurt, or milk. They may also contain other ingredients such as ice, sugar, honey, cereal grains, herbs, vegetables, dairy proteins, vitamins, and minerals. These products usually have a higher viscosity than milk or fruit juice due to the content of solids or stabilizers (Mehta et al. 2017).

Consumers usually take smoothies fresh or stored for a short time in refrigeration after pasteurization or freezing. There are different methodologies for producing milk smoothies. However, industrially produced drinks are commonly prepared using a high-speed mixer to disperse powdered products and grind solid foods, mainly supplemented frozen, followed by homogenization and heat treatment (Rani et al. 2016).

On the other hand, milkshakes are considered cold beverages based on ice cream, food additives (emulsifiers and stabilizers), and other ingredients (fruit or chocolate syrup). Additionally, it is normally served with whipped cream as a topping. For the production of milkshakes, it is required to homogenize the previously pasteurized ingredients. In the case of automatic machines, standard in fast-food restaurants, a milkshake base is mixed with the flavors and additives and consumed at the moment (Mehta et al. 2017). Like milk smoothies, the milkshake industrial processing steps include mixing, homogenization, and pasteurization (usually high-temperature short-time pasteurization, HTST). It is important to mention that the air emulsion in these products is essential to have good sensory characteristics and acceptance (McCrickerd et al. 2015).

17.4.3 Functional Milk Beverages

Functional milk beverages are produced from milk or flavored milk supplemented with certain functional ingredients with proven health benefits such as proteins, fiber, minerals, bioactive components, plant extracts, polyphenols, and ω-3 FA. Additionally, low-caloric beverages are manufactured by replacing sugar or fats (Chandrapala and Zisu 2016).

The consumption of functional milk beverages is rapidly increasing because of their nutritional and functional properties. The functional milk segment has been considered the fastest-growing segment with various products (Pinto et al. 2018). The current demand for nutritious beverages by health-conscious consumers involves functional milk beverages classified as low calorie, zero-calorie, low fat/lactose/carbohydrates, high protein, and low cholesterol. Specifically, the major innovations seen in this segment are a variety of functional or bioactive components and the packaging section.

17.4.3.1 Low-Sugar Functional Milk Beverages

Sugar plays a fundamental role in the flavor of the product and further affects texture and even color. Low-sugar functional milk beverages can be prepared by sugar replacements with natural and artificial nonnutritive sweeteners. In general, consumers select the "no sugar added" claim to the "low-sugar" claim, most likely because they assume that a certain amount of sugar and other sugar replacements are added. However, another procedure to increase the sweetness of low-sugar milk beverages is through enzymes, such as lactase, which hydrolyzes lactose into glucose and galactose, which comparatively possess higher sweetening power (McCain et al. 2018). Moreover, these products would be suitable for lactose-intolerant people. In this regard, the sweetness generated in milk by lactose hydrolysis is not very appreciated in countries where milk is traditionally consumed such as Northern Europe and North America (Dekker et al. 2019).

17.4.3.2 Low-Lactose Functional Milk Beverages

The lactose-free claim is gaining more popularity among consumers. These beverages are becoming a preferred choice due to the wide availability of good-quality lactose-free dairy products. Low lactose or lactose-free milk beverages can positively influence lactose-intolerant people's health, as they can continue having the nutritional benefits of dairy products. The products are treated either by neutral lactase or by supplementing acid lactase (Dekker et al. 2019).

17.4.3.3 Protein-Altered Functional Milk Beverages

The native composition of dairy proteins can be altered for enhanced functionality or nutrition purposes, e.g., adding more dairy/nondairy proteins, bioactive peptides, and protein hydrolysates. Proteins like caseinates and whey are sometimes added to products like protein-enriched snacks, yogurts, nutrition bars, and milk beverages. There is an increased demand for high-protein milk items supplemented with additional amounts of casein or whey proteins among sport and health-conscious consumers (Banovic et al. 2018).

Milk proteins can act as precursors of bioactive peptides with biological activity, especially by the inhibition of the angiotensin-converting enzyme (ACE-I). These bioactive peptides lower blood pressure and are generated during fermentation processes by starter/probiotic microorganisms. Therefore, it is foreseeable that the peptide content and its bioactivity will depend on the microorganism used and the type of milk (Vincenzetti et al. 2017).

On the other hand, milk protein hydrolysates are also used to prepare protein-based functional milk beverages to decrease allergenicity, increase adsorption, and improve functional properties (Ha et al. 2019). In general, hydrolysis is carried out with proteases. The degree of hydrolysis can impact the final application since highly hydrolyzed proteins provide a bitter taste. In that case, debittering agents such as specific peptidases are applied. These hydrolysates may be appropriate to replace native milk proteins in beverages for children and adults who experience allergy problems. The higher the degree of hydrolysis, the lesser the allergenicity. Furthermore, the hydrolysates have better intestinal absorption and release potential bioactive peptides (Li-Chan 2015).

17.4.3.4 Fat-Altered Functional Milk Beverages

The consumer trend is to opt for low-fat beverages with a low percentage of saturated FA, and hence skimmed milk is often used. However, the consistency and mouthfeel are not comparable to full-cream milk. Therefore, manufacturers add hydrocolloids, such as gums and modified starches, to restore the removed fat's effects.

Another alternative is to replace the milk fat with fat with specific health benefits, such as unsaturated FA from plant and marine sources. The FA with the most significant interest in functional milk beverages are α-linoleic acid (C18:3 ω-3, ALA), eicosapentaenoic acid (C20:5 ω-3, EPA), docosahexaenoic acid (C22:6 ω-3, DHA), and conjugated linoleic acid (CLA), due to their antioxidative role and prevention of chronic diseases (Corbo et al. 2014). Some examples of these products are Natrel Omega-3® (Natrel, Canada) and Natural Linea® (Corporacion Alimentaria Penanata S.A., Spain).

Nowadays, consumers are also very much concerned about the consumption of high levels of cholesterol. In this way, manufacturers have developed skimmed milk functional beverages supplemented with plant sterols (phytosterols and phytostanols), which diminish the absorption of all sterols in the digestive tract and, therefore, serum cholesterol levels (Özer and Kirmaci 2010). Usually, phytosterols occur naturally in most vegetable oils but are partially removed during deodorization, which is the last stage of the conventional refining process (Chapter 8). Commercial phytosterols are obtained from tall oil, a by-product of the wood pulp industry, or from the deodorizer distillate (Moreau 2015). Thus, the consumption of drinks containing plant sterols could be an excellent strategy to reduce cholesterol, although the final beverage matrix and bioavailability-related aspects should be considered (Vaghini et al. 2016). Some examples of these products are Benecol® (Raisio Benecol Ltd., Finland) and Danacol® (Danone S.A., Spain).

17.4.3.5 Vitamin- and Mineral-Fortified Functional Milk Beverages

Vitamins and minerals are generally added to milk beverages to compensate for possible losses during processing and optimize the recommended intakes. Among the minerals most used to fortify milk beverages are calcium, iron, and magnesium. Calcium can be added to milk beverages as inorganic salts such as calcium chloride, calcium lactate, calcium gluconate, or milk-based calcium. Calcium salts cause a pH reduction and therefore affect product stability. To solve this problem, sodium phosphate salts, which aid to adjust the pH, are often used. Nevertheless, the low level of calcium salts' solubility often leads to sedimentation in the beverage due to calcium-induced protein instability (Özer and Kirmaci 2010). In this sense, micronized tricalcium citrate improves dispersion and avoids sedimentation. This additive keeps particles in suspension longer and improves mouthfeel (McCarthy et al. 2017). Regarding iron, although there are many reports on the benefits of iron fortification, current scientific evidence on the effects of iron in fortified milk beverages is somewhat limited. Iron absorption appears to be positively affected by microbial dairy fermentation since acidic products such as lactic acid are generated (Silva et al. 2008). The major challenges for the iron-fortified milk beverages are the low level of solubility of this metal, the induction of protein instability, and the appearance of undesirable colors. On the other hand, fortified milk beverages with magnesium, selenium, and zinc are less common on the market. In all cases, the limiting factors are the solubility or loss of minerals after thermal processes (Minj et al. 2020).

The fortification of milk beverages with vitamins has been practiced for decades, especially of fat-soluble vitamins (A, D, and E) that are partially removed with the fat fraction during skimmed milk production. In any case, the fortification of milk beverages with vitamin D would be the most relevant commercially because it improves calcium absorption and strengthens the immune system. Additionally, during milk processing, some of the B-group vitamins are degraded to varying levels. In this regard, folic acid and vitamins B6 and B12 can be degraded up to 20% at sterilization temperatures (Özer and Kirmaci 2010). In this case, the strategy would be to either fortify with these vitamins directly or produce them directly in products through microbial fermentation (Smithers 2015).

17.4.3.6 High-Fiber and Prebiotic Functional Milk Beverages

Dietary fiber (DF) addition in milk beverages is a recent topic of interest because its consumption has been associated with preventing chronic diseases such as cardiovascular disease or diabetes (Chapter 21). However, only soluble dietary fiber (SDF) can be used in beverages due to its hydrosolubility (Gunenc et al. 2017). Examples of SDF include pectins, β-glucans, inulin, fructooligosaccharides (FOS), galactooligosaccharides (GOS), or galactomannans. Inulin, FOS, and β-glucans are the most used fibers to fortify functional milk beverages due to their prebiotic effect or enhancement of the hind gut bacteria activity. These sources have important technofunctional properties such as increased water-holding capacity improving viscosity and texture or mouthfeel. Moreover, SDF are used as fat substitutes for the production of low-calorie beverages (da Silveira et al. 2015, Hurtado-Romero et al. 2020). Several DF and prebiotic ingredients have been used in fortified dairy products, the most recent ones being extracted from agrifood by-products like pectins from artichoke or β-glucans from oats (Hurtado-Romero et al. 2020). In this regard, it has been shown that the consumption of milk fortified with 1.4% oat β-glucan is associated with decreased serum cholesterol levels (Lazaridou et al. 2014).

17.4.3.7 Phytochemical-Fortified Functional Milk Beverages

There are many research studies focused on formulations of functional dairy drinks that incorporate plant extracts rich in phytochemicals; however, few of these drinks reach the market or remain there, in part due to the restrictions that the authorities (EFSA in the European Union or FDA in the United States) in terms of health claim. Plant extracts can contain different bioactive phytochemicals, including phenolic compounds, saponins, carotenoids, betalains, and isoflavones. For example, soy extracts containing genistein, daidzein, and glycitein are commonly used to formulate soy isoflavones-fortified functional milk beverages (Seyhan et al. 2016). However, more data on the doses and duration of consumption should be generated to get evidence regarding the positive effect of soy isoflavones-enriched milk, including a lowered risk of osteoporosis, heart disease, or menopausal symptoms (Farid et al. 2019).

17.4.4 Functional Whey-Based Beverages

Whey is a liquid by-product formed during the process of cheese production rich in α-lactalbumin, β-lactoglobulin, and serum albumin. Depending on pH, whey can be classified as acidic (pH < 5) and sweet (pH 6–7). Since at acid pH, milk proteins can sediment after heat treatment, sweet whey would be the most appropriate to make whey-based beverages. In general, these beverages provide many nutrients at a low cost and can be consumed by a wide range of consumers, including children and elderly people. All the variants using milk can be alternatively produced with whey (see Section 17.5.3).

Among all the functional dairy beverages present on the market, those made from whey combined with fruits or fruit extracts would be the most successful and innovative. In most cases, the whey-based beverage health-promoting features will depend on the type of fruit since each possesses certain types and contents of bioactive compounds. For instance, apple juice rich in polyphenols, phenolic acids, and flavonoids, known to reduce incidence or risk of chronic diseases, is supplemented in various fruit whey-based formulations (Goudarzi et al. 2015).

In addition to fruit, another innovative idea for this type of whey-based beverages is the addition of prebiotic ingredients since they improve the physicochemical characteristics of the product and enhance functional properties in terms of intestinal health. A whey-based beverage was formulated with strawberries and the prebiotic xylooligosaccharide (XOS) at a concentration of 1.25/100 mL, finding an improvement in the functional properties in terms of antioxidant, antidiabetic, and antihypertensive activities (Souza et al. 2019).

On the other hand, there are other carbonated drinks (thirst quenching–type beverages) where the water portion is replaced by deproteinated whey. These drinks usually have other minor components, such as sweeteners, colorants, flavors, and carbon dioxide.

17.4.5 Functional Buttermilk-Based Beverages

Buttermilk is obtained from the production of butter and is considered a by-product. Its composition is similar to skimmed milk, although it contains a high phospholipid content, which are separated from fat globules during butter-making. Therefore, buttermilk can be used to develop functional beverages due to the presence of proteins, phospholipids, and minerals and can be useful in preventing diseases such as diabetes or cancer (Mudgil et al. 2016). Buttermilk-based beverages can be fortified with bioactive compounds like DF, prebiotics, fruit juices, and bioactive peptides. For instance, cultured buttermilk was prepared with or without fortification of partially hydrolyzed guar gum as SDF source, improving the product's physicochemical and sensory properties while increasing the healthy potential via the prevention of digestive disorders and diabetes (Mudgil and Barak 2016).

17.4.6 Traditional and Novel Processing Techniques

In general terms, the main processes involved in processing milk-based beverages and the type of beverage are homogenization, heat treatment, acidification, enzymatic processing, and stabilization. The homogenization improves the mouthfeel or textural quality of milk beverages. The acidification to pH ranging from 3.6 to 4.6 positively affects the beverage in terms of viscosity, shelf life, and organoleptic properties. The heat treatment is associated with an increase in the product's shelf life. Inactivation of microorganisms can be carried out by pasteurization at 72–75°C/15–20 s or by UHT sterilization at 135–150°C for a few seconds. Certain enzymes can be utilized during processing, as in the case of lactase or enzymes to obtain protein hydrolysates. Finally, to achieve a good texture and viscosity, hydrocolloids are usually used as stabilizers (Minj et al. 2020).

As stated above, a heat treatment is used to ensure the quality and shelf life of the product, in addition to using appropriate packaging. For example, dairy-based beverages submitted to a UHT treatment using Tetra Brick® technology have a shelf life of four to six months. However, some of the heat-labile compounds such as vitamins, minerals, and amino acids can be compromised. Besides, the energy of the thermal processing treatment is comparatively more expensive. In this sense, innovative nonthermal or cold processing technologies are appearing. The advantages of these technologies are that they destroy pathogenic microorganisms, yield more nutritious products, and products are perceived with better freshness and organoleptic properties (de Toledo Guimarães et al. 2018). Moreover, these processes are more energy-efficient and sustainable. Some of these nonthermal technologies are high hydrostatic pressure, ultrasound, or electrical pulses discussed in Chapter 2. Nevertheless, more research is required to apply these technologies to create novel dairy-based beverages at the industrial level.

In this section, the production and the role of unfermented dairy-based beverages as functional snacks were presented. Additional nutraceutical properties of milk-based snacks are discussed in Chapter 21.

17.5 DAIRY-BASED DESSERTS

Dairy-based desserts are considered snacks and are defined as semisolid foods based on milk mixed with sugar, thickeners (starch, flours, and cereals), eggs, and hydrocolloids (alginate or carrageenan). The category includes ice cream, pudding, custard, chocolate, mousse, crème brûlée, cheesecake, and flan. Their process includes cooking, baking, or steaming when the final structure is set derived from the interaction of milk casein with carrageenan and where viscosity increased because of the gelatinization of starch-based ingredients (Schoenfuss and Chandan 2011, Chandan and Kilara 2016). In ice cream, a freezing step is also included, and other details are described in the following sections. The final texture of dairy-based desserts ranges from soft, creamy, and gelled to firm and moldable (Schoenfuss and Chandan 2011).

Dairy-based desserts have a clear production and development trend toward low-sugar, less fat, fewer additives, more fruit, ready to eat, and high quality tasting natural home-made alternatives (Tetra Pak International 2020, Dussimon 2014). Outside these trends, the category decreased in North America by 200 million packaged units during 2013–2018. Other alternatives considered healthier as spoonable yogurt (see Section 17.4) increased around 700 million units in the same period (Dussimon 2014).

17.5.1 Ice Cream

17.5.1.1 Definitions and Categories

Frozen desserts appeared most probably in China, where the first blends of ice with fruits or fruit juice were reported. In Europe, these desserts were introduced in the 13th century, and massive ice cream production started in 1851, after the introduction of mechanical refrigeration and the development of electric equipment (Varnam and Sutherland 2001). Frozen dairy desserts can be described as products containing milk solids (with or without milk fat), aerated and consumed frozen. This category includes frozen custard, frozen yogurt, sherbet, and ice cream, all in hard frozen or soft frozen presentations, popular especially during hot weather, temperature being an important component of the pleasure associated with ingestion of these desserts (Goff 2006, Eccles et al. 2013, Goff and Hartel 2013). Within the frozen dairy desserts category, ice cream is the most widely consumed. Its definition varies globally because of the different traditions and national regulations (Goff and Hartel 2013). However, a general description is "ice cream is whipped dairy product that is manufactured by freezing and consumed frozen, including products that contain either dairy or non-dairy fats" (Goff 2006). In some countries, nondairy fats are not allowed in ice cream, so these must be commercialized with different or alternatives names. In the United States, frozen desserts with vegetable fats are sold as "mellorine." In Europe, the only countries where the use of nondairy fats are allowed are Ireland, Portugal, and the United Kingdom, where the word "dairy" is used before "ice cream" when the formulation includes dairy fats (Varnam and Sutherland 2001, Schoenfuss and Chandan 2011).

In the United States, the allowable ingredients for frozen desserts and ice creams are defined by the FDA, where the "frozen products with milk, include ice cream, frozen custard, goat's milk ice cream, sherbet and mellorine." The *custard*, also named "French ice-cream," can be defined as ice cream with egg or at least 1.4% of solids from the yolk (Varnam and Sutherland 2001, FDA 2020).

The fat percentage (dairy) in an ice cream must be at least 10%, and the milk solids nonfat (MSNF) also not less than 10% (Schoenfuss and Chandan 2011). In Mexico, ice cream manufacture is regulated in NOM-243, where ice cream definition is "a food produced with or without agitation of a pasteurized mix through freezing and composed by a combination of dairy ingredients with or without non-dairy fats, fruits, egg, and its derivatives, flavors, sweeteners, and other additives" (SSA 2010). In the European Union, standards for ice cream are less restrictive than in the United States and can include a product comprised of "water and/or milk, edible fats (from non-dairy sources), proteins and sugars, with no minimum requirements." In the European definition of ice cream, gelato is typically lower in fat (4–8%) and higher in sugar (up to 25%) than regular formulations. These differences yield a softer and pliable product (Schoenfuss and Chandan 2011, Goff and Hartel 2013).

Regarding ice milk categories, these products can be classified according to the point of manufacture/consumption as (a) take-home products, defined as grocery store purchases and home consumption. This specific category represents more than 90% of the US sales; (b) impulse products are handheld, single-serving eat-on-the-spot products, also defined as novelty products; and (c) artisanal/parlor sales or the ice cream manufactured at the site of the purchase and consumed or sold directly by the manufacturer to the final consumer (Goff and Hartel 2013, Gunzburg 2020).

In terms of fat content, the ice cream categories are (a) high-fat premium-type products generally containing 14–18% fat; (b) regular ice cream, with fat contents between 10% and 14%, and (c) low-fat or nonfat versions, with <10% fat. The last category may not meet the usual definitions of ice cream because, in the United States, ice cream must have a minimum of 10% fat (Goff 2006, Goff and Hartel 2013). Considering freezing as a parameter for classification, two broad categories exist: (a) soft-serve ice creams, which are served and consumed immediately after freezing and without the hardening step. These products are usually lower in fat and are commonly supplemented with carbohydrates and protein-based fat substitutes. The MSNF varies from 10% to 16%, and the total solids content varies from 30% to 35%. The second (b) category consists of hard ice cream, which is frozen again to harden the final product (Schoenfuss and Chandan 2011, Goff and Hartel 2013).

In the United States, sales of ice cream and frozen desserts ascended to US$17 billion in 2020, and US$18.3 billion are projected for 2025. The ice cream represents 94% of all ice cream and frozen dessert categories in retail sales. Impulse ice cream presentations reduced sales during 2020 attributed to the coronavirus lockdown, while bulk dairy ice cream benefited from this event, reaching US$7.3 billion in sales. Ice cream and frozen desserts reached 2.9 million tons in 2020 production (an annual increase of 3%). Ninety-one percent of ice cream and frozen desserts in sales are distributed in store-based retailing and 9% in vending machines, e-commerce, and direct selling, with a 6.4% increase in e-commerce sales (Gunzburg 2020). In the United States, the most important ice cream brands in sales share are Häagen-Dazs, Ben & Jerry's, and Blue Bell, with 5.8%, 5.3%, and 5.3% of the total sales, respectively. The most important companies are Dreyer's Grand Ice Cream Holdings Inc., Unilever Ice Cream, and Ben & Jerry Homemade Ltd., with 20.2%, 12.3%, and 5.3% of total US sales, followed by Blue Bell Creameries, Inc. with 5.3% (Gunzburg 2020).

In summary, ice cream is the most popular dairy frozen dessert. Its definition varies according to countries, but, in general, it can be defined as a dairy frozen mix made from fat, MSNF, sweeteners, stabilizers, and emulsifiers flavor and color, where the air has been incorporated during freezing.

17.5.1.2 Ingredients and Typical Formulations

Ice cream and most dairy frozen desserts contain seven categories of ingredients: fat, MSNF, sweeteners, stabilizers, emulsifiers, water, and flavorings (Goff and Hartel 2013). Fat and MSNF are chosen depending on availability, cost, legislation, and desired product quality (Marshall et al. 2003). The source of dairy and nondairy fat is cream, butter, or vegetable fats, whereas the most common MSNF products are condensed milk, whole milk, skimmed milk powder, whey powder, or whey protein products. The most used sweeteners are sucrose, high-fructose corn syrup, glucose, and nonnutritive sweeteners, whereas, in terms of stabilizers, locust bean gum, guar gum, CMC (carboxymethyl cellulose), and carrageenan usually are utilized. Regarding emulsifiers, the most typical are mono- and diglycerides and polysorbate 80. All ingredients are balanced with milk or water to an average of 37% solids in the mix to be frozen (Marshall et al. 2003, Goff 2006). Some typical formulations are depicted in Table 17.4.

17.5.1.3 Ice Cream Manufacturing Process and Trends

Frozen dairy desserts have a similar manufacturing process, all involving the critical and key operations of mixing and freezing (Figure 17.6).

TABLE 17.4
Typical Formulation of Several Grades of Ice Cream and Frozen Desserts[a]

Ingredients (% by Weight)	Nonfat Ice Cream	Economy Ice Cream	Standard Ice Cream	Premium Ice Cream	Superpremium Ice Cream
Milk fat	<0.5	10	10–12	12–14	14–18
Milk solids nonfat	12–14	10–11	9–10	8–10	5–8
Sweeteners[b]	18–22	15–17[(c)]	14–17[(d)]	13–16	14–17[(e)]
Stabilizers and emulsifiers	1.0	0.4	0.2–0.4	0.2–0.4	0.0–0.2
Total solids	28–32	35–36	36–38	38–40	40–42

[a] Based on information of Goff and Hartel (2013) and Schoenfuss and Chandan (2011).
[b] In this category of ingredients are sucrose, glucose, corn syrup solids, maltodextrins, and other bulking agents.
[c] Around 9.5% sugar and 7.8% corn syrup solids.
[d] Around 12% sugar and 4% corn syrup solids.
[e] All as sugar.

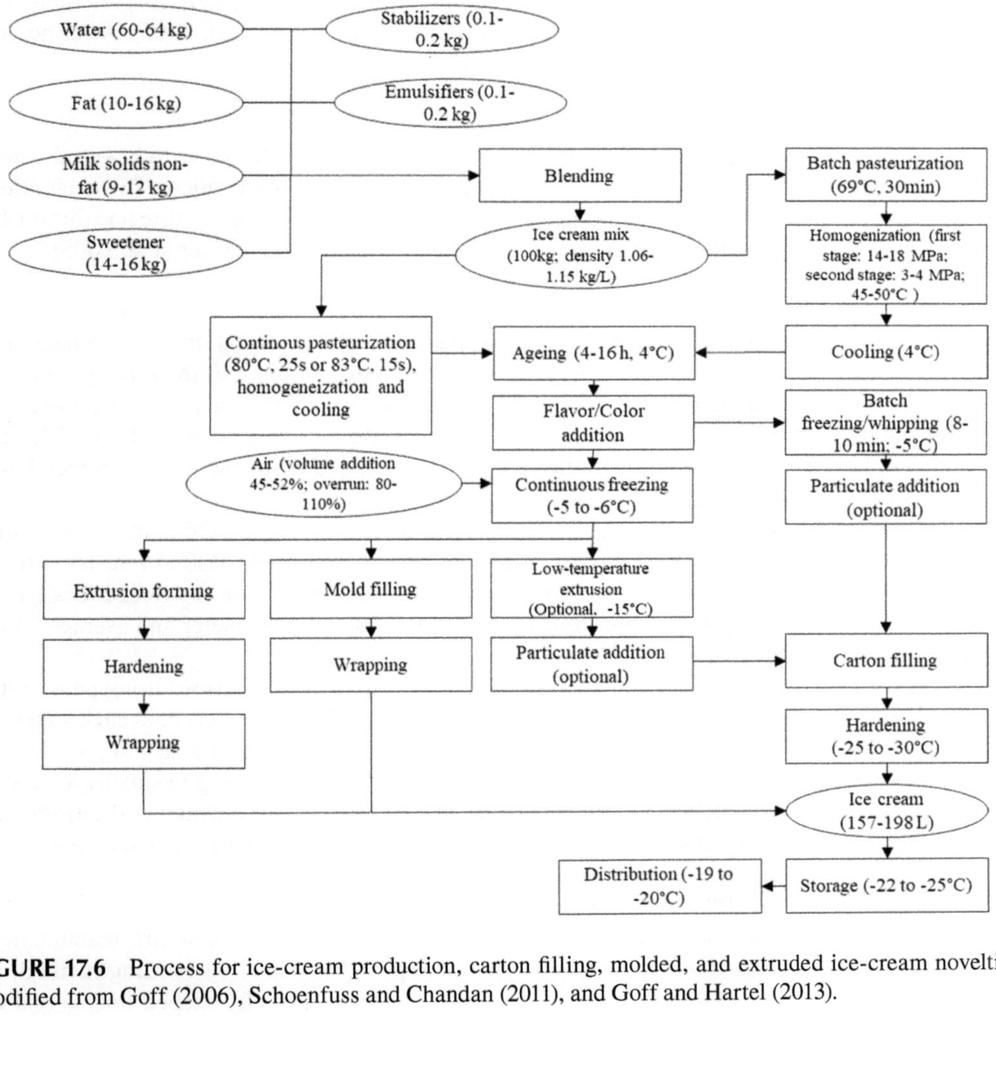

FIGURE 17.6 Process for ice-cream production, carton filling, molded, and extruded ice-cream novelties. Modified from Goff (2006), Schoenfuss and Chandan (2011), and Goff and Hartel (2013).

17.5.1.3.1 Mix Preparation and Blending

Ice cream manufacture starts with liquid and dry ingredients standardized to a percentage of solids and fat (Table 17.4). Stabilizers are premixed with other dry ingredients, mainly sugar, and added to liquid ingredients. The blend is made in-line or through recirculation in an agitated tank. This step is important to minimize air incorporation and avoid pasteurization problems (Schoenfuss and Chandan 2011).

17.5.1.3.2 Pasteurization

Batch or continuous pasteurization schemes have several objectives, but the most relevant is to ensure the destruction of pathogenic microorganisms. Besides, it has a positive impact on flavor development, viscosity, and the efficiency of the next step of homogenization. The time–temperature profile for mix pasteurization is higher than for fluid milk (12–12.5% solids) because of the higher solid contents. The most common parameters for batch pasteurization are 69°C for 30 min and for HTST 80°C/25 s or 83°C for 15 s (Goff 2006, Goff and Hartel 2013, FDA 2015, Gharibzahedi et al. 2018). Temperature is reduced to 45–50°C after pasteurization (Gharibzahedi et al. 2018).

17.5.1.3.3 Homogenization

During homogenization, the hot mix is treated first at pressures of 14–18 MPa (first stage) and then to 3–4 MPa (second stage). After homogenization, the diameter of the fat droplets is reduced to 2 μm or less (Cook and Hartel 2010), enhancing the droplet surface area, which reacts preferably with emulsifiers over proteins because of their relatively smaller size (Eisner et al. 2005). After homogenization, the mix is cooled down to 4°C before aging.

17.5.1.3.4 Aging

In the stage of aging or ripening, the ice cream mix is maintained at 4°C for 4–16 h with the main goals of (a) hydration of milk protein and stabilizers and (b) fat crystallization. This step allows stabilizers to perform their function, producing a mix with smooth body and texture and diminishes ice and lactose crystal growth during storage (Bahramparvar et al. 2011, Eisner et al. 2005).

17.5.1.3.5 Freezing

It is a stage with many physical changes for the mix. It starts with the formation and stabilization of foam, partial coalescence of fat, water freezing, and freeze concentration of the mix by removing water from the solution in the form of ice (Goff 2006). This step is performed in a scraped-surface batch or continuous freezer, which incorporates air and produces small ice crystals at −5°C. The batch process lasts 8–10 min, aerates at atmospheric pressure, and chills the ice cream blend until 30–35% of the water is frozen.

On the other hand, continuous freezers incorporate air under 3.5–5.0 atm (760 mmHg) and chills the product until 35–55% of the water is frozen (Mitten 1983, Adapa et al. 2000). In the latter freezing technology, the residence time is measured in seconds and depends on the freezer design and desired overrun. The ice crystals range from 1 μm to over 150 μm in diameter and average 35 μm (Cook and Hartel 2010).

Some novel technologies to improve freezing performance have been described using power ultrasound to shorten the processing time, yielding better quality foods. Ultrasound generates cavitation leading to the production of gas bubbles and the occurrence of micro-streaming. Cavitation promotes ice nucleation, accelerates heat and mass transfer, and fractures ice crystals, leading to smaller ice crystal distribution (Zheng and Sun 2006). The amount of air whipped into the ice cream mix determines the overrun percentage or the increase in ice cream volume (Eisner et al. 2005, Zheng and Sun 2006).

17.5.1.3.6 Forming (Extrusion, Mold, and Carton Filling)

Ice cream is deposited in carton packages, and novelty or impulse products are manufactured by molding soft-frozen ice cream or extrusion and cutting. In the extrusion forming method, ice cream from a continuous freezer is pumped at −6°C through a specialty-shaped nozzle and cut upon exiting with a wire. Ice cream sandwiches and frozen bars are examples of this technology. For molded products, the ice cream mix is pumped into molds immersed in chilled brine or glycol. After partial freezing, sticks are inserted and the product further frozen (Schoenfuss and Chandan 2011, Goff and Hartel 2013).

The low-temperature extrusion (Figure 17.6) follows conventional freezing. In this step, the product lowers from –5°C up to –15°C inside a piece of equipment with a slowly rotating screw that prevents large ice crystals and reduces the size of dispersed phase components. The process provides smoother texture and higher resistance to recrystallization during storage (Bolliger et al. 2000). Among the advantages of this technology is maintaining the ice cream microstructure, useful in reduced-fat products where a desirable creamy mouthfeel is desired (Schoenfuss and Chandan 2011, Goff and Hartel 2013). Low-temperature extrusion is also known as the "low-churned process" (Goff and Hartel 2013).

17.5.1.3.7 Hardening

This second freezing stage is made in forced-air convection freezers as air blast tunnels reduce the internal temperature to –25°C to –30°C. The aim is to freeze the rest of the ice cream water because only 30–55% was frozen after the first freezing stage. During hardening, some significant changes can occur in air bubbles, ice crystals, and fat in terms of distribution, size, and morphology (Homayouni et al. 2018). Hardening must be then made as fast as possible to minimize changes and maintain an adequate ice structure in finished products (Goff and Hartel 2013).

17.5.1.3.8 Flavor, Color, and Particulate Ingredients

Flavorings and coloring agents are added in the "white mix" before the first freezing step, and particulate or "bulky" ingredients as fruits, nuts, or bakery pieces are generally added after freezing.

17.5.2 Rice Pudding

Rice pudding is a dessert made mainly of cooked starch generally associated with polished white rice (Chapter 7) mixed with milk, egg, cinnamon, and other flavorings. The starches are responsible for soft and consistent texture (Budinoff 2003). A general formula for rice pudding is presented in Table 17.5.

As reported by Lo et al. (2003), rice pudding production includes the following steps: (a) *Selection of rice*: mostly rice pudding is made with white-polished rice kernels. Rice with high amylopectin

TABLE 17.5
General Formulation Used for the Production of Rice Pudding[a]

Ingredient	%
Water	62.7
Heavy cream	12.0
Nonfat dry milk	5.4
Sugar	11.2
Rice	7.0
Flavor/color	0.6
Carrageenan	0.5
Egg yolk	0.34
Sodium stearoyl lactylate (SSL)	0.2
Trisodium pyrophosphate	0.04
Sodium acid pyrophosphate	0.02

[a] From Schoenfuss and Chandan (2011).

must be avoided because it yields an undesirable sticky texture in the final product; (b) *Preheating and concentration*: water is added to a batch tank and heated at 65°C and then carrageenan and sugar are added and thoroughly dispersed. After the temperature is reduced to 50°C and the nonfat dry milk is added, sodium stearoyl lactylate is included with the ingredients and mixed at 65°C. Heating the mixture continues until the rice softens and gelatinizes into a thicker mixture; (c) *Homogenization*: the mixture is homogenized using 13.8 and 3.5 MPa in the first stage and second stage, respectively; (d) *Sterilization*: the mixture is pumped into a scraped-surface heat exchanger and heated to 115°C for 26 min. It is cooled afterward to 80°C in another cylinder. The sterilized mixture is transferred to a sterile buffer tank for further storage; (e) *Packaging*: the rice pudding is transferred to a form-and-fill plastic container under aseptic conditions; (f) *Cooling*: finally, the rice pudding is rapidly cooled below 15°C, and the texture is maintained for 90 days at refrigerated storage conditions.

17.5.3 Chocolate

17.5.3.1 Chocolate Milk

Chocolate milk is a sweetened chocolate-flavored beverage that is very popular and widely commercialized worldwide, with strong and continuous growth since 2015. The chocolate milk market is projected to reach US$9.31 billion by 2024, registering a CAGR of 3.6% during 2019–2024 (Toma and Săsean 2020). The milk industry has positioned chocolate milk as an ideal beverage to drink, especially for children and sports people, not only as a refreshment or snack but also as a food with high nutritional attributes (Deshpande et al. 2019).

Chocolate milk is considered a good source of nutrients that come from milk and cocoa, such as proteins, carbohydrates, fat, calcium, vitamin D, riboflavin, K, and P. Chocolate milk also contains smaller amounts of Zn, Se, I, Mg, and vitamins A, B1, B6, and B12. It is also important to mention that chocolate is a source of polyphenols associated with neuroprotective, antioxidant, antimicrobial, and cardioprotective activities (Chapter 21, Kardum and Gilbetic 2018). On the other hand, since it is a sweetened drink, a cup of chocolate milk can contain up to three teaspoons of added sugar. Thus, drinking too much can easily exceed the recommendation of limiting added sugars to less than ten teaspoons of added sugar per day for the average adult (USDA Dietary Guidelines). Additionally, chocolate milk has been discovered recently as a potential recovery after a grueling workout. Chocolate milk is rich in carbohydrates and proteins and also contains beneficial compounds to replenish sugars, fluids, and electrolytes lost during exercise (Moore 2015).

Depending on the manufacturer, different formulations to make chocolate milk are normally used. In general, the basic chocolate milk formulation consists of 93% milk, 6.3% sugar, 0.65% cocoa powder, and 0.05% stabilizer (usually carrageenan), in addition to other minor ingredients such as phosphate salts and flavorings. The product is often standardized to 1–2% fat, meaning there is 1–2% fat in the milk before adding the other ingredients. Milk is the ingredient that provides nutrients and flavor, whereas cocoa powder (often processed with alkali) provides the chocolate flavor. Sugar or artificial sugar substitutes provide the sweetness required to balance the flavor profile of the beverage. Carrageenan or other stabilizers help to suspend the cocoa powder in the beverage, imparts thickness or body, and avoids the separation of ingredients during storage.

Furthermore, formulations generally contain dipotassium phosphate, which acts as buffering agent, pH regulator, and enhancer of the salty and chocolate flavor. In general, processing for chocolate milk is as follows: (a) standardized and homogenized milk is preheated to 35–40°C; (b) cocoa powder, sugar, and other ingredients are slowly added to the warm milk and mixed to dissolve them properly; (c) the milk mixture is then pasteurized or sterilized at 71°C for 30 min and at 108–111°C for 25–30 min, respectively; (d) the milk mixture is bottled, although this step can be done before the thermal treatment; and (e) the final product is kept under refrigeration (pasteurized bottles) or at room temperature (sterilized bottles) (Guneser et al. 2019). Although thermal treatments are still the most widely used technology for stabilizing chocolate milk, nonthermal emerging technologies appeared as promising ways to improve the product's benefits. For instance, cold plasma technology seems to be an excellent alternative for chocolate milk processing because it improves functional properties, especially in terms of concentration of polyphenols which are known to be potent antioxidants that improve health indices (Coutinho et al. 2019).

Regarding future trends in chocolate milk, it must be considered that consumers are looking for options that have lower sugar content while maintaining the typical chocolate flavor. The reduction in sugar is achieved with artificial sweeteners, the current trend being the usage of natural plant extracts such as stevia or monk fruit. On the other hand, manufacturers already offer varieties that are low in fat or high in protein and other micronutrients. Additionally, many combinations with chocolate, including chocolate cherry, chocolate banana, chocolate mint, or peanut butter, are opening the market to new indulgent flavors. Ultimately, consumer choices' evolution will be marked based on preferred tastes and the desired nutritional content. In this sense, kids often prefer a sweeter flavor profile, while a product for the adult market might provide a stronger and less-sweet chocolate profile.

17.5.3.2 Milk-Based Beverage Mixes

Many chocolate-flavored milk-based beverage mixes are known. The dry mixes that can be diluted before consumption to form a chocolate-flavored milk-based beverage are trendy among these formulations. These beverage mixes can be reconstituted with milk, powdered milk diluted in water, or plain water.

The formulations contain milk solids to provide the nutritional and sensory properties of the reconstituted product. However, milk solids are relatively expensive and add extra cost to the product. These costs can be reduced by partially or fully replacing milk solids with whey solids. Whey can provide protein fortification to the dry product, being a reasonably inexpensive ingredient. However, whey solids may generate an objectionable off-flavor to beverages or impart a different texture than the milk provides. Consumers often reject mixes that contain whey solids and therefore add whole milk or powdered milk to the product to enhance its sensory properties (Aribah et al. 2020). Accordingly, manufacturers are improving their formulations to make a dry chocolate milk–based beverage mix that can be reconstituted with plain water and tastes like chocolate milk without requiring large amounts of milk or whey solids. The key to the dry beverage mix is the ideal combination of sugar (40–60%), cocoa powder (3–13%), milk solids, which can include fat (10–32%), texturizing agents (<6%), a protein source (0–20%), and other ingredients such as vitamins, minerals, and flavorings in variable amounts, depending on the product. In many cases, the dairy ingredient is substituted for a powdered nondairy creamer (Dolan et al. 1996).

The production of dry mixes or blends require mixing the necessary quantities of each ingredient. Usually, cocoa powder and sugar are mixed beforehand in order to produce the first premix. The cocoa powder and sugar must be mixed well so that the cocoa particles adhere to the sugar particles so as to improve the dry beverage mix's dispersibility when reconstituted with water. Lecithin can be added to this premix as a processing aid. Afterward, the rest of the ingredients are mixed so as to obtain a second premix. Subsequently, the two premixes are brought together to produce the final product. The dry chocolate-flavored beverage mix can further comprise other ingredients to provide flavor, texture, stability, or nutrition benefits. Such optional ingredients can include, for example, natural and artificial colors and flavors as vanilla, coffee, caramel, or nuts.

17.5.3.3 Milk Chocolate Bars

Milk chocolate is a confectionery product made from cocoa, sugar, milk, and other minor ingredients. The most popular choice is the traditional bar due to its availability, convenience, and cost-friendliness. Chocolate bars are now available in various choices, such as chocolate coated with nuts, candies, chocolate chips, and others. The main milk chocolate producers worldwide are Ferrero, Hershey, Mondelez, Mars, and Nestlé. The global milk chocolate market is expected to show substantial growth in the coming years. According to the Zion Market Research (https://www.zionmarketresearch.com/report/milk-chocolate-market), the global milk chocolate market accounted for US$57.95 billion in 2018 and is expected to reach around US$90.65 billion by 2025, at a CAGR of 6.6% between 2019 and 2025.

The processing of milk chocolate begins with the cocoa fruit, which is usually grown and harvested in Southeast Asia, South America, and West Africa. Once the cocoa pod is harvested, the cocoa beans are removed, fermented, and dried. These beans are then cleaned, roasted, and ground, causing cocoa butter to melt down to form cocoa liquor. The liquefied cocoa is then separated into the cocoa solids and fat or butter, which are mixed in different proportions. From here, the other two key ingredients, milk and sugar, come into play. Milk is added in powdered form since an excess of water could damage the properties of the chocolate bars. Whole milk powder is commonly used, although skimmed milk options can also be used.

In some cases, it is likely that condensed milk will be used, although it will depend on the type of chocolate bar. On the other hand, sugar, either from sugarcane or sugar beet, is added at the same time as the powdered milk. This mixture must be refined for particle size reduction required to improve chocolate quality. Then, the conching process is carried out, which can be defined as the working of chocolate flake and crumb into a fluid phase, coupled with flavor modification. For milk chocolate, temperatures below 50°C are recommended to avoid Maillard reactions.

Additionally, tempering involving a thermal treatment is applied to produce homogeneous fat crystals and yield crunchy and shiny chocolate bars. Finally, the liquid chocolate is poured into molds and formed into bars (Beckett 2019). As another category of chocolate, it is important to mention the case of "white chocolate," made with cocoa butter, sugar, and milk solids. Despite being high in cocoa butter, it does not contain cocoa solids.

Current trends in the milk chocolate market include the use of alternative lipids or sweeteners. To replace cocoa butter, cocoa butter equivalents (CBE) are often used. CBE is selected due to the similarity in fatty acid and triglyceride composition, resulting in similar physical characteristics, although in some cases differences in flavor and aroma can be detected. On the other hand, maltitol and tagatose have shown excellent qualities as sugar replacements (Engeseth and Pangan 2018). Due to consumer concerns, certain ingredients have been included in the chocolate formulation to improve its nutritional content. Prebiotics, probiotics, and antioxidants are examples of food ingredients that have been investigated. For instance, a prebiotic low-sugar milk chocolate with good sensory properties was developed with inulin, polydextrose, and maltodextrin as sugar substitutes (Farzanmehr and Abbasi 2009). Finally, a milk chocolate variant has gained significant importance in recent years. The bars contain dried fruits, caramel, walnuts, or cookies. Brands such as Kit Kat, Snickers, or Mars Bars are some examples.

17.5.4 Others

17.5.4.1 Mousse

Mousse is defined as a chilled air dessert made from mainly four basic ingredients: whipped cream, beaten eggs, gelatin, or sugar meringue combined with whipped cream (Clafin 2015). Since this dessert is aerated, the density is positively related to the air-trapping and the overrun. The volume of mousses vary from 100% to 150% in order to impart a light texture and the desired foam-like consistency (Schoenfuss and Chandan 2011). There are two additional ingredients in the formulation (Table 17.6): emulsifier and gelatin. These are critical to create and maintain fine air bubbles that greatly affect product texture.

The general manufacturing process of mousse includes the following: (a) *Ingredients mix*: it is an essential step to avoid air incorporation that could make pasteurization and homogenization problematic; (b) *Heat treatment:* the mix is subjected to HTST treatment at 80–85°C for 20–30 s. During this process, the serum proteins denature improving the water-binding capacity; (c) *Homogenization:* this step can take place upstream or downstream the thermal treatment. Homogenization can control the mousse's structure since it enhances the production of more stable emulsions; (c) *Heat treatment:* the gelatin is dissolved in hot water (80°C); (d) *Aeration and blending:* the gelatin is added

TABLE 17.6
General Formulation of Chocolate Mousse[a]

Ingredient	Quantity
Cream 35% fat (L)	16.0
Sugar (kg)	3.0
Gelatin (kg)	0.22
Cocoa, alkalized 10–12% fat (kg)	0.6
Chopped chocolate (kg)	3.0
Cream cheese (kg)	1.2
Water (L)	1.2

[a] From Schoenfuss and Chandan (2011).

TABLE 17.7
General Formulation of Baked Flan[a]

Ingredient	Usage Level (kg)
Skim milk	65.0
Cream cheese	15.0
Liquid eggs	20.0
Sugar	14.0
Flavor (cinnamon)	0.125
Lemon rind	0.55
Liquid caramel	6.0

[a] From Schoenfuss and Chandan (2011).

to the mixture while continuing the whipping process for 10–15 s at 20–30°C; (e) *Packaging and storage*: the final dessert is stored under frozen conditions.

17.5.4.2 Flan

Flan is a popular dessert obtained by mixing milk, eggs, and sugar, which results in a moldable gel that can be easily removed from the container. The flan texture should be firm, creamy, and cohesive so as to impart a good mouthfeel. Table 17.7 details a typical formulation for this product.

The flan manufacturing process includes mixing the ingredients (skimmed milk, cinnamon, lemon rind, cream cheese, eggs, and sugar) for 30 min. The second step is filling containers, where liquid caramel can be added (optional). If used, it is placed in the bottom of the container before addition of the blend. Once containers are filled, they are vacuum-sealed and heat-treated at 70–105°C for 25–35 min. Finally, the product is cooled down at 0–8°C and stored (Salmon Garcia de los Salmones 2006).

REFERENCES

Adapa, S., Schmidt, K.A., Jeon, I.J., Herald, T.J., and Flores, R.A. 2000. Mechanisms of ice crystallization and recrystallization in ice cream: A review. *Food Reviews International*, 16(3):259–271. https://doi.org/10.1081/FRI-100100289.

Ahrens, G., Kriszio, H., and Langer, G. 2006. Microwave vacuum drying in the food processing industry. In: *Advances in Microwave and Radio Frequency Processing*, ed. M. Willert-Porada, 426–435. Berlin: Springer. https://doi.org/10.1007/978-3-540-32944-2.

Aribah, S.A., Sanjaya, A.P., Muhammad, D.R.A., and Praseptiangga, D. 2020. Sensorial and physical properties of chocolate beverage prepared using low fat cocoa powder. In: *AIP Conference Proceedings* (Vol. 2219, No. 1, p. 070007). Solo, Indonesia: AIP Publishing LLC.

Aryana, K.J., and Douglas, W.O. 2017. A 100-year review: Yogurt and other cultured dairy products. *Journal of Dairy Science*, 100(12):9987–10013. https://doi:10.3168/jds.2017-12981.

Asgar, M.A., Yamauchi, R., and Kato, K. 2003. Modification of pectin in Japanese persimmon fruit during the sun-drying process. *Food Chemistry*, 81(4):555–560. https://doi.org/10.1016/S0308-8146(02)00489-2.

Azami, T., Niakousari, M., Hashemi, S.M.B., and Torri, L. 2018. A three-step sensory-based approach to maximize consumer acceptability for new low-sugar licorice-chocolate-flavored milk drink. *LWT-Food Science and Technology*, 91:375–381.

Bahramparvar, M., and Mazaheri Tehrani, M. 2011. Application and functions of stabilizers in ice cream. *Food Reviews International*, 27(4):389–407. https://doi.org/10.1080/87559129.2011.563399.

Banovic, M., Arvola, A., Pennanen, K., Duta, D.E., Brückner-Gühmann, M., Lähteenmäki, L., and Grunert, K.G. 2018. Foods with increased protein content: A qualitative study on European consumer preferences and perceptions. *Appetite*, 125:233–243. https://doi.org/10.1016/j.appet.2018.01.034.

Beckett, S.T. 2019. *The Science of Chocolate*. Cambridge: RSC Publishing.

Berk, Z. 2009. Frying, baking and roasting. In: *Food Process Engineering and Technology*, 3rd edition, 525–531. New York: Academic Press.

Bernier, R., and Labbe, R. 2019. *Research Report: Specialty Cheese Market*. https://agriculture.vermont.gov/sites/agriculture/files/documents/AgDevReports/SpecialtyCheeseMarketResearchReport.pdf.

Bochnak-Niedźwiecka, J., and Świeca, M. 2020. Quality of new functional powdered beverages enriched with lyophilized fruits: Potentially bioaccessible antioxidant properties, nutritional value, and consumer analysis. *Applied Sciences*, 10(11):3668. https://doi.org/10.3390/app10113668.

Bolliger, S., Kornbrust, B., Goff, H.D., Tharp, B.W., and Windhab, E.J. 2000. Influence of emulsifiers on ice cream produced by conventional freezing and low-temperature extrusion processing. *International Dairy Journal*, 10(7):497–504. https://doi.org/10.1016/S0958-6946(00)00071-6.

Budinoff, D.B. 2003. Rice pudding method and composition. US Patent No. US 2003/0099757.

Bylund, G. 2015. *Dairy Processing Handbook*. 3rd edition. Lund, Sweden: Tetra Pak Processing Systems AB.

Chandan, R.C., and Kilara, A. 2016. Pudding and dairy-based desserts. In: *Dairy Processing and Quality Assurance*, eds. R.C. Chandan, A. Kilara, and N.P. Shah, 397–427. Sussex, UK: Wiley Blackwell.

Chandrapala, J., and Zisu, B. 2016. Novel trends in engineered milk products. *The Journal of Dairy Research*, 83(3):268. https://doi.org/10.1017/S0022029916000479.

Clafin, K.W. 2015. Mouse. In: *The Oxford Companion to Sugar and Sweets*, ed. G. Darra, 464. New York: Oxford University Press.

Cook, K.L.K., and Hartel, R.W. 2010. Mechanisms of ice crystallization in ice cream production. *Comprehensive Reviews in Food Science and Food Safety*, 9(2):213–222. https://doi.org/10.1111/j.1541-4337.2009.00101.x.

Corbo, M.R., Bevilacqua, A., Petruzzi, L., Casanova, F.P., and Sinigaglia, M. 2014. Functional beverages: The emerging side of functional foods: Commercial trends, research, and health implications. *Comprehensive Reviews in Food Science and Food Safety*, 13(6):1192–1206. https://doi.org/10.1111/1541-4337.12109.

Coutinho, N.M., Silveira, M.R., Fernandes, L.M., Moraes, J., Pimentel, T.C., Freitas, M.Q., Silva, M.C., Raices, R.S.L., Senaka-Ranadheera, C., Borges, F.O., Neto, R.P.C., Tavares, M.B.I., Fernandes, F.A.N., Fonteles, T.V., Nazzaro, F., Rodrigues, S., and Ctuz, A.G. 2019. Processing chocolate milk drink by low-pressure cold plasma technology. *Food Chemistry*, 278:276–283. https://doi.org/10.1016/j.foodchem.2018.11.061.

da Silveira, E.O., Neto, J.H.L., da Silva, L.A., Raposo, A.E., Magnani, M., and Cardarelli, H.R. 2015. The effects of inulin combined with oligofructose and goat cheese whey on the physicochemical properties and sensory acceptance of a probiotic chocolate goat dairy beverage. *LWT – Food Science and Technology*, 62(1):445–451. https://doi.org/10.1016/j.lwt.2014.09.056.

Dairy Management. 2010. *Snacking, Identifying a World of Opportunities for Dairy*. Dairy Management Inc. Prepared for the Innovation Center for U.S. Dairy. pp 1–40.

Danone Institute International. 2015. What is yogurt? Process, history and frequently asked questions. https://www.yogurtinnutrition.com/what-yogurt/.

De Toledo Guimarães, J., Silva, E.K., de Freitas, M.Q., de Almeida Meireles, M.A., and da Cruz, A.G. 2018. Non-thermal emerging technologies and their effects on the functional properties of dairy products. *Current Opinion in Food Science*, 22:62–66. https://doi.org/10.1016/j.cofs.2018.01.015.

Dekker, P.J., Koenders, D., and Bruins, M.J. 2019. Lactose-free dairy products: Market developments, production, nutrition and health benefits. *Nutrients*, 11(3):551. https://doi.org/10.3390/nu11030551.

Deshpande, S., Singh, S., Panneerselvam, A., and Rajeswari, V.D. 2019. Nutrients in caffeinated beverages—An overview. In: *Caffeinated and Cocoa Based Beverages*, eds. A.M. Grumezescu and A.M. Holban, 367–389. Woodhead Publishing.

Dolan, K.M., Hughes, D.L., Leavell, K.A., Swaine Jr, R.L., and Hayes, M.Z. 1996. *Dry Chocolate Flavored Beverage Mix*. U.S. Patent Application No. 08/465042.

Dussimon, K. 2014. Premium look prevails for dairy desserts in Western Europe and the US. https://0-www-portal-euromonitor-com.biblioteca-ils.tec.mx/portal/analysis/tab.

Eccles, R., Du-Plessis, L., Dommels, Y., and Wilkinson, J.E. 2013. Cold pleasure. Why we like ice drinks, ice-lollies and ice cream. *Appetite*, 71:357–360. https://doi.org/10.1016/j.appet.2013.09.011.

Eisner, M.D., Wildmoser, H., and Windhab, E.J. 2005. Air cell microstructuring in a high viscous ice cream matrix. *Colloids and Surfaces A: Physicochemical and Engineering Aspects*, 263(1–3):390–399. https://doi.org/10.1016/j.colsurfa.2004.12.017.

Engeseth, N.J., and Pangan, M.F. 2018. Current context on chocolate flavor development—A review. *Current Opinion in Food Science*, 21:84–91. https://doi.org/10.1016/j.cofs.2018.07.002.

FAO/WHO. 2011. Milk and milk products. http://www.fao.org/3/i2085e/i2085e00.pdf.

Farid, M., Kodama, K., Arato, T., Okazaki, T., Oda, T., Ikeda, H., and Sengoku, S. 2019. Comparative study of functional food regulations in Japan and globally. *Global Journal of Health Science*, 11(6):132–145. https://doi.org/10.5539/gjhs.v11n6p132.

Farzanmehr, H., and Abbasi, S. 2009. Effects of inulin and bulking agents on some physicochemical, textural and sensory properties of milk chocolate. *Journal of Texture Studies*, 40(5):536–553. https://doi.org/10.1111/j.1745-4603.2009.00196.x.

FDA. 2015. Grade a pasteurized milk ordinance (PMO). https://www.fda.gov/media/99451/download.

FDA. 2020. Code of federal regulations. Title 21. https://www.accessdata.fda.gov/scripts/cdrh/cfdocs/cfcfr/CFRSearch.cfm?fr=131.200#:~:text=Sec.,131.200 Yogurt.&text=Yogurt%2C before the addition of,percent%2C expressed as lactic acid.

Fisberg, M., and Machado, R. 2015. History of yogurt and current patterns of consumption. *Nutrition Reviews*, 73(1):4–7. https://doi.org/10.1093/nutrit/nuv020.

FSANZ. 2015. *Australia New Zealand Food Standard Code*. Australian Government. https://www.legislation.gov.au/Details/F2015L00413.

Gharibzahedi, S.M.T., Koubaa, M., Barba, F.J., Greiner, R., George, S., and Roohinejad, S. 2018. Recent advances in the application of microbial transglutaminase crosslinking in cheese and ice cream products: A review. *International Journal of Biological Macromolecules*, 107(B):2364–2374. https://doi.org/10.1016/j.ijbiomac.2017.10.115.

Goff, H.D. 2006. Ice cream. In: *Advanced Dairy Chemistry, Volume 2 Lipids*, eds. P.F. Fox and P.L.H. McSweeney. Boston: Springer. https://doi.org/10.1007/0-387-28813-9_12.

Goff, H.D., and Hartel, R.W. 2013. *Ice Cream*, 7th edition. New York: Springer. https://doi.org/10.1007/978-1-4614-6096-1.

Goudarzi, M., Madadlou, A., Mousavi, M.E., and Emam-Djomeh, Z. 2015. Formulation of apple juice beverages containing whey protein isolate or whey protein hydrolysate based on sensory and physicochemical analysis. *International Journal of Dairy Technology*, 68(1):70–78. https://doi.org/10.1111/1471-0307.12155.

Gunenc, A., Hosseinian, F., and Oomah, B.D. 2017. Dietary fiber-enriched functional beverages in the market. In: *Dietary Fiber Functionality in Food and Nutraceuticals*, eds. F. Hosseinian, B.D. Oomah, and R. Campos-Vega, 45–75. Wiley Online Books. https://doi.org/10.1002/9781119138105.ch3.

Guneser, O., Hosoglu, M.I., Guneser, B.A., and Yuceer, Y.K. 2019. Engineering of milk-based beverages: Current status, developments, and consumer trends. In: *Milk-Based Beverages*, eds. A.M. Grumezescu and A.M. Holban, 1–37. Cambridge: Woodhead Publishing.

Gunzburg, S. 2020. Ice cream and frozen desserts in the US. https://0-www-portal-euromonitor-com.biblioteca-ils.tec.mx/portal/ResultsList/Index.

Ha, H.K., Rankin, S.A., Lee, M.R., and Lee, W.J. 2019. Development and characterization of whey protein-based nano-delivery systems: A review. *Molecules*, 24(18):3254. https://doi.org/10.3390/molecules24183254.

Hess, J.M., Jonnalagadda, S.S., and Slavin, J.L. 2016. What is a snack, why do we snack, and how can we choose better snacks? A review of the definitions of snacking, motivations to snack, contributions to dietary intake, and recommendations for improvement. *Advances in Nutrition*, 7(3):466–475. https://doi.org/10.3945/an.115.009571.

Homayouni, A., Javadi, M., Ansari, F., Pourjafar, H., Jafarzadeh, M., and Barzegar, A. 2018. Advanced methods in ice cream analysis: A review. *Food Analytical Methods*, 11(11):3224–3234. https://doi.org/10.1007/s12161-018-1292-0.

Huang, L.L., and Zhang, M. 2012. Trends in development of dried vegetable products as snacks. *Drying Technology*, 30(5):448–461. https://doi.org/10.1080/07373937.2011.644648.
Hurtado-Romero, A., Del Toro-Barbosa, M., García-Amezquita, L.E., and García-Cayuela, T. 2020. Innovative technologies for the production of food ingredients with prebiotic potential: Modifications, applications, and validation methods. *Trends in Food Science and Technology*, 104:117–131. https://doi.org/10.1016/j.tifs.2020.08.007.
Ibrahim, F.S., Dabiza, N.M., Abd El-Lattief, O.A., and Abd EL Razik, S.T. 1999. Microstructure, amino acids content and microflora of farm house concentrated fermented milk "Laban Zeer". *Egyptian Journal of Dairy Science*, 27(2):291–300.
Ilao, S.M.M., Panabang, M.B., Depositario, D.P.T., and Baticados, G.N. 2017. Product attribute analysis for latte drinks: Implications for a coffee-flavored carabao milk-based drink. *International Food Research Journal*, 24(1):142–147. http://ifrj.upm.edu.my/24%20(01)%202017/(17).pdf.
Kardum, N., and Glibetic, M. 2018. Polyphenols and their interactions with other dietary compounds: Implications for human health. *Advances in Food and Nutrition Research*, 84:103–144. https://doi.org/10.1016/bs.afnr.2017.12.001.
Kolakowski, P., and Pawlikowska, K. 2012. Kefir as a probiotic delivery vehicle. *Milchwissenschaft-Milk Science International*, 67(2):159–162.
Lazaridou, A., Serafeimidou, A., Biliaderis, C.G., Moschakis, T., and Tzanetakis, N. 2014. Structure development and acidification kinetics in fermented milk containing oat β-glucan, a yogurt culture and a probiotic strain. *Food Hydrocolloids*, 39:204–214. https://doi.org/10.1016/j.foodhyd.2014.01.015.
Li-Chan, E.C. 2015. Bioactive peptides and protein hydrolysates: Research trends and challenges for application as nutraceuticals and functional food ingredients. *Current Opinion in Food Science*, 1:28–37. https://doi.org/10.1016/j.cofs.2014.09.005.
Lo, C.Y.G., Vellucci, D.J., Messick, D.L., Peet, H.L., Sole, M.M., McPherson, A.E., MacDonald, J.L., and Leshik, R.R. 2003. Refrigerated ready to eat aseptically produced rice pudding. US Patent No. US. 2003/6596335.
Mahomud, M.S., Katsuno, N., and Nishizu, T. 2017. Role of whey protein-casein complexes on yoghurt texture. *Reviews in Agricultural Science*, 5:1–12. https://doi.org/10.7831/ras.5.1.
Marshall, R.T., Goff, H.D., and Hartel, R.W. 2003. *Ice Cream*. New York: Kluwer Academic Publishers.
McCain, H.R., Kaliappan, S., and Drak, M.A. 2018. Invited review: Sugar reduction in dairy products. *Journal of Dairy Science*, 101(10):8619–8640. https://doi.org/10.3168/jds.2017-14347.
McCarthy, N.A., Power, O., Wijayanti, H.B., Kelly, P.M., Mao, L., and Fenelon, M.A. 2017. Effects of calcium chelating agents on the solubility of milk protein concentrate. *International Journal of Dairy Technology*, 70(3):415–423. https://doi.org/10.1111/1471-0307.12408.
McCrickerd, K., Lensing, N., and Yeomans, M.R. 2015. The impact of food and beverage characteristics on expectations of satiation, satiety and thirst. *Food Quality and Preference*, 44:130–138. https://doi.org/10.1016/j.foodqual.2015.04.003.
Mehta, D., Kumar, M.S., and Sabikhi, L. 2017. Development of high protein, high fiber smoothie as a grab-and-go breakfast option using response surface methodology. *Journal of Food Science and Technology*, 54(12):3859–3866. https://doi.org/10.1007/s13197-017-2841-z.
Minj, J., Sudhakaran, V.A., and Kumari, A. 2020. *Dairy Processing: Advanced Research to Applications*. Singapore: Springer. https://doi.org/10.1007/978-981-15-2608-4.
Mitten, H.L. 1983. Proper hardening ensures quality. *Dairy Field*, 166(6):58–61.
Moreau, R.A. 2015. Composition of plant sterols and stanols in supplemented food products. *Journal of AOAC International*, 98(3):685–690. https://doi.org/10.5740/jaoacint.SGEMoreau.
Moore, D.R. 2015. Nutrition to support recovery from endurance exercise: Optimal carbohydrate and protein replacement. *Current Sports Medicine Reports*, 14(4):294–300. https://doi.org/10.1249/JSR.0000000000000180.
Mordor Intelligence. 2020. Yogurt market – Growth, trends and forecasts (2020–2025). https://www.mordorintelligence.com/industry-reports/yogurt-market.
Mudgil, D., and Barak, S. 2016. Development of functional buttermilk by soluble fibre fortification. *Agro Food Industry Hi-Tech*, 27(2):44–47.
Mudgil, D., Barak, S., and Darj, I.P. 2016. Development and characterization of functional cultured buttermilk utilizing *Aloe vera* juice. *Food Bioscience*, 15:105–109. https://doi.org/10.1016/j.fbio.2016.06.001.
Mukhopadhyay, S., and Ukuku, D.O. 2018. The role of emerging technologies to ensure the microbial safety of fresh produce, milk and eggs. *Current Opinion in Food Science*, 19:145–154. https://doi.org/10.1016/j.cofs.2018.01.013.

Nazir, M., Arif, S., Khan, R.S., Nazir, W., Khalid, N., and Maqsood, S. 2019. Opportunities and challenges for functional and medicinal beverages: Current and future trends. *Trends in Food Science and Technology*, 88:513–526. https://doi.org/10.1016/j.tifs.2019.04.011.

Özer, B.H., and Kirmaci, H.A. 2010. Functional milks and dairy beverages. *International Journal of Dairy Technology*, 63(1):1–15. https://doi.org/10.1111/j.1471-0307.2009.00547.x.

Pinto, L.P., Silva, H., Kuriya, S., Maçaira, P.M., Cyrino, F.L., Cruz, A.G., Esmerino, E.A., and Freitas, M.Q. 2018. Understanding perceptions and beliefs about different types of fermented milks through the application of projective techniques: A case study using Haire's shopping list and free word association. *Journal of Sensory Studies*, 33(3):123–126. https://doi.org/10.1111/joss.12326.

Radas, P., and Pawel, R. 2017. Method to produce cheese chips and raised cheese chips. US Patent No. US 2017/0020155.

Rafiq, L., Zahoor, T., Sagheer, A., Khalid, N., Rahman, U., and Liaqat, A. 2020. Augmenting yogurt quality attributes through hydrocolloidal gums. *Asian-Australasian Journal of Animal Sciences*, 33(2):323. https://doi.org/10.5713/ajas.18.0218.

Rakcejeva, T., Zagorska, J., Dukalska, L., Galoburda, R., and Eglitis, E. 2009. Physical-chemical and sensory characteristics of Cheddar cheese snack produced in vacuum microwave dryer. *Chemine Technologija*, 3(52):16–20.

Rani, R., Kumar, M.H.S., and Sabikhi, L. 2016. Process optimization for a ready-to-serve breakfast smoothie from a composite milk–sorghum base. *International Journal of Dairy Technology*, 69(3):372–379. https://doi.org/10.1111/1471-0307.12269.

Routray, W., and Mishra, H.N. 2011. Scientific and technical aspects of yogurt aroma and taste: A review. *Comprehensive Reviews in Food Science and Food Safety*, 10(4):208–220. https://doi.org/10.1111/j.1541-4337.2011.00151.x.

Salampessy, J., and Kailasapathy, K. 2011. Fermented dairy ingredients. In *Dairy Ingredients for Food Processing*, eds. R.C. Chandan and A. Kilara, 335–356. Ames: Wiley-Blackwell.

Salmon Garcia de los Salmones, C. 2006. Method of preparing a cheese flan. US Patent No. US 2006/0280853.

Savaiano, D.A. 2014. Lactose digestion from yogurt: Mechanism and relevance. *The American Journal of Clinical Nutrition*, 99(5):1251S–1255S. https://doi.org/10.3945/ajcn.113.073023.

SCFI/SAGARPA. 2018. Yogurt-denominación, especificaciones fisicoquímicas y microbiológicas, información comercial y métodos de prueba. CDMX: Secretaría de Gobernación. https://www.dof.gob.mx/nota_detalle.php?codigo=5549317&fecha=31/01/2019.

Schoenfuss, T.C., and Chandan, R.C. 2011. Dairy ingredients in dairy food processing. In *Dairy Ingredients for Food Processing*, eds. R.C. Chandan and A. Kilara, 421–72. Ames: Wiley-Blackwell. https://doi.org/10.1002/9780470959169.ch17.

Schouten, R. 2020. Slideshow: The trends culturing yogurt innovation. https://www.foodbusinessnews.net/articles/16346-slideshow-the-trends-culturing-yogurt-innovation.

Seyhan, E., Yaman, H., and Özer, B. 2016. Production of a whey-based functional beverage supplemented with soy isoflavones and phytosterols. *International Journal of Dairy Technology*, 69(1):114–121. https://doi.org/10.1111/1471-0307.12229.

Shah, N.P., and Champagne, C.P. 2016. Cultured milk and yogurt. In *Dairy Processing and Quality Assurance*, eds. R.C. Chandan, A. Kilara, and N.P. Shah, 235–265. Oxford: Wiley Blackwell.

Silva, M.R., Dias, G., Ferreira, C.L., Franceschini, S.C., and Costa, N.M. 2008. Growth of preschool children was improved when fed an iron-fortified fermented milk beverage supplemented with *Lactobacillus acidophilus*. *Nutrition Research*, 28(4):226–232. https://doi.org/10.1016/j.nutres.2008.02.002.

Smithers, G.W. 2015. Wheying up the options – Yesterday, today and tomorrow. *International Dairy Journal*, 48:2–14. https://doi.org/10.1016/j.idairyj.2015.01.011.

Souza, F.P., Celso, F.B., Guimarães, J.T., Pimentel, T.C., Esmerino, E.A., Freitas, M.Q., Raices, R., Silva, M.C., and Cruz, A. 2019. The addition of xyloligoosaccharide in strawberry-flavored whey beverage. *LWT- Food Science and Technology*, 109:118–122. https://doi.org/10.1016/j.lwt.2019.03.093.

SSA. 2010. Productos y servicios. Leche. fórmula láctea, producto lácteo combinado y derivados lácteos. Disposiciones y especificaciones sanitarias. Métodos de Prueba. Mexico. http://dof.gob.mx/nota_detalle.php?codigo=5160755&fecha=27/09/2010.

Tamime, A.Y., and Robinson, R.K. 2007. *Yogurt Science and Technology*. Cambridge: Woodhead.

Tasneem, M., Siddique, F., Ahmad, A., and Farooq, U. 2014. Stabilizers: Indispensable substances in dairy products of high rheology. *Critical Reviews in Food Science and Nutrition*, 54(7):869–879. https://doi.org/10.1080/10408398.2011.614702.

Technomic. 2019. Trends in dairy: Snacking. realcaliforniamilk.com/foodservice.

Tetra Pak International. 2019a. Introducing clean label yogurt. Designing alternatives to additives. https://www.tetrapak.com/insights/cases-articles/introducing-clean-label-yoghurt.

Tetra Pak International. 2019b. Understanding ambient yoghurt challenges and opportunities. https://www.tetrapak.com/insights/cases-articles/understanding-ambient-yoghurt.

Tetra Pak International. 2020. Best-practice lines for desserts. https://www.tetrapak.com/solutions/processing/applications/prepared-food/desserts.

Thompson, P. 2006. Unique method of baking parmesan cheese and creating a variety of resulting products. US Patent No. US 2006/0110518.

Toma, S.G., and Săseanu, A.S. 2020. Chocolate, a global business. 6th BASIQ *International Conference on New Trends in Sustainable Business and Consumption*, eds. R Pamfilie, V. Dinu, L. Tăchiciu, D. Pleşea and C. Vasiliu, 1233–1240. Bucharest: ASE.

Tzia, C., Giannou, V., and Varzakas, T. 2015. Snack foods. In: *Handbook of Food Processing Food Safety, Quality, and Manufacturing Processes*, eds. T. Varzakas and C. Tzia, 573–583. Boca Raton: CRC Press.

Vaghini, S., Cilla, A., Garcia-Llatas, G., and Lagarda, M.J. 2016. Bioaccessibility study of plant sterol-enriched fermented milks. *Food and Function*, 7(1):110–117. https://doi.org/10.1039/C5FO00458F.

Varnam, A.H., and Sutherland, J.P. 2001. *Milk and Milk Products. Technology, Chemistry and Microbiology.* Gaithersburg: Aspen Publishers, Inc.

Vedamuthu, E.R. 2011. Functional ingredients from dairy fermentations. In *Dairy Ingredients for Food Processing*, eds. R.C. Chandan and A. Kilara, 357–374. Ames: Wiley Blackwell.

Vincenzetti, S., Pucciarelli, S., Polzonetti, V., and Polidori, P. 2017. Role of proteins and of some bioactive peptides on the nutritional quality of donkey milk and their impact on human health. *Beverages*, 3(3):34. https://doi.org/10.3390/beverages3030034.

Walker, K.Z., Woods, J., Ross, J., and Hechtman, R. 2010. Yoghurt and dairy snacks presented for sale to an Australian consumer: Are they becoming less healthy? *Public Health Nutrition*, 13(7):1036–1041. https://doi.org/10.1017/S1368980009992965.

Walstra, P., Wouters, J.T.M., and Geurts, T.J. 2006. *Dairy Science and Technology.* Boca Raton: Taylor and Francis.

Warde, A., and Yates, L. 2017. Understanding eating events: Snacks and meal patterns in Great Britain. *Food, Culture and Society*, 20(1):15–36. https://doi.org/10.1080/15528014.2016.1243763.

Zheng, L., and Sun, D.W. 2006. Innovative applications of power ultrasound during food freezing processes—A review. *Trends in Food Science and Technology*, 17(1):16–23. https://doi.org/10.1016/j.tifs.2005.08.010.

18 Snacks from Animal, Poultry, and Sea Foods

Sergio O. Serna-Saldivar

CONTENTS

18.1 INTRODUCTION

The development of meat- and marine-based snacks have their origin in meat and fish drying practices that were developed in order to prevent spoilage and foodborne illnesses and their convenience when refrigeration was not available. Historically, it is documented that the ancient Egyptian culture about 3000 BC preserved quails or other birds with salting and later on North American and African cultures produced different versions of dried meats. Likewise, smoking was developed as a meat preservation method by the Celts and Romans and fermentation of meats as a common practice among antique Western civilizations. These old preservation techniques are commonly used today to produce shelf-stable animal-based snacks that are perceived as highly nutritional and suited for people adopting low-carbohydrate regimens (Leroy et al. 2013).

Raw meats and marine products mostly consist of water, protein, fat, and minerals and therefore are practically free of carbohydrates, especially when post-*rigor mortis* meats are employed (Serna-Saldivar 2008). From the nutritional viewpoint, muscle tissues from animal and marine resources are considered as one of the good quality proteins because they are highly digested and more importantly contain all essential amino acids required for optimum growth. Additionally, they are excellent sources of trace minerals iron and zinc and most B-vitamins, including B_{12}. The heme iron associated with meat is considered the best bioavailable source for the human system. Unfortunately, the amount of total and saturated fat and cholesterol associated with animal meats have negatively influenced consumption.

Nutritionist and dietitians usually limit the consumption of meat products, especially in patients suffering from gout disease, hypercholesterolemia/lipidemia, and arteriosclerosis. A recent boom in sales and production of meat-based snacks has been observed, especially in developed countries

DOI: 10.1201/9781003129066-18

due to the new low-carbohydrate and high-protein diets. In the United States, sales of puffed and extruded pork rinds totaled US$380 million in 2020 (Wunsch 2020).

The major concern related to animal-based snacks is that they are susceptible to developing pathogenic bacteria that cause foodborne illnesses, which is still considered as the main food security problem worldwide. Fresh meat is an ideal substrate for bacteria due to its chemical composition, high water activity (A_w), and a pH of 5.2–6.5. These factors readily decompose the product and favor the growth of putrefactive and pathogenic bacteria when meats or marine products are not properly stored at low temperatures. Other major concerns about smoked and cured products are the carcinogenic polycyclic aromatic hydrocarbons (PAH) and nitrosamines discussed in Chapter 20.

In order to produce shelf-stable snacks, most products are simply dehydrated or treated with additives to reduce A_w to safe levels. Shelf-stable snacks can also be produced by combined preservation methods, the most common being the combination of A_w by dehydration and pH by the use of acidulants or acid forming bacteria cultures. In addition, meat-based snacks can be purposely treated with antimycotic agents (propionates, sorbates, benzoates), antioxidants (BHA, BHT), and vacuum/modified atmosphere packaged to avoid growth of xerophilic molds which enhance fat rancidity (Lawrie and Ledward 2006).

Due to the traditional role of salting, smoking, and fermentation in the shelf-life of red meats, ready-to-eat meat products such as dry-cured hams, pastrami, and fermented sausages have commonly been used for centuries (Leroy et al 2013). The packaging of these products play an important role to keep their organoleptic properties and microbial shelf life.

Recently, the focus of the meat snack industry has been to improve the quality and stability of traditional snacks or to launch novel products with reduced fat, cholesterol, and sodium and the development of new snacks supplemented with prebiotic fiber sources and probiotics (Lawrie and Ledward 2006, De Vuyst et al 2008, Decker and Park 2010, Kumar et al 2019).

The objective of this chapter is to describe the most common industrial processes to obtain shelf-stable and ready-to-eat snacks from domestic and game animals, poultry, and marine sources.

18.2 MICROBIAL STABILITY OF MEATS AND MARINE PRODUCTS

Raw and processed meats, fish, and crustaceans are considered as highly perishable foods because of their high moisture, adequate pH, and proteins that are very prone to microbial degradation. The high moisture make these raw products to contain an A_w higher than 0.98, which practically favors the growth of any type of microorganism. Besides, practically all of these products have a neutral pH which also favors the growth of most microorganisms. This is the main reason why these foods are considered among the most prone to develop putrefactive and pathogenic bacteria. The strategies to preserve these products are via dehydration, storage at refrigeration or freezing temperatures, and storage under vacuum or controlled atmospheres. In addition, products can be fermented with natural or starter cultures in order to increase acidity or lower the pH and add antimicrobial agents like organic acids, preservatives, and depressants of water activity. Figure 18.1 summarizes the three most relevant intrinsic food and three extrinsic factors that greatly affect the susceptibility of meat and marine products to microbial deterioration. Table 18.1 summarizes the main types of pathogenic bacteria with their optimum range of pH, A_w, and temperature responsible for the majority of the foodborne illnesses associated with animal, poultry, and seafood products. These bacteria can be inactivated at acidic pH, A_w lower than 0.85 and most when exposed to refrigeration temperatures, and storage under low-oxygen concentrations. Most meat-based snacks are expected to be stable at ambient temperature and therefore are produced with the combination of factors mentioned before. There are clear synergistic effects among pH, A_w, oxygen availability, and use of antimicrobial agents like nitrates and preservatives (propionates, benzoates, sorbates); so product developers employ

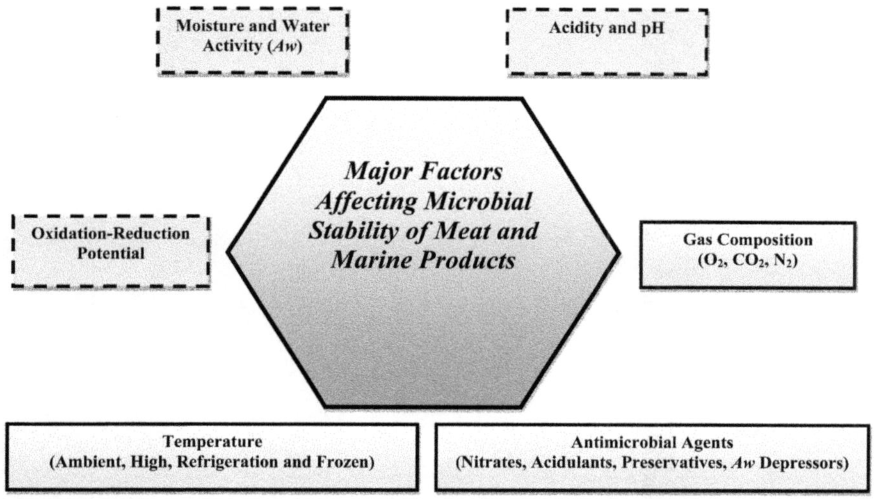

FIGURE 18.1 Major intrinsic (dashed rectangles) and extrinsic (solid rectangles) factors that affect the stability, shelf life, and susceptibility to pathogenic bacteria of fresh and processed meats and marine products.

hurdle technologies to prevent microbial growth at ambient temperature and optimize their sensorial properties. Most snacks are produced via dehydration to lower the moisture and A_w to safe levels or by the combination of partial dehydration with the use of salt, curing agents, acids (intentionally added or naturally produced via fermentation), and preservatives.

Recently, *Listeria monocytogenes* has received special attention because this pathogenic bacterium (Table 18.1) can grow at refrigeration temperature and in intermediate-moisture acidic meats and even dehydrated meats (Gormley et al 2010). The survival of this bacterium was evaluated by Ingham et al (2004) on 15 RTE meat products obtained via drying, fermentation, and/or smoking. The products included beef jerky, snack stick, summer sausage, smoked cured beef, and pork rinds. The A_w of the products ranged from 0.27 (pork rinds) to 0.98 (smoked cured beef slices). Products were inoculated with *L. monocytogenes,* repackaged under either vacuum or air, and then stored either at room temperature (21.8°C) or under refrigeration (5.8°C) for 4–11 weeks. Counts of *L. monocytogenes* fell for all products during storage, ranging from a decrease of 0.8 log CFU on smoked cured beef slices during 11 weeks under vacuum at refrigeration to a decrease of 3.3 log CFU on a pork rind product stored for five weeks under air at room temperature. The smoking, cooking, and drying processes used to make the products can be considered effective antimicrobial processes under the USDA regulations mandating control *of L. monocytogenes* on RTE meat and poultry products.

Raw marine and crustacean tissues are more prone to bacterial decomposition and contamination with foodborne pathogens than red meat and poultry products. These tissues are ideal substrate for bacteria due to its chemical composition, high moisture, A_w, and a pH near neutral. These factors readily decompose marine product and can favor the growth of pathogenic bacteria, especially when it is not properly stored under refrigeration or suffer from temperature abuse (Nilsson and Gram 2002).

18.3 MEATS AND POULTRY SNACK PRODUCTS

18.3.1 Fresh Composition of Meats and Poultry Products

The typical composition of different types of raw meats from domestic and game animals and poultry/fowl are depicted in Tables 18.2 and 18.3 (USDA 2021). The average composition of raw beef

TABLE 18.1
Characteristics and Growth Conditions of the Main Pathogenic Bacteria that Can Contaminate Different Types of Meat, Poultry, and Marine Products and Cause Food-Borne Illnesses[a,b]

Pathogenic Bacteria	Classification of Bacteria	A_w	pH		Temperature °C	
			Minimum	Maximum	Minimum	Maximum
Bacillus cereus	Facultative anaerobe	0.92–0.93	4.3–5	8.8–9.3	4.3	55.0
Campylobacter jejuni	Microaerophile	0.98	4.9	9.5	3.0	45.0
Clostridium botulinum (nonproteolytic)	Anaerobe	0.97	5.0	9.0	3.3	45.0
Clostridium botulinum (proteolytic)	Anaerobe	0.93–0.94	4.6	9.0	10.0	48.0
Clostridium perfringens	Anaerobe	0.93–0.97	5.0	8–9	10.0	50–52
Escherichia coli	Facultative anaerobe	0.95	4.4	9.0	7.0	44–49.4
E. coli O157:H7	Facultative anaerobe		4.5		8.0	44–45
Listeria monocytogenes	Facultative anaerobe	0.92	4.3	9.4	–0.4	45.0
Salmonella spp.	Facultative anaerobe	0.94–0.95	3.7	9.5	5.2	46.2
Shigella sp.	Facultative anaerobe	0.96	4.8	9.3	6.1	47.1
Staphylococcus aureus	Facultative anaerobe	0.83–0.90	4.0	10.0	7.0	49.0
Yersinia enterocolitica	Facultative anaerobe	0.94	4.2	10.0	–1.3	42.0
Vibrio cholera	Facultative anaerobe	0.97	5	10.0	10.0	43.0
Vibrio parahemoliticus	Facultative anaerobe	0.94	4.8	11.0	5.0	45.3

[a] From National Advisory Committee on Microbiological Criteria for Foods (2010) and FDA (2021).
[b] All bacteria are strongly inhibited by nitrates and nitrites used as curing salts in hams and many other cured products.

TABLE 18.2
Nutrient Composition of Unprocessed Red Meat Products Used as Raw Materials for Production of Snacks

Nutrient	Beef, Carcass, Choice, Raw	Beef, Chuck, Arm Pot Roast, Raw	Beef, Brisket, Separable Lean and Fat, Raw	Beef, Ground, 75% Lean Meat/25% Fat, Raw	Pork, Carcass, Separable Lean and Fat, Raw	Horse Meat, Raw	Goat, Raw	Lamb, Composite of Trimmed Cuts, Separable Lean and Fat, Raw	Bison, Meat, Ground, Raw	Deer, Meat, Ground, Raw
Water, g	57.26	57.92	59.07	58.16	49.83	72.63	75.84	64.93	64.25	71.15
Energy, kcal	291	294	277	293	376	133	109	229	223	157
Protein, g	17.32	17.75	17.94	15.76	13.91	21.39	20.6	17.84	18.67	21.78
Tryptophan, g	0.201	0.199	0.127	0.065	0.16	0.265	0.306	0.208	0.141	0.192
Lysine, g	1.443	1.476	1.627	1.296	1.23	1.823	1.532	1.576	1.732	1.756
Methionine, g	0.43	0.454	0.501	0.394	0.347	0.473	0.552	0.457	0.505	0.505
Fat, g	24.05	24.16	22.18	25	35.07	4.6	2.31	16.97	15.93	7.13
Saturated, g	9.75	10.31	8.95	9.48	12.44	1.44	0.71	8.19	6.80	3.36
Monounsaturated, g	10.47	10.73	9.50	10.96	15.93	1.61	1.03	6.91	6.22	1.34
Polyunsaturated, g	0.92	0.96	0.84	0.608	3.80	0.65	0.17	0.72	0.75	0.394
Cholesterol, mg	74	70	81	75	74	52	57	92	70	80
Calcium, mg	8	7	16	21	19	6	13	13	11	11
Iron, mg	1.83	2.01	1.59	1.79	0.69	3.82	2.83	1.43	2.6	2.92
Sodium, mg	59	57	59	67	42	53	82	74	66	75
Thiamin, mg	0.08	0.104	0.065	0.044	0.595	0.13	0.11	0.133	0.13	0.547
Riboflavin, mg	0.16	0.169	0.13	0.144	0.207	0.1	0.49	0.261	0.227	0.287
Niacin, mg	3.54	3.106	4.041	3.804	3.846	4.6	3.75	4.934	4.91	5.7
Folate, mcg	7	7	10	8	4	0	5	0	11	4
Vitamin E, mg	0	0	0.44	0.45	0	0	0	0	0.24	0.45

Source: USDA (2021).

TABLE 18.3
Nutrient Composition of Unprocessed Poultry/Bird Products Used as Raw Materials for Production of Snacks

Nutrient	Chicken, Broilers or Fryers, Meat and Skin, Raw	Turkey, All Classes, Meat and Skin, Raw	Ostrich, Ground, Raw	Duck, Domesticated, Meat and Skin, Raw
Water, g	65.99	70.4	71.07	48.5
Energy, kcal	215	160	165	404
Protein, g	18.6	20.42	20.22	11.49
Tryptophan, g	0.207	0.226	0.18	0.144
Lysine, g	1.509	1.857	1.785	0.912
Methionine, g	0.493	0.574	0.565	0.291
Fat, g	15.06	8.02	8.7	39.34
Saturated, g	4.31	2.26	2.177	13.22
Monounsaturated, g	6.24	2.9	2.654	18.69
Polyunsaturated, g	3.23	1.98	0.91	5.08
Cholesterol, mg	75	68	71	76
Calcium, mg	11	15	7	11
Iron, mg	0.9	1.43	2.91	2.4
Sodium, mg	70	65	72	63
Thiamin, mg	0.06	0.064	0.182	0.197
Riboflavin, mg	0.12	0.155	0.267	0.21
Niacin, mg	6.801	4.085	4.377	3.934
Folate, mcg	6	8	7	13
Vitamin E, mg	0.3	0.35	0.24	0.7

Source: USDA (2021).

(*Bos taurus*, *Bos indicus*), pork (*Sus scrofa*), lamb (*Ovis aries*), goat (*Capra hircus*), and chicken (*Gallus gallus*) meats is about 65–75% moisture, 15–20% crude protein, 10–35% fat, and 1% ash. The composition is affected by the type of species, age, breed, type of muscle, degree of fattening, and other less relevant factors. Meats from turkey (*Meleagris gallopavo*), ostrich (*Struthio camelus*), horse (*Equus caballus*), and game animals such as deer (*Cervus elaphus*) and bison (*Bison bison*) are popular because of their comparatively lower fat contents than pork and beef meats (Tables 18.2 and 18.3).

18.3.2 Dehydrated Snacks Derived from Meats and Poultry

The principle of drying meat is the reduction of the water activity (A_w) to levels of less than 0.7, which do not support the growth of putrefactive and pathogenic microorganisms. The most economical ways to dehydrate meat products is by solar or artificial drying with hot convective air of small or thin slices to speed up the process and save energy. Dehydration aimed to prevent microbial growth and contamination increases fat oxidation because this reaction is faster when high temperatures are used and due to the exposure of meat slices to oxygen and ultraviolet light (Chapter 7). In order to prevent fat oxidation, artificial or natural antioxidants such as butylated hydroxy anisole (BHA), butylated hydroxy toluene (BHT), tocopherols, or oleoresins can be used (Chapter 7). In general, most spoilage and pathogenic bacteria do not grow at A_w values below 0.92, and spoilage yeasts and molds at <0.90 and 0.80, respectively (Bechtel 2001, Serna-Saldivar 2008, Santchurn et al 2012).

The drying phenomenon is a well-understood physical process (Chapter 2) where the meat product is normally exposed to hot air, which evaporates moisture from the surface by convective mass transfer. The evaporation rate is primarily governed by the temperature, relative humidity of the air stream, and air velocity. The water evaporation will proceed at a constant rate while there is free water at the surface. As water is removed from the surface, there will be established a moisture gradient from the interior to the exterior through a combination of liquid diffusion, vapor diffusion, molecular motion, and osmosis. As the meat gradually dehydrates through the movement of the interior water to the surface, the moisture concentration gradient or drying rate will diminish. This part of the drying process is known as "diffusion governed" or "falling rate." Drying will proceed at a progressively decreasing rate until equilibrium is reached, at which stage the meat product is said to have reached the "*equilibrium moisture content*" (Doe 2002). These principles should be considered in order to optimize the properties of dried meat products.

An array of different dehydrated meat snacks with long shelf life at room temperature are available on the market as premium quality products. Table 18.4 summarizes the main types of dehydrated shelf-stable meat products produced in different parts of the world. Most of these products are manufactured from beef, followed by poultry and game meat. These products are popular among several segments of the population because they contain high amounts of good quality protein, iron, and B-vitamins, including cyanocobalamin or B_{12} (Chapter 20). The sources vary in total fat, saturated to unsaturated fatty acids (FA), and cholesterol. Interestingly, among game meat, ostrich is gaining acceptance because it is recognized as a dietetic and tasty type of meat (Bechtel 2001, Serna-Saldivar 2008, Santchurn et al 2012). Zdanowska-Sąsiadek et al (2018) recently compared dehydrated meat snacks made from ostrich, beef, and chicken meat in relation to their nutrients, including anserine, carnosine, heme iron, and FA profile (Chapter 20).

Typical formulations for manufacturing several types of dehydrated meat products are depicted in Table 18.5. The meat is dehydrated enough to lower the A_w to a level where most microorganisms cannot grow. Generally, muscles from the beef round, shoulder, and knuckle are preferred to manufacture dried meat. After two to three days of salting and curing, the meat is dehydrated for two to three days at temperatures 30–40°C.

The Native Americans called dried meat *charqui* that was later anglicized into jerky (Table 18.4). Jerky is the most common ready-to-eat snack from animal origin and one of the oldest snack items. Today different types of jerkies are produced: beef, venison, buffalo, bear (*Ursus* sp.), emu (*Dromaius novaehollandiae*), elk (*Cervus elaphus*), poultry, fish, and beef-soy (*Glycine max*) protein concentrate. Commercially packed jerky can be kept for one year in storage at ambient temperature. Modern jerky is a nutrient-dense dehydrated meat product that is shelf-stable and normally prepared following five sequential steps: strip preparation, marinating, and interventions such as acid drips, lethality treatments, and drying (Figure 18.2). The USDA Food Safety and Inspection Service (FSIS) determined that meat and poultry jerky might not be adequately processed to achieve the lethality necessary to produce a safe product. FSIS has identified two critical control points where producers need to focus and ensure the wholesomeness of the product. First, jerky should be adequately heat-treated to meet the lethality performance standards if the requirement for moist cooking is not achieved. Second, FSIS became aware that some processors rely upon the maximum moisture:protein ratio rather than A_w for determining whether the process adequately dries the product to yield a shelf-stable item. While a moisture:protein ratio of 0.75:1 or less remains part of the standard of identity, A_w should be assessed to verify if jerky was properly dried. Needless to say, the jerky process must eliminate potential microbial hazards that can produce foodborne illnesses. The main pathogens of concern listed in Table 18.1 are *L. monocytogenes*, *Salmonella* sp., *Staphylococcus aureus*, and *Escherichia coli*. *Listeria* is of particular concern because of recent outbreaks from contaminated jerky. The HACCP (Hazard Analysis and Critical Control Points) and SOP (Standard Operating Procedures) programs should be aimed toward the control of the growth of *L. monocytogenes* (American Meat Institute Foundation 1994, Serna-Saldivar 2008)

TABLE 18.4
Features of Traditional Dehydrated and Shelf-Stable Meat Products Produced Throughout the World

Traditional Name	Region	Processing and Product Characteristics	References
Charque or cherqui	The United States	Charqui or jerky is a beef product that should have less than 45% moisture and 15% of mineral residue with A_w ranging between 0.70 and 0.75. Manufacturing cuts for charque must be carefully prepared after deboning to produce muscle pieces of uniform thickness. The wet salting or brining operations is carried out in vats of approximately 80 cm depth. A brine strength should be between 95° and 100° and should be maintained between 15° and 20°C. The immersed meat pieces are agitated vigorously, by the use of paddles, for a period of 50 min. Then, the slices of meat are dry-salted on concrete floors covered with a 1-cm thick layer of salt. The floor should be slightly inclined, falling away into lateral channels to carry away the meat juice expressed during salting. The meat pieces are stacked into piles of less than 1.5 m, separated from each other by layers of coarse marine salt (1 mm thick). In addition, each pile should be capped with a 2-cm covering of salt. The pile is restacked after 8 h applying additional layers of coarse salt between each successive meat layer every 16 additional hours before being repositioned into its original order. The piles are remade every 24 h during four to five days in order to ensure even salt penetration and water loss. Before sun or artificial drying, the meat pieces are washed with acidified water (pH 5.5) to remove excess salt adhering to the surface. For sun drying, the slices of charque are commonly placed on wooden rails positioned parallel in runs for a maximum period of initial drying of 4–8 hours preferably at temperatures that do not exceed 40°C on the meat surface. The meat pieces are exposed to the sun each day over a period of four to five days	Serna-Saldivar 2008, Huang, and Nip, 2001,
Bitlong	South Africa	Biltong is a dried uncooked meat product. The basic formulation shown in Table 18.4 consists of preferably low-fat or lean beef, salt, spices, acetic acid or vinegar, salt pepper or curing agents, and antimycotic agents such as sorbates and/or benzoates. Biltong is dehydrated to lower its moisture to about 25% and generally sold in sticks or slices	Serna-Saldivar 2008,
Kilishi	Nigeria	Prepared from the hindquarters of beef carcasses, but mutton and goat meat are also employed. Lean cuts of meat are sliced into 17–20 mm thickness and an average length of 140 cm in preparation for effective sun drying at 30–36°C on a raised bed for 4 h. Then the slices are immersed in a slurry of groundnut cake powder and seasonings for 1 h and redried in the sun under similar conditions for 3–12 h to lower the moisture to 20–25%. The product is finally roasted over a fire for 3–5 min with final moisture of 12–15%	Huang and Nip, 2001
Balanqu	Africa	This dehydrated meat product is produced from boneless slabs of lean/organ meat from beef, mutton, or goat seasoned with salt, spices, groundnut flour, and oil, and roasted over a low-burning smokeless fire. The meat is boned and cut into chunks that are sliced into slabs of 4–6 cm. The meat slabs are dusted on both surfaces with the seasoning mixture before the roasting that last for 30–60 min	Huang and Nip, 2001

(Continued)

TABLE 18.4 (CONTINUED)
Features of Traditional Dehydrated and Shelf-Stable Meat Products Produced Throughout the World

Traditional Name	Region	Processing and Product Characteristics	References
Dendeng	Indonesia	This intermediate-moisture meat product is prepared from beef, chicken, pork, or fish, containing coconut sugar, salt, and spices. There are two major versions: the Dendeng sayat prepared from thin slices of meat and the dendeng giling prepared from ground meat. Dendeng has a sweet taste due to its high sugar content and a characteristic flavor. The preparation consists of first slicing the meat to about 3-mm thick followed by soaking in a blend of coconut or palm sugar, salt, and spices consisting of coriander, tamarind, garlic, and galangal root, for up to 12 h at ambient temperature. The cured meat slices are placed on trays to be sun-dried (dendeng sayat) or artificially dried to lower the A_w to 0.52–0.67 or 26% moisture. For dendeng giling, the minced meat is mixed with ground spices, coconut sugar, and salt and then rolled into thin sheets 3-mm thick, and put on bamboo trays to be sun-dried	
Bak kua, rou pu	Malaysia	This spicy pork meat product is sliced to 2–3 mm and supplemented with a high sugar content. Traditionally, this product is made from pork ham or picnic shoulder. The meat is first sliced and then soaked in a curing mixture made of salt, sugar, soy sauce, monosodium glutamate, spices, and coloring agents. After curing for about 2 h, the slices are placed on slightly oiled bamboo trays at 40–70°C for several hours until the moisture content drops to 30–40%. After drying, the slices are stacked up and cut into squares. The product is not stable at room temperature, so it should be refrigerated or frozen	Huang and Nip, 2001
Rougan and roupu	China	These are prepared from sliced and cured pork meat that is cut into very thin slices (2 mm) of lean meat in preparation for curing with a mixture of sugar, salt, soy sauce, monosodium glutamate, and spices for 24–36 h at 4°C. Thereafter, the thin slices are placed on an oiled bamboo basket or wire rack and dried at 50–60°C until they lose approximately 50% of their original weight. The partially dehydrated meat is further dried by roasting over charcoal or deep fat fried to give a final A_w of about 0.6. The major difference between roupu and rougan is that the curing of the first product is made with more, sugar so it is sweeter in taste	Huang and Nip, 2001
Shafu	China	The preparation of shafu is based on the principles of rougan, but it is produced with nitrite curing salts. The final shelf-stable product has comparatively higher A_w, lighter color, softer texture, and less sweet	Huang and Nip, 2001
Tasajo	Cuba and Latin America	Intermediate-moisture product made from beef sheets that are soaked in brine, drained, rubbed in salt and layered with salt in vats for several days, washed, and sun-dried	Huang and Nip, 2001
Cecina	Spain, Mexico, and Latin America	Spanish Cecina is salted, cured, dried, and smoked for about 20 days at 15°C and then aged for approximately two months in a cold dry environment. The Mexican counterpart is made from thin slices of lean meat that are sun-dehydrated and covered with a thin layer of safflower (*Carthamus tinctorius*) oil and vinegar	Serna-Saldivar, 2008

TABLE 18.5
Typical Formulations of Dried Meats Commonly Used as Snacks[a,b]

	Dehydrated Meat Snacks		
Ingredient	**Jerkey[c]**	**Beef[c]**	**Bitlong**
Beef, lean	100	100	100
Water		3	
Salt	4	3	4
Sucrose		1	
Vinegar			1
Black pepper	0.50	0.40	
Spices		0.15	
Curing agents (nitrate, nitrite)		0.06	

[a] From Ockerman (1989).
[b] Products are usually dehydrated to lower their moisture content to 15–20% moisture and A_w lower than 0.8. Sometimes formulations are also supplemented with antimycotic agents like potassium sorbate, sodium benzoate, and/or calcium propionate.
[c] Beef meat can be substituted or mixed with elk, venison, buffalo, bear, emu, poultry, and soybean protein concentrate.

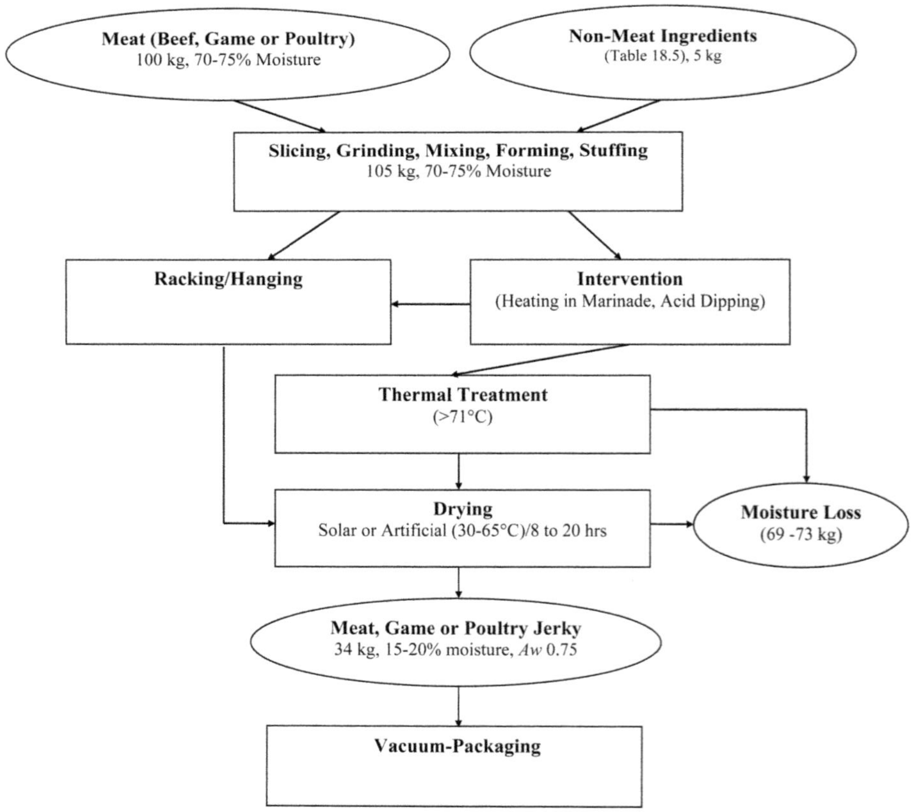

FIGURE 18.2 Diagram of the typical process for production of dehydrated meat snacks like jerky.

The main control points to ensure the elimination of microbial hazards are the lethality and drying steps. The lethality treatment destroys existing pathogens, whereas the drying step reduces A_w to a level that avoids pathogen growth and/or recontamination. The USDA current recommendation for making jerky safely is to heat meat to 71°C before drying. This thermal treatment will inactivate any contaminating pathogenic bacteria. The US Code of Federal Regulations has established an A_w of 0.85 as the critical value for shelf stability of jerky. At this A_w, none of the potentially hazardous pathogens (Table 18.1) can grow and the only remaining microbial issue is molds. The aerobic molds can be controlled with antimycotic agents, vacuum packaging, or preferably by the combination of both. It is worth mentioning that the minimum A_w for *Staphylococcus*, *Listeria*, *E. coli*, and *Salmonella* are 0.86, 0.92, 0.95, and 0.95, respectively (Serna-Saldivar 2008).

Drying of meat has been done for many centuries because it preserves the meat for long periods while maintaining its nutritional attributes. The meat dehydration process is usually combined with smoking and salting, food processes practiced since prehistoric times. There are basically three types of jerky products: strips, ground/formed, and stuffed into sausage configurations. The key of the manufacturing process is to use lean meat and carefully dehydrate the meat to lower the A_w to a safe level; meat usually contains 65–75% moisture. Approximately 3 kg of fresh meat cuts yield about 1 kg of dehydrated meat. Jerky and related snacks are usually expensive because the final product usually contains a moisture:protein ratio of 0.75:1 or less. Generally, jerky products are salted and can in some instances treated with curing agents (nitrates), liquid or natural smoke, flavorings, and humectants that lower the A_w and improve texture and chewiness (Table 18.5).

The typical flowchart of the jerky manufacturing procedure is summarized in Figure 18.2 by the selection of preferably lean meat high in protein in order to manufacture good quality jerky. The selection of mature animals or game animals fed on natural grasslands that yield carcasses with a higher lean:fat ratio is practiced. The fresh meat is normally cut in long strips or ground/comminuted. Whole muscle is sliced, ground, and formed into strips and then marinated in a solution that often contains salt, sugar, flavorings, spices, and curing agents before the subsequent steps of intervention, lethality, and drying. The lethality treatment is a process aimed toward the reduction of the number of pathogenic microorganisms to make the product safe and fit for human consumption. Examples of lethality treatments are dry or wet heating or the use of known antimicrobial agents. Examples of such interventions are preheating the meat or poultry strips in the marinate to achieve a minimum internal temperature of 71°C. Because heating in the marinate may produce unacceptable flavors, sometimes this operation is done in water. The other way to reduce pathogens is dipping the meat strips into a 5% acetic acid solution for 10 min before placing them into the marinate. This intervention may also result in undesirable flavors (Serna-Saldivar 2008).

For jerky strips, meat is first sliced into 2–3 mm-thick strips, which are placed on wire racks in preparation for drying/smoking. For formed jerky, the ground meat is placed in molds, heated and cut into the desired shape before dehydration; while for the sausage type, the meat is stuffed in natural or artificial casings. Dryers or smoke houses are set to a temperature of approximately 55–80°C. According to Hamm and Deatherage (1960), the exposure of fresh meat to temperatures of 20–30°C has no apparent effect on the water-holding capacity (WHC) of meat proteins. Heat treatments at temperatures above 40°C decrease WHC due to the net disappearance of acidic groups at pH values greater than the isoelectric point due to decreased repulsion between negatively charged peptide chains. At pH values less than the isoelectric point, the decrease in negative groups results in greater WHC. The dehydration time varies according to the dimensions of the strips or formed jerky and settings of the dryer, but in most instances lasts more than 20 h. During the drying process, meat undergoes extensive physical and chemical changes. The process is fundamentally a diffusion of water to the meat surface, which is driven by the difference in water vapor partial pressure between the air and the meat surface. Both continuous belt oven and fluidized bed driers are commercially employed to dry meat. The removal of water from the muscle cells as well as between muscle fibers results in stronger hydrophobic interactions of the myofibrils, thereby hardening the texture of meat for desirable chewiness (Xiong and Benjy Mikel 2001). According to Figure 18.2, 100 kg of fresh meat yields approximately 34 kg of 17.5% moisture-based jerky.

Formed jerky is manufactured by blending and massaging chunks of preferably lean meat, salt, various spices, and curing agents (Table 18.5). In some instances, soybean concentrate, which acts as a binder, is included in the formulation before molding. If the level of binder exceeds 3.5%, based on the raw product weight, it is mandatory to declare it on the food label. The resulting piece of formed meat is then cut into strips and dehydrated as explained before. Another version of this snack is the so-called sausage jerky made by mixing the formulation and stuffing it into small-diameter casings before drying. The long shelf life of jerky products is achieved mainly by reducing the A_w by dehydration and addition of salt and other water binders (i.e., glycerin, propylene glycol), by lowering the pH with lactic acid cultures or acidulants such as lactic and acetic acids, and adding antimicrobial agents such as propionates, sorbates, and/or benzoates. The microbial safety of jerky is the utmost important factor in its production because the snack is stored at room temperature and is prone to microbial cross-contamination with pathogens associated with spices and other ingredients. The safety of the product is enhanced with the use of curing agents because most pathogenic bacteria cannot grow in a medium containing nitrates/nitrites (Serna-Saldivar 2008).

Lyophilization or freeze-drying can be alternatively used to dehydrate meats with the main advantage of reducing microbial hazards because the drying process is performed with frozen meat configurations. However, the technology is seldom used because in most instances it is prohibitively expensive (Xiong and Benjy Mikel 2001, Barbut 2001). The comparatively major advantages of freeze-drying are that the original meat configuration is maintained and products do not shrink and maintain a porous internal honeycomb texture, so they can rehydrate more easily.

Another category of dried meat products is those containing higher moisture content but a low A_w achieved by the addition of agents that depress water activity like glycerin, propylene glycol, polysorbates, and other approved water binders. These binders are used in levels of approximately 3% and change the flavor of the finished meat snack. Major concerns in these products are that consumers usually complain about the metallic and sweet flavor imparted by these additives and the possibility of mold growth. Therefore, these intermediate moist meat snacks are frequently treated with antimycotic agents like sorbates, propionates, or benzoates. Moreover, according to Xiong and Benjy Mikel (2001), intermediate-moisture meat products tend to have higher rate of lipid oxidation and rancidity.

18.3.3 Intermediate-Moisture Sausages and Dry-Cured Hams

Intermediate-moist sausages and different types of dry-cured hams are also increasing in popularity because of their high-protein and low-carbohydrate contents. Semidried sausages are products based on ground meat supplemented with spices, curing agents, and in many instances with starter cultures that will drop the pH to levels of 5 or less. These products would have moisture contents of less than 50%, a relatively low A_w, an acidic pH, and a moisture:protein ratio of 2:1. Manufacturing of semidry or intermediate-moisture sausage and dry-cured hams are based on historical traditions of the art and science of food preservation. These products are classified according to texture, ethnic origin, meat source or formulation, and chemical composition (Tables 18.6 and 18.7). The art of dry salt curing is a method dating back to prehistoric times and was the first curing method practiced by humans. The principle of this practice is that the moisture is gradually drawn out of the meat by the salt and curing agents leaving the meats drier and harder with a briny and cured flavor (Martin 2012). The partial removal of the moisture that diminishes A_w and the preservative effects of the sodium nitrates and nitrites allow the long preservation of these products without refrigeration.

18.3.3.1 Hams

Dry-cured hams are normally made with pork hams without the shank bone and are not thermally treated. Since dry cured hams are not cooked, the raw meat should be free of the roundworm

TABLE 18.6
Main Features and Key Processing Characteristics of Dry-Cured Hams and Semidry Sausages

Major Process	Products	Manufacturing Characteristics
Dry-cured hams		
Curing	Parma	This popular Italian ham is made from 1-year-old pigs weighing roughly 180 kg raised in northern Italy and processed for a total curing time between 10 and 15 months. After slaughter, the carcasses are frozen and trimmed of subcutaneous fat into a drumstick shape, salted, seasoned, and frozen for eight days. Then, they are massaged to squeeze out the juice, tenderize the meat, and are salted again before deep freezing for 18 days. The hams are allowed to thaw in fresh air for one to two days and placed for one month chilling at 0°C. Then hams are placed in the drying room for approximately three months. Finally, the hams are scoured with warm water, sandpapered to an attractive outer finish
Curing	Virginia (Country)	The Virgina ham is prepared by dry curing. Fresh hams are first thoroughly rubbed with salt and packed in barrels for seven weeks in a dry-cool place. Then hams are rubbed with molasses, brown sugar, black pepper, cayenne pepper, and saltpeter or curing agents. The hams are placed in a cool room for two weeks and finally hanged in a meat house for one-month aging
Salting	Serrano and Iberian (Spanish)	Serrano and Iberian hams are preferably produced with the hind legs of pigs, although some hams are also produced from the front legs (*paleta* or *paletilla*). The Iberian is cured for comparatively longer times and contains more fat that imparts a different flavor profile. The manufacturing process consists of first burning or covering the hams with sea salt or salt for about two weeks to enhance their partial dehydration and prevent the growth of putrefactive and pathogenic microorganisms. Then, the legs are taken out of the salting bins, washed, and treated again with salt. The hams are then placed in a cellar with strict controls of temperature and humidity for 1.5–3 months. Finally, the serrano hams are ripened for 15–18 months, whereas the Iberia hams for up to 36 months. Within this category of hams, the *bellota* or acorn hams are produced from special breeds of pasture pigs that are fed a diet mainly based on acorns and wild herbs
Curing, smoking	Westphalian or German hams	The ham is prepared by rubbing with salt and curing agents. After two weeks of curing, hams are placed in a 22% brine and allowed to cure for 18 more days. The hams are taken out of the brine for a one-month ripening period in a dry cellar. The final step of their preparation consists of smoking with beech wood for one week
Curing	Scotch	Scotch hams are made from fresh-skinned and deboned and manually defatted hams that are treated with a mild curing salts. The cured ham is rolled, tied, and placed in a cellulose casing
Curing and smoking	Chinese	Chinese hams are dry-cured with an ingredient blend of salt, sugar, and sodium nitrate which are rubbed on the ham surface. Approximately 78 g of the curing blend is used per kilogram of meat. The curing mix is applied at three intervals during 30–40 days at refrigeration temperature in order to enhance the penetration of the components. During the second phase, hams are hung in the smokehouse and subjected to cold smoking (21–32.2°C) for two to three days. After smoking, the third or ripening phase consists of aging for six to nine months at temperatures 20–30°C. The characteristic flavor of this ham is the result of the combination of curing, smoking, and prolonged aging

(Continued)

TABLE 18.6 (CONTINUED)
Main Features and Key Processing Characteristics of Dry-Cured Hams and Semidry Sausages

Major Process	Products	Manufacturing Characteristics
Sausages		
Fermented-dried	Italian sausages: salami pepperoni	Cured with nitrites, pre-drying 18–48 h in green room, one to two months in dry room. Final moisture 30–36% with a pH of 4.8–5.0. The moisture:protein ratio ranges from 1.6–1.9:1
Fermented semidried smoked	Beef Lebanon	Contain salt, sugar, and curing agents. Smoking at 40°C for approximately six days. Final moisture 50–56%. pH 4.7. Moisture:protein ratio 2–3.7:1
Cured, fermented and flavored with fenugreek, salt, and spices	Pastrami	Preferably manufactured from the whole *longissimus dorsi* muscle obtained from beef, although it is also produced from pork, mutton, and even turkey meat. The combination of curing and drying makes this product shelf stable. The whole production process generally lasts one month
Fermented semidried cooked/smoked	German: Summer Thuringer Cervelat	Drying at 30–40°C for 18–48 h, cooking to an internal temperature of about 60°C. Final moisture and pH around 50% and 4.9, respectively. Moisture:protein ratio 1–3.7:1
Acidified semidried	Mexican Chorizo	Fresh sausage produced from coarsely ground pork and beef that is highly condimented and red-colored with pepper paste, oregano, garlic, and other spices. The product has an acidic pH due to vinegar addition

Source: Huang and Nip (2001), Xiong and Benjy Mikel (2001), Serna-Saldivar (2008), Gou (2012), Herring and Smith (2012), Martin (2012), Obuz et al. (2012).

Trichinae spiralis. According to Martin (2012), the nitrates or curing salts have many functionalities in cured products, the main ones being the prevention of bacterial contamination, the delay of lipid oxidation, and the generation of the typical color and flavor compounds that characterize these products.

The dry curing process is especially time-consuming and labor-intensive because the curing agents are rubbed onto the meat surface. During the long curing time, the meat gradually dehydrates in a house or site that has controlled or adequate environmental conditions. The production of these dry-cured hams is long. For instance, the total curing time for Parma hams (Table 18.6) lasts 10–15 months.

The old and renowned Virgina ham is also prepared by dry curing. Fresh hams are first thoroughly rubbed with salt and packed in barrels for seven weeks in a dry-cool place. Thereafter, hams are rubbed with molasses, brown sugar, black and cayenne peppers, and saltpeter or curing agents before placing hams in a cold room for two weeks and finally hanged in a meat house for about one month ripening (Huang and Nip 2001, Gou 2012, Martin 2012).

The traditional Serrano and Iberian hams have been produced for many centuries in the Iberian Peninsula by curing only with salt (Table 18.6), although nitrates are authorized as preservatives in the curing process (Morales et al 2006, Lorido, et al 2015). The process starts by the selection of hind legs and their classification in terms of size or weight. The hams are placed in salting bins where they are completely covered in sea salt. Following salting, the hams are washed and hung in post-salting sheds for about two months at temperatures less than 6°C and high air humidity. During this phase, the salt equilibrates within the ham. Then, the Serrano hams enter the curing phase of four to five months where temperatures rise slowly and humidity decreases, enhancing the partial dehydration or shrinkage of the ham pieces. During ripening, proteolysis, lipolysis, chemical or enzymatic oxidation, Maillard reactions, and Strecker degradation take place, contributing

TABLE 18.7
Typical Formulations of Dried-Cured Hams and Various Intermediate-Moisture Sausages Consumed Throughout the World

	Hams				Salami					
Ingredient	Dry Cured	Westphalian	Pepperoni	Cervelat	Cotto	Genoa	Summer Sausage	Lebanon Bologna	Thuringer	Chorizo
Beef, lean			50	35	30.0	35	60	100	60	25
Pork ham	100	100								
Pork trim			50		60.0	30	40		25	50
Pork backfat						35				
Lard				25						25
Beef heart/head/cheek				40	10.0		–		15	–
Water		20			25.0		–			–
Salt	5.1	4	3.4	3.25	2.75	3	2.75	3.5	2.5	2.30
Sucrose	0.9				–		0.50		0.5	
Corn syrup					2.00		–			
Dextrose			0.1	0.50	1.00		1.00	0.75		
Natural red pepper paste										10.00
Paprika			0.25						0.25	
Vinegar										10.0
Oregano										0.50
Fresh garlic			0.03						0.15	1.50
Garlic, powder						0.06				
Mustard					1.00		0.20	0.06	0.25	
Black pepper			0.2		0.20		0.25	0.37		0.20
White pepper						0.25			0.30	
Spices			0.3		0.50		0.20	0.06	0.30	0.20
Commercial starter					–		0.13			–
Curing agents (nitrate, nitrite)	0.07	0.06	0.18		0.25	0.015	0.13	0.045	0.015	–
Sodium erythorbate					0.05		0.05			–

[a] From Ockerman (1989).

to the development of the characteristic flavor and aroma. Intramuscular fat is essential for flavor formation due to the generation of volatile compounds, especially during the long ripening stage (Rivas-Cañedo et al 2021).

The Westphalian or German hams are sliced and eaten raw. Hams are prepared by rubbing with salt and curing agents, placed in a concentrated brine, ripened in a dry house and finally smoked (Table 18.6, Huang and Nip 2001, Serna-Saldivar 2008).

The production of country hams is based on dry curing of uncooked meat and the final products normally shrink a minimum of 18% and contain a total salt content of 4.5–5.5%. Consequently, the flavor, texture, and appearance of country hams differ from common pickle-cured hams. The country hams are salty and more flavorful as they contain lower moisture and a harder and chewy texture. The curing mixture with salt, nitrite, and/or nitrate is divided into three equal portions, which are applied by rubbing around the entire cut at days 1, 7, and 14. It is essential to ensure that the ham surface is adequately rubbed with the cure mixture because the cure penetrates into the ham through the flesh and not through the skin. In order to avoid contamination with pathogenic bacteria and obtain best product quality, country hams are cured under refrigeration with air at a relative humidity of 70–90%. Hams are refurbished two to three times during the 30–50-day total curing time. The optimum curing time depends on the size and weight of hams and normally takes approximately 1.5–2 days per pound (0.454 kg). Following curing, country hams are kept for about 20 more days for salt equalization at an average temperature of 10°C and air relative humidity of 80%. Finally, the hams are hung by the shank or alternatively placed in stockinettes for ripening or aging for six to nine additional months. The temperature of the maturing room varies from 21° to 35°C and the air humidity from 50% to 60% to facilitate gradual dehydration of the country ham for the development of the characteristic firm texture and flavor (Xiong and Benjy Mikel 2001, Gou et al 2012, Martin 2012).

The lysosomal enzymes cathepsins and lipases play an important role in flavor development of dry-cured hams. The first proteolytic enzymes degrade muscle proteins into smaller peptides which impart characteristic flavors, whereas lipases generate free fatty acids and lipoxygenases oxidize fats into an array of volatile and flavorful compounds. Frequently, molds will grow on the surface of dry-cured hams, especially during the salt equalization and aging stages. These molds are generally innocuous or harmless and are removed by rubbing with a fabric dampened with edible oil. The matured hams normally contain moisture 50–60% and an average salt content of 5% (Chapter 20). There are other sorts of dry-rubbed hams made in other countries that are more or less similar to country hams made in the United States like the popular Spanish serrano, Chinese Jinhua, Italian prosciutto, and German Black Forest (Xiong and Benjy Mikel 2001).

According to Martin (2012), the aforementioned dry curing process may also be commingled with brine injection. Hams are normally injected with about 10% of saturated brine and then the dry curing blend rubbed onto the product.

18.3.3.2 Sausages

The first use of fermentation in meat is dated back to the Babylonian culture around 1500 BC. Today, fermented sausages are broadly divided into dry and semidry categories based on the processing method and product features. For both types, lactic acid bacteria grow and ferment products generating an acidic product. Dry and semidry sausages usually contain 30–40% and 40–50% moisture, respectively. The most consumed products are pepperoni and salami (Xiong and Benjy Mikel 2001).

The most critical steps in the fermented sausage production is the time to lower the pH of fresh meat so as to prevent the growth of spoilage and pathogenic microorganisms. The final pH of fermented sausages typically ranges from 4.8 to 5.2 (Xiong and Benjy Mikel 2001, Cocolin and Rantsiou 2012). Lactic acid bacteria can develop and multiply either by "chance or natural inoculation" or by purposely inoculating with a starter culture. In natural fermentation, these bacteria are present in the environment and in many cases a portion of already fermented meat from a previous batch is inoculated to a new lot. The natural fermentation procedure normally takes longer time compared to the inoculation procedure and its main disadvantage is that the bacteria population and type are difficult to control. Therefore, if the initial population of lactic bacteria is small, the meat pH cannot be rapidly lowered, increasing the probability of microbial spoilage and possible growth of pathogens (Table 18.1). Many lactic bacteria from chance inoculation are heterofermentative, and yield acetic acid, alcohol, and gases that harm the organoleptic properties of end-products. This is the main reason why nowadays most commercial producers employ selected starter cultures that are procured in two major forms: frozen concentrate and freeze-dried. These commercial cultures are typically blends of two or more different microorganisms. The most common examples are *Lactobacillus*, *Pediococcus*, *Lactococcus* (all three are homofermentative), and *Micrococcus* (used to reduce nitrate to nitrite).

Organic acidulants, such as encapsulated glucono-delta-lactone and lactic acid, are sometimes mixed with fresh meat to prevent the growth of other undesirable microorganisms and stimulate the growth of the starter culture. In order to produce lactic acid during fermentation, sugars like sucrose or dextrose must be present in the meat formulation. The amount of sugar added to dry or semidry sausages are typically in the range of 0.5–2.0%. The lower the desired pH, the more the sugar needed.

Fermented sausages are salted and cured, and both nitrates and nitrites are used. Salt is needed to facilitate dehydration and impart flavor. Fermented sausages are usually heavily spiced, making the product particularly palatable, whereas nitrates and nitrites stabilize red meat color, contribute to the typical cured flavor development, and inhibit food-poisoning microorganisms. Nitrosamines, which are known carcinogens to laboratory animals, are formed when these curing salts react with secondary amines (Chapter 20). Smoke is also an important component of the preservation of cured meats. In addition, it contributes to flavor and color. Smoke contains a wide variety of organic compounds, including antibacterial and antioxidants such as phenolics and formaldehyde. However, many of the polycyclic aromatic hydrocarbons formed are known to be carcinogens. Smoke may be applied by natural combustion of woods or by the application of liquid smoke by direct adding, spraying, or dipping.

After ground meat is blended with all ingredients, including the starter culture, the mixture is stuffed and subsequently incubated in a closed chamber (sometimes a smoke house) to facilitate fermentation. Temperature of maturing rooms is typically maintained at 21–24°C with a 75–80% relative humidity for dry sausages, and 30–37°C with a 75–80% relative humidity for semidry sausages. The fermentation time, however, is longer for dry sausages (one to three days) than for semidry counterparts (8–20 h). Semidry sausages are normally smoked and end up the process with a relatively high-moisture content (A_w of about 0.95) and therefore require refrigeration to prevent microbial spoilage (Xiong and Benjy Mikel 2001, Serna-Saldivar 2008).

For dry sausage production, the fermented meat is placed in a drying room to allow further dehydration and flavor development. As a general recommendation, the temperature of the drying room should be controlled to 7–13°C and the relative humidity of about 70%. The air of the drying room should be changed periodically to ensure air quality and prevent moisture buildup on the surface of the sausage. The drying time varies considerably, depending on the size (diameter) and type of product. Most dry sausages are aged for somewhere between ten days and three months. Dry sausages are not cooked and most are not smoked. They do not require refrigeration after manufacture because of their low-moisture content ($A_w < 0.91$) and acidic conditions, which act synergistically (Figure 18.1) to preserve the product. Semidry sausages, however, are generally cooked to an internal temperature of at least 68°C following fermentation to ensure the destruction of pathogenic bacteria.

The manufacturing procedures for dry and semidry sausage operations vary among regions, but generally consist of freezing of incoming meats before breaking, grinding, or chopping (Figure 18.3). Freezing is needed to ensure uniform particle size. The optimum grinding temperature is generally of –3°C. The ground and frozen meats are blended with salt, curing agents, sugar, and spices in paddle mixers. Starter cultures are normally added to the blend after the salt, sugar, and spices

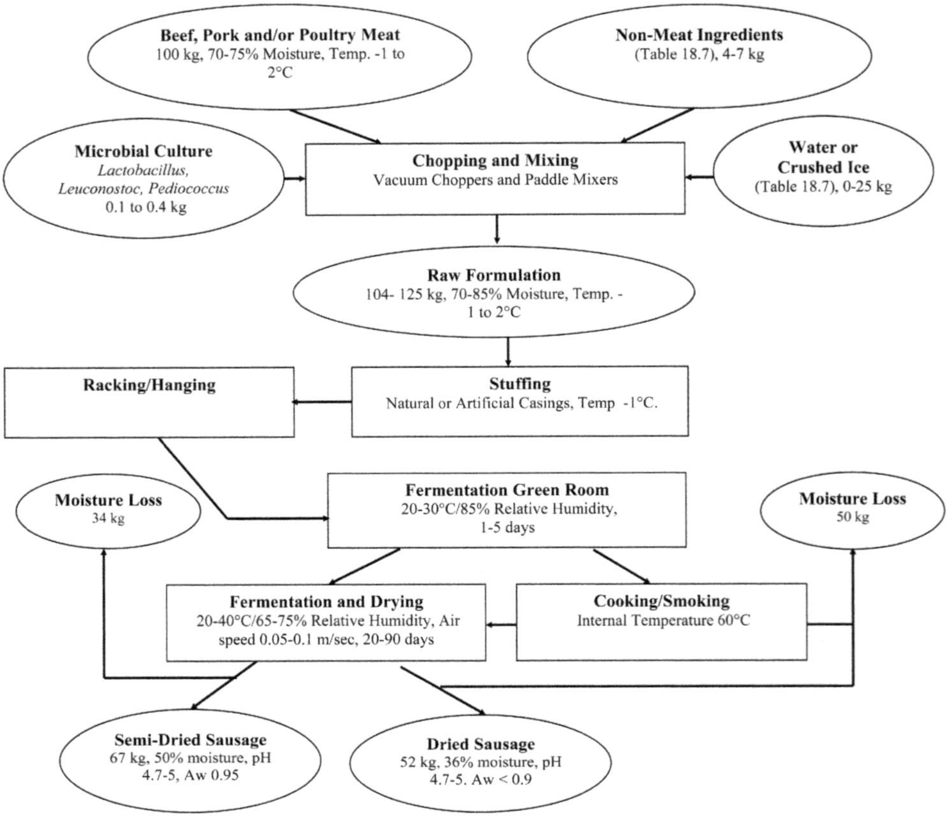

FIGURE 18.3 Diagram of typical processes for production of semidried and dried sausage products.

have been incorporated. After mixing, the product is placed in a green room to initiate fermentation. The product is then stuffed at a temperature of about –1°C to prevent fat smearing. It is important to prevent overmixing and loss of particle size. The use of vacuum mixers helps to minimize both oxidative changes in fats and the loss of color. Sausages have traditionally been stuffed into natural gut casings, although the new trend is to use synthetic casings. Fibrous casings should be soaked in water before stuffing. After stuffing, sausages are hung on sticks and racked in a green or maturing room for fermentation. The product is allowed to ferment for three to five days at temperatures 20–40°C. The length of fermentation depends on the initial culture concentration, room temperature, and formulation (i.e., use of sugar) and products are fermented until achieving a certain pH or acidity level. The control of relative humidity during fermentation is required to prevent excessive case hardening and shriveling. Some products such as summer sausages are smoked and cooked at 65°C until the internal temperature reaches 60°C. For noncooked Italian dry sausages, the product moves from stuffing into the drying room set at the temperature of about 15°C and 70% relative humidity for one to three months. Control of proper moisture loss is accomplished by control of air speed and air relative humidity. At the pre-drying stage, sausages should lose about 1–1.5% and during the drying stage, at less than 0.7% weight daily. Air velocity in the pre-drying room is normally set at 0.5–0.8 m/s, whereas in the drying room at only 0.05–0.1 m/s (Xiong and Benjy Mikel 2001). Yields of semidried and dry sausages vary according to formulations, but in general terms 100 kg of raw meat(s) yield 76 and 56 kg, respectively (Figure 18.3).

Intermediate-moisture sausages can be classified according to the particle size or degree of grinding into three categories: coarsely ground, emulsion type, and muscle cube. Most of the sausages belong to the emulsion type. Emulsion type sausages greatly depend on the functionality of the myofibrillar proteins that in the presence of salt produce binding and a good and stable emulsion. Typical formulations of popular semidried sausages around the world are presented in Table 18.6 (Ockerman 1989).

Sausages from Germany are normally smoked and sometimes cooked, Italian types are not cooked or smoked, and Lebanon is a raw sausage that is heavily smoked. Generally, San Francisco- and Hungarian-style salamis are not cooked and identified by the presence of surface molds. The Lebanon bologna is a semidry coarse-ground naturally fermented product produced with starter cultures (Tables 18.5 and 18.6). Beef is salted and held for eight to ten days at temperatures of 1–3°C. The meat is ground and mixed with salt, sugar, curing agents, and spices and stuffed into casings in preparation for cold smoking for up to seven days. The product is extremely stable due to the high-salt content and acidic pH (<5).

Pastrami (Table 18.6) from the Romanian word "*pǎstra*," meaning "to preserve," is a traditional shelf-stable meat product normally obtained from the whole *longissimus dorsi* muscle of beef, although pastrami is also produced from pork, mutton, and even turkey meat. The key for its stability at room temperature is the combination of curing and drying which lowers the A_w to levels 0.85–0.91. The first step of the process consists of dry curing the raw meat pieces with a curing blend of nitrate/nitrite, sucrose, and glucose. After dry curing, the meat pieces are washed to remove the curing mixture from the surface and air-dried until the targeted moisture content. The meat product is covered with fenugreek (*Trigonella foenum-graecum*) paste and dried again in preparation for ripening. The spreadable fenugreek paste containing red pepper, garlic, salt, and water is the most relevant ingredient because it imparts the characteristic taste and aroma and also protects the product from undesirable microorganisms. The whole production process of pastrami generally lasts one month (Obuz et al. 2012, Karabıyıklı et al. 2015).

Semidry and dry sausages contain moisture contents of about 50% and 30–36%, respectively (Figure 18.3). The pH range of these products range from 4.7 to 5 and titratable acidity (lactic acid) may range from 0.4% to 1.5%. Most of these products are shelf-stable due to the concomitant effect of acids produced by fermenting lactic acid forming bacteria, addition of salt and in some instances curing agents, and the partial dehydration that lowers the A_w. The starter culture requirements for use in these products include salt tolerance and fast growth in 6% brine, ability to grow in the presence of curing agents or nitrates, optimum growth at 30–40°C, lack of lipolytic and proteolytic activities, and inactivation at temperatures of approximately 60°C. The lactic acid helps to protect the product, especially during the initial phase of drying. As acid is being produced, the pH of the system gradually drops near the isoelectric point of meat proteins when they exhibit the lowest water-holding capacity.

The new trend of the industry is to produce sausages with special strains of bacteria that also produce significant amounts of bacteriocins such as nisin, enterocin, pediocin, pentocin, and sakacin. Bacteriocins are ribosomal-synthesized antimicrobial peptides that act against Gram-positive and Gram-negative pathogenic and/or deteriorating bacteria. The bacteria produced carry specific immune mechanisms that protect it from its own bacteriocins. The most studied bacteriocin-producing microorganisms are lactic acid bacteria (*Lactococcus*, *Enterococcus*, *Pediococcus*, and *Lactobacillus* genera), as they have great potential application in food biopreservation, since the majority have *Generally Recognized as Safe* or GRAS status. The bacteriocins have strong inhibitory effects which act against *Staphylococcus* sp., *E. coli*, *L. monocytogenes*, *Bacillus cereus*, *Clostridium perfringens*, and *Salmonella typhimurium* (De Martinis et al 2002, Nilsson and Gram 2002, Woraprayote et al 2016, da Costa et al 2019).

18.3.4 Pork Rinds and Cracklings

Pork rinds and cracklings have been manufactured for many years and recently have regained popularity due to the adoption of low-carbohydrate, keto, and Atkins diets. This unique type of snack category grew 24.3% in dollar sales in the US market for the 52 weeks ended June 2019 (https://www.bakingbusiness.com/articles/49992-the-proliferation-of-pork-rinds).

The flavor of pork puffed or expanded pork rinds is rather bland and is affected by the tap of frying oil during the process, whereas the texture of the fried products is crisp and friable. The pork rinds practically free of starch and other carbohydrates are not hygroscopic as many other cereal-based expanded snacks. Pork rinds are rich in protein, but the protein quality is low because most proteins are from connective tissue (collagen and elastin) practically free of essential amino acids lysine and tryptophan (Mertz 1993, Serna-Saldivar 2008).

18.3.4.1 Production

The raw materials for pork rind production are green belly skins, green fat hog skins, and green harn skins from any type of hog. Belly skins are considered the most adequate feedstock for production of quality end-products. The raw pork skins containing an average of 63% moisture, 25% protein, and 10% fat are mechanically removed from the carcass with a Townsend Derinding Skinner machine and normally dipped for 30 s in brine at near boiling temperature. The brine containing about 9% (w/w) salt in water is usually mixed with 1.5% glucose and 1.5% sucrose. The thermal treatment ensures the destruction of both pathogenic and putrefactive bacteria (Matz 1993, Serna-Saldivar 2008).

The skin is then cut or diced into small squares. The dicer will cut frozen skin into uniform pieces in automatic cuber or chipper. The dimensions of the resulting pellets are controlled by the spacing of the cutting blades. Next, the skins are rendered with lard at 110–120°C in a steam-jacketed kettle equipped with propeller-type stirrers for about 4 h. Rendering decreases the dimensions of the squares in about 50% and moisture and fat contents concentrating the connective tissue. It is important to keep the pork skins submersed during the heat-rendering step. Then, pellets are removed from the hot lard and allowed to drain excess lard and cool down. The resulting pellets are scalped or sized by screens in order to remove smaller pellets. The sized pellets or half products are stored in plastic bags and maintained under refrigeration. Pork rinds are obtained after rehydrating the pellets in a solution containing flavorings followed by deep-fat frying for approximately 60 s at temperatures near 200°C (Figure 18.4). Pork lard is the preferred frying fat because it allows the setting of the fryer to a higher temperature when compared with vegetable oils. During the first stage of frying, the pellets expand to form a light-density product that entraps air and tends to float on the fryer. The higher the fat temperature, the higher the expansion rate. The exiting pork rinds are drained and immediately flavored in a tumbler with salt-based flavorings based on hot pepper, citric acid, barbecue, or powdered cheese. The yield of pork rinds based on 100 kg of raw pork skins is about 40% (Serna-Saldivar 2008). Resulting pork rinds are mainly composed of protein and fat absorbed during the frying step. The packaging of the pork rinds should protect the product from breaking, moisture absorption, and fat oxidation (Chapter 4). Kõrge and Laos (2019) recently concluded that the packaging of pork rinds in PET/PETmet/PE of 74 µm and PP/metPP of 50 µm with nitrogen in the headspace prevented rancidity and textural changes during 120 days of storage.

A low-fat, high protein, carbohydrate-free, and easy-to-process microwavable pork rind snack has been launched on the market. Pellets are packaged in similar bags with susceptors used for microwavable popcorn (Chapter 4). The low fat pork rinds are expanded in the microwave after 2–3 min cooking at full power. Resulting expanded pork rinds contain 70% protein and only about 10% fat and 70 mg cholesterol/100 g serving. The detailed composition of this product is covered in Chapter 20.

18.4 MARINE SNACK PRODUCTS

18.4.1 Fresh Composition of Seafoods

The average composition of different types of raw fish and shrimp depicted in Table 18.8 (USDA 2021) clearly indicates that the main constituents of these products is water (70–80% moisture) followed by protein (15–20% protein). Therefore, the flesh from crustaceans and fish is highly susceptible

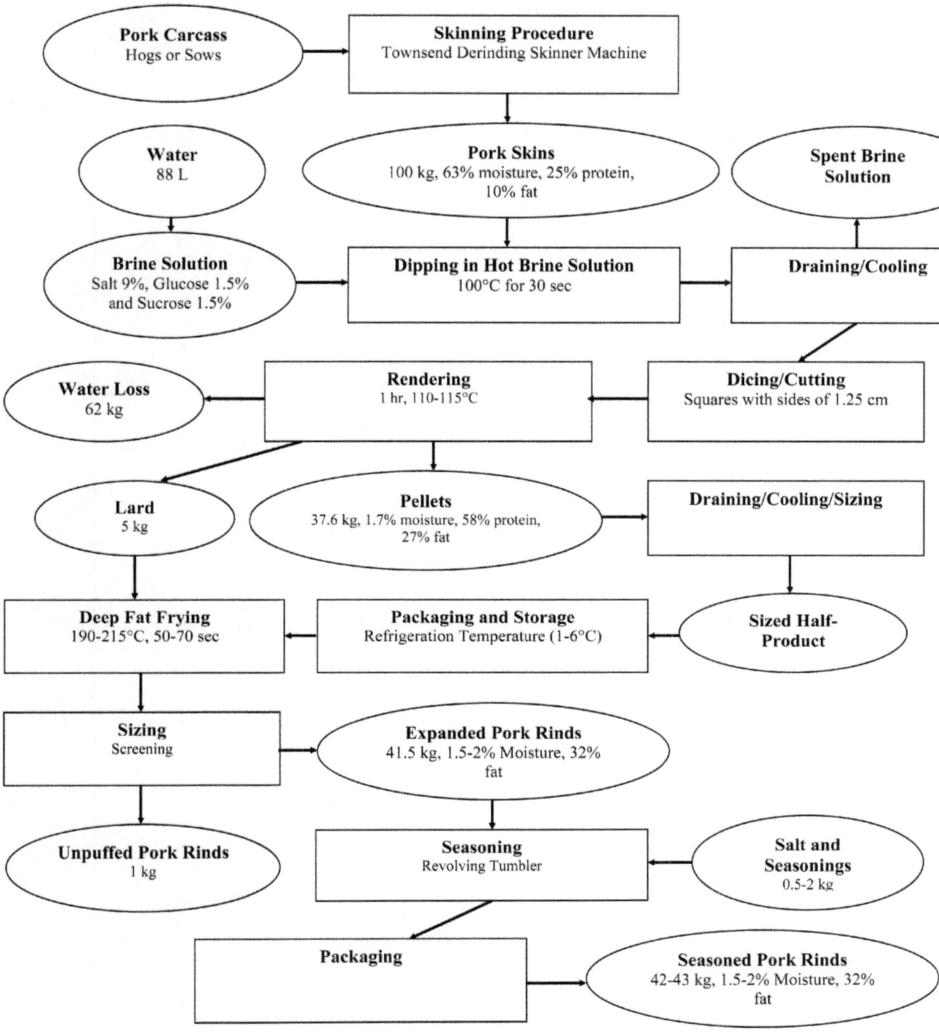

FIGURE 18.4 Diagram of the process for production of puffed and seasoned pork rinds.

to microbial contamination, including troublesome pathogenic bacteria (Table 18.1). One relevant aspect related to marine products is that they are practically carbohydrate-free and contain lower lipids compared to most meat products. Species that contain low fat in their tissues are preferred for production of snacks. This is to reduce rancidity and off-flavors during prolonged storage. More importantly, the lipids contain a favorable PUFA profile that protect humans against cardiovascular diseases. Some species like salmon contain significant amounts of ω-3-dodecahexahecanoic acid known as DHA considered as highly nutraceutical (Chapter 21).

One major difference between marine fats and red meat fats is the amount of polyunsaturated fatty acids. Marine sources contain higher amounts of polyunsaturated fatty acids and are the major source of long-chain PUFA such as EPA and DHA (Chapter 7).

18.4.2 Dehydrated Marine Products

Dried marine products are considered important snacks in many coastal areas around the globe, primarily in Japan, China, Korea, England, Scandinavian countries, and the United States. The

TABLE 18.8
Nutrient Composition of Unprocessed Marine Products Used as Raw Materials for Production of Snacks

	Crustaceans	Fish												
Nutrient	**Shrimp**	**Anchovy**	**Bluefish**	**Cod**	**Whitefish**	**Turbot**	**Spot**	**Snapper**	**Salmon**	**Pout**	**Pollock**	**Mackerel**	**Halibut**	**Herring**
Water, g	75.86	73.37	70.86	81.22	72.77	76.95	75.95	76.87	70.24	81.36	81.56	63.55	77.92	72.05
Energy, kcal	106	131	124	82	134	95	123	100	168	79	81	205	110	158
Protein, g	20.31	20.35	20.04	17.81	19.09	16.05	18.51	20.51	21.30	16.64	17.18	18.60	20.81	17.96
Tryptophan, g	0.283	0.228	0.224	0.199	0.214	0.180	0.207	0.230	0.239	0.186	0.192	0.208	0.233	0.201
Lysine, g	1.768	1.869	1.840	1.635	1.753	1.474	1.700	1.883	1.956	1.528	1.578	1.708	1.911	1.650
Methionine, g	0.572	0.602	0.593	0.527	0.565	0.475	0.548	0.607	0.630	0.492	0.509	0.551	0.616	0.532
Lipids, g	1.73	4.84	4.24	0.67	5.86	2.95	4.90	1.34	8.56	0.91	0.80	13.69	2.29	9.04
Saturated, g	0.328	1.282	0.915	0.131	0.906	0.750	1.450	0.285	1.495	0.320	0.164	3.257	0.325	2.040
Monounsaturated, g	0.253	1.182	1.793	0.094	1.996	0.610	1.330	0.251	4.125	0.329	0.125	5.456	0.750	3.736
Polyunsaturated, g	0.669	1.637	1.060	0.231	2.149	0.880	1.090	0.459	1.879	0.032	0.412	3.350	0.730	2.133
Cholesterol, mg	152	60	59	43	60	48	60	37	62	52	71	70	32	60
Carbohydrates, g	0.91	–	–	–	–	–	–	–	–	–	–	–	–	–
Calcium, mg	52	147	7	16	26	18	14	32	6	10	5	12	47	57
Iron, mg	2.41	3.25	0.48	0.38	0.37	0.36	0.32	0.18	0.47	0.28	0.23	1.63	0.84	1.10
Sodium, mg	148	104	60	54	51	150	29	64	47	61	99	90	54	90
Thiamine, mg	0.028	0.055	0.058	0.076	0.140	0.066	0.160	0.046	0.203	0.080	0.065	0.176	0.060	0.092
Riboflavin, mg	0.034	0.256	0.080	0.065	0.120	0.080	0.220	0.003	0.149	0.060	0.058	0.312	0.075	0.233
Niacin, mg	2.552	14.024	5.950	2.063	3.000	2.200	7.000	0.284	5.780	2.100	1.290	9.080	5.848	3.217
Folic acid, mcg	–	–	–	–	–	–	–	–	–	–	–	–	–	–
Vitamin E, mg	1.10	0.57	–	0.64	0.20	–	–	0.50	–	–	0.64	1.52	0.85	1.07

Source: USDA (2021).

production of different versions of dried fish is one of the oldest methods of processing practiced since prehistoric ages. Salting was introduced in the Middle Ages and nowadays products are normally artificially dried, although solar drying is still done in many developing countries.

Nowadays many consumers pay attention to fishery products as their main source of protein because they contain a highly digestible protein well balanced in terms of essential amino acids and many of these products are rich in ω-3-FA including DHA.

The safety of shelf-stable dried marine product is achieved at A_w of 0.85 or below, because at these A_w's the lack of free water will prevent the growth and toxin production of all pathogenic bacteria, including *S. aureus* and *Clostridium botulinum* (Table 18.1). *S. aureus*, which is capable of growing at a lower A_w than other pathogenic bacteria, is considered the target pathogen in the dehydration of fish products (FDA 2021, Nilsson and Gram 2002). The combination of low A_w (<0.85) with acidic pH (<4.6) and the addition of significant amounts of salt, curing salts, or preservatives act synergistically ensuring the inactivation of foodborne bacteria. In addition, processors should manage the time the fish is exposed to temperatures that are favorable for bacteria growth (Burt 1988, Doe 2002, Nilsson and Gram 2002, Serna-Saldivar 2008, Wang et al 2011).

Practically all sorts of fish can be dehydrated, but pelagic species such as anchovies (*Engraulis encrasicolus*), sardines (*Clupea harengre*), herring (*Clupea harengus*), and mackerel (*Scomber scombrus*) are preferred because of their comparatively longer shelf life. One of the most popular fish for drying is cod (*Gadus macrocephalus*) that is simply dried to produce stockfish or salted into a product named Klipfish (Shaviklo 2012, Table 18.9). Fatty fish is less frequently used because they become rancid and therefore have shorter expectations of shelf life. Qui et al (2019) recently studied the lipid oxidation during the processing and storage of dried seafood products and concluded that various factors, including pretreatment, drying method, and storage conditions, have an impact on the oxidative stability of these seafood products. Salting before drying significantly diminishes microbial contamination and problems because it rapidly lowers the A_w below the range for pathogen growth (Table 18.1).

The simplest form of drying is exposing the fish placed on racks, either whole or split, directly to the sun for one to three days. However, this ancient method make products more prone to be contaminated by dust and fly larvae, and develop fat oxidation or rancidity. This method is preferred for drying small fish species like anchovies. Salting the fish by dipping in brine helps to reduce the incidence of fly larvae infestation, as does raising the fish off the ground onto racks. Having the fish on racks allows air to circulate, enhancing the rate of drying. If solar drying is employed, the facilities should provide covering during rainfall. Some solar dryers are equipped with glass or plastic coverings that increases the temperature of the air around the fish (Doe 2002)

Dehydrated fish products can be divided into fully or partially dehydrated products. The fully dried products are dehydrated to drop the A_w below 0.75 and have prolonged shelf life between one week and several months provided that they are properly packaged and stored. These intermediate-moisture fish typically have a shelf life of up to one week and are usually kept under refrigeration before consumption (Doe 2002).

Dehydrated fish products are also categorized into hard- and soft-cured. Hard curing consists of drying at high temperatures to lower the A_w to safe levels, whereas soft-cured products contain higher moisture that favors organoleptic and textural properties. Soft curing usually combines salting, light drying, and in some instances smoking. A soft-cured fish product usually contains 46% moisture. Regardless of the type of product, fish snacks should be packaged in airtight containers or pouches and kept preferably at low temperatures during distribution to minimize rancidity and growth of xerophilic molds (Burt 1988, Doe 2002, Serna-Saldivar 2008, Wang et al 2011).

Fish jerky is popular in some regions of the world. Fish for jerky must be extremely fresh to start with a raw product low in bacterial counts. Species like tuna (*Thunnus thynnus*), catfish (*Ictalurus punctatus*), trout (*Salmonidae* sp.), salmon (*Salmo salar*), and whitefish (*Coregonus* spp.) should be avoided because they contain more fat than other species (Table 18.8). Several marine species like anchovies, sardines, herrings, saury, shellfish, small shrimps, and fish by-products (fish frames,

TABLE 18.9
Features of Traditional Dehydrated and Shelf-Stable Marine Products Produced Throughout the World[a]

Seafood Product/Region	Process and Product Characteristics
Lutefisk/Nordic countries	Dehydrated whitefish produced from aged stockfish, or dried and salted cod that are pickled in lye to develop a gelatinous texture after rehydration
Stockfish-cod (*Gadus morrhua*)/ Norway	Cod is split lengthwise but remains joined at the tail, rinsed in freshwater, then hung to dry for 1½–2 months in a relative humidity of 70–80%; post-drying in well-aired storerooms (A_w = 0.74)
Anchovy (*Anchola mitchilli*)/ Malaysia	Whole anchovies are boiled in 10% brine in bamboo baskets for 1 min with agitation and then removed from the brine and allowed to drain and cool down. Fish are spread in preparation for solar drying for one day (A_w = 0.79)
Herring kippers (*Clupea* sp.)/ Norway and Isle of Men	Herrings are split in a butterfly fashion from tail to head along the dorsal ridge, gutted, washed, soaked in 80% brine for 15 min, hung to drain for 1 h, smoked at maximum temperature of 29°C in dryers for 4 h
Tuna (*Thunnus* sp.) Cubes/Japan	Tuna is steam-cooked, cooled, and trimmed. The flesh is then cut into small cubes about 1.0–1.2 cm in preparation for 1 h, steam-cooking in the presence of seasonings such as salt, sugar, soy sauce, and spices. The cooked cubes are then dehydrated for approximately 1 h in dryers
Skipjack tuna (*Katsuwonus pelamis*)/Sri Lanka	Skipjack is dry-salted for approximately 30 min or cooked in brine for a few minutes. Then the fish is smoked and dried on coals for 10–12 h or alternatively sun-dried for six to seven days (A_w = 0.64–0.67)
Bonito (*Sarda sarda*)/Japan	Bonito (Japan) fillets without skin are boiled for 60–90 min, inoculated with molds (*Aspergillus glaucus*, *A. melleus*, and *Penicillium glaucum*), broiled in a smoking furnace, and then solar-dried for about 1 month to decrease the A_w = 0.76
Mackerel (*Scomber scomber*)/ Indonesia	Mackerel is first treated in a saline saturated brine containing 30% salt by weight. Then the fish is solar-dried for 48 h to lower the A_w to 0.89
Sand lance or eels (*Ammodytidae* sp.) /Japan	The whole fish is boiled for 10 min in 10–12% brine, the solution drained, and the partially cooked fish dehydrated for two to three days by sun-drying (A_w = 0.98)
Clupeid fish bones/ Southeast Asia	Clupeids that include herrings, sardines, and menhadens are seasoned fish bones and are a popular crunchy and seasoned snack prepared from fish frames from small clupeids that are fried for 2–3 min
Catfish (*Wallago attu*)/Malaysia	Fish are split, gutted, and the snout cut off. They are placed in vats for salt-pickling for one to two days. After pickling, they are solar-dried for two to three days to lower the A_w to about 0.82
	Shrimp East Asia – small-sized shrimps are cooked in 2% brine for 30 min at boiling temperature followed by peeling and drying
Ebi-senbei (shrimp crackers)/ Japan	This traditional Japanese snack is prepared by first removing shells and deveining shrimp. Shrimps are flattened, minced, sprayed with a dilute salt solution mixed with cooked and mashed glutinous rice, tapioca flour, or starch to form a dough that is flattened and dehydrated by baking. Alternatively, the crackers can be deep-fat fried to develop different flavor and the desired crispiness
Keropok (seafood crackers)/ Southeast Asia	These popular crackers are made from a dough prepared from surimi, minced shellfish, or shrimp mixed with tapioca starch, salt, baking powder, and seasonings. The dough is normally flattened into thin disks and fried to develop the characteristic crispiness and flavor

[a] From Burt (1988) and Shaviklo (2012).

cutoffs) are mainly used for developing shelf-stable snacks. Table 18.9 summarizes the main types of dehydrated fish products found in different countries around the globe.

One of the key steps in fish jerky production is curing consisting in placing the fish strips in cold brine followed by dry curing with a mixture of salt and seasonings at refrigeration temperature. After 6–10 h curing, the product is ready for dehydration at a temperature of 60–65°C until the moisture drops to a level of 15–20%. In Japan, it is common to treat whole small fish or fillets of larger species with a seasoning mix consisting of soy sauce, sugar, sake, and monosodium glutamate for 1–2 h before drying and smoking (Burt 1988, Doe 2002, Serna-Saldivar 2008, Wang et al 2011).

Shrimps (*Penaeidae* or *Pandalidae* sp.) are frequently dried in Japan. The crustaceans are boiled in 2% brine for 30 min and then sun-dried for one to two days or dehydrated in an artificial dryer. The resulting shrimps are manually peeled and sized before packaging.

These products are available as fresh, frozen, canned, smoked, or dried. Snacks from these marine sources can be obtained by single or combined methods like salting, drying, smoking, marinating, and extrusion (Shaviklo 2012). Most snacks are obtained from low-value fish or even fish by-products.

Drying of marine products improves sensory attributes such as flavor and texture and enables products to be stored for long periods without refrigeration. Fish jerky is commonly prepared from sliced fish, which has been marinated and dried. Prior to drying, the fish is soaked in a strong marinade. Smoking in many instances is used during dehydration to impart different flavors, prevent fat oxidation, and inhibit microbial growth. The smoking operation can be applied via cold, hot, or dry smoking. In the first, the smoke is applied at temperatures of about 18–25°C for relatively long periods of time (one to three days), whereas hot smoking cooks the fish at 55–75°C. Smoke drying cooks and dehydrates the fish at temperatures 70–150°C for only 1–4 h. In Europe, fish is usually brined before smoking small species like sardines and anchovies (Shaviklo 2012).

The nutritional values of an array of processed marine products are covered in Chapter 20.

REFERENCES

American Meat Institute Foundation. 1994. *The Hazard Analysis and Critical Control Points System of the Meat and Poultry Industry*. Washington: AMI Publishing & Resources.

Barbut, S. 2001. Preservation by chilling, heating and other means. In: *Poultry Products Processing: An Industry Guide*, ed. S. Barbit, 197–238. Boca Raton: CRC Press.

Bechtel, P.J. 2001. Snack foods of animal origin. In: *Snack Foods Processing*, eds. E.W. Lusas and L.W. Rooney, 421–437. Lancaster: Technomic Publishing Co.

Burt, J.R. 1988. *Fish Smoking and Drying: The Effect of Smoking and Drying on the Nutritional Properties of Fish*, 1–166. Barking: Elsevier Science Publishers Ltd.

Cocolin, L., and Rantsiou, K. 2012. Meat fermentation. In: *Handbook of Meat and Meat Processing*, ed. Y.H. Hui, 557–572. Boca Raton: CRC Press.

da Costa, R.J., Voloski, F.L., Mondadori, R.G., Duval, E.H., and Fiorentini, Â.M. 2019. Preservation of meat products with bacteriocins produced by lactic acid bacteria isolated from meat. *Journal of Food Quality*, Article ID 4726510, 12 pages.

Decker, E.A., and Park, Y. 2010. Healthier meat products as functional foods. *Meat Science*, 86(1):49–55.

De Martinis, E.C.P., Alves, V.F., and Franco, B.D.G.M. 2002. Fundamentals and perspectives for the use of bacteriocins produced by lactic acid bacteria in meat products. *Food Reviews International*, 18(2–3):191–208.

De Vuyst, L., Falony, G., and Leroy, F. 2008. Probiotics in fermented sausages. *Meat Science*, 80(1):75–78.

Doe, P.E. 2002. Fish drying. In: *Safety and Quality Issues in Fish Processing*, ed. H.A. Bremner, 350–359. Cambridge: CRC Press, Woodhead Publishing Limited.

FDA (Food and Drug Administration). 2021. Appendix 4: Bacterial pathogen growth and inactivation. In: *Fish and Fishery Products Hazards and Controls Guidance*, SGR 129, Fourth Edition, 417–438, Department of Health and Human Services, Public Health Service, Center for Food Safety and Applied Nutrition, Food and Drug Administration.

Gormley, F.J., Little, C.L., Grant, K.A., De Pinna, E., and McLauchlin, J. 2010. The microbiological safety of ready-to-eat specialty meats from markets and specialty food shops: A UK wide study with a focus on *Salmonella* and *Listeria monocytogenes*. *Food Microbiology*, 27(2):243–249.

Gou, P., Arnau, J., Garcia-Gil, N., Fulladosa, E., and Serra, X. 2012. Dry cured ham. In: *Handbook of Meat and Meat Processing*, ed. Y.H. Hui, 673–688. Boca Raton: CRC Press.

Hamm, R., and Deatherage, F.E. 1960. Changes in hydration, solubility and charges of muscle proteins during heating of meat. *Food Research*, 25(5):587–610.

Herring, J.L., and Smith, B.S. 2012. Meat smoking technology. In: *Handbook of Meat and Meat Processing*, ed. Y.H. Hui. 547–556. Boca Raton: CRC Press.

Huang, T.H., and Nip, W.K. 2001. Intermediate-moisture meat and dehydrated meat. In: *Meat Science and Applications*, eds. Y.H. Hui, W.K. Nip, R.W. Rogers and O.A. Young. 403–442. New York: Marcel Dekker Inc.

Ingham, S.C., Buege, D.R., Dropp, B.K., and Losinski, J.A. 2004. Survival of *Listeria monocytogenes* during storage of ready-to-eat meat products processed by drying, fermentation, and/or smoking. *Journal of Food Protection*, 67(12):2698–2702.

Karabıyıklı, Ş., Öncül, N., and Cevahiroğlu, H. 2015. Microbiological safety of pastrami: A traditional meat product. *LWT – Food Science and Technology*, 64(1):1–5.

Kõrge, K., and Laos, K. 2019. The influence of different packaging materials and atmospheric conditions on the properties of pork rinds. *Journal of Applied Packaging Research*, 11(3):1.

Kumar, P., Verma, A.K., Kumar, D., Umaraw, P., Mehta, N., and Malav, O.P. 2019. Meat snacks: A novel technological perspective. In: *Innovations in Traditional Foods*, ed. C.M. Galanakis, 293–321. Dusford: Elsevier (Woodhead Publishing).

Lawrie, R.A., and Ledward, D.A. 2006. The storage and preservation of meat: II Moisture control. In *Lawrie's Meat Science*, 235–263. Cambridge: CRC, Woodhead Publishing Limited.

Leroy, F., Geyzen, A., Janssens, M., De Vuyst, L., and Scholliers, P. 2013. Meat fermentation at the crossroads of innovation and tradition. A historical outlook. *Trends in Food Science and Technology*, 31(2):130–137.

Lorido, L., Estévez, M., Ventanas, J., and Ventanas, S. 2015. Comparative study between Serrano and Iberian dry-cured hams in relation to the application of high hydrostatic pressure and temporal sensory perceptions. *LWT – Food Science and Technology*, 64(2):1234–1242.

Martin, J.M. 2012. Meat curing technology. In: *Handbook of Meat and Meat Processing*, ed. Y.H. Hui. 531–548. Boca Raton: CRC Press.

Matz, S.A. 1993. *Snack Food Technology*. 3rd edition. New York: AVI-Van Nostrand.

Morales, P., Calzada, J., and Nunez, M. 2006. Effect of high-pressure treatment on the survival of *Listeria monocytogenes* Scott A in sliced vacuum-packaged Iberian and Serrano cured hams. *Journal of Food Protection*, 69(10):2539–2543.

National Advisory Committee on Microbiological Criteria for Foods. 2010. Parameters for determining inoculated pack/challenge study protocols. *Journal of Food Protection*, 73(1):140–202.

Nilsson, L., and Gram, L. 2002. Improving the control of pathogens in fish products. In: *Safety and Quality Issues in Fish Processing*, ed. H.A. Bremner, 54–84. Cambridge: CRC Press, Woodhead Publishing Limited.

Obuz, E., Akkaya, L., and Gok, V. 2012. Turkish pastirma: A dry-cured beef product. In: *Handbook of Meat and Meat Processing*, ed. Y.H. Hui, 689–700. Boca Raton: CRC Press.

Ockerman, H.W. 1989. *Sausage and Processed Meat Formulations*, 1–603. New York: Van Nostrand Reinhold.

Qiu, X., Chen, S., and Lin, H. 2019. Oxidative stability of dried seafood products during processing and storage: A review. *Journal of Aquatic Food Product Technology*, 28(3):329–340.

Rivas-Cañedo, A., Martínez-Onandi, N., Gaya, P., Nuñez, M., and Picon, A. 2021. Effect of high-pressure processing and chemical composition on lipid oxidation, aminopeptidase activity and free amino acids of Serrano dry-cured ham. *Meat Science*, 172:108349.

Santchurn, S.J., Arnaud, E., Zakhia-Rozis, N., and Collignan, A. 2012. Drying principles and applications. In: *Handbook of Meat and Meat Processing*, ed. Y.H. Hui, 505–522. Boca Raton: CRC Press.

Serna-Saldivar, S.O. 2008. Snacks of animal origin. In: *Industrial Manufacture of Snack Foods*, ed. S.O. Serna-Saldivar, 297–327. London: Kennedys Publications Ltd.

Shaviklo, A.R. 2012. Fish snacks from value addition of low value fish and processing by-products. *INFOFISH International*, 5:42–44.

USDA. 2021. Nutrient data laboratory. Agricultural Research Service. Electronic page: http://www.nal.usda.gov/fnic/foodcomp.

Wang, Y., Zhang, M., and Mujumdar, A.S. 2011. Trends in processing technologies for dried aquatic products. *Drying Technology*, 29(4):382–394.

Woraprayote, W., Malila, Y., Sorapukdee, S., Swetwiwathana, A., Benjakul, S., and Visessanguan, W. 2016. Bacteriocins from lactic acid bacteria and their applications in meat and meat products. *Meat Science*, 120:118–132.

Wunsch, N.-G. 2020. Top puffed/extruded pork rind vendors by dollar sales in U.S. 2020. Statista. https://www.statista.com/statistics/916649/puffed-extruded-pork-rind-brand-dollar-sales-us/
Xiong, Y.L., and Benjy Mikel, W. 2001. Meat and meat products. In: *Meat Science and Applications*, eds. Y.H. Hui, W.K. Nip., R.W. Rogers and O.A. Young. New York: Marcel Dekker, Inc.
Zdanowska-Sąsiadek, Ż., Marchewka, J., Horbańczuk, J.O., Wierzbicka, A., Lipińska, P., Jóźwik, A., Atanasov, A.G., Huminiecki, Ł., Sieroń, A., Sieroń, K., Strzałkowska, N., Stelmasiak, A., De Smet, S., Van Hecke, T., and Hoffman, L.C. 2018. Nutrients composition in fit snacks made from ostrich, beef and chicken dried meat. *Molecules*, 23(6):1267.

19 Quality Control and Assurance in the Snack Food Industry

Sergio O. Serna-Saldivar and Daniel A. Jacobo-Velázquez

CONTENTS

19.1 INTRODUCTION

The postharvest endeavors and manufacture of snack foods are subjected to many quality control measurements in the sequential stages of production. There are tests for raw materials, intermediate processed products, and finished products. There are numerous quality tests commonly used, varying from simple to sophisticated. The quality control of the snack industries is of utmost importance because it relates to yields and quality of end-products. In addition, tests can predict relevant processing parameters and adjustments in formulations. Normally, these industries require more sophisticated and specialized equipment that measures the factor quickly and even automatically on the processing line. These tests are interrelated and in most instances complement each other.

Most snacks are obtained after frying (Chapter 3) or by applying oil and therefore usually contain low-protein and low-moisture contents. Therefore, shelf-life expectations greatly depend on the susceptibility of the oil to rancidity and deterioration. In addition, snacks are frequently analyzed in terms of textural properties related to residual moisture and oil contents, flavor, color, and amount of salt and flavorings (Serna-Saldivar 2010).

DOI: 10.1201/9781003129066-19

The goal of the integrated quality program in the industry is to produce snacks that will meet the customer's expectations without posing any health hazards or risks. An effective quality control program should prevent defects and chronic waste, keep good quality records, establish a trace recall program, and ensure the expected shelf life. Responsibilities of the quality assurance and control programs also include production forms, returns, receiving, food safety and allergen programs, labeling, and supplier certification. The quality assurance program plays a key role and function to meet the customer's expectations through the implementation of documentation, practices, and tests that ensure the quality of snacks.

The aim of this chapter is to discuss the importance of quality assurance programs and describe the main quality control measurements applied to raw materials, intermediate products, and end-products of the different categories of snacks.

19.2 QUALITY ASSURANCE PROGRAMS

There are various major quality assurance programs used by the food industry, particularly by snack food processors. These systems complement each other and are aimed to certify the quality and wholesomeness of products. Good manufacturing practices (GMP) are the basis of all quality programs for producing legal and safe products. GMP refers to personal hygiene, sanitation, food processing conditions and practices, storage conditions, pest control, and even design of facilities. On the other hand, the hazard analysis and critical control points (HACCP) program is designed to complement GMP or prerequisites with the aim to ensure the safety in foods through definitive guidelines (Waniska 2001).

19.2.1 Good Manufacturing Practices

GMP are the chores required in order to conform to the guidelines recommended by regulatory agencies that control the manufacturing and sale of foods and beverages. These guidelines provide the minimum requirements that the food manufacturer should meet to ensure that their products are consistently high in quality. The ultimate aim of GMP is to prevent harm and foodborne illnesses to the consumer.

The GMP program is key for the implementation of a reliable quality assurance program. This program is supported by qualified and specialized personnel in quality control matters. Figure 19.1 depicts the major components of an adequate GMP program. An acceptable program starts with the strong support from the top management of the food company and assignment of the necessary resources to implement an effective platform. The key is to establish a quality management department or section, which will dictate all actions toward the implementation of the hygiene of personnel and sanitation of production facilities to minimize microbial, physical, and chemical risks, including toxic compounds. In order to facilitate this process, food companies apply GMP related to the construction of new facilities or modification of existing facilities, especially in terms of sanitation stations, plant design, and control of pests. One of the main functions of GMP programs today is to implement an effective traceability and recall programs to prevent potential risks to customers. The overall day-to-day GMP program is based on documentation and the training of all personnel with the aim to ensure internal quality standards and customer satisfaction. Other tenets include ensuring the finished food products are free from contamination (physical, chemical, or microbial) and that its production has been consistent and well-documented.

19.2.2 Hazard Analysis and Critical Control Points

The HACCP is a management system designed to ensure that snacks are safe for the consumers. The main goal of the program is to detect physical, biological, and chemical hazards from raw material production, procurement, and handling to manufacturing, packaging, and distribution of end–food

FIGURE 19.1 Key elements for the adequate implementation of good manufacturing practices.

products. The HACCP program designs measures to diminish these risks to a safe level, so it attempts to avoid hazards rather than to inspect finished products for the effects of those hazards (US Food and Drug Administration 1997, Mortimore 2000). The current GMPs are an essential foundation for the operation of effective HACCP plans. For successful implementation of a HACCP plan, the upper management must be strongly committed to the program and related activities and responsibilities properly documented. One of the main objectives of the HACCP as part of the Total Quality Management program is the traceability for the recall of defective foods. Table 19.1 summarizes the seven key fundamental principles for the implementation of a HACCP program (US Food and Drug Administration 1997). The different segments of the snack industry highly depend on HACCP programs to ensure product quality and safety. In terms of potential foodborne illnesses, animal, poultry, marine, and dairy products are of more concern because raw materials are highly susceptible to microbial contamination. On the other hand, grain- and fruit-based snacks focus their quality assurance programs toward quality issues in terms of physical (metals, glass, plastic, and stones) and chemical hazards (pesticides, non-GRAS food additives), although snacks based on maize and peanuts are frequently monitored in terms of mycotoxins (Chapter 20).

19.2.3 International Organization for Standardization (ISO)

ISO 22000 is the most popular current international program related to food safety. This system can be applied independent of other management system standards or preferably integrated with other elements like the GMP and HACCP programs mentioned before (Surak 2007). According to Psomas and Kafetzopoulos (2015), the effectiveness of the HACCP program described above (Table 19.1) is significantly better when companies adopt the ISO 22000. The overall goal is to ensure that all relevant food safety hazards are identified and adequately controlled at each step of the food chain to protect consumers from risks or hazards and to build and maintain their trust. The ISO 22000 emphasizes the communication between organizations both upstream and downstream in the food chain. Communication with customers and suppliers about identified hazards and

TABLE 19.1
The Seven Key Principles for the Implementation of a Hazard Analysis and Critical Control Points (HACCP) Program

Principle	Description
Conduct hazard analysis	When conducting a hazard analysis, both quality and safety concerns must be clearly differentiated. Develop a list of hazards which are more related to affect product quality and likely to cause injury or illness if not effectively controlled. Hazards are divided into physical, chemical, and biological that are reasonably likely to cause illness or harm to the consumer. It is important to consider raw ingredients and products generated in each stage of the manufacturing process. This analysis should identify associated control measurements, which will be critically important to establish or determine the control points
Establish critical control points	In this second step, the HACCP team should decide which potential hazards must be addressed based on the severity of the potential hazard and its likely occurrence. A critical control point is defined as a step at which control can be applied and is essential to prevent or eliminate a food quality or safety hazard or reduce them to acceptable levels. Upon completion of this exercise, the hazards associated with each step in the food production should be listed along with quality control measurements used to control the risks. The critical control points must be carefully developed and documented
Establish critical limits	After establishing the critical control points, the HACCP team should set the critical limits (maximum and/or minimum values) to which a physical, chemical, or biological parameters must be controlled with the aim to prevent, eliminate, or diminish risks to acceptable levels. Each control point should have one or more control measures to ensure that the identified hazards are properly monitored. These critical limits may be based upon factors such as temperature, time, physical dimensions, moisture level, water activity (a_w), pH, titratable acidity, and concentrations of salt, preservatives, and nitrates. The critical limits may be derived from scientific sources such as regulatory standards and guidelines, experimental results, and experts
Establish monitoring procedures	Monitoring serves several purposes and is essential for food safety management, tracking the sequential food operations, and control, i.e., exceeding or not meeting a critical limit. When a deviation occurs, an appropriate corrective action must be taken. Values and data of the monitoring procedures should be documented and used for verification. Most procedures need to be reliable and rapid because they relate to online processes. Examples of these activities include visual observations and measurements of temperature, cook times, pH, moisture, color, and a_w. Ideally, monitoring should be continuous, and instrumental equipment must be carefully calibrated
Establish corrective actions	The next goal is to establish strategies to prevent, eliminate, or reduce the occurrence of hazards. Corrective actions are necessary where there is a deviation from the established critical limit. The corrective actions should include the following elements: (1) determine and correct the cause of noncompliance; (2) determine the disposition of noncompliant product, and (3) document the corrective actions. It is very important to assign a trained expert to oversight and decide corrective actions and take care of disposition of noncompliant food products
Establish verification procedures	Verification validates the HACCP program and ensures that the plan is operating correctly. The information needed to validate the program include (1) expert advice and consulting of scientific studies; and (2) in-plant observations, measurements, and evaluations. The verification activities should be preferably carried out by a trained expert within a company, third party consultants, and regulatory agencies
Establish documentation procedures	The final and very relevant step of the implementation of an adequate HACCP program is the documentation of the HACCP plan and HACCP team clearly indicating their responsibilities. The HACCP plan should clearly describe the hazards or risks of concern, the critical control limits, the monitoring with their corrective actions, and the verification procedures and schedule. In addition, the food, intended use, process flow diagram, and distribution should be documented

Source: US Food and Drug Administration (1997).

control measures will assist in clarifying customer and supplier requirements. The most effective food safety systems are established, operated, and updated within the framework of a structured management system and incorporated into the overall management activities of the company. The more recent ISO 22000 has been aligned with ISO 9001 in order to enhance the compatibility of the two standards. In comparison with the ISO 9001, the ISO 22000 makes more emphasis on procedural-orientated industrial-specific risks for any type of food, which includes farming, processing, manufacturing, catering, storage, and distribution. Food companies that seek the standard certification are evolving toward integrated management systems, including Occupational Health & Safety (ISO 45001) and Environmental (ISO 14001), with the idea to improve the safety of the workforce and integrate relevant environmental issues related to the food organization (Surak 2007, Varzakas and Arvanitoyannis 2008, Psomas et al. 2011, International Organization for Standardization (ISO 22000:2018) 2018).

19.3 QUALITY CONTROL OF GRAINS AND THEIR MILLED FRACTIONS

The postharvest endeavors and manufacture of grain-based snacks are subjected to many quality control measurements in the sequential stages of production. There are tests for whole grains, other non-grain and functional additives or raw materials, intermediate processed items (i.e., refined and whole flours, semolina, grits, starches, etc.), and finished food products. The accurate determination of grain properties and characteristics play a key role in the selection of raw materials for further processing into milled fractions and end-products and to ensure their sanitary conditions. For the specific case of cereals (Chapters 6 and 8), pulses, and pseudocereals (Chapter 13), the quality control of the various types of milled products is of utmost relevance because it relates to yields and quality of end-snack foods. In addition, tests can predict important processing parameters and adjustments in formulations and processing. Generally, these industries require more sophisticated and specialized equipment. These tests are interrelated and, in most instances, complement each other (Serna-Sadlivar 2010, Serna-Saldivar 2012)

19.3.1 Cereals

Most snacks are produced from whole cereals, dry-milled fractions, and/or derived starches covered in Chapter 6. The quality control for milling industries is divided into two major categories: quality of cereals used as raw materials and quality of dry or wet-milled products such as flours, refined grits, and starches. As in most other food industries, the quality of raw material greatly affects the quality of the end-products. The parameters usually performed to incoming lots of grains are moisture, test weight, dockage or foreign material, broken kernels, damaged kernels, endosperm texture, hardness, average kernel weight, and aflatoxins (Table 19.2). For the specific case of wheat, which upon water addition and mixing yields a viscoelastic dough, the protein, starch damage (falling number), and dough rheological properties are frequently assessed to differentiate classes of wheats with different end uses.

Dry and wet millers normally rely on laboratory milling procedures to assess the quality of grains and obtain critical information for in-plant processing. Rice is normally dehulled and decorticated to obtain white rice. The yield of whole rice, second heads, and brokens is normally determined and white rice further analyzed for color, amount of fissures, and chemical characterization (Chapter 6). On the other hand, the maize D-T dry-milling procedure described in Chapter 6 yields an array of grits and flours that are usually analyzed for color, particle size distribution, ash, starch damage, and fat and fiber contents. There are several laboratory or pilot plant equipment used to mill wheat, rice, and maize into the typical intermediate products widely used and demanded by snack processors. For the specific case of refined and other wheat flours, the dough rheological instruments like mixolab, farinograph, alveograph, extensigraph, and mixograph are normally employed to determine gluten strength and elasticity (Serna-Saldivar 2010, Serna-Saldivar 2012).

TABLE 19.2
Quality Control Methods and Parameters Commonly Used to Determine the Quality of Cereals, Pulses, and Pseudocereals and Their Milled Fractions

Quality Control	Equipment Instrument	Importance
Whole Grains		
Moisture	NIRA, electric conductivity moisture meters, moisture balances, air oven method (AACC 44-15A)	Moisture is one of the most important parameters because it affects yields and stability during storage. Grains stored out of their optimum moisture are more prone to damage (heat, insect/mold damage) and normally have high enzymatic activity that compromise functionality and end quality
GMO maize or soybean	PCR (polymerase chain reaction)	Molecular biology techniques detect genetically modified organisms using PCR assay. A primer discriminates between the intended and unintended DNA sequence so that only the intended sequence is amplified. Many countries around the world have established mandatory labeling of food products containing GMO despite the fact that they are GRAS for human consumption
Foreign material	Dockage test meter	The amount of foreign material affects yields. Lots of grains with a high amount of extraneous material are usually more prone to insect and mold damage and require costly precleaning actions
Test or volumetric weight	Winchester Bushel meter (AACC Method 55-10)	It is one of the most important measurements for grading cereals and legume seeds. Test weight is important because it is closely related to the grain condition. Insect-, mold-, and heat-damaged kernels have lower test weight compared to healthy counterparts
Kernel weight	Automatic seed counter	The kernel weight gives an indirect indication of grain size and correlates with milling yields. Large cereal grains have higher and lower proportions of endosperm and pericarp, respectively, whereas large pulses have a higher proportion of cotyledons to testa or hulls
Density	Pycnometer (nitrogen displacement)	The density is related to the endosperm texture of the kernel and correlates well with dry and wet milling yields
Broken kernels and splits	Sieving	Broken kernels of cereal grains and splits of legumes are more prone to insect and mold damage and deterioration
Endosperm texture of cereals	Grain viewer (light box) or dissecting kernels	The observation of grain samples positioned on the top of a light box or dissected samples allows determination of the proportion of floury to vitreous endosperm. This is very important for popcorn and cereals that are selected for dry milling and wet milling processes. In wheat, the endosperm texture is closely related to class or use (durum, hard, and soft)
Hardness of cereals	TADD mill	Hardness is indirectly determined by subjecting samples to decortication with the tangential abrasive dehulling device (TADD) for a given amount of time. The higher the dry matter removed, the softer the grain
Fissures or stress cracks	Grain viewer (light box)	The observation of grain samples positioned on the top of a light box allows determination of stress cracks and fissures on the endosperm caused by defective drying or grain mishandling. This test is critically important for rice millers
Heat-damaged kernels	Visual inspection	Heat-damaged kernels are recognized because of their darker coloration. The heat damage is due to the generation of heat from an active grain during storage. A heat-damaged kernel has important alterations in their chemical constituents. In cereals, heat damage is closely related to starch damage due to amylase activity

(Continued)

TABLE 19.2 (CONTINUED)
Quality Control Methods and Parameters Commonly Used to Determine the Quality of Cereals, Pulses, and Pseudocereals and Their Milled Fractions

Quality Control	Equipment Instrument	Importance
Insect and mold damage	Visual inspection	Insect-damaged kernels are recognized by observing the punctures or holes in the grain or the presence of a web-like material. Mold damage grains are recognized because of the darker coloration and dirty appearance. Molds have potent amylolytic and lipolytic enzymes that damage starch and germ lipids, respectively. Several types of molds can produce high levels of mycotoxins that are harmful for humans and domestic animals
Starch damage	AACC Method 76-30A, Falling Number (AACC 56-81A) or viscoamylograph	Starch damage due to faulty storage or sprouting can be determined via hydrolysis with alpha amylase or indirectly determined by a measurement of the viscosity of a slurry prepared from the ground grain or flour. The lower the viscosity, the greater the starch damage
Protein content	NIRA	The amount of protein can be rapidly determined using an infrared analyzer that was previously calibrated with a set of standards. This is particularly important to determine the class of wheat
Mycotoxins	Black light test equipped with ultraviolet light bulbs (365 nm)	It is an extremely convenient screening test for the detection of mycotoxins in grains. The detection of a bright greenish gold fluorescence is a tip-off that a mold is present and that the sample may contain mycotoxins
	Aflatest, Fumonitest, Ochratest, and other related analyses	The utilization of ELISA columns to selectively isolate mycotoxins followed by their quantification via fluorescence allows a quick and reliable determination of these important metabolites regulated by most regulatory agencies. Test can be performed in less than 10 min. The assessment of aflatoxins in maize and peanuts is mandatory and generally performed with ELISA tests
Laboratory rice milling	Sample sheller for rough rice and McGill rice miller	The equipment provides a rapid and efficient method of shelling and milling rough rice into white rice. They accurately test rice samples for milling yield by automatically controlling both the milling pressure and the time cycle
Laboratory maize milling	Laboratory maize milling procedure (Brekke et al. 1971, Peplinski et al. 1984)	Maize is initially conditioned to 16% moisture for 16 h, then to 21% for about 2 h, and finally to 24% for 15 min before degerming. The tempered maize is degermed in an experimental horizontal drum equipped with a steel screen with 7.9-mm diameter round holes. The degermed stock is dried, classified, and air aspirated, and further reduced in particle size by passing through three pairs of independent corrugated rolls and classified/sized with sieves again. The yield of dry-milled fractions and co-products is calculated based on the original grain
Laboratory wheat milling	Quadrumat Junior or Senior Brabender Mill, Chopin Wheat Roller Mill, Buhler Wheat Roller Mill (AACC Method 26-20)	These milling equipment consist of break and reduction rolls and sieves to separate dry-milled fractions and calculate flour extraction or milling yield. The resulting refined flours can be further characterized or used to determine rheological properties
Milled Fractions (grits, semolinas, meals, flours)		
Particle size	Rotap equipped with different sieves	The particle size determination of dry-milled fractions greatly affects extrusion and baking processes and quality of end-snacks

(Continued)

TABLE 19.2 (CONTINUED)
Quality Control Methods and Parameters Commonly Used to Determine the Quality of Cereals, Pulses, and Pseudocereals and Their Milled Fractions

Quality Control	Equipment Instrument	Importance
	Laser ray diffraction analysis (Hareland 1994)	This quick method allows the determination of the particle size distribution of flours, meals, dry-milled fractions (grits) and starches. The commercial equipment are capable of determining from sub-nanometer to millimeters in particle size. Results are usually given in a diagram which clearly indicate the amounts or population of different particle sizes
Specks	Visual determination	The amount of specks (i.e., black specks in maize grits, bran specks in flours) can negatively affect the appearance of processed snack foods. Specks are present in dry-milled fractions due to broken sieves or the improper calibration of purifiers
Extraneous material, filth, insect fragments, roden hairs, roden excreta	AACC Methods 28-00, 28-19, 28-40, 28-41-A, 28-50, 28-51	Dry-milled samples are digested with diverse chemical reagents and the resulting hydrolysate treated with mineral oil and other chemical compounds to separate extraneous material, insect fragments, etc. The number and types of extraneous material are viewed in a wide field microscope
Rodent feces	Microscope technique (AACC Method 28-50)	Dry-milled samples are treated with trichloromethane and carbon tetrachloride to separate feces. Excreta fragments are viewed in a microscope
Color score	Color meters (Hunter Lab, Agtron [AACC Method 14-30] and others)	The color of dry-milled fractions affects color of end-products. Most methods are quickly performed without destroying the sample
Ash content	Muffle furnace, AACC Method 08-01	This is one of the most relevant parameters associated with milling efficiency of cereal grains. Most refined dry-milled products from cereals should contain less than 1% ash
Fat content	NIRA, Goldfisch or Soxhlet apparatus (AACC Method 30-25)	This is one of the most important parameters associated with milling efficiency. Most refined dry-milled products should contain less than 1% fat. The lower the fat content, the better the expansion rate in direct extrusion processes. The fat content of legume seeds and pseudocereals is very important because it is related to susceptibility to oxidation
Starch content	NIRA, analytical determination of total starch (AACC 76-11)	It is the most important single component of cereal dry-milled fractions and greatly affects functionality
Starch gelatinization	Birefringence with microscope equipped with polarized filters	Microscopic technique used to determine the relative amount of gelatinized vs. native starch granules (starch granules show Maltese cross)
Starch damage	AACC Method 76-11, Falling Number (AACC 56-81A) or viscoamylograph	Starch damage due to faulty storage or sprouting can be determined via hydrolysis with alpha amylase or indirectly determined by a measurement of the viscosity of slurry prepared from the ground grits or flour. The lower the viscosity, the greater the starch damage

(*Continued*)

TABLE 19.2 (CONTINUED)
Quality Control Methods and Parameters Commonly Used to Determine the Quality of Cereals, Pulses, and Pseudocereals and Their Milled Fractions

Quality Control	Equipment Instrument	Importance
Starch properties	Viscoamylograph AACC Method 22-10, (Shuey and Tipples 1982, Crosbie and Ross 2007)	It is one of the most important functional tests because after analyzing the viscoamylograph curve, it could be determined temperature at the start of gelatinization, peak viscosity, shear thinning, and set back viscosity (retrogradation). These properties greatly influence processing characteristics and quality of finished products. The most common instruments are the regular and micro Brabender Visco-Amylo-Graph and the Rapid Visco Analyzer
Protein content	NIRA (AACC Method 39-10)	The amount of protein can be rapidly determined using an infrared analyzer that was previously calibrated with a set of standards. This is particularly important to determine the class of wheat flour (soft, intermediate, or all-purpose and hard wheat flours) and protein content of feedstocks produced from legumes, pulses, and pseudocereals
	Kjeldahl methods (AACC 46-10, 46-12 or 46-13) Udy dye method (AACC 46-14A)	These are the most widely used analytical procedures to determine nitrogen and protein. Their main disadvantage is that they take a long analysis time
Gluten content of Wheat	Handwashing method (AACC 38-10) or Glutomatic, AACC (AACC Method 38-11)	Wheat gluten is determined after producing a dough that is washed with water or a salt solution to selectively remove starch and soluble protein. The wet residue is weighed and then dried to determine wet and dry gluten, respectively. The Glutomatic rapidly determines gluten content automatically. These tests are used to differentiate classes of wheat
Rheological properties of wheat flour	Mixolab (Dubat et al. 2013)	This instrument was designed based on the principles of farinograph and amylograph because it first determines dough-mixing characteristics and gluten strength and then pasting, viscosity, and retrogradation properties of the same wheat dough system. Therefore, it is the only instrument enabling complete characterization of the gluten protein and starch functionality in a single assay. The flour is first mixed with the optimum amount of water, next its viscosity is tested similarly as in the farinograph. Thereafter, the temperature of the dough is gradually increased to test changes in viscosity through heating and cooling cycles. The instrument has an integrated simulator that automatically calculates water absorption, dough development time, and dough stability and viscosities. The instrument is suited to test gluten properties, starch damage, amylase activity, and ingredient functionality
	Farinograph (AACC Methods 54-20 or 54-21) (D'Appolonia and Kunerth 1984)	It is an instrument that records the resistance of a wheat dough to mixing and evaluates the gluten properties of a wheat flour. After the complete analysis of the farinograph curve, processors obtain optimum water absorption, optimum dough development time, dough stability, and mixing tolerance index. The instrument is widely used to determine the strength of the flour and also to determine if the flour is soft, all-purpose, or hard
	Alveograph. (Dubois et al. 2008)	It is an instrument that records strength or resistance and extensibility of a dough that was optimum mixed with a salt solution. These properties are determined after introducing air at a constant rate to a preformed piece of dough until the dough bubble bursts. The instrument records the strength of the flour (P), extensibility (L or G), and work required to inflate the bubble. The P/L ratio and work (W) determine if the flour is soft, all purpose or hard

(*Continued*)

TABLE 19.2 (CONTINUED)
Quality Control Methods and Parameters Commonly Used to Determine the Quality of Cereals, Pulses, and Pseudocereals and Their Milled Fractions

Quality Control	Equipment Instrument	Importance
	Extensigraph (AACC Method 54-10) (Rasper and Preston (1991)	It is an instrument that records strength or resistance and extensibility of a dough that was optimum mixed. The extensigraph is widely used to determine the strength of the flour and determine if the flour is soft, all-purpose, or hard. The extensigraph curve yields important parameters such as maximum resistance (R_{max}), resistance after 50 mm of dough stretching (R_{50}), dough extensibility (E), and work (area under the curve)
	Mixograph (AACC Method 54-40)	It is an instrument that records the resistance of a wheat dough to mixing and evaluates the gluten properties of a wheat flour. After the complete analysis of the mixograph curve, processors obtain optimum dough development time, dough stability, and the strength of the flour (soft, all-purpose or hard). This instrument is widely used as an screening test by plant breeders

Source: AACC (2000), Serna-Saldivar (2008), Serna-Saldivar (2010), Serna-Saldivar (2012).

19.3.2 Legumes, Pseudocereals, and Proteins

The quality control assays for legume seeds and pseudocereals are alike to those employed to assess cereal grains (Table 19.2). However, raw legume seeds are additionally assessed in terms of presence of antinutritional factors (Chapter 20). Most processes of legumes or pulses pay special attention to the concentrations of trypsin inhibitor, urease, and lectins, which are partially or totally inactivated by thermal treatments, sprouting, and fermentation. Protein is normally the major chemical component in these grains that are also used as raw materials to produce protein concentrates and isolates. Therefore, from the functionality viewpoint, there are many assays that are practiced in both animal- and vegetable-based proteins. The main functionality tests are water-holding capacity, water solubility index, fat-binding capacity, emulsification capacity, protein and nitrogen dispersibility indexes, free amino nitrogen, heat coagulation, and foaming properties that are critically important in the formulation of snacks. Table 19.3 depicts the main quality control tests applied to assess antinutritional factors, protein functionality of flours, concentrates and isolates, and protein quality from the nutritional viewpoint.

19.4 QUALITY CONTROL OF CRUDE AND REFINED OILS

The quality control of oilseeds as raw materials and during oil crushing, refining, hydrogenation, fractionation, and esterification plays a key role in the value of end-products and dictates processing adjustments. The condition of oilseeds is usually monitored by physical and chemical properties summarized in Tables 19.2 and 19.3, particularly for the amount, characteristics, and condition of the extracted oil (Table 19.4). O'Brien (2009) details practically all the physical, chemical, functionality, and sensory analyses of crude, refined, and modified fats and oils.

The quality control of crude oil is important because it greatly affects processing conditions during the refining steps covered in Chapter 7. The official methods of AOCS (2009) are the analytical assays most commonly used. The most relevant assays for oil refiners is the total refining loss because it very highly correlates with the refined oil yield. The mechanically and/or solvent-extracted crude oil is generally tested for phosphorus related to the amount of phospholipids and

TABLE 19.3
Quality Control Methods and Parameters Commonly Used to Determine the Quality of Flours and Protein Concentrates and Isolates Derived from Legumes

Quality Control	Equipment Instrument	Importance
Antinutritional Factors		
Trypsin inhibitor	AOCS Official Method Ba 12-75, Ba 12ª-2020.Kakade et al. (1974)	The assay is based on measurement of the difference in absorbance obtained in the absence and the presence of an inhibitor, where a synthetic substance is used as the substrate for bovine trypsin. It is used to assess trypsin inhibitors in soybeans, pulses, and other legumes and products processed from these sources. The activity is negatively correlated with the application of thermal treatments
Urease activity	AOCS Official Method Ba. 9-58. Katz (1964)	This assay is based on the measurement of pH change or differential and therefore is simpler compared to the trypsin inhibitor test. That is the reason it is widely used to indirectly determine the extent of heat treatments and the presence of trypsin inhibitors
Lectins	Liener (1955), Jaffe et al. (1974), Rizzi et al. (2003)	Lectins are proteins that bind to carbohydrate-containing molecules and are considered as antinutritional factors because they cause intestinal problems and agglutination of red blood cells. Lectins in raw soybeans can inhibit growth and cause death in animals. These glycoproteins have been assayed since the 1950s using trypsinized rabbit red blood cells and a photometer. The dilution equivalent to settling of 50% of cells is calculated and expressed as hemagglutinating units/mg protein (HU). Lectin levels can vary from 37 to 323 HU and are deactivated with thermal treatments. New ELISA kits and instrumental analysis have been developed for the faster and more accurate quantitation of lectins
Oligosaccharides	Megazyme colorimetric kit K-RAFGL raffinose/sucrose/D-glucose assay	Grain legumes contain significant amounts of flatus-producing oligosaccharides (raffinose, stachyose, and verbascose) which are considered as antinutritional factors, although recent studies indicate that they act as important prebiotics favoring the growth of intestinal microbiota and human health
Functional Tests		
Water-holding capacity and solubility indexes	Zayas (1997)	These tests are widely applied because they are closely related to the ability of proteins to hold water or to solubilize in water. Samples are dispersed in distilled water and mixed with a stirrer for 20 min at 25°C and then centrifuged at 1080 ×g for 30 min. The amount of protein in the supernatant indicates solubility, whereas the WHC is calculated as the difference in weight of the hydrated flour (pellet) and the weight of the original flour or as grams of water per gram of protein
Fat-binding capacity	Lin and Zayas (1987)	The fat-binding capacity is determined by stirring a diluted sample solution with corn oil for 30 min followed by allowing this mixture to stand at room temperature for an additional 30 min (25°C). The mixture is then centrifuged for 30 min at approximately 1,080 ×g. After the disposal of excess oil, the FBC was calculated as the weight of the residue divided by the original weight or alternatively expressed as grams of oil per gram of protein

(*Continued*)

TABLE 19.3 (CONTINUED)
Quality Control Methods and Parameters Commonly Used to Determine the Quality of Flours and Protein Concentrates and Isolates Derived from Legumes

Quality Control	Equipment Instrument	Importance
Emulsifying capacity, emulsification activity index, and emulsification stability index	Pearce and Kinsella (1978), McWatters and Holmes (1979)	The emulsification capacity is determined in protein suspensions at 25°C. The suspension is continuously blended at high speed with a given amount of soybean oil, which is gradually incorporated at a flow rate of 0.5 g/s. This mixture is continuously blended until the inversion point (water-in-oil) is observed. Emulsification capacity is expressed as the maximum amount of oil emulsified per gram of protein. The emulsification activity and stability indexes test 2% protein suspension at 25°C. The suspension is blended with soybean oil for 1 min using a Waring Blender equipped with a microcontainer at low speed. The emulsion is immediately diluted with 0.1% SDS at a 500× dilution factor, and the absorbance measured at 500 nm. The diluted emulsion is then incubated at 95°C in a water bath, and the absorbance measured at time zero and at 10 min
Free (FAN) or alpha amino nitrogen (AAN)	AOAC ninhydrin colorimetric procedure 945.30 L	The FAN or AAN tests measure the concentration of individual amino acids and small peptides in samples and therefore are related to the rate of hydrolysis and solubility of proteins. The assay is based in the extraction of samples in water or a buffered solution in preparation with the colorimetric reaction with ninhydrin. Results are calculated after comparing with a standard curve prepared from glycine
Protein dispersibility index (PDI)	AOCS Official Method Ba 10-65. High stir method, centrifugation and determination of protein in supernatant	The assay determines the dispersible protein in soybean products with better and faster results compared to the NSI or slow stir method. The method is highly useful to determine the degree of thermal treatment or protein denaturation of soybean proteins. Low and high PDI values indicate high and low heat treatments
Nitrogen solubility index (NSI)	AOCS Official Method Ba 11-65. Low stir and determination of nitrogen in supernatant after centrifugation	This method determines the dispersible nitrogen related to solubility in soybean products. In contrast to the alternative fast-stir PDI method, the NSI will give generally lower results
Foaming capacity, density, and stability	Patel et al. (1988), Tsutsui (1988), Sorgentini et al. (1995)	Protein solutions adjusted to different pH are whipped or blended at high speed in a graduated container for a certain period of time. The foam capacity is calculated as the foam volume after blending/original volume, whereas foam stability by assessing the drop of volume after 30 min
Heat coagulation test	Regenstein and Regenstein (1984)	This test of protein is performed to determine the capacity of test proteins dispersed in water to coagulate when exposed to heat. Coagulation is the capacity of the protein to form a clot or coagulum after an initial denaturation normally caused by heat. It involves the rupture of hydrogen bonds within peptide chains, and when an advanced state is reached, denaturation becomes irreversible. The principle of the assay is to prepare a protein solution at pH 7, which is centrifuged in order to recuperate the supernatant. An aliquot of the supernatant is heat-treated at 98–100°C for 30 min in a water bath and then the coagulum is separated via centrifugation and filtration from the new supernatant. The uncoagulated protein is assayed in the supernatant and the coagulated protein calculated by difference. The albumins and globulins are the protein fractions that tend to have higher coagulation. The heat coagulation of proteins occurs due to denaturation and agglutination or the separation of the denatured protein in a particular form and occurs at higher extent at the isoelectric point of proteins

(Continued)

TABLE 19.3 (CONTINUED)
Quality Control Methods and Parameters Commonly Used to Determine the Quality of Flours and Protein Concentrates and Isolates Derived from Legumes

Quality Control	Equipment Instrument	Importance
Protein Nutritional Quality		
In vitro digestibility	pH drop of proteins after protease hydrolysis. Hsu et al. (1977)	The pH drop *in vitro* technique correlates significantly with *in vivo* protein digestibility values and only takes 10 min. The principle is that proteins lower the pH, depending on their rate of digestion. The assay is compared with casein
Amino acid composition	HPLC with ion exchange or reverse-phase chromatography and mass spectroscopy. Otter (2012)	Most methods first hydrolyze the protein into individual amino acids with acid, basic, or enzyme hydrolyses. The amino acids are quantified after ninhydrin reaction or after derivatization and dual wavelength detection. Tryptophan is lost during acid hydrolysis and therefore is quantified after basic or enzyme hydrolyses. The limiting amino acid, chemical score, can be calculated after assessing the quantity of essential amino acids based on the requirement (generally for preschool infants)
C-PER	Satterlee et al. (1982)	Calculation of the protein efficiency ratio by determining *in vitro* protein digestibility and the amino acid composition. The C-PER highly correlates with *in vivo* PER bioassays

free fatty acids (FFA) or acidity in order to determine the amount and strength of alkali used during neutralization. The color of crude oil is also a relevant parameter because it affects the amount of bleaching earths required to achieve a certain color value in refined oils (Table 19.4).

The quality control of refined oils consists of more determinations because these products are fit for human consumption or further processing via blending, fractionation, hydrogenation, or esterification (Chapter 7). The FA composition of refined oils is usually tested via gas chromatography – flame ionization detector. The test consists of the hydrolysis of the triglycerides to yield FFA that are methylated or derivatized with the aim of making these compounds nonpolar. The sample is injected, and the FA eluted according to their molecular weight and degree of instauration. The FA profile is used to calculate the ratio of saturated:monounsaturated PUFA required for food labeling purposes (Chapter 20). A similar chromatography technique is used to determine the amount of *cis*- and *trans*-FA. The major modification of the procedure is the use of other gas carriers such as helium, long capillary columns, and *trans*-FA standards. This is important because most countries around the world are enforcing the declaration of *trans*-FA in food labels and some are even prohibiting the use of *trans*-fats in foods (Chapter 20). The stability or susceptibility of refined oils to oxidation is the most relevant functionality test. The peroxide value is used to determine the amount of peroxides, and the active oxygen method (AOM) is the stability test most commonly used by the industry. The AOM method is rapid and determines the oxidative stability of processed oils. The accelerated instrumental method predicts the resistance of the heated oil to oxidation by bubbling air into the sample. The AOM test quantifies via peroxide value the time or induction period where the oil starts forming peroxides at a constant rate or the number of hours required to reach 100 mEq peroxides/kg oil. Other important test for refined oils are fat acidity, color, iodine value, and the smoke/ignition/fire points, although these last tests are more used to test quality of used oils (Table 19.4).

Several criteria have been proposed to determine the state or condition of the frying oils but none has been entirely satisfactory. The acid value is one of the most relevant criteria to test the condition of the oil, although it is not necessarily correlated to sensory properties. The iodine and refractive index values are used to follow blending, fractionation, and hydrogenation processes. Both tests are

TABLE 19.4
Quality Control Parameters Most Commonly Used to Determine the Quality of Mechanically and/or Solvent-Extracted Crude and Refined Oils

Quality Control Parameter	Equipment or Technique	Importance
Crude Oil		
Free fatty acid or acidity	AOCS Ca5a-40, AACC 58-15 Titration with alkali Kits for quick tests	This is the most important crude oil test because it closely relates to refining losses and quality of the raw material. Extracted fat is titrated with a dilute alkali solution to neutrality and the free fatty acids expressed as oleic equivalents
Phosphorus (phospholipids)	AOCS Ca 12–55 Ashing followed by molybdate reaction	The total phosphorus content in crude oils is determined by first ashing the oil sample in the presence of ZnO followed by the spectrophotometric measurement as a blue phosphomolybdic acid complex and the conversion of values to phospholipid equivalents. The test is relevant because the phospholipids or gums are removed during the first stage of the oil-refining scheme and these compounds are important nutraceuticals and emulsifiers in the food industry
Peroxide value	AOCS Cc8-53, AACC 58-16 Amount of iodine liberated by its reaction with potassium iodide	This is the most used method to determine the stability or oxidation of the crude oil. The main oxidation products of oils are hydroperoxides, the quantity of which can be quantitatively measured by the amount of iodine liberated by its reaction with potassium iodide. Values are expressed on mEq/kg of oil
Refining loss	AOCS Ca9-52 Laboratory refining test with caustic soda	It is considered the most ideal quantitative criterion of quality of crude oils. A crude or degummed oil sample is treated with caustic soda and the amount of refined oil obtained. Refining losses vary among kinds of oils, the values range from 5% to 15%
Iodine value	AOCS Cd 1-25, AACC 58-30 Wijs titration	Most common test to determine the degree of instauration of the extracted oil. The higher the iodine values the higher the degree of instauration and susceptibility to oxidation
Saponification Value	AOCS Cd 3-25, AACC 58-50 Grams of fat/oil saponified by 1 mol (56.104 g) of KOH	Analytic test used to determine the average molecular weight of fatty acids defined as the mg of KOH required to completely reacting with 1 g sample
Color	Lovibond (AOCS Cc 13b-45, AACC 58-12) or any type of colorimeter	The color of the crude oil is a legitimate factor in determining its value, inasmuch as dark oils require more bleaching earths to obtain acceptable light colored oil
Fatty acid profile	Gas chromatography with flame ionization detector (AACC 58-18)	This is the method of choice for determining the fatty acid profile of fats and oils and very useful to determine ratios of saturated:monounsaturated: polyunsaturated fatty acids for labeling purposes. The fatty acids are derivatized (methyl esters) prior to the analysis. New techniques to separate and quantify *cis*- and *trans*-isomers have been developed
Refined Oils		
Peroxide value	AOCS Cc8-53, AACC 58-16 Amount of iodine liberated by its reaction with potassium iodide	This is the most used method to determine the stability of fresh refined and used oils. The main oxidation products of oils are hydroperoxides the quantity of which can be quantitatively measured by the amount of iodine liberated by its reaction with potassium iodide. Values are expressed on milliequivalents/kg oil

(*Continued*)

TABLE 19.4 (CONTINUED)
Quality Control Parameters Most Commonly Used to Determine the Quality of Mechanically and/or Solvent-Extracted Crude and Refined Oils

Quality Control Parameter	Equipment or Technique	Importance
Active oxygen method (AOM)	AOCS Cd12-57, AACC 58-54	Most common rapid method to determine oxidative stability of processed oils. The method predicts the resistance of the heated oil (97.8°C) to oxidation by bubbling air into the sample. The AOM method determines the number of hours required to reach 100 mEq peroxides/kg oil
Color	Lovibond tintometer (AOCS Cc 13b-45, AACC 58-12) or any type of colorimeter	The color of the refined oil is one of the most important factors affecting its value. The color after refining and bleaching is a more reliable index than the color of the crude oil or the color after neutralization. The most important index of the Lovibond color test is the red value
Smoke, fire, and flash point	AOCS Ce 9a-48 or Cc 9b-55, AACC 58-82	The smoke, fire, and flash points are measures of its thermal stability when heated in the presence of air. The smoke point is the temperature at which smoking is first detected in a laboratory apparatus, whereas the flash point is the temperature at which the volatile products are evolved at such a rate that they ignite but do not support combustion. The fire point is the temperature at which the volatile products support continued combustion. Free fatty acids are much more volatile than triglycerides; therefore, these points are closely related to the free fatty acid content and the condition of used oils
Electronic nose	Shen et al. (2001), Gan et al. (2005), Xu et al. (2016), Upadhyay et al. (2017), Majchrza et al. (2018)	The novel electronic nose chromatography (mass spectrometry and gas chromatography) technology is increasingly being used for the quality control of vegetable crude and refined oils, fats, and snacks with the advantages of short time of analysis, adequate sensitivity, and relatively low cost. The main advantage of this technology is that it effectively quantifies important volatile compounds which correlate positively with organoleptic and olfactory tests

Source: Firestone (1998), AACC (2000), Serna-Saldivar (2008), AOCS (2009), O'Brien (2009).

aimed toward the determination of the degree of instauration of the oil. The main advantage of the refractive index is that it is performed in a short time and do not destroy the sample and therefore is more used for in-plant rapid testing. The functionality of oleins, stearines, partially hydrogenated oils, shortening, and esterified shortening are usually tested via the determination of melting point and solid fat index (SFI). The method is based on dilatometry and is ideally suited to test functionality of plastic shortenings and tropical oils used for confectionary purposes. The SFI is used to determine the temperature range in which a fat/oil melts and relates to the fatty acid composition and distribution. The SFI is based on the dilatometry behavior of the fat, which is considered as a fingerprint of these products (Table 19.4).

As explained in Chapter 3 related to frying operations, during deep-fat frying of snacks, the oil is kept at elevated temperatures (165–210°C) and exposed to air, water, and chemical compounds associated with the snack being processed. These conditions lead to thermal, oxidative, and hydrolytic breakdown reactions of the oil. The frying oil gradually oxidizes, polymerizes, smokes, and produces volatiles that affect negatively the sensory properties of the snacks, and some organic products are also implicated in producing adverse or toxicological effects covered in Chapter 20. According to Weisshaar (2014), several official assays such as total polar materials (TPM) and polymerized triacylglycerides (PTG) are the most related to sensory properties of snack foods. When oxidative alterations strongly predominate over thermal alterations, sensory defects can appear before TPM and PTG. In that case, additional parameters like anisidine value, carbonyl value, and epoxy FA should be considered (Bou et al. 2012). A unique and powerful tool for the simultaneous detection of multiple parameters corresponding to thermal and oxidative alterations of oils is near-infrared spectroscopy (NIRA), which will become more important because of the short assay time and convenience.

The quality control of frying oil operations (Chapter 3) generally consists of rapid in-plant determinations. Undoubtedly, the most common assay is the FFA content or acid value because the frying oil gradually increases its value as it is heated and used (Table 19.4). The FFA in fresh oil is generally less than 0.05% and abused oils may contain levels higher than 1.5%. There are many strip tests used to follow the condition of the frying oil and spot tests that monitor free fatty acids by a colorimetric procedure. A high FFA or acid value increases dielectric constant, oil color values, foaming, smoking, and production of polar compounds and decreases oil stability. As the oil degrades and forms polar compounds, the dielectric constant increases. The advantage of the dielectric value is that it is quick and easy to perform, but it has the disadvantage that the reading is influenced by many outside and uncontrollable factors such as water and oil type. The peroxide value is the most common assay used to determine the susceptibility of the oil to oxidation and the AOM is a derived method widely used by the industry. The AOM method requires more time, but it is very accurate to determine and predict the oxidative stability and shelf life of new and processed oils. The method predicts the resistance of the heated oil to oxidation by bubbling air into the sample and quantifies the peroxide value through the accelerated assay (generally 100 mEq peroxides/kg oil). There are also strips that can give rough indications of the quality of the oil from the peroxide value viewpoint. The quick RAU test consists of a colorimetric kit that contains Redox indicators that react with the total amount of oxidized compounds in a frying oil. The condition of the oil is determined by comparing the color developed to a four-color scale. Other important tests for frying oils are polar compounds, color, and smoke/ignition/fire points (Table 19.4). Recently, the rate of oil oxidation, adulteration, and properties of used oils have been successfully assessed with the electronic nose technology. The major advantages of this instrumental analysis is the short assay time and the strong correlation with many oxidative indexes discussed above and sensory properties of oils and oil-rich snack foods (Shen et al. 2001, Gan et al. 2005, Xu et al. 2016, Upadhyay et al. 2017, Majchrza et al. 2018).

19.5 QUALITY CONTROL OF SNACK FOODS

19.5.1 Grain-Based Snacks

19.5.1.1 Popcorn

The quality control for popcorn is divided into two major types: quality of raw kernels and quality of flakes. The quality of raw kernels greatly affects processing and quality of the burst flakes. The parameters usually performed to incoming lots of kernels are moisture, test weight, dockage or foreign material, broken/damaged kernels, thousand kernel weight, and analysis of mycotoxins, mainly aflatoxins. These tests are detailed in Table 19.2. The two most important quality control parameters for raw kernels are moisture and kernel damage because both greatly affect expansion volume and percent of unpopped kernels. In 1975, the Grades and Standards Committee of the Popcorn Institute developed the Metric Weight Volume Tester (MWVT) manufactured by C. Cretors & Co. (Chicago, USA) as the official testing method for the industry. The device measures popping expansion as cubic centimeters of popped flakes per gram of raw kernels. The power is set to 1,500 W and temperature at 260°C. When the pan temperature is reached, 250 g popcorn and 63 g oil are added. At the completion of popping, the volume is measured in a cylinder 13 cm in diameter (Snack Food Association 1992). The main advantage of this equipment is the consistency of popping expansion volume and the low variability regardless of the operator. Expansion volume is calculated as the total popped volume (mL) divided by the original sample weight (g), and the flake size or individual kernel expansion is measured by the popped volume (mL)/number of popped kernels. The percent unpopped kernels is determined by measuring the number of unpopped kernels/original number of kernels) × 100 (Mohamed et al. 1993). The number of kernels can be effectively accurately and quickly determined in an automatic seed counter. A similar volume/expansion test developed by the Technical Committee of the Popcorn Institute has been especially devised for microwave popcorn (Popcorn Institute 1989).

The different types of popped corn products are additionally tested for total oil, saturated fat, and salt or sodium because these nutrients should be declared along with other mandatory nutrients in the food label (Chapter 20). The stability and oxidative condition of the oil associated with the packaged popcorn products are mainly assessed in terms of peroxide value, AOM, hexanal, and other oil stability tests summarized in Table 19.4.

19.5.1.2 Wheat-Based Snacks

The quality of wheat-based snacks discussed in Chapter 10 greatly depends on the characteristics of the wheat flour and other supplemented ingredients. Table 19.5 depicts the chief quality control parameters used to assess properties of wheat-based snacks. The critical rheological assays used to determine dough gluten strength, extensibility, and functionality are summarized in Table 19.2. Flours are generally characterized by moisture, protein, wet or dry gluten, starch damage (falling number), and color. The rheological properties of wheat flour can be determined with the mixolab, farinograh, extensigraph, and alveograph (D'Appolonia and Kunerth 1984, Dubois et al. 2008, Carson and Edwards 2009, Rasper and Preston 1991, Serna-Saldivar 2012).

Standardized laboratory methods have been devised to study yeast activity (Rubenthaler et al. 1980) and functionality of flour and other raw materials used to produce table bread (straight dough and sponge procedures), French bread, crackers, and pretzels (Pizzinatto and Hoseney 1980, AACC 2000, Serna-Saldivar 2012). These laboratory procedures are also used for product development and research purposes and can easily be adopted in manufacturing plants to adjust processing parameters and conditions and produce uniform products with top quality. Other analytical approaches commonly used to study wheat-based snacks are textural properties with the texture analyzer or Instron, which normally determines shear force or compression force. Moisture, protein, and fat of raw materials and finished products can be rapidly determined with a standardized near-infrared analyzer.

Sensory evaluation tests are commonly used to determine the consumer's preference of finished snacks. Tests more widely practiced are the triangular and preference tests (Table 19.5). Triangular tests are ideally suited to identify the consumer's preference between two samples because panelists try to identify the odd and paired samples among the three samples in a set. The triangular test is widely used for product development purposes. In the preference test, panelists evaluate the color, aroma, flavor, texture, and overall acceptability of the products generally using a 5- or 9-point hedonic scale (Serna-Saldivar 2012). This test is mainly used for product development, shelf-life studies, and determining functionality of additives or ingredients.

19.5.1.3 Lime-Cooked Snacks

The quality of nixtamalized or lime-cooked snacks and related products depends on the soundness and characteristics of the raw maize (Table 19.2) and other ingredients and on manufacturing practices discussed in Chapter 11. The major quality parameters controlled during the whole tortilla chip manufacturing process are summarized in Table 19.6.

Lately, the issue of the utilization of non-GMO maize genotypes has become important in many countries around the world. Testing for GMO is usually performed via polymerase chain reaction (PCR). The processing of maize into snacks or other food products involves steps that extract and may fragment or otherwise compromise DNA molecules resulting in smaller DNA yields for

TABLE 19.5
Quality Control Parameters Most Commonly Used to Assess Quality of Wheat-Based Snack Foods

Quality Control	Equipment Instrument	Importance
Yeast activity	Pressurometer, gaseograph (Rubenthaler et al. 1980)	Tests ideally suited to test yeast activity. They record gas pressure generated during fermentation
Baking procedures	Straight dough process (AACC Method 10-10A) (Finney 1984)	100 g flour is mixed with yeast, sugar, shortening, malt, salt, and different amounts of oxidizing agents in order to produce an optimally developed dough. The dough is fermented, punched, or degassed with sheeting rolls, paned in standardized baking pans, and proofed in preparation for baking. From this laboratory procedure, important processing parameters and bread quality characteristics are obtained. The major parameters are optimum water absorption, optimum mix time, proof height, oven spring, bread volume, bread density, and crumb texture
	Sponge dough procedure (AACC Method 10-11)	Bread baking test for assessing quality by the sponge dough method using 700 g flour. The sponge is made by mixing part of the total flour with part of the total water and all yeast and yeast food. The sponge is fermented for 4 h and then mixed with the rest of the flour and ingredients. Resulting doughs are fermented, panned, and proofed following a time schedule before baking for 25 min at 225°C. The volume, density, proof height, oven spring, crumb texture, and color are evaluated
French bread	Serna-Saldivar (2012)	Wheat flour is mixed with yeast, salt, malt, and different amounts of oxidizing agents or other additives in order to produce optimally developed dough. The dough is fermented for a short time, punched or degassed with sheeting rolls, formed into the characteristic form, and proofed in preparation for baking. From this laboratory procedure, important processing parameters and quality characteristics are obtained. The major parameters are optimum water absorption, optimum mix time, proof height, oven spring, bread volume, bread density, and crumb texture
Lab. cracker procedure	Pizzinatto and Hoseney (1980)	The yield and quality of crackers can be determined in the laboratory using a standardized procedure where flour is divided into two lots: sponge and dough, mixed to attain optimum dough properties, fermented, and the cracker dough shaped into the typical rectangular configuration. Saltines could be salted before baking
Lab. pretzel procedure	Serna-Saldivar (2012)	The yield and quality of pretzels can be determined in the laboratory using a standardized procedure where flour is mixed to obtain dough that is manually shaped into pretzel forms, fermented, NaOH washed, salted, and baked
Extruded products (crispbread)	Expansion rate	Test widely used to monitor in-plant expansion of extrudates. Many factors affect the expansion rate: quality of raw materials (see Chapter 2), tempering, extrusion conditions (temperature–pressure gradient), and pressure release valves or vents prior to die forming
Water activity (a_w)	Water activity meter	The determination of a_w is critically important for low-moisture snack foods. These products are manufactured to reach a certain or given a_w level. Water activity affects microbial growth, enzyme activity, and product texture
Color value	Color meters (Hunter Lab, NIRA, and others)	The color of the end-products is one of the most important attributes affecting the acceptability of these snacks. The color of the extrudate is affected by the color of the raw materials, addition of fiber sources, % reducing sugars, and baking conditions
Hardness	Instron or TAXT2 texturometer (Serna-Saldivar 2012)	Products are compressed at a constant rate in order to obtain a force–deformation curve. The hardness affects the texture and overall acceptability of the extrudates
Organoleptic tests	Triangular tests (Serna-Saldivar 2012)	Ideally suited for comparison of two different samples. The trained or untrained panelist tries to identify the odd and paired samples among the three samples in the set and then is asked why she/he considered the odd sample different
	Hedonic preference test (Serna-Saldivar 2012)	Trained or untrained panelists evaluate the color, aroma, flavor, texture, and overall acceptability of the products in a sensory evaluation laboratory. Products are usually evaluated using a 9-point hedonic scale where 4 = like extremely, 0 = neither like nor dislike, and –4 dislike extremely

TABLE 19.6
Quality Control Parameters Most Commonly Used to Assess Quality of Nixtamal, *Masa*, Tortillas and Chips

Quality Control	Equipment Instrument	Importance
Nixtamal		
Moisture	NIRA, microwave, moisture meters/balances, air oven method (AACC 44-15A)	The moisture content of the cooked-steeped nixtamal is correlated to the extent of cooking
pH	Potentiometer calibrated with buffers	The pH of the nixtamal correlates to the lime absorbed during cooking and steeping and the amount of washing. The pH affects the flavor, texture, and color of end-products
Pericarp removal test	Use of May- Gruenwald dye (Serna-Saldivar et al. 1991)	Maize kernels are placed in nylon bags for lime-cooking for a given amount of time (20 min) and then stained with a May Gruenwald dye containing methylene blue and eosin Y. After washing excess dye, kernels are subjectively evaluated by determining the blue-greenish stained pericarp. Insufficient removal of the pericarp causes darkening and off colors in the snacks
Optimum lime-cooking time and dry matter loss	Mini-cooking trial (Serna-Saldivar et al. 1993)	Maize kernels are placed in a nylon bag and lime-cooked for three different times and then steeped in the cooking liquor overnight. Optimum cooking time is calculated using regression equations solving the time required to achieve a certain nixtamal moisture level (50% for table tortillas and 48% for tortilla chips) and dry matter loss is calculated by estimating the loss incurred during the cooking process. Both parameters are of utmost importance for production of fresh *masa* and snacks
Dough or *Masa*		
Moisture	NIRA, microwave, moisture meters/balances, air oven method (AACC 44-15A)	The moisture content of the masa greatly affects machinability and its rheology
Particle size distribution	Use of sieves to fractionate wet *masa*. Method of Pflugfelder et al. (1988)	A *masa* sample is fractionated in a set of sieves (US No. 35, 60, 100 and filter paper) with the aid of water sprayed with a squeeze bottle. The amount of sample recovered from each sieve is weighed and dried. The particle size distribution can be expressed on wet or dry matter basis. The method is fast and ideally suited to adjust the grinders in processing plants
Color value	Color meters (Hunter Lab and others)	The color of the *masa* is greatly affected by the color of the raw grain, the amount of lime used during cooking/steeping, and the amount of lime removed during nixtamal washing. The color of the *masa* correlates with the color of the raw and fried tortilla chips
Starch gelatinization	Birefringence with microscope equipped with polarized filters	Microscopic technique used to determine the relative amount of gelatinized vs. native starch granules (starch granules show Maltese cross)
	Glucoamylase method with subsequent measurement of glucose (AACC Method 76-16)	Chemical analysis in which the *masa* is subjected to amyloglucosidase hydrolysis followed by the determination of glucose with glucose oxidase. The ESS assay correlates with degree of cooking and starch gelatinization

(*Continued*)

TABLE 19.6 (CONTINUED)
Quality Control Parameters Most Commonly Used to Assess Quality of Nixtamal, *Masa*, Tortillas and Chips

Quality Control	Equipment Instrument	Importance
Viscoamylograph starch properties	Viscoamylograph (Shuey and Tipples 1982, Crosbie and Ross 2007)	It is one of the most important functional tests because after analyzing the viscoamylograph curve, temperature at the start of gelatinization, peak viscosity, shear thinning, and setback viscosity (retrogradation) could be determined. These properties greatly influence processing characteristics and quality of end-products. Peak viscosity is closely related to the amount of starch gelatinized during lime-cooking and grinding processes
Water absorption and solubility indexes	Centrifugation of slurry (Anderson et al. 1969)	These parameters are greatly affected by the degree of starch gelatinization or extent of cooking
Masa hardness and texture	Instron or TAXT2 compression test	A given piece of *masa* is compressed at a constant rate in order to obtain a force–deformation curve
Masa texture and rheology	Penetrometer	The force required to deform *masa* obtained after hydration with a given amount of water gives an indication of amount of cooking or starch gelatinization
	Adhesiveness test (Ramirez-Wong et al. 1993).	The adhesiveness of *masa* can be rapidly evaluated with a mechanical stickiness device consisting of two rectangular bars in a parallel arrangement. A block of *masa* is formed between the bars and partially split with a cutter and the adhesiveness measured by raising the movable bar until the *masa* detaches from it. The degree of adhesion of *masa* to the upper bar gives an adhesiveness index
	Texture analyzer or Instron. TPA analysis (Rooney and Suhendro 2001)	The texture profile analysis or TPA consists of a two-bite compression assay developed to determine adhesivity, cohesiveness, firmness, springiness, and other rheological parameters of *masa* using a texture analyzer. The technique was developed for quality control purposes and is sensitive to different degrees of lime-cooking, *masa* particle size, other processing conditions, and addition of gums and other additives
	Mixolab assay adapted to *masa* (Espinosa-Ramírez et al. 2020)	The Mixolab protocol was successfully used for the analysis of rheological properties of hydrated *masa* and proved to predict water absorption requirements, mixing conditions, *masa* stability, gelatinized starch properties, and retrogradation. The mixolab parameters also correlated with tortilla quality and texture
Tortillas		
Moisture	NIRA, microwave, moisture meters/balances, air oven method (AACC 44-15A)	The moisture content of baked and/or equilibrated tortilla affects oil absorption during frying
Pilot plant alkaline cooking	Tortilla pilot plant	Procedure to evaluate alkaline cooking properties of 3–1,000 kg lots of maize. Kernels are placed in a perforated nylon bag, lime-cooked, steeped, washed, and stone-ground into *masa*. The *masa* is sheeted and formed into tortillas or tortilla pieces in a small-scale three-tier baking oven. Yields of tortillas are calculated by knowing the original moisture content of the grain and the amount of tortillas obtained. Resulting tortillas can be used for texture, color, organoleptic, and shelf life stability tests

(*Continued*)

TABLE 19.6 (CONTINUED)
Quality Control Parameters Most Commonly Used to Assess Quality of Nixtamal, *Masa*, Tortillas and Chips

Quality Control	Equipment Instrument	Importance
Texture tests	Texture analyzer or Instron equipped with Kramer shear tests (Suhendro et al. 1998b)	The tortillas bending technique is sensitive to changes in tortilla texture throughout storage or equilibration
	Rollability test (Suhendro et al. 1998a)	The objective rollability test uses a custom-designed rollability fixture consisting of an acrylic cylindrical dowel and a metal chain that connects the cylinder dowel to the texture analyzer arm. The force and work required to pull the axle is determined. The technique is fast, simple, and sensitive to changes in tortilla texture
Color value	Color meters (Hunter Lab and others)	The color of the tortilla correlates with the color of the fried product
Fried Snacks		
Moisture	NIRA, microwave, moisture meters/balances, air oven method (AACC 44-15A)	The moisture content of the fried chip affects its texture and sensory attributes
Oil content	NIRA	It is the most important chemical determination for snacks. This is generally determined via near-infrared analysis. When properly calibrated, NIRA instruments determine fat and other chemical components in several seconds. However, instrument calibration and recalibration are critical for reliable analyses
	Hydraulic press (Serna-Saldivar 2008)	The hydraulic or Carver press is widely used to determine oil for quick quality control purposes. A ground sample is placed in a chamber for mechanical-pressing extraction. After the pressing cycle is completed, the free oil or the partially defatted snack sample is weighed. The instrument should be calibrated for each type of snack because they leave different amounts of residual oil (around 12% oil). Values should be correlated to solvent extraction values
	Solvent extraction Goldfisch or Soxhlet extractor (AACC Method 30-25)	The ether extraction technique is the official method of fat analysis. It consists of extracting oil using petroleum ether, followed by removal of the solvent and weighing of the oil. The main disadvantage of ether extraction is that it takes several hours to perform the test
Peroxide value	AOCS Cc8-53 AACC 58-16	It is the most widely used quality control parameter to estimate rancidity of snack foods. The main oxidation products of oils are hydroperoxides, the quantity of which can be quantitatively measured by the amount of iodine liberated by its reaction with potassium iodide. Values are expressed on mEq/kg oil
Active oxygen method (AOM)	AOCS Cd12-57 AACC 58-54	It is the most common rapid method to determine oxidative stability of oil-laden products. The method predicts the resistance of the extracted heated oil (97.8°C) to oxidation by bubbling air into the sample. The AOM method determines the number of hours required to reach 100 mEq peroxides/kg oil
Hexanal content	Gas chromatography with mass spectroscopy (Nzekoue et al. 2019)	Hexanal is a volatile closely related to extent of lipid oxidation and rancidity. It is determined in the headspace with solid-phase microextraction of the package and correlates with peroxide values and organoleptic properties

(*Continued*)

TABLE 19.6 (CONTINUED)
Quality Control Parameters Most Commonly Used to Assess Quality of Nixtamal, *Masa*, Tortillas and Chips

Quality Control	Equipment Instrument	Importance
Blistering or pillowing	Visual observation (Serna-Saldivar 2008)	The number and frequency of blisters or pillows in the fried tortilla chips are of upmost importance. Blistering is controlled by the extent of cooking, moisture content, and particle size of the *masa* and also by the temperature profiles during baking and frying operations
Hardness	Instron or TAXT2 texturometer	Chips are compressed at a constant rate in order to obtain a force–deformation curve. Values are greatly affected by the thickness of the chip
Color value	Color meters (Hunter Lab and others)	The color of the chip is one of the most important attributes affecting the acceptability of chips. The color of the chip is affected by the type of maize used as raw material, % residual lime or *masa* pH, and baking and frying temperatures
Salt	AACC Methods 40-60 ($AgNO_3$ Method) or 40-61 (total chloride method)	Added salt can be determined by estimating the amount of sodium or chloride. The salt is extremely important in finished products because it is closely related to the amount added as flavoring or as an indicator of the percentage of covering. Sodium is also important from the nutritional label
Sensory evaluation	Triangular test (Serna-Saldivar 2012)	It is ideally suited for comparison of two different samples. The trained or untrained panelist tries to identify the odd and paired samples among the three samples in the set and then is asked why she/he considered the odd sample different
	Hedonic preference test (Serna-Saldivar 2012)	Trained or untrained panelists evaluate the color, aroma, flavor, texture, and overall acceptability of the products in a sensory evaluation laboratory. Products are usually evaluated using a 5- or 9-point hedonic scale

testing (Lipp et al. 2005). Good quality maize begins with the farmer and ends in the packaged product. The quality of raw maize is critically important because it also affects processing conditions (Rooney and Suhendro 2001, Rooney and Serna-Saldivar 2015). In addition, a useful quality control guide to produce nixtamalized foods is included in the Corn Quality Assurance Manual of the Snack Food Association (Snack Food Association 1992).

The quality of the grain is so important that the industry recognizes "food grade maizes" as the best raw materials for production of both corn and tortilla chips (Rooney and Serna-Saldivar 2015). In the United States, in order to classify grain hybrids or varieties as "food grade," they should pass minimum standards of physical properties and other tests. The optimum kernels are selected based on bulk and true density values, floater kernels, endosperm hardness, presence of mycotoxins, and degree of insect/mold/heat damage (Table 19.2). The color of tortillas and snacks are influenced by the color of the kernel and cob, the amount of lime used during cooking, the extent of washing, and the final pH. Kernels from red cobs can yield off-colored tortillas because the tissue that remains after threshing is high in pigments (Rooney and Serna-Saldivar 2015).

The most important factor to control during alkaline cooking is the extent of cooking or optimum cooking time (Table 19.6). Most industries subjectively evaluate cooking by observing the condition of the cooked-steeped nixtamal. Analytical approaches that have been applied include tests for enzyme-susceptible starch, loss of birefringence, viscosity with consistometers, amylograph peak viscosity, and Instron shear force. The viscoamylograph (Shuey and Tipples 1982) and rapid viscoanalyzer (RVA, Crosbie and Ross 2007) are effective and fast ways to determine extent of nixtamal cooking (Rooney and Suhendro 2001, Rooney and Serna-Saldivar 2015).

Optimum cooking and steeping are determined subjectively by evaluating the extent of pericarp removal (Serna-Saldivar et al. 1991), kernel softening, and overall appearance of the nixtamal. There are strong correlations between the degree of cooking and water uptake in nixtamal, the best and most practical way to control this parameter is to monitor the amount of moisture absorbed by the lime-cooked and steeped nixtamal (Serna-Saldivar et al. 1993). Plasticity, cohesiveness, and stickiness are some of the subjective rheological properties of *masa* used to judge optimum cooking conditions (Snack Food Association 1992). Ramirez-Wong et al. (1993) developed a simple and rapid test to assess the stickiness of *masa*. The simple instrument consists of two rectangular bars in a parallel arrangement. A block of *masa* is formed between the bars and partially split with a wire cutter and the stickiness measured by raising the movable bar until the *masa* detaches from it. The degree of adhesion of *masa* to the upper bar gave an adhesiveness index.

Two important considerations in stone-grinding are the degree of grinding (*masa* particle size) and the amount of water supplied per unit of time. Pflugfelder et al. (1988) developed a simple method to determine the particle size distribution of wet *masa*.

A texture profile analysis (TPA) consisting of a two bite compression test was developed to determine adhesivity, cohesiveness, firmness, springiness, and other rheological parameters of *masa* using a texture analyzer. The technique was developed for quality control purposes and is sensitive to different degrees of lime cooking, *masa* particle size, other processing conditions, and addition of gums and other additives (Rooney and Suhendro 2001). Recently, Espinosa-Ramírez et al. (2020) developed an assay to assess rheological properties of hydrated *masas* with the novel Mixolab. The assay proved to predict water absorption requirements and mixing conditions of dry *masa* flours and other important parameters like dough stability, starch gelatinization, starch gel stability, and retrogradation. More importantly, this instrument was efficient to discern among the quality of different samples and parameters correlated adequately with tortilla moisture and texture.

Objective bending, rollability, and extensibility techniques were developed to measure tortilla texture with the texture analyzer (Suhendro et al. 1998a,b). The bending technique detects changes in tortilla texture throughout storage and correlates with subjective rollability and flexibility test scores. Likewise, the rollability technique, measured as the force and work required to pull an axle that caused a tortilla to roll around a dowel, was more sensitive to changes in texture and tortilla thickness than subjective evaluations. These techniques can be used to evaluate the effect of formulation and processing changes on fresh and stored tortillas.

Near-infrared or NIRA instruments can be used to monitor in-plant oil, moisture, and color of fried chips. Color, crispness, and texture of corn and tortilla chips and the amount of oil absorbed are influenced by *masa* moisture content, thickness of the chips, frying temperature, and residence time. A hydraulic press can be used to measure oil content in a short period of time. Chips are placed in the press and the oil pressed out is related to a calibration chart standardized for a given product. The amount of salt and the percentage of seasoning in fried products can be analyzed by procedures suggested by the Snack Food Association (1992). Chips are usually analyzed for peroxide value, active oxygen stability, and hexanal because they are excellent indicators of lipid oxidation and correlate with the consumer's preference and shelf life of the packaged products.

Sensory evaluation tests are widely used to determine the consumer's preference of finished products. Tests are usually performed in a sensory evaluation laboratory furnished with individual evaluation booths. Tests more widely practiced are the triangular and preference tests (Table 19.6). Triangular tests are ideally suited to identify the consumer's preference between two samples because panelists try to identify the odd and paired samples among the three samples in a set. The triangular test is widely used for product development purposes. In the preference test, trained or untrained panelists evaluate the color, aroma, flavor, texture, and overall acceptability of the products generally using a 5- or 9-point hedonic scale (Serna-Saldivar 2012). This test is used for product development purposes, shelf-life studies, and to study functionality of additives/technologies.

19.5.1.4 Extruded Snacks

The quality of extruded snacks covered in Chapter 12 greatly depends on the characteristics of the raw materials and on the various manufacturing steps used throughout processing. The quality of cereal-based grits, meals, and flour and refined raw and modified starches mostly described in Chapter 6 is critically important because it also affects processing conditions. Moisture, ash, particle size distribution, starch damage, and color are important quality attributes of raw materials. For starches, the viscoamylograph assay is very useful because it predicts pasting, retrogradation, and other important functionalities imparted by the starch (Shuey and Tipples 1982, Crosbie and Ross 2007). For oils, the stability, acidity, iodine value, and FA composition are important factors to consider (Table 19.7).

One of the most relevant factors to control during extrusion cooking is the extrudate density of direct-extruded puffs or expanded pellets (Table 19.7). There are different ways to assess the density of these second- or third-generation snacks. Most processors subjectively evaluate the bulk density by determining the weight of a certain volume of extrudates. The air displacement pycnometer can be used to determine more accurately the density. Analytical approaches that are commonly used to study textural properties of extrudates are the Instron or texture analyzer shear force or compression tests. As in most snack food products, the final moisture content and fat contents are continuously monitored. Moisture can be quickly determined with Near Infrared Instruments and the gravimetric methods using the gravimetric oven technique. The amount of fat can be rapidly determined via the use of Near Infrared Instruments or the Carver press and analytically assayed via solvent extraction in a Soxhlet or Goldfisch apparatus. The Near Infrared Instruments or colorimeters can also be used to monitor in-plant color of extruded products.

Frequent monitoring of the extrusion process is required to adjust settings for product uniformity. The most significant factors to control are moisture in the tempered raw materials, feed rate, temperatures/pressure in the different zones of the extruder and the die, and the extrudate expansion rate and moisture content.

The amount of salt and the percentage of seasoning in fried products can be analyzed by procedures suggested by the Snack Food Association (Snack Food Association 1992). Sensory evaluation tests are used to determine the consumer's preference of finished products. Tests more widely practiced are the triangular and preference tests (Table 19.7). Triangular tests are ideally suited to identify the consumer's preference between two samples because panelists try to identify the odd and paired samples among the three samples in a set. This particular test is frequently used for product development purposes. In the preference test, trained or untrained panelists evaluate the color, aroma, flavor, texture, and overall acceptability of the products using a hedonic scale (Serna-Saldivar 2012). This test is used for product development purposes, shelf-life studies, and to study functionality of ingredients, additives, and new processing technologies.

19.5.2 Potato, Fruit, and Vegetable Snacks

19.5.2.1 Potato Chips

The quality control of potatoes and chips or crisps has been thoroughly reviewed by Yee and Bussell (2007) and Gould (1999). Some of methods regularly used by the industry to assess the suitability of raw potatoes for chip processing are synopsized in Table 19.8. These methods are used to assess the quality of a particular lot of tubers during reception at the plant level, or may be performed for the selection of a new potato cultivars for processing (Chapter 14). Common tests used for raw tuber evaluation include potato size distribution, specific gravity, reducing sugar content, and slicing and trimming losses. A frying test is sometimes commonly performed by quality control and the moisture content, color, oil absorption or uptake, percent of defects, harsh, and bulking and yield are attributes measured on fried potato chips. The major potential toxicological compound associated with fried potato chips is acrylamide. The concentration of this important potential carcinogenic

TABLE 19.7
Quality Control Parameters Most Commonly Used to Assess Quality of Extruded Products

Quality Control	Equipment Instrument	Importance
Moisture	NIRA, microwave, moisture meters/ balances, air oven method (AACC 44-15A)	The moisture content of raw materials, tempered–equilibrated blends, half products or pellets, extrudates and finished products affect processing, product quality, and texture of snacks
Bulk density	Winchester Bushel meter (AACC Method 55-10) or measurement of weight/volume	Bulk density is important because it is closely related to the extrudate expansion rate and extrusion conditions. The advantage of this simple test is that it can be rapidly determined
Density	Pycnometer (nitrogen or air displacement)	True density is important because it is closely related to the quality of pellets or half-products. The advantage of this simple test is that it can be rapidly determined
Expansion rate or radial expansion	Diameter or extrudate/ diameter of die orifice	This test is widely used to monitor in-plant expansion of extrudates. Many factors affect the expansion rate: quality of raw materials (see Chapter 2), tempering, extrusion conditions (temperature–pressure gradient), and pressure release valves or vents prior to die forming
Starch content	NIRA, analytical determination of total starch (AACC 76-11)	It is the most important single component of dry-milled fractions, extrudates, and finished products. Starch greatly affects functionality and expansion rate
Starch gelatinization	Birefringence with microscope equipped with polarized filters	Microscopic technique used to determine the relative amount of gelatinized vs. native starch granules (starch granules show Maltese cross)
	Glucoamylase method with subsequent measurement of glucose (AACC Method 76-16)	Chemical analysis in which the *masa* is hydrolyzed with amyloglucosidase followed by the determination of glucose with the glucose oxidase technique. The assay correlates with degree of cooking and starch gelatinization
Starch properties	Viscoamylograph (Shuey and Tipples 1982, Crosbie and Ross 2007)	It is one of the most relevant functional tests because after analyzing the viscoamylograph curve, temperature at the start of gelatinization, peak viscosity, shear thinning, and setback viscosity (retrogradation) could be determined. These properties greatly influence processing characteristics and quality of end-products. Peak viscosity is related to the extent of starch gelatinization and starch damage
Water absorption index and water solubility index	Centrifugation of slurry (Anderson et al. 1969)	These parameters are greatly affected by the degree of starch gelatinization affected by extrusion conditions
Oil content	Near-infrared analyzers (NIRA)	It is the most important chemical determination for snacks. This is generally determined via near-infrared analysis. When properly calibrated, NIRA instruments determine fat and other chemical components in several seconds. However, instrument calibration and recalibration are critical for reliable analyses
	Hydraulic press	The hydraulic or Carver press is widely used to determine oil for quick quality control purposes. A ground sample is placed in a chamber for mechanical-pressing extraction. After the pressing cycle is completed, the free oil or the partially defatted snack sample is weighed. The instrument should be calibrated for each type of snack because they leave different amounts of residual oil (around 12% oil). Values should be correlated to solvent extraction values
	Solvent extraction Goldfisch or Soxhlet extractor (AACC Method 30-25)	The ether extraction technique is the official method of fat analysis. It consists of extracting oil using petroleum ether, followed by removal of the solvent and weighing of the oil. The main disadvantage of ether extraction is that it takes several hours to perform the test

(Continued)

TABLE 19.7 (CONTINUED)
Quality Control Parameters Most Commonly Used to Assess Quality of Extruded Products

Quality Control	Equipment Instrument	Importance
Peroxide value	AOCS Cc8-53 AACC 58-16	It is the most widely used quality control parameter to estimate rancidity of snack foods. The main oxidation products of oils are hydroperoxides, the quantity of which can be quantitatively measured by the amount of iodine liberated by its reaction with potassium iodide
Active oxygen method (AOM)	AOCS Cd12-57, AACC 58-54	It is the most common rapid method to determine oxidative stability of oil-laden products. The method predicts the resistance of the extracted heated oil (97.8°C) to oxidation by bubbling air into the sample
Hexanal content	Gas chromatography with mass spectroscopy (Nzekoue et al. 2019).	Hexanal is a volatile closely related to extent of lipid oxidation and rancidity. The volatile is normally assessed in the headspace of the package via solid-phase microextraction and correlates with peroxide values and sensory tests
Hardness	Instron or TAXT2 texturometer	Extrudates are compressed at a constant rate in order to obtain a force–deformation curve. The hardness affects the texture and overall acceptability of the extrudates
Color scores	Color meters (Hunter Lab, NIRA, and others)	The color of the extrudates is one of the most important attributes affecting the acceptability of these snacks. The color of the extrudate is affected by the color of the raw materials, % reducing sugars, baking and frying temperatures, and % covering
Salt	AACC Methods 40-60 ($AgNO_3$ method) and 40-61(total chloride method)	Added salt can be determined by estimating the amount of sodium or chloride. The salt is extremely important in extruded products because it is closely related to the amount added as flavoring or as an indicator of the percentage of covering. Sodium is also important from the nutritional label viewpoint
Sensory evaluation	Triangular tests (Serna-Saldivar 2012)	It is ideally suited for comparison of two different samples. The trained or untrained panelist tries to identify the odd and paired samples among the three samples in the set and then is asked why she/he considered the odd sample different
	Hedonic preference test (Serna-Saldivar 2012)	Trained or untrained panelists evaluate the color, aroma, flavor, texture, and overall acceptability of the products in a sensory evaluation laboratory. Products are normally evaluated using a 5- or 9-point hedonic scale

compound is normally determined with gas chromatography (Table 19.8). The generation and toxicology of acrylamide is discussed in Chapter 20.

19.5.2.2 Fruits and Vegetables

The quality control of fruits and vegetables has been exhaustively detailed by other authors (Ranganna 1986, Aked 2002, Perera and Perera 2005). The quality control applied to fruits and vegetables is broadly divided into the quality of fresh produce, especially in terms of product quality and cross-contamination with pathogenic bacteria and pesticides, and quality of processed products, including shelf-stable snacks. The quality of processed fruit and vegetable products is determined by the properties and conditions of the raw materials utilized (type of cultivar, maturity, cultural practices) and the efficiency and care taken during handling, storage, distribution, and processing (Chapter 16). As with fresh horticultural commodities, it is key to establish quality standards and to ensure their fulfilment through the use of a quality assurance program in terms of Good Agricultural Practices,

TABLE 19.8
Quality Control Determinations for the Selection of Raw Potato Tubers and Fried Potato Chips

Quality Test	Principle and Relevance
Raw Potatoes	
Size distribution (Gould and Plimpton 1985)	The immediate axis length is determined by passing the tubers through ring sizers designed at 10.8, 8.3, and 6.4 cm. Approximately, 4.5 kg of potatoes are used as one replicate for the determination of size distribution. It is best to perform the measurement in triplicate. The size range should be reported as the percentage of tubers, which fall into each size designation (small, medium, large, and extra-large). Large and medium tubers are preferred for potato chip production
Specific gravity (Gould and Plimpton 1985).	Determined by the weight in air/weight in water method at 20°C. The test should be preferably performed with 60–100 individual tubers
Reducing sugar analysis (AOAC Method 906.1-1906, Gusakov et al. 2011)	The most commonly used methods to determine amounts of reducing sugars in foods is the Somogyi-Nelson, 2,5 dinitrosalicylic acid (DNS) colorimetric assay (Gusakov et al. 2011) and the Munson-Walker methods (AOAC 906.1-1906). An aqueous extract from potatoes is prepared and filtered before the determinations. In addition, there are commercial analytical kits used for routine analysis of potato chip companies. Reducing sugars should be below an upper limit range of 2,875–3,300 ppm fresh weight (equivalent to 0.29–0.33% w/w) and is of upmost importance because it affects color of fried chips
Peeling and trimming losses (Gould and Plimpton 1985, Gould 1999)	Peeling and trimming losses are determined by differential weighing before and after each unit operation. Peeling and trimming losses are calculated by the following equations: *Percent peeling losses* = [(initial weight – peeled weight)/initial weight] × 100 *Percent trimming losses* = [(peeled weight – trimmed weight)/peeled weight] × 100 It is desirable to lose <10% of the initial potato weight during peeling and trimming losses should preferably be below 10%
Slicing (Gould and Plimpton 1985, Gould 1999)	To perform the slice thickness evaluation, the blades of the slicer must be sharp. Variations in slice thickness may be recorded by measuring the thickness in various positions of each slice with a micrometer. The processing facility can also keep records of the average number of slices per centimeter and of the standard deviation in slice thickness for a single and a group of slices. The thickness of the slices are normally 1.27–1.78 mm for straight-cut slices and 2.54–3.05 mm for crinkle-cut counterparts
Frying test (Gould and Plimpton 1985)	A frying test is frequently performed to evaluate a particular lot of tubers during reception at the plant level, or for the selection of a new potato cultivar for processing. The frying conditions should be consistent every time a new lot is evaluated as well the slice thickness (1.27–1.78 mm). The fryer should be thermostatically controlled and should hold sufficient good quality oil to allow a standard size sample (250 g minimum) a temperature drop of less than 11°C from the time the chips are placed in the oil until they are fried. When the fryer temperature falls lower than 177°C, excessively light colored and possibly soft centered/oily chips may be produced. After frying for 2 min, the sample is evaluated for yield, residual moisture content, color, oil uptake, percent of defects, harsh, and bulking
Fried Potato Chips	
Potato chip yield (Gould and Plimpton 1985, Gould 1999)	The calculation of potato chip yield is obtained after weighing the raw potato slices and the resulting fried potato chips using the following equation: [(weight of raw potato slices – weight of fried potato chips)/initial weight of raw slices] × 100. Normally, yields of 28–36% are considered optimum for commercial potato chip operations
Defective chips (Gould and Plimpton 1985)	The evaluation consists of first obtaining a representative sample of potato chips. Then, chips are visually examined in order to sort and separate defective chips which are classified into minor or major defects. For instance, potato chips with major defects contain a defective area of 3.35 cm^2, whereas a minor defect of 1.61 cm^2. The other important defect is the amount of blisters which are determined by separating and counting blistered or pillowed chips and expressed as percentage

(*Continued*)

TABLE 19.8 (CONTINUED)
Quality Control Determinations for the Selection of Raw Potato Tubers and Fried Potato Chips

Quality Test	Principle and Relevance
Percent harsh (Gould and Plimpton 1985)	A package of potato chips containing a minimum of 100 g is opened and the sample separated using sieves into chips, crumbs, and harsh (defined as anything smaller than a set area of 6.5 cm^2). The crumbs and harsh are weighed and the percentage of each in the original sample is calculated
Bulking test (specific volume) (Gould and Plimpton 1985)	For the evaluation of bulking, a sample of fried chips (100 g) is weighed and then placed into a 500-mL beaker and the volume recorded. Specific volume is calculated and reported as cm^2/g (Gould and Plimpton 1985)
Moisture NIRA, moisture meters/balances, air oven method (AACC 44-15A)	Moisture after potato chip frying is important because it is related to crispiness. Normally the target moisture content is 1–2%
Oil uptake or absorption NIRA, Soxhlet solvent extractor (AACC Method 30-25), hydraulic Carver press (Snack Food Association 1992)	The critical determination of oil content is generally made via a properly calibrated near-infrared analyzer or by the Soxlet or ether extraction official methods. The hydraulic or Carver press is used to determine oil for quick quality control purposes. The press should be calibrated for potato chips comparing with official assays. The advantage of the NIRA and hydraulic press methods is that they are performed in short times. Tubers with specific gravities above 1.085 produce chips with oil contents in the range of 35–40%. When a de-oiling step is used, the oil content of potato chips can be reduced to 18–25%
Color Colorimeters and comparison charts	Potato chip color may be determined subjectively by rating the color of a sample of 20 chips by the use of potato chip reference charts published by the Snack Food Association (1992). Instrumental color determinations may also be obtained by crushing the potato chip sample into fine pieces using a mortar and pestle, and packing the crumbles in an appropriate glass cylindrical container. Instrumental lightness (L), chroma, and hue angles can be recorded using an instrument such as the Hunter Lab colorimeter
Texture (hardness) TAXT2 texture analyzer	The texture analyzer compresses chips at a constant rate in order to obtain a force–deformation curve. The hardness affects the texture and overall acceptability of the potato chips
Proxide value AOCS Cc8-53, AACC 58-16	It is the most widely used quality control parameter to estimate rancidity of snack foods. The main oxidation products of oils are hydroperoxides, the quantity of which can be quantitatively measured by the amount of iodine liberated by its reaction with potassium iodide: less than 100 mEq/kg
Active oxygen method AOCS Cd12-57, AACC 58-54	It is the most common rapid method to determine oxidative stability of oil-laden products. The method predicts the resistance of the extracted heated oil (97.8°C) to oxidation by bubbling air into the sample
Salt AACC Methods 40-60 ($AgNO_3$ method) and 40-61(total chloride method)	Added salt can be determined by estimating the amount of sodium or chloride. The salt is extremely important in extruded products because it is closely related to the amount added as flavoring or as an indicator of the percentage of covering. Sodium is also important from the nutritional label

(*Continued*)

TABLE 19.8 (CONTINUED)
Quality Control Determinations for the Selection of Raw Potato Tubers and Fried Potato Chips

Quality Test	Principle and Relevance
Hexanal Gas chromatography with mass spectroscopy (Nzekoue et al. 2019)	Hexanal is a volatile closely related to extent of lipid oxidation and rancidity. The volatile is normally assessed in the headspace of the package via solid-phase microextraction and correlates with peroxide values and sensory tests. Hexanal levels above 62 ppm may be related to undesirable rancid flavors (Li et al. 2000)
Electronic nose (Shen et al. 2001, Gan et al. 2005, Xu et al. 2016, Upadhyay et al. 2017, Majchrza et al. 2018)	The novel electronic nose chromatography (mass spectrometry and gas chromatography) technology is increasingly being used for the quality control of snacks with the advantages of short time of analysis, adequate sensitivity, and relatively low cost. The main advantage of this assay is that it effectively measures the relevant volatile compounds that correlate positively with sensory evaluation and olfactory tests
Organoleptic tests (Serna-Saldivar 2012)	These are triangular or hedonic tests. Triangular tests are suited for comparison of two different samples, whereas hedonic tests conducted with trained or untrained panelists evaluate the color, aroma, flavor, texture, and overall acceptability of the products in a sensory evaluation laboratory. Products are normally evaluated using a 5- or 9-point hedonic scale
Acrylamide (Ghiasvand and Hajipour 2016)	Acrylamide is a potentially toxic and carcinogenic compound that can be produced in significant amounts in fried potato chips due to Maillard reactions. The toxicology of this compound is covered in Chapter 20. The most common determination of acrylamide in potato chips is by using headspace solid-phase microextraction coupled with gas chromatography–flame ionization detection or via quantification with mass spectroscopy

Good Manufacturing Practices, and HACCP programs described before. The selection of appropriate cultivars for particular products and determination of the degree of maturation are of utmost relevance. The quality of the irrigation water and the harvesting and postharvest practices are aimed toward the minimization of the respiration process and cross-contamination with important bacteria like *Listeria*, *Escherichia coli 0157*, and other coliforms, which if not removed can persist in the food chain and eventually cause foodborne illness to the consumers. Fresh fruits are not considered to be high-risk products with respect to food safety as they normally have an acidic pH and become completely undesirable for consumption long before any hazardous microorganisms or toxins might develop. On the other hand, fresh vegetables which normally contain a comparatively higher pH compared to fruits are also low-risk items because of their relatively high and low fiber and sugar contents, respectively (Ranganna 1986, Aked 2002, Perera and Perera 2005).

The fresh produce is normally evaluated in terms of size and shape, defects (underripe and overripe, insect and mold-damaged), density or specific gravity, maturity, firmness, color, soluble solids, and titratable acidity. In most fruits, the brix to acidity ratio is used as a quick indicator of maturity and flavor.

Enzymes catalyze desirable or undesirable reactions in fruits and vegetables and some are indicators of the efficiency of thermal treatments. Therefore, there are simple kits or assays to determine enzymes which catalyze color, flavor, and textural changes associated with ripening of fruits and horticultural crops. The PPO (polyphenol oxidase) catalyzes browning of senescent and cut items, whereas the polygalacturonase (PG) is an important pectolytic enzyme heavily implicated in the softening or loss of texture during ripening of fruits and horticultural commodities. On the other hand, LPO (lipoxygenase) and peroxidases catalyze the formation of off-flavor and off-odor metabolites and color changes. All these enzymes are denatured after blanching and application of other thermal treatments to prevent loss of quality of end-products. Additionally, these enzymes are inhibited using certain food additives and/or exclusion of oxygen. The instrumental analysis of activities normally involves the extraction of the enzymes with water or buffers followed by a colorimetric determination.

Most snacks produced from fruits or vegetables are processed via regular or osmotic dehydration (Chapter 16) and therefore contain lower moisture and higher solids and soluble sugar concentrations, which significantly lowers the a_w to levels that impede growth of putrefactive and pathogenic bacteria, molds, and yeasts. The most relevant quality control measurements are moisture, pH, titratable acidity, a_w, reducing sugars, color, aroma, texture, and flavor sensory tests previously discussed for other types of snacks. From the sanitary viewpoint, snacks are usually tested in terms of insect fragments and aerobic plate counts, coliforms, and other microbial plate counts. Finally, the residual amounts of different pesticides are assessed to protect consumers from their toxicological effects (Ranganna 1986, Aked 2002, Perera and Perera 2005).

19.6 QUALITY CONTROL OF DAIRY, MEAT, AND SEAFOOD

19.6.1 Dairy Products

The quality of dairy products is assessed in terms of a wide range of criteria, including, chemical, physical, functional, microbiological, and nutritional characteristics. The ultimate goal is to offer high-quality and uniform products that are safe to the consumers. Milk and processed dairy products are one of the most susceptible foods for microbial contamination due to the high moisture, a_w, adequate pH, and mainly because of the intrinsic nutrients that favor microbial growth.

Similar to other food groups, the quality control of dairy products is broadly divided into quality of raw milk and processed products like yoghurt, fresh and fermented cheeses, powdered milk, dairy candies, and many other related foods. Table 19.9 summarizes the main physicochemical, microbial, organoleptic, and toxicological quality tests applied to raw milk and processed dairy products. The importance and description of these methods are covered by Tamime and Robinson (2000) and Mostert ad Jooste (2012). In terms of raw milk, the most important parameters are total solids, fat content, titratable acidity, and somatic cell counts, whereas in processed related products a_w, pH, acidity, color, texture, organoleptic assays, and toxicological tests especially in terms of aflatoxins M and residual antibiotics, pesticides, and hormones. One of the most important tests are those related to the presence of pathogenic bacteria that can severely harm human health. In this sense, the industry routinely inspects products in terms of aerobic plate counts and pathogens like *Staphylococcus*, *Listeria*, *Salmonella*, *Yersinia*, and coliforms. For the particular case of yoghurt, cheese, and milk, confectionary products widely used as snacks, the pH, titratable acidity, a_w, % salt, % fat, % protein, protein:fat ratio, color, texture, viscosity, sensory analysis, and pathogenic and probiotic microbial counts are of paramount importance.

19.6.2 Meat- and Marine-Based Products

The quality of animal, poultry, and marine based snacks is more rigorous than the grain- or fruit-based counterparts because of the greater possibilities of contamination with pathogenic bacteria. Table 19.9 summarizes main quality control parameters used for raw meats and finished products. Raw materials are generally characterized for moisture, protein, and fat because formulations are generally adjusted according to the moisture:protein:fat ratio. The type of protein is critically important especially for semidried sausages because the muscle proteins (myofibrillar) act as binder and fat emulsifier. In contrast, connective tissue proteins (collagen and elastin) do not impart functionality and counteract the positive effects of the myofibrillar proteins. Moisture is generally tested via near-infrared analyzer (NIRA), microwave oven, or infrared dehydration and fat with the NIRA, Fosslet assay, or modified Bobcock procedure.

TABLE 19.9
Quality Control Parameters Most Commonly Used to Assess Quality of Dairy and Meat/Poultry/Marine Raw Products and Related Snacks

Quality Control	Equipment Instrument	Importance
Moisture and total solids	NIRA, microwave (AOAC 985.14) moisture meters, moisture balances, air oven method (AACC 44-15A), vacuum drying (AOAC 950.46) Pre-dry method for milk (AOAC 16.032)	Moisture and total solids are one of the most relevant parameters to determine in dairy and meat/marine raw materials, intermediate, and finished products. In milk, the direct forced air method, and the AOAC pre-dry method 16.032 (Clark et al. 1989) are normally used. The determination of moisture in processed meats is essential to determine the moisture:protein ratio in finished products
Moisture protein ratio	Moisture divided by protein contents	The moisture:protein ratio of raw and processed meats are widely used for quality assurance and product control purposes. The ratio is a controlling factor in determining added water and is specified as a finished product standard for processed meat products. Carcass fresh meats usually contain 3.6–4:1 ratios, whereas organ meats contain 4.0–6.3:1 and processed meats from 0.75:1 (jerky, Chapter 18) and 1.6-1.9:1 (dry salami and pepperoni, Chapter 18)
Protein content	NIRA (AACC Method 39-10)	The amount of protein can be rapidly determined using an infrared analyzer that was previously calibrated with a set of standards. This is particularly important to determine the moisture:protein ratio. The myofibrillar or muscle proteins impart functionality to meat products when treated with salt
	Kjeldahl methods (AACC 46-10, 46-12, or 46-13) Udy dye method (AACC 46-14A)	These are the most widely analytical procedures to determine nitrogen and protein. The main disadvantage is that these analytical methods take a long analysis time
Fat content	Solvent extraction Goldfisch or Soxhlet extractor (AOAC 960.39, AACC Method 30-25), Fosslet technique (AOAC 976.21), rapid microwave solvent extraction method (AOAC 985.15)	The ether extraction technique is the official method of fat analysis. It consists of extracting oil using petroleum ether, followed by removal of the solvent and weighing of the oil. The main disadvantage of ether extraction is that it takes several hours to perform the test. The fat content of raw meats and finished products is extremely important because it affects sensory properties, shelf life, and functionality. The rapid techniques for determination of fat in meat products are the Fosslet and rapid microwave solvent methods
	NIRA	When properly calibrated, NIRA instruments determine fat and other chemical components in several seconds. However, instrument calibration for each type of meat and recalibration are critical for reliable analyses
	Butyrometer or Gerber test for milk (AOAC method 2000.18, Kleyn et al. 2001)	The Gerber method is used worldwide as a simple and rapid method for determining fat in raw and processed milks. Samples are acid digested with sulfuric acid and then isoamyl alcohol is added in order to release the fat which is measured by the graduated Gerber butyrometer
	Modified Babcock test	Rapid fat test commonly used for meats. Samples are acid digested in Paley test bottles and centrifuged. The floating fat is determined after reading the top graduation of the bottle

(Continued)

TABLE 19.9 (CONTINUED)
Quality Control Parameters Most Commonly Used to Assess Quality of Dairy and Meat/Poultry/Marine Raw Products and Related Snacks

Quality Control	Equipment Instrument	Importance
Somatic cells counts in milk (DMSCC – Single Strip Procedure FDA-2400d and Instrumental tests)		The somatic cell count (leukocytes or white blood cells) in milk is the main biomarker for assessing its quality and counts increase as an immune response to a mastitis-causing pathogens. It is traditionally performed by exhaustive methods consisting of the visual observation of stained cells in smears through a microscope. A healthy cow produces milk with less than 100,000 cells, whereas infected cows normally have 300,000 or more. Over the last years, new methods for the identification of somatic cells were developed – Scepter™ 2.0 Cell Counter, California Mastitis Test, Fossomatic and portable deLaval cell counter (flow cytometry) which are less dependent on the operator and therefore provide precise and reliable measures
Connective tissue	Determination of praline and hydroxyproline	The amount of connective tissue protein is important because it has a detrimental effect on protein functionality (decreases binding). The connective tissue is indirectly determined via the analysis of praline and hydroxyproline, amino acids present in high amounts in connective tissue and negligible amounts in myofibrillar proteins
pH	Potentiometer calibrated with buffers	The pH of the raw meat is an important quality control measurement because it is different for fresh meat and post rigor mortis meat. The pH also affects the susceptibility of meat to microbial attack, the water-holding capacity of meat proteins, and flavor, texture, and color of end-products
Acidity	Acidity in milk (AOAC International Method 947.05)	A milk sample is first boiled and cooled with distilled water. Then phenolphthalein is added in preparation for titration with standardized NaOH until the color changes to pink for 30 s
	Titratable-indicator method	This is an important test for animal-based snacks because it greatly affects shelf life. It is also important for fermented processed meats because fermentation is closely related to acidity (usually expressed as lactic acid). An aqueous extract obtained from the product is titrated with a dilute alkali solution to neutrality and the acidity expressed as lactic acid equivalents
Water activity (a_w)	Water activity meter	The determination of a_w is critically important for processed milk products (Chapter 17) and low-moisture meat-based snacks like jerky and semidry or dried sausages (Chapter 18). These products are manufactured to reach a certain a_w value because it affects microbial growth, shelf life, and product texture
Color score	Color meters (Hunter Lab, NIRA, and others)	The color of processed milk products and cured and regular meat products is one of the most important attributes affecting the acceptability of these snacks. The color and pigmentation of raw meats, the % myoglobin, the use of curing agents, the pH/acidity, and heat treatments applied during processing affect the color of meat based snacks

(Continued)

TABLE 19.9 (CONTINUED)
Quality Control Parameters Most Commonly Used to Assess Quality of Dairy and Meat/Poultry/Marine Raw Products and Related Snacks

Quality Control	Equipment Instrument	Importance
Hardness	Instron or TAXT2 texturometer	Dairy and processed meat products are compressed at a constant rate in order to obtain a force–deformation curve. The hardness affects the texture and overall acceptability of the snacks
Salt	Volhard method, AACC Methods 40-60 ($AgNO_3$ Method), AOAC 983.18 for meats, 937.09 for seafood, and/or AACC 40-61(total chloride method), indicating strip method (AOAC 971.19)	Sodium chloride contents of processed dairy products and animal-based snacks are determined officially by the Volhard method. Samples are acid digested and chloride ion released and reacted with $AgNO_3$. The AgCl is back titrated to determine NaCl. Added salt can be determined by estimating the amount of sodium or chloride. The salt is extremely important in snack meat/marine-based products because it acts as preservative and affects flavor, a_w, and organoleptic properties. Sodium is also important because it should be declared on the nutritional label
Nitrates/nitrites	Xylenol method AOAC 935.48. NED colorimetric method for cured meats (AOAC 973.31)	Nitrates and nitrites are responsible for the excellent public health record of cured meat products because they prevent growth of pathogens. Nitrites react with hemoglobin and myoglobin iron and upon heating result in cured meat color. Most cured meat products should contain less than 156 ppm of nitrates
Aerobic plate counts	AOAC 988.18	Total aerobic counts are widely used to determine microbial populations in fresh and processed dairy and meat/poultry/marine products. In meats, most of these bacteria are *Pseudomonas* which decrease shelf life
Pathogenic bacteria	*Listeria* (ELISA tests), *Coliforms*, *Escherichia coli* (AOAC 978.23 and 983.25), *Staphylococcus* (AOAC 980.37, and 987.09), *Clostridium perfringens* (AOAC 976.30), *Salmonella* (AOAC 967.25, 967.26, 967.27, 967.28, 975.54 986.35, 989.12), *Vibrio* (AOAC 988.20), and others	Pathogenic bacteria can be identified via preparation of specific culture media, diagnostic kits, serological tests, and colorimetric clonal enzyme immunoassays. ELISA tests have gained popularity because they are very practical in terms of time consumed and accuracy
Trimethylamine nitrogen in seafood	AOAC 971.13	In seafood products, this test is highly related to the degree of deterioration. TMA is determined via nitrogen determination of a supernatant
Fat oxidation	TBA (Fernandez et al. 1997)	It is the most widely used quality control parameter to estimate oxidation of meat-based foods. The colorimetric technique is based in the reaction of thiobarbituric acid with malondialdehydes
Organoleptic assays	Triangular and hedonic tests (Serna-Saldivar 2012)	Triangular tests are normally used to compare two samples, whereas hedonic tests conducted with trained or untrained panelists evaluate the color, aroma, flavor, texture, and overall acceptability of several similar products in a sensory evaluation laboratory. Products are usually evaluated using a 5- or 9-point hedonic scale

(*Continued*)

TABLE 19.9 (CONTINUED)
Quality Control Parameters Most Commonly Used to Assess Quality of Dairy and Meat/Poultry/Marine Raw Products and Related Snacks

Quality Control	Equipment Instrument	Importance
Nitrosamines	AOAC 984.18, 980.22	Nitrosamines mainly formed in cured meat products exposed to harsh thermal treatments are known carcinogens (Chapter 20). Some regulatory agencies have strict controls on maximum levels of nitrosamines allowed in processed meats
Aflatoxins in milk (Vaz et al. 2020)	Aflatest and chromatography	The utilization of ELISA columns to selectively isolate aflatoxins followed by their quantification via fluorescence allows a quick and reliable determination of aflatoxins M regulated by regulatory agencies (maximum levels vary from 0.05 to 5 μg/L in fluid milk, 0.25–0.5 μg/kg in chesses and other processed dairy products, and 5 μg/kg in dry milk). Test can be performed in less than 10 min. The more accurate determination of aflatoxin M in milk is also performed by high-performance liquid chromatography with photochemical post-column derivatization and fluorescence and mass spectroscopy detection
Heavy metals	Atomic absorption spectrophotometer or inductively coupled plasma-mass spectrometry (ICP-MS)	The concentrations of bioaccumulated heavy metals (Pb, Cd, and Hg) in meats, fish, milk, or other food matrixes are normally assessed after an acid hydrolysis of the food in a microwave oven followed by the assessment of heavy metals using atomic absorption spectroscopy furnished with the proper lamps or by ICP-MS. The last instrumental analysis is capable of determining heavy or toxic metals in ppq (parts per quadrillion)
Antibiotics	Immunological (ELISA) or microbial inhibition screening tests, GC, and liquid chromatography technologies (Schenck and Callery 1998, Jayalakshmi et al. 2017), UPLC-quadrupole MS (Cepurnieks et al. (2015)	There are various approved analytical methods for screening and confirmation of antibiotic residues in animal products. The main ones are microbial inhibition test, biosensor, ELISA, HPLC, GC, liquid chromatography with mass spectrometry (LC–MS), liquid chromatography–tandem mass spectrometry (LC–MS/MS), and ultra-performance liquid chromatography–mass spectrometry (UPLC–MS). More recently, the quadrupole-MS instrumental analysis is capable of determining 26 antibiotic residues (including sulfonamides, macrolides, tetracyclines, penicillins, and quinolones) in both meat and milk. The analytical procedure subjects a representative sample of meat or milk to simple extraction procedures followed by UPLC-quadrupole-MS
Hormones	Gas or liquid chromatographies. GC–mass spectrometry is especially employed (Seo et al. 2005)	Important growth hormones like estradiol, testosterone, progesterone, zeranol, and diethylstilbestrol were determined simultaneously in meat. The extraction is carried out using methanol–water and ultrasonication. The hormones in the extract are effectively separated from lipids by freezing-lipid filtration, followed by C_8-solid-phase extraction. The quantitation is performed using isotope-labeled internal standards by GC–MS in the selected ion-monitoring mode

(*Continued*)

TABLE 19.9 (CONTINUED)
Quality Control Parameters Most Commonly Used to Assess Quality of Dairy and Meat/Poultry/Marine Raw Products and Related Snacks

Quality Control	Equipment Instrument	Importance
Pesticides	GC or HPLC systems with different detectorsLeDoux (2011)	Pesticide residues (mainly organochlorine and organophosphorus) from meat and poultry products, milk and processed dairy products, fish and shellfish, and eggs are first extracted and then quantified by different chromatographic techniques. Most pesticide residue detection methods for food samples comprise two key preparation steps prior to identification/quantification: extraction of target analytes from the bulk of the matrix, and partitioning of the residues in an immiscible solvent and/or cleanup of analytes from matrix coextractives, especially fat which interferes with assays. Following the extraction/ purification procedures, pesticides are separated either by GC or liquid chromatography and then identified and quantified using different types of detectors like electron-capture, flame photometric, nitrogen–phosphorus, fluorescence, and diode-array

Standardized laboratory methods are widely used to test meat pH, acidity, salt, and nitrate/nitrite contents. The meat acidity is typically used to follow the production of bacteria-acidified sausages discussed in Chapter 18 because it affects susceptibility to growth of pathogenic bacteria, organoleptic properties, and water-holding capacity of the product. The antimicrobial salt and nitrates are extremely important to quantify because they affect bacterial growth, water activity, and the organoleptic properties of the finished products (flavor, color, and texture). In addition, nitrates are usually added to cured products to a maximum level of 156 ppm.

Freshness and rancidity are usually tested with the trimethylamine (TMA) and thiobarbituric (TBA) assays, respectively. TMA is widely used to test fish and shellfish sources and determines the amount of volatile nitrogen, so it strongly correlates with the aerobic microbial contamination. The TBA assay is preferred for testing animal and marine products because it correlates better with rancidity. The TBA assay determines the amount of malondialdehyde present in a product that had lipid decomposition.

The analytical approaches that are commonly used to study textural properties of animal-based snacks are the Instron or texture analyzer shear force or compression tests.

From the food safety viewpoint, animal, poultry, fish, and crustacean based products should be frequently monitored in terms of putrefactive and pathogenic microbial counts (Table 19.9) to ensure the wholesomeness of the products and prevent foodborne illnesses. The US Food Safety and Inspection Service (FSIS) published a mandatory HACCP system of process control for all inspected meat and poultry plants. The HACCP system that regulates shelf-stable meat and poultry products is aimed toward the prevention of biological, physical, and chemical contamination. There are basically two kinds of microbial hazards: infectious bacteria such as *Listeria monocytogenes*, *Salmonella* sp., and *Clostridium perfringens* that must be inactivated by an effective lethal treatment and establish measurements to prevent recontamination. The other type are toxigenic bacteria such as *Clostridium botulinum* and *Staphylococcus aureus* that must be inhibited from growing and producing their toxins during processing and during the shelf life of the finished product. These bacteria can pose a hazard when the drying process is too low or insufficient. The hazard of these

TABLE 19.10
Quality Control Tests Most Commonly Used to Assess Quality of Packaging Materials

Quality Control	Equipment Instrument	Importance
Film dimensions, thickness, and weight/area	Measurements of dimensions, weight, and thickness	These three tests are regularly applied as quick quality control measurements of plastic laminates. Thickness is determined with a micrometer with a precision of 0.002 mm and the weight of a certain area of the film determined with an analytical balance
Tensile properties	Universal Instron or texture analyzer	This is one of the most important tests for plastic films. The plastic film is subjected to a constant rate of deformation; so the resistance, deformation, elongation, and rupture of the material is obtained via the analysis of a force–extensibility graph
Film resistance	Falling dart method	It is a standard test in which a free falling dart impacts a film with a given thickness fixed to a 12.7-cm diameter ring. The height of the test and weight of the dart are varied in order to rupture 50% of the tested films
Tear resistance and thin sheeting	Universal Instron testing, texture analyzer, Elmendorf apparatus	Tear resistance is an important characteristic of flexible packaging materials because it affects the mechanical performance and machinability. The test is also useful to determine the force necessary to open plastic bags or pouches. A force deformation graph is obtained
Peel resistance of adhesives	Texture analyzer or Universal Instron machine	Standard test applied to see the resistance of films glued together
Seal strength	Static method Dynamic method	These are tests widely used to determine strength of thermoplastic materials fin- or lap-sealed. Values are expressed in terms of force/thermo-sealed film area. In the static method, the thermo-sealed area of the material is horizontally fixed to a support. A certain weight applies the force to separate the thermo-sealed films. The dynamic method uses a texture analyzer or dynamometer to measure the force required to separate the sealed plastic materials
Hot tack	Packforsk equipment	The hot tack value expresses the resistance of the thermo-sealed area right after hot sealing and its effectiveness is influenced by the seal temperature and pressure, characteristics of the thermo-sealed materials, the chemical composition of the plastic film, and film thickness, among others. Normally, the hot tack values increases with temperature until reaching a maximum value. Excess temperatures usually diminish values due to the loss of viscosity
Type of plastic film or chemical composition	Infrared absorption spectra	The infrared absorption spectra is like a fingerprint for the different polymers that constitute packaging materials. This spectroscopy technique can be applied directly on the film, or in chemically dissolved films. For nontransparent films, potassium bromate is used
Headspace gas composition	Gas chromatography-thermal conductivity detector	The composition of gases in the headspace is important for snacks because they could affect rate of product oxidation/rancidity and shelf life. The test is also important in packaging operations where inert gases are flushed into the plastic bag. The test consists of sampling the headspace and then the injection of the sample into a GC-thermal conductivity for determination of the gas composition
Oxygen permeability test	Coulometric sensor	It is a standard test widely used to study the oxygen transmission rate through a determined plastic film. The plastic material is placed in a diffusion cell where oxygen or oxygen–nitrogen is flushed in order to determine permeability. The oxygen that passed the film is determined via a coulometric sensor

(Continued)

TABLE 19.10 (CONTINUED)
Quality Control Tests Most Commonly Used to Assess Quality of Packaging Materials

Quality Control	Equipment Instrument	Importance
Gas permeability test		This method determines the permeability rate of various gases through two packaging films located in a diffusion cell. The chemically characterized gas is applied at a constant flow rate in the external chamber and permeates into the intermediate part of the cell that contains the two sealed plastic films. The permeability is determined by analyzing the composition of the gas contained between the two films
Vapor water permeability test	Gravimetric method	This test determines the permeability of vapor water through a plastic material placed in a capsule where the rim of the film is glued with wax and a hygroscopic material placed inside the capsule. The area of the plastic sample and thickness of the film are obtained. The capsule is placed in an environmental room with strict temperature and relative humidity controls (two conditions are used 25°C/75% RH and 38°C/90% RH). The test is based on the weight or water gain of a hygroscopic material (generally calcium chloride) located inside the chamber determined with an analytical balance. Values are commonly expressed in g/m^2/day. The assay is important for water-sensible or hygroscopic foods such as snacks
Fat or grease permeability		This rapid test allows the determination of the speed of fat permeation in flexible packaging materials. The assay consists of placing samples of the plastic film some with the internal face up and others with the external face up on a glass. The fat-containing food is placed in direct contact with the surface of the film and a weight placed on top of the whole arrangement. The test is performed at a fixed temperature (40 or 60°C). The time required for the glass surface to get fat contamination (light dispersion in the area where the weight is placed) is registered
Haze and luminous transmittance	Spectrophotometer UV	The light transmittance is determined as the % incident light and the haze as the intensity of light transmitted with dispersion. A high haze value hinders visibility and color distinction. These values affect the optical properties of the films and the visual appearance of the packaged product. More importantly, a high light transmission rate catalyzes oxidation reactions that affect rancidity, availability of vitamins, and pigments responsible for color

Source: Serna-Saldivar (2008), ASTM (2018), AOAC (2019).

bacteria is controlled by time–temperature profiles during drying, reducing the a_w to a level that ensures the inactivation of pathogens, and packaging of the finished product.

Beef, pork, and poultry meats are also inspected in terms of presence of residual anabolic hormones, antibiotics, heavy metals, and pesticides. The steroid or growth hormones are used to improve feed conversion efficiency and promote better growth rates in livestock production, but, unfortunately, they can negatively affect humans who are the last consumer of the food chain. The known toxicological and harmful effects of these compounds have propitiated the ban of contaminated meats in the European Union or the strict use and control of maximum levels in many other countries. As a result, governmental regulatory agencies have dedicated resources and efforts to develop analytical methods for monitoring hormones in meats and their products. On the other hand, veterinary antibiotics that are frequently used to treat and prevent a variety of bacterial diseases of both monogastric and polygastric or ruminant animals can persist in products like meat, liver, kidney, and milk. Residual antibiotics are especially important in the milk of dairy cows that have been treated for mastitis. Antibiotics might pose health risks to consumers, especially those who are allergic to the different types of commercial antibiotics. In order to quantify these antimicrobial agents, sensitive, reliable, and selective assays are used in order to monitor compliance to maximum residue limits (MRL) established by public health agencies (Table 19.9).

The presence of significant amounts of heavy metals, mainly lead, arsenic, and chromium, and organochlorine or organophosphate pesticides in meats and milk can also harm human health. The bioaccumulable lead can cause saturnism and neurological disorders, whereas pesticides the risk of developing cancer. For these and other reasons, the regulatory agencies pay much attention to the quantification of these toxicological compounds (Table 19.9)

19.7 QUALITY CONTROL OF PACKING MATERIALS

All snack food processors pay attention to the effectiveness of their packaging operations because it affects shelf life of the packaged product and cost. The characteristics and properties of synthetic and natural packaging materials are covered in Chapter 4. The quality of the packaging materials should be carefully monitored to protect the product and the image of the manufacturing company. As explained before, most snacks are packaged in flexible materials, which are formed, filled, and sealed in one operation (Chapter 4). The type of laminate or film is determined effectively with an infrared analyzer. Each type of polymer yields a characteristic spectra or pattern like a fingerprint. This spectroscopy technique can be applied directly on the film or in chemically dissolved materials (Table 19.10).

Most packaging materials used for snacks should meet strict specifications of physical properties (thickness, weight/area, tensile strength, film resistance, tear resistance, seal resistance, hot tack, etc.) and chemical composition (Table 19.10). Seal resistance and hot tack are important because most snacks are packaged with inert gases. The headspace composition of the package could be determined via gas chromatography equipped with a solid-phase microextraction or SPME. The quality control methods widely employed to test packaging materials are published and detailed by the American Society for Testing and Materials (ASTM 2018).

REFERENCES

Aked, J. 2002. Maintaining the post-harvest quality of fruits and vegetables. In: *Fruit and Vegetable Processing*, ed. W. Jongen. Boca Raton: Woodhead Publishing (CRC Press).

American Association of Cereal Chemists (AACC). 2000. *AACC Approved Methods of Analysis*. 10th edition. St. Paul: American Association of Cereal Chemists.

Anderson, R.A., Conway, H.F., Pfeifer, V.F., and Griffin, E.I. 1969. Roll and extrusion-cooking of grain sorghum grits. *Cereal Science Today*, 14(11):373–381.

AOAC. 2019. *Official Methods of Analysis*. 21st edition. Washington: AOAC International.

AOCS. 2009. *Official Methods and Recommended Practices of the AOCS*. 6th edition. Danvers: The Society.

ASTM. 2018. *Standard Guide for Packaging Test Method Validation*. West Conshohochen: American Society for Testing and Materials.

Bou, R., Navas, J.A., Tres, A., Codony, R., and Guardiola, F. 2012. Quality assessment of frying fats and fried snacks during continuous deep-fat frying at different large-scale producers. *Food Control*, 27(1):254–267.

Carson, G.R., and Edwards, N.M. 2009. Criteria of wheat and flour quality. In: *Wheat Chemistry and Technology*, eds. K. Khan and P.R. Shewry, 97–118. St. Paul: American Association of Cereal Chemists.

Cepurnieks, G., Rjabova, J., Zacs, D., and Bartkevics, V. 2015. The development and validation of a rapid method for the determination of antimicrobial agent residues in milk and meat using ultra performance liquid chromatography coupled to quadrupole–Orbitrap mass spectrometry. *Journal of Pharmaceutical and Biomedical Analysis*, 102:184–192.

Clark, J.L., Barbano, D.M., and Dunham, C.E. 1989. Comparison of two methods for determination of total solids content of milk: Collaborative study. *Journal of the Association of Official Analytical Chemists*, 72(5):712–718.

Crosbie, G.B., and Ross, A.S. 2007. *The RVA Handbook*. St. Paul: American Association of Cereal Chemists.

D'Appolonia, B.L., and Kunerth, H. 1984. *The Farinograph Handbook.* 3rd edition. St. Paul: American Association of Cereal Chemists.

Dubat, A., Rosell, C.M., and Gallagher, E. 2013. *Mixolab: A New Approach to Rheology*, 1–112. St. Paul: AACC-International.

Dubois, M., Dubat, A., and Launay, B. 2016. *The AlveoConsistograph Handbook*, 1–78. St. Paul: AACC-International.

Espinosa-Ramírez, J., Rosell, C.M., Serna-Saldivar, S.O., and Perez-Carrillo, E. 2020. Evaluation of the quality of nixtamalized maize flours for tortilla production with a new Mixolab protocol. *Cereal Chemistry*, 97(2):527–539.

Fernandez, J., Perez-Alvarez, J.A., and Fernandez-Lopez, J.A. 1997. Thiobarbituric acid test for monitoring lipid oxidation in meat. *Food Chemistry*, 59(3):345–353.

Firestone, D. 1998. *Official and Recommended Methods of the AOCS.* 5th edition. Champaign: American Chemical Society.

Gan, H.L., Tan, C.P., Man, Y.C., NorAini, I., and Nazimah, S.A.H. 2005. Monitoring the storage stability of RBD palm olein using the electronic nose. *Food Chemistry*, 89(2):271–282.

Ghiasvand, A.R., and Hajipour, S. 2016. Direct determination of acrylamide in potato chips by using headspace solid-phase microextraction coupled with gas chromatography-flame ionization detection. *Talanta*, 146:417–422.

Gould, W.A. 1999. *Potato Production, Processing and Technology*, 111–178. Timonium: CTI Publications, Inc.

Gould, W.A., and Plimpton, S.L. 1985. Quality evaluation of potato cultivars for processing. *Research Bulletin*, Ohio-Agricultural-Research and Development Center. No. 1172.

Gusakov, A.V., Kondratyeva, E.G., and Sinitsyn, A.P. 2011. Comparison of two methods for assaying reducing sugars in the determination of carbohydrase activities. *International Journal of Analytical Chemistry*, Article ID 283658.

Hareland, G.A. 1994. Evaluation of flour particle size distribution by laser diffraction, sieve analysis and near-infrared reflectance spectroscopy. *Journal of Cereal Science*, 20(2):183–190.

Hsu, H.W., Vavak, D.L., Satterlee, L., and Miller, G.A. 1977. A multienzyme technique for estimating protein digestibility. *Journal of Food Science*, 42(5):1269–1273.

International Organization for Standardization (ISO 22000:2018). 2018. *Food Safety Management Systems – Requirements for any Organization in the Food Chain.* Athens: Hellenic Body for Standardisation (ELOT).

Jaffe, W.G., Levy, A., and González, D.I. 1974. Isolation and partial characterization of bean phytohemagglutinins. *Phytochemistry*, 13(12):2685–2693.

Jayalakshmi, K., Paramasivam, M., Sasikala, M., Tamilam, T.V., and Sumithra, A. 2017. Review on antibiotic residues in animal products and its impact on environments and human health. *Journal of Entomology and Zoology Studies*, 5(3):1446–1451.

Kakade, M.L., Rackis, J.J., McGhee, J.E., and Puski, G. 1974. Determination of trypsin inhibitor activity of soy products: A collaborative analysis of an improved procedure. *Cereal Chemistry*, 51(3):376–382.

Katz, S.A. 1964. Direct potentiometric determination of urease activity. *Analytical Chemistry*, 36(13):2500–2501.

Kleyn, D.H., Lynch, J.M., Barbano, D.M., Bloom, M.J., and Mitchell, M.W. 2001. Determination of fat in raw and processed milks by the Gerber method: Collaborative study. *Journal of AOAC International*, 84(5):1499–1508.

LeDoux, M. 2011. Analytical methods applied to the determination of pesticide residues in foods of animal origin. A review of the past two decades. *Journal of Chromatography A*, 1218(8):1021–1036.

Li, J., Li, H., and Ho, C.T. 2000. *Static Headspace Solid-Phase Microextraction Analysis (SPME) of Volatile Aldehydes and Fatty Acids in Potato Chips.* Book of Abstracts: Ann. Mtg. July 22–26. Dallas, TX: Institute of Food Technologists.

Liener, I. 1955. The photometric determination of the hemagglutinaing activity of soyin and crude soybean extracts. *Archives of Biochemistry and Biophysics*, 54(1):223–231.

Lin, C.S., and Zayas, J.F. 1987. Functionality of defatted corn germ proteins in a models system: Fat-binding capacity and water retention. *Journal of Food Science*, 52(5):1308–1311.

Lipp, M., Shillito, R., Giroux, R., Spiegelhalter, F., Charlton, S., Pinero, D., and Song, P. 2005. Polymerase chain reaction as an analytical tool in agricultural biotechnology. *Journal of AOAC International*, 88(1):136.

Majchrzak, T., Wojnowski, W., Dymerski, T., Gębicki, J., and Namieśnik, J. 2018. Electronic noses in classification and quality control of edible oils: A review. *Food Chemistry*, 246:192–201.

McWatters, K.H., and Holmes, M.R. 1979. Influence of pH and salt concentration on nitrogen solubility and emulsification properties of soy flour. *Journal of Food Science*, 44(3):770–773, 781.

Mohamed, A.A., Ashman, R.B., and Kirelis, A.W. 1993. Pericarp thickness and other kernel physical characteristics relate to microwave popping quality of popcorn. *Journal of Food Science*, 58(2):342–346.

Mortimore, S. 2000. An example of some procedures used to assess HACCP systems within the food manufacturing industry. *Food Control*, 11(5):403–413.

Mostert, J.F., and Jooste, P.J. 2012. Quality control in the dairy industry. In *Dairy Microbiology Handbook*, 3rd edition, ed. R.K. Robinson, 655–736. New York: Wiley Interscience.

Nzekoue, F.K., Caprioli, G., Fiorini, D., Torregiani, E., Vittori, S., and Sagratini, G. 2019. HS-SPME-GC-MS technique for FFA and hexanal analysis in different cheese packaging in the course of long term storage. *Food Research International*, 121:730–737.

O'Brien, R.D. 2009. Fats and oil analysis. In: *Fats and Oils. Formulating and Processing for Applications*, 3rd edition. Boca Raton: CRC Press.

Otter, D.E. 2012. Standardised methods for amino acid analysis of food. *British Journal of Nutrition*, 108(S2):S230–S237.

Patel, P.D., Stripp, A.M., and Fry, J.C. 1988. Whipping test for the determination of foaming capacity of protein: A collaborative study. *International Journal of Food Science and Technology*, 23(1):57–63.

Pearce, K.N., and Kinsella, J.E. 1978. Emulsifying properties of proteins: Evaluation of turbidimetric technique. *Journal of Agriculture and Food Chemistry*, 26(3):716–723.

Peplinski, A.J., Anderson, R.A., and Eckhoff, S.R. 1984. A dry milling evaluation of trickle sulfur dioxide treated corn. *Cereal Chemistry*, 61:289.

Perera, C.O., and Perera, A.D. 2005. Quality assurance, quality control, inspection, and sanitation. In: *Processing Fruits Science and Technology*, eds. D.M. Barrett, L. Somogyi and H. Ramaswamy. Boca Raton: CRC Press.

Pflugfelder, R.L., Rooney, L.W., and Waniska, R.D. 1988. Fractionation and composition of commercial corn masa. *Cereal Chemistry*, 65:262–266.

Pizzinatto, A., and Hoseney, R.C. 1980. A laboratory method for saltine crackers. *Cereal Chemistry*, 57(4):249–252.

Popcorn Institute. 1989. *Test for Evaluating Popcorn Hybrids in a Microwave Oven*. Chicago: Technical Committee Popcorn Institute.

Psomas, E., Fotopoulos, C., and Kafetzopoulos, D. 2011. Core process management practices, quality tools and quality improvement in ISO 9001 certified manufacturing companies. *Business Process Management Journal*, 17(3):437–460.

Psomas, E.L., and Kafetzopoulos, D.P. 2015. HACCP effectiveness between ISO 22000 certified and non-certified dairy companies. *Food Control*, 53:134–139.

Ramirez-Wong, B., Sweat, V.E., Torres, P.I., and Rooney, L.W. 1993. Development of two instrumental methods for corn masa texture evaluation. *Cereal Chemistry*, 70(3):286–290.

Ranganna, S. 1986. *Handbook of Analysis and Quality Control for Fruit and Vegetable Products*. 2nd edition. New Delhi: Tata McGraw HillPublished Company Limited.

Rasper, V.F., and Preston, K.R. 1991. *The Extensigraph Handbook*, 1–50. St. Paul: American Association of Cereal Chemists.

Regenstein, J.M., and Regenstein, C.E. 1984. Protein functionality for food scientists. In: *Food Protein Chemistry. An Introduction for Food Scientists*, eds. J. Regenstein and C.E. Regenstein, 274–332. Orlando: Academic Press.

Rizzi, C., Galeoto, L., Zoccatelli, G., Vincenzi, S., Chignola, R., and Peruffo, A.D.B. 2003. Active soybean lectinin foods: quantitative determination by ELISA using immobilized asialofetuin. *Food Research International*, 36(8):815–821.

Rooney, L.W., and Serna-Saldivar, S.O. 2015. Food grade corn quality for lime cooked tortillas and snacks. In: *Tortillas: Wheat Flour and Corn Products*, eds. L.W. Rooney and S.O. Serna-Saldivar, 227–246. Saint Paul: American Association of Cereal Chemists.

Rooney, L.W., and Suhendro, E.L. 2001. Food quality of corn. In: *Snack Foods Processing*, 1st edition, eds. E.W. Lusas and L.W. Rooney, 37–72. Boca Raton: CRC Press.

Rubenthaler, G.L., Finney, P.L., Demaray, D.E., and Finney, K.F. 1980. Gasograph: Design, construction, and reproducibility of a sensitive 12-channel gas recording instrument. *Cereal Chemistry*, 57(3):212–216.

Satterlee, L.D., Kendrick, J.G., Marshall, H.F., Jewell, D.K., Ali, R.A., Heckman, M.M., Steinke, H.F., Sarward, G., and Slump, P. 1982. In vitro assay for predicting protein efficiency ratio as measured by rat bioassay: Collaborative study. *Journal of the Association of Official Analytical Chemists*, 65(4):798–809.

Schenck, F.J., and Callery, P.S. 1998. Chromatographic methods of analysis of antibiotics in milk. *Journal of Chromatography A*, 812(1–2):99–109.
Seo, J., Kim, H.Y., Chung, B.C., and Hong, J. 2005. Simultaneous determination of anabolic steroids and synthetic hormones in meat by freezing-lipid filtration, solid-phase extraction and gas chromatography–mass spectrometry. *Journal of Chromatography. Part A*, 1067(1–2):303–309.
Serna-Saldivar, S.O. 2008. *Industrial Manufacture of Snack Foods*. London: Kennedys Publications Ltd.
Serna-Saldivar, S.O. 2010. Quality control of cereal grains and their products. In *Cereal Grains: Properties, Processing and Nutritional Attributes*, ed. S.O. Serna-Saldivar, 463–533. Boca Raton: CRC Press (Taylor & Francis Group).
Serna-Saldivar, S.O. 2012. *Cereal Grains: Laboratory Reference and Procedures Manual*, 1–368. Boca Raton: CRC Press (Taylor & Francis Group).
Serna-Saldivar, S.O., AlmeidaDominguez, H.D., Rooney, L.W., Gomez, M.H., and Bockholt, A.J. 1991. Method to evaluate ease of pericarp removal on lime-cooked corn kernels. *Crop Science*, 31(3):842–844.
Serna-Saldivar, S.O., Gomez, M.H., AlmeidaDominguez, H.D., Islas Rubio, A., and Rooney, L.W. 1993. A method to evaluate the lime cooking properties of corn (*Zea mays*). *Cereal Chemistry*, 70:762–764.
Shen, N., Moizuddin, S., Wilson, L., Duvick, S., White, P., and Pollak, L. 2001. Relationship of electronic nose analyses and sensory evaluation of vegetable oils during storage. *Journal of the American Oil Chemists' Society*, 78(9):937–940.
Shuey, W.C., and Tipples, K.H. 1982. *The Amylograph Handbook*. St. Paul: American Association of Cereal Chemists.
Snack Food Association. 1992. *Corn Quality Assurance Manual*, 1–79. Alexandria: The Snack Food Association.
Sorgentini, D.A., Wagner, J.R., and Anon, M.C. 1995. Effects of thermal treatment of soy protein isolate on the characteristics and structure–function relationship of soluble and insoluble fractions. *Journal of Agricultural and Food Chemistry*, 43(9):2471–2479.
Suhendro, E.L., Almeida-Dominguez, H., Rooney, L.W., and Waniska, R.D. 1998a. Objective rollability method for corn tortilla texture measurement. *Cereal Chemistry Journal*, 75(3):320–324.
Suhendro, E.L., Almeida-Dominguez, H., Rooney, L.W., Waniska, R.D., and Moreira, R.G. 1998b. Tortilla bending technique: An objective method for corn tortilla texture measurement. *Cereal Chemistry Journal*, 75(6):854–858.
Surak, J.G. 2007. A recipe for safe food: ISO 22000 and HACCP. *Quality Progress*, October:21–27.
Tamime, A.Y., and Robinson, R.K. 2000. Quality control in yoghurt manufacturing. In: *Yoghurt Science and Technology* eds. Tamime, A.Y., and Robinson, R.K. Cambridge: Woodhead Publishing Limited (CRC Press).
Tsutsui, T. 1988. Functional properties of heat-treated egg yolk low density lipoprotein. *Journal of Food Science*, 53(4):1103–1106.
Upadhyay, R., Sehwag, S., and Mishra, H.N. 2017. Frying disposal time of sunflower oil using hybrid electronic nose-fuzzy logic approach. *LWT – Food Science and Technology*, 78:332–339.
US Food and Drug Administration. 1997. HACCP principles & application guidelines. National Advisory Committee on Microbiological Criteria for Foods. https://www.fda.gov/food/hazard-analysis-critical-control-point-haccp/haccp-principles-application-guidelines.
Varzakas, T.H., and Arvanitoyannis, I.S. 2008. Application of ISO 22000 and comparison to HACCP for processing of ready to eat vegetables: Part I. *International Journal of Food Science and Technology*, 43(10):1729–1741.
Vaz, A., Cabral Silva, A.C., Rodrigues, P., and Venancio, A. 2020. Detection methods for aflatoxin M1 in dairy products. *Microorganisms*, 8(2):246.
Waniska, R.D. 2001. Evaluation methods and quality control for snacks. In: *Snack Foods Processing*, eds. E. Lusas and L.W. Rooney, 607–629. Lancaster: Technomic Publishing Co.
Weisshaar, R. 2014. Quality control of used deep-frying oils. *European Journal of Lipid Science and Technology*, 116(6):716–722.
Xu, L., Yu, X., Liu, L., and Zhang, R. 2016. A novel method for qualitative analysis of edible oil oxidation using an electronic nose. *Food Chemistry*, 202:229–235.
Yee, N.G., and Bussell, W.T. 2007. Good potatoes for good potato crisps, a review of current potato crisp quality control and manufacture. *Global Science Books*, 1:271–286.
Zayas, J.F. 1997. Water holding capacity of proteins. In: *Functionality of Proteins in Food*, 76–133. Berlin: Springer.

20 Nutritional Composition and Toxic Compounds of Snack Foods

Sergio O. Serna-Hernandez and Sergio O. Serna-Saldivar

CONTENTS

20.1 INTRODUCTION

In the present-day societies where the number of obese people continues to surge, regular diets commonly exceed the recommended allowances for energy, fat, sugars, and sodium. Globally, in 2016, about 1,900 million people were overweight and at least 650 million obese with body mass index (BMI) of 25–30 and >30, respectively. More worrisome is the fact that, according to the World Health Organization (WHO), the worldwide prevalence of obesity nearly tripled between 1975 and 2016. In terms of childhood and adolescents, in 2019, an estimated 38.2 million children aged <5 years and over 340 million children and adolescents aged 5–19 were overweight or obese in 2016.

DOI: 10.1201/9781003129066-20

Once considered a high-income country problem, overweight and obesity are now on the rise in low- and middle-income countries (www.who.int/news-room/fact-sheets/detail/obesity-and-overweight).

Obesity and overweight pose a major risk for chronic diseases, including type-2 diabetes, cardiovascular disease (CVD), hypertension, and certain forms of cancers (Chapter 21). The key causes are increased consumption of energy-dense foods high in saturated fats and sugars accompanied by a reduced physical activity. In order to stop the rising obesity, the public health agencies throughout the globe recommend the consumption of no more than 30% of the total energy intake from fats and oils. In addition, the general recommendation is to consume about one-third of saturated, monounsaturated, and polyunsaturated fatty acids (FAs). The FA composition of fats and oils should be balanced and for this case, the lipids should contain approximately 22 g of each FA class. A healthful balance of saturated: monounsaturated:polyunsaturated FA, especially in terms of the ω-6:ω-3 ratio, is being highly considered for the production of novel snacks. These health-associated problems are exacerbated when the source of the fat/oil contains significant amounts of *trans*-FAs or the oil is highly oxidized/polymerized due to continuous usage or is thermally abused especially during deep fat frying. The consumption of *trans*-FA increases the problems of arteriosclerosis, cancer, and other chronic diseases to a higher degree compared to saturated FA (Hunter et al. 2009). This is the reason why most regulatory agencies force processors to declare the amount of these harmful fats on the food label. In fact, several countries around the globe are embracing the production of foods with zero *trans*-fats. The controversy of *trans*-fats started in the 1980s after a scientific study comparing diets of Scandinavians to those of Americans. The conclusion of the investigation was that Scandinavians had relatively low levels of CVD even though they consumed large amounts of saturated fats, whereas the American group with lower intakes of saturated fats consumed higher levels of *trans*-FA and had higher incidence of CVD.

Most snacks are salted or flavored with salt-based powdered or liquid seasonings (Chapter 8); therefore, these products normally contain large quantities of sodium. This essential electrolyte is considered the single nutrient most abused by modern people because an average diet usually provides at least ten times of the recommended intake. The high consumption of sodium is the main reason why hypertension is responsible for more than 50% of the cardiovascular diseases, which are regarded as the main cause of death today.

Nowadays food scientists are including in their research and product development programs new efforts toward the development of novel snacks with minimal fat calories and fat decomposition products related to adverse effects on human health. More challenging is the production of healthy snacks with similar sensory properties (flavor, texture, and overall acceptability) compared to the regular counterparts. Needless to say, this presents a huge challenge for food developers because it is difficult to substitute the flavor- and texture-imparting properties of frying oils and salt that act as flavor enhancers.

The objective of this chapter is to review the nutritional value of all categories of snack foods covered in this book and the new industrial trends in terms of nutritional attributes. Moreover, the chapter also covers main toxicological compounds most frequently associated with raw materials widely used for production of snacks and related compounds produced during processing. The other big trend is the production of nutraceutical or functional snacks with proven positive health effects, especially in terms of preventing oxidative stress, snacks aimed toward the special diet systems like low carbohydrate, keto, and others.

20.2 FOOD LABELING

Most regulatory agencies around the world have mandatory nutrition labeling regulations applied for almost all processed foods sold in packages or containers. Nutritional labeling differs in countries around the globe, although the spirit is to provide useful information to customers related to the food ingredients and composition. In most instances, the guidelines include the amount of food and number of servings in a package, its common name, the list of ingredients, the address and information of the processors, and the nutritional information that in many instances can include health claims especially related to chronic diseases and obesity. The aim of the nutritional label is to inform consumers of the raw materials used to manufacture the food item and the most relevant nutrients affecting human well-being. It is relevant to inform consumers of the raw materials because segments of the general population are prone to food allergies, the most common being lactose, gluten, eggs, almonds, peanuts, soybeans, and additives such as sulfites and monosodium glutamate. Emphasis is laid on the relationship between nutrients and dietary concerns.

In the United States, chronic diseases such as heart disease and cancer are the leading cause of death and disability. Nearly one in three adults have high blood pressure, which is the leading cause of heart disease and strokes. In addition, almost 70% of the adults are overweight (BMI: 25–30) and 40% are obese (BMI > 30). Almost 20% of the children and adolescent population in the United States are obese. In order to prevent these alarming statistics, the FDA has adopted new strategies aimed to modernize health claims, ingredient labels, and standards of identity plus the implementation of nutritional facts label and nutritional education. The most worrisome nutrients that negatively affect public health are sodium, sugars, calories provided by fats, and both saturated and *trans* fats. On the other hand, several nutrients such as dietary fiber, vitamins, and minerals are being promoted because they prevent oxidative stress, metabolic syndrome, obesity, and chronic diseases. The other important change is that the FDA plans to reevaluate the ingredient list on food packages to make it more consumer-friendly because new consumers pay attention to "clean labels" which should be more readable and understandable. The FDA also changed regulations related to nutrition facts and issued several technical guidance documents on issues such as fiber and serving sizes to help manufacturers meet the requirements of the new Nutrition Facts rule. Consumers are starting to see updated labels based on current science that provide more information to empower them to choose healthful diets. The regulatory agencies also state that when a label contains the term "Light" or "Lite," it indicates that a food has at least one-third fewer calories or 50% less fat, or 50% less sodium compared to the regular counterpart. This is critically important in the labeling of snacks. The other common terminologies are "Low Fat" "Reduced Fat," and "Fat Free." Reduced-fat foods contain at least 25% less fat per serving than a serving of the same food that contains the regular amount of fat, whereas low-fat foods contain more fat than fat-free foods and provide 3 g or less of fat per serving. Finally, The FDA permits foods to be labeled "Fat-free" if they contain less than 0.5 g of fat per serving. Normally, in order to meet this stringent requirement, some types of snacks such as baked products are not supplemented and/or treated with fats, frying oils, and fat-based seasonings. For animal- and marine-based snacks, nutrient claims can include lean and extra lean statements. In order to claim that a product is lean, the seafood or meat should contain less than 10 g fat, 4.5 g saturated fat, and 95 mg cholesterol per 100 g. For extra-lean items, the 100 g serving should contain less than 5 g fat, 2 g saturated fat, and 95 mg cholesterol (https://ask.usda.gov/s/article/What-does-lean-and-extra-lean-beef-mean-on-a-nutrition-label). The FDA allows the labeling of health claims for the following nutrients: calcium and osteoporosis, sodium and hypertension, dietary fat and cancer, dietary saturated fat and risk of CVD, cholesterol and CVD, fiber and cancer, soluble fiber and CVD, folate and neural tube defects, and sugar alcohol and dental caries.

Consumers have gradually learned of the strong relationship between the abuse of some nutrients and the incidence of chronic diseases such as CVD, hypertension, hypercholesterolemia, diabetes, and cancer, discussed in Chapter 21. Table 20.1 depicts the mandatory and optional nutrients with their corresponding RDA considering a 2,000 kcal diet and the reason why they are relevant from the nutrition and human health viewpoints. The new change in the food labeling of the United States is the inclusion of the amount of sugars purposely added to the food, so the sugar content is divided into the sugar naturally contained in the food product plus sugars (sucrose, glucose, syrups, etc.) added to it during production. In other parts of the world, the regulatory agencies now demand the inclusion of "seals" which indicate that a given food source exceeds amounts of sugars, sodium, fat, and calories. All these strategies are aimed to facilitate communication through the food label and

TABLE 20.1
Mandatory and Optional Nutrients of US Food Labels with Their Reference to Daily Intake (RDI) Values and Positive or Negative Relationships with Human Health

Nutrient	RDI	Nutritional and Health Implications
Mandatory nutrients		
Total calories	2,000 kcal	The total calorie intake is closely related with overweight, obesity, and health
Fat	65–78 g	The fat intake should be limited in people suffering from obesity and chronic diseases. Fats should provide less than 30% of the total caloric intake
Saturated fat	<20 g	The high intake of saturated fats is one of the main factors that increase cardiovascular disease
trans-Fats	0	*trans*-Fatty acids have detrimental effects on human health because they increase the risk of cardiovascular disease, hypercholesterolemia, and cancer. Most new food labeling regulations mandate processors to declare the amount of *trans*-fats. Some regulatory agencies recently prohibited the use of *trans*-fats because they are more harmful than saturated fats
Cholesterol	<300	The high intake of cholesterol from animal sources is associated with higher risk of cardiovascular disease and hypercholesterolemia
Total carbohydrate	275–300 g	The total carbohydrate content is constituted by complex and simple carbohydrates. Complex (starch) are digested more slowly than simple carbohydrates and have lower glycemic index
Sugar	<50 g	The consumption of high amounts of sugars increases risk of diabetes and caries or tooth decay. The modern human diets tend to abuse the consumption of simple sugars considered as empty foods. Many countries recently enacted the mandatory declaration on food labels of added sugars to inform consumers of exogenous addition of simple sugars
Total dietary fiber	25–28 g	The consumption of both insoluble and soluble dietary fiber lowers the caloric density of foods and helps to prevent cardiovascular disease, colon cancer, constipation, and hemorrhoids and lower the glycemic index of foods. Most dietary fibers contain important amounts of nutraceutical compounds that have positive health implications
Protein	50 g	Protein is especially important to sustain growth in infants, children, and adolescents or teenagers. It also helps to prevent diseases via production of antibodies
Calcium	1,000–1,300 mg	Calcium is the most important macro-mineral in human nutrition and the main component of bones and teeth. Its absorption is enhanced by vitamin D. This mineral also plays fundamental roles in cell signaling, blood clotting, muscle contraction, and nerve function. Growing children especially requires calcium for proper growth. Its deficiency is the main cause of osteoporosis, especially among postmenopausal women and geriatric people
Sodium	2,300–2,400 mg	The main function of sodium is to regulate the total amount of water in the body, the blood volume, and its critical role in muscle contraction and brain functioning. However, this electrolyte is the most abused nutrient associated with Western diets and the most responsible for hypertension
Potassium	3,500–4,700 mg	The main function of potassium is as an electrolyte, so its deficiency in combination with excessive sodium leads to hypertension and improper function of the heart, renal system, muscles, and the nervous system. The deficiency of this electrolyte can produce fatigue, muscle weakness, constipation, and abdominal cramping
Iron	18 mg	Iron is the most important micro-mineral in human nutrition. It is deficient especially among low socioeconomic children and pregnant women. Iron helps to prevent anemia and strengths the immune system

(*Continued*)

TABLE 20.1 (CONTINUED)
Mandatory and Optional Nutrients of US Food Labels with Their Reference to Daily Intake (RDI) Values and Positive or Negative Relationships with Human Health

Nutrient	RDI	Nutritional and Health Implications
Vitamin D	10–20 mcg	It has gained more relevance because it is essential for the proper absorption of calcium and muscle contraction for the prevention of osteoporosis. In addition, it plays a key role in preventing inflammation and keeping the heart, lungs, kidney, and brain healthy. Clinical studies have shown that this vitamin lowers blood pressure and prevents cognitive impairment and Alzheimer's disease
Major optional nutrients		
Calories from fat	<585 kcal	The calories from fat should be limited, especially in overweight, obese, and superobese people (BMI > 35) and people suffering from chronic diseases such as cardiovascular diseases and arteriosclerosis
Monosaturated fat		Monounsaturated fatty acids like oleic and palmitoleic are comparatively healthier compared to saturated fats. The recommendation is to consume about one-third of the total fat with these fatty acids.
Polyunsaturated fat		These fats might contain ω-6 linoleic acid considered as essential in human nutrition. The consumption of polyunsaturated fatty acids, mainly those classed as ω-3, have been associated with many health benefits like lowering blood cholesterol, triglycerides, and therefore cardiovascular disease, and improving brain functioning. The long-chain polyunsaturated fatty acids EPA and DHA are considered highly nutraceutical because of their implication in blood coagulation and prevention of neurodegenerative diseases and age-related macular degeneration
Insoluble dietary fiber		Insoluble dietary fibers mainly constituted of cellulose and lignin prevent constipation and cancer and lower glycemic index of foods
Soluble dietary fiber		Most soluble dietary fibers constituted by pectins, B glucans, arabinoxylans, inulin, and soluble hemicelluloses are prebiotic, modulate the population of beneficial or probiotic bacteria in the hindgut, and diminish blood cholesterol levels and colon cancer
Sugar alcohols		These are monosaccharide and disaccharide derivatives like sorbitol, xylitol, lactitol, and mannitol used as sweeteners, which have limited digestion and absorption and do not provide the same energy as sugars. These compounds are classified as carbohydrates or artificial sweeteners. Besides helping diabetic people, sugar alcohols do not promote the formation of cavities
Phosphorus	1,000 mg	The main function of phosphorus is the formation of bones and teeth along with calcium and produce ATP that stores energy. Therefore, it plays a relevant role in the use of carbohydrates, fats, and proteins required for the growth, maintenance, and repair of cells. Phosphorus is also required for regular renal function, nerve signaling, muscle contraction, and heartbeat. Moreover, phospholipids play a key role in cell membrane structure and function
Magnesium	400 mg	This mineral is required for hundreds of metabolic reactions in the body, including regulation of blood glucose levels and production of energy and proteins. In practical terms, it helps to maintain normal nerve and muscle and heart function, a healthy immune system, and bones. It also helps adjust blood glucose levels. It aids in the production of energy and protein
Iodine	150 mcg	This trace mineral is required for production of energy through synthesis of thyroid hormones and correct thyroid function. These hormones influence proper bone and brain development during pregnancy and infancy. Severe deficiency can cause goiter, stunted fetus growth, intellectual disability, and delayed sexual development

(*Continued*)

TABLE 20.1 (CONTINUED)
Mandatory and Optional Nutrients of US Food Labels with Their Reference to Daily Intake (RDI) Values and Positive or Negative Relationships with Human Health

Nutrient	RDI	Nutritional and Health Implications
Vitamin A	5,000 IU	Vitamin A has multiple roles in human nutrition and health. It is the most important liposoluble vitamin in human nutrition and considered within the three most important antioxidant vitamins, so it prevents cardiovascular disease and cancer. One of the major roles of this vitamin is the prevention of night blindness, age-related macular degeneration, and infectious diseases through a stronger immune system
Vitamin C	60 mg	Ascorbic acid is an antioxidant that protects cells against free radicals and is also required to form blood vessels, collagen, muscle, and antibodies. It is also involved in the absorption and storage of iron and wound healing. Vitamin C prevented scurvy in the Middle Ages and today it is used to prevent oxidative stress, cancer, and cardiovascular disease
Vitamin E	12 mg	Vitamin E or tocopherol is a potent antioxidant which protects cells from oxidative damage caused by free radicals and plays a key role in aging. This vitamin helps to keep the immune system strong against bacteria and viruses and is involved in the formation of red blood cells. For these reasons, vitamin E can prevent cancer, cardiovascular diseases, dementia, and hepatic disease
Vitamin K	80 mcg	This fat-soluble vitamin is required for proper blood clotting or coagulation and plays a relevant role in preventing osteoporosis
B_1	1.5 mg	It is a cofactor in the metabolism of carbohydrates for the efficient conversion into energy. It prevents beriberi, characterized by weakness, complications in the nervous system, brain, muscles, heart, stomach, and intestines. It is also involved in the flow of electrolytes into and out of muscle and nerve cells
B_2	1.7 mg	It is also involved in the metabolism of carbohydrates and required to maintain normal skin and mucous membranes, red blood cells, vision, and proper functioning of the nervous system
Niacin	20 mg	It is a cofactor in the metabolism of carbohydrates via NADH for production of energy. It prevents pellagra characterized by dermatitis, diarrhea, and dementia
B_6	2 mg	Pyridoxine is key in the protein, carbohydrate, and fat metabolism and the production of red blood cells, neurotransmitters, and antibodies to strengthen the immune system
Folate	400 mcg	It is essential for the synthesis of DNA and RNA. The deficiency is associated with anemia, swelling of the tongue or glossitis, diarrhea, and mental disorders. It is very important in pregnant women to prevent malformations in newborns (neural tube defects and spina bifida)
B_{12}	6 mcg	Cobalamin is crucial for nerve tissue health, brain function, and the formation of red blood cells and prevents macrocytic anemia. It is also involved in the production and regulation of DNA. Its deficiency can result in irreversible and potentially severe damage of the nervous system and brain, fatigue, depression, constipation, and weight loss
Biotin	300 mcg	It helps the body as cofactor to metabolize proteins, fats, and carbohydrates. The symptoms of biotin deficiency include hair thinning and loss, red rash, conjunctivitis, brittle nails, and neurological symptoms, which might include depression and hallucinations
Pantothenic acid	10 mg	This vitamin is needed for proper growth and is required for the synthesis of coenzyme A, fatty acids, adrenal hormones, and endogenous cholesterol. Severe deficiency might cause depression, vomiting, and numbing and tingling of the hands and feet

consumers of important nutritional parameters that influence caloric intake, obesity, hypertension, and/or chronic diseases.

The FDA also has made important efforts to inform and protect Americans prone to food allergies. These allergies affect millions of Americans and their families and reactions vary in severity from mild symptoms involving hives and lip swelling to severe, life-threatening anaphylaxis that may involve fatal respiratory problems and shock. To protect those with food allergies and other food hypersensitivities, the FDA enforces regulations requiring companies to list ingredients on packaged foods and beverages. For certain foods or substances that cause allergies or other hypersensitivity reactions, there are more specific labeling requirements. The FDA conducts inspections and sampling to check if major food allergens are properly labeled on products and if the food plants implement controls to prevent allergen cross-contact during manufacturing and packaging. The major eight food allergens are milk, eggs, fish, shellfish, tree nuts, peanuts, wheat, and soybeans. These food sources are responsible for 90% of food allergies in the United States and food labels should declare that these foods contain a "major food allergen" with the name of the allergen source (www.fda.gov/food/food-labeling-nutrition/food-allergies).

20.3 NUTRITIONAL COMPOSITION OF SNACK FOODS

Snacks are produced from the four major groups of foods – cereals/grains, fruits/vegetables, dairy, and meat/poultry/fish products – and therefore have contrasting nutritional properties. Table 20.2 summarizes advantages, disadvantages, and main potential toxicological compounds associated with the wide array of snacks covered in this book. Most snacks are perceived as empty and high-energy foods because they are, in most instances, produced from refined flours or grits rich in starch that are cooked in preparation for deep fat frying and sodium-rich seasonings. These snacks are high in lipids, energy from fat, carbohydrates, and sodium and low in dietary fiber, micronutrients, and phytochemicals. This is the main reason why snack food processors are changing their food product development strategies toward the utilization of composite flours that in many instances include whole grains, raw materials rich in soluble and insoluble dietary fiber, and new processes aimed to produce zero added fat products or alternatively diminish the oil uptake during frying. In addition, processors are trying to reduce the sodium content with the usage of salt substitutes and flavor enhancers. The industry is rapidly evolving with the product development of new snacks from animal and dairy origin, legumes, pseudocereals, and tree nuts. Animal and dairy snacks contain high amounts of good quality proteins and are low in soluble sugars, whereas snacks rich in legumes, pseudocereals, almonds, and tree nuts normally provide high protein, dietary fiber, and lipids with high amounts of mono- and polyunsaturated FA.

20.3.1 Popcorn

Popcorn is one of the best nutritionally snack foods because it is rich in starch and dietary fiber and contains important amounts of vitamins present in this special type of flinty maize. In fact, popcorn has a long history of being a staple for the ancient Mesoamerican cultures. The American Diabetes Association and the American Dietetic Association allow popcorn as a bread exchange on weight control diets (Serna-Saldivar 2008).

Most of the popcorn sold for microwaving or in cinemas, amusement parks, and grocery stores is oil-popped or flavored with fat-based flavorings and therefore is more energy-dense compared to air-popped products. Starch is the main nutrient in both types of popcorns and this complex carbohydrate is generally regarded as nutritionally superior to simpler carbohydrates because the starch polymer is hydrolyzed or digested more slowly, thereby producing less pronounced blood glucose elevation or glycemic index and less insulin requirement. This is especially important for diabetic people, but it is also advantageous for nondiabetics because starchy foods prolong the feeling of satiety (Serna-Saldivar 2008). Oil-popped popcorn contains more than 25% oil and provides

TABLE 20.2
General Advantages and Disadvantages and Potential Toxicological Compounds Associated with the Different Broad Categories of Snacks

Category of Snack Food	Main Nutritional Advantages	Main Nutritional Disadvantages	Main Health Risks
Popcorn	Good source of energy, low in sugars, relatively high in dietary fiber	Most products contain high levels of sodium. Medium protein quality in terms of essential amino acid profile. Wet-popped and flavored products tend to contain high amounts of fats, mainly saturated from tropical or hydrogenated oils. Caramel-coated popcorn is also rich in soluble sugars	Potential risk of aflatoxins
Wheat-based	Good source of energy, low in sugars	Medium protein quality in terms of essential amino acid profile. Most products contain high levels of sodium	Contain gluten that is allergic to celiac patients
Lime-cooked	Good source of energy, low in sugars, relatively high in dietary fiber	High levels of oil which increases energy density and sodium due to salt and seasoning addition. Medium protein quality in terms of essential amino acid profile	Potential risk of lysinoalanine production
Extruded	Good source of energy, practically free of sugars	High levels of oil which increases energy density and sodium due to salt and seasoning addition. Low levels of micronutrients. Medium protein quality in terms of essential amino acid profile. The starch has high glycemic index	Products containing wheat, rye, barley, oat, or triticale may cause allergic reactions against celiac patients
Legumes and pseudocereals	Good source of protein with adequate essential amino acid profile, energy, dietary fiber, micronutrients and phytochemicals with low levels of sugars. Contain oligosaccharides and soluble fiber that act as prebiotic. The starch and sugars have low glycemic index	Legumes and pseudocereals may contain antinutritional compounds like trypsin inhibitor, urease, and hemagglutinins that need to be deactivated via thermal processing, sprouting, or fermentation	Risk of aflatoxins and food allergens (mainly soybeans and peanuts)
Potato	Good source of energy, low in sugars	Potato-based snacks are practically devoid of protein and dietary fiber and normally contain very high levels of oil sodium. The starch has high glycemic index	The major risk of potato products is potential generation of acrylamide during frying
Almonds and tree nuts	Good source of natural oils rich in mono- or polyunsaturated fatty acids, energy, protein, dietary fiber, micronutrients, and lipophilic and hydrophilic phytochemicals with low levels of sugars	Almonds and tree nuts are very energy-dense due to the high-lipid content	Risk of aflatoxins, other toxins, and some may trigger allergic reactions in susceptible people

(*Continued*)

TABLE 20.2 (CONTINUED)
General Advantages and Disadvantages and Potential Toxicological Compounds Associated with the Different Broad Categories of Snacks

Category of Snack Food	Main Nutritional Advantages	Main Nutritional Disadvantages	Main Health Risks
Fruits and vegetables	Most products contain very low levels of fat and are rich in both insoluble and soluble dietary fibers, phytochemicals, vitamin C, and other important micronutrients	Products are naturally rich in soluble sugars with high glycemic index and low levels of protein	Products treated with sulfites may trigger allergic reactions in susceptible people
Dairy	Excellent source of good quality protein, calcium, magnesium, and other important micronutrients. May be a source of bioactive peptides	Milkfat is rich in saturated fatty acids and is devoid of dietary fiber	May cause allergic reactions in lactose-intolerant people
Animal- and poultry-based	Excellent source of protein with excellent essential amino acid profile. The minerals iron and zinc are highly bioavailable. Contains high levels of B vitamins	Snacks based on animal products are rich in saturated fat, cholesterol, and salt and are devoid of dietary fiber	Cured products that are thermally abused might contain high levels of nitrosamines
Marine-based	Excellent source of protein with excellent essential amino acid profile. The minerals iron and zinc are highly bioavailable. Contains high levels of B vitamins. Fish from cold regions tend to have high levels of DHA and EPA considered as highly nutraceuticals	Snacks based on marine products are normally rich in salt and cholesterol and are devoid of dietary fiber	Some products rich in histamines might cause allergic reactions

approximately 500 kcal/100 g serving (Table 20.3). In contrast, air-popped popcorn only contains the intrinsic oil mainly associated with the germ (4%) and 380 kcal. Air-popped popcorn also contains comparatively higher amounts of dietary fiber and protein. One 100-g serving supplies about 15 g of dietary fiber and 10–12 g protein. The protein quality is not considered good because even though it is highly digested, it does not contain sufficient amounts of lysine and tryptophan for optimum infant growth. Most of the dietary fiber fraction of popcorn is insoluble, so it prevents constipation, hemorrhoids, and diverticulosis and protects against colorectal cancer by binding dietary carcinogens. In addition, the insoluble dietary fiber acts as a physical barrier against the rapid absorption of glucose by the epithelial gastrointestinal cells.

Dentists have long endorsed popcorn as an alternative to sweet-flavored snacks because most products are free of sugars and rich in dietary fiber. The pericarp-rich fiber helps to clean between the teeth reducing the chances of caries. The exception is the caramel-coated popcorn and related confectionary products that contain significant amounts of sugar or sweeteners (more than 50%,

TABLE 20.3
Nutritional Composition of Regular and Sugar-Coated Popcorn Products (100 g)

Nutrient	Popcorn, Air-Popped	Popcorn, Oil-Popped	Popcorn, Cakes	Popcorn Microwave, Low Fat and Sodium	Popcorn, Caramel-Coated, with Peanuts	Popcorn, Sugar Syrup or Caramel, Fat-Free	Cracker Jack Popcorn: Caramel Coated with Peanuts	Cracker Jack Popcorn: Chocolate Peanut Butter
Water, g	4.1	2.8	5	2.8	3.3	6	ND	ND
Energy, kcal	382	500	384	429	400	381	423	357
Protein, g	12	9	9.7	12.6	6.4	2	7.1	3.6
Lipids, g	4.2	28.1	3.1	9.5	7.8	1.4	7.1	1.8
Carbohydrates, g	77.9	57.2	80.1	73.4	80.7	90.1	81.1	85.2
Sugars, g	0.7	0.5	0.7	0.5	47.8	74.9	52.9	60.7
Dietary fiber, g	15.1	10	2.9	14.2	3.8	2.5	3.5	3.6
Calcium, mg	10	10	9	11	66	18	0	0
Iron, mg	2.66	2.78	1.87	2.28	3.91	0.80	2.54	0
Sodium, mg	4	884	288	490	295	286	247	571
Thiamine, mg	0.203	0.134	0.075	0.350	0.051	0.038	ND	ND
Riboflavin, mg	0.283	0.136	0.178	0.110	0.126	0.058	ND	ND
Niacin, mg	1.944	1.550	6.006	2.070	1.990	0.342	ND	ND
Folate mcg	23	17	18	17	16	4	ND	ND
Vitamin E, mg	0.29	5.01	0.29	5.01	0.85	0.13	ND	ND
Type of fat, g								
Saturated	0.570	4.890	0.480	1.410	1.040	0.200	0	0
Monounsaturated	1.100	8.170	0.920	4.085	2.730	0.260	ND	ND
Polyunsaturated	1.900	13.420	1.350	3.572	3.270	0.634	ND	ND

Source: USDA (2021).

Table 20.3) that generally stick to the dentures increasing the chances of getting dental caries and are obviously contraindicated for diabetic patients (Serna-Saldivar 2008).

20.3.2 Wheat-Based Snacks

Among salted snacks, wheat-based items are considered the best in terms of calories. The nutritional value of these products resembles more bakery products than snacks. The main difference is that since these products are low in moisture, they are usually more energy-dense than fresh bread counterparts.

Pretzels are considered low energy because most formulations do not contain any added fat. A 100-g serving of regular and whole-wheat pretzels provides approximately 380 and 360 kcal, respectively (Table 20.4). In addition, whole-wheat pretzels contain significant amounts of dietary fiber because a 100 g serving provides approximately 30% of the recommended dietary fiber intake of 25 g (Table 20.1). This snack is ideally suited for diabetic people because it does not contain significant amounts of sugars, most of the calories are provided by the complex carbohydrate starch, and the fiber retards glucose absorption. Most commercially available pretzels are high in sodium and therefore are not recommended for patients with hypertension. One 100-g serving of regular pretzels provides more than 200 mg sodium. Saltines, crackers, and crisp breads are also considered healthful snacks because they do not contain high amounts of fats. The nutritional value of crackers resembles more bakery products than snacks. The main difference is that, since these products are low in moisture, they are usually more energy-dense than other bakery products. Saltines and crackers are also considered healthful bakery items because they do not contain high amounts of fats. Whole-wheat or Graham crackers contain important quantities of dietary fiber and reduced calories. Among crackers, the oil-sprayed or oil-filled items normally contain higher amounts of lipids and calories. Some whole-wheat products contain relevant quantities of dietary fiber and comparatively lesser amounts of calories (Table 20.4).

Recently, the popularity of pita chips is increasing because customers think of this snack as healthy. These products resemble the composition of pita bread, but they are more energy-dense due to their comparatively lower moisture content. Besides regular pita chips, there are an array of derived items available on the market like rye, garlic-seasoned, cheese-flavored, and multigrain. One-hundred grams of this last item can provide more than 40% of the recommended dietary fiber intake (Table 20.4).

Wheat-based snacks should be avoided for the population segment that is gluten intolerant. The wheat protein or gluten causes important allergic reactions and gastrointestinal problems such as malnutrition, diarrhea, bleeding, bloating, and bowel inflammation. The label of these snacks should contain a warning indicating that products were manufactured with gluten-containing ingredients.

20.3.3 Granola and Nutritional Bars

One of the snacks that have recently gained more market and sales are the category of granola and nutritional bars because most consumers consider them healthy. These products have been tailor-made in order to fulfill several markets like the 40–30–30 (indicating the calories provided by carbohydrates, protein, and fat respectively), granola, high protein, and low carbohydrate/high dietary fiber. Particularly, the demand for high-protein snack bars has grown significantly by the people engaged in sports activities and dieting and as meal substitutes. These snack bars provide a healthy alternative to conventional snacks due to its high content of protein (15–35%) and other nutritionally beneficial ingredients (Constantin and Istrati 2019). However, not all of these bars are necessarily highly nutritional because some contain large amounts of calories, soluble sugars, and carbohydrates. The most nutritionally acceptable bars are those produced with significant amounts of oats, pulses, soybeans, peanuts, almonds, tree nuts, and dried fruits. In particular, groats either in laminated or meal form provide β-glucans that according to numerous clinical trials diminish total and LDL cholesterol levels related to risk of CVD. In addition, this type of soluble fiber is highly prebiotic and helps to correct the dysbiosis of the microbiome of the hindgut that is more common in obese individuals. From the data presented in Table 20.5, the chemical composition of the snack

TABLE 20.4
Nutritional Composition of Main Categories of Wheat-Based Snacks (100 g)

Nutrient	Pretzels				Crackers/Saltines						
	Chips, Hard Plain	Hard, Multigrain	Hard, Whole Wheat	Hard, Chocolate-Coated	Saltines	Nabisco Graham	Nabisco Ritz Crackers	Nabisco Wheat Thins, Baked	Whole Wheat	Cheese, Regular	Cheese, Sandwich-Type Cheese Filling
Water, g	3.1	3.3	3.9	3.1	5.1	3.6	3.1	2.5	2.7	3.1	4.8
Energy, kcal	375	381	362	467	418	424	492	470	443	503	490
Protein, g	9.8	10.2	11.1	7.1	9.5	7	7.2	8.3	8.8	10.1	8.9
Fat, g	2.9	3	2.6	17.6	8.6	10	22.9	20.0	17.2	25.3	24.4
Carbohydrates, g	78.4	80.1	81.2	70.1	74.1	76.2	64.2	69.1	68.6	58.2	58.8
Dietary fiber, g	3.3	4.2	7.7	3.5	2.8	3.4	1.9	3	10.5	2.4	1.9
Calcium, mg	28	28	28	141	19	79	147	80	50	151	89
Iron, mg	4.47	4.70	2.69	2.54	5.57	4.13	4.05	3.70	3.08	4.77	2.13
Sodium, mg	2154	1216	203	494	941	659	776	578	659	995	878
Thiamine, mg	0.414	0.426	0.440	0.360	0.702	0.228	0.253	0.300	0.201	0.570	0.508
Riboflavin, mg	0.324	0.337	0.291	0.317	0.487	0.240	0.306	0.300	0.100	0.428	0.457
Niacin, mg	5.141	5.433	6.537	2.398	6.442	3.780	3.810	4	4.519	4.671	3.478
Folate mcg	177	179	54	74	216	46	60	42	28	152	82
Vitamin E, mg	0.46	0.49	ND	0.62	1.15	ND	ND	ND	0.86	0.06	1.99
Type of fat, g											
Saturated	0.381	0.396	0.560	8.611	1.653	1.560	3.920	3.190	3.393	9.371	4.810
Monounsaturated	0.709	0.725	1.030	3.293	1.986	3.600	17.950	7	5.880	12.105	14.307
Polyunsaturated	1.260	1.309	0.830	2.820	4.835	0.460	1.770	1.240	6.596	2.472	2.327

(*Continued*)

TABLE 20.4 (CONTINUED)
Nutritional Composition of Main Categories of Wheat-Based Snacks (100 g)

Nutrient		Crispbread				Pita chips		
	Bread Sticks	Sesame Sticks, Wheat-Based, Salted	Rye Crispbread	Rye and Sesame	Sesame and Sea Salt	Pita Chips	Baked	Multigrain
Water, g	6.1	2	6.1	ND	ND	2	2	2
Energy, kcal	412	541	366	350	333	457	429	464
Protein, g	12	10.9	7.9	10	6.7	11.8	10.7	14.3
Fat, g	9.5	36.7	1.3	7.5	10	15.2	8.9	14.3
Carbohydrates, g	68.4	46.5	82.2	60	66.7	68.3	75	67.8
Dietary fiber, g	3	2.8	16.5	20	6.7	3.8	3.6	10.7
Calcium, mg	22	170	31	400	73	17	0	0
Iron, mg	4.28	0.74	2.43	1.8	ND	4.59	3.86	3.90
Sodium, mg	657	1488	2.39	350	533	854	536	500
Thiamine, mg	0.589	0.123	0.145	ND	ND	0.560	ND	ND
Riboflavin, mg	0.553	0.063	1.040	ND	ND	0.307	ND	ND
Niacin, mg	5.281	1.552	0.676	ND	ND	7.043	ND	ND
Folate mcg	162	22	25	ND	ND	135	ND	ND
Vitamin E, mg	1.01	3.85	0.81	ND	ND	6.93	ND	ND
Type of fat, g								
Saturated	1.410	6.480	0.145	2.500	ND	1.342	ND	0
Monounsaturated	3.568	10.910	0.165	ND	ND	10.842	ND	8.930
Polyunsaturated	2.090	17.420	0.557	ND	ND	2.130	ND	1.800

Source: USDA (2021).

TABLE 20.5
Nutritional Composition of Different Types of Snack Bars (100 g)

Nutrient	Kellogg's Nutri-Grain Cereal-Based	Quaker Chewy Granola Bar	General Mills Nature Valley Crunchy Granola Bar	Breakfast Bars, Oats, Sugar, Raisins, Coconut	Snack, Granola Bar, Soft, Uncoated, Nut and Raisin	Formulated Bar, High Fiber, Chewy, Oats and Chocolate	Balance Original Bar (40:30:30)	Cereal or Granola Bar, Fruit and Nut	Kellogg's Nutri Grain Fruit and Nut Bar	Nutrition Bar, High Protein
Water, g	16.7	6.5	3.9	4.1	6.1	12	7.8	7.9	11.3	12.2
Energy, kcal	365	418	471	464	454	389	415	487	403	412
Protein, g	4.2	5.7	10.1	9.8	8	5	28	14	9.4	30.3
Lipids, g	8.7	16.6	19.8	17.6	20.4	10	12	31.9	10.9	15.2
Carbohydrates, g	67.6	70.2	64.4	66.7	63.6	69.8	48.7	44.3	66.7	38.4
Sugars, g	29	28.9	28.6	27.5	ND	25	34.9	22.6	32.5	15.2
Dietary fiber, g	3	3.8	5.3	3.1	5.6	22.5	3.1	6.6	7.5	7.3
Calcium, mg	576	41	61	60	84	250	543	115	65	322
Iron, mg	6.14	2.19	2.95	3.18	2.18	0.90	10.90	3.09	5.65	5.40
Sodium, mg	354	292	294	251	254	237	253	45	195	436
Thiamine, mg	1.153	0.169	0.264	0.280	0.190	0.688	1.628	0.188	0.465	0.176
Riboflavin, mg	2.067	0.082	0.119	0.110	0.190	0.091	1.708	0.462	0.616	0.437
Niacin, mg	19.267	0.765	1.581	1.750	2.610	7.903	8.991	3.056	6.442	1.268
Folate, mcg	14	16	23	81	30	35	126	54	127	184
Vitamin E, mg	0.93	0.31	2.09	0.82	ND	1.47	5.84	10.20	3.72	4.61
Type of fat, g										
Saturated	1.442	6.061	2.370	12.690	9.540	3.397	1.149	3.834	1.930	3.384
Monounsaturated	1.989	6.967	4.380	1.933	4.220	3.522	8.115	19.004	5.705	4.290
Polyunsaturated	4.544	1.576	12.050	1.634	5.520	2.334	1.969	7.487	2.581	5.326

bars varies greatly, mainly depending on major ingredients present in the formulation. From the nutritional perspective, snack bars are commonly categorized as high-energy, high-protein, high-fiber, multigrain, 40:30:30, cereal-based, fruit-based, nut-based, sweetened, and chocolate coated, among others.

20.3.4 Nixtamalized or Lime-Cooked Snacks

The nixtamalization process significantly affects the nutritional profile of tortillas and related snacks (Serna-Saldivar 2015). Deep-fat-fried corn and tortilla chips thoroughly covered in Chapter 11 are considered as energy-dense foods due to their high starch and oil contents and recently have also gained popularity because they are an excellent choice for gluten-intolerant people. As most cereals, nixtamalized products lack good quality protein due to the deficiency of two essential amino acids: lysine and tryptophan. Besides, they lack adequate levels of key micronutrients such as iron, zinc, and vitamins A, D, E, and B_{12}. The supplementation of tortilla chips with legumes, oilseeds, and pseudocereals greatly improve the overall nutritional value of nixtamalized snacks. Regular corn and tortillas chips contain at least twice as much calories compared to regular soft tortillas because of the oil uptake and due to the removal of practically all moisture during deep fat frying. From the negative side, these snacks contain high levels of sodium provided by salt and seasonings (Table 20.6). Corn chips are higher in calories and oil than tortilla chips. The average oil content of regular corn and tortilla chips is 36% and 24%, respectively. Starch is the single chemical component present in the highest amount in lime-cooked snack foods and this carbohydrate comprises more than 50% of their total weight. The starch fraction in nixtamalized snacks is gelatinized and therefore is more rapidly digested in the gastrointestinal tract when compared with native starch. Snacks with lower glycemic index can be produced by supplementing whole grains, insoluble and/or soluble dietary fiber sources, and low glycemic starches. During the sequential thermal treatments applied to snacks, one small portion of the starch becomes resistant. The major type of resistant starch is the type 3, which is also known as retrograded.

The second major chemical compound is the oil. The balance among saturated, unsaturated, and polyunsaturated fatty acids present in the finished product is greatly influenced by the chemical composition of the frying oil, although flavored chips also contain other fat sources used as carriers for seasonings. A 100-g serving of regular corn and tortilla chips provides more than 540 and 500 kcal, respectively. In this sense, processors have launched many light and low-fat or baked tortilla chips that normally contain less than 15% and 6% fat, respectively. The low-fat tortilla chip contains about 60 less kcal per 100 g serving (Table 20.6). In terms of minerals, corn and tortilla chips contain moderate to high quantities of calcium due to lime cooking/steeping. There is overwhelming evidence that the calcium contained in lime is highly bioavailable to humans (Serna-Saldivar 2015). In terms of vitamins, it is well-known that the nixtamalization process has a positive effect on the availability of niacin in the diet.

Regarding specialty nixtamalized snacks, the most promising are those supplemented with other grains like amaranth, flaxseed, sesame, whole wheat, brown rice, and/or oatmeal and fried with polyunsaturated commercial oils aimed to improve protein content per serving, protein quality, dietary fiber, and polyunsaturated fats. The other popular category is the blue maize tortilla chips rich in anthocyanins because these products are viewed as healthy by the consumers.

20.3.5 Extruded Snacks

Consumer preference for extruded foods covered in Chapter 12 is mainly due to convenience, attractive appearance, and texture. This versatile technology allows processors to combine different raw materials to potentially enhance the nutritional profile. The nutritional composition of main types of extruded snacks is provided in Table 20.7. Direct expanded and third-generation snacks are normally manufactured from refined grits and starches and contain high quantities of oil added during

TABLE 20.6
Nutritional Composition of Major Nixtamalized Snack Foods (100 g)

	Corn Nuts		Corn Chips		Tortilla Chips					
Nutrient	**Plain**	**Barbecue Flavor**	**Plain**	**Barbecue Flavor**	**Plain**	**Regular Ranch Flavor**	**Light (Baked with Less Oil)**	**Low Fat, Baked**	**Multigrain Sun Chips ™**	**Tortilla Chips, Takis Fuego ™**
Water, g	1.3	1.6	1	1.2	1.8	1.7	1.3	1.7	2	ND
Energy, kcal	446	436	539	523	501	490	465	415	491	500
Protein, g	8.5	9	6.6	7	7	7.6	8.7	11	8	7.1
Lipids, g	15.6	14.3	33.4	32.7	26.2	23.8	15.6	5.7	21.1	28.6
Carbohydrates, g	71.9	71.7	56.9	56.2	62.9	64.6	73.4	80	67.3	57.1
Dietary fiber, g	6.9	8.4	4.9	5.2	6.5	3.9	5.7	5.3	8.8	7.1
Calcium, mg	9	17	127	131	154	141	159	159	20	143
Iron, mg	1.67	1.70	1.32	1.54	1.52	1.46	1.63	1.60	1.74	1.29
Sodium, mg	549	976	630	763	528	612	1003	419	423	1393
Thiamine, mg	0.040	1.510	0.030	0.080	0.080	0.110	0.220	0.220	0.140	ND
Riboflavin, mg	0.130	0.370	0.140	0.210	0.180	0.240	0.280	0.280	0.02	ND
Niacin, mg	1.690	0.190	1.180	1.650	1.280	1.460	0.420	0.420	2.100	ND
Folate mcg	0	0	20	39	10	17	16	16	25	ND
Vitamin E, mg	1.98	ND	1.36	ND	3.53	ND	3.53	0.77	7.04	ND
Type of fat, g										
Saturated	2.430	2.580	4.550	4.460	5.020	4.560	2.840	0.850	2.166	8.930
Monounsaturated	9.460	7.360	9.660	9.480	15.450	14.050	6.340	1.660	11.501	ND
Polyunsaturated	3.050	3.220	16.480	16.170	3.630	3.300	5.020	2.870	6.513	ND

Source: USDA (2021).

TABLE 20.7
Nutritional Composition of Direct-Expanded and Third-Generation Extruded Snacks (100 g)

	Direct-Extruded Snacks						Third Generation Snacks	
Nutrient	**Puffs or Twist, Cheese Flavor**	**Puffs or Twist, Cheese Flavor**	**Cheese Puffs and Twist, Low Fat**	**Cones, Plain**	**Cones, Nacho Flavor**	**Onion Flavor**	**Puffed Wheat, Sabritones ™ Chile and Lime Flavored**	**Wagon Wheels, Wheat Snack**
Water, g	1.5	1.5	3	2	1.9	2	ND	ND
Energy, kcal	554	554	432	510	536	500	500	367
Protein, g	7.6	7.6	8.5	5.8	6.5	7.7	7.1	6.7
Fat, g	34.4	34.4	12.1	26.9	31.7	22.6	28.6	20
Carbohydrate, g	53.8	53.8	72.4	62.9	57.3	65.1	57.1	40
Dietary fiber, g	1.1	1.1	10.7	1.1	1	3.9	3.6	3.33
Calcium, mg	58	58	357	3	36	29	0	7
Iron, mg	2.35	1.10	1.28	2.54	1.27	3.72	2.57	6.67
Sodium, mg	1050	1050	1285	1022	952	981	2214	667
Thiamin, mg	0.264	0.260	0.540	0.320	0.210	0.220	ND	ND
Riboflavin, mg	0.354	0.200	0.610	0.240	0.120	0.310	0.243	ND
Niacin, mg	3.232	1.180	7.140	1.410	1.410	3.180	4.286	ND
Folate, mcg	120	23	97	3	5	110	ND	ND
Vitamin E, mg	4.25	ND	4.25	ND	ND	2.69	ND	ND
Type of fat, g								
Saturated	6.590	6.590	2.100	22.750	26.70	4.340	7.140	8.330
Monounsaturated	20.280	20.280	3.500	1.700	2.140	13.360	ND	ND
Polyunsaturated	4.760	4.760	5.740	0.780	0.840	3.140	ND	ND

Source: USDA (2021).

seasoning or absorbed by the pellet or extrudate during deep fat frying. Most extruded snacks contain more than 30% oil and 500 kcal/100 g serving. Therefore, extruded snacks are a rich source of calories and oil, but do not contain significant quantities of other nutrients such as protein, dietary fiber, micronutrients, and phytochemicals (Camire 2012). These nutrients associated with the germ, aleurone, and bran were removed during the various types of dry milling processes discussed in Chapter 6.

The sequential heat treatments that the starch-rich feedstock suffers within the extruder barrel and then throughout deep fat frying cause starch gelatinization and dextrinization. The degree of starch gelatinization is greatly influenced by the heat, pressure, and residence time in the extruder and the oil temperature and frying time in the deep fat fryer. Gelatinized starch and dextrins are more rapidly degraded and transformed into glucose in the gastrointestinal tract, so extruded snacks are considered as high glycemic. Overheating or temperature abuse during the high-shear extrusion and other processes can create resistant starch type 3 that acts as prebiotic in the hindgut. Since most extruded snacks are not major sources of protein, the digestibility and protein quality have little impact on the overall nutritional value. Several changes in proteins occur during extrusion. The main one is denaturation, improved digestibility, and reduced lysine bioavailability, especially when the feedstock contains significant amounts of reducing sugars or alkali that lowers lysine bioavailability (Camire 2012). The nutritional significance of *Maillard* reactions is of utmost relevance from the nutritional viewpoint. Protein nutritional value is dependent on the digestibility as well as the level and availability of essential amino acids, mainly lysine. On the other hand, extruded snacks are generally high in sodium. A 100-g serving of extruded snacks generally provides more than 500 mg of this electrolyte (Table 20.7).

20.3.6 Snacks from Legumes, Pseudocereals, and Other Grains

Grain legumes, pulses, and pseudocereals play an important role in human nutrition, especially in the dietary pattern of low-income groups inhabiting in developing countries. The more affluent consumer is also increasing the consumption of these whole grains because they look for new natural alternatives to improve health. Legumes, pulses, and pseudocereals (Chapter 13) are considered nutritionally superior compared to whole-grain cereals, mainly due to their higher protein content and quality and higher concentrations of dietary fiber, phytochemicals, and micronutrients (Serna-Saldivar 2020).

Recently, the daily recommendations for the United States and Canada indicate that whole grains should provide at least 50% of the total daily grain intake (US Department of Health 2015). Likewise, the recommendations for people inhabiting in the European community and Australia recommend the consumption of four to nine daily servings of whole grains. The goal of these recommendations is to improve health by preventing obesity, metabolic syndrome, diabetes, CVD, oxidative stress, and cancer (Chen et al. 2016, Serna-Saldivar 2020).

Legumes should be consumed after thermal processing, sprouting, or fermentation because in raw forms they contain trypsin inhibitors, urease, and hemagglutinins. Therefore, the thermal processes of roasting, extrusion cooking, and deep fat frying inactivate these antinutritional compounds and normally improve both protein digestibility and resistant starch formation.

Most consumers perceive legumes and pseudocereals as a vegan source of high-quality protein and dietary fiber (Chapter 13). In addition, the carbohydrates associated with these grain sources are slow releasing and considered low glycemic. Starch, the major carbohydrate associated with pulses, greatly differs from cereals or tuber starches because they are slow releasing and tend to become more resistant after thermal processing. The slow and reduced starch digestibility has been attributed to its high-molecular-weight amylose. These resistant starches are highly prebiotic when reaching the large intestine and increase the fecal bulk and short-chain fatty acid production, mainly butyrate. One of the major advantages of the consumption of whole-grain legumes and pseudocereals is that they are excellent sources of both insoluble and soluble dietary fibers that lower caloric density and act as important prebiotics. Additionally, the dietary fiber is associated with an array of relevant phytochemicals that exert many health benefits especially related to prevention of oxidative stress (Tharanathan and Mahadevamma 2003).

Table 20.8 summarizes the nutritional composition of main snacks produced from legumes and pseudocereals. Peanuts are a high-caloric food item because of the high-oil and low-moisture contents. The caloric density increases when peanuts are oil-roasted and decreases when the naturally occurring oil is removed from the product. Peanuts contain about 26% protein that is considered good quality protein because it is mainly constituted by lysine-rich globulins. The sulfur-containing amino acids methionine and cystine are the ones that limit protein quality. Peanuts naturally contain 47–50% oil and contain a favorable ratio of saturated to polyunsaturated fatty acids. The oil contains 40–45% oleic acid (18:1 Δ 9) and 30–35% ω-6 linoleic acid (18:2 Δ 9, 12). Peanut oil is the only vegetable oil that contains trace levels of ω-6 arachidonic acid (20:4 Δ 5, 8, 11, 14). Both linoleic and arachidonic acids are relevant because they can be metabolized into prostaglandins and other important eicosanoids with positive health implications (Brenna and Diau 2007, Gibson et al. 2011).

Different types of soybeans are commonly used to increase the protein content of snacks like bars and extruded products. The soybeans can be used for the production of roasted nuts (Table 20.8), sprouted for the manufacture of natural meals with low levels of trypsin inhibitor (Maetens et al. 2017), or alternatively processed into full-fat or defatted meals or flours. The virtue of these products is that they contain at least twice the amount of protein compared to fresh meat and an array of phytochemicals, including isoflavones (Xu and Chang 2008). The defatted soybean, concentrate, and isolates contain 50%, 65%, and 88% protein, respectively, with excellent digestibility and amino acid scores comparable with the best animal sources. The other important attribute of soybean proteins is that they effectively improve protein quality of cereal-based products. Generally, the inclusion of 6–10% of soybean meal, concentrate, or isolates increase about 20% the protein content of cereal-based snacks and greatly improves protein quality or biological value.

Pulses are good sources of protein, slow-digesting starch, insoluble and soluble dietary fibers, oligosaccharides, and relevant antioxidants mainly of phenolic nature (Chapter 13). In addition, they commonly contain less than 10% lipids rich in polyunsaturated fatty acids. These nutrients and phytochemicals have been related to relevant health benefits, especially in terms of prevention of chronic diseases and cancers. Among pulses, chickpeas have recently gained attention and are nowadays the third most produced pulse worldwide. The value of chickpeas is due to the relatively high amounts of good quality protein (18–22%), resistant starch, and other nondigestible carbohydrates, dietary fiber, and phytochemicals compared with other grains. Chickpea protein consumption has gained importance due to its high protein quality, amino acid profile, and bioactive peptides, which are known to benefit human health, and because it is comparatively less allergenic compared to peanuts or soybeans. The carbohydrates in chickpea constitute more than 60% of the total seed weight and are mainly constituted by starch (41–50% of the total seed weight). The chickpea starch is rich in slowly digestible and resistant fractions. In chickpea seeds, the resistant starch is 20–30% of total starch. As a result, the glycemic index of chickpea foods is comparatively very low, favoring diabetic patients and the microbiome of the hindgut. Besides starch, chickpeas are rich in oligosaccharides, insoluble dietary fiber, and soluble dietary fiber with significant amounts of prebiotic β-glucans. These nondigestible carbohydrates play a relevant role in cholesterolemia, glucose homeostasis, cancer prevention, and the modulation of microbiota composition (Guardado-Felix et al. 2019, Milan-Norris et al. 2019).

Buckwheat is considered a functional pseudocereal with high nutritional value. The roasted dry groats (Table 20.8) contain protein of high nutritional value, dietary fiber, resistant starch, the flavonoids (rutin and quercetin), and vitamins and minerals. In addition, the oil contains a high level of polyunsaturated FA such as linoleic. Sprouting of whole buckwheat enhances its nutritional and functional properties by increasing protein digestibility and the bioavailability of other nutritional and functional components (Wijngaard and Arendt 2006).

TABLE 20.8
Nutritional Composition of Snack Products Based on Legumes, Pulses, and Pseudocereals (100 g)

Nutrient	Peanuts				Peanuts Mixed with Nuts, Dry Roasted, Salted	Peanuts Mixed with Nuts, Oil Roasted, Salted	Soybean Nuts	Chickpea Snacks	
	All Types Dry Roasted, Salted	All Types Oil Roasted, Salted	Japanese Style	Chocolate-Coated				Roasted Chickpea Snack	Plain Hummus
Water, g	1.6	1.5	1.5	1.9	1.8	3.2	2	ND	51.5
Energy, kcal	585	599	500	519	594	615	469	464	260
Protein, g	23.7	28	17.9	13.1	17.3	15.5	38.6	17.9	8.2
Lipids, g	49.7	52.5	28.6	33.5	51.5	56.2	25.4	21.4	17.8
Carbohydrates, g	21.5	15.3	46.4	49.4	25.4	22.3	30.2	57.1	19.9
Sugars, g	4.9	4.2	7.1	37.6	4.8	4.4	4.2	10.7	2.9
Dietary fiber, g	8.0	9.4	7.1	4.7	9.0	5.5	17.7	10.70	5.90
Calcium, mg	54	61	71	104	70	106	138	71	113
Iron, mg	2.26	1.52	2.86	1.31	3.70	2.57	3.9	3.86	3.36
Sodium, mg	813	320	571	41	669	306	163	411	481
Thiamine, mg	0.438	0.085	ND	0.116	0.200	0.504	0.100	ND	0.305
Riboflavin, mg	0.098	0.089	ND	0.174	0.200	0.486	0.145	ND	0.129
Niacin, mg	13.520	13.820	ND	4.250	4.700	1.964	1.410	ND	1.363
Folate mcg	145	120	ND	8	50	56	211	ND	113
Vitamin E, mg	7.80	6.94	ND	3.46	10.94	8.20	0.91	ND	1.09
Type of fat, g									
Saturated	6.890	8.670	4	14.600	6.900	9.090	3.674	1.790	2.429
Monounsaturated	24.640	25.940	12	12.920	31.400	33.140	5.610	ND	8.554
Polyunsaturated	15.690	15.270	8	4.330	10.770	11.450	14.339	ND	5.845

(Continued)

Nutrient	Buckwheat Groats, Roasted, Dry	Amaranth Grain, Uncooked	Amaranth Grain, Cooked	Crunchy Amaranth Snack	Quinoa, Cooked	Chia Seeds, Dried	Sesame Seed, Toasted, with Added Salt	Flaxseed	Roasted Pumpkin Seeds
Water, g	8.4	11.3	75.2	ND	71.6	5.8	5	7	7.1
Energy, kcal	346	371	102	567	120	486	567	534	522
Protein, g	11.7	13.6	3.8	13.3	4.4	16.5	17	18.3	33
Lipids, g	2.7	7	1.6	36.7	1.9	30.7	48	42.2	42.1
Carbohydrates, g	75	65.3	18.7	43.3	21.3	42.1	26	28.9	13.4
Sugars, g	ND	1.7	ND	13.3	0.9	ND	0.5	1.6	ND
Dietary fiber, g	10.3	6.7	2.1	11.7	2.8	34.3	16.9	27.3	3.9
Calcium, mg	17	159	47	450	17	631	131	255	43
Iron, mg	2.47	7.61	2.10	2.33	1.49	7.72	7.78	5.73	14.94
Sodium, mg	11	4	6	317	7	16	588	30	575
Thiamine, mg	0.224	0.116	0.015	ND	0.107	0.620	1.205	1.644	0.210
Riboflavin, mg	0.271	0.200	0.022	ND	0.110	0.170	0.466	0.161	0.319
Niacin, mg	5.135	0.923	0.235	ND	0.412	8.830	5.438	3.080	1.742
Folate mcg	42	82	22	ND	42	49	96	87	57
Vitamin E, mg	ND	1.19	0.19	ND	0.63	0.50	0.25	0.31	ND
Type of fat, g									
Saturated	0.591	1.459	ND	8.330	0.231	3.330	6.722	3.663	7.970
Monounsaturated	0.828	1.685	ND	ND	0.528	2.309	18.127	7.527	13.100
Polyunsaturated	0.828	2.778	ND	ND	1.078	23.665	21.039	28.730	19.210

Source: USDA (2021).

Amaranth is a highly nutritional ancient pseudocereal with a superior amount of proteins when compared to true cereals. It is a reasonably well-balanced food with functional properties that have been shown to provide health benefits. The protein content of amaranth of 13–17% is in between the amount provided by cereals and pulses. The protein is considered good quality because of the balanced essential amino acids and relatively high protein digestibility value. The amaranth is rich in dietary fiber, monounsaturated and polyunsaturated fatty acids, resistant starch, vitamins, and minerals (Caselato-Sousa and Amaya-Farfan 2012, Serna-Saldivar and Sanchez-Hernandez 2020).

The ancient American pseudocereals quinoa and chia are gaining popularity throughout the world. These gluten-free pseudocereals are valued because of their nutritional value and bioavailability in regard to proteins, dietary fiber, complex carbohydrates, fatty acids, and micronutrients (Chapter 13). The oil of chia is rich in ω-3 linolenic acid and an array of antioxidants that prevent cardiovascular diseases, hypertension, diabetes, and nervous system disorders (Muñoz et al. 2013, Navruz-Varli and Sanlier 2016, Pereira et al. 2019).

Sesame is highly valued for its protein, carbohydrates, oil rich in polyunsaturated FA, micronutrients, and phytochemicals like lignans, tocopherols, and phytosterols. The presence of these substances enhances the keeping quality of the oil and health benefits, including antioxidant, anticancer, and antitumoral (Pathak et al. 2014).

Among oilseeds, flaxseed or linseed is the one that contains the highest amounts of ω-3 linolenic acid. Besides, these seeds are viewed as excellent sources of proteins, dietary fiber, carbohydrates, micronutrients, and phytochemicals.

New product containing pumpkin or squash seeds are being introduced into the snack market because the seed kernels are rich in polyunsaturated oil, protein (24%), iron, zinc, vitamin A, lysine, and tryptophan. More than 40% of the lipid fraction is composed of linoleic (18:2 Δ 9,12) acid. A 100-g portion of roasted and salted seeds provide more than 520 kcal (Table 20.8). The squash seeds are also valued as health promoting because of their phenolics, sterols, and phytoestrogens (secoisolariciresinol and lariciresinol) that prevent hyperlipidemia and osteoporosis of menopausal women and reduction of hormone-dependent cancer tumors. The pumpkin seeds are especially rich in tocopherols that act as free radical scavengers, antiaging, and antioxidants (Fu et al. 2006, Saavedra et al. 2015).

20.3.7 Potato Chips

The category of potato chips is still considered the most popular savory snack in the world. The nutrient composition of different types of potato chips is presented in Table 20.9. Potato chips are high caloric because of the high starch and oil contents and low moisture, which is removed during the deep fat frying operation. The caloric density of potato chips varies according to the oil content. Regular chips usually contain 36% oil absorbed during the frying process. Needless to say, the balance among saturated, unsaturated, and polyunsaturated fatty acids present in the finished product is greatly influenced by the chemical composition of the frying oil, although flavored chips also contain other fat sources used as carriers for flavorings. A 100-g serving of regular chips provides more than 550 kcal equivalent to 25% of the recommended energy intake for a 2,000 kcal diet (Table 20.9). There are versions of low-fat and fat-free potato chips that contain 480 and 380 kcal, respectively (Table 20.9). Potato chips also contain high amounts of sodium because salt is the most common flavoring. One 100-g serving of regular salted chips provides more than 25% of the RDI of sodium, considering a 2,000 kcal diet with a daily sodium requirement of 2,200 mg. Potato chips are regarded as a poor source of dietary fiber, essential minerals such as calcium, iron, and zinc, and vitamins. These products practically do not contain any vitamin C because of the heat treatments the potato suffers during processing, although raw potatoes contain approximately 20 mg ascorbic acid/100 g. These are the main reasons why these items are viewed as junk or empty foods and have had a bad image among dietitians and physicians.

TABLE 20.9
Nutritional Composition of Different Types of Potato Chips (100 g)

	Potato Chips									
Nutrient	**Plain**	**Plain and Salted**	**Sour Cream and Onion-Flavor**	**BarbecueFlavor**	**Cheese Flavor**	**Restructured Baked**	**Potato Sticks**	**Fat-Free, Salted**	**Light**	**Reduced Fat**
Water, g	1.4	1.9	1.8	1.9	1.8	1.4	2.2	2	1.4	1
Energy, kcal	558	536	531	491	496	469	522	379	501	471
Protein, g	5.9	7	8.1	7.7	8.5	5	6.7	9.6	5.6	7.1
Fat, g	38.4	34.6	33.9	32.4	27.2	18.2	34.4	0.6	25.7	20.8
Carbohydrates, g	51	52.9	51.5	52.8	57.7	71.4	53.3	83.8	64.9	66.9
Dietary fiber, g	3.6	4.8	5.2	4.4	5.2	4.8	3.4	7.5	3.6	5.9
Calcium, mg	24	24	72	50	72	125	18	35	34	21
Iron, mg	1.50	1.63	1.60	1.94	1.84	0.80	2.27	3.57	1.51	1.35
Sodium, mg	656	594	625	750	793	917	250	643	428	492
Thiamin, mg	0.210	0.170	0.190	0.220	0.160	0.340	0.100	1.080	0.190	0.210
Riboflavin, mg	0.120	0.200	0.200	0.220	0.160	0.070	0.120	0.120	0.060	0.270
Niacin, mg	3.150	3.830	4.030	4.690	5	4.090	4.790	6.450	4.200	7
Folate, mcg	7	45	62	83	0	0	40	45	27	27
Vitamin E, mg	2.19	ND	ND	ND	ND	2.18	9.11	0.04	2.18	5.47
Type of fat, g										
Saturated	9.450	5.430	8.890	8.050	8.600	2.630	8.880	0.150	5.130	4.160
Monounsaturated	7.270	18	6.120	6.540	7.720	9.910	6.160	0.015	5.920	4.800
Polyunsaturated	19.980	9.160	17.420	16.370	9.560	4.230	17.900	0.260	13.490	10.940

Source: USDA (2021).

20.3.8 Almonds and Tree Nuts

The popularity of almonds and tree nuts covered in Chapter 15 has increased because many epidemiological and clinical studies have shown that consumption of these snacks are associated with reduced risk of CVD because they contain high amounts of polyunsaturated fats and many hydrophilic and lipophilic phytochemicals. Table 20.10 depicts the comparative nutritional attributes of various types of almonds and dry- or oil-roasted tree nuts. Almonds and tree nuts are considered as one of the more energy-dense snacks because of the high-oil and low-moisture contents. For instance, almonds contain about 50% oil, 20% protein, more than 10% dietary fiber, and approximately 600 kcal/100 g serving. Interestingly, the almond oil contains less than 10% saturated fat. The major fatty acid is oleic (18:1 ω-9). On the other hand, the tree nuts with the highest fat and energy contents are macadamia and pecans because they contain approximately 75% oil. The lipids of these nuts are associated with natural-occurring phospholipids, phytosterols, and vitamin E. Vitamin E has received attention lately because it is a potent antioxidant and considered the second line of defense against free radicals that cause oxidative stress and cancer.

Yang et al. (2009) analyzed the antioxidant and antiproliferative activities of ten sorts of nuts (almonds, Brazil nuts, cashews, hazelnuts, macadamia, peanuts, pecans, pistachios, walnuts, and pine nuts) and concluded that walnuts exerted the highest antioxidant activity as measured with the total oxygen radical scavenging capacity or TOSC assay followed by pecans, peanuts, and pistachios. Walnuts and pecans exhibited the highest antiproliferative cancer activity against hepatic $HEPG_2$ and colon Caco-2 cells.

20.3.9 Snacks from Fruits and Vegetables

Most of the fruits and vegetables employed as snacks or as ingredients are dehydrated or used as intermediate-moisture items (Chapter 16). Therefore, they contain higher energy and nutrient density compared to their fresh counterparts. These food products have relatively low amounts of proteins and lipids, so they are mainly constituted by carbohydrates rich in dietary fiber. The simple and complex carbohydrates account for 50–80% of the dry solids of these items. The main difference between fruits and vegetables is that the first category is richer in soluble sugars, whereas the second category in dietary fiber. The most abundant sugars are glucose, fructose, and sucrose which are contraindicated for diabetic people. In terms of dietary fiber, fruits and vegetables are considered excellent sources.

Overall, the amount and profile of dietary fiber associated with fruits and vegetables is highly related to the type, variety, ripening stage, growth conditions, and processing. Most of the fruits contain higher fiber concentration than vegetables; however, depending on the factors mentioned above, the fiber concentration and composition in each product will vary, as well as their functional properties such as water- and oil-holding capacities. Both fruits and vegetables represent an excellent alternative for snacks, especially in order to reach the recommended dietary fiber intake for health wellness and their association with phytochemicals or bioactives that act synergistically to improve health (Morales Peña et al. 2020). The main fiber components are cellulose, cross-linking glycans, pectin, lignin, resistant starch, and nondigestible oligosaccharides. The fiber content in fresh fruits and vegetables ranges from 1% to 3% and cellulose is the most abundant fraction. The consumption of soluble fibers like pectins modulates intestinal function and are readily fermented into acetic, propionic, and butyric acids in the colon, which prevent CVD, obesity, cancer, and blood cholesterol and pressure (Vincente et al. 2014, Morales de la Peña et al. 2020).

Regarding micronutrients, fruits and vegetables are a good source of vitamins and minerals. The most relevant vitamins are A (retinol), B_1 (thiamine), B_2 (riboflavin), B_3 (niacin), B_5 (pantothenic acid), B_6 (pyridoxine), B_9 (folic acid), biotin, choline, C, E, and K, although the specific concentrations and distribution change according to the sort of product, environmental conditions, cultivation practices, and processing. On the other hand, the most abundant minerals are potassium, calcium, magnesium, and phosphorus (Vincente et al. 2014). Table 20.11 depicts the most popular

TABLE 20.10
Nutritional composition of Nut-Based Snack Products (100 g)

	Almonds			Cashews		Hazelnuts	Macadamia	Pecans		Pistachio
Nutrient	**Dry Roasted, Salted**	**Honey Roasted, Unblanched**	**Oil Roasted, Salted**	**Dry Roasted Salted**	**Oil Roasted, Salted**	**Dry Roasted, Unsalted**	**Dry Roasted, Salted**	**Dry Roasted, Salted**	**Oil Roasted, Salted**	**Dry Roasted, Salted**
Water, g	2.6	1.7	2.8	1.7	2.3	2.5	1.6	1.1	1.1	2
Energy, kcal	597	594	607	574	581	646	716	710	715	568
Protein, g	22.1	18.2	21.2	15.3	16.8	15	7.8	9.5	9.2	21.4
Lipids, g	52.8	49.9	55.2	46.4	47.8	62.4	76.1	74.3	75.2	46
Carbohydrates, g	19.3	27.9	17.7	32.7	30.2	17.6	12.8	13.6	13	26.8
Dietary fiber, g	11.8	13.7	10.5	3	3.3	9.4	8	9.4	9.5	10.3
Calcium, mg	266	263	291	45	43	123	70	72	67	110
Iron, mg	4.51	2.83	3.68	6.00	6.05	4.38	2.65	2.80	2.47	4.20
Sodium, mg	339	130	339	640	308	0	265	383	393	405
Thiamine, mg	0.074	0.114	0.092	0.200	0.363	0.338	0.710	0.450	0.473	0.840
Riboflavin, mg	0.859	0.953	0.781	0.200	0.218	0.123	0.087	0.107	0.110	0.158
Niacin, mg	3.850	2.819	3.665	1.400	1.736	2.050	2.274	1.167	1.200	1.425
Folate mcg	33	32	27	69	25	88	10	16	15	50
Vitamin E, mg	26	ND	25.97	0.92	0.92	15.28	0.57	1.30	2.53	1.93
Type of fat, g										
Saturated	4.050	4.730	4.210	9.160	8.480	4.510	11.950	6.240	7.240	5.560
Monounsaturated	33.660	32.400	34.790	27.320	25.920	46.610	59.280	43.960	40.970	24.220
Polyunsaturated	12.650	10.470	13.520	7.840	8.550	8.460	1.500	20.570	23.580	13.900

Source: USDA (2021).

TABLE 20.11
Nutritional Composition of Snacks Based on Fruits and Vegetables (100 g)

Nutrient	Raisins	Cranberries, Dried	Blueberries, Dried	Strawberry, Raw	Mango, Dried	Apple, Dried	Banana, Dehydrated	Banana Chips	Pineapple, Dried	Snacks, Fruit Leather	Fruit Snacks Candy
Water, g	15.5	15.8	14.8	91	16.6	31.8	3	4.3	16	10.2	11.8
Energy, kcal	299	308	317	32	319	243	346	519	313	371	352
Protein, g	3.3	0.2	2.5	0.7	2.5	0.9	3.9	2.3	1.4	0.1	0.1
Lipids, g	0.3	1.1	2.5	0.3	1.2	0.3	1.8	33.6	0.5	3	0
Carbohydrates, g	79.3	82.8	80	7.7	78.6	65.9	88.2	58.4	81.5	85.8	88
Sugars, g	65.2	72.6	67.5	4.9	66.3	57.2	47.3	35.3	72.9	49.2	68.2
Dietary fiber, g	4.5	5.3	7.5	2	2.4	8.7	9.9	7.7	3.7	0	0
Calcium, mg	62	9	19	16	0	14	22	18	35	32	0
Iron, mg	1.79	0.39	0.90	0.41	0.23	1.40	1.15	1.25	0.79	1.01	0
Sodium, mg	26	5	3	1	162	87	3	6	3	317	23
Thiamine, mg	0.106	0.013	0.092	0.024	0.062	0	0.180	0.085	0.147	0.084	0.025
Riboflavin, mg	0.125	0.028	0.127	0.022	0.085	0.159	0.240	0.017	0.085	0.02	0.008
Niacin, mg	0.766	0.548	1.154	0.386	2	0.927	2.800	0.710	1.195	0.1	0.008
Folate, mcg	5	0	12	24	68	0	14	14	24	2	1
Vitamin E, mg	0.12	2.10	2.35	0.29	4.02	0.53	0.39	0.24	0.05	0.56	0
Type of fat, g											
Saturated	0.094	0.088	0.200	0.015	0.287	0.052	0.698	28.970	0.024	0.672	0
Monounsaturated	0.024	0.298	0.405	0.043	0.439	0.013	0.153	1.950	0.035	1.480	0
Polyunsaturated	0.053	0.182	1.300	0.155	0.222	0.093	0.337	0.630	0.106	0.552	0

(Continued)

TABLE 20.11 (CONTINUED)
Nutritional Composition of Snacks Based on Fruits and Vegetables (100 g)

Nutrient	Carrot, Dehydrated	Tomato, Sun-Dried	Onion, Dehydrated, Flakes	Chili Pepper, Raw	Kale Chips	Sweet Potato Chips	Mixed Vegetable Chips	Dill Pickles	Sweet Pickles
Water, g	4	14.6	3.9	89.9	ND	4.5	5.4	94.7	76.2
Energy, kcal	341	258	349	34	553	532	495	12	91
Protein, g	8.1	14.1	9	1.4	25.9	2.9	4.7	0.5	0.6
Lipids, g	1.5	3	0.5	0.4	22.4	32.4	26.6	0.4	0.4
Carbohydrates, g	79.6	55.8	83.3	7.7	85.3	56.8	59.2	2	21.2
Sugars, g	38.8	37.6	37.4	4.7	1.8	8.8	6.9	1.3	18.3
Dietary fiber, g	23.6	12.3	9.2	2.2	15.9	8.8	7.8	1	1
Calcium, mg	212	110	257	13	1053	59	66	54	61
Iron, mg	3.93	9.09	1.55	0.64	12.92	2.12	1.6	0.23	0.25
Sodium, mg	275	107	21	6	2629	35	329	808	457
Thiamine, mg	0.534	0.528	0.500	0.056	2	0.088	0.180	0.047	0.025
Riboflavin, mg	0.417	0.489	0.100	0.078	2.350	0.161	0.095	0.057	0.030
Niacin, mg	6.567	9.050	0.990	1.262	7.647	2.088	2.268	0.109	0.115
Folate, mcg	55	68	166	25	238	37	52	8	1
Vitamin E, mg	5.45	0.01	0.18	2.13	ND	9.82	6.5	0.12	0.36
Type of fat, g									
Saturated	0.256	0.426	0.078	0.067	15.290	2.941	1.824	0.079	0.067
Monounsaturated	0.076	0.487	0.064	0.027	ND	9.155	14.988	0.005	0.004
Polyunsaturated	0.730	1.115	0.180	0.176	ND	12.224	6.766	0.122	0.106

Source: (USDA 2021).

intermediate-moisture and dehydrated fruits and vegetables used for the preparation of fruit bars and mixes with peanuts and tree nuts. Among these items, intermediate-moisture raisins, cranberries, blueberries, and strawberries have gained market because of their high antioxidant properties, microbial stability, and flavor. In addition, the table also contains relevant information related to shelf-stable products employed to produce snacks.

20.3.10 Dairy-Based Snacks

Milk and dairy food products are characterized for providing a variety of essential and high-quality nutrients, such as proteins, sugars, and lipids, and micronutrients, such as minerals and vitamins. Specifically, these products contain essential minerals like calcium, phosphorus, potassium, magnesium, selenium, and zinc that are required for appropriate growth and bone development and essential vitamins like riboflavin (B_2), cianocobalamin (B_{12}), and vitamins A and D. This last vitamin acts synergistically with calcium to favor its absorption and deposition in bones (Rozenberg et al. 2015, Gasmalla et al. 2017, Górska-Warsewicz et al. 2019).

The main carbohydrate in dairy products is lactose, whereas most of the triglycerides are constituted by short-chain saturated fatty acids. In addition, milk contains significant amounts of phospholipids and cholesterol. The main virtue of dairy products is the protein quality of both whey and casein, which constitute around 20% and 80% of the total protein content, respectively. These two protein fractions, individually or combined, are considered as very complete in terms of essential amino acids required for optimum growth of babies and infants.

Lactose is one of the most relevant allergens and should be declared in food labels. Lactose intolerance is a condition with more than 3 million annual cases in which individuals are unable to digest this sugar due to the lack of intestinal lactase. The allergic reaction can be developed by both children and adults and is more common in Asian Americans, African Americans, Mexican Americans, and Native Americans. As a result, there are lactose-free dairy products obtained by different technologies, which include the removal of the disaccharide from milk or the hydrolysis of the sugar with lactase. The quantity of lactose in dairy products can be diminished by removing the water-soluble components, for example, when processing cheeses, especially those categorized as ripened or fermented (Chapter 17), and butter. Additionally, the fermentation of milk for production of yogurts and related products (Chapter 18) with lactic acid bacteria leads to the breakdown of lactose into glucose and galactose; this is the reason why fermented dairy products (e.g., yogurts) have decreased amounts of lactose (Dekker et al. 2019). Milk can also be skimmed to remove most of its fat, which can then be utilized to generate additional dairy snack products such as cheeses and yogurts. Milk and other dairy products, such as creams and butters, can be used as raw materials for the generation of more specialized snack-based food products. Some examples of these include chocolate, milkshakes, ice creams, mousses, and many others discussed in Chapter 18. In many cases, these types of snack products have increased amounts of sugar and fats in their formulations.

The popularity of fermented dairy products has gained interest because many products are designed as functional, especially in terms of probiotics. The *Lactobacillus* and *Bifidobacteria* species ferment milk and generate bioactive compounds that have been proven to have positive effects on health. These should normally provide a minimum count of 10^9 CFU/mL; however, these minimum populations might differ among countries around the globe (Nyanzi et al. 2021). These bioactive compounds include short-chain fatty acids, bioactive peptides, and oligosaccharides that, alongside the prebiotic microorganisms, have positive effects in human health by preventing diseases like hypertension, obesity, cancer, and diabetes (Gasmalla et al. 2017).

Even though milk- and snack-based dairy products are rich in essential and desirable nutrients, many products contain significant amounts of sugar, saturated fats, and cholesterol, components that can impact human health negatively when excessively consumed (Rogelj 2000). Consuming these types of snacks need to be limited and moderated to avoid and prevent the incidence of obesity and CVD.

Table 20.12 summarizes many of the dairy-based products thoroughly covered in Chapter 18. Most of these products have extended shelf stability at refrigeration or even at room temperature because they contain more solids, high amounts of salt, or added sugars that lower water activity (A_w) and many are fermented, so they have significantly lower pHs. Among these dairy-based snacks, the most popular are different sorts of yogurts, cheeses, and dairy-based candies.

20.3.11 Snacks from Animal, Poultry, and Marine Sources

From the nutritional viewpoint, meat is considered as one of the best sources of good quality protein because it contains all essential amino acids required for optimum growth, an excellent source of trace minerals iron and zinc and most B vitamins, including B_{12} or cianocobalamin. The heme iron from meat is considered the most bioavailable source for the human system. Unfortunately, the amount of fat, type of fatty acid composition, and cholesterol associated with meat have negatively influenced consumption. Nutritionist and dietitians usually limit the consumption of meat products especially in patients suffering from chronic diseases and elevated uric acid or gout. The fat associated with meat products is rich in saturated fatty acids that are known to exacerbate hypercholesterolemia. Saturated fatty acids inhibit the regeneration of the LDL receptor commonly named B-100. Therefore, the lipoproteins take longer to internalize into the cells (Clarke et al. 1997). Dehydration of meats and marine products under the best possible conditions results in little or no loss of vitamins or amino acids with the exception of thiamin or vitamin B_1 (about 40% loss).

Although pork rinds and cracklings have a high protein concentration, the protein quality is deficient because it is composed of connective tissue rich in collagen and elastin. These proline-rich proteins are practically devoid of lysine and therefore have a very low protein quality or balance of essential amino acids.

Normally, the different types of fresh meats contain more than 70% protein, so processed and dehydrated meats usually have higher concentrations of all nutrients. The different types of jerkies are normally dehydrated to less than 20% moisture (Table 20.13), normally with the addition of substantial amounts of regular salt and/or curing salts in order to lower the A_w and prevent spoilage at ambient temperature and growth of pathogenic microorganisms. Therefore, these snacks contain higher amounts of proteins, fat, and sodium. Zdanowska-Sąsiadek et al. (2018) compared three dried meat snacks made from ostrich, beef, and chicken meat in relation to their nutrients content, including fat, fatty acids, heme iron, and peptides, like anserine and carnosine. The composition of dried ostrich, beef, and chicken meats with and without spices was compared. Independent of the type of meat, all contained high concentration of proteins, minerals (heme iron especially in ostrich), and biologically active peptides (carnosine – in beef, anserine – in ostrich and chicken meat). However, the n-3 fatty acids levels differed significantly among types. The ostrich jerky contained four times less fat as compared to beef and half of that assayed in the dehydrated chicken. Both poultry meats contained more polyunsaturated FA and more favorable proportion of PUFA:SFA. Table 20.13 also depicts typical composition of pork rinds and processed and cured meat products that are frequently used as snacks.

Marine products are viewed as more nutritious compared to animal counterparts because they usually contain more protein and an oil composition containing a more favorable ratio of unsaturated to saturated fatty acids. More importantly, several marine sources contain significant amounts of nutraceutical fatty acids DHA (22:6) and EPA (20:5). These long-chain fatty acids are synthesized in the body from linolenic acid but the conversion is very inefficient. On the contrary, the conversion of EPA into DHA is very effective. DHA is considered as the most nutraceutical fatty acid because it forms part of cell membranes and positively impact their functionality. Therefore, DHA helps to prevent heart attacks and improve brain and vision function. It is especially important in the growing fetus and during infancy. The cold-temperature or pelagic fishes are the ones that contain the highest amounts of these nutraceutical fatty acids. Unfortunately, these highly unsaturated fatty acids are very prone to oxidation and rancidity, so in many products with expected long shelf lives, it is necessary to supplement natural or synthetic antioxidants and mineral sequestrants. Table 20.14

TABLE 20.12
Nutritional Composition of Dairy-Based Snack Foods (100 g)

Nutrient	Plain, Whole Milk, Yogurt	Plain, Greek, Whole Milk, Yogurt	Cheddar Cheese	Mozzarella Cheese	Cheese Sauce	Crunchy Cheddar Cheese Snack	Parmesan Cheese Chips	Probiotic Dairy-Based Beverage	Rice Pudding	Mousse	Vanilla Milkshake	Candy, Milk Chocolate
Water, g	87.9	81.3	36.6	50	74.1	ND	ND	ND	75.2	63.6	74.5	1.5
Energy, kcal	61	97	408	299	160	583	566	75	108	209	112	535
Protein, g	3.5	9	23.3	22.2	3.2	41.7	50	1.3	3.2	4.4	3.9	7.7
Lipids, g	3.3	5	34	22.1	13	41.7	38.9	0.1	2.2	15	3	29.7
Cholesterol, mg	13	13	100	79	8	167	111	ND	12	120	12	23
Carbohydrates, g	4.7	4	2.4	2.4	7.7	8.3	0	17	18.4	16.3	17.8	59.4
Sugars, g	4.7	4	0.3	0	0.3	0	0	ND	11.6	15.1	17.8	51.5
Dietary fiber, g	0	0	0	0	0.3	0	0	0	0.3	0.6	0	3.4
Calcium, mg	121	100	707	505	74	975	1333	50	95	103	146	189
Iron, mg	0.05	0	0.16	0.44	0.53	3.33	0	15	0.11	0.54	0.1	2.35
Sodium, mg	46	35	654	486	651	1167	1278	0	97	35	95	79
Thiamine, mg	0.029	0.020	0.030	0.030	0.080	ND	ND	ND	0.030	0.050	0.030	0.112
Riboflavin, mg	0.142	0.280	0.440	0.280	0.090	ND	ND	ND	0.180	0.190	0.200	0.298
Niacin, mg	0.075	0.210	0.050	0.100	0.580	ND	ND	ND	0.220	0.150	0.150	0.386
Folate, mcg	7	5	21	7	4	ND	ND	ND	4	13	7	12
Vitamin E, mg	0.06	0.01	0.75	0.19	1.23	ND	ND	ND	0.07	0.46	0.05	0.51
Type of fat, g												
Saturated	2.096	2.400	19.200	13.900	3.100	25	19.440	0.100	1.210	8.670	1.890	18.509
Monounsaturated	0.893	2.140	7.440	6.570	5.070	ND	ND	ND	0.560	4.410	0.880	7.186
Polyunsaturated	0.092	0.470	1.180	0.770	4.340	ND	ND	ND	0.130	0.840	0.110	1.376

Source: USDA (2021).

TABLE 20.13
Nutritional Composition of Animal-Based Snacks (100 g)

	Beef Snacks			Pork Rinds/Skins			Shelf-Stable, Cured Meat Products				
Nutrient	**Jerky**	**Sticks, Smoked**	**Pork, Oriental Style, Dehydrated**	**Plain**	**Barbecue Flavor**	**Low-Fat Microwavable**	**Ham, Prosciutto**	**Pepperoni, Beef, and Pork**	**Salami, Italian, Pork**	**Dry, Cured, Serrano Ham**	**Pastrami, Beef, Cured**
Water, g	23.4	19.1	21.8	1.8	2.1	2	55.9	28.6	34.6	ND	69.5
Energy, kcal	410	550	615	545	538	414	195	504	425	286	147
Protein, g	33.2	21.5	11.8	61.3	57.9	78.6	27.8	19.3	21.7	25	21.8
Fat, g	25.6	49.6	62.4	31.3	31.8	10.7	8.3	46.3	37	21.4	5.8
Cholesterol. mg	48	133	67	95	115	ND	70	97	80	89	68
Carbohydrate, g	11	5.4	1.4	0	1.6	0	0.3	1.2	1.2	0	0.4
Dietary fiber, g	1.8	ND	0	0	0	0	0	0	0	0	0
Calcium, mg	20	68	10	30	43	ND	10	19	10	0	10
Iron, mg	5.42	3.40	0.50	0.88	1.04	ND	1.11	1.33	1.52	1.29	2.22
Sodium, mg	1785	1531	685	1838	2667	2571	2695	1582	1890	2286	1078
Thiamin, mg	0.154	0.141	0.450	0.099	0.084	ND	0.567	0.271	0.930	ND	0.052
Riboflavin, mg	0.142	0.436	0.140	0.283	0.432	ND	0.242	0.257	0.330	ND	0.161
Niacin, mg	1.732	4.540	2.100	1.549	3.355	ND	3.881	4.987	5.600	ND	4.260
Folate, mcg	134	0	2	0	31	ND	5	5	2	ND	6
Vitamin E, mg	0.49	ND	0.36	0.53	ND	ND	0.28	0	ND	ND	0.12
Type of fat, g											
Saturated	10.850	20.800	23.060	11.370	11.560	3.57	2.780	17.708	13.100	12.5	2.681
Monounsaturated	11.305	20.470	28.560	14.780	15.030	ND	3.821	20.770	18.200	ND	2.118
Polyunsaturated	1.011	4.420	7.320	3.640	3.450	ND	0.970	4.458	3.600	ND	0.145

Source: USDA (2021).

TABLE 20.14
Nutritional Composition of Marine-Based Snacks (100 g)

	Dried Products							
Nutrient	**Cod, Dried, Salted**	**Herring, Dried, Packed in Oil**	**Whitefish, Dried**	**Chum Salmon, Dried**	**Shrimp, Dried**	**Salmon Jerky**	**Snack, Shrimp Cracker**	**Herring Pickled**
Water, g	16.1	18.2	20.6	22.8	38.9	ND	13.6	55.2
Energy, kcal	290	489	371	378	253	286	426	262
Protein, g	62.8	44.5	62.4	62.1	51.7	39.3	7.1	14.2
Fat, g	2.4	34.6	13.4	14.4	3.4	5.4	17.9	18
Cholesterol. mg	152	ND	266	214	638	89	2	13
Carbohydrate, g	0	0	0	0	0	21.4	59.1	9.6
Dietary fiber, g	0	0	0	0	0	0	5.6	0
Calcium, mg	160	ND	810	155	367	0	20	77
Iron, mg	2.50	ND	4.10	2.20	5.39	0	1.91	1.22
Sodium, mg	7027	ND	200	190	2203	1750	571	870
Thiamin, mg	0.268	0.010	0.053	0.356	0.018	ND	0.264	0.036
Riboflavin, mg	0.240	ND	0.438	0.463	0.038	ND	0.091	0.139
Niacin, mg	7.500	2.200	11.200	14.650	1.413	ND	2.631	3.300
Folate, mcg	25	ND	11	12	23	ND	23	2
Vitamin E, mg	2.84	ND	0	0.82	2.78	ND	3.28	1.71
Type of fat, g								
Saturated	0.462	ND	2.850	2.395	0.476	0	5.357	2.381
Monounsaturated	0.342	ND	4.380	5.139	0.273	1.790	7.997	11.947
Polyunsaturated	0.804	ND	2.160	2.323	1.732	0	3.971	1.679

Source: USDA (2021).

describes the nutritional composition of several types of marine products used as snacks. Several of these snacks contain more than 50% protein.

20.4 TOXICOGENIC COMPOUNDS ASSOCIATED WITH SNACKS

Snack foods may contain toxic compounds that can harm human health. These compounds can be intrinsic, exogenous, and generated during processing. In the first category, there are several antinutritional factors like the ones contained in raw legume seeds. The main types are trypsin inhibitor, urease, and hemagglutinin that synthesize these grains as defense mechanisms against biotic agents (Rodriguez-Sifuentes et al. 2020). Exogenous toxic compounds are mainly produced by mycotoxin-producing molds in common raw materials used for the manufacturing of snacks. Undoubtedly, peanuts and maize or corn are the most prone to contamination in the field or storage, although other seeds and tree nuts may also contain these harmful secondary metabolites that cause acute and chronic toxicities. The third way to produce harmful toxicogenic compounds is via processing, especially during deep fat frying.

20.4.1 Antinutritional Factors

Legume oilseeds and pulses in their raw forms contain these antinutritional factors that are subdivided into three major classes: trypsin inhibitor, urease, and lectins. These factors are capable of killing laboratory animals fed with large amounts of raw legume seeds in a couple of weeks. This is the reason why legume seeds need to be thermally treated, sprouted, or fermented for the production of human foods.

20.4.1.1 Trypsin Inhibitors (TIs)

Legume TIs are classified into two families according to their molecular size: Kunitz, the larger moieties with molecular weights around 20 kDa, and the lower molecular weight Bowman-Birk of approximately 8 kDa. There are legumes containing both types like soybean, whereas others only synthesize one type like common beans and lentils, which produced the Bowman-Birk family. It has been reported that not only the legume type but other factors, like cultivar and growing location, influence concentrations of TIs in common beans (De Mejía et al. 2003). Among raw legumes, soybeans are the type that contain the highest TI's activity and toxicological effects.

TIs strongly inhibit the activity of key pancreatic enzymes trypsin and chymotrypsin, thereby reducing digestion and absorption of dietary proteins by the formation of complexes that are indigestible even in the presence of high amounts of digestive enzymes. TIs enhance poor protein assimilation, digestive problems, pancreatic enlargement, muscle wasting, and depression (Aviles Gaxiola et al. 2018 a,b). In general, the most effective treatments to deactivate TIs are thermal such as cooking, frying, microwaving, extrusion, and infrared irradiation. The processes with medium effectiveness are ultrasound, gamma irradiation, germination, and high pressure applied alone (Aviles-Gaxiola et al. 2018a,b).

20.4.1.2 Urease Activity

Urease is a metalloenzyme rich in methionine common to most raw legume seeds, which decomposes urea to ammonia and carbamate and carbon dioxide. Soybean is the legume that contains the highest amounts of ureases, which exert entomotoxic and fungitoxic protective effects. The reported LD_{50} for mammals is 2 mg/kg. Fortunately, the inactivation of urease is almost identical to that of TIs and, therefore, it is easier to perform. Urease activity is comparatively easily determined by pH differential tests. Therefore, urase activity has been used as an indirect method to determine levels of trypsin inhibitor and heat processing of soybean meals and products (Prachayawarakorn et al. 2004, Real-Guerra et al. 2013, Yalcin and Basman 2015).

20.4.1.3 Lectins (Hemagglutinins)

The antinutritional lectins also known as hemagglutinins are a diverse group of proteins of nonimmune origin, which reversibly bind to specific glycans. One class of plant lectins groups all carbohydrate-binding proteins that are constitutively expressed in high amounts, especially in legume seeds. They have a major role in plant defense against insects and predators. The classical soybean lectin is a tetrameric glycoprotein of four identical 30 kDa units (Van Holle and Van Damme 2015). Lectins are of special importance in soybeans, kidney beans, common beans, and other legume seeds. Similar to TIs and ureases, lectins are effectively deactivated after thermal treatments. However, the incomplete inactivation of lectins can result in serious health problems because they have high affinity to intestinal epithelial cells forming glycoconjugates causing acute inflammation and higher intestinal permeability. Moreover, they interfere with nutrient digestion and absorption, stimulate shifts in the microbiome, and modulate the immune state of the digestive tract. Systemically, they can disrupt lipid, carbohydrate, and protein metabolism, promote enlargement and/or atrophy of key internal organs like the pancreas, and are also known to alter the hormonal and immunological status (Vasconcelos and Oliveira 2004).

20.4.2 Mycotoxins

Mycotoxicoses are today considered as the second most relevant public health problem in the planet. Mycotoxigenic molds can contaminate cereal grains, legumes, pulses, pseudocereals, and tree nuts in the field or during faulty storage. In general terms, the growth of molds and production of mycotoxins is greatly affected by the moisture content of the grains or fruits and the environmental conditions, especially in terms of temperature. Generally speaking, molds damage grains and fruits containing 0.85 or more of A_w and when the environmental relative humidity and temperature are higher than 70% and 25°C, respectively. The most relevant genus of mycotoxigenic molds are *Fusarium*, *Aspergillus*, and *Penicillum*, commonly known as pink, black, and green molds, respectively (Serna-Saldivar 2010). By far, the most relevant concern related to mold infestations is the risk of mycotoxicoses in products obtained from peanuts and corn or maize.

Mycotoxins are secondary metabolites secreted as defense mechanisms by fungi and unfortunately cause severe diseases known for hundreds of years. It is well-documented cases of ergotism in the 17th century and ochratoxicosis during the Middle Ages. Nevertheless, it has been only 60 years since mycotoxicosis became relevant and a serious public health hazard. It was until the 1960s when the strong effects of aflatoxins secreted by *Aspergillus flavus* were recognized when contaminated peanut flour caused the death of approximately 100,000 turkeys in England. Later on, aflatoxins were recognized as one of the most potent occurring carcinogens in nature and one of the major food security problems worldwide. Aflatoxins and the other sorts of mycotoxins summarized in Table 20.15 have received attention and their presence monitored by regulatory agencies around the globe. During the past two decades, fumonisins have received special attention because of their potent harmful health effects on equines and humans.

There are several types of aflatoxins, the most important and toxic is the B_1. Other important metabolites are B_2, G_1, G_2, M_1, and M_2. The last two occur in the milk of lactating mammals that consumed mycotoxin-contaminated foods. These metabolites are of upmost importance because they cause toxicity at concentrations of 10 ppb. The most common aflatoxins in peanuts and maize are B_1 and B_2. Both are potent carcinogens and mutagenic agents, although the G toxins are approximately ten times less toxic compared to B counterparts (Serna-Saldivar 2010). Undoubtedly, peanuts and maize are the most susceptible grains to mold infestation and mycotoxins. This is because the pods of peanuts develop and mature under the soil and the maize kernels mature in a cob that is covered with husks. In most countries, the maximum level of total aflatoxins allowed is 20 ppb for humans and 200 ppb for animals. In milk, the maximum permitted level is only 0.5 ppb.

TABLE 20.15
Toxicological Effects of the Main Mycotoxins that Occur in Grains, Tree Nuts, and Fruits Used for Production of Snacks

Mycotoxin	Causal Agent	Main Occurrence	Toxicological Effects on Humans and Domestic Animals
Aflatoxins	*Aspergillus flavus, Aspergillus parasiticus*	Maize and peanuts	There are several types of aflatoxins, the most relevant and toxic in grains being B_1. Other important metabolites are B_2, G_1, G_2, M_1, and M_2. There are some aflatoxins that are partially modified during metabolism, the typical examples are the M_1 and M_2 present in milk of lactating mammals. These metabolites are of great importance because they cause toxicity at very low concentrations (10 ppb). The most common aflatoxins in cereals and legumes (peanuts) are B_1 and B_2. Aflatoxins produce acute hepatitis, widespread hemorrhages, poor immunological response, and are potent carcinogens and mutagenic agents. B aflatoxins are approximately ten times more toxic than G types
Ochratoxin	*Aspergillus ochraceus, Penicillum verbicosum*	Maize, peanuts, tree nuts, and dried fruits	Ochratoxicoses have been documented for several centuries in regions of the Balkans where maize is staple and the crop was exposed to unusual rainfalls during maturation. Chemically, they are isocoumarin derivatives bound to phenylalanine, the ochratoxin A (OTA) being the most relevant. Most ochratoxicosis are related to field-contaminated maize due to excess rainfall before harvesting or due to storage at high moisture. Peanuts are also very prone to develop the mold and mycotoxin. The toxicosis involves progressive renal failure and atrophia, anemia, polyuria, anorexia, headaches, and uremia. In most instances, the disease is fatal. Laboratory rats fed with high concentration of ochratoxins develop cancer and the progeny teratogenic defects
Fumonisin	*Fusarium moniliforme, Fusarium verticilloides, Fusarium proliferatum*	Maize	Fumonisins were recently discovered and are the group of toxins that have received more attention lately. Fumonisins are toxic because they interfere with the synthesis of sphingomyelins, inhibit the enzyme ceramide synthetase, and modify cell proliferation, increasing the production of cytokines. Fumonisins are highly toxic to livestock, mainly horses or equines, and have harmful effects on humans and other monogastrics. In humans, fumonisins have been related to esophageal cancer and interference with folic acid metabolism. Therefore, they can exacerbate fetal malformation such as neural tube defects. Contaminated feeds can cause death in horses several hours after ingestion. Equines show the classic symptoms of "Leukoncephalomalacia" characterized by disorientation and death. In swine, fumonisins cause pulmonary edema and hepatotoxicity
Zeralenone	*Fusarium graminearum*	Corn and wheat	Chemically, it is an acid lactone with a phenolic resorcinol configuration. These toxins have caused problems in humans and other monogastric animals. Swine are the most affected animals. It causes the "Estrogenic syndrome" or animal feminization, characterized by vulvovaginitis, vaginal prolapse and infertility in sows and testicular atrophy, infertility and swelling of the mammary glands in boars. In high concentrations, zeralenone lowers fertility, ovulation, fetal development, and litter survival rate and induce thymic atrophy. Most cases are related to contaminated maize or feed produced from maize, although the presence of zeralenone in other cereal grains has also been documented

(*Continued*)

TABLE 20.15 (CONTINUED)
Toxicological Effects of the Main Mycotoxins that Occur in Grains, Tree Nuts, and Fruits Used for Production of Snacks

Mycotoxin	Causal Agent	Main Occurrence	Toxicological Effects on Humans and Domestic Animals
Trichothecenes (T-2 toxin)	*Fusarium*	Barley Wheat Corn	Approximately 180 trichotecenes are known. High-moisture cereals are frequently contaminated with these toxins considered as potent protein synthesis inhibitors and immune suppression. Trichothecenes can produce the Fatal Syndrome named "Alimentary Toxemia," characterized by leucopenia (reduced white blood cell counts), multiple hemorrhages, loss of bone marrow, and esophageal cancer. In addition, the consumption of these toxins cause vomiting and reduce the efficiency of feed conversion in swine due to the considerably lower feed intake
Deoxynivalenol (DON) or vomitoxin	*Fusarium graminearum*	Corn Wheat Barley	Vomitoxin is a deoxynivalenol derivative. The toxin causes feed refusal, vomiting, and lower feed efficiency. The brewing industry is especially concerned about these toxins because they may contaminate beer when infested barley or malt is used
Ergot alkaloids (ergotamine, ergotine)	*Claviceps purpurea*	Rye Wheat Barley	Historically, there is evidence to indicate that ergot poisonings have affected human and domestic animals for centuries. Ergotism is no longer prevalent due to improvements in preventing contaminated grains from entering the food chain. Two types of ergotisms are recognized: convulsive and gangrenous. Convulsive is characterized by a tingling sensation under the skin, itching, numbness of the hands and feet, twitching and muscle cramps, and convulsions and hallucinations. Postmortem findings include bleeding and softening of the brain and spinal cord lesions. The full mental recovery from affected victims is rare. In gangrenous ergotism, the fingers and toes become necrotic and in the most severe cases the feet and entire limbs shriveled and dropped off

Source: Serna-Saldivar (2010).

Ochratoxins produced the Balkan endemic nephropathy first described in Bulgaria. Most ochratoxicosis are related to field-contaminated maize, although it also occurs in peanuts, tree nuts, and dried fruits such as raisins. The main ochratoxin known as OTA produce progressive nephropathy or renal dysfunction, cancer, and in severe cases death (Serna-Saldivar 2010). Fumonisins were recently discovered and are the group of mycotoxins that have received more attention in recent times. This is due to its high toxicity to livestock, mainly horses, and the toxicological effects to humans. In humans, fumonisins have been related to esophageal cancers in China and South Africa and interference with folic acid metabolism. Therefore, these toxins have been associated with neural tube defects or fetal malformations (Marasas et al. 2004, Scott 2012). Other relevant mycotoxins are zeralenone, T-2, vomitoxin, and ergotin/ergotamine produced by different types of *Fusarium* (Desjardins 2006). The characteristic and toxic effects of these toxins are summarized in Table 20.15.

20.4.3 Toxicogenic Compounds Generated during Processing

Snacks are normally produced after the application of one or several thermal treatments like baking, roasting, thermoplastic extrusion, and especially deep fat frying. The studies on the nutritional and physiological effects of frying oils indicate that only overheated and overused oils may have detrimental effects. These treatments are known to enhance the formation of several types of toxic compounds like acrylamide, nitrosamines, polycyclic aromatic hydrocarbons (PAH), and cyclic fatty acid monomers. The consumption of snacks with significant levels of these toxic compounds pose health risks related to cancer, hypertension, Alzheimer, Parkinson, and thymus necrosis (Soriguer et al. 2003, Hosseini et al. 2016).

20.4.3.1 Acrylamide

It has been almost 20 years when acrylamide formation in heated snack foods was categorized as potential carcinogenic and neurotoxic. Since then, the most worrisome snack has been fried potato chips. Acrylamide is a degradation compound generated during frying, toasting, and roasting at temperatures higher than 120°C at low-humidity conditions. The generation of acrylamide in foods is related to *Maillard* reactions through the condensation of the amino acid asparagine with reducing sugars such as fructose or glucose (Carrieri et al. 2010, Hosseini et al. 2016). Therefore, the control of the main causes of *Maillard* reactions (temperature, water activity, pH, and the content of precursors) can be modulated to diminish its concentration and toxicity. Since the content of reducing sugars appears to be a strong determinant for acrylamide formation in potatoes, therefore, the selection of cultivars low in reducing sugars and storage at moderate temperatures might present an option for reduction (Amrein et al. 2003). For potato chips, the concentration of acrylamide increases with higher frying time and with the contact surface area. In order to reduce the concentrations of acrylamide in snacks, the raw materials should be carefully selected in terms of reducing sugars and asparagine contents with the implementation of several pretreatments like blanching, fermentation, and even application of asparaginase (Matthaus and Haase 2014, Hosseini et al. 2016).

20.4.3.2 Nitrosamines

Nitrosamines which are classified as carcinogens by the International Agency for Research on Cancer (IARC), mutagenic, teratogenic, and hepatotoxic are formed by reaction of nitrogen oxide and secondary amines. *N*-Nitrosodimethylamine and *N*-nitrosodiethylamine are considered the most carcinogenic to humans (Al-Kaseem et al. 2014). Foods which have been shown to contain nitrosamines include cured or nitrite-treated meats like bacon, salami, smoked sausages, and hams (Chapter 18); beer and cheeses (Chapter 17); and hot smoked and fried fish products (Chapter 18). Starting materials for nitrosamine formation in meat products are nitrate, nitrite, primary, secondary, and tertiary amines, amides, proteins, peptides, and amino acids or precursors of these, which are transformed into precursors by microbial action. The temperature and time of cooking, nitrite concentration, and storage conditions of processed meats have a significant effect on their concentration. For this reason, the use of these nitrosating agents for curing meat is strictly controlled in many countries around the globe. Moreover, some known inhibitors of the nitrosation reaction, such as ascorbic acid and sulfur dioxide, are used in the processing of various sorts of snacks (Yurchenko and Molder 2006, Yurchenko and Molder 2007, Al-Kaseem et al. 2014, Park et al. 2015).

20.4.3.3 Polycyclic Aromatic Hydrocarbons (PAHs)

PAHs are lipophilic aromatic hydrocarbons with two or more fused benzene rings in various structural configurations (Purcaro et al. 2006). These toxic compounds may cause several types of cancers such as skin, pulmonary, bladder, and gastrointestinal. Besides, they have also been related to hepatic and renal damage, mutations, and increase in cardiopulmonary mortality (Lawal 2017). The major contributors to PAH intake in the average diet are oils and fats and snacks processed at high temperatures (frying, grilling, roasting, and smoking). For this reason, the Commission Regulation (2005) has recently set a maximum level of 2 ppb for benzo[*a*]pyrene in oils and fats intended for direct consumption or for use as an ingredient in foods. Lawal (2017) recently reviewed the occurrence and toxicological effects of PAHs and concluded that the degradation of the frying oil involves the propagation of radical species catalyzed by elevated temperatures. The optimal temperatures for the formation of PAHs are 660–740°C; however, Li et al. (1994) concluded that the fumes of cooking oil (180–270°C) contain significant amounts of PAH. Heterocyclic aromatic amines and volatile decomposition products such as 1,3 butadiene, benzene, acrolein, and formaldehyde have been suggested to be responsible for the mutagenic properties of the fumes from cooking oils (Shields et al. 1995, Purcaro et al. 2006). Chen and Chen (2003) studied the generation of different vegetable oils (soybean, canola, and sunflower oils) and concluded that the degree of unsaturation of fatty acids affected the variety and amounts of PAHs formed in the fumes. Among these oils, the soybean oil, which had the highest degree of unsaturation, produced the highest amounts of PAHs. According to de Oliveira Galvao et al. (2014), roasted cashew nuts (Chapter 15) contained 12 different types of PAHs, the main ones being benzo[*k*]fluoranthene, indene[1,2,3-*c,d*]pyrene, benzo[*ghi*]perylene, phenanthrene, and benzo[*b*]fluoranthene.

20.4.3.4 Cyclic Fatty Acid Monomers (CFAM)

CFAMs are among the most toxic compounds formed during deep fat frying of snacks because they could be easily absorbed by the epithelial cells of the gastrointestinal tract. These fatty acid monomers are mainly produced from C18 polyunsaturated FA linoleic and α-linolenic acids during heating of vegetable oils at temperatures of ≥200°C. Thermal-abused oils can contain up to 0.66% CFAM of total fatty acids. The main toxic effects of CFAM are the alteration of hepatic enzymes which induce hepatomegaly and steatosis. In laboratory animals, CFAM also increase their blood triglyceride and cholesterol levels. Among monomeric products, CFAM originating from C18 PUFA have also been associated with cancer (Frankel 1984, Romero et al. 2000, Mboma et al. 2018). Rojo and Perkins (1987) assayed two- to fivefold increase in the total amount of CFAM in partially hydrogenated soybean oil (736–1,803 ppm) when it was intermittently heated for 80 h. Apparently, a significant inhibition of the cyclization mechanism occurs when the frying oil contains natural antioxidants like tocopherols (Hosseini et al. 2016). Frankel et al. (1984) researched the generation of CFAM during frying processes of fast foods with vegetable shortening, animal–vegetable shortening, partially hydrogenated vegetable oils, and regular vegetable oils. The highest amounts of CFAM were assayed in cottonseed oil used for the production of french fries.

REFERENCES

Al-Kaseem, M., Al-Assaf, Z., and Karabeet, F. 2014. Determination of seven volatile N-nitrosamines in fast food. *Pharmacology and Pharmacy*, 5(2):Article ID: 42998.

Amrein, T.M., Bachmann, S., Noti, A., Biedermann, M., Barbosa, M.F., Biedermann-Brem, S., Grob, G., Keiser, A., Realini, P., Escher, F., and Amado, R. 2003. Potential of acrylamide formation, sugars, and free asparagine in potatoes: A comparison of cultivars and farming systems. *Journal of Agricultural and Food Chemistry*, 51(18):5556–5560.

Aviles-Gaxiola, S., Chuck-Hernandez, C., del Refugio Rocha-Pizaña, M., García-Lara, S., López-Castillo, L.M., and Serna-Saldivar, S.O. 2018a. Effect of thermal processing and reducing agents on trypsin inhibitor activity and functional properties of soybean and chickpea protein concentrates. *LWT – Food Science and Technology*, 98:629–634.

Aviles-Gaxiola, S., Chuck-Hernandez, C., and Serna-Saldívar, S.O. 2018b. Inactivation methods of trypsin inhibitor in legumes: A review. *Journal of Food Science*, 83(1):17–29.

Brenna, J.T., and Diau, G.Y. 2007. The influence of dietary docosahexaenoic acid and arachidonic acid on central nervous system polyunsaturated fatty acid composition. *Prostaglandins, Leukotrienes, and Essential Fatty Acids*, 77(5–6):247–250.

Camire, M.E. 2012. Nutritional changes during extrusion cooking. In: *Advances in Food Extrusion Technology*, eds. M. Maskan and A. Altan, 87–102. Boca Raton: CRC Press (Taylor and Francis Group).

Carrieri, G., Anese, M., Quarta, B., De Bonis, M.V., and Ruocco, G. 2010. Evaluation of acrylamide formation in potatoes during deep frying: The effect of operation and configuration. *Journal of Food Engineering*, 98(2):141–149.

Caselato-Sousa, V.M., and Amaya-Farfán, J. 2012. State of knowledge on amaranth grain: A comprehensive review. *Journal of Food Science*, 77(4): R93–R104.

Chen, G.-C., Tong, X., Xu, J.-Y., Han, S.-F., Wan, Z.-X., Qin, J.-B., and Qin, L.Q. 2016. Whole-grain intake and total, cardiovascular, and cancer mortality: A systematic review and meta-analysis of prospective studies. *American Journal of Clinical Nutrition*, 104(1):164–172.

Chen, Y.C., and Chen, B.H. 2003. Determination of polycyclic aromatic hydrocarbons in fumes from fried chicken legs. *Journal of Agriculture and Food Chemistry*, 51(14):4162–4167.

Clarke, R., Frost, C., Collins, R., Appleby, P., and Peto, R. 1997. Dietary lipids and blood cholesterol: Quantitative meta-analysis of metabolic ward studies. *The British Medical Journal*, 314(7074):112.

Constantin, O.E., and Istrati, D. 2019. Functional foods of snack bars. In: *Functional Foods*, ed. V. Lagouri, 47–60. London: Intech Open.

De Mejía, E.G., Guzmán-Maldonado, S.H., Acosta-Gallegos, J.A., Reynoso-Camacho, R., Ramírez-Rodríguez, E., Pons-Hernandez, J.L., Gonzalez-Chavira, M.M., Castellanos, J.Z., and Kelly, J.D. 2003. Effect of cultivar and growing location on the trypsin inhibitors, tannins, and lectins of common beans (*Phaseolus vulgaris* L.) grown in the semiarid highlands of Mexico. *Journal of Agriculture and Food Chemistry*, 51(20):5962–5966.

de Oliveira Galvão, M.F., de Melo Cabral, T., de André, P.A., de Fátima Andrade, M., de Miranda, R.M., Saldiva, P.H.N., de Castro Vsconcellos, P., and Batistuzzo de Medeiros, S.R. 2014. Cashew nut roasting: Chemical characterization of particulate matter and genotocixity analysis. *Environmental Research*, 131:145–152.

Dekker, P., Koenders, D., and Bruins, M. 2019. Lactose-free dairy products: Market developments, production, nutrition and health benefits. *Nutrients*, 11(3):551. doi:10.3390/nu11030551.

Desjardins, A.E. 2006. *Fusarium Mycotoxins. Chemistry, Genetics and Biology.* Saint Paul: American Phytopathological Society.

Frankel, E.N., Smith, L.M., Hamblin, C.L., Creveling, R.K., and Clifford, A.J. 1984. Occurrence of cyclic fatty acid monomers in frying oils used for fast foods. *Journal of the American Oil Chemists´ Society*, 61(1):87–90.

Fu, C.L., Shi, H., and Li, Q.H. 2006. A review on pharmacological activities and utilization technologies of pumpkin. *Plant Foods for Human Nutrition*, 61(2):70–77.

Gasmalla, M.A.A., Tessema, H.A., Salaheldin, A., Alagmad, K., Hassanin, H.A.M., and Aboshora, W. 2017. Health benefits of milk and functional dairy products. *MOJ Food Processing & Technology*, 4(4):108–111. doi:10.15406/mojfpt.2017.04.00099.

Gibson, R.A., Muhlhausler, B., and Makrides, M. 2011. Conversion of linoleic acid and alpha-linolenic acid to long-chain polyunsaturated fatty acids (LCPUFAs), with a focus on pregnancy, lactation and the first 2 years of life. *Maternal and Child Nutrition*, 7(Supplement 2):17–26.

Górska-Warsewicz, H., Rejman, K., Laskowski, W., and Czeczotko, M. 2019. Milk and dairy products and their nutritional contribution to the average polish diet. *Nutrients*, 11(8):1771. doi:10.3390/nu11081771.

Guardado-Felix, D., Milan-Noris, A.K., Serrano-Sandoval, S.N., and Serna-Saldivar, S.O. 2019. Chickpea proteins and peptides in health. In: *Handbook of Chickpeas: Nutritional Value, Health Benefits and Management*, eds. A.T. Lund and N.D. Schultz. New York: Nova Science Publishers.

Hosseini, H., Ghorbani, M., Meshginfar, N., and Mahoonak, A.S. 2016. A review on frying: Procedure, fat, deterioration progress and health hazards. *Journal of the American Oil Chemists' Society*, 93(4):445–466.

Hunter, J.E., Zhang, J., and Kris-Etherton, P.M. 2009. Cardiovascular disease risk of dietary stearic acid compared with trans, other saturated, and unsaturated fatty acids: A systematic review. *American Journal of Clinical Nutrition*, 91(1):46–63.

Lawal, A.T. 2017. Polycyclic aromatic hydrocarbons. A review. *Cogent Environmental Science*, 3(1):1339841.

Li, S.G., Pan, D.F., and Wang, G.X. 1994. Analysis of polycyclic aromatic hydrocarbons in cooking oil fumes. *Archives of Environmental Health*, 49(2):119–122.

Maetens, E., Hettiarachchy, N., Dewettinck, K., Horax, R., Moens, K., and Moseley, D.O. 2017. Physicochemical and nutritional properties of a healthy snack chip developed from germinated soybeans. *LWT-Food Science and Technology*, 84:505–510.

Marasas, W.F.O., Riley, R.T., Hendricks, K.A., Stevens, V.L., Sadler, T.W., Gelineau-van Waes, J., Missmer, S.A., Cabrera, J., Torres, O., Gelderblom, W.C.A., Allegood, J., Martinez, C., Maddox, J., Miller, J.D., Starr, L., Sullards, M.C., Roman, A.V., Voss, K.A., Wang, E., and Merrill, A.H. 2004. Fumonisins disrupt sphingolipid metabolism, folate transport, and neural tube development in embryo culture and in vivo: A potential risk factor for human neural tube defects among populations consuming fumonisin-contaminated maize. *Journal of Nutrition*, 134(4):711–716.

Matthaus, B., and Haase, N.U. 2014. Acrylamide – Still a matter of concern for fried potato food? *European Journal of Lipid Science and Technology*, 116(6):675–687.

Mboma, J., Leblanc, N., Wan, S., Jacobs, R.L., Tchernof, A., Dubé, P., Angers, P., and Jacques, H. 2018. Liver and plasma lipid changes induced by cyclic fatty acid monomers from heated vegetable oil in the rat. *Food Science and Nutrition*, 6(8):2092–2103.

Milan-Noris, A.K., Montoya-Rodríguez, A., De la Rosa-Millan, J., and Serna-Saldivar, S.O. 2019. Health effects of starch and undigested carbohydrates associated to chickpea based foods. In: *Handbook of Chickpeas: Nutritional Value, Health Benefits and Management*, eds. A.T. Lund and N.D. Schultz. New York: Nova Science Publishers.

Morales de la Peña, M., Odriozola-Serrano, I., Oms-Oliu, G., and Martín-Belloso, O. 2020. Dietary fiber in fruits and vegetables. In: *Science and Technology of Fibers in Food Systems*, eds. J. Welti-Chanes, S.O. Serna-Saldivar, O. Campanella and V.A. Tejada-Ortigoza, 123–152. New York: Springer.

Muñoz, L.A., Cobos, A., Diaz, O., and Aguilera, J.M. 2013. Chia seed (*Salvia hispanica*): An ancient grain and a new functional food. *Food Reviews International*, 29(4):394–408.

Navruz-Varli, S., and Sanlier, N. 2016. Nutritional and health benefits of quinoa (*Chenopodium quinoa* Willd.). *Journal of Cereal Science*, 69:371–376.

Nyanzi, R., Jooste, P.J., and Buys, E.M. 2021. Invited review: Probiotic yogurt quality criteria, regulatory framework, clinical evidence, and analytical aspects. *Journal of Dairy Science*, 104(1):1–19.

Park, J.E., Seo, J.E., Lee, J.Y., and Kwon, H. 2015. Distribution of seven N-nitrosamines in food. *Toxicological Research*, 31(3):279–288.

Pathak, N., Rai, A.K., Kumari, R., Thapa, A., and Bhat, K.V. 2014. Sesame crop: An underexploited oilseed holds tremendous potential for enhanced food value. *Agricultural Sciences*, 5(6):Article ID:46023.

Pereira, E., Encina-Zelada, C., Barros, L., Gonzales-Barron, U., Cadavez, V., and Ferreira, I.C.F.R. 2019. Chemical and nutritional characterization of *Chenopodium quinoa* Willd (quinoa) grains: A good alternative to nutritious food. *Food Chemistry*, 280:110–114.

Prachayawarakorn, S., Prachayawasin, P., and Soponronnarit, S. 2004. Effective diffusivity and kinetics of urease inactivation and color change during processing of soybeans with superheated-steam fluidized bed. *Drying Technology*, 22(9):2095–2118.

Purcaro, G., Navas, J.A., Guardiola, F., Conte, L.S., and Moret, S. 2006. Polycyclic aromatic hydrocarbons in frying oils and snacks. *Journal of Food Protection*, 69(1):199–204.

Real-Guerra, R., Stanisçuaski, F., and Carlini, C.R. 2013. Soybean urease: Over a hundred years of knowledge. In: *A Comprehensive Survey of International Soybean Research-Genetics, Physiology, Agronomy and Nitrogen Relationships*. London: IntechOpen.

Rodríguez Sifuentes, L., Elzbieta Marszalek, J., Chuck-Hernandez, C., and Serna-Saldivar, S.O. 2020. Legumes protease inhibitors as biopesticides and their defense mechanisms against biotic factors. *International Journal of Molecular Sciences*, 21(9):3322. doi:10.3390/ijms21093322.

Rogelj, I. 2000. Milk, dairy products, nutrition and health. *Food Technology and Biotechnology*, 38(2):143–147.

Rojo, J.A., and Perkins, E. 1987. Cyclic fatty acid monomer formation in frying fats. I. Determination and structural study. *Journal of the American Oil Chemists´ Society*, 64(3):414–421.

Romero, A., Cuesta, C., and Sanchez-Muniz, F.J. 2000. Cyclic fatty acid monomers and thermoxidative alteration compounds formed during frying of frozen foods in extra virgin olive oil. *Journal of the American Oil Chemists´ Society*, 77(11):1169–1175.

Rozenberg, S., Body, J.J., Bruyère, O., Bergmann, P., Brandi, M.L., Cooper, C., Devogelaer, J.P., Gielen, E., Goemaere, S., Kaufman, J.M., Rizzoli, R., and Reginster, J.Y. 2015. Effects of dairy products consumption on health: Benefits and beliefs – A commentary from the Belgian bone club and the European society for clinical and economic aspects of osteoporosis, osteoarthritis and musculoskeletal diseases. *Calcified Tissue International*, 98(1):1–17. doi:10.1007/s00223-015-0062-x.

Saavedra, M.J., Aires, A., Dias, C., Almeida, J.A., De Vasconcelos, M.C.B.M., Santos, P., and Rosa, E.A. 2015. Evaluation of the potential of squash pumpkin by-products (seeds and shell) as sources of antioxidant and bioactive compounds. *Journal of Food Science and Technology*, 52(2):1008–1015.

Scott, P.M. 2012. Recent research on fumonisins: A review. *Food Additives and Contaminants: Part A*, 29(2):242–248.

Serna-Saldivar, S.O. 2008. Nutritional value of snack foods. In: *Industrial Manufacture of Snack Foods*, ed. S.O. Serna-Saldivar, 371–408. London: Kennedys Publications Ltd.

Serna-Saldivar, S.O. 2010. Grain storage pest. In: *Cereal Grains: Properties, Processing and Nutritional Attributes*, ed. S.O. Serna-Saldivar, 149–176. Boca Raton: CRC Press (Taylor & Francis Group).

Serna-Saldivar, S.O. 2015. Nutrition and fortification of corn and wheat tortillas. In *Tortillas: Wheat Flour and Corn Products*, eds. L.W. Rooney and S.O. Serna-Saldivar, 29–63. Saint Paul: American Association of Cereal Chemists.

Serna-Saldivar, S.O. 2020. Relación del consumo de granos enteros con la salud pública. In: *Papel de los cereales de grano entero en la salud*, 197–214. Granada: ILSI Nor-Andino y Fundación Iberoamericana de Nutrición (FINUT).

Serna-Saldivar, S.O., and Sanchez-Hernandez, D. 2020. Dietary fiber in cereals, legumes, pseudocereals and other seeds. In: *Science and Technology of Fibers in Food Systems*, eds. J. Welti-Chanes, S.O. Serna-Saldivar, O. Campanella and V.A. Tejada-Ortigoza, 87–122. New York: Springer.

Shields, P.G., Xu, G.X., Blot, W.J., Fraumeni, J.F., Trivers, G.E., Pellizzari, D.E., Qu, Y.H., Gao, Y.T., and Harris, C.C. 1995. Mutagens from heated Chinese and U.S. cooking oils. *Journal of the National Cancer Institute*, 87(11):836–841.

Soriguer, F., Rojo-Martinez, G., Dobarganes, M.C., Almeida, J.M.G., Esteva, I., Beltran, M., Ruiz De Adana, M.S., Tinahones, F., Gómez Zumaquero, J.M., García-Fuentes, E., and González-Romero, S. 2003. Hypertension is related to the degradation of dietary frying oils. *American Journal of Clinical Nutrition*, 78(6):1092–1097.

Tharanathan, R.N., and Mahadevamma, S. 2003. Grain legumes—A boon to human nutrition. *Trends in Food Science and Technology*, 14(12):507–518.

US Department of Health and Human Services; US Department of Agriculture. 2015–2020. *Dietary Guidelines for Americans.* 8th edition. Washington, DC: US Dept of Health and Human Services; December 2015. http://www.health.gov/DietaryGuidelines.

USDA. 2021. Nutrient data laboratory. Agricultural Research Service. Electronic page: http://www.nal.usda.gov/fnic/foodcomp.

Van Holle, S., and Van Damme, E.J. 2015. Distribution and evolution of the lectin family in soybean (*Glycine max*). *Molecules*, 20(2):2868–2891.

Vasconcelos, I.M., and Oliveira, J.T.A. 2004. Antinutritional properties of plant lectins. *Toxicon*, 44(4):385–403.

Vincente, A.R., Manganaris, G.A., Ortiz, C.M., Sozzi, G.O., and Crisosto, C.H. 2014. Nutritional quality of fruits and vegetables. In: *Postharvest Handling, a Systems Approach*, eds. W. Florkowski, R. Shewfelt, S. Prussia, and N. Banks, 69–121. San Diego: Academic Press.

Wijngaard, H., and Arendt, E.K. 2006. Buckwheat. *Cereal Chemistry Journal*, 83(4):391–401.

Xu, B., and Chang, S.K. 2008. Total phenolics, phenolic acids, isoflavones, and anthocyanins and antioxidant properties of yellow and black soybeans as affected by thermal processing. *Journal of Agricultural and Food Chemistry*, 56(16):7165–7175.

Yalcin, S., and Basman, A. 2015. Effects of infrared treatment on urease, trypsin inhibitor and lipoxygenase activities of soybean samples. *Food Chemistry*, 169:203–210.

Yang, J., Liu, R.H., and Halim, L. 2009. Antioxidant and antiproliferative activities of common edible nut seeds. *LWT – Food Science and Technology*, 42(1):1–8.

Yurchenko, S., and Molder, U. 2006. Volatile N-nitrosamines in various fish products. *Food Chemistry*, 96(2):325–333.

Yurchenko, S., and Mölder, U. 2007. The occurrence of volatile N-nitrosamines in Estonian meat products. *Food Chemistry*, 100(4):1713–1721.

Zdanowska-Sąsiadek, Ż., Marchewka, J., Horbańczuk, J.O., Wierzbicka, A., Lipińska, P., Jóźwik, A., Atanasov, A.G., Huminiecki, Ł., Sieroń, A., Sieroń, K., Strzałkowska, N., Stelmasiak, A., De Smet, S., Van Hecke, T., and Hoffman, L.C. 2018. Nutrients composition in fit snacks made from ostrich, beef and chicken dried meat. *Molecules*, 23(6):1267.

21 Nutraceutical and Health Implications of Snack Foods

Sergio O. Serna-Saldivar and Daniel A. Jacobo-Velázquez

CONTENTS

21.1 INTRODUCTION

The strong association between foods and health has been recognized since more than 2,500 years. Hippocrates, the Greek philosopher and father of medicine, postulated the saying "Allow your food to be your medicine and your medicine to be your food." Unfortunately, among the different types of foods, the consumption of snacks have been considered as detrimental in terms of human health mainly due to their high caloric, fat, and sodium contents and low amounts of dietary fiber, proteins, and micronutrients. For this reason, the current snack food industry is looking for alternatives to upgrade the nutrient and nutraceutical properties of new developments aimed to improve health and quality of life.

It is well-known that the diet is one of the main factors affecting the incidence of overweight, obesity, chronic diseases, and life expectancy. The prevention of cancer and chronic diseases such as hypertension, coronary heart disease, type-2 diabetes, and hyperlipidemia through the diet and adequate food habits is the best way to improve life expectancy and quality. Recent statistics clearly

DOI: 10.1201/9781003129066-21

indicate that the length of life keeps gradually improving despite the COVID pandemic. The prediction for inhabitants born in 2020 in more developed countries is 80.5 years. The other important aspect to ponder is that the population dynamics is shifting toward higher numbers of geriatric people. For instance, the United States geriatric population of year 2020 estimated is 54 million which will increase to 71 million by 2030. Moreover, according to the World Health Organization, the proportion of the world's population over 60 years will nearly double from 12% to 22% between 2015 and 2050 (https://www.who.int/news-room/fact-sheets/detail/ageing-and-health).

Snack food processors are challenged with these new food development trends and have to focus on the consumer's needs if they want to survive in the competitive market. The bad image of snacks among conscientious consumers, dieticians, and medical doctors have to be reverted through new functional product developments and marketing. The greatest potential is in the development of low-fat foods high in fiber, protein, and essential nutrients required by infants, millennials, and geriatric people. These products rich in hydrophilic and lipophilic antioxidants and further enriched or fortified with vitamins such as A, E, thiamin, riboflavin, niacin, folic acid, and B_{12} and/or essential minerals such as calcium, iron, zinc, and selenium may gradually revert the "junk food" image.

The FDA and other health regulatory agencies around the globe are allowing food processors to include claims of some functional compounds because they have proven positive effects on human health. Most nutraceuticals are potent antioxidants that block free radicals and consequently lower oxidative stress. Therefore, these natural phytonutrients prevent cancer, fibrosis, high triglycerides and cholesterol, cardiovascular diseases (CVD), and aging.

The production of nutraceutical or health-promoting snacks is not easy, especially those produced from cereal grains and tubers. The product development of novel snacks rich in whole grains, dietary fiber sources, and phytochemicals is problematic because these ingredients lower functionality and yield products with different colors, textures, and especially flavors. For instance, the addition of whole-grain meals from cereals, pulses, or pseudocereals rich in dietary fiber and lipids and fiber by-products negatively affect expansion rate of second- and third-generation extruded products (Chapter 12) and properties of wheat-based and nixtamalized foods covered in Chapters 10 and 11, respectively. The higher fiber acts as a physical barrier interfering with important functional compounds such as starch and gluten. The major opportunity is to enrich snacks with soluble fiber sources like inulin and β-glucans and probiotics because this combination improves health in many aspects.

The objectives of this chapter are to review the different types of nutrients, phytochemicals, and probiotic microorganisms that can be incorporated into the different categories of snacks foods covered in this book and can favor human health, especially by preventing obesity, chronic or non-communicable diseases, and neurological diseases.

21.2 ASSOCIATION BETWEEN CONSUMPTION OF SNACK FOODS AND HEALTH

Consumption of snacks is commonly associated with the ingestion of unhealthy foods, since in most cases they are calorie-dense foods with high-lipid contents. In this context, there is an association between the consumption of snacks and obesity and the development of different chronic-degenerative diseases, mainly those associated with the metabolic syndrome. Therefore, there is a need to develop healthy snacks, with lower calories and higher protein and dietary fiber and that also provide bioactive compounds and prebiotics. According to a recent study, when a Dutch community ($N = 1{,}087$) was asked to generate a concept about an ideal snack, its two main characteristics included "tasty" and "healthy." Therefore, these two features should be considered when designing next-generation snacks that are palatable and attractive to customers that make food choices (Schlinkert et al. 2020). The addition of phytochemicals and other nutraceutical compounds to snacks is an appropriate strategy to complement basic nutrition provided by foods. The new trends of the snack industry is to use whole grains from cereals, pseudocereals and legume-pulses, vegetable proteins, oils free of *trans*-FA and with a more favorable polyunsaturated profile, natural dehydrated fruits and vegetables, insoluble and soluble dietary sources or by-products, resistant starches (RS), extracts rich in antioxidants, and probiotics that individually and synergistically favor human health. In the following sections, the main health benefits of lipophilic and hydrophilic nutraceuticals and prebiotics are discussed.

21.3 THE WHOLE GRAIN CONCEPT

Whole grains are defined as products consisting of the entire grain, and are being promoted worldwide as more healthful because they offer nutrients that are lost during the dry-milling processes aimed to obtain the more popular, functional, and used refined products. Whole grains of cereals contain the pericarp or bran, germ, and the aleurone layer, which are rich in dietary fiber, micronutrients, and phytochemicals. The whole grains of legume-pulses constituted by the testa or hull, cotyledons, and hilum and pseudocereals also contain high dietary fiber, vitamins, minerals, and an array of nutraceutical compounds that exert positive effects on human health. A large body of evidence on whole grains in relation to human health outcomes has accumulated over the past 20 years (Slavin et al. 1997, Serna-Saldivar 2020).

Epidemiological and clinical trials clearly indicate the inverse significant relationship between whole grain intake and obesity, metabolic syndrome, CVD, type-2 diabetes, oxidative stress, inflammation, and cancer. Most public health agencies recommend the consumption of at least three servings of whole grains for the control and management of chronic diseases and cancer that are responsible for approximately 76% of the deaths (60% attributable to chronic diseases, mainly CVD, and 16% to cancer). There are many mechanisms of why whole grains improve human health. The most relevant are that whole grains normally diminish caloric density of foods, the rate of glucose absorption, serum cholesterol levels, and important markers of inflammation. Moreover, whole grains improve satiety, the viscosity of the intestinal contents, the fecal volume, and binding of bile acids and dietary carcinogenic compounds. Importantly, whole grains contain higher amounts of both hydrophilic and lipophilic phytochemicals and antioxidants that are known to prevent oxidative stress and regulate key genes involved with chronic diseases and cancer (Slavin et al. 1997, Behallm et al. 2006, Van de Vijver et al. 2009, Ye et al. 2012, Giacco et al. 2014, Chen et al. 2016, Serna-Saldivar 2020).

The other important contribution of whole grains is in terms of prebiotic compounds that improve the function of the hindgut through the enhancement of growth and activity of probiotic microorganisms correcting dysbiosis. The microbiota generates short-chain FA and other metabolites, which prevent cancer and pathogenic and chronic disorders (Lu et al. 2004, El Khoury et al. 2012, Zeng et al. 2014, Senes-Guerrero et al. 2020).

21.4 DIETARY FIBER AND RESISTANT STARCHES

21.4.1 Total Dietary Fiber

Dietary fiber is the portion of food that cannot be digested by human digestive enzymes (Yangilar 2013). It is composed of nondigestible carbohydrates, lignin, cellulose, hemicellulose, pectic substances, arabinoxylans, β-glucans, mannans, and fructans and other associated substances of plant origin like resistant starches (Serna-Saldivar and Ayala-Soto 2020). The dietary fiber is broadly classified as insoluble or soluble, based on whether it forms a solution when mixed with water (soluble) or not (insoluble). Insoluble fiber includes cell wall components, such as lignin, cellulose, and hemicelluloses, while soluble fiber includes noncellulosic polysaccharides, such as pectins, β-glucans, galactomannan gums, and a large range of nondigestible oligosaccharides, including inulin and resistant starch (Figure 21.1, Serna-Saldivar and Ayala-Soto 2020). Whole cereals, legume pulses, and pseudocereals (Serna-Saldivar and Sanchez-Hernandez 2020) are the main sources of cellulose, lignin, and hemicelluloses, including the relevant β-glucans, whereas fruits and vegetables (Yangilar 2013, Morales de la Peña et al. 2020) are the primary sources of pectin, gums, and

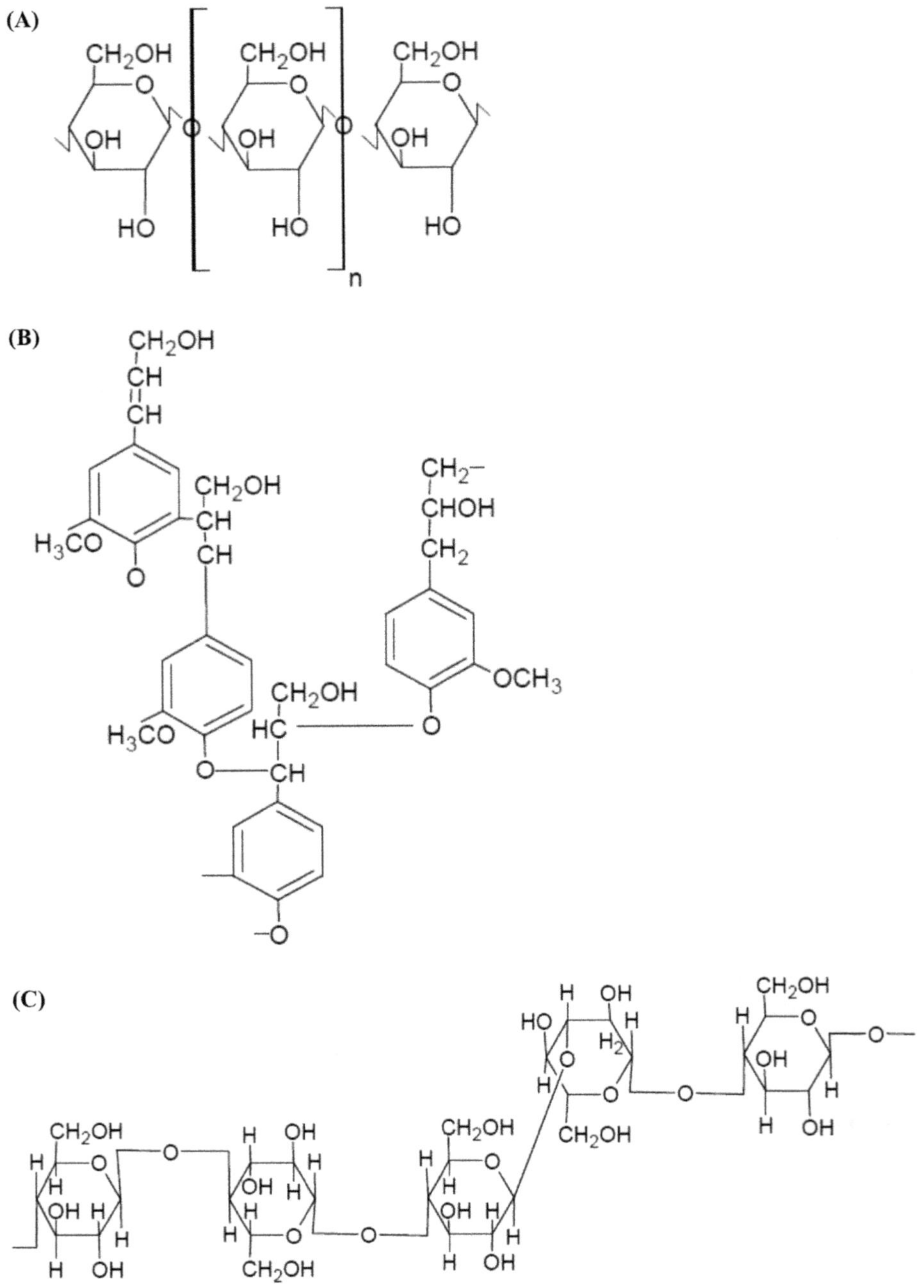

FIGURE 21.1 Chemical structures of the main insoluble and soluble dietary fiber components generally associated with cell walls of plant materials: (a) cellulose, (b) lignin, (c) hemicellulose, (d) arabinoxylans, (e) β-glucans, (e) fructooligosaccharides, and (g) pectin.

mucilages. The detailed compositions of the dietary fiber of cereals, pulses, pseudocereals, major fruits, and horticultural commodities were recently reviewed (Morales de la Peña et al. 2020, Serna-Saldivar and Sanchez-Hernandez 2020). Among cereals, oat has been recognized as the best source of dietary fiber and the cereal that provides significant quantities of soluble dietary fiber mainly consisting of prebiotic β-glucans (Serna-Saldivar and Sanchez- Hernandez 2020). Rolled or flaked

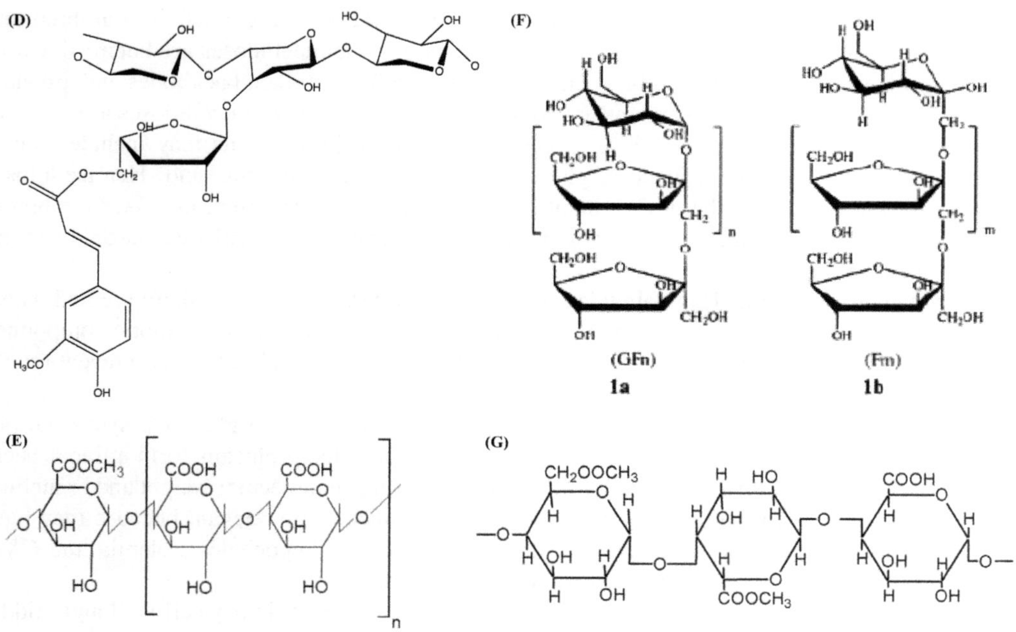

FIGURE 21.1 Continued.

and oat meals normally contain from 10–12% total dietary fiber of which 4–5% is soluble. The dry milling processes applied to cereal grains considerably diminish the amounts of total dietary fiber in decorticated grains, grits, meals, and flours. Needless to say, the fiber-rich by-products obtained by these industries (wheat and maize brans, rice bran and polishings) are concentrated forms that are used to increase fiber contents of many snacks. Legume-pulses, pseudocereals, and oilseeds are normally used as whole grains and therefore are good sources of both insoluble and soluble dietary fibers. As in cereals, most of the dietary fiber is concentrated in the hulls. Whole legumes normally provide 10–20% dietary fiber: common beans, cowpeas, and kidney beans being the best sources. Amaranth, quinoa, chia, buckwheat, whole sesame, and flaxseed or linseed (Chapter 13) are also good sources of dietary fiber (Serna-Saldivar and Sanchez-Hernandez 2020).

Health benefits of dietary fiber include laxation and reduction of blood cholesterol and blood glucose, thus preventing hyperlipidemia, CVD, diabetes, obesity, and some forms of cancer, such as colon (Yangilar 2013, Ye et al. 2012, Antunes-Ricardo et al. 2020, Torres et al. 2020, Tovar et al. 2020). According to most regulatory agencies, the recommended daily intake of total dietary fiber for a 2,000-kcal diet is 25 g. Of this total amount, it is recommended to consume half insoluble and half soluble dietary fractions.

Due to the health benefits of dietary fiber, the food industry has incorporated it into foods. Cereal brans are the most common sources of dietary fiber used by the food industry. However, cereal brans have generally negative effects on quality and organoleptic properties of the end-product, and have higher content of phytic acid, which is a compound that can block the absorption of trace minerals and cause malnutrition (Turksoy and Ozkaya 2011). On the other hand, fruit and vegetable fibers are of better quality due to their higher soluble dietary fiber contents, better water- and oil-holding capacities, and colonic fermentability (Eim et al. 2008).

21.4.2 Soluble and Prebiotic Fiber Components

The soluble dietary fiber components of the diet have strong implication for health mainly due to their prebiotic effects that stimulate the growth and development of beneficial microbiota in the hindgut. The major types of soluble constituents are hemicelluloses associated with arabino and glucoxylans, β-glucans and fructooligosaccharides, or inulins. Hemicelluloses are hetero-linear or branched polysaccharides found in plant cell walls that have β-1–4-linked backbones and are classified into xyloglucans, xylans, mannans, glucomannans, and glucans. Hemicelluloses are of lower molecular weight compared to cellulose. Component monosaccharide units may include xylose, arabinose, galactose, mannose, glucose, glucuronic acid, and galacturonic acid. Hemicelluloses are normally categorized by the predominant carbohydrate monomer present in xylans (arabinoxylans and glucoxylans), β-glucans, and xyloglucans (Scheller and Ulvskov 2010, Serna-Saldivar and Ayala-Soto 2020).

Arabinoxylans (Figure 21.1) are abundant in cereal endosperm cell walls, where rye and wheat contain the highest contents followed by barley, maize, rice, and oats. These prebiotic compounds have the potential to lower the glycemic index of cereals (Broekaert et al. 2011) and prevent type-2 diabetes (Van Den Abbeele et al. 2011).

On the other hand, β-glucans (Figure 21.1) comprise other group of β-D-glucose polysaccharides inherent to cell walls of cereals, algae, yeast, and molds. Normally, β-glucans form a linear backbone, primarily with 1–3- or 1–4-β-glycosidic linkages and vary in molecular weight and branching which affects solubility, viscosity, gelation, and health. β-Glucans are relevant because they have strong prebiotic and immune-modulatory properties and prevent hypocholesterolemia and CVD (Zekovic 2005, Bacic et al. 2009, Brown and Williams 2009).

Pectic polysaccharides or pectins (Figure 21.1) are present in the primary cell wall and middle lamella of all seed-bearing plants and are especially important in fruits. These soluble dietary fractions have relevant technological and nutraceutical properties mainly because of their prebiotic properties and ability to form gels. Chemically, pectins are defined as a family of covalently linked galacturonic acid–rich polysaccharides. The galacturonic acid linked at the *O*-1 and the *O*-4 positions comprises approximately 70% of pectins. Pectins are usually extracted from citrus byproducts in order to produce pure pectins widely used to produce marmalades or gels. The solubility of pectins increases with increasing extent of esterification (Serna-Saldivar and Ayala-Soto 2020).

Fructans are defined as polyfructosylfructose polymers (Figure 21.1) of four to several hundred residues that possess a sucrose core. After starch, fructans are the most abundant nonstructural polysaccharides found in nature (Franck 2006) associated with many plants and in some instances with fungi and bacteria. The degree of polymerization of fructans is variable depending on the source, the climate and growing conditions, the harvesting time, and storage conditions. Fructans also include oligofructose, which contain a degree of polymerization lower than 10 (Roberfroid 2007). In branched fructans, the branching linkages are usually β-2–6. The two major sources of inulins are chicory and agave and these are chiefly texture enhancer, sugar and fat replacers in chocolate, yogurt, and ice cream preparations, and a source of prebiotics in bakery goods (Chapter 10) and dairy-based foods (Chapter 17). The sweetness level of inulin is about 10% of the sucrose (Shoaib et al. 2016). Bornet et al. (2002) concluded that the short-chain fructooligosaccharides escape digestion in the human upper intestine and reach the colon where they are totally fermented mostly to lactate and short-chain FA acetate, propionate, and butyrate. More importantly, the fructooligosaccharides stimulate the growth of Bifidobacteria, which also suppress the growth of potentially pathogenic bacteria like *Clostridium perfringens* in the colon. Because of their fermentation, their caloric value is approximately half of the caloric value of sugars (2 kcal/g). Moreover, the fructooligosaccharides enhance the fecal bulking effect and diminish colon tumor development by enhancing both colon butyrate concentrations and local immune system effectors.

21.4.3 Resistant Starches

RS comprise those forms of starch which are not accessible to human digestive enzymes and therefore are fermented in the hindgut producing short-chain FA. The concept of RS has evoked new interest in the bioavailability of starch and in its use as a source of prebiotic dietary fiber (Sajilata

TABLE 21.1
Classification of Resistant Starches and Their Main Sources

Type of Resistant Starch	Description	Sources	Resistance Minimized
RS1	Physically inaccessible or protected starch	Whole or partially milled cereal grains, pulses, legumes, and vegetables	Milling, chewing
RS2	Ungelatinized resistant starch granules with type B crystallinity	Raw potatoes, green bananas, legumes (chickpeas, beans), high amylose maize	Thermal treatments or cooking
RS3	Retrograded or crystalline starch	Cooked and cooled potatoes, bread, snacks, breakfast cereals subjected to gelatinization and cooling which enhance retrogradation	Processing and amylase treatments
RS4	Chemically modified starches (cross-linked and distarch phosphate-esterified starches)	Modified starches intentionally used for bakery items and snack bars	
RS5	Amylose-lipid complexed starch	Formed from high AM starches that require higher temperatures for gelatinization and are more susceptible to retrograde	

Source: Sajilata et al. (2006), Fuentes-Zaragoza et al. (2011), and Ashwar et al. (2016).

et al. 2006, Serna-Saldivar 2010, Fuentes-Zaragoza et al. 2011, Ashwar et al. 2016). These starches have great potential to produce healthy snacks because most contain or are processed from starch feedstocks. The main potential uses is in extruded snacks, snack bars, and processed dairy products. The resistant starches are divided into five current categories named RS1 to RS5 (Table 21.1).

Briefly, RS1 represents starch that is resistant because it is in a physically inaccessible form such as partly milled grains and seeds and some in very dense types of processed foods. RS1 is heat-stable in most normal thermal operations and enables its use as an ingredient in a wide variety of conventional foods. RS2 represents starch that is in a certain granular form and is resistant to enzyme digestion. The starch in these raw granules is tightly packed in a radial pattern and is relatively dehydrated. This compact structure limits the accessibility of amylolytic enzymes. The most relevant RS3 is mainly constituted of retrograded amylose formed during cooking and cooling phases as it normally occurs in the production of many food, including snacks (bakery, tortilla chips, and extruded items covered in Chapters 10–12). The RS4 and RS5 are chemically modified and are currently viewed as low-caloric or a prebiotic dietary source. These are not natural and processed via chemical reactions like cross-linking (Sajilata et al. 2006, Fuentes-Zaragoza et al. 2011, Ashwar et al. 2016).

Commercial preparations of RS are now available and can be added to foods as an ingredient for lowering the energy density and improving textural and organoleptic characteristics along with increasing the amount of prebiotic dietary fiber. Among the health benefits of the different RS include improvement of colonic health and microflora, management of diabetes, lower glycemic index and blood cholesterol levels, reduced bile stone formation, increased mineral absorption, and potential to modify fat oxidation. These nutraceutical properties are closely related to the rate and extent of digestion and absorption in the duodenum. Among these different physiological roles, the prebiotic effects are of great interest for nutritionists and health specialists because RS have high fermentability which results in gut-associated immune and microbiota modulation as well as a significant production of short-chain FA (Sajilata et al. 2006, Fuentes-Zaragoza et al. 2011, Ashwar et al. 2016).

21.5 NUTRACEUTICAL OILS AND LIPOPHILIC NUTRACEUTICALS

Fat is in most instances a major chemical component of processed snack foods, especially those obtained after deep fat frying and seasoning operations like potato chips (Chapter 14), extruded snacks (Chapter 12), and corn and tortilla chips (Chapter 11). The amount of lipids greatly affects the caloric density because fat and oils provide 2.2 times more energy compared to carbohydrates and proteins (9 kcal/g of energy instead of 4 kcal/g). More importantly, the ratio between saturated: monounsaturated:polyunsaturated (PUFA) fatty acids greatly affects their effects on human health. The most relevant and nutraceutical PUFA are linoleic, linolenic, EPA, and DHA. On the other hand, the most detrimental FA for human health are *trans*-FA followed by saturated FA.

Crude fats and oils are almost always associated with other minor lipids like phospholipids, phytosterols, and lipophilic nutraceuticals like tocopherols, carotenoids/xanthophylls, and policosanols. The chemical structures of these minor compounds are depicted in Figures 21.2 and 21.3. These minor components are considered important nutraceuticals because they help to prevent oxidative stress, CVD, cancer, and neurological disorders. The xanthophylls lutein and zeaxanthin and the provitamin A- or B-carotenes also help to prevent night blindness and eye disorders like age-related macular degradation or AMD (Lee et al. 2009).

21.5.1 Nutraceutical Oils (Ω-3 and Ω-6 Fatty Acids and DHA/EPA)

There is an opportunity to upgrade the nutritional quality of snacks since most contain high amounts of fats/oils. Snacks should be free of *trans*-FA because they have negative effects on human health. In order to avoid these fats, processors should avoid partially hydrogenated oils frequently used in deep fat frying operations and plastic vegetable shortenings (Chapter 7). The control of the saturated:PUFA ratio is also of utmost importance. New snacks should have high contents of ω-3 FA, although the production of snacks with high amounts of PUFA is difficult because they are more prone to oxidation and rancidity. This could be prevented by the addition of natural antioxidants such as tocopherols and/or improved packaging technologies like vacuum, controlled atmosphere, and active and intelligent packaging (Chapter 4).

ω-3 FA are PUFA of 18–22 carbons with a double bond at the n-3 terminal position. Mammalian cells cannot synthesize these compounds, so they should be obtained through the diet. These nutraceutical ω-3 oils have several positive effects on human health. These long-chain PUFAs lower activity of key enzymes that synthesize cholesterol and fatty acids and play an important and key role in the synthesis of prostaglandins.

Linoleic acid (ω-6, 18:2) is the most common PUFA of the diet because it occurs as the main FA in many highly consumed foods and oils like soybean, cottonseed, maize, peanut, sunflower, and safflower (Chapter 7). The ω-6 linoleic acid is considered essential in human nutrition because it is necessary for the synthesis of important physiological compounds, including hormones like prostaglandins and other FA like EPA and DHA with high implications in human health.

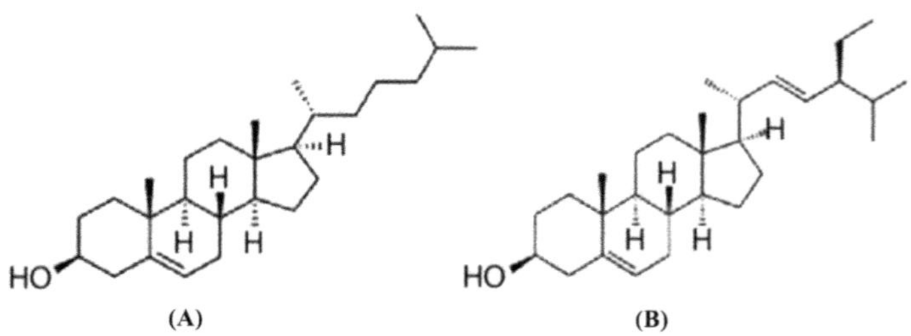

FIGURE 21.2 Comparison of the chemical structures of cholesterol with a plant sterol: (a) cholesterol and (b) stigmasterol.

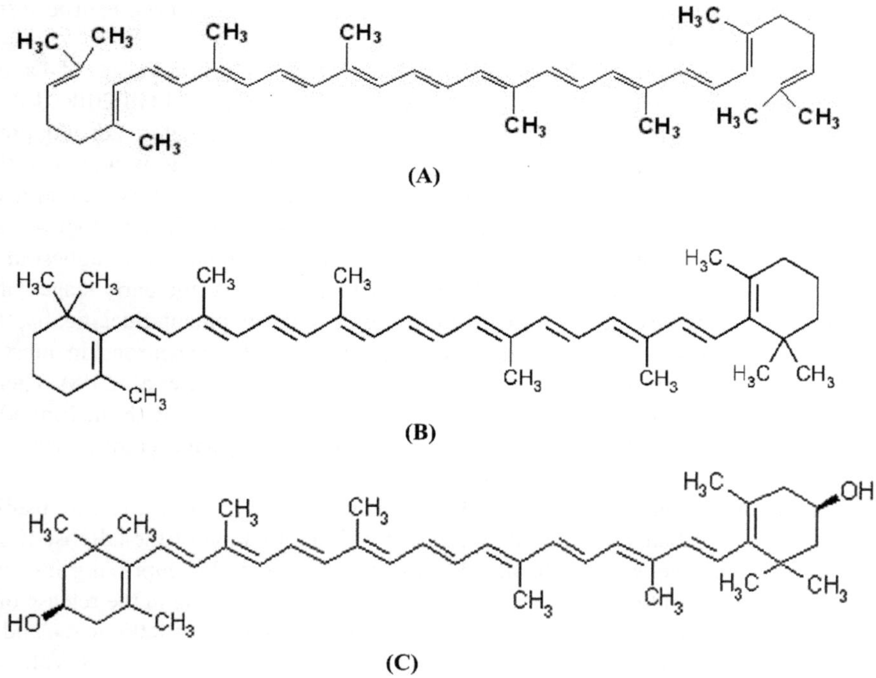

FIGURE 21.3 Chemical structure of important carotenes and a xanthophyll: (a) lycopene, (b) β-carotene, and (c) lutein.

Linolenic acid (ω-3, 18:3) is after linoleic acid the most common PUFA found in foods. The major sources of this FA are flaxseed or linseed and soybean oil, which contain more than 50% and approximately 7% of this ω-3 FA, respectively (Chapter 7). The consumption of linolenic acid spares the requirement of linoleic and facilitates the conversion into other nutraceutical fatty acids used to synthesize anti-inflammatory eicosanoids and hormones.

The other essential bioactive ω-3 PUFAs are eicosapentaenoic acid (EPA, ω-3, C20:C5) and docosahexaenoic acid (DHA, ω-3, C22:C6). Fish or algae oils harvested from species adapted to cold temperatures are the most common commercial sources of these ω-3 PUFAs. The World Health Organization (2008) recommends the daily consumption of 250–500 mg of combined EPA and DHA for healthy adults.

EPA and DHA are key for the production of the phospholipid bilayer membrane and lipid rafts, functioning to protect cell membrane integrity, fluidity, and structure (De Cicco et al. 2019). Also, some studies found that these two FAs are inhered to fight against degenerative and chronic diseases (Tacon and Metian 2013, Punia et al. 2019, Uauy et al. 2001, 2003). Previous studies report that EPA and DHA have anti-inflammatory activity and are attributable to its effect on cell membrane reorganization and on specific receptors. They are known to modulate inflammatory gene expression through binding to intracellular receptors. EPA is used as a substrate by cyclooxygenase and lipoxygenase enzymes to produce eicosanoids and, in combination with DHA, inhibits the arachidonic acid metabolism in the cell membrane, thereby decreasing the activation of cyclooxygenase-1,which is a gene involved in inflammation (Punia et al. 2019).

These long-chain FAs have been found to play a crucial role in maintaining cognitive function in individuals through their neuroprotective properties (Robinson et al. 2010, Dyall 2015). The ω-3 PUFAs are the major constituents of neuronal membranes, and particularly DHA is present in higher proportions (up to 60%). DHA is more rapidly incorporated into gray matter, where neuron cell bodies are located, so it is essential to develop useful brain function. Therefore, the adequate intake of DHA during childhood is important to have normal brain function, neurotransmission, vision, and synapsis plasticity (Robinson et al. 2010).

Previous investigations showed that the supplementation of ω-3 PUFAs at the age of 4 and 8 years generated improved academic performance evaluation (Rangel-Huerta and Gil 2018). Likewise, it has been found that the accumulation of DHA in the brain begins in the uterus during pregnancy. This is typically generated during the second stage when the growth of gray matter in the brain accelerates (Rangel-Huerta and Gil 2018). It has also been found that there is a rapid accumulation of DHA in a child's brain in the last trimester of pregnancy and in the first two years of life. Likewise, other researchers found that the brain's gray matter accumulates the highest amount of DHA up to 5 years of age. Therefore, ω-3 FA supplementation during the early stages of a child and during pregnancy may be crucial for their future cognitive development. Specifically, the chain length and high unsaturation of DHA provide fluidity and flexibility to the neurons' membrane. This allows and facilitates the transduction of signals inside the cell (Orellana et al. 2018). Therefore, it has been found that the imbalance or deficiency of these FAs is associated with the limited or poor development of a child. This can be translated into deficiencies in language, communication, motor development, verbal fluency, and problem-solving.

Other investigations suggest that dietary ω-3 PUFA found in fish oil may improve insulin sensitivity or reduce type-2 diabetes through the inhibition of inflammation of the adipose tissue (Gao et al. 2017). PUFAs have shown beneficial effects in these patients by improving insulin action through several metabolic pathways, such as hepatic lipogenesis, reduction in the release of triacylglycerols from the liver, improvement in ketogenesis, hepatic oxidation of fatty acids, and skeletal muscle, and by the stimulation of relevant antioxidant enzymes (Kumari et al. 2018, Valdes-Ramos et al. 2019). Moreover, the consumption of oils rich in PUFAs has been related to G protein-coupled receptor (FFAR1), which responds to medium and long unsaturated fatty acids, resulting in an increase of insulin secretion during elevated glucose levels (Eleazu et al. 2018). Other studies using supplementation of fish oil demonstrated an association with a lower risk with the important mechanism of insulin resistance (Farsi et al. 2014).

21.5.2 Phospholipids

Like triglycerides, the backbone of phospholipids is glycerol but only the primary and secondary residues are esterified to FA, whereas the third site is esterified to a phosphate group which in turn is linked to a head of choline, ethanolamine, serine, inositol, or glycerol. Crude vegetable oils contain 1–3% phospholipids, of which the majority is dominated by phosphatidyl choline, ethanolamine, and inositol. Most of the original concentrations of phospholipids are removed during the first stage or degumming step of the refining process detailed in Chapter 7. In the food industry, the phospholipids recovered after degumming are extracted and refined in order to use them as natural emulsifiers, which are ordinarily known as lecithins. In foods, phospholipids increase the oxidative stability of fats and oils and high-fat foods and act synergistically with tocopherols and phenolic antioxidants (Pokorny 2001).

The phospholipids are important polar lipids, which contribute to the structure of cell membranes forming bilayers (Liu and Deber 1997, Dixit et al. 2011). Therefore, the several types of phospholipids exert many positive effects on human health. There is overwhelming scientific evidence that the dietary supplementation has a positive impact in the prevention of inflammation, CVD, and cancer. One of the major positive effects is that phospholipids diminish hepatic lipid levels because they interfere with sterol absorption in the intestinal lumen. Phospholipids stimulate bile acid and cholesterol secretion and have the ability to increase plasma HDL levels (Cohn et al. 2008). Phospholipid isolates could be used as hepatoprotective nutraceuticals. Particularly, phosphatidyl choline helps to keep the proper functioning of the liver and its deficiency is related to increased susceptibility to hepatic cancer (Serna-Saldivar 2010). Lecithin and choline lower the risk of CVD and positively affect brain and mental development of both the fetus and infants. In geriatrics, the chronic inadequacy has been related to Alzheimer's disease. Choline released from lecithin during metabolism is considered as one of the most relevant neurotransmitters and is commonly supplemented to geriatrics in order to maintain proper brain function. Both phosphatidyl inositol and serine reduce blood triglycerides, fatty liver, bipolar disorders, and neurodegenerative diseases. Inositol, from phosphatidyl inositol or from the enzymatic hydrolysis of phytic acid by phytases, is known as a potent nutraceutical because it is essential for brain functioning and therefore is especially recommended for newborns and geriatrics. Inositol is considered a potent biological active compound and by some nutritionists as part of the B-vitamin complex because it helps to maintain cell membrane integrity and other relevant metabolic functions such as part of phosphatidyl inositol required for the proper functioning of the cerebrum and heart. Furthermore, it helps in the synthesis of RNA and to transport lipids and cholesterol. As a result, inositol is considered hypocholesterolemic, cardioprotective, and anticarcinogenic (Canty and Zeisel 1994, Orthoefer and List 2006, Serna-Saldivar 2010, Küllenberg et al. 2012).

21.5.3 Lipophilic Nutraceuticals

Oils associated with cereals, legumes, almonds, tree nuts, and other plant-based foods contain significant amounts of lipophilic nutraceuticals. Likewise, the extracted vegetable oils contain concentrated amounts of these minor components especially in the unrefined or crude forms. The main categories are phytosterols, carotenoids, tocols, and policosanols.

21.5.3.1 Phytosterols

Vegetable oils contain 0.2–0.4% of plant sterols or phytosterols. The major moieties are β-sitosterol, campesterol, stigmasterol, avenasterol, and brassicasterol. These phytosterols differ from cholesterol only in the structure of their side chains. The sterols differ from stanols in being unsaturated versus saturated at the C5–C6 double bond in their B-ring. Figure 21.2 compares the structure of cholesterol with stigmasterol. Phytosterols inhibit the absorption of cholesterol from the small intestine of human, thus effectively lowering plasma total and LDL cholesterol (Weststrate and Meijer 1998, Hendriks et al. 1999, Normen et al. 2000, Moreau et al. 2002, Ostlund 2004). A typical dietary intake level of phytosterols is only 200–400 mg/day, significantly below the effective level for cholesterol-lowering. Human feeding studies suggest sterol and stanol are similarly effective (Weststrate and Meijer 1998, Law 2000, Serna-Saldivar 2010, Dixit et al. 2011).

Phytosterols include sterols and stanols which constitute part of the plant cell membranes. Plant sterols are C-28 or C-29 sterols differing from cholesterol (C-27) due to presence of an extra methyl or ethyl groups on the cholesterol side chain (Figure 21.2). Stanols are saturated sterols, having no double bonds in the ring structure. There has been much interest in the health benefits of consuming phytosterols, especially since the FDA has issued a health claim for their use. Phytosterols compete for cholesterol absorption and therefore are considered hypocholesterolemic and preventive against CVD. A daily intake of 2 g of plant sterols inhibits the absorption of cholesterol from the small intestine, thus effectively lowering total blood cholesterol by 5–20%. In addition, the intake decreases oxidized LDL by 10–15% (Brufau et al. 2008, Serna-Saldivar 2010). Moreover, the supplementation of 0.2% β-sitosterol decreases the occurrence of chemically induced colon tumors. Stigmasterol (Figure 21.2) inhibits several pro-inflammatory and matrix degradation mediators typically involved in osteoarthritic-induced cartilage degradation through the inhibition of the NF-κB pathway (Gabay et al. 2010).

Pszczola (2005) described a snack bar that combines cocoa flavonols and phytosterols. Clinical trials have shown that consumers who ate two bars a day had a significant reduction in blood cholesterol and that these compounds acted synergistically and exerted multiple and positive cardiovascular benefits.

21.5.3.2 Carotenes and Xanthophylls

Carotenoids are lipophilic natural compounds synthesized by photosynthetic organisms like plants and algae and also by some nonphotosynthetic bacteria and fungi (Clotault et al. 2008). They serve two essential functions in plants: photoprotection and accessory pigments in photosynthesis (Mayne 1996). In green tissues, carotenoids are located in chloroplasts and function as light-harvesting antennae and photooxidation protectants, whereas in flowers and fruits they give rise to colors that attract insects and animals to enhance pollen and seed dispersal. These compounds are also involved in the production of hormones, flavors, aromas, defense compounds, and plant development (Clotault et al. 2008). In terms of foods, they are mainly present in fruits, vegetables, some cereal grains, pulses, and pseudocereals. On the other hand, animals (including humans) cannot synthesize carotenoids, so they need to acquire them from dietary sources (Amorim-Carrilho et al. 2014).

Carotenoids are considered as antioxidants with enormous potential to improve health through the protection against oxidative stress. These lipid-soluble pigments are classified into two groups: carotenes, which are nonoxygenated molecules, such as α-carotene, β-carotene and lycopene; and xanthophylls, which are molecules that contain oxygen or OH groups, such as lutein, zeaxanthin, and cryptoxanthin (Amorim-Carrilho et al. 2014). Figure 21.3 depicts the chemical structures of two important carotenoids and a xanthophyll (lutein). Lycopenes (Figure 21.3) are one of the most valuable nutraceuticals for the prevention of prostate cancer, whereas lutein and zeaxanthin are very helpful to maintain eye health and vision, especially through the prevention of age macular degeneration in geriatric people (Fullmer and Shao 2001). Of the 700 different carotenoids found in nature, about 50 have pro-vitamin A activity, meaning once consumed by humans, they can be converted to retinol or vitamin A. The three most relevant ones are α-carotene, β-carotene (Figure 21.3), and β-cryptoxanthin (Amorim-Carrilho et al. 2014).

Vitamin A is considered the most multifunctional vitamin as it serves many functions in the human body, such as in vision, immunity, reproduction, lipid metabolism, insulin response, energy balance, and the nervous system (Tanoury et al. 2013). Besides pro-vitamin A activity, carotenoids have been found to exert other health benefits. Carotenoids, to a greater or lesser degree, show scavenging properties against excessive numbers of free radicals that may be produced over the course of a cell's cycle (Santocono et al. 2007). This antioxidant capacity has been the most investigated, and has been suggested as their main mechanism of action (Amorim-Carrilho et al. 2014). The ability of carotenoids to absorb light and to quench singlet oxygen and free radicals has been associated with their polyene structure (Mayne 1996).

Although carotenoids are present in many common human foods, deeply pigmented fruits, vegetables, and juices constitute the major dietary sources (Amorim-Carrilho 2014). In general, carotenoids are mostly found in red-, orange-, and yellow-colored fruits and vegetables, with carrot and yellow pumpkin being one of the richest sources of these pigments.

Several types of grains may also contain significant amounts of carotenoids and xanthophylls. In cereals, yellow endosperm genotypes like yellow maize, durum wheat, and yellow sorghum contain significant amounts of carotenoids. In fact, plan breeders have developed genotypes that express high quantities of β-carotenes and other carotenoids in order to alleviate the deficiency of vitamin A in many developing countries around the globe. Examples are the development of high-carotenoid maize and the genetically modified golden rice (Serna-Saldivar 2010).

Instrumental HPLC analyses performed by Hart and Scott (1995) showed that good sources of lycopene were tomatoes and their processed products, whereas of β-carotenes were broccoli, carrots, butterhead lettuce, mixed vegetables, parsley, spinach, and watercress. The best sources of lutein (>1,000 μg/100 g) are broccoli, butterhead lettuce, parsley, peas, peppers, spinach, and watercress. According to these authors, there was little or no loss of carotenoids on cooking, and green vegetables showed an average increase in lutein levels of 24% and in β-carotene levels of 38%.

Yellow endosperm maize (*Zea mays*) is recognized as the cereal which contains the highest levels (up to 63 μg/g) of carotenoids. Most of the cultivars present zeaxanthin as the major moiety followed by β-cryptoxanthin, α-carotene, and β-carotene. Among commercial wheats, the durum or pasta tetraploid wheats with yellow endosperm contain the highest levels of carotenoids (around 5–6 μg/g). Unfortunately, the white polished rice and oats are practically devoid of these compounds.

For instance, the contents of lutein and carotenes (α- and β-carotenes) in oats are on the order of only 0.2 and 0.01 mg/kg, respectively (Mellado Ortega and Hornero Mendez 2015)

Among legume pulses, french, field, and green beans, lentils, and green peas are rich sources of carotenes.

The marigold (*Tagetes erecta*) flower, *Spirulina fusiformis*, and lucerna (*Medicago sativa*) are unconventional food sources that contain extremely high levels of carotenes and xanthophylls (Kandlakunta et al. 2008, Siriamornpun et al. 2012) with potential to be used as supplements for the production of snacks with high levels of these nutraceuticals.

21.5.3.3 Tocopherols and Tocotrienols

The chemical family of the tocols comprises four (α, β, γ, and δ) tocopherols and four (α, β, γ, and δ) tocotrienols. These lipophilic compounds are synthesized by photosynthetic organisms and play a critical role in the functioning of biological membranes. α-Tocopherol is the primary moiety responsible for vitamin E activity in mammalians (Kim et al. 2007) and is regarded as having the highest vitamin activity because it is readily absorbed and retained by mammalian cells (Kamal-Eldin and Appelqvist 1996, Dutta and Dutta 2003). The main physiological role function of tocols is to prevent PUFA oxidation.

Tocopherols and tocotrienols are known for their antioxidant activity and protection against oxidative stress. Needless to say, the α-tocopherol or vitamin E is essential for humans because it plays critical roles in terms of protecting oxidative stress, vision, reproduction, and the proper functioning of blood, brain, and skin. From the nutraceutical viewpoint, tocopherols are considered the second most relevant mechanism of defense against free radicals and oxidative stress, which diminish risk of several degenerative diseases (Serna-Saldivar 2010). Furthermore, the increased tocotrienol consumption has been associated with reduced serum cholesterol levels and the strengthening of the immune system by T lymphocytes (Bender 1992, Berdanier 2002). Nesaretnam et al. (2007) further indicate that there is a growing interest in the benefits of tocotrienols due to their recently reported activities beyond that of tocopherols in suppressing tumors and inhibiting platelet aggregation.

The major natural sources of tocols are tree nuts (Kornsteiner et al. 2006), almonds, (Maguire et al. 2004), avocado (Lu et al. 2009), and oilseeds like peanuts, soybeans, sesame, flax, safflower, and sunflower (Karunanandaa et al. 2005). In terms of vegetable oils, practically all contain significant levels with the exception of coconut oil (Desei et al. 1988). The chemical or physical refining processes described in Chapter 7 cause significant losses of these compounds, especially during the deodorization step. A comparison among vegetable oils indicate that refined soybean and sunflower oils contain higher levels of tocols than virgin olive oil (Desei et al. 1988, Psomiadou et al. 2000).

The cereal grains contain most of the tocopherols in the germ and the major contribution of these nutraceuticals is to the oil extraction from the germ of maize obtained after dry or wet milling (Chapter 6).

Tocopherols can be extracted and purified from the deodorizer distillate by-product obtained during oil refining (Chapter 7) and the value-added ingredients used as natural antioxidant and nutrient. About 20–50% of the total tocopherols present in crude vegetable oils are lost during the overall refining process, particularly during the deodorization step (Naz et al. 2011, Desei et al. 1988). However, their antioxidant activity greatly diminishes when fresh oils are subjected to the high temperatures of frying. This antioxidant is increasingly being employed to produce green labels and high-antioxidant foods. Tocopherols can effectively donate hydrogen atoms to lipid peroxyl radicals to interrupt the propagation of lipid oxidation.

21.5.3.4 Policosanols

Policosanols are wax-like materials consisting of a mixture of long-chained alcohols containing mostly hexacosanol (26:0), octacosanol (28:0), triacontanol (30:0), and docatriacontanol (32:0). These compounds are commercially obtained from by-products of sugarcane or from the pericarp and/or germ of maize or other cereal grains (Acosta-Estrada et al. 2019). According to Hwang et al.

(2005), the yield of this wax-like material from maize was 10 mg/100 g of dry kernels. Policosanols, as determined by HPLC in the wax-material, were only 2% of the total, with docosanol (22:0) being the major one found in maize, rice, and wheat (Hwang et al. 2005, Serna-Saldivar 2010). Kawaniski et al. (1991) reported octacosanol (28:0) contents of germs of maize, rice, and wheat of 0.82, 0.46, and 0.22 mg/100 g, respectively. A considerable amount of policosanols can be extracted from sorghum grain or its dried-distilled grains obtained from biorefineries (Hwang et al. 2004). The policosanol consisted primarily of octacosanol (28:0) and triacontanol (30:0) that comprised more than 80% of the total policosanols present in whole sorghum kernels and its dried-distilled grains (Hwang et al. 2004).

Policosanols reportedly exert physiological activities such as reducing serum lipid levels and platelet aggregation (Gouni-Berthold and Berthold 2002, Hargrove et al. 2004, Acosta-Estrada et al. 2019). In laboratory animals, the administration of policosanols diminished the risk of atheroma formation by reducing platelet aggregation, endothelial damage, and foam cell formation. Additionally, policosanols have been shown to lower total and LDL cholesterol levels by 13–23% and 19–31%, respectively, while increasing HDL cholesterol 8–29%. Policosanols are thought to improve lipid profiles by reducing hepatic cholesterol biosynthesis while enhancing LDL clearance (Varady et al. 2003). Sanchez et al. (2010) treated human patients with 20 mg/day of policosanols plus 125 mg/day of aspirin after hospital discharge and concluded that the intervention produced good results in the neurological recovery, recurrence of events, and cholesterol control in patients with ischemic stroke and previous transient ischemic attack.

21.6 BIOACTIVE PEPTIDES

Bioactive peptides have been defined as specific protein fragments that have a positive influence on physiological and metabolic functions or condition of the body and may have ultimate beneficial effects on human health. Most of the physiological activities of proteins are performed by peptide sequences encrypted in the parent protein, which become active when cleaved intact. Bioactive peptides are released during enzymatic proteolysis in the gastrointestinal tract or via *in vitro* hydrolysis with selected proteases and also during food processes like fermentation, sprouting, and cooking. Bioactive peptides can be generated from plants, milk, eggs, animal products, fungi, and microorganisms. The most important ones are associated with fermented dairy products and hydrolyzed legume proteins from soybeans and common beans. The enzymatic hydrolysis occurring during digestion or fermentation can liberate an enormous amount of BP whose activities span from antimicrobial, antithrombotic, antihypertensive, opioid, immunomodulatory, mineral binding, antioxidative, and even anticarcinogenic. The bioactive peptides are commonly recuperated from protein hydrolysates, which are used to supplement functional foods. These peptides could also be further separated and purified into nutraceuticals for nonpharmacological therapies. Bioactive peptides from foods are valuable functional agents in healthy diets that can prevent and treat chronic diseases, other diseases, and even cancer (Shahidi and Zhong 2008, Daliri et al. 2017, Sanchez and Vázquez 2017).

21.6.1 Bioactive Peptides in Dairy Products

Most of the research related to bioactive peptides has been performed with dairy products, covered in Chapter 17, because digested or fermented milk proteins exert a wide range of nutritional, functional, and biological activities. Increasing attention has been given to their physiologically active peptides that are liberated in the human gastrointestinal tract after digestion or alternatively produced from fermented milk products treated with proteolytic starter cultures or proteolytic enzymes. A variety of naturally formed bioactive peptides have been found and identified in fermented dairy foods such as yogurt, sour milk, and cheeses (Korhonen and Pihlanto 2006). These bioactive peptides exert physiological effect on the digestive, cardiovascular, immune, and nervous systems. Several investigations have identified a great number of peptide sequences with specific bioactivities derived from the hydrolysis of caseins and whey proteins (lactalbumins and lactoglobulins). The most functional bioactive peptides are tripeptides IPP and VPP which have strong antihypertensive properties and positive effects in human trials. In addition, both caseins and whey proteins produce bioactive peptides that have antioxidant, antithrombotic, hypocholesterolemic, immunomodulatory, cytomodulatory, opioid, antiappetizing, and antimicrobial properties. The myriad of innate bioactive peptides derived from casein and whey proteins present an excellent source of natural ingredients for different applications in functional snacks. In the near future, several breakthrough products based on these ingredients will be launched for infants, elderly, and immunocompromised people as well as to improve performance and prevent diet-related chronic diseases and cancer (Korhonen and Pihlanto 2006, Park and Nam 2015).

21.6.2 Bioactive Peptides in Plant-Based Sources

Two excellent reviews related to bioactive peptides associated with vegetal sources are published by Malaguti et al. (2014) and Rizzello et al. (2016). Among plant sources, legumes and pulses (Lopez-Barrios et al. 2014, Tak et al. 2021) discussed in Chapter 13 are the most promising due to their relatively high-protein content that upon digestion or protein hydrolysis generate an array of bioactive peptides with different physiological effects. Bioactive peptides produced during digestion of soybean seeds and fermented soybean milk with *Lactobacillus plantarum* were investigated by Singh and Vij (2018). The results indicated that soybean proteins experimented degradation during protein digestion generating a large number of bioactive peptides with antimicrobial activity and antioxidative and antihypertensive activities. These studies also concluded that the angiotensin-inhibitory peptides had good stability against temperature of processing and gastrointestinal digestion and therefore may be useful for their application as ingredients for the development of functional foods. Likewise, the pulses chickpeas, cowpeas, field peas, common bean, mung bean, and kidney bean among others yield after protein digestion hydrolysates with significant inhibition of angiotensin-converting enzyme related to hypertension. Moreover, many hydrolysates obtained from pulses contain bioactive peptides with antioxidant properties and are preventive of cancer, inflammation, and hypercholesterolemia. The phaseolin, vicilin, and legumin protein fractions are the ones identified with best potential to generate bioactive peptides (Lopez Barrios et al. 2014).

Although cereal grains contain less than one half of the protein of legumes and pulses, they also produce bioactive peptides after digestion or hydrolysis with proteases. Digested wheat and oat showed the presence of angiotensin-converting enzyme inhibitory peptides and dipeptidyl peptidase inhibitor, as well as peptides with antithrombotic, antioxidant, hypotensive, and opioid activities. On the other hand, wheat and rice have proteins with peptidic sequences showing anticancer activity. Wheat and barley showed the greatest diversity and abundance of peptides with potential biological activity among the cereal proteins (Malaguti et al. 2014).

The encapsulation of extracted or purified bioactive peptides has emerged as an effective strategy to improve stability and generate a controlled release of these bioactive compounds for application in food matrices. Spray-drying is the most common technology to produce microencapsulated products (Lopez Barrios et al. 2014).

21.7 HYDROPHILIC PHYTOCHEMICALS

Most studied hydrophilic antioxidants belong to the family of phenolics which are subdivided into simple phenols, flavonoids/anthocyanins, and tannins. These compounds normally produce colors, which are greatly influenced by the pH, and astringent flavors to foods. Most of these compounds are associated with fruits and vegetables, although some special types of whole cereals and especially legume pulses contain significant quantities.

21.7.1 Phenolics and Polyphenols

Phenolic compounds or polyphenols are the most widely distributed secondary metabolites present in the plant kingdom (Dicko et al. 2006). These compounds are a wide range of chemical compounds characterized by a benzene ring with one or more hydroxyl groups attached to it. The structure of different types of phenolics are depicted in Figure 21.4. Phenolics are normally derived from benzoic or cinnamic acids, whereas flavonoids and anthocyanins are built from two units: a C6-C3 unit from cinnamic and a C6 fragments from malonyl–CoA. Tannins are polymers of five to seven units or more of flavan-3-ol or catechins (Figure 21.4). They are widely present in the plant kingdom, where more than 8,000 are known, with different chemical structures and activities (Hsu et al. 2006). They are crucial for the growth, reproduction, and defense of plants (Ghasemzadeh and Ghasemzadeh 2011).

Common sources of phenolic compounds include cereal grains, pulses, fruits, vegetables, wine, tea, and coffee (Dos Santos et al. 2006). Most phenolics associated with cereal grains are present in the pericarp, testa or seed coat, aleurone, and glumes or husks (Serna-Saldivar 2010, Acosta-Estrada et al. 2019). Ferulic acid (Figure 21.4) in its free, conjugated, and bound forms is the most relevant simple phenolic present in cereal kernels. It is a hydroxycinnamic acid associated with cell walls. The insoluble-bound diferulic forms are covalently linked to polysaccharides mainly in the pericarp and aleurone layers. The function of these compounds is to cross-link the strength of cell walls, therefore playing an important role in the lignification process that influences physical and textural attributes of plants and foods (Acosta-Estrada et al. 2019).

Although phenolic compounds do not have any known nutritional function, they have received considerable attention because of their role in human health. Most of their biological activities have been associated with their antioxidant activity, meaning they are able to scavenge reactive nitrogen and oxygen species, which are responsible for damaging the body through oxidative stress (d'Ischia et al. 2006; Phuong-Lien et al. 2016). However, these compounds are also able to protect the body through nonantioxidant mechanisms (Srivastava et al. 2000). Regardless of the ferulic acid form, it is considered as a potent antioxidant and a nutraceutical that prevents inflammation, cancer, LDL oxidation, and neuron degeneration (Kanski et al. 2002, Lin et al. 2016). Generally, phenolic compounds possessing *ortho*-hydroxyl groups possess higher antioxidant capacity than others.

Among cereals, maize and sorghum possess higher antioxidant activity compared to wheat, oats, and rice. Several types of sorghums, which include high-tannin genotypes, are known to contain the highest phenolics and antioxidant capacity. In fact, some sorghums exert higher antioxidant capacity compared to the most significant sources of antioxidants such as blackberries (Awika and Rooney 2004, Awika et al. 2004). The dry milling and nixtamalization procedures described in Chapters 6 and 11 diminish very significantly the amount of phenolics and antioxidants associated with the refined milled fractions.

Flavonoids are a type of polyphenolic compounds, a class of secondary plant metabolites, which have been reported to exert antioxidant and are considered anti-inflammatory, antimutagenic, and anticarcinogenic agents with the capacity of modulating cellular enzyme expression such as xanthine oxidase, cyclooxygenase, and lipoxygenase among others that increase reactive oxygen species concentration. There are more than 6,000 flavonoids reported and subclassified into anthocyanin, chalcone, flavanones, flavones, flavonols, and isoflavonoids according to their chemical structure (Panche et al. 2016).

Given the diversity of flavonoids and the novelty of their study, it has been complicated to establish a daily recommended index; nevertheless, a recent study has proposed an acceptable daily intake of 400 mg based in different studies that showed occasional effects with 250 mg/day dose, but become more consistent at 400 dosage. A minimum intake cannot be established contrary to vitamins or other micronutrients given that there are no symptoms of deficiency (Peluso and Palmery 2015).

Most fruits, vegetables, pulses, cereals, and pseudocereals contain significant amounts of flavonoids such as flavanols, flavanones, flavones, and anthocyanins. The six major subclasses of

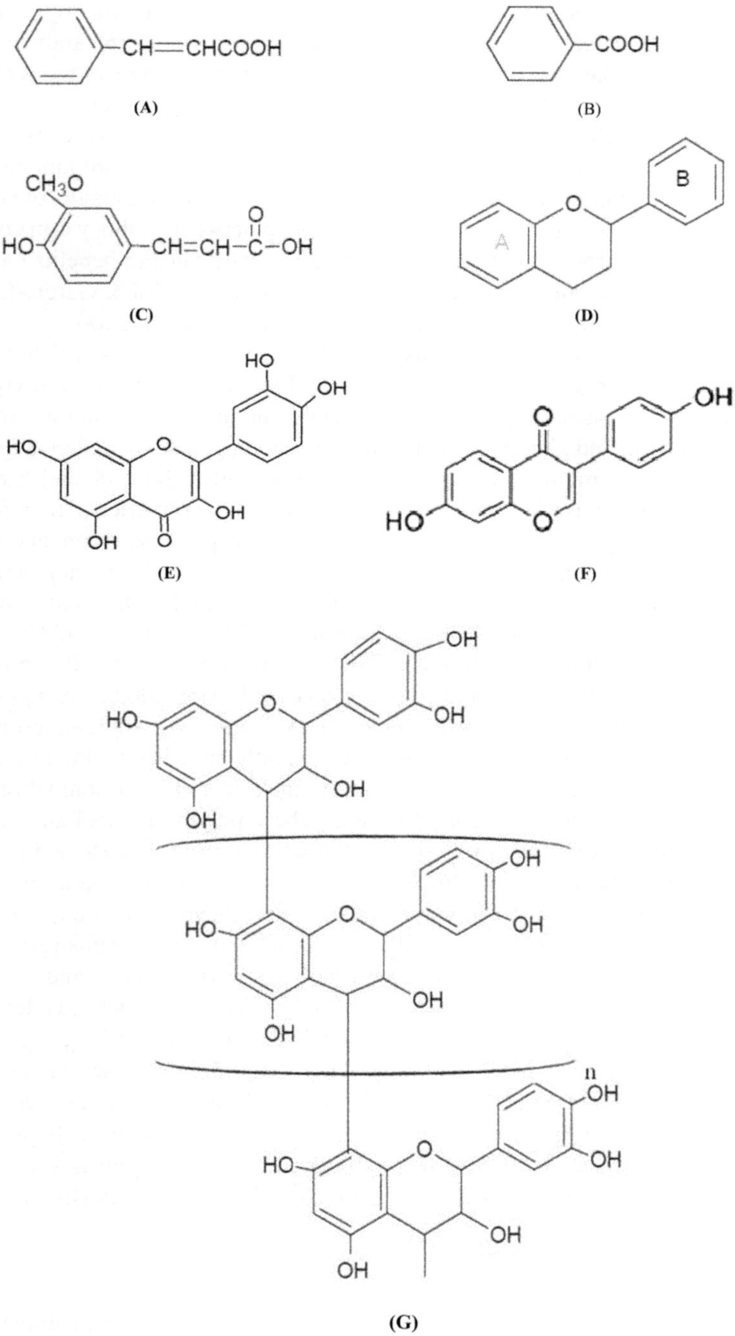

FIGURE 21.4 Chemical structures of main types of phenolic compounds: (a) simple phenolic (cinnamic acid), (b) simple phenolic (benzoic acid), (c) simple phenolic (ferulic acid), (d) anthocyanidin, (e) flavonoid (quercetin), (f) isoflavone (daidzein), and (g) condensed tannins.

flavonoids include the flavones (e.g., apigenin, luteolin), flavonols (quercetin, myricetin), flavanones (naringenin, hesperidin), catechins or flavanols (epicatechin, gallocatechin), anthocyanidins (cyanidin, pelargonidin), and isoflavones (genistein, daidzein). Most of the flavonoids are present in plants attached to sugars or oligosaccharides (glycosides) (Ross and Kasum 2002).

The pigmented pulses like black beans, Desi chickpeas, and black soybeans and cereals like blue- or purple-pigmented maize and Shawaya sorghum contain significant amounts of anthocyanins similar to the ones found in many berries and red wine. Cumulative results of epidemiological, *in vitro*, and *in vivo* research suggest an inverse relationship between anthocyanin consumption and the incidence of various chronic and degenerative diseases. Their health benefits have been related to their high antioxidant and antiradical activities (Fimognari et al. 2004, Garcia-Lara et al. 2012, Gutierrez-Uribe and Serna-Saldivar 2014, Oomah et al. 2010, Wu et al. 2006).

Isoflavones, also known as phytoestrogens, is a subgroup of flavonoids which contain a phenyl ring B connected at position 3 of 1,4-benzopyrone ring (Figure 21.4). In nature, soybean is the most abundant source of isoflavones (up to 3 mg/g dry weight) and normally contain three types of glycosidic (daidzin, genistin, and glycitin) or aglycones (daidzein, genistein, and glycitein). Isoflavones are structurally similar to mammalian estradiol and can bind to both α- and β-estrogen receptors. As most flavonoids, the isoflavones exert many beneficial health effects; therefore, they are of immense help in preventing several diseases, including cancer (prostate, colon, and breast cancers), hypercholesterolemia, CVD, atherosclerosis, osteoporosis, and relief of menopausal symptoms in certain women. The last two positive effects are partially attributed to the weak estrogenic activity of isoflavones (Ross and Kasum 2002, Faraj and Vasanthan 2004, Setchell and Cassidy 1999).

Due to their ample availability, whole and defatted soybeans have potential to produce nutraceutical snacks containing significant amounts of isoflavones (Messina 2002, Messina and Lane 2007, Messina and Hilakivi-Clarke 2009). For example, snacks containing soy derived plant sterols for lowering cholesterol levels and isoflavones have been recently launched on the market.

Condensed tannins (Figure 21.4), also known as proanthocyanidins or nonhydrolyzable tannins, are polymers formed by the condensation of flavans. These polymeric yield anthocyanidins when depolymerized under oxidative conditions. Tannins affect the color of processed products and normally impart an astringent taste. Even though they are considered as potent antioxidants, some nutritionists consider them as antinutritional factors because they bind dietary proteins and inhibit digestive enzymes lowering protein digestibility (Serna-Sadivar 2010). Fruits (grapes, persimmon, pomegranates, blueberries, blackberries, cherries, cranberries), tree nuts (peanuts, walnuts, pecans, cashew), pulses (black soybeans and common beans, red beans, black chickpeas, lentils), vegetables (squash), tea, spices, red wine, chocolate, and cocoa powder are foods or beverages, which contain significant amounts of tannins. Among commercial cereals, sorghum is the only one that may contain significant amounts of condensed tannins. The tannins have nutraceutical benefits due to their strong antioxidant capacity. Experimental studies have demonstrated that the high antioxidant tannins prevent oxidative stress and the growth of pathogenic bacteria and viruses and decrease colon and mammary cancers in induced animals (Serrano et al. 2009, Serna-Saldivar et al. 2010).

21.7.2 Terpenes

Terpenes are a group of secondary metabolites present in fruits, vegetables, legumes, molds, yeast, and microorganisms. Terpenes are classified according to the number of isoprenes (five carbon organic compound with the formula $CH_2{=}C(CH_3){-}CH{=}CH_2$) units that comprise their molecules. They are divided into monoterpenes, sesquiterpenes, diterpenes, triterpenes, and tetraterpenes. Figure 21.5 shows examples of monoterpenes, sesquiterpenes, and triterpenes that are found in food matrices. These compounds have received much attention due to their quimiopreventive effects attributed to the inhibition of free radicals and induction of important antioxidant and detoxifying enzymes that protect humans against oxidative stress. The most studied monoterpenes are limonene classified as cyclic, perillyl alcohol, carvona, and geraniol, whereas the most studied diterpenes are

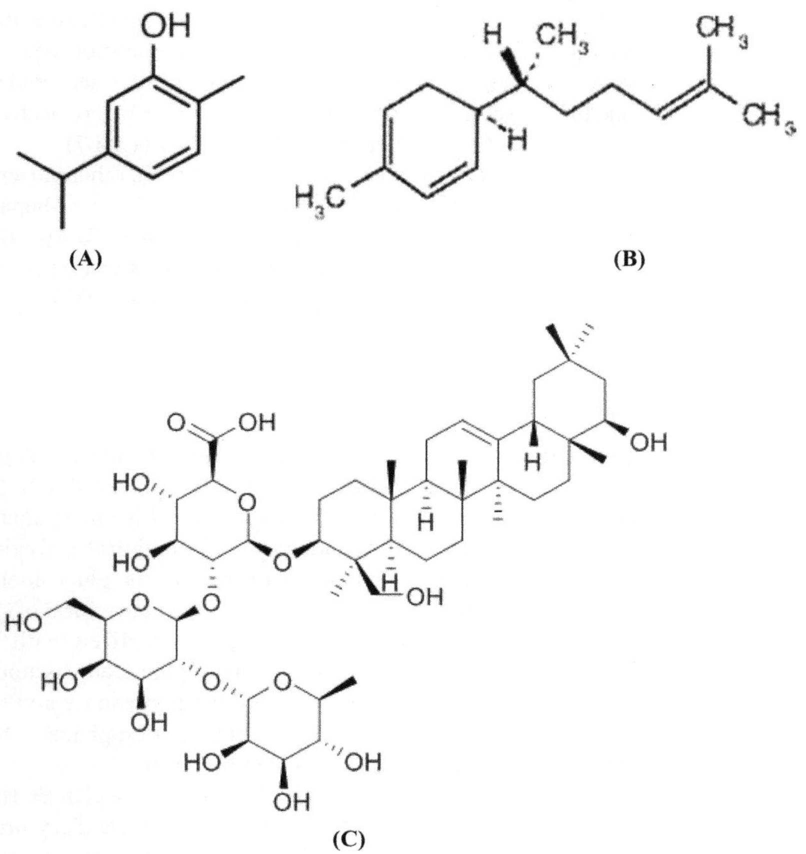

FIGURE 21.5 Chemical structures of main types of terpenes: (a) monoterpene (carvacrol), (b) sesquiterpene (zingiberene), and (c) triterpene (glycosylated soyasaponin).

carnosic acid, carnosol, and rosmanol. The two most relevant types of terpenes are sesquiterpenes (15 C) and triterpenes (30 C) also known as saponins. The most common saponins are derivatives of betulinic, ursolic, and oleanolic acids (Wagner and Elmadfa 2003).

These phytochemicals are known to induce apoptosis and affect expression of key genes involved with oxidative stress and cancer (Crowell and Elson 2000, Kerwin 2004). In foods, sesquiterpenes are found in essential oils extracted from celery, eucalyptus, and orange. The more naturally occurring triterpenes are associated with legumes-pulses, alfalfa, and ginseng (root of plants belonging to the genus *Panax*). The popular supplements of ginseng used in traditional Chinese medicine for more than 4,000 years have many benefits in human health. The active compounds known as ginsenosides (glycosidic saponins) are primarily responsible for health-beneficial effects in laboratory animals and humans. In animals, it is reported to improve the cognitive function in a mouse model of Alzheimer's disease and display anxiolytic activity and neuroprotective effects against neuronal damage resulting from ischemic stroke. Administration of ginseng inhibited the hypertrophy in heart failure by regulation of reactive oxygen species in mice as well as depletion of cardiac contractile function in rats. It also has an antidiabetic and antiobesity potential as it increases insulin sensitivity and inhibits formation of adipose tissue. Likewise, the consumption of ginseng by humans improves brain and cognitive function and immunological activity and prevents inflammation, type-2 diabetes, cancer, and viral infections (Ernst 2010, Szczuka et al. 2019).

Legumes seeds like soybean, common beans, peanuts, and chickpeas also contain significant amounts of triterpenes glycosides or soysaponins (Figure 21.5), which are currently attracting lot of scientific attention. They can be classified according to the type of aglycon, the moiety attached at the C-22 position on the aglycon, and the carbohydrate sequences at the C-3 position on the aglycon. Approximately 30 different soysaponins are reported, but their presence and quantity differ from genetic and agronomical variation. Soysaponins are found to have several biological activities such as hepatoprotective, antihyperlipidemic, anticancer, antioxidative, and antiviral (Guclu-Ustündag and Mazza 2007).

Saponins extracted from sprouted black beans and their hulls, chickpeas, other pulses, and Agave have concluded that these phytochemicals prevent oxidative stress, obesity, and hepatic steatosis and modulate lipid metabolism and biliary cholesterol secretion (Guajardo-Torres 2013, Chavez-Santoscoy et al. 2014, Leal Diaz et al. 2016, Singh et al. 2017) and the successful incorporation of extracts in nixtamalized and bakery products (Chavez-Santoscoy et al. 2016 a,b), covered in Chapters 10 and 11.

21.7.3 Isothiocyanates and Sulforaphanes

Isothiocyanates are defense-related molecules present in plants, mainly those belonging to the Brassicaceae family, including watercress, kale, cabbage, Brussel sprouts, cauliflower, and broccoli (Palliyaguru et al. 2018). Sulforaphane (Figure 21.6) is the most studied isothiocyanate, due to its potential therapeutic effect, including antimicrobial, anticancer, and antiviral activities (Abbaoui et al. 2018). Sulforaphane is generated after the enzymatic hydrolysis of the glucosinolate glucoraphanin (Figure 21.6). The enzyme responsible for its conversion into the active form of sulforaphane is the myrosinase (a thioglucosidase), which in intact cells is compartmentalized in different organelles (Sivapalan et al. 2018). Sulforaphane synthesis occurs during plant-cell disruption induced by insect attack or during mastication process by humans where the myrosinase and the substrate (glucoraphanin) get in contact. An additional mechanism to obtain sulforaphane is by microbial thioglucosidase activity produced by the gut microbiota (Tian et al. 2018).

The mechanism of action of sulforaphane is through sulfur binding with Kelch ECH-associating protein 1 (KEAP1), leading to conformational modification and release of nuclear erythroid 2-related factor 2 (Nrf2), which accumulates and increases transcription of the genes that have antioxidant response elements (AREs) (Dinkova-Kostova et al. 2017). AREs genes include those that codify for critical antioxidant enzymes such as quinone oxidoreductase, glutathione reductase, leukotriene B4 dehydrogenase, and other transcription factors for related antioxidant and antiinflammation pathways (Lubelska et al. 2016).

Sulforaphane exhibits more reliable results than glucoraphanin when ingested; metabolites at urinary excretion show a 70–90% assimilation for sulforaphane versus 2–15% for the glucosinolate

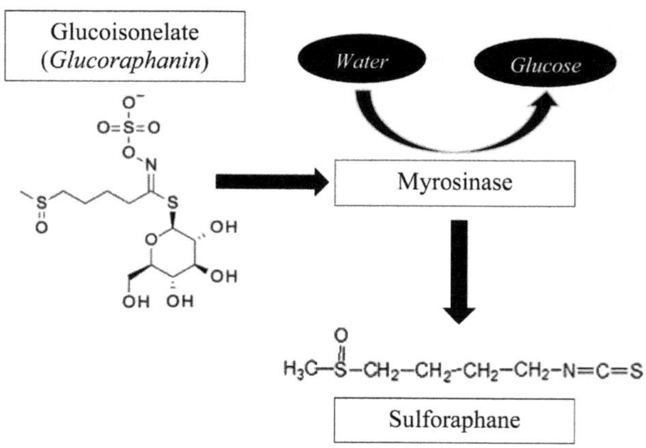

FIGURE 21.6 Enzyme-mediated conversion of a glucosinolate (glucoraphanin) into the bioactive form of sulforaphane.

form. However, there are technical complications for isolating and formulating supplements for direct dosing of sulforaphane. Raw and dry vegetables remain the most important sources of sulforaphane in the glucoraphanin form, and current efforts are focused on increasing the concentration of glucoraphanin (Yagishita et al. 2019). The potential of synergistics effects of sulforaphanes with other antioxidants is promising since the sulfur-containing phytochemicals protect organisms from oxidative stress with a different physiological mechanism.

21.8 PROBIOTICS

Probiotics are live microorganisms that have been extensively studied and explored commercially, especially in developed countries around the globe. In order to be categorized as probiotic, it has to exert different benefits to the consumer (Soccol et al. 2010). According to Kerry et al. (2018), humans are a reservoir of heterogeneous and vivacious group of microbes, which together forms the human–microbiome organism. The gastrointestinal tract serves as a home to 100–1,000 microbial species, which primarily modulate the host internal environment and thereby play a major role in health. This remarkable symbiotic relationship has recently attracted extensive investigations in nutrition and health. The benefits of probiotics to human health have been demonstrated in hundreds of scientific research studies. *Lactobacillus* and *Bifidobacterium* are the main probiotic groups; however, there are reports of the probiotic potential of *Pediococcus*, *Lactococcus*, *Bacillus*, and yeasts like *Saccharomyces boulardii*. The main probiotic lactobacillus are *L. plantarum*, *L. paracasei*, *L. acidophilus*, *L. casei*, *L. rhamnosus*, *L. crispatus*, *L. gasseri*, *L. reuteri*, and *L. bulgaricus*, whereas the chief bifidobacteria are *B. longum*, *B. catenulatum*, *B. breve*, *B. animalis*, and *B. bifidum*. These strains exhibit potent anti-inflammatory, antiallergic, anticarcinogenic and are useful to combat practically all chronic diseases. Probiotics are mainly acquired from the diet through the consumption of dairy foods (Chapter 17), although recently many other foods are supplemented with these encapsulated microorganisms. Probiotic microrganisms consist of a wide variety of species and subspecies, which have the ability to adhere, colonize, and modulate the gastrointestinal tract, especially in the large intestine. The viability of probiotics is key for developing probiotic food products. For that reason, probiotic cultures are processed mainly via encapsulation (chitosan, alginates, dextrins, biodegradable polysaccharides) with spray-drying technologies (Anselmo et al. 201). According to Huang et al. (2017), the spray-drying of probiotics enables a larger production scale than the freeze-drying currently used with the concurrent savings in energy and production costs. The aim is to prolong their viability with different types of protective matrices to ensure their improved resistance during storage and processes and from the harsh conditions of the digestive system.

Probiotics are designed to defy several challenges like oral mastication and delivery, the acidic conditions of the stomach, the exposure to gastrointestinal enzymes, and the bile salts in the duodenum. In addition, probiotics should rapidly adhere to a phenomenon named mucoadhesion in order to effectively colonize the intestines (Anselmo et al. 2016).

Supplementation with probiotics, prebiotics, and synbiotics has shown promising results against various enteric pathogens due to their unique ability to compete with them for adhesion sites or to stimulate and modulate the host's immune response via activation of specific genes in and outside the gastrointestinal tract.

REFERENCES

Abbaoui, B., Lucas, C.R., Riedl, K.M., Clinton, S.K., and Mortazavi, A. 2018. Cruciferous vegetables, isothiocyanates, and bladder cancer prevention. *Molecular Nutrition and Food Research*, 62(18):e1800079.

Acosta-Estrada, B., Gutierrez-Uribe, J., and Serna-Saldivar, S.O. 2019. Minor constituents and phytochemicals of the kernel. In: *Corn: Chemistry and Technology*, 3rd edition, ed. S.O. Serna-Saldivar, 369–403. Cambridge: Elsevier (Woodhead Publishing), AACC-International.

Amorim-Carrilho, K.T., Cepeda, A., Fente, C., and Regal, P. 2014. Review of methods for analysis of carotenoids. *TrAC Trends in Analytical Chemistry*, 56:49–73.

Anselmo, A.C., McHugh, K.J., Webster, J., Langer, R., and Jaklenec, A. 2016. Layer-by-layer encapsulation of probiotics for delivery to the microbiome. *Advanced Materials*, 28(43):9486–9490.

Antunes-Ricardo, M., Villela-Castrejón, J., Gutiérrez-Uribe, J.A., and Serna-Saldivar, S.O. 2020. Dietary fiber and cancer. In: *Science and Technology of Fibers in Food Systems*, eds. J. Welti-Chanes, S.O. Serna-Saldivar, O. Campanella, and V. Tejada Ortigoza, 241–276. New York: Springer.

Ashwar, B.A., Gani, A., Shah, A., Wani, I.A., and Masoodi, F.A. 2016. Preparation, health benefits and applications of resistant starch. A review. *Starch-Stärke*, 68(3–4):287–301.

Awika, J.M., and Rooney, L.W. 2004. Sorghum phytochemicals and their potential aspects on human health. *Phytochemistry*, 65(9):1199–1221.

Awika, J.M., Rooney, L.W., and Waniska, R.D. 2004. Properties of 3-deoxyanthocyanins from sorghum. *Journal of Agricultural and Food Chemistry*, 52(14):4388–4394.

Bacic, A., Fincher, G.B., and Stone, B.A. 2009. *Chemistry, Biochemistry, and Biology of (1–3)-[Beta]Glucans and Related Polysaccharides*. Amsterdam: Academic Press.

Behallm, K.M., Scholfield, D.J., and Hallfrisch, J. 2006. Whole-grain diets reduce blood pressure in mildly hypercholesterolemic men and women. *Journal of the American Dietetic Association*, 106(9):1445–1449.

Bender, D.A. 1992. Vitamin E: Tocopherols and tocotrienols. In: *Nutritional Biochemistry of the Vitamins*, ed. Bender, D.A., 87–105. Cambridge: Cambridge University Press.

Bernadier, C.D. 2002. *Handbook of Nutrition and Food*. New York: CRC Press.

Bornet, F.R.J., Brouns, F., Tashiro, Y., and Duvillier, V. 2002. Nutritional aspects of short-chain fructooligosaccharides: Natural occurrence, chemistry, physiology and health implications. *Digestive and Liver Disease*, 34:S111–S120.

Broekaert, W.F., Courtin, C.M., Verbeke, K., Van de Wiele, T., Verstraete, W., and Delcour, J.A. 2011. Prebiotic and other health-related effects of cereal-derived arabinoxylans, arabinoxylan-oligosaccharides, and xylooligosaccharides. *Critical Reviews in Food Science and Nutrition*, 51(2):178–194.

Brown, G.D., and Williams, D.L. 2009. (1,3)-β-glucans in innate immunity: Mammalian systems. In: *Chemistry, Biochemistry and Biology of (1–3) β-Glucans and Related Polysaccharides*, eds. A. Bacic, G.B. Fincher, and B.A. Stone, 579–620. Burlington: Academic Press.

Brufau, G., Canela, M.A., and Rafecas, M. 2008. Phytosterols: Physiologic and metabolic aspects related to cholesterol-lowering properties. *Nutrition Research*, 28(4):217–225.

Canty, D., and Zeisel, S.H. 1994. Lecithin and choline in human health and disease. *Nutrition Reviews*, 52(10):327–339.

Chavez-Santoscoy, R.A., Gutiérrez-Uribe, J.A., Serna-Saldivar, S.O., and Perez-Carrillo, E. 2016a. Production of maize tortillas and cookies from nixtamalized flour enriched with anthocyanins, flavonoids and saponins extracted from black bean (*Phaseolus vulgaris*) seed coats. *Food Chemistry*, 192:90–97.

Chavez-Santoscoy, R.A., Gutierrez-Uribe, J.A., Torre-Villalvazo, I., Granados, O., Serna-Saldivar, S.O., Torre, N., Palacios-Gonzalez, B., and Tovar, A.R. 2014. Flavonoids and saponins extracted from black bean (*Phaseolus vulgaris* L.) seed coats modulate lipid metabolism and biliary cholesterol secretion in C57BL/6 mice. *British Journal of Nutrition*, 112(6):886–899.

Chavez-Santoscoy, R.A., Lazo-Velez, M.A., Serna-Saldivar, S.O., and Gutiérrez-Uribe, J.A. 2016b. Delivery of flavonoids and saponins black bean (*Phaseolus vulgaris*) seed coats incorporated into whole wheat bread. *International Journal of Molecular Sciences*, 17(2):222.

Chen, G.C., Tong, X., Xu, J.Y., Han, S.F., Wan, Z.X., Qin, J.B., and Qin, L.Q. 2016. Whole-grain intake and total, cardiovascular, and cancer mortality: A systematic review and meta-analysis of prospective studies. *The American Journal of Clinical Nutrition*, 104(1):164–172.

Clotault, J., Peltier, D., Berruyer, R., Thomas, M., Briard, M., and Geoffriau, E. 2008. Expression of carotenoid biosynthesis genes during carrot root development. *Journal of Experimental Botany*, 59(13):3563–3573.

Cohn, J.S., Wat, E., Kamili, A., and Tandy, S. 2008. Dietary phospholipids, hepatic lipid metabolism and cardiovascular disease. *Current Opinion in Lipidology*, 19(3):257–262.

Crowell, P., and Elson, C. 2000. Isoprenoids, health and disease. In: *Handbook of Nutraceuticals and Functional Food*, ed. R. Wildman, 31–53. Boca Raton: CRC Press.

Daliri, E.B.M., Oh, D.H., and Lee, B.H. 2017. Bioactive peptides. *Foods*, 6(5):32.

De Cicco, P., Catani, M.V., Gasperi, V., Sibilano, M., Quaglietta, M., and Savini, I. 2019. Nutrition and breast cancer: A literature review on prevention, treatment and recurrence. *Nutrients*, 11(7):1514.

Desai, I.D., Bhagavan, H., Salkeld, R., and de Oliveira, J.E.D. 1988. Vitamin E content of crude and refined vegetable oils in Southern Brazil. *Journal of Food Composition and Analysis*, 1(3):231–238.

Dicko, M.H., Gruppen, H., Traore, A.S., Voragen, A.G.J., and van Berkel, W.J.H. 2006. Phenolic compounds and related enzymes as determinants of sorghum for food use. *Biotechnology and Molecular Biology Review*, 1(1):21–38.
Dinkova-Kostova, A.T., Fahey, J.W., Kostov, R.V., and Kensler, T.W. 2017. KEAP1 and done? Targeting the NRF2 pathway with sulforaphane. *Trends in Food Science and Technology*, 69(B):257–269.
d'Ischia, M., Panzella, L., Manini, P., and Napolitano, A. 2006. The chemical basis of the antinitrosating action of polyphenolic cancer chemopreventive agents. *Current Medicinal Chemistry*, 13(26):3133–3144.
Dixit, A.K., Antony, J.I.X., Sharma, N.K., and Tiwari, R.K. 2011. Soybean constituents and their health benefits. *Opportunity, Challenge and Scope of Natural Products in Medicinal Chemistry*, 2011:367–383.
Dos Santos, M.D., Almeida, M.C., Lopes, N.P., and de Souza, G.E.P. 2006. Evaluation of the anti-inflammatory, analgesic and antipyretic activities of the natural polyphenol chlorogenic acid. *Biological and Pharmaceutical Bulletin*, 29(11):2236–2240.
Dutta, A., and Dutta, S. 2003. Vitamin E and its role in the prevention of atherosclerosis and carcinogenesis: A review. *Journal of the American College of Nutrition*, 22(4):258–268.
Dyall, S.C. 2015. Long-chain omega-3 fatty acids and the brain: A review of the independent and shared effects of EPA, DPA, and DHA. *Frontiers in Aging Neuroscience*, 7:52.
Eim, V.S., Simal, S., Rossello, C., and Femenia, A. 2008. Effects of addition of carrot dietary fibre on the ripening process of a dry fermented sausage (sobrassada). *Meat Science*, 80(2):173–182.
El Khoury, D., Cuda, C., Luhovyy, B.L., and Anderson, G.H. 2012. Beta glucan: Health benefits in obesity and metabolic syndrome. *Journal of Nutrition and Metabolism*, Article ID 851362.
Eleazu, C., Charles, A., Eleazu, K., and Achi, N. 2018. Free fatty acid receptor 1 as a novel therapeutic target for type 2 diabetes mellitus-current status. *Chemico-Biological Interactions*, 289:32–39.
Ernst, E. 2010. Panax ginseng: An overview of the clinical evidence. *Journal of Ginseng Research*, 34(4):259–263.
Faraj, A., and Vasanthan, T. 2004. Soybean isoflavones: Effects of processing and health benefits. *Food Reviews International*, 20(1):51–75.
Farsi, P.F., Djazayery, A., Eshraghian, M.R., Koohdani, F., Saboor-Yaraghi, A.A., Derakhshanian, H., Zarei, M., Javanbakht, M.H., and Djalali, M. 2014. Effects of supplementation with omega-3 on insulin sensitivity and non-esterified free fatty acid (NEFA) in type 2 diabetic patients. *Arquivos Brasileiros de Endocrinologia & Metabologia*, 58(4):335–340.
Fimognari, C., Berti, F., Nüsse, M., Cantelli-Forti, G., and Hrelia, P. 2004. Induction of apoptosis in two human leukemia cell lines as well as differentiation in human promyelocytic cells by cyanidin-3-*O*-glucopyranoside. *Biochemical Pharmacology Journal*, 67(11):2047–2056.
Franck, A. 2006. Inulin. In: *Food Polysaccharides and Their Applications*, eds. A.M. Stephen, G.O. Phillips, and P.A. Williams, 335–351. Boca Raton: Taylor & Francis.
Fuentes-Zaragoza, E., Sanchez-Zapata, E., Sendra, E., Sayas, E., Navarro, C., Fernández-López, J., and Pérez-Alvarez, J.A. 2011. Resistant starch as prebiotic: A review. *Starch-Stärke*, 63(7):406–415.
Fullmer, L.A., and Shao, A. 2001. The role of lutein in eye health and nutrition. *Cereal Foods World*, 46(9):408–413.
Gabay, O., Sanchez, C., Salvat, C., Chevy, F., Breton, M., Nourissat, G., Wolf, C., Jacques, T.F., and Berenbaum, F. 2010. Stigmasterol: A phytosterol with potential anti-osteoarthritic properties. *Osteoarthritis and Cartilage*, 18(1):106–116.
Gao, H., Geng, T., Huang, T., and Zhao, Q. 2017. Fish oil supplementation and insulin sensitivity: A systematic review and meta-analysis. *Lipids in Health and Disease*, 16(1):131.
Garcia-Lara, S., Gutierrez-Uribe, J.A., and Serna-Saldivar, S.O. 2012. Nutracetical properties of blue maize. In: *Maize: Cultivation, Uses and Health Benefits*, ed. J.C. Jimenez-Lopez, 113–135. New York: Nova Publishers.
Ghasemzadeh, A., and Ghasemzadeh, N. 2011. Flavonoids and phenolic acids: Role and biochemical activity in plants and human. *Journal of Medicinal Plants Research*, 5(31):6697–6703.
Giacco, R., Costabile, G., Della Pepa, G., Anniballi, G., Griffo, E., Mangione, A., Ciprinao, P., Viscovo, D., Clemente, G., Landberg, R., and Pacini, G. 2014. A whole grain cereal-based diet lowers postprandial plasma insulin and triglyceride levels in individuals with metabolic syndrome. *Nutrition, Metabolism, and Cardiovascular Diseases*, 24(8):837–844.
Gouni-Berthold, I., and Berthold, H.K. 2002. Policosanol: Clinical pharmacology and therapeutic significance of a new lipid lowering agent. *American Heart Journal*, 143(2):356–365.
Guajardo-Flores, D., Serna-Saldivar, S.O., and Gutiérrez-Uribe, J.A. 2013. Evaluation of the antioxidant and antiproliferative activities of extracted saponins and flavonols from germinated black beans (*Phaseolus vulgaris* L.). *Food Chemistry*, 141(2):1497–1503.

Guclu-Ustündag, O., and Mazza, G. 2007. Saponins: Properties, applications and processing. *Critical Reviews in Food Science and Nutrition*, 47(3):231–257.

Gutierrez-Uribe, J.A., and Serna-Saldivar, S.O. 2014. El Frijol como alimento functional. In: *Los alimentos funcionales*, eds. G.A. Gonzales Aguilar, A.F. Gonzalez Cordova, B. Vallejo Cordova, E. Alvarez Parrilla and H.S. Garcia, 211–232. Mexico: AGT Editor.

Hargrove, J.L., Greenspan, P., and Hartle, D.K. 2004. Nutritional significance and metabolism of very long chain fatty alcohols and acids from dietary waxes. *Experimental Biological and Medicine*, 229(3):215–226.

Hart, D.J., and Scott, K.J. 1995. Development and evaluation of an HPLC method for the analysis of carotenoids in foods, and the measurement of the carotenoid content of vegetables and fruits commonly consumed in the UK. *Food Chemistry*, 54(1):101–111.

Hendriks, H.F.J., Weststrate, J.A., van Vliet, T., and Meijer, G.W. 1999. Spreads enriched with three different levels of vegetable oil sterols and the degree of cholesterol lowering in normocholesterolaemic and mildly hypercholesterolaemic subjects. *European Journal of Clinical Nutrition*, 53(4):319–327.

Hsu, P.K., Chien, P.J., Chen, C.H., and Chau, C.F. 2006. Carrot insoluble fiber-rich fraction lowers lipid and cholesterol absorption in hamsters. *LWT – Food Science and Technology*, 39(4):338–342.

Huang, S., Vignolles, M.L., Chen, X.D., Le Loir, Y., Jan, G., Schuck, P., and Jeantet, R. 2017. Spray drying of probiotics and other food-grade bacteria: A review. *Trends in Food Science and Technology*, 63:1–17.

Hwang, K.T., Kim, J.E., and Weller, C.L. 2005. Policosanol contents and composition in wax like materials extracted from selected cereals of Korean origin. *Cereal Chemistry Journal*, 82(3):242–245.

Hwang, K.T., Weller, C.L., Cuppett, S.L., and Hanna, M.A. 2004. Policosanol contents and composition of grain sorghum kernels and dried distilled grains. *Cereal Chemistry Journal*, 81(3):345–349.

Kamal-Eldin, A., and Appelqvist, L.A. 1996. The chemistry and antioxidant properties of tocopherols and tocotrienols. *Lipids*, 31(7):671–701.

Kandlakunta, B., Rajendran, A., and Thingnganing, L. 2008. Carotene content of some common (cereals, pulses, vegetables, spices and condiments) and unconventional sources of plant origin. *Food Chemistry*, 106(1):85–89.

Kanski, J., Aksenova, M., Stoyanova, A., and Butterfield, D.A. 2002. Ferulic acid antioxidant protection against hydroxyl and peroxyl radical oxidation in synaptosomal and neuronal cell culture systems in vitro: Structure-activity studies. *The Journal of Nutritional Biochemistry*, 13(5):273–281.

Karunanandaa, B., Qi, Q., Hao, M., Baszis, S.R., Jensen, P.K., Wong, Y.H.H., Jiang, J., Venkatramesh, M., Gruys, K.J., Moshiri, F., Post-Beittenmiller, D., Weiss, J.D., and Valentin, H.E. 2005. Metabolically engineered oilseed crops with enhanced seed tocopherol. *Metabolic Engineering*, 7(5–6):384–400.

Kawaniski, K., Aoki, K., Hashimoto, Y., and Matsunobu, A. 1991. Free primary alcohols in oils and waxes from germs and kernels and other components of nuts, seeds, fruits and cereals. *Journal of the American Oil Chemist Society*, 68(11):869–872.

Kerry, R.G., Patra, J.K., Gouda, S., Park, Y., Shin, H.S., and Das, G. 2018. Benefaction of probiotics for human health: A review. *Journal of Food and Drug Analysis*, 26(3):927–939.

Kerwin, S.M. 2004. Soy saponins and the anticancer effects of soybeans and soy-based foods. *Current Medicinal Chemistry – Anti-Cancer Agents*, 4(3):263–272.

Kim, H., Lee, H., and Min, D. 2007. Effects and prooxidant mechanisms of oxidized alpha tocopherol on the oxidative stability of soybean oil. *Journal of Food Science*, 72(4):C223–C230.

Korhonen, H., and Pihlanto, A. 2006. Bioactive peptides: Production and functionality. *International Dairy Journal*, 16(9):945–960.

Kornsteiner, M., Wagner, K.H., and Elmadfa, I. 2006. Tocopherols and total phenolics in 10 different nut types. *Food Chemistry*, 98(2):381–387.

Küllenberg, D., Taylor, L.A., Schneider, M., and Massing, U. 2012. Health effects of dietary phospholipids. *Lipids in Health and Disease*, 11(1):1–16.

Kumari, A., Arora, S., Choudhary, S., Singh, A.K., and Tomar, S.K. 2018. Comparative stability of aspartame and neotame in yoghurt. *International Journal of Dairy Technology*, 71(1):81–88.

Law, M. 2000. Plant sterol and stanol margarines and health. *British Medical Journal*, 320(7238):861–864.

Leal-Díaz, A.M., Noriega, L.G., Torre-Villalvazo, I., Aleman, G., Lopez, P., Sanchez-Tapia, M., Velazquez-Villegas, L.A., Avila-Nava, A., Ordaz, G., Gutierrez-Uribe, J.A., Serna-Saldivar, S.O., Torres, N., and Tovar Palacio, A. 2016. Aguamiel concentrate from *Agave salmiana* and its extracted saponins prevents obesity, hepatic steatosis and microbiota dysbiosis in C57BL6 mice. *Scientific Reports*, 6:34242.

Lee, J.D., Shannon, J.G., So, Y.S., Sleper, D.A., Nelson, R.L., Lee, J.H., and Choung, M.G. 2009. Environmental effects on lutein content and relationship of lutein and other seed components in soybean. *Plant Breeding*, 128(1):97–100.

Lin, D., Xiao, M., Zhao, J., Li, Z., Xing, B., Li, X., Kong, M., Li, L., Zhang, Q., Liu, Y., Chen, H., Qin, W., Wu, H., and Chen, S. 2016. An overview of plant phenolic compounds and their importance in human nutrition and management of type 2 diabetes. *Molecules*, 15(10):E1374.

Liu, L.-P., and Deber, C.M. 1997. Anionic phospholipids modulate peptide insertion into membranes. *Biochemistry*, 36(18):5476–5482. https://doi.org/10.1021/bi970030n

Lopez-Barrios, L., Gutierrez-Uribe, J.A., and Serna-Saldivar, S.O. 2014. Bioactive peptides and hydrolysates from pulses and their potential use as functional ingredients. *Journal of Food Science*, 79(3):R273–R283.

Lu, Q.Y., Zhang, Y., Wang, Y., Wang, D., Lee, R.P., Gao, K., Byrns, R., and Heber, D. 2009. California Hass avocado: Profiling of carotenoids, tocopherol, fatty acid, and fat content during maturation and from different growing areas. *Journal of Agricultural and Food Chemistry*, 57(21):10408–10413.

Lu, Z.X., Walker, K.Z., Muir, J.G., and O'Dea, K. 2004. Arabinoxylan fibre improves metabolic control in people with Type II diabetes. *European Journal of Clinical Nutrition*, 58(4):621–628.

Lubelska, K., Wiktorska, K., Mielczarek, L., Milczarek, M., Zbroińska-Bregisz, I., and Chilmonczyk, Z. 2016. Sulforaphane regulates NFE2L2/Nrf2-dependent xenobiotic metabolism phase II and phase III enzymes differently in human colorectal cancer and untransformed epithelial colon cells. *Nutrition and Cancer*, 68(8):1338–1348.

Maguire, L.S., O'sullivan, S.M., Galvin, K., O'connor, T.P., and O'brien, N.M. 2004. Fatty acid profile, tocopherol, squalene and phytosterol content of walnuts, almonds, peanuts, hazelnuts and the macadamia nut. *International Journal of Food Sciences and Nutrition*, 55(3):171–178.

Malaguti, M., Dinelli, G., Leoncini, E., Bregola, V., Bosi, S., Cicero, A.F., and Hrelia, S. 2014. Bioactive peptides in cereals and legumes: Agronomical, biochemical and clinical aspects. *International Journal of Molecular Sciences*, 15(11):21120–21135.

Mayne, S.T. 1996. Beta-carotene, carotenoids, and disease prevention in humans. *The FASEB Journal*, 10(7):690–701.

Mellado-Ortega, E., and Hornero-Méndez, D. 2015. Carotenoids in cereals: An ancient resource with present and future applications. *Phytochemistry Reviews*, 14(6):873–890.

Messina, M., and Hilakivi-Clarke, L. 2009. Early intake appears to be the key to the proposed protective effects of soy intake against breast cancer. *Nutrition and Cancer*, 61(6):92–798.

Messina, M., and Lane, B. 2007. Soy protein, soybean isoflavones and coronary heart disease risk: Where do we stand? *Future Lipidology*, 2(1):55–74.

Messina, M.J. 2002. Soy foods and soybean isoflavones and menopausal health. *Nutrition in Clinical Care*, 5(6):272–282.

Morales de la Peña, M., Odriozola-Serrano, I., Oms-Oliu, G., and Martín-Belloso, O. 2020. Dietary fiber in fruits and vegetables. In: *Science and Technology of Fibers in Food Systems*, eds. J. Welti-Chanes, S.O. Serna-Saldivar, O. Campanella, and V. Tejada Ortigoza, 123–152. New York: Springer.

Moreau, R., Whitaker, B., and Hicks, K. 2002. Phytosterols, phytostanols, and their conjugates in foods: Structural diversity, quantitative analysis, and health-promoting uses. *Progress in Lipid Research*, 41(6):457–500.

Naz, S., Sherazi, S.T.H., and Talpur, F.N. 2011. Changes of total tocopherol and tocopherol species during sunflower oil processing. *Journal of the American Oil Chemists' Society*, 88(1):127–132.

Nesaretnam, K., Yew, W.W., and Wahid, M.B. 2007. Tocotrienols and cancer: Beyond antioxidant activity. *European Journal of Lipid Science and Technology*, 109(4):445–452.

Normen, L., Dutta, P., Lia, A., and Andersson, H. 2000. Soy sterol esters and beta-sitostanol ester as inhibitors of cholesterol absorption in human small bowel. *American Journal of Clinical Nutrition*, 71(4):908–913.

Oomah, B.D., Corbe, A., and Balasubramanian, P. 2010. Antioxidant and anti-inflammatory activities of bean (*Phaseolus vulgaris* L.) hulls. *Journal of Agricultural and Food Chemistry*, 58(14):8225–8230.

Orellana, P., Valenzuela, R., Valenzuela, A., Morales, G.I., Orellana, P., Valenzuela, R., and Morales, G.I. 2018. Efectos neuro protectores del ácido araquidónico y del ácido docosahexaenoico en las etapas extremas de la vida: Una visión integradora. *Revista Chilena de Nutrición*, 45(1):80–88.

Orthoefer, F.T., and List, G.R. 2006. Phospholipids/lecithin: A class of nutraceutical lipids. In: *Nutraceutical and Specialty Lipids and Their Co-Products*, ed. F. Shahidi, 509–530. Boca Raton: CRC Press.

Ostlund, R. 2004. Phytosterols and cholesterol metabolism. *Current Opinion in Lipidology*, 15(1):37–41.

Palliyaguru, D.L., Yuan, J.M., Kensler, T.W., and Fahey, J.W. 2018. Isothiocyanates: Translating the power of plants to people. *Molecular Nutrition and Food Research*, 62(18):1700965. https://doi.org/10.1002/mnfr.201700965

Panche, A.N., Diwan, A.D., and Chandra, S.R. 2016. Flavonoids: An overview. *Journal of Nutritional Science*, 5:e47.

Park, Y.W., and Nam, M.S. 2015. Bioactive peptides in milk and dairy products: A review. *Korean Journal for Food Science of Animal Resources*, 35(6):831–840.

Peluso, I., and Palmery, M. 2015. Flavonoids at the pharma-nutrition interface: Is a therapeutic index in demand? *Biomedicine and Pharmacotherapy*, 71:102–107.

Phuong-Lien, D.I., Bich-Tram, P.T., and Toan, H.T. 2016. Antioxidant properties of food natural phenolic compounds – A review. *Innovations in Food Research*, 2:1–5.

Pokorny, J. 2001. Natural antioxidant functionality during food processing. In: *Antioxidants in Food, Practical Applications*, eds. J. Pokorny, N. Yanishlieva, and M. Gordon, 335–351. Cambridge: CRC Press, Woodhead Publishing Ltd.

Psomiadou, E., Tsimidou, M., and Boskou, D. 2000. α-Tocopherol content of Greek virgin olive oils. *Journal of Agricultural and Food Chemistry*, 48(5):1770–1775.

Pszczola, D.E. 2005. From soup to soynuts: The broadening uses of soy. *Food Technology*, 59(2):44–55.

Punia, S., Sandhu, K.S., Siroha, A.K., and Dhull, S.B. 2019. Omega 3-metabolism, absorption, bioavailability and health benefits–A review. *PharmaNutrition*, 10:100162.

Rangel-Huerta, O.D., and Gil, A. 2018. Effect of omega-3 fatty acids on cognition: An updated systematic review of randomized clinical trials. *Nutrition Reviews*, 76(1):1–20.

Rizzello, C.G., Tagliazucchi, D., Babini, E., Rutella, G.S., Saa, D.L.T., and Gianotti, A. 2016. Bioactive peptides from vegetable food matrices: Research trends and novel biotechnologies for synthesis and recovery. *Journal of Functional Foods*, 27:549–569.

Roberfroid, M. 2007. Inulin-type fructans; functional food ingredients. *The Journal of Nutrition*, 137(11):2493S–2502S.

Robinson, J.G., Ijioma, N., and Harris, W. 2010. Omega-3 fatty acids and cognitive function in women. *Women's Health*, 6(1):119–134.

Ross, J.A., and Kasum, C.M. 2002. Dietary flavonoids: Bioavailability, metabolic effects, and safety. *Annual Review of Nutrition*, 22(1):19–34.

Sajilata, M.G., Singhal, R.S., and Kulkarni, P.R. 2006. Resistant starch–A review. *Comprehensive Reviews in Food Science and Food Safety*, 5(1):1–17.

Sanchez, A., and Vázquez, A. 2017. Bioactive peptides: A review. *Food Quality and Safety*, 1(1):29–46.

Sanchez, J., Mas, R., Mendoza, S., Fernández, J.C., and Ruiz, D. 2010. Effects of policosanol on patients with ischemic stroke with previous transient ischemic attack: A long-term follow-up. *Revista CENIC Ciencias Biológicas*, 41(1):23–29.

Santocono, M., Zurria, M., Berrettini, M., Fedeli, D., and Falcioni, G. 2007. Lutein, zeaxanthin and astaxanthin protect against DNA damage in SK-N-SH human neuroblastoma cells induced by reactive nitrogen species. *Journal of Photochemistry and Photobiology B: Biology*, 88(1):1–10.

Scheller, H.V., and Ulvskov, P. 2010. Hemicelluloses. *Annual Review of Plant Biology*, 61:263–289.

Schlinkert, C., Gillebaart, M., Benjamins, J., Poelman, M., and de Ridder, D. 2020. The snack that has it all: People's associations with ideal snacks. *Appetite*, 152:104722.

Senes-Guerrero, C., Gradilla-Hernandez, M.S., Garcia-Gamboa, R., and Garcia-Cayuela, T. 2020. Dietary fiber and gut microbiota. In: *Science and Technology of Fibers in Food Systems*, eds. J. Welti-Chanes, S.O. Serna-Saldivar, O. Campanella, and V. Tejada Ortigoza, 277–298. New York: Springer.

Serna-Saldivar, S.O. 2010. Role of cereals in human nutrition and health. In: *Cereal Grains: Properties, Processing and Nutritional Attributes*, ed. S.O. Serna-Saldivar, 565–628. Boca Raton: CRC Press (Taylor & Francis Group).

Serna-Saldivar, S.O. 2020. Relación del consumo de granos enteros con la salud pública. In: *Papel de los cereales de grano entero en la salud*, 197–214. Granada: ILSI Nor-Andino y Fundación Iberoamericana de Nutrición (FINUT).

Serna-Saldivar, S.O., and Ayala-Soto, F. 2020. Chemical composition and biosynthesis of dietary fiber components. In: *Science and Technology of Fibers in Food Systems*, eds. J. Welti-Chanes, S.O. Serna-Saldivar, O. Campanella, and V. Tejada Ortigoza, 15–44. New York: Springer.

Serna-Saldivar, S.O., and Sanchez-Hernandez, D. 2020. Dietary fiber in cereals, legumes, pseudocereals and other seeds. In: *Science and Technology of Fibers in Food Systems*, eds. J. Welti-Chanes, S.O. Serna-Saldivar, O. Campanella, and V. Tejada Ortigoza, 87–122. New York: Springer.

Serrano, J., Puupponen-Pimiä, R., Dauer, A., Aura, A.M., and Saura-Calixto, F. 2009. Tannins: Current knowledge of food sources, intake, bioavailability and biological effects. *Molecular Nutrition and Food Research*, 53:S310–S329.

Setchell, K.D.R., and Cassidy, A. 1999. Dietary isoflavones: Biological effects and relevance to human health. *The Journal of Nutrition*, 129(3):758S–767S.

Shahidi, F., and Zhong, Y. 2008. Bioactive peptides. *Journal of AOAC International*, 91(4):914–931.

Shoaib, M., Shehzad, A., Omar, M., Rakha, A., Raza, H., Sharif, H.R., Shakeel, A., Ansari, A., and Niazi, S. 2016. Inulin: Properties, health benefits and food applications. *Carbohydrate Polymers*, 147:444–454.

Singh, B., Singh, J.P., Singh, N., and Kaur, A. 2017. Saponins in pulses and their health promoting activities: A review. *Food Chemistry*, 233:540–549.

Singh, B.P., and Vij, S. 2018. In vitro stability of bioactive peptides derived from fermented soy milk against heat treatment, pH and gastrointestinal enzymes. *LWT-Food Science Technology*, 91:303–307. https://doi.org/10.1016/j.lwt.2018.01.066

Siriamornpun, S., Kaisoon, O., and Meeso, N. 2012. Changes in colour, antioxidant activities and carotenoids (lycopene, β-carotene, lutein) of marigold flower (*Tagetes erecta* L.) resulting from different drying processes. *Journal of Functional Foods*, 4(4):757–766.

Sivapalan, T., Melchini, A., Saha, S., Needs, P.W., Traka, M.H., Tapp, H., Dainty, J.R., and Mithen, R.F. 2018. Bioavailability of glucoraphanin and sulforaphane from high-glucoraphanin broccoli. *Molecular Nutrition and Food Research*, 62(18):1700911.

Slavin, J., Jacobs, D., and Marquart, L. 1997. Whole-grain consumption and chronic disease: Protective mechanisms. *Nutrition and Cancer*, 27(1):14–21.

Soccol, C.R., de Souza Vandenberghe, L.P., Spier, M.R., Pedroni Medeiros, A.B., Yamaguishi, C.T., Lindner, J.D.D., Pandey, A., and Thomaz-Soccol, V. 2010. The potential of probiotics: A review. *Food Technology and Biotechnology*, 48(4):413–434.

Srivastava, R.C., Husain, M.M., Hasan, S.K., and Athar, M. 2000. Green tea polyphenols and tannic acid act as potent inhibitors of phorbol ester-induced nitric oxide generation in rat hepatocytes independent of their antioxidant properties. *Cancer Letters*, 153(1–2):1–5.

Szczuka, D., Nowak, A., Zakłos-Szyda, M., Kochan, E., Szymańska, G., Motyl, I., and Blasiak, J. 2019. American ginseng (*Panax quinquefolium* L.) as a source of bioactive phytochemicals with pro-health properties. *Nutrients*, 11(5):1041.

Tacon, A.G., and Metian, M. 2013. Fish matters: Importance of aquatic foods in human nutrition and global food supply. *Reviews in Fisheries Science*, 21(1):22–38.

Tak, Y., Kaur, M., Amarowicz, R., Bhatia, S., and Gautam, C. 2021. Pulse derived bioactive peptides as novel nutraceuticals: A review. *International Journal of Peptide Research and Therapeutics*. https://doi.org/10.1007/s10989-021-10234-8.

Tanoury, Z.A., Piskunov, A., and Rochette-Egly, C. 2013. Vitamin A and retinoid signaling: Genomic and nongenomic effects. *Journal of Lipid Research*, 54(7):1761–1775.

Tian, S., Liu, X., Lei, P., Zhang, X., and Shan, Y. 2018. Microbiota: A mediator to transform glucosinolate precursors in cruciferous vegetables to the active isothiocyanates. *Journal of the Science of Food and Agriculture*, 98(4):1255–1260.

Torres, M., Avila-Nava, A., Medina-Vera, I., and Tovar, A.R. 2020. Dietary fiber and diabetes. In: *Science and Technology of Fibers in Food Systems*, eds. J. Welti-Chanes, S.O. Serna-Saldivar, O. Campanella, and V. Tejada Ortigoza, 201–218. New York: Springer.

Tovar, A.R., Guevara-Cruz, M., Serralde-Zuñiga, A.E., and Torres, M. 2020. Dietary fiber and hyperlipidemia and cardiovascular disease. In: *Science and Technology of Fibers in Food Systems*, eds. J. Welti-Chanes, S.O. Serna-Saldivar, O. Campanella, and V. Tejada Ortigoza, 219–240. New York: Springer.

Turksoy, S., and Ozkaya, B. 2011. Pumpkin and carrot pomace powders as a source of dietary fiber and their effects on the mixing properties of wheat flour dough and cookie quality. *Food Science and Technology Research*, 17(6):545–553.

Uauy, R., Hoffman, D.R., Mena, P., Llanos, A., and Birch, E.E. 2003. Term infant studies of DHA and ARA supplementation on neurodevelopment: Results of randomized controlled trials. *Journal of Pediatrics*, 143:S17–S25.

Uauy, R., Hoffman, D.R., Peirano, P., Birch, D.G., and Birch, E.E. 2001. Essential fatty acids in visual and brain development. *Lipids*, 36(9):885–895.

Valdes-Ramos, R., Guadarrama-López, A.L., Martínez-Carrillo, B.E., and Harbige, L.S. 2019. n-3 polyunsaturated fatty acids in type-2 diabetes mellitus. In: *Bioactive Food as Dietary Interventions for Diabetes*, eds. R.R. Watson and V.R. Preddy, 193–209. London: Academic Press (Elsevier).

Van den Abbeele, P., Van De Wiele, T., and Possemiers, S. 2011. Prebiotic effect and potential health benefit of arabinoxylans. *Agro Food Industry Hi-Tech*, 22(2):9–12.

Van de Vijver, L.P.L., van den Bosch, L.M., van den Brandt, P.A., and Goldbohm, R.A. 2009. Whole-grain consumption, dietary fibre intake and body mass index in the Netherlands cohort study. *European Journal of Clinical Nutrition*, 63(1):31–38.

Varady, K.A., Wang, Y., and Jones, P.J. 2003. Role of policosanols in the prevention and treatment of cardiovascular disease. *Nutrition Reviews*, 61(11):376–383.

Wagner, H.K., and Elmadfa, I. 2003. Biological relevance of terpenoids: Overview focusing on mono-, di- and tetraterpenes. *Annals of Nutrition and Metabolism*, 47(3–4):95–105.

Weststrate, J.A., and Meijer, G.W. 1998. Plant sterol-enriched margarines and reduction of plasma total-and LDL-cholesterol concentrations in normocholesterolaemic and mildly hypercholesterolaemic subjects. *European Journal of Clinical Nutrition*, 52(5):334.

Wu, X., Beecher, G.R., Holden, J.M., Haytowitz, D.B., Gebhardt, S.E., and Prior, R.L. 2006. Concentrations of anthocyanins in common foods in the United States and estimation of normal consumption. *Journal of Agricultural and Food Chemistry*, 54(11):4069–4075.

Yagishita, Y., Fahey, J.W., Dinkova-Kostova, A.T., and Kensler, T.W. 2019. Broccoli or sulforaphane: Is it the source or dose that matters? *Molecules*, 24(19):3593.

Yangilar, F. 2013. The application of dietary fiber in food industry: Structural features, effects on health and definition, obtaining and analysis of dietary fiber: A review. *Journal of Food and Nutrition Research*, 1(3):13–23.

Ye, E., Chacko, S.A., Chou, E.L., Kugizaki, M., and Liu, S. 2012. Greater wholegrain intake is associated with lower risk of type 2 diabetes, cardiovascular disease, and weight gain. *The Journal of Nutrition*, 142(7):1304–1313.

Zekovic, D.B., Kwiatkowski, S., Vrvić, M.M., Jakovljević, D., and Moran, C.A. 2005. Natural and modified (1→3)-β-D-glucans in health promotion and disease alleviation. *Critical Reviews in Biotechnology*, 25(4): 205–230.

Zeng, H., Lazarova, D., and Bordonaro, M. 2014. Mechanisms linking dietary fiber, gut microbiota and colon cancer prevention. *World Journal of Gastrointestinal Oncology*, 6(2): 41–51.

Index

For Product Safety Concerns and Information please contact our EU
representative GPSR@taylorandfrancis.com
Taylor & Francis Verlag GmbH, Kaufingerstraße 24, 80331 München, Germany

www.ingramcontent.com/pod-product-compliance
Ingram Content Group UK Ltd.
Pitfield, Milton Keynes, MK11 3LW, UK
UKHW050056200926
475702UK00019B/2875

* 9 7 8 0 3 6 7 6 5 3 5 1 4 *